THE COASTAL OCEAN

A DERIVATIVE OF ENCYCLOPEDIA OF OCEAN SCIENCES, 2ND EDITION

Editor

KARL K. TUREKIAN

AMSTERDAM • BOSTON • HEIDELBERG • LONDON • NEW YORK • OXFORD
PARIS • SAN DIEGO • SAN FRANCISCO • SINGAPORE • SYDNEY • TOKYO
Academic Press is an imprint of Elsevier

ELSEVIER

ACADEMIC
PRESS

Academic Press is an imprint of Elsevier
32 Jamestown Road, London NW1 7BY, UK
30 Corporate Drive, Suite 400, Burlington, MA 01803, USA
525 B Street, Suite 1900, San Diego, CA 92101-4495, USA

Material in the work originally appeared in *Encyclopedia of Ocean Sciences* (Elsevier Ltd., 2001) and *Encyclopedia of Ocean Sciences*, 2nd Edition (Elsevier Ltd., 2009), edited by John H. Steele, Steve A. Thorpe and Karl K. Turekian.

Notice
No responsibility is assumed by the publisher for any injury and/or damage to persons or property as a matter of products liability, negligence or otherwise, or from any use or operation of any methods, products, instructions or ideas contained in the material herein, Because of rapid advances in the medical sciences, in particular, independent verification of diagnoses and drug dosages should be made

British Library Cataloguing in Publication Data
A catalogue record for this book is available from the British Library

Library of Congress Cataloging-in-Publication Data
A catalog record for this book is available from the Library of Congress

ISBN: 978-0-12-3813954

For information on all Academic Press publications
visit our website at www.elsevierdirect.com

Working together to grow
libraries in developing countries

www.elsevier.com | www.bookaid.org | www.sabre.org

ELSEVIER   BOOK AID International   Sabre Foundation

The Coastal Ocean

Editor-in-Chief

John H. Steele

Marine Policy Center, Woods Hole Oceanographic Institution, Woods Hole,
Massachusetts, USA

Editors

Steve A. Thorpe

National Oceanography Centre, University of Southampton,
Southampton, UK
and
School of Ocean Sciences, Bangor University, Menai Bridge, Anglesey, UK

Karl K. Turekian

Yale University, Department of Geology and Geophysics, New Haven,
Connecticut, USA

Subject Area Volumes from the Second Edition

Climate & Oceans edited by Karl K. Turekian
Elements of Physical Oceanography edited by Steve A. Thorpe
Marine Biology edited by John H. Steele
Marine Chemistry & Geochemistry edited by Karl K. Turekian
Marine Ecological Processes edited by John H. Steele
Marine Geology & Geophysics edited by Karl K. Turekian
Marine Policy & Economics guest edited by Porter Hoagland, Marine Policy Center,
Woods Hole Oceanographic Institution, Woods Hole, Massachusetts
Measurement Techniques, Sensors & Platforms edited by Steve A. Thorpe
Ocean Currents edited by Steve A. Thorpe
The Coastal Ocean edited by Karl K. Turekian
The Upper Ocean edited by Steve A. Thorpe

CONTENTS

WAVES, TIDES, SURGES, TSUNAMI AND SEICHES

TOPOGRAPHY AND SEA LEVEL

PLANKTON AND BENTHOS

MANAGEMENT, MARICULTURE AND FISHERIES

POLLUTION

SEDIMENTS, SLIDES, SLUMPS AND CYCLING

CIRCULATION AND MODELS

REMOTE SENSING BY ACOUSTICS, AIRCRAFT AND SATELLITES

RIGS, STRUCTURES AND SHIPPING

INDEX

THE COASTAL ZONE: INTRODUCTION

The coastal zone, where the land meets the sea, has unique qualities affecting both the ocean and the land. The interactions of the two domains result in impacts on ocean circulation and chemical and salinity modifications of sea water on the one hand and the erosion and modification of the coast line on the other. The coastal boundary ocean is sometimes referred to as the "brown" ocean in temperate and arctic environs. In oligotrophic tropical and subtropical coastal zones the characterization is more clearly one of clear blue waters.

The oceanic boundaries control the surface current trajectories of the world oceans and the oceans behave as transmitters of the energy of the sea from climate impacts and earthquakes to the continental margins. The resulting impact on a region heavily occupied by humans makes for important stories of suffering and property loss.

The ocean boundaries with land take many forms ranging from estuarine, lagoonal and deltaic features to ice controlled features on the one hand to coral reefs on the other. These varied environments each have their own special properties requiring a detailed analysis in order to understand natural and human induced variations.

The human engagement with the coastal zone ranges from the response to intense storms and tsunamis to the utilization for transportation and recreation. The coast also provides an arena of resource exploitation. These activities include off-shore drilling for oil, mariculture and the use of sand and gravel resources for building materials.

Humans have impacted the coastal zone not only by their construction of cities with their harbors but also by impacting the surrounding waters of these cities with the pollution of human activity. Not only the direct impact of the usual human input of sewage and industrial waste but also the more remotely derived transport of contaminants from mining, manufacturing and agriculture can impact the coastal ocean by causing eutrophication and metal pollution.

The range of processes, both natural and human induced, at the coastal zone are engaged in the articles in this collection. Theoretical, geographical and practical issues are addressed for the various environments characteristic of the varied coastal zones of the planet.

Karl K. Turekian
Editor

RIVERS, ESTUARIES AND FIORDS

RIVER INPUTS

J. D. Milliman, College of William and Mary,
Gloucester, VA, USA

Introduction

Rivers represent the major link between land and the ocean. Presently rivers annually discharge about $35\,000\,km^3$ of freshwater and $20–22 \times 10^9$ tonnes of solid and dissolved sediment to the global ocean. The freshwater discharge compensates for most of the net evaporation loss over the ocean surface, groundwater and ice-melt discharge accounting for the remainder. As such, rivers can play a major role in defining the physical, chemical, and geological character of the estuaries and coastal areas into which they flow.

Historically many oceanographers have considered the land–ocean boundary to lie at the mouth of an estuary, and some view it as being at the head of an estuary (or, said another way, at the mouth of the river). A more holistic view of the land–ocean interface, however, might include the river basins that drain into the estuary. This rather unconventional view of the land–sea interface is particularly important when considering the impact of short- and medium-term changes in land use and climate, and how they may affect the coastal and global ocean.

Uneven Global Database

A major hurdle in assessing and quantifying fluvial discharge to the global ocean is the uneven quantity and quality of river data. Because they are more likely to be utilized for transportation, irrigation, and damming, large rivers are generally better documented than smaller rivers, even though, as will be seen below, the many thousands of small rivers collectively play an important role in the transfer of terrigenous sediment to the global ocean. Moreover, while the database for many North American and European rivers spans 50 years or longer, many rivers in Central and South America, Africa and Asia are poorly documented, despite the fact that many of these rivers have large water and sediment inputs and are particularly susceptible to natural and anthropogenic changes.

The problem of data quality is magnified by the fact that the available database spans the latter half of the twentieth century, some of the data having been collected $>20–40$ years ago, when flow patterns may have been considerably different to present-day patterns. In many cases, more recent data can reflect anthropogenically influenced conditions, often augmented by natural change. The Yellow River in northern China, for example, is considered to have one of the highest sediment loads in the world, 1.1 billion tonnes y^{-1}. In recent years, however, its load has averaged <100 million tons, in response to drought and increased human removal of river water. How, then, should the average sediment load of the Yellow River be reported – as a long-term average or in its presently reduced state?

Finally, mean discharge values for rivers cannot reflect short-term events, nor do they necessarily reflect the flux for a given year. Floods (often related to the El Niño/La Niña events) can have particularly large impacts on smaller and/or arid rivers, such that mean discharges or sediment loads may have little relevance to short-term values. Despite these caveats, mean values can offer sedimentologists and geochemists considerable insight into the fluxes (and fates) from land to the sea.

Because of the uneven database (and the often difficult access to many of the data), only in recent years has it been possible to gather a sufficient quantity and diversity of data to permit a quantitative understanding of the factors controlling fluvial fluxes to the ocean. Recent efforts by Meybeck and Ragu (1996) and Milliman and Farnsworth (2002) collectively have resulted in a database for about 1500 rivers, whose drainage basin areas collectively represent about 85% of the land area emptying into the global ocean. It is on this database that much of the following discussion is based.

Quantity and Quality of Fluvial Discharge

Freshwater Discharge

River discharge is a function of meteorological runoff (precipitation minus evaporation) and drainage basin area. River basins with high runoff but small drainage area (e.g., rivers draining Indonesia, the Philippines, and Taiwan) can have discharges as great as rivers with much larger basin areas but low runoff (**Table 1**). In contrast, some rivers with low

Table 1 Basin area, runoff, and discharge for various global rivers

River	Basin area ($\times 10^3 km^2$)	Runoff (mm y^{-1})	Discharge (km^{3}y^{-1})
Amazon	6300	1000	6300
Congo	3800	360	1350
Ganges/Bramaputra	1650	680	1120
Orinoco	1100	1000	1100
Yangtze (Changjiang)	1800	510	910
Parana/Uruguay	2800	240	670
Yenisei	2600	240	620
Mekong	800	690	550
Lena	2500	210	520
Mississippi	3300	150	490
Choshui (Taiwan)	3.1	1970	6.1
James (USA)	20	310	6.2
Cunene (Angola)	100	68	6.8
Limpopo (Mozambique)	380	14	5.3

The first 10 rivers represent the highest discharge of all world rivers, the Amazon having discharge equal to the combined discharge of the next seven largest rivers. Basin areas of these rivers vary from 6300 (Amazon) to 800 (Mekong) $\times 10^3$ km^2, and runoffs vary from 1000 (Amazon) to 150 mm y^{-1} (Mississippi). The great variation in runoff can be seen in the example of the last four rivers, each of which has roughly the same mean annual discharge (5.3–6.8 km^3 y^{-1}), but whose basin areas and runoffs vary by roughly two orders of magnitude.

runoff (such as the Lena and Yenisei) have high discharges by virtue of their large drainage basin areas. The Amazon River has both a large basin (comprising 35% of South America) and a high runoff; as such its freshwater discharge equals the combined discharge of the next seven largest rivers (**Table 1**). Not only are the coastal waters along north-eastern South America affected by this enormous discharge, but the Amazon influence can be seen as far north as the Caribbean > 2000 km away. The influence of basin area and runoff in controlling discharge is particularly evident in the bottom four rivers listed in **Table 1**, all of which have similar discharges (5.3–6.2 km^3 y^{-1}) even though their drainage basin areas and discharges can vary by two orders of magnitude.

Sediment Discharge

A river's sediment load consists of both suspended sediment and bed load. The latter, which moves by traction or saltation along the river bed, is generally assumed to represent 10% (or less) of the total sediment load. Suspended sediment includes both wash load (mostly clay and silt that is more or less continually in suspension) and bed material load (which is suspended only during higher flow); bed material includes coarse silt and sand that may move as bed load during lower flow. A sediment-rating curve is used to calculate suspended sediment concentration (or load), which relates measured concentrations (or loads) with river flow. Sediment load generally increases exponentially with flow, so that a two-order of magnitude increase in flow may result in three to four orders of magnitude greater suspended sediment concentration.

In contrast to water discharge, which is mainly a function of runoff and basin area, the quantity and character of a river's sediment load also depend on the topography and geology of the drainage basin, land use, and climate (which influences vegetation as well as the impact of episodic floods). Mountainous rivers tend to have higher loads than rivers draining lower elevations, and sediment loads in rivers eroding young, soft rock (e.g., siltstone) are greater than rivers flowing over old, hard rock (e.g., granite) (**Figure 1**). Areas with high rainfall generally have higher rates of erosion, although heavy floods in arid climates periodically can carry huge amounts of sediment.

The size of the drainage basin also plays an important role. Small rivers can have one to two orders of magnitude greater sediment load per unit basin area (commonly termed sediment yield) (**Figure 1**) than larger rivers because they have less flood plain area on which sediment can be stored, which means a greater possibility of eroded sediment being discharged directly downstream. Stated another way, a considerable amount of the sediment eroded in the headwaters of a large river may be stored along the river course, whereas most of the sediment eroded along a small river can be discharged directly to the sea, with little or no storage.

Small rivers are also more susceptible to floods, during which large volumes of sediment can be transported. Large river basins, in contrast, tend to be self-modulating, peak floods in one part of the basin are often offset by normal or dry conditions in another part of the basin. The impact of a flood on a small, arid river can be illustrated by two 1-day floods on the Santa Clara River (north of Los Angeles) in January and February 1969, during which more sediment was transported than the river's cumulative sediment load for the preceding 25 years! The combined effect of high sediment yields from small rivers can be seen in New Guinea, whose more than 250 rivers collectively discharge more sediment to the ocean than the Amazon River, whose basin area is seven times larger than the entire island.

Our expanded database allows us to group river basins on the basis of geology, climate, and basin area, thereby providing a better understanding of how these factors individually and collectively influence sediment load. Rivers draining the young,

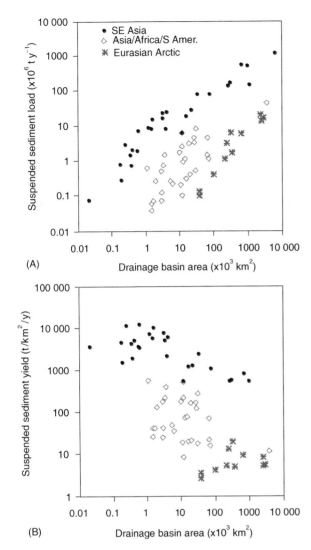

(A)

(B)

Figure 1 Suspended sediment discharge from rivers with various sized basins draining wet mountains in south-east Asia with young (assumed to be easily erodable) rocks (dots); wet mountains in south Asia, north-eastern South America, and west Africa with old (assumed to be less erodable) rocks (open diamonds); and semi-arid to humid mountainous and upland rivers in the Eurasian Arctic. Sediment loads generally increase with basin size, and are one to two orders of magnitude greater for south-east Asian rivers than for rivers with equal rainfall but old rocks, which in turn are greater than for rivers with old rocks but less precipitation (A). Sediment yields for smallest south-east Asian rivers are as much as two orders of magnitude greater than for the largest, whereas the yield for Eurasian Arctic rivers shows little change with basin size (B).

easily eroded rocks in the wet mountains of south-east Asia, for instance, have one to two orders of magnitude greater sediment loads (and sediment yields) than similar-sized rivers draining older rocks in wet Asian mountains (e.g., in India or Malaysia), which in turn have higher loads and yields than the rivers from the older, drier mountains in the Eurasian Arctic (**Figure 1**).

Because of the many variables that determine a river's sediment load, it is extremely difficult to calculate the cumulative sediment load discharged from a land mass without knowing the area, morphology, geology, and climate for every river draining that land mass. This problem can be seen in the last four rivers listed in **Table 2**. The Fitzroy-East drains a seasonally arid, low-lying, older terrain in north-eastern Australia, whereas the Mad River in northern California (with a much smaller drainage basin), drains rainy, young mountains. Although the mean annual loads of the two rivers are equal, the Mad has 112-fold greater sediment yield than the Fitzroy-East. The Santa Ynez, located just north of Santa Barbara, California, has a similar load and yield, but because it has much less runoff, its average suspended sediment concentration is 13 times greater than the Mad's.

Dissolved Solid Discharge

The amount of dissolved material discharged from a river depends on the concentration of dissolved ions and the quantity of water flow. Because dissolved concentrations often vary inversely with flow, a river's dissolved load often is more constant throughout the year than its suspended load.

Dissolved solid concentrations in river water reflect the nature of the rock over which the water flows, but the character of the dissolved ions is controlled by climate as well as lithology. Rivers with different climates but draining similar lithologies can have similar total dissolved loads compared with rivers with similar climates but draining different rock types. For example, high-latitude rivers, such as those draining the Eurasian Arctic, have similar dissolved solid concentrations to rivers draining older lithologies in southern Asia, north-eastern South America and west Africa, but lower concentrations than rivers draining young, wet mountains in south-east Asia (**Figure 2**).

The concentration of dissolved solids shows little variation with basin area, small rivers often having concentrations as high as large rivers (**Figure 2B**). At first this seems surprising, since it might be assumed that small rivers would have low dissolved concentrations, given the short residence time of flowing water. This suggests an important role of ground water in both the dissolution and storage of river water, allowing even rivers draining small basins to discharge relatively high concentrations of dissolved ions.

Four of the 10 rivers with highest dissolved solid discharge (Salween, MacKenzie, St Lawrence, and Rhine) in **Table 3** are not among the world leaders in

Table 2 Basin area, TSS (mg l⁻¹), annual sediment load, and sediment yield (t km² per year) for various global rivers

River	Basin area ($\times 10^3 km^2$)	TSS (gl^{-1})	Sediment load ($\times 10^6 t\, y^{-1}$)	Sediment yield ($t\, km^{-2} y^{-1}$)
Amazon	6300	0.19	1200	190
Yellow (Huanghe)	750	25	1100	1500
Ganges/Bramaputra	1650	0.95	1060	640
Yangtze (Changjiang)	1800	0.51	470	260
Mississippi	3300	0.82	400 (150)	120 (45)
Irrawaddy	430	0.6	260	600
Indus	980	2.8	250 (100)	250 (100)
Orinoco	1100	0.19	210	190
Copper	63	1.4	130(?)	2100
Magdalena	260	0.61	140	540
Fitzroy-East (Australia)	140	0.31	2.2	16
Arno (Italy)	8.2	0.69	2.2	270
Santa Ynez (USA)	2	23	2.3	1100
Mad (USA)	1.2	1.7	2.2	1800

Loads and yields in parentheses represent present-day values, the result of river damming and diversion. The first 10 rivers represent the highest sediment loads of all world rivers, the Amazon, Ganges/Brahmaputra and Yellow rivers all having approximately equal loads of about 1100×10^6 t y⁻¹. No other river has an average sediment load greater than 470×10^6 t y⁻¹). Discharges for these rivers vary from 6300 (Amazon) to $43 \times$ km³ y⁻¹ (Yellow), and basin areas from 6300 (Amazon) to 63×10^3 km² y⁻¹ (Copper). Corresponding average suspended matter concentrations and yields vary from 0.19 to 25 and 190 to 2100, respectively. The great variation in values can be seen in the example of the last four rivers, which have similar annual sediment loads ($2.2–2.3 \times 10^6$ t y⁻¹), but whose average sediment concentrations and yields vary by several orders of magnitude.

terms of either water or sediment discharge (**Tables 1** and **2**). The prominence of the Rhine, which globally can be considered a second-order river in terms of basin area, discharge, and sediment load, stems from the very high dissolved ionic concentrations (19 times greater than the Amazon), largely the result of anthropogenic influence in its watershed (see below).

The last four rivers in **Table 3** reflect the diversity of rivers with similar dissolved solid discharges. The Cunene, in Angola, is an arid river that discharges about as much water as the Citandy in Indonesia, but drains more than 20 times the watershed area. As such, total dissolved solid (TDS) values for the two rivers are similar (51 vs 62 mg l⁻¹), but the dissolved yield (TDS divided by basin area) of the Cunene is < 1% that of the Citandy. In contrast, the Ems River, in Germany, has a similar dissolved load to the Citandy, but concentrations are roughly three times greater.

Fluvial Discharge to the Global Ocean

Collectively, the world rivers annually discharge about 35 000 km³ of fresh water to the ocean. More than half of this comes from the two areas with highest precipitation, south-east Asia/Oceania and north-eastern South America, even though these two areas collectively account for somewhat less than 20% of the total land area draining into the global ocean. Rivers from areas with little precipitation,

such as the Canadian and Eurasian Arctic, have little discharge (4800 km³) relative to the large land area that they drain (> 20 × 10⁶ km²).

A total of about $20–22 \times 10^9$ t y⁻¹ of solid and dissolved sediment is discharged annually. Our estimate for suspended sediment discharge (18×10^9 t y⁻¹) is less accurate than our estimate for dissolved sediment (3.9×10^9 t y⁻¹) because of the difficulty in factoring in both basin area (see above) and human impact (see below). Given the young, wet mountains and the large anthropogenic influence in south-east Asia, it is perhaps not surprising that this region accounts for 75% of the suspended sediment discharged to the global ocean. In fact, the six high-standing islands in the East Indies (Sumatra, Java, Celebes, Borneo, Timor, and New Guinea) collectively may discharge as much sediment (4.1×10^9 t y⁻¹) as all non-Asian rivers combined. Southeast Asia rivers are also the leading exporters of dissolved solids to the global ocean, 1.4×10^9 t y⁻¹ (35% of the global total), but Europe and eastern North America are also important, accounting for another 25%.

Another way to view river fluxes is to consider the ocean basins into which they empty. While the Arctic Ocean occupies < 5% of the global ocean basin area (17×10^6 km²), the total watershed draining into the Arctic is 21×10^6 km², meaning a land/ocean ratio of 1.2. In contrast, the South Pacific accounts for one-quarter of the global ocean area, but its land/ocean ratio is only 0.05 (**Table 4**). The greatest fresh water input occurs in the North Atlantic (largely

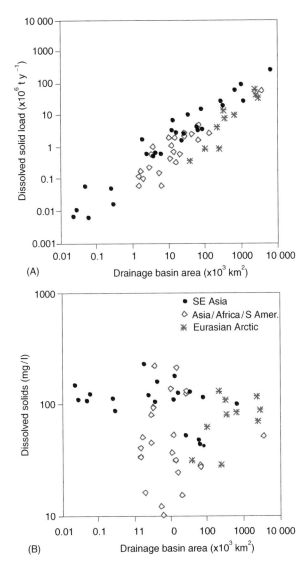

Figure 2 Dissolved solid load versus basin area for rivers draining south-east Asian rivers (solid dots), south Asia, west Africa, north-east South America (open diamonds – high rainfall, old rocks), and the Canadian-Eurasian Arctic (asterisks – low precipitation, old rocks). Note that while the dissolved loads relative to basin size are much closer than they are for suspended load, the difference seems to increase with decreasing basin area (A). The lack of correlation between dissolved concentrations and basin area (B) suggests that residence time of a river's surface water may play less of a role than ground water in determining the quantity and character of the dissolved solid fraction.

because of the Amazon and, to a lesser extent, the Orinoco and Mississippi), but the greatest input per unit basin area is in the Arctic ($28\,\mathrm{cm\,y^{-1}}$ if evenly distributed over the entire basin). The least discharge per unit area of ocean basin is the South Pacific ($4.5\,\mathrm{cm\,y^{-1}}$ distributed over the entire basin). In terms of suspended sediment load, the major sinks are the Pacific and Indian oceans (11.1 and $4 \times 10^9\,\mathrm{t\,y^{-1}}$, respectively), whereas the North Atlantic is the major

sink for dissolved solids ($1.35 \times 10^9\,\mathrm{t\,y^{-1}}$; **Table 4**), largely due to the high dissolved loads of European and eastern North American rivers.

This is not to say, of course, that the fresh water or its suspended and dissolved loads are evenly distributed throughout the ocean basins. In fact, most fluvial identity is lost soon after the river discharges into the ocean due to mixing, flocculation, and chemical uptake. In many rivers, much of the sediment is sequestered on deltas or in the coastal zone. In most broad, passive margins, in fact, little sediment presently escapes the inner shelf, and very little reaches the outer continental margin. During low stands of sea level, on the other hand, most fluvial sediment is discharged directly to the deep sea, where it can be redistributed far from its source(s) via mass wasting (e.g., slumping and turbidity currents). This contrast between sediment discharge from passive margins during high and low stands of sea level is the underlying basis for sequence stratigraphy. However, in narrow active margins, often bordered by young mountains, many rivers are relatively small (e.g., the western Americas, East Indies) and therefore often more responsive to episodic events. As such, sediment can escape the relatively narrow continental shelves, although the ultimate fate of this sediment is still not well documented.

Changes in Fluvial Processes and Fluxes – Natural and Anthropogenic

The preceding discussion is based mostly on data collected in the past 50 years. In most cases it reflects neither natural conditions nor long-term conditions representing the geological past. While the subject is still being actively debated, there seems little question that river discharge during the last glacial maximum (LGM), 15 000–20 000 years ago, differed greatly from present-day patterns. Northern rivers were either seasonal or did not flow except during periodic ice melts. Humid and sub-humid tropical watersheds, on the other hand, may have experienced far less precipitation than they do at present.

With the post-LGM climatic warming and ensuing ice melt, river flow increased. Scattered terrestrial and marine data suggest that during the latest Pleistocene and earliest Holocene, erosion rates and sediment delivery increased dramatically as glacially eroded debris was transported by increased river flow. As vegetation became more firmly established in the mid-Holocene, however, erosion rates apparently decreased, and perhaps would have remained relatively low except for human interference.

Table 3 Basin area, TDS ($mg\,l^{-1}$), annual dissolved load, and dissolved yield (t/km^2 per year) for various global rivers

River	Basin area ($\times 10^3 km^2$)	TDS ($mg\,l^{-1}$)	Dissolved load ($\times 10^6 t\,y^{-1}$)	Dissolved yield ($t\,km^{-2}y^{-1}$)
Amazon	6300	43	270	43
Yangtze (Changjiang)	1800	200	180	100
Ganges/Bramaputra	1650	130	150	91
Mississippi	3300	280	140	42
Irrawaddy	430	230	98	230
Salween	320	310	65	200
MacKenzie	1800	210	64	35
Parana/Uruguay	2800	92	62	22
St Lawrence	1200	180	62	52
Rhine	220	810	60	270
Cunene (Angola)	110	51	0.35	3
Torne (Norway)	39	30	0.37	9
Ems (Germany)	8	180	0.34	42
Citandy (Indonesia)	4.8	62	0.38	79

The first 10 rivers represent the highest dissolved loads of all world rivers, only the Amazon, Yangtze, Ganges/Brahmaputra, and Mississippi rivers having annual loads $>100 \times 10^6\,t\,y^{-1}$. Discharges vary from 6300 (Amazon) to $74 \times km^3\,y^{-1}$ (Rhine), and basin areas from 6300 (Amazon) to $220 \times 10^3\,km^2$ (Rhine). Corresponding average dissolved concentrations and yields vary from 43 to 810 and 43 to 270, respectively. The bottom rivers have similar annual dissolved loads ($0.34-0.38 \times 10^6\,t\,y^{-1}$), but their average sediment concentrations and yields vary by factors of 6–26 (respectively), reflecting both natural and anthropogenic influences.

Table 4 Cumulative oceanic areas, drainage basin areas, and discharge of water, suspended and dissolved solids of rivers draining into various areas of the global ocean

Oceanic area	Basin area ($\times 10^6\,km^2$)	Drainage basin ($\times 10^6\,km^2$)	Water discharge ($km^3\,y^{-1}$)	Sediment load ($\times 10^6\,t\,y^{-1}$)	Dissolved load ($\times 10^6\,t\,y^{-1}$)
North Atlantic	44	30	12 800	2500	1350
South Atlantic	46	12	3300	400	240
North Pacific	83	15	6000	7200	660
South Pacific	94	5	4300	3900	650
Indian	74	14	4000	4000	520
Arctic	17	21	4800	350	480
Total	358	98	35 200	18 000	3900

For this compilation, it is assumed that Sumatra and Java empty into the Indian Ocean, and that the other high-standing islands in Indonesia discharge into the Pacific Ocean. Rivers discharging into the Black Sea and Mediterranean area are assumed to be part of the North Atlantic drainage system. (Data from Milliman and Farnsworth, 2001.)

Few terrestrial environments have been as affected by man's activities as have river basins, which is not surprising considering the variety of uses that humans have for rivers and their drainage basins: agriculture and irrigation, navigation, hydroelectric power, flood control, industry, etc. Few, if any, modern rivers have escaped human impact, and with exception of the Amazon and a few northern rivers, it is difficult to imagine a river whose flow has not been strongly affected by anthropogenic activities. Natural ground cover helps the landscape resist erosion, whereas deforestation, road construction, agriculture, and urbanization all can result in chan-nelized flow and increased erosion. Erosion rates and corresponding sediment discharge of rivers draining much of southern Asia and Oceania have increased substantially because of human activities, locally as much as 10-fold. While land erosion in some areas of the world recently has been decreasing (e.g., Italy, France, and Spain) due to decreased farming and increased reforestation; deforestation and poor land conservation practices elsewhere, particularly in the developing world, have led to accelerated increases in both land erosion and fluvial sediment loads.

While increased river basin management has led to very low suspended loads for most northern European rivers, mining and industrial activities have resulted in greatly elevated dissolved concentrations.

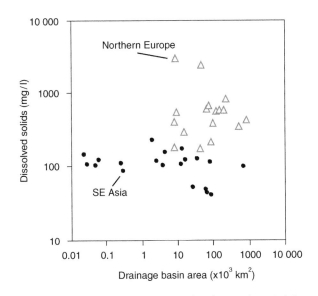

Figure 3 Dissolved solid concentration for south-east Asian rivers (also shown in **Figure 2**) compared with the much higher concentrations seen in northern European rivers (excluding those in Scandinavia). The markedly higher levels of dissolved solids in northern European rivers almost certainly reflects greater mining and industrial activity.

Compared to south-east Asian rivers, whose waters generally contain 100–200 mg l^{-1} dissolved solids, some European rivers, such as the Elbe, can have concentrations greater than 1000 mg l^{-1} (**Figure 3**), in large part due to the mining of salt deposits for potassium. Since the fall of the Soviet empire, attempts have been made to clean up many European rivers, but as of the late 1990s, they still had the highest dissolved concentrations in the world.

Interestingly, while sediment loads of rivers may be increasing, sediment flux to the ocean may be decreasing because of increased river diversion (e.g., irrigation and flood protection through levees) and stoppage (dams). As of the late 1990s there were >24 000 major dams in operation around the world. The nearly 200 dams along the Mississippi River, for example, have reduced sediment discharge to the Gulf of Mexico by >60%. In the Indus River, construction of irrigation barrages in the late 1940s led to an 80% reduction in the river's sediment load. Even more impressive is the almost complete cessation of sediment discharge from the Colorado and Nile rivers in response to dam construction. Not only have water discharge and sediment flux decreased, but high and low flows have been greatly modulated; the effect of this modulated flow, in contrast to strong seasonal signals, on the health of estuaries is still not clear.

Decreased freshwater discharge can affect the circulation of shelf waters. Many of the nutrients utilized in coastal primary production are derived from upwelled outer shelf and slope waters as they are advected landward to offset offshore flow of surface waters. Decreased river discharge from the Yangtze River resulting from construction of the Three Gorges Dam might decrease the shoreward flow of nutrient-rich bottom shelf waters in the East China Sea, which could decrease primary production in an area that is highly dependent on its rich fisheries.

Retaining river water within man-made reservoirs also can affect water quality. For example, reservoir retention of silicate-rich river water can lead to diatom blooms within the man-made lakes and thus depletion of silicate within the river water. One result is that increased ratios of dissolved nitrate and phosphate to dissolved silica may have helped change primary production in coastal areas from diatom-based to dinoflagellates and coccolithophorids. One result of this altered production may be increased hypoxia in coastal and shelf waters in the north-western Black Sea and other areas off large rivers.

Taken in total, negative human impact on river systems seems to be increasing, and these anthropogenic changes are occurring faster in the developing world than in Europe or North America. Increased land degradation (partly the result of increased population pressures) in northern Africa, for instance, contrasts strongly with decreased erosion in neighboring southern Europe. Dam construction in Europe and North America has slackened in recent years, both in response to environmental concerns and the lack of further sites of dam construction, but dams continue to be built in Africa and Asia.

Considering increased water management and usage, together with increased land degradation, the coastal ocean almost certainly will look different 100 years from now. One can only hope that the engineers and planners in the future have the foresight to understand potential impacts of drainage basin change and to minimize their effects.

See also

Estuarine Circulation.

Further Reading

Berner RA and Berner EK (1997) Silicate weathering and climate. In: Ruddiman RF (ed.) *Tectonic Uplift and Climate Change*, pp. 353–365. New York: Plenum Press.

Chen CTA (2000) The Three Gorges Dam: reducing the upwelling and thus productivity in the East China Sea. *Geophysics Research Letters* 27: 381–383.

Douglas I (1996) The impact of land-use changes, especially logging, shifting cultivation, mining and urbanization on sediment yields in tropical Southeast Asia: a review with special reference to Borneo. *Int. Assoc. Hydrol. Sci. Publ.* 236: 463–472.

Edmond JM and Huh YS (1997) Chemical weathering yields from basement and orogenic terrains in hot and cold climates. In: Ruddiman RF (ed.) *Tectonic Uplift and Climate Change*, pp. 329–351. New York: Plenum Press.

Humborg C, Conley DJ, Rahm L, *et al.* (2000) Silicon retention in river basins: far-reaching effects on biogeochemistry and aquatic food webs in coastal marine environments. *Ambio.*

Lisitzin AP (1996) *Oceanic Sedimentation, Lithology and Geochemistry.* Washington, DC: American Geophysics Union.

Meybeck M (1984) Origin and variable composition of present day river-borne material. In: *Material Fluxes on the Surface of the Earth, National Research Council Studies in Geophysics*, pp. 61–73. Washington: National Academy Press

Meybeck M and Ragu A (1996) *River Discharges to the Oceans. An Assessment of Suspended Solids, Major Ions and Nutrients.* GEMS-EAP Report.

Milliman JD (1995) Sediment discharge to the ocean from small mountainous rivers: the New Guinea example. *Geo-Mar Lett* 15: 127–133.

Milliman JD and Farnsworth KL (2002) *River Runoff, Erosion and Delivery to the Coastal Ocean: A Global Analysis.* Oxford University Press (in press).

Milliman JD and Meade RH (1983) World-wide delivery of river sediment to the oceans. *J Geol* 91: 1–21.

Milliman JD and Syvitski JPM (1992) Geomorphic/ tectonic control of sediment discharge to the ocean the importance of small mountainous rivers. *J Geol* 100: 525–544.

Milliman JD, Ren M-E, Qin YS, and Saito Y (1987) Man's influence on the erosion and transport of sediment by Asian rivers: the Yellow River (Huanghe) example. *J Geol* 95: 751–762.

Milliman JD, Farnworth KL, and Albertin CS (1999) Flux and fate of fluvial sediments leaving large islands in the East Indies. *Journal of Sea Research* 41: 97–107.

Thomas MF and Thorp MB (1995) Geomorphic response to rapid climatic and hydrologic change during the late Pleistocene and early Holocene in the humid and sub-humid tropics. *Quarterly Science Review* 14: 193–207.

Walling DE (1995) Suspended sediment yields in a changing environment. In: Gurnell A and Petts G (eds.) *Changing River Channels*, pp. 149–176. Chichester: John Wiley Sons.

ESTUARINE CIRCULATION

K. Dyer, University of Plymouth, Plymouth, UK

Introduction

Estuaries are formed at the mouths of rivers where the fresh river water interacts and mixes with the salt water of the sea. Even though there is only about a 2% difference in density between the two water masses, the horizontal and vertical gradients in density causes the water circulation, and the mixing created by the tides, to be very variable in space and time, resulting in long residence times for pollutants and trapping of sedimentary particles. Both salinity and temperature affect density, but the salinity changes are normally of greatest influence.

Most estuaries are the result of a dramatic rise in sea level of about 100 m during the last 10 000 years, following the end of the Pleistocene glaciation. The river valleys were flooded by the sea, and the valleys infilled with sediment to varying extents. The degree of infilling provides a wide range of topographic forms for the estuaries. Those estuaries that have had large inputs of sediment have been filled, and may have been built out into deltas, where the sediment flux is extreme. These are typical of tropical and monsoon areas. Where the sediment discharge was less, the estuary may still have many of the morphological attributes of river valleys: a sinuous, meandering outline, a triangular cross-sectional form with a deep central channel, and wide shallow flood plains. These are termed drowned river valleys, and are typical of higher latitudes. Where the land mass was previously covered by the glaciers, the river valleys may have been drastically overdeepened, and the U-shaped valleys became fiords.

Because of the shallow water, the sheltered anchorages, and the ready access to the hinterland, estuaries have become the centers of habitation and of industrial development, and this has produced problems of pollution and environmental degradation. The solution to these problems requires an understanding of how the water flows and mixes, and how the sediment accumulates.

Definition

Estuaries can be defined and classified in many ways, depending on whether one is a geologist/geographer, a physicist, an engineer, or a biologist. The most comprehensive physical definition is that

> An estuary is a semi-enclosed coastal body of water that has free connection to the open sea, extending into the river as far as the limit of tidal influence, and within which sea water is measurably diluted with fresh water derived from land drainage.

There are normally three zones in an estuary: an outer zone where the salinity is close to that of the open sea, and the horizontal gradients are low; a middle zone where there is rapid change in the horizontal gradients and where mixing occurs; and an upper or riverine zone, where the water may be fresh throughout the tide despite there being a tidal rise and fall in water level.

Tides in Estuaries

The tidal rise and fall of sea level is generated in the ocean and travels into the estuary, becoming modified by the shoreline and by shallow water. The tidal context is microtidal when the range is less than 2 m; mesotidal when the range is between 2 m and 4 m; macrotidal when it is between 4 m and 6 m; and hypertidal when it is greater than 6 m. A narrowing of the estuary will cause the range of the tide to increase landward. However, this is counteracted by the friction of the seabed on the water flow, and the fact that some of the tidal energy is reflected back toward the sea. The result is that the time of high and low water is later at the head of the estuary than at the mouth, and the currents turn some tens of minutes after the maximum and minimum water level. This phase difference means that the tidal wave is a standing wave with a progressive component. In high tidal range estuaries, the currents and the range of the tide generally increase toward the head, until in the riverine section the river flow becomes important. These estuaries are often funnel-shaped with rapidly converging sides. The volume of water between high and low tide, known as the tidal prism, is large compared with the volume of water in the estuary at low tide. Since the speed of progression of the tidal wave increases with water depth, the flooding tide rises more quickly than the ebbing tide, leading to an asymmetrical tidal curve and currents on the flood tide larger than those on the ebb tide. These are flood-dominant estuaries. The high tidal range leads to high current velocities, which can produce considerable sediment transport, also dominant in the flood direction.

When the estuary is relatively shallow compared with the tidal range, the wet cross-sectional area is greater at high water than at low water, and the discharge of water per unit of velocity is greater also. There is thus a greater discharge landward near high water than seaward around low water. This creates a Stokes Drift towards the head of the estuary that has to be compensated for by an extra increment of tidally averaged flow toward the mouth in the deeper areas. The Eulerian mean flow measured at a fixed point, the nontidal drift, is therefore greater than the flow due to the river discharge. Additionally, in the upper estuary where intertidal areas are extensive, the deeper channel changes from being flood-dominant to ebb dominance. This produces an obvious location for siltation and the need for dredging.

In microtidal estuaries, friction exceeds the effects of convergence and the tidal range diminishes toward the head. These estuaries are generally fan-shaped, with a narrow mouth and extensive shallow water areas. The currents at the mouth are larger than those inside, and the ebb current velocities are larger than the flood currents, i.e., the estuary would be ebb-dominant.

It has been observed that the ratio of certain estuarine dimensions frequently appears to be constant, leading to the concept of an equilibrium estuary. In particular, the variations of breadth and depth along the estuary are often exponential in form. The O'Brien relationship shows that the tidal prism volume is related to the cross-sectional area of the estuary. As the tidal prism increases, the increased currents through the cross-section cause erosion and the area increases until equilibrium is reached. This implies that there is ultimately a balance between the amount of sediment carried landward on the flood tide and that carried seaward on the ebb, leading to zero net accumulation. This is a widely used concept for the prediction of morphological change.

Circulation Types

From a morphological point of view, estuaries can be classified into many categories in terms of their shape and their geological development. These reflect the degree of infilling by sediment since the ice age. Present-day processes are also important in distinguishing those estuaries dominated by tidal currents and those whose mouths are drastically affected by the spits and bars created by wave-induced longshore drift of sediment. However, though important, these classifications do not tell us much about the water and how its movement affects the discharge of pollutants and of sediment. The classifications, originally proposed by Pritchard in 1952, describe the tidally averaged differences in vertical and longitudinal salinity structure, and the mean water velocity profiles. These have been extended to consider the processes during the tide that contribute to the averages.

Salt Wedge Estuaries

Where the river discharges into a microtidal estuary, the fresher river water tends to flow out on top of the slightly denser sea water that rests almost stationary on the seabed and forms a wedge of salt water penetrating toward the head of the estuary. The salt wedge has a very sharp salinity interface at the upper surface – a halocline. There is a certain amount of friction between the two layers and the velocity shear between the rapidly flowing surface layer and the almost stationary salt wedge produces small waves on the halocline; these can break when the velocity shear between the layers is large enough. The breaking waves inject some of the salty lower layer water into the upper layer, thus enhancing its salinity. This process, known as entrainment, is equivalent to an upward flow of salt water. It is not a very efficient process, but requires a small compensating inflow in the salt wedge. Salt wedge estuaries have almost fresh water on the surface throughout, and almost pure salt water near the bed. The slight amount of tidal movement of the water does not create much mixing, except in some of the shallower water areas, but does act to renew the salty water trapped in hollows on the bed. This process is shown in **Figure 1**. During the tide, the stratification remains high, but the halocline becomes eroded from its top surface during the ebb tide and rises during the flood tide because of new salt water intruding along the bottom. Fiords often have a shallow rock sill at their mouth that isolates the deeper interior basins. The renewal of bottom water is infrequent, often only every few years. Because fiords are so deep the lower layer is effectively stationary, and the mixing is dominated by entrainment.

Partially Mixed Estuaries

In mesotidal and macrotidal estuaries the tidal movement drives the whole water mass up and down the estuary each tide. The current velocities near to the seabed are large, producing turbulence, and creating mixing of the water column. This is a much more efficient process for mixing than entrainment, but it is greatest near the bed, whereas entrainment is maximum in mid-water. However, both processes

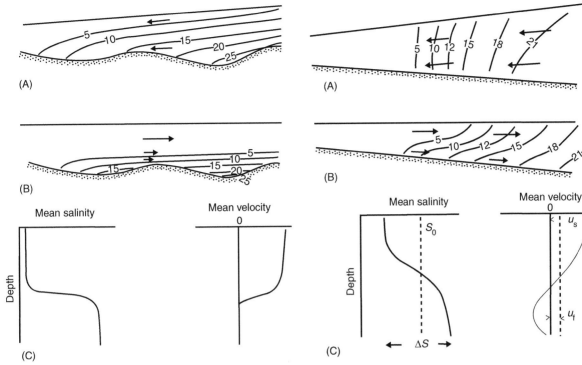

Figure 1 Diagrammatic longitudinal salinity and velocity distributions in a salt wedge estuary in PSU. (A) At mid flood tide. (B) At mid ebb tide. Contours show representative salinities; arrows show relative strength of currents. (C) Tidally averaged mean salinity and velocity profiles.

Figure 2 Diagrammatic longitudinal salinity and velocity distributions in a partially mixed estuary in PSU. (A) At mid flood tide. (B) At mid ebb tide. Contours show representative salinities; arrows show relative strength of currents. (C) Tidally averaged mean salinity and velocity profiles, including the definition of parameters for **Figure 4**: S, salinity; ΔS, salinity difference normalized by S_0 and circulation parameter u_s/u_f; u_s, surface mean flow; u_f, depth mean flow.

can be active together. Turbulent mixing is a two-way process, mixing fresher water downward and salty water upward. As a result, the mean vertical profile of salinity has a more gentle halocline (**Figure 2**). The salt intrusion is now a much more dynamic feature, changing its structure regularly with a tidal periodicity. During the flood tide, the surface water travels faster up the estuary than the water nearer the bed, and the salinity difference between surface and bed is minimized. Conversely, during the ebb, the fresher surface water is carried over the near-bottom salty water, and the stratification is enhanced. This process is known as tidal straining. It is to be expected that entrainment will be large during the ebb tide, and turbulent mixing will be dominant on the flood tide. Since each tide must discharge an amount of fresh water equivalent to an increment of the river inflow, and the water involved in this is now salty, there must be a significant mean advection of salt water up the estuary near to the bed to compensate for that discharged. There is thus a mean outflow of water and salt on the surface, and a mean inflow near the bed. The latter must diminish toward the head of the salt intrusion, and there is a level of no net motion near to the halocline where the sense of the mean flow velocity changes. At the landward tip of the salt intrusion there is a convergence, a null zone, where the tidally averaged near-bed flow diminishes to zero. The vertical turbulent mixing acting in combination with the mean horizontal advection is known as vertical gravitational circulation. Because of the meandering shape of the estuary, the flow is not straight but tends to spiral on the bends. This leads to the development of lateral differences in salinity and velocity, the shallower regions generally being better mixed and ebb-dominated.

Well-mixed Estuaries

When the tidal range is macrotidal or hypertidal, the turbulent mixing produced by the water flow across the bed is active throughout the water column, with the result that there is very little stratification. Nevertheless, there can be considerable lateral differences across the estuary in salinity and in mean flow velocity, as if the vertical circulation were turned on its side. This results in the currents on one side of the channel being flood-dominated while those on the other side are ebb-dominated, often separated by a mid-channel bank (**Figure 3**). As a consequence, during the tide the water tends to flow preferentially landward up one side of the bank, cross the channel at high water, and ebb down the other side. The bed sediment also is driven to circulate around the bank in the same way.

The above descriptive classification shows the general relationship between the structure of an estuary, the tidal range, and the river flow. Quantified classifications depend on producing numerical criteria that relate these variables. When the flow ratio - the ratio of river flow per tidal cycle to the tidal prism volume – is 1 or greater, the estuary is highly stratified. When it is about 0.25, the estuary is partially mixed; and for less than 0.1 it is well mixed. Alternatively, the ratio of the river discharge velocity (R/A, where R is the river volume discharge, and A is the cross-sectional area) divided by the root-mean-square tidal velocity gives values of less than 10^{-2} for well-mixed conditions and greater than 10^{-1} for stratified conditions. There are many alternative proposals that depend on describing the processes controlling the tidally averaged vertical profiles of salinity and of current velocity. One example uses a stratification parameter, which relates the surface to near-bed salinity difference (ΔS) normalized by the depth mean salinity (S_0), and a circulation parameter that expresses the ratio of the surface mean flow (u_s) to the depth mean flow (u_f) (**Figure 2**). Use of these classification schemes shows that estuaries plot as a series of points depending on position in the estuary, on river discharge, and on tidal range. They form part of a continuous sequence, rather than falling into distinct types, and an estuary can change in both space and time. Near the head of the estuary the river flow will be more important than nearer the mouth, and consequently the structure may be more stratified. Alternatively, if the tidal currents rise toward the head of the estuary, better-mixed conditions may prevail. **Figure 4** shows the classification scheme based on the stratification–circulation parameters, together with indication of the direction in which an estuary would plot for different circumstances.

Changes in river discharge of water force the estuarine circulation to respond. An increase will push the salt intrusion downstream and increase the stratification. Conversely, a decrease will allow the salt intrusion to creep further landward, and the stratification will decrease because the tidal motion will become relatively more important than the stabilizing effect of the fresh water input. Because of the drastic variation of tidal elevation and currents between spring and neap tides, well-mixed conditions

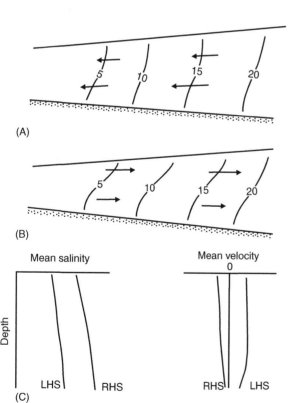

(A)

(B)

Mean salinity   Mean velocity

(C)

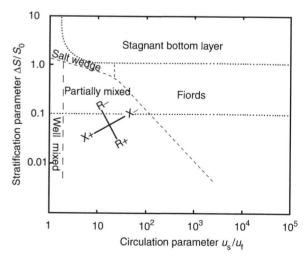

Figure 3 Diagrammatic plan views of the horizontal distribution of salinity in a well-mixed estuary (PSU). (A) At mid flood tide. (B) At mid ebb tide. Contours show representative salinities; arrows show relative strengths of currents. (C) Vertical mean profiles for left-hand side and right-hand side of the estuary looking down estuary.

Figure 4 Quantified estuarine classification scheme based on stratification and circulation parameters. R − and R + show the direction of movement of the plot of an estuarine location with increase or decrease in river discharge; X − and X + with movement toward the estuarine head or mouth. Parameters are defined in **Figure 2**. From Dyer (1972) with permission. © John Wiley & Sons.

may occur at spring tides and partially mixed at neaps. This must occur by a relatively stronger mixing on the flood tide during the increasing tidal range, and by increased stabilization on the ebb during decreasing tidal range. The change between the two conditions may occur very rapidly at a critical tidal range as the effects of mixing rapidly break down the stratification.

Additionally, a wind blowing landward along the estuary will tend to restrict the surface outflow, and may even reverse the sense of the vertical gravitational circulation. A down-estuary wind would enhance the stratification. Also, atmospheric pressure changes will cause a total outflow or inflow of water, as the mean water level responds. Thus large variations in the mixing and water circulation are likely on the timescale of a few days, the progression rate of atmospheric depressions.

Flushing

Within the estuary there is a volume of fresh water that is continually being carried out to sea and replaced by new inputs of river water. At the mouth, not all of the brackish water discharged on the ebb tide returns on the flood, the proportion being very variable depending on the coastal circulation close to the mouth, and fresh water comprises a fraction of that lost. The flushing or residence time is the time taken to replace the existing fresh water in the estuary at a rate equal to the river discharge. From direct measurements of the salinity distribution it is possible to calculate the volume of accumulated fresh water and the flushing time for various river flows. The flushing time changes rapidly with discharge at low river discharge, but slowly at high river flow, being buffered to a certain extent by the consequent changes in stratification. Flushing times are generally of the order of days to tens of days, rising with the size of the estuary.

Turbidity Maximum

A distinctive feature of partially and well-mixed estuaries is the turbidity maximum. This is a zone of high concentrations of fine suspended sediment, higher than in the river or lower down the estuary, that depends upon the estuarine circulation for its existence. The maximum is located near the head of the salt intrusion (**Figure 5**), and the maximum concentrations tend to rise with the tidal range, from of the order of 100 ppm to 10 000 ppm, though there are variations depending on the availability of sediment. Often the total amount of suspended sediment

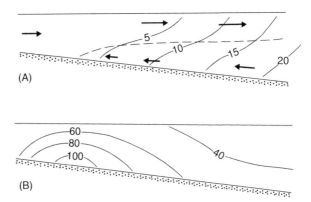

Figure 5 (A) Diagrammatic longitudinal distribution of tidally averaged salinity and current velocities in a partially mixed estuary. Contours show salinities; arrows show relative current velocities. Dashed line shows level of no net motion. (B) Diagrammatic distribution of mean suspended sediment concentration (in ppm), showing the turbidity maximum.

exceeds several million tonnes. The location of the maximum coincides with a mud-reach of bed sediment, and with muddy intertidal areas. The continual movement of the turbidity maximum and the sorting of the sediment appears to create a gradient of sediment grain size along the estuary. Closer to the mouth, the bed is generally dominated by more sandy sediment. Because of the affinity of the fine particles for contaminants, prediction of the characteristics of the turbidity maximum is important. The position of the maximum changes with river discharge and with tidal range, but with a time lag behind the variation in the salt intrusion position. During the tide there are drastic changes in concentration. At high water the maximum is located well up the estuary and concentrations are relatively low because of settling. During the ebb the current re-entrains the sediment from the bed, and advects it down the estuary. Settling again occurs at low water and there is further reentrainment and advection on the flood tide. Thus there is active cycling of sediment between the water column and the bed, and the whole mass becomes very well sorted in the process. The behavior of the sediment will depend on the settling velocity and the threshold for erosion. The settling velocity varies with concentration and with the intensity of the turbulence because flocculation of the particles occurs, forming large, loose but fast-settling floes. Settling causes high-concentration layers to build up on the bed, and at neap tides these may persist throughout the tide, forming layers often seen on echo-sounders as fluid mud.

The turbidity maximum is maintained by a number of trapping mechanisms. Vertical gravitational circulation carries the fine particles and flocs downstream on the surface, and they settle toward the bed because the turbulent mixing is reduced by the stratification. Near the bed the landward mean flow carries the suspended particles toward the head of the salt intrusion, together with new material coming in from the sea. There they meet other particles carried downstream by the river flow and are suspended by the high currents. Additionally, the asymmetry of the currents during the tide leads to tidal pumping of sediment, the flood-dominant currents ensuring that the sediment transport on the flood exceeds that on the ebb. The sediment is thus pumped landward until the asymmetry of the currents is reversed by the river flow – somewhere landward of the tip of the salt intrusion. The pumping process is affected by the settling of sediment to the bed and its reerosion, which introduces phase lags between the concentration and current velocity variations and enhances the asymmetry in the sediment transport rate during the tide. Measurements have shown that all of these processes are important, but tidal pumping is dominant in well-mixed estuaries. Both vertical gravitational circulation and tidal pumping are important in partially mixed estuaries. As a result the sediment in the turbidity maximum is derived from the coastal sea as much as from the rivers.

In microtidal salt wedge estuaries the surface fresh water carries sediment from the river seaward throughout the estuary, discharging into the sea at high river flow. The clearer water in the underlying salt wedge is often revealed in the wakes of ships.

Estuarine Modeling

The water flows transport salt, thus affecting the density distribution, and the densities affect the longitudinal pressure gradients that drive the water flow. Turbulent eddies produced by the flows cause exchanges of momentum as well as of salt, and produce frictional forces that help to resist the flow. Estuarine models attempt to predict the salt distribution, the water circulation and the sediment transport through application of the fundamental equations for mass, momentum, and sediment continuity, which formalize the above interactions. In principle, it is possible to measure directly most of the terms in these equations, apart from those that involve the turbulent exchanges. In practice, approximations are required to solve the equations. Either tidally averaged or within-tide conditions can be modeled. For sediment, the definition of the settling velocity and the threshold are required.

One-dimensional (1D) models assume that the estuarine characteristics are constant across the cross-section, and that the estuary is of straight prismatic shape. For tidally average conditions the assumption is that the seaward advection of salt on the mean flow is balanced by a landward dispersion of salt at a rate determined by the horizontal salinity gradient and a longitudinal eddy dispersion coefficient. This coefficient obviously incorporates the effects of turbulent mixing, the tidal oscillation and any actual nonuniformity of conditions.

Two dimensional (2D) models can either assume that conditions are constant with depth but vary across the estuary (2DH), or are constant across the estuary but vary with depth (2DV). The eddy dispersion and eddy viscosity coefficients will be different for the two situations. Using the 2DH model one would not be able to explore the effects of vertical gravitational circulation, whereas with the 2DV model the effects of the shallow water sides of the channel would be missing. In many cases a further approximation may be made by assuming the water is of constant density.

Three-dimensional models are obviously the ideal as they represent fully the vertical and horizontal profiles of velocity, density, and suspended sediment concentration. The only unknown terms are those relating to the turbulent mixing, the diffusion terms. Although computationally 3D models are becoming less costly to run, it is still difficult to obtain sufficient data to calibrate and validate them.

Once realistic flow models have been constructed, they can be used to explore the transport of sediment and estuarine water quality. However, the limitations on the field data restricts their use for prediction. For instance, the majority of the sediment travels very near to the bed, and it is difficult to measure the concentrations of layers only a few centimeters thick.

An active topic for research and modeling at present is the prediction of morphological change in estuaries resulting from sea level rise. This involves integrating the sediment transport over the tide, for many months or years, in which case the accumulated errors can become overriding. These results can then be compared with the simple empirical relationships stemming from considerations of estuarine equilibrium.

See also

Coastal Circulation Models. Fiord Circulation. Fiordic Ecosystems. Tides.

Further Reading

Dyer KR (1986) *Coastal and Estuarine Sediment Dynamics*. Chichester: Wiley.

Dyer KR (1997) *Estuaries: A Physical Introduction, 2nd edn*. Chichester: Wiley.

Kjerfve BJ (1988) *Hydrodynamics of Estuaries*, vol. I: *Estuarine Oceanography*; vol. II: *Hydrodynamics*. Boca Raton, FL: CRC Press.

Lewis R (1997) *Dispersion in Estuaries and Coastal Waters*. Chichester: Wiley.

Officer CB (1976) *Physical Oceanography of Estuaries (and Associated Coastal Waters)*. New York: Wiley.

Perillo GME (1995) *Geomorphology and Sedimentology of Estuaries*. Amsterdam: Elsevier.

GAS EXCHANGE IN ESTUARIES

M. I. Scranton, State University of New York, Stony Brook, NY, USA

M. A. de Angelis, Humboldt State University, Arcata, CA, USA

Introduction

Many atmospherically important gases are present in estuarine waters in excess over levels that would be predicted from simple equilibrium between the atmosphere and surface waters. Since estuaries are defined as semi-enclosed coastal bodies of water that have free connections with the open sea and within which sea water is measurably diluted with fresh water derived from land drainage, they tend to be supplied with much larger amounts of organic matter and other compounds than other coastal areas. Thus, production of many gases is enhanced in estuaries relative to the rest of the ocean. The geometry of estuaries, which typically have relatively large surface areas compared to their depths, is such that flux of material (including gases) from the sediments, and fluxes of gases across the air/water interface, can have a much greater impact on the water composition than would be the case in the open ocean. Riverine and tidal currents are often quite marked, which also can greatly affect concentrations of biogenic gases.

Gas Solubility

The direction and magnitude of the exchange of gases across an air/water interface are determined by the difference between the surface-water concentration of a given gas and its equilibrium concentration or gas solubility with respect to the atmosphere. The concentration of a specific gas in equilibrium with the atmosphere (C_{eq}) is given by Henry's Law:

$$C_{eq} = p_{gas}/K_H \qquad [1]$$

where p_{gas} is the partial pressure of the gas in the atmosphere, and K_H is the Henry's Law constant for the gas. Typically as temperature and salinity increase, gas solubility decreases. (Note that Henry's Law and the Henry's Law constant also may be commonly expressed in terms of the mole fraction of the gas in either the gas or liquid phase.) For gases that make up a large fraction of the atmosphere (O_2, N_2, Ar), p_{gas} does not vary temporally or spatially. For trace atmospheric gases (carbon dioxide (CO_2), methane (CH_4), hydrogen (H_2), nitrous oxide (N_2O), and others), p_{gas} may vary considerably geographically or seasonally, and may be affected by anthropogenic activity or local natural sources.

Gas Exchange (Flux) Across the Air/Water Interface

The rate of gas exchange across the air/water interface for a specific gas is determined by the degree of disequilibrium between the actual surface concentration of a gas (C_{surf}) and its equilibrium concentration (C_{eq}), commonly expressed as R:

$$R = C_{surf}/C_{eq} \qquad [2]$$

If $R = 1$, the dissolved gas is in equilibrium with the atmosphere and no net flux or exchange with the atmosphere occurs. For gases with $R < 1$, the dissolved gas is undersaturated with respect to the atmosphere and there is a net flux of the gas from the atmosphere to the water. For gases with $R > 1$, a net flux of the gas from the water to the atmosphere occurs.

Models of Gas Exchange

The magnitude of the flux (F) in units of mass of gas per unit area per unit time across the air/water interface is a function of the magnitude of the difference between the dissolved gas concentration and its equilibrium concentration as given by Fick's First Law of Diffusion:

$$F = k(C_{surf} - C_{eq}) \qquad [3]$$

where k is a first order rate constant, which is a function of the specific gas and surface water conditions. The rate constant, k, also known as the transfer coefficient, has units of velocity and is frequently given as

$$k = D/z \qquad [4]$$

where D is the molecular diffusivity (in units of cm^2 s^{-1}), and z is the thickness of the laminar layer at the air/water interface, which limits the diffusion of gas across the interface.

In aquatic systems, C_{surf} is easily measured by gas chromatographic analysis, and C_{eq} may be calculated readily if the temperature and salinity of the water are known. In order to determine the flux of gas (F) in or out of the water, the transfer coefficient (k) needs to be determined. The value of k is a function of the surface roughness of the water. In open bodies of water, wind speed is the main determinant of surface roughness. A number of studies have established a relationship between wind speed and either the transfer coefficient, k, or the liquid laminar layer thickness, z (**Figure 1**). The transfer coefficient is also related to the Schmidt number, Sc, defined as:

$$Sc = v/D \qquad [5]$$

where v is the kinematic viscosity of the water. In calmer waters, corresponding to wind speeds of <5 m s^{-1}, k is proportional to Sc$^{-2/3}$. At higher wind speeds, but where breaking waves are rare, k is proportional to S$^{-1/2}$. Therefore, if the transfer coefficient of one gas is known, the k value for any other gas can be determined as:

$$k_1/k_2 = Sc_1^n/Sc_2^n \qquad [6]$$

where n = the exponent. For short-term steady winds, the transfer coefficient for CO_2 has been derived as

$$k_{CO_2} = 0.31(U_{10})^2(Sc/600)^{-0.5} \qquad [7]$$

where U_{10} is the wind speed at a height of 10 m above the water surface. Eqns [6] and [7] can be used to estimate k for gases other than CO_2. In most estuarine studies wind speeds are measured closer to the water surface. In such cases, the wind speed measured at 2 cm above the water surface can be approximated as $0.5U_{10}$.

In restricted estuaries and tidally influenced rivers, wind speed may not be a good predictor of wind speed due to limited fetch or blockage of prevailing winds by shore vegetation. Instead, streambed-generated turbulence is likely to be more important than wind stress in determining water surface roughness. In such circumstances, the large eddy model may be used to approximate k as:

$$k = 1.46(D*u*l^{-1})^{1/2} \qquad [8]$$

where u is the current velocity, and l is equivalent to the mean depth in shallow turbulent systems. Much of the reported uncertainty (and study to study variability in fluxes) is caused by differences in assumptions related to the transfer coefficient rather than to large changes in concentration of the gas in the estuary.

Direct Gas Exchange Measurements

Gas exchange with the atmosphere for gases for which water column consumption and production processes are known can be estimated using a dissolved gas budget. Through time-series measurements of biological and chemical cycling, gas loss or gain across the air/water interface can be determined by difference. For example, in the case of dissolved O_2, the total change in dissolved O_2 concentration over time can be attributed to air/water exchange and biological processes. The contribution of biological processes to temporal changes in dissolved O_2 may be estimated from concurrent measurements of phosphate and an assumed Redfield stoichiometry, and subtracted from the total change to yield an estimate of air/water O_2 exchange.

Gas fluxes also may be measured directly using a flux chamber that floats on the water surface. The headspace of the chamber is collected and analyzed over several time points to obtain an estimate of the net amount of gas crossing the air/water interface enclosed by the chamber. If the surface area enclosed

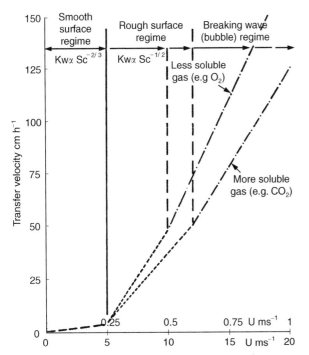

Figure 1 Idealized plot of transfer coefficient (Kw) as a function of wind speed (u) and friction velocity (u*). (Adapted with permission from Liss PS and Merlivat L (1985) Air–sea gas exchange rates. In: Buat-Menard P (ed.) The Role of Air–sea Exchange in Geochemical Cycling, p. 117. *NATO ASI Series C, vol. 185*. Dordrecht: Reidel.)

by the chamber is known, a net gas flux can be determined. Some flux chambers are equipped with small fans that simulate ambient wind conditions. However, most chambers do not use fans and so do not take account of the effects of wind-induced turbulence on gas exchange. Despite this limitation, flux chambers are important tools for measuring gas exchange in environments (such as estuaries or streams) where limited fetch or wind breaks produced by shoreline vegetation make wind less important than current-induced turbulence in shallow systems. While flux chambers may alter the surface roughness and, hence, gas/exchange rates via diffusion, flux chambers or other enclosed gas capturing devices also are the best method for determining loss of gases across the air/water interface due to ebullition of gas bubbles from the sediment.

Measurement of radon (Rn) deficiencies in the upper water column can be used to determine gas exchange coefficients and laminar layer thickness, which can then be applied to other gases using eqn [3]. In this method, gaseous ^{222}Rn, produced by radioactive decay of ^{226}Ra, is assumed to be in secular equilibrium within the water column. ^{222}Rn is relatively short-lived and has an atmospheric concentration of essentially zero. Therefore, in near-surface waters, a ^{222}Rn deficiency is observed, due to flux of ^{222}Rn across the air/water interface. The flux of ^{222}Rn across the air/water interface can be determined by the depth-integrated difference in measured ^{222}Rn and that which should occur based on the ^{226}Ra inventory. From this flux, the liquid laminar layer thickness, z, can be calculated.

Other volatile tracers have been used in estuaries to determine gas exchange coefficients. These tracers, such as chlorofluorocarbons (CFCs) or sulfur hexafluoride (SF_6), are synthetic compounds with no known natural source. Unlike ^{222}Rn, these gases are stable in solution. These tracers may be added to the aquatic system and the decrease of the gas due to flux across the air/water interface monitored over time. In some estuaries, point sources of these compounds may exist and the decrease of the tracer with distance downstream may be used to determine k or z values for the estuary.

Individual Gases

Methane (CH₄)

Atmospheric methane plays an important role in the Earth's radiative budget as a potent greenhouse gas, which is 3.7 times more effective than carbon dioxide in absorbing infrared radiation. Despite being present in trace quantities, atmospheric methane plays an important role controlling atmospheric chemistry, including serving as a regulator of tropospheric ozone concentrations and a major sink for hydroxyl radicals in the stratosphere. Methane concentrations have been increasing annually at the rate of approximately 1–2% over the last two centuries. While the contribution of estuaries to the global atmospheric methane budget is small because of the relatively small estuarine global surface area, estuaries have been identified as sources of methane to the atmosphere and coastal ocean and contribute a significant fraction of the marine methane emissions to the atmosphere.

Surface methane concentrations reported primarily from estuaries in North America and Europe range from 1 to >2000 nM throughout the tidal portion of the estuaries. Methane in estuarine surface waters is generally observed to be supersaturated (100% saturation ~2–3 nM CH_4) with R-values ranging from 0.7 to 1600 (**Table 1**). In general, estuarine methane concentrations are highest at the freshwater end of the estuary and decrease with salinity. This trend reflects riverine input as the major source of methane to most estuaries, with reported riverine methane concentrations ranging from 5 to 10 000 nM. Estuaries with large plumes have been observed to cause elevated methane concentrations in adjacent coastal oceans. In addition to riverine input, sources of methane to estuaries include intertidal flats and marshes, ground-water input, runoff

Table 1 Methane saturation values (R) and estimated fluxes to the atmosphere for US and European estuaries

Geographical region[a]	R[b]	Flux CH₄ ($\mu mol\,m^{-2}\,h^{-1}$)
North Pacific coast, USA	3–290	6.2–41.7[c]
North Pacific coast, USA	1–550	3.6–8.3[c]
Columbia River, USA	78	26.0[c]
Hudson River, USA	18–376	4.7–40.4[c]
Tomales Bay, CA, USA	2–37	17.4–26.3[c]
Baltic Sea, Germany	10.5–1550	9.4–15.6[c]
European Atlantic coast	0.7–1580	5.5[c]
Atlantic coast, USA	n.a.	102–1107[c]
Pettaquamscutt Estuary, USA	81–111	0.8–14.2[c]
		541–3375[d]

[a]For studies that report values for a single estuary, the major river feeding the estuary is provided. For studies that report values for more than one estuary, the oceanic area being fed by the estuaries is given.
[b]R = degree of saturation = measured concentration/atmospheric equilibrium concentration. n.a. indicates values not available in reference.
[c]Diffusive flux.
[d]Ebullition (gas bubble) flux.

from agricultural and pasture land, petroleum pollution, lateral input from exposed bank soils, wastewater discharge and emission from organic-rich anaerobic sediments, either diffusively or via ebullition (transport of gas from sediments as bubbles) and subsequent dissolution within the water column. Anthropogenically impacted estuaries or estuaries supplied from impacted rivers tend to be characterized by higher water-column methane concentrations relative to pristine estuarine systems. Seasonally, methane levels in estuaries are higher in summer compared with winter, primarily due to increased bacterial methane production (methanogenesis) in estuarine and riverine sediments.

Methane can be removed from estuarine waters by microbial methane oxidation and emission to the atmosphere. Methane oxidation within estuaries can be quite rapid, with methane turnover times of $<2\,h$ to several days. Methane oxidation appears to be most rapid at salinities of less than about 6 (on the practical salinity units scale) and is strongly dependent on temperature, with highest oxidation rates occurring during the summer, when water temperatures are highest. Methane oxidation rates decrease rapidly with higher salinities.

Methane diffusive fluxes to the atmosphere reported for estuaries (**Table 1**) fall within a narrow range of $3.6–41.7\,\mu mol\,m^{-2}\,h^{-1}$ ($2–16\,mg\,CH_4\,m^{-2}\,day^{-1}$). Using a global surface area for estuaries of $1.4 \times 10^6\,km^2$ yields an annual emission of methane to the atmosphere from estuaries of $1–8\,Tg\,y^{-1}$. Because the higher flux estimates given in **Table 1** generally were obtained close to the freshwater end-member of the estuary, the global methane estuarine emission is most likely within the range of $1–3\,Tg\,y^{-1}$, corresponding to approximately 10% of the total global oceanic methane flux to the atmosphere, despite the much smaller global surface area of estuaries relative to the open ocean.

Methane is also released to the atmosphere directly from anaerobic estuarine sediments via bubble formation and injection into the water column. Although small amounts of methane from bubbles may dissolve within the water column, the relatively shallow nature of the estuarine environment results in the majority of methane in bubbles reaching the atmosphere. The quantitative release of methane via this mechanism is difficult to evaluate due to the irregular and sporadic spatial and temporal extent of ebullition. Where ebullition occurs, the flux of methane to the atmosphere is considerably higher than diffusive flux (**Table 1**), but the areal extent of bubbling is relatively smaller than that of diffusive flux and, except in organic-rich stagnant areas such as tidal marshes, probably does not contribute

significantly to estuarine methane emissions to the atmosphere. Methane emission via ebullition has been observed to be at least partially controlled by tidal changes in hydrostatic pressure, with release of methane occurring at or near low tide when hydrostatic pressure is at a minimum.

Nitrous Oxide (N$_2$O)

Nitrous oxide is another important greenhouse gas that is present in elevated concentrations in estuarine environments. At present, N_2O is responsible for about 5–6% of the anthropogenic greenhouse effect and is increasing in the atmosphere at a rate of about 0.25% per year. However, the role of estuaries in the global budget of the gas has only been addressed recently.

Nitrous oxide is produced primarily as an intermediate during both nitrification (the oxidation of ammonium to nitrate) and denitrification (the reduction of nitrate, via nitrite and N_2O, to nitrogen gas), although production by dissimilatory nitrate reduction to ammonium is also possible. In estuaries, nitrification and denitrification are both thought to be important sources. Factors such as the oxygen level in the estuary and the nitrate and ammonium concentrations of the water can influence which pathway is dominant, with denitrification dominating at very low, but non-zero, oxygen concentrations. Nitrous oxide concentrations are typically highest in the portions of the estuary closest to the rivers, and decrease with distance downstream. A number of workers have reported nitrous oxide maxima in estuarine waters at low salinities ($<5–10$ on the PSU scale), but this is not always the case. The turbidity maximum has been reported to be the site of maximum nitrification (presumably because of increased residence time for bacteria attached to suspended particulate matter, combined with elevated substrate (oxygen and ammonium)).

Table 2 presents a summary of the data published for degree of saturation and air–estuary flux of nitrous oxide from a variety of estuaries, all of which are located in Europe and North America. Concentrations are commonly above that predicted from air–sea equilibrium, and estimates of fluxes range from $0.01\,\mu mol\,m^{-2}\,h^{-1}$ to $5\,\mu mol\,m^{-2}\,h^{-1}$. Ebullition is not important for nitrous oxide because it is much more soluble than methane. Researchers have estimated the size of the global estuarine source for N_2O based on fluxes from individual estuaries multiplied by the global area occupied by estuaries to range from $0.22\,Tg\,N_2O\,y^{-1}$ to $5.7\,Tg\,y^{-1}$ depending on the characteristics of the rivers studied. Independent estimates based on budgets of nitrogen input to rivers, assumptions about the fraction of

Table 2 Nitrous oxide saturation values (R) and estimated fluxes to the atmosphere for US and European estuaries

Estuary	R	Flux[a] N_2O ($\mu mol\,m^{-2}\,h^{-1}$)
Europe		
Gironde River	1.1–1.6	n.a.
Gironde River	$\approx$ 1.0–3.2	n.a.
Oder River	0.9–3.1	0.014–0.165
Elbe	2.0–16	n.a.
Scheldt	$\approx$ 1.0–31	1.27–4.77
Scheldt	$\approx$ 1.2–30	3.56
UK		
Colne	0.9–13.6	1.3
Tamar	1–3.3	0.41
Humber	2–40	1.8
Tweed	0.96–1.1	$\approx$ 0
Mediterranean		
Amvraikos Gulf	0.9–1.1	0.043 $\pm$ 0.0468
North-west USA		
Yaquina Bay	1.0–4.0	0.165–0.699
Alsea River	0.9–2.4	0.047–0.72
East coast USA		
Chesapeake Bay	0.9–1.4	n.a.
Merrimack	1.2–4.5	n.a.

[a]All fluxes given are for diffusive flux to the atmosphere. n.a. indicates that insufficient data were given to permit calculation of flux.

Table 3 Fluxes of carbon dioxide from estuaries in Europe and eastern USA

Estuary	R	Flux[a] CO_2 ($mmol\,m^{-2}\,h^{-1}$)
European rivers		
Northern Europe	0.7–61.1	1.0–31.7
Scheldt estuary	0.35–26.2	4.2–50
Portugal	1.6–15.8	10–31.7
UK	1.1–14.4	4.4–10.4
Clyde estuary	$\approx$ 0.7–1.8	n.a.
East coast USA		
Hudson River (tidal freshwater)	1.2–5.4	0.67–1.54
Georgia rivers	Slight supersaturation to 22.9	1.7–23

[a]n.a., insufficient data were available to permit calculation of this value.

inorganic nitrogen species removed by nitrification or denitrification, and the fractional 'yield' of nitrous oxide production during these processes indicate that nitrous oxide fluxes to the atmosphere from estuaries is about 0.06–0.34 Tg $N_2O\,y^{-1}$.

Carbon Dioxide (CO₂) and Oxygen (O₂)

Estuaries are typically heterotrophic systems, which means that the amount of organic matter respired within the estuary exceeds the amount of organic matter fixed by primary producers (phytoplankton and macrophytes). Since production of carbon dioxide then exceeds biological removal of carbon dioxide, it follows that estuaries are likely to be sources of the gas to the atmosphere. At the same time, since oxidation of organic matter to CO_2 requires oxygen, the heterotrophic nature of estuaries suggests that they represent sinks for atmospheric oxygen. In many estuaries, primary productivity is severely limited by the amount of light that penetrates into the water due to high particulate loadings in the water. In addition, large amounts of organic matter may be supplied to the estuary by runoff from agricultural and forested land, from ground water, from sewage effluent, and from organic matter in the river

itself. There are many reports of estuarine systems with oxygen saturations below 1 (undersaturated with respect to the atmosphere), but few studies in which oxygen flux to the estuary has been reported. However, estuaries are often dramatically supersaturated with respect to saturation with CO_2, especially at low salinities, and a number of workers have reported estimates of carbon dioxide flux from these systems (**Table 3**).

Dimethylsulfide (DMS)

DMS is an atmospheric trace gas that plays important roles in tropospheric chemistry and climate regulation. In the estuarine environment, DMS is produced primarily from the breakdown of the phytoplankton osmoregulator 3-(dimethylsulfonium)-propionate (DMSP). DMS concentrations reported in estuaries are generally supersaturated, ranging from 0.5 to 22 nM, and increase with increasing salinity. DMS levels in the water column represent a balance between tightly coupled production from DMSP and microbial consumption. Only 10% of DMS produced from DMSP in the estuarine water column is believed to escape to the atmosphere from estuarine surface water, since the biological turnover of DMS (turnover time of 3–7 days) is approximately 10 times faster than DMS exchange across the air/water interface. A large part of the estuarine DMS flux to the atmosphere may occur over short time periods on the order of weeks, corresponding to phytoplankton blooms. The DMS flux for an estuary in Florida, USA, was estimated to be on the order of <1 nmoles $m^{-2}\,h^{-1}$. Insufficient data are available to determine reliable global DMS air/water exchanges from estuaries.

Hydrogen (H₂)

Hydrogen is an important intermediate in many microbial catabolic reactions, and the efficiency of hydrogen transfer among microbial organisms within an environment helps determine the pathways of organic matter decomposition. Hydrogen is generally supersaturated in the surface waters of the few estuaries that have been analyzed for dissolved hydrogen with R-values of 1.5–67. Hydrogen flux to the atmosphere from estuaries has been reported to be on the order of 0.06–0.27 nmol m^{-2} h^{-1}. The contribution of estuaries to the global atmospheric H$_2$ flux cannot be determined from the few available data.

Carbon Monoxide (CO)

Carbon monoxide in surface waters is produced primarily from the photo-oxidation of dissolved organic matter by UV radiation. Since estuarine waters are characterized by high dissolved organic carbon levels, the surface waters of estuaries are highly supersaturated and are a strong source of CO to the atmosphere. Reported R-values for CO range from approximately 10 to >10 000. Because of the highly variable distributions of dissolved CO within surface waters (primarily as the result of the highly variable production of CO), it is impossible to derive a meaningful value for CO emissions to the atmosphere from estuaries.

Carbonyl Sulfide (OCS)

Carbonyl sulfide makes up approximately 80% of the total sulfur content of the atmosphere and is the major source of stratospheric aerosols. Carbonyl sulfide is produced within surface waters by photolysis of dissolved organosulfur compounds. Therefore, surface water OCS levels within estuaries exhibit a strong diel trend. Carbonyl sulfide is also added to the water column by diffusion from anoxic sediments, where its production appears to be coupled to microbial sulfate reduction. Diffusion of OCS from the sediment to the water column accounts for ~75% of the OCS supplied to the water column and is responsible for the higher OCS concentrations in estuaries relative to the open ocean. While supersaturations of OCS are observed throughout estuarine surface waters, no trends with salinity have been observed. Atmospheric OCS fluxes to the atmosphere from Chesapeake Bay have been reported to range from 10.4 to 56.2 nmol m^{-2} h^{-1}. These areal fluxes are over 50 times greater than those determined for the open ocean.

Elemental Mercury (Hg⁰)

Elemental mercury is produced in estuarine environments by biologically mediated processes. Both algae and bacteria are able to convert dissolved inorganic mercury to volatile forms, which include organic species (monomethyl- and dimethyl-mercury) and Hg^0. Under suboxic conditions, elemental mercury also may be the thermodynamically stable form of the metal. In the Scheldt River estuary, Hg^0 correlated well with phytoplankton pigments, suggesting that phytoplankton were the dominant factors, at least in that system. Factors that may affect elemental mercury concentrations include the type of phytoplankton present, photo-catalytic reduction of ionic Hg in surface waters, the extent of bacterial activity that removes oxygen from the estuary, and removal of mercury by particulate scavenging and sulfide precipitation. Fluxes of elemental mercury to the atmosphere have been estimated for the Pettaquamscutt estuary in Rhode Island, USA, and for the Scheldt, and range from 4.2–29 pmol m^{-2} h^{-1}, although the values are strongly dependent on the model used to estimate gas exchange coefficients.

Volatile Organic Compounds (VOCs)

In addition to gases produced naturally in the environment, estuaries tend to be enriched in by-products of industry and other human activity. A few studies have investigated volatile organic pollutants such as chlorinated hydrocarbons (chloroform, tetrachloromethane, 1,1-dichloroethane, 1,2-dichloroethane, 1,1,1-trichloroethane, trichloroethylene and tetrachloroethylene) and monocyclic aromatic hydrocarbons (benzene, toluene, ethylbenzene, o-xylene and m- and p-xylene). Concentrations of VOCs are controlled primarily by the location of the sources, dilution of river water with clean marine water within the estuary, gas exchange, and in some cases, adsorption onto suspended or settling solids. In some cases (for example, chloroform) there also may be natural biotic sources of the gas. Volatilization to the atmosphere can be an important 'cleansing' mechanism for the estuary system. Since the only estuaries studied to date are heavily impacted by human activity (the Elbe and the Scheldt), it is not possible to make generalizations about the importance of these systems on a global scale.

Conclusions

Many estuaries are supersaturated with a variety of gases, making them locally, and occasionally regionally, important sources to the atmosphere. However, estuarine systems are also highly variable in the amount of gases they contain. Since most estuaries studied to date are in Europe or the North American continent, more data are needed before global budgets can be reliably prepared.

Glossary

Air–water interface The boundary between the gaseous phase (the atmosphere) and the liquid phase (the water).

Catabolic Biochemical process resulting in breakdown of organic molecules into smaller molecules yielding energy.

Denitrification Reduction of nitrate via nitrite to gaseous endproducts (nitrous oxide and dinitrogen gas).

Ebullition Gas transport by bubbles, usually from sediments.

Estuary Semi-enclosed coastal body of water with free connection to the open sea and within which sea water is measurably diluted with fresh water derived from land drainage.

Gas solubility The amount of gas that will dissolve in a liquid when the liquid is in equilibrium with the overlying gas phase.

Henry's Law Constant Proportionality constant relating the vapor pressure of a solute to its mole fraction in solution.

Liquid laminar thickness The thickness of a layer at the air/water interface where transport of a dissolved species is controlled by molecular (rather than turbulent) diffusion.

Molecular diffusivity (D) The molecular diffusion coefficient.

Nitrification Oxidation of ammonium to nitrite and nitrate.

Practical salinity scale A dimensionless scale for salinity.

Redfield stoichiometry Redfield and colleagues noted that organisms in the sea consistently removed nutrient elements from the water in a fixed ratio (C : N : P=106 : 16 : 1). Subsequent workers have found that nutrient concentrations in the sea typically are present in those same ratios.

Transfer coefficient The rate constant which determines the rate of transfer of gas from liquid to gas phase.

Further Reading

Bange HW, Rapsomanikis S, and Andreae MO (1996) Nitrous oxide in coastal waters. *Global Biogeochemical Cycles* 10: 197–207.

Bange HW, Dahlke S, Ramesh R, Meyer-Reil L-A, Rapsomanikis S, and Andreae MO (1998) Seasonal study of methane and nitrous oxide in the coastal waters of the southern Baltic Sea. *Estuarine, Coastal and Shelf Science* 47: 807–817.

Barnes J and Owens NJP (1998) Denitrification and nitrous oxide concentrations in the Humber estuary, UK and adjacent coastal zones. *Marine Pollution Bulletin* 37: 247–260.

Cai W-J and Wang Y (1998) The chemistry, fluxes and sources of carbon dioxide in the estuarine waters of the Satilla and Altamaha Rivers, Georgia. *Limnology and Oceanography* 43: 657–668.

de Angelis MA and Lilley MD (1987) Methane in surface waters of Oregon estuaries and rivers. *Limnology and Oceanography* 32: 716–722.

de Wilde HPJ and de Bie MJM (2000) Nitrous oxide in the Schelde estuary: production by nitrification and emission to the atmosphere. *Marine Chemistry* 69: 203–216.

Elkins JW, Wofsy SC, McElroy MB, Kolb CE, and Kaplan WA (1978) Aquatic sources and sinks for nitrous oxide. *Nature* 275: 602–606.

Frankignoulle M, Abril G, Borges A, *et al.* (1998) Carbon dioxide emission from European estuaries. *Science* 282: 434–436.

Frost T and Upstill-Goddard RC (1999) Air–sea exchange into the millenium: progress and uncertainties. *Oceanography and Marine Biology: An Annual Review* 37: 1–45.

Law CS, Rees AP, and Owens NJP (1992) Nitrous oxide: estuarine sources and atmospheric flux. *Estuarine, Coastal and Shelf Science* 35: 301–314.

Liss PS and Merlivat L (1986) Air–sea gas exchange rates: introduction and synthesis. In: Baut-Menard P (ed.) *The Role of Air–Sea Exchange in Geochemical Cycling*, pp. 113–127. Dordrecht: Riedel.

Muller FLL, Balls PW, and Tranter M (1995) Processes controlling chemical distributions in the Firth of Clyde (Scotland). *Oceanologica Acta* 18: 493–509.

Raymond PA, Caraco NF, and Cole JJ (1997) Carbon dioxide concentration and atmospheric flux in the Hudson River. *Estuaries* 20: 381–390.

Robinson AD, Nedwell DB, Harrison RM, and Ogilvie BG (1998) Hypernutrified estuaries as sources of N_2O emission to the atmosphere: the estuary of the River Colne, Essex, UK. *Marine Ecology Progress Series* 164: 59–71.

Sansone FJ, Rust TM, and Smith SV (1998) Methane distribution and cycling in Tomales Bay, California. *Estuaries* 21: 66–77.

Sansone FJ, Holmes ME, and Popp BN (1999) Methane stable isotopic ratios and concentrations as indicators of methane dynamics in estuaries. *Global Biogeochemical Cycles* 463–474.

Scranton MI, Crill P, de Angelis MA, Donaghay PL, and Sieburth JM (1993) The importance of episodic events in controlling the flux of methane from an anoxic basin. *Global Biogeochemical Cycles* 7: 491–507.

Seitzinger SP and Kroeze C (1998) Global distribution of nitrous oxide production and N input in freshwater and coastal marine ecosystems. *Global Biogeochemical Cycles* 12: 93–113.

CHEMICAL PROCESSES IN ESTUARINE SEDIMENTS

W. R. Martin, Woods Hole Oceanographic Institution, Woods Hole, MA, USA

Introduction

The physical and chemical environments of estuarine sediments vary over wide ranges. The sediments may lie beneath nearly fresh water or may have salinities near that of the open ocean. They may be intertidal or always under several meters of water. They may be covered by grasses, by algal or bacterial mats, or may be free of surface biological cover. They may be permeable sands, affected by groundwater flow, or may be impermeable muds. Nonetheless, common features determine their chemical environments. They exist in productive, coastal waters and therefore experience large fluxes of particulate organic matter. They experience large inputs of terrestrial material that supply Fe and Mn oxides. Most estuarine sediments lie underneath oxic water columns, and support abundant macrofauna that both mix sedimentary particles and irrigate the seabed. Common features such as oxic bottom water, abundant supplies of organic matter, Fe and Mn oxides, a supply of sulfur from seawater (except for freshwater end members of estuarine systems), and active macrofauna produce the set of biogeochemical processes that determines the chemical environment and its effects on carbon, nutrient, and contaminant cycling in the coastal ocean.

The upper sediment column in estuarine and coastal environments can be regarded as a slowly stirred reactor to which substrates are added at the top and mixed downward. Transport processes, both in the dissolved and solid phases, distribute reactants within the reactor, while at the same time biogeochemical processes alter the particles and the pore waters surrounding them. This combination of transport and reaction drives solute exchanges between the sediments and overlying waters that are important to coastal carbon, nutrient, and contaminant cycles. It transforms the 'reactive' component of sedimentary particles so that the accumulating sediments differ in important ways from the particles that fall through the water column to the seafloor. These processes are outlined in **Figure 1**.

Because the rates of chemical reactions in this system are faster than mixing rates, the reactor is not chemically uniform; rather, roughly speaking, it has a layered structure. In all but a few locations, there is a thin, oxic layer at the sediment surface. As particles pass through this layer and through the underlying anoxic sediments, they are altered by heterotrophic respiration: organic matter is gradually oxidized, and a series of oxidants – O_2, N(V) in NO_3^-, Mn(IV) and Fe(III) in particulate oxides, and S(VI) in SO_4^{2-} – are consumed. Some of the energy originally stored in organic matter remains in the reduced products of these reactions: Fe(II), and S(−II), and, to some extent, Mn(II). This energy is available to chemolithotrophic bacteria, and the result is rapid internal cycling between oxidized and reduced forms of Mn, Fe, and S. Reduced Mn, Fe, and S that escape bacterial oxidation may be oxidized abiotically by O_2 in the surface oxic layer. The result of these processes is sediments with an oxic 'cap', in which precipitation of newly formed Fe and Mn oxides may limit benthic/water column exchange of many solutes, and an underlying anoxic layer in which rapid cycling between oxidized and reduced forms limits the rate of burial of products of heterotrophic respiration. The chemical processes defining this system are oxidation–reduction reactions, including 'heterotrophic respiration', by which organic matter is oxidized, 'chemolithotrophic respiration', most importantly oxidizing Fe(II) and S(−II), and 'abiotic oxidation'; and 'authigenic mineral formation', which converts dissolved products of these reactions to solid phases. The stirring within the sedimentary reactor, which is vital to its operation, is accomplished by solute diffusion and particle mixing, accomplished predominantly by the feeding and burrowing activities of sedimentary fauna.

In the following paragraphs, oxidation/reduction reactions in sediments and authigenic mineral formation are discussed. The effects of these reactions on the sedimentary environment are illustrated using profiles of solutes in sedimentary pore waters at a near-shore location. Then, there is a brief discussion of the role that sedimentary chemical processes play in nutrient cycles and the cycling of anthropogenic contaminants.

It is important to note that this article is not a comprehensive review, but instead focuses on the range of chemical reactions occurring in coastal sediments

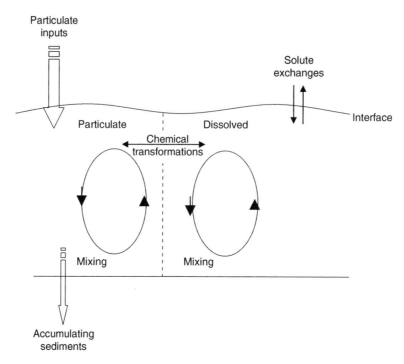

Figure 1 Biogeochemical cycles in estuarine and coastal sediments. Particles that fall to the sediment–water interface simultaneously undergo chemical reactions and are mixed through the upper few centimeters of the sediment column. Chemical transformations alter the particles and the sedimentary pore waters surrounding them, driving solute exchange between sediments and the water column and determining the composition of accumulating sediments.

and their role in creating sedimentary environments that are important in coastal carbon, nutrient, and contaminant cycles. Therefore, a single type of coastal sediment – muddy, impermeable sediments underlying oxic bottom water – is discussed. Other sediment types are important: for example, permeable, sandy sediments are widespread, and sediments underlying permanently or seasonally anoxic bottom water can be important. In some high-energy areas, the effects of sediment resuspension can dominate sediment–water-column exchange. Although this article emphasizes sediments as sites of respiration, primary production at the surface of sediments in shallow waters can be important. There are important differences between sedimentary environments that determine their involvement in coastal biogeochemical processes. Nonetheless, similar chemical processes occur in all of these environments. The goal of this article is to illustrate these fundamental chemical processes, so that their role in a range of sedimentary environments can be examined.

Oxidation/Reduction Reactions

The distinctive chemical environment of estuarine and coastal sediments is determined to a large extent by electron transfer, or 'redox' reactions, in which a chemical species with a greater affinity for electrons accepts these carriers of negative charge from a species with a lesser affinity. In the process, energy is released. Although these reactions are often slow, they are catalyzed by organisms – in particular, microbes – that can harness the released energy. To understand the chemical environment of sediments, it is necessary to know which redox reactions will occur.

Oxidants and Reductants

In chemical terms, we can consider some constituents of sediments as oxidants, capable of accepting electrons under conditions extant in the sediments, and reductants, capable of donating them. The primary sources of reductants are the organic matter, formed by production in the water column or on the sediment surface or by anthropogenic activities, that reaches the sediments, and the reduced products of sedimentary reactions. Oxidants diffuse from bottom water (O_2, NO_3^-, and SO_4^{2-}), arrive to the sediments with the particulate flux (Fe and Mn oxides), or are formed by sedimentary processes (NO_3^- and authigenic Fe and Mn oxides). The principal electron donors and acceptors are listed in **Table 1**.

pe: A Measure of the Affinity to Accept Electrons

Free electrons do not exist in aqueous solution. Nonetheless, since each electron acceptor (or donor)

Table 1 Oxidants and reductants in shallow-water sediments

e^- acceptor/donor	Formal oxidation state	Chemical form	Main sources
Oxidants			
O	0	Dissolved O_2	Solute transport from bottom water
N	$+V$	NO_3^-	Sedimentary nitrification
Mn	$+IV, +III$	Mn oxyhydroxides	Terrigenous material
			Cycling within sediments
Fe	$+III$	Fe oxyhydroxides	Terrigenous material
			Cycling within sediments
S	$+VI$	SO_4^{2-}	Solute transport from bottom water
			Cycling within sediments
Reductants			
C	A range of oxidation states	Organic matter	Marine production and anthropogenic
N			
S			
Mn	$+II$	Dissolved Mn^{2+}	Product of microbial respiration
Fe	$+II$	Dissolved Fe^{2+} FeS	Product of microbial respiration
S	$-II$	Dissolved HS^- FeS	Product of microbial respiration

can, in principle, react with several different electron donors (or acceptors), it is useful to consider 'half-reactions' of the sort

$$\tfrac{1}{4}O_2 + H^+ + e^- \Leftrightarrow \tfrac{1}{2}H_2O \qquad [1]$$

In this reaction, the oxygen atom starts in the 0 oxidation state and ends in the $-II$ state. To undergo this change, each O atom 'accepts' two electrons (from an unspecified source). Thus, 1/4 of a mole of O_2 molecules accepts 1 mol of electrons. In formal terms, we can write an 'equilibrium constant' for the half-reaction:

$$K = \frac{(H_2O)^{1/2}}{(O_2)^{1/4}(H^+)(e^-)} \qquad [2]$$

in which parentheses denote activities of the species. Using the definitions, $pH = -\log(H^+)$ and $pe = -\log(e^-)$, we can rewrite this equation:

$$pe = \log\{K\} - \log\left\{\frac{(H_2O)^{1/2}}{(O_2)^{1/4}}\right\} - pH \qquad [3]$$

The equilibrium constant, K, can – in principle – be determined from the free energies of formation of the reactants and products through the relationship between ΔG^0 and K. However, since free electrons do not exist in aqueous solution, chemists have devised a way of eliminating the free electron from the reaction. Half-reactions of the sort shown in eqn [1] are combined with with the oxidation of $H_2(g)$ to $H^+(aq)$. This reaction releases one electron and is 'defined' to have a free energy change of 0. Some algebraic manipulation will show that this combination of

half-reactions allows the calculation of ΔG^0 for reaction [1] as

$$\Delta G^0 = \tfrac{1}{2}G_f^0(H_2O(l)) - \tfrac{1}{4}G_f^0(O_2(aq)) \qquad [4]$$

Then,

$$\log(K) = -\frac{\Delta G^0}{2.303RT} \equiv pe^0 \qquad [5]$$

can be calculated from tabulated thermodynamic data. In these equations, ΔG^0 is the Gibbs free energy yield of the reaction, when all reactants and products are in their standard states, R is the gas constant, and T is temperature (in K). Now, eqn [3] can be rewritten as

$$pe = pe^0 - \log\left\{\frac{(H_2O)^{1/2}}{(O_2)^{1/4}}\right\} - pH \qquad [6]$$

A similar equation can be derived for any half-reaction, written as a reduction of the reactant involving a single electron. Then, three quantities can be calculated:

1. pe^0, with all reactants and products having activity $= 1$;
2. $pe(sw)$, which is applicable to reactions in seawater, and therefore is calculated with $pH = 7.5$ and $[HCO_3^-] = 2$ mM, but other species with activity $= 1$;
3. pe, with all species at concentrations that are observed in the environment.

pe^0 (calculation (1)) is not very useful for environmental calculations, as *in situ* activities are far from 1.

Table 2   Redox couples important to shallow water marine sediments.

Reaction	pe			
	pe^0	pe(sw)	pe	
$\frac{1}{4}O_2 + H^+ + e^- \rightarrow \frac{1}{2}H_2O$	21.49	13.99	$O_2 = 350\,\mu M$	13.1
			$= 1\,\mu M$	12.5
$\frac{1}{5}NO_3^- + \frac{6}{5}H^+ + e^- \rightarrow \frac{1}{10}N_2 + \frac{3}{5}H_2O$	20.81	11.81	$N_2 = 500\,\mu M$	
			$NO_3^- = 1\,\mu M$	10.94
			$= 30\,\mu M$	11.24
$\frac{1}{2}MnO_2 + 2H^+ + e^- \rightarrow \frac{1}{2}Mn^{2+} + H_2O$	21.54	6.54	$Mn^{2+} = 1\,\mu M$	9.54
			$= 50\,\mu M$	8.69
$FeOOH + 3H^+ + e^- \rightarrow Fe^{2+} + 2H_2O$	15.98	-6.52	$Fe^{2+} = 1\,\mu M$	-0.52
			$= 400\,\mu M$	-3.12
$\frac{1}{8}SO_4^{2-} + \frac{9}{8}H^+ + e^- \rightarrow \frac{1}{8}HS^- + \frac{1}{2}H_2O$	4.21	-4.23	$HS^-/SO_4^{2+} = 0.1\,\mu M/28\,mM$	-4.12
			$= 1\,mM/1\,mM$	-4.23
$\frac{1}{4}HCO_3^- + \frac{5}{4}H^+ + e^- \rightarrow \frac{1}{4}CH_2O + \frac{1}{2}H_2O$	1.43	-7.95	$HCO_3^- = 2\,mM$	-8.62
$\frac{1}{8}NO_3^- + \frac{5}{4}H^+ + e^- \rightarrow \frac{1}{8}NH_4^+ + \frac{3}{8}H_2O$	14.88	5.41		

The reactions are written as one-electron half-reactions. pe values are calculated from thermodynamic data from Stumm and Morgan (1996). Note that the reduced form of carbon is shown as 'CH$_2$O'. This commonly used shorthand denotes a carbohydrate (not formaldehyde), and its standard state free energy is taken to be one-sixth that of glucose ($C_6H_{12}O_6$, or, in this notation, '$(CH_2O)_6$').

pe(sw) is often used because the difference between this calculation (2) and pe (3) is often small.

pe^0, pe(sw), and pe values are listed in **Table 2** for the electron acceptors and donors in **Table 1**, and they are shown graphically in **Figure 2**. The order of decreasing pe (**Figure 2**) is the order of decreasing affinity of the oxidant species for electrons. All reactions in **Table 2** and **Figure 2** are written as reductions (a species accepts an electron) involving the transfer of one electron. Thus, complete redox reactions, in which one chemical species is reduced while a second is oxidized, can be constructed by combining the reaction as written in **Table 2** for the oxidant (which is reduced) with the reverse of a reaction that is below it on the pe scale. The resulting reaction will have $\Delta G < 0$:

$$\Delta G_{redox} = -2.303RT(pe_{reduction} - pe_{oxidation}) \quad [7]$$

When the species that is oxidized lies below the species that is reduced on the pe scale, the reaction will release energy and can occur either abiotically or via microbial catalysis.

The Oxidation of Organic Matter in Estuarine Sediments

Most of the organic matter falling to the seafloor in estuarine and coastal environments is formed by local primary production. A useful (but not quite accurate) representation of marine organic matter is

$$(CH_2O)_{106}(NH_3)_{16}(H_3PO_4) \quad [8]$$

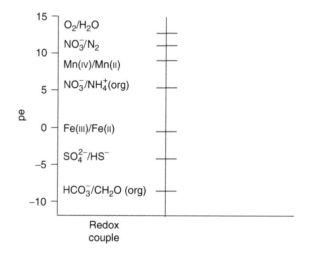

Figure 2   A graphical depiction of the pe of several environmentally important reduction reactions. A reduction as depicted in this figure, when combined with the oxidation of the reduced member of a couple listed lower on the figure, will yield a $\Delta G < 0$ under the conditions at which the pe values were calculated.

In this simple model, organic C is assumed to be in the form of a carbohydrate, and organic N in the form of ammonia (or a primary amine). In actuality, organic C has a lower average oxidation state than shown in this representation, but the model is still useful and widely used. We have used this organic matter stoichiometry, and neglected P since it is not oxidized during organic matter breakdown, to calculate the relative free energy yield for the oxidation of the C and N in organic matter by the different

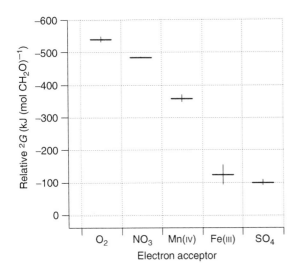

Figure 3 The relative free energy yield of the oxidation of a 'model' organic compound by the different electron acceptors. The model compound was $(CH_2O)(NH_3)_{16/106}$. ΔG values were calculated using the data in **Table 2**. NH_3 was assumed to be oxidized to NO_3^- by O_2 and to N_2 by NO_3^-.

electron acceptors. CH_2O is oxidized to CO_2 (converted to HCO_3^- in the marine environment) and NH_3 to NO_3^- (oxidation by O_2), N_2 (oxidation by NO_3^-), or NH_4^+ ($Mn(IV)$, $Fe(III)$, $S(VI)$). The result is shown in **Figure 3**. The vertical bars shown for each electron acceptor represent the range in ΔG expected over the ranges concentrations of reactants and products observed in sediments. There is a distinct order of ΔG. In order of free energy yield per mole of CH_2O, the order of yield by reaction with the different electron acceptors is

$$O_2 > NO_3^- > Mn(IV) > Fe(III) \geq SO_4^{2-} \quad [9]$$

Observations in a variety of sedimentary environments confirm that this order of electron acceptor use is followed. Apparently, organisms that can obtain a higher energy yield per mole of organic matter oxidized are favored over those obtaining less energy per mole. When an electron acceptor with a higher free energy yield is depleted, the next in the order is used.

Under the circumstances depicted in **Figure 1**, with reactants supplied at the sediment surface and mixed downward, the sequence of reactions by which organic matter is oxidized leads to a predictable sequence of solute concentrations. Starting from the sediment surface and moving downward: first, O_2 is removed from the pore waters; over the depth range where O_2 is removed, NO_3^- is added; it is then removed as NO_3^- is used as an electron acceptor; when NO_3^- is depleted, Mn^{2+} is added to the pore waters, followed by Fe^{2+}; then, SO_4^{2-} is removed and H_2S is added.

An example of this sequence of concentrations is shown in pore waters from a coastal site in **Figure 4**. The site from which the pore water profiles shown were taken lies under a 5-m water column with salinity ~ 31 psu and bottom water dissolved O_2 concentrations ranging from 240 to 390 μM during the year. The annually averaged organic carbon oxidation rate is high, $\sim 800\,\mu mol\,C\,cm^{-2}\,yr^{-1}$. Clearly, under these conditions, O_2 is very rapidly consumed in the sediments. The absence of dissolved Fe^{2+}, which is very rapidly oxidized by O_2, in the upper 2 mm of the sediments indicates that the O_2 flux from overlying water is consumed within the upper 2 mm of the sediment column. There is little NO_3^- in the water overlying the sediments, but, as for O_2, its supply to the sediments is consumed in the upper 2 mm. Dissolved Mn^{2+} appears in the upper 2 mm, indicating that its production by oxidation of organic matter (and potentially other reduced phases) occurs very near the sediment/water interface. The shape of the dissolved Fe^{2+} profile shows that its release begins within 3–7 mm of the sediment–water interface. Finally, the SO_4^{2-} profile shows that sulfate reduction begins by ~ 20 mm below the interface. In these high-carbon-flux environments, the redox zones are compressed into a very small region near the sediment–water interface; nonetheless, the pore water data are consistent with the order of use suggested by ΔG calculations. This order has been observed at many locations. The site shown falls in the class of 'sulfide-dominated' sediments, in which the sulfate reduction rate is rapid enough to lead to a buildup of H_2S in the pore waters. Other sediments, either with lower salinity (hence a smaller source of SO_4^{2-}) or lower organic matter oxidation rates (hence a lower sulfate reduction rate), do not show this sulfide buildup. These sites are often considered 'iron-dominated' sediments because the pore water dissolved Fe^{2+} concentration never drops to near 0 and H_2S never increases to high levels.

Authigenic Mineral Formation

Reduced Fe and Mn Phases

The dissolved SO_4^{2-} profile in **Figure 4** clearly shows that sulfate reduction, which produces H_2S, occurs in the region of the sediments where there are very large concentrations of Fe^{2+} in the pore waters. However, no H_2S is present in the pore waters in this depth region, and H_2S does not appear until the dissolved Fe^{2+} concentration begins its rapid decline to values $< 1\,\mu M$. The explanation for both the absence of H_2S and the decline of Fe^{2+} is the formation of solid-phase FeS. It is believed that the solid phase

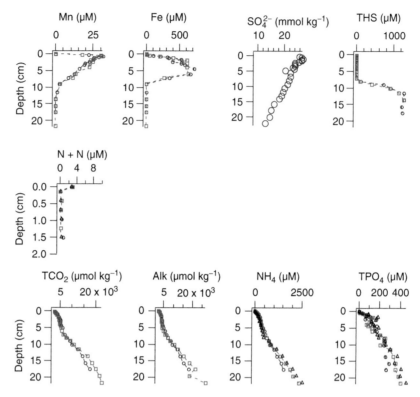

Figure 4 Pore water concentration vs. depth profiles, measured at a site in Hingham Bay in Boston Harbor, Massachusetts, in October 2001. Different Symbols represent replicate cores taken at the site, all at the same time. The depth ranges over which chemical reactions occur in the sediments and the rates of these reactions can be inferred from the solute profiles. The methods used for making these inferences are outlined in a classic paper by Froelich *et al.* Two important assumptions underlying the interpretation of the profiles are: (1) They are approximately in 'steady state'. That is, the rates of the transport and reaction processes whose balance sets the shape of the profiles are rapid relative to the rates at which environmental variability causes them to fluctuate over time. (2) The most important transport process in the sediments is vertically oriented molecular (or ionic) diffusion in the pore waters. Possible deviation from these simplifications can be seen in the pore water TCO_2, alkalinity ('Alk'), and NH_4^+ profiles, which all have inflections near the depth at which H_2S builds up in pore waters. These inflections could be caused either by transient changes in solute concentrations or by removal of the solutes from the sediments by 'sediment irrigation', a solute transport process due to the activities of animals living in the sediments. Data from Morford JL, Martin WR, Kalnejais LH, François R, Bothner M, and Karle I-M (2007) Insights on geochemical cycling of U, Re, and Mo from seasonal sampling in Boston Harbor, Massachusetts, USA. *Geochimica et Cosmochimica Acta* 71: 895–917.

formed is a fine-grained, poorly crystalline form of mackinawite (which we will call FeS_m). Studies of the dissolution of mackinawite have shown that, in alkaline solutions (such as seawater), FeS_m may be in equilibrium with either dissolved Fe^{2+} and HS^- or with the uncharged species, FeS_{aq}^0. In either case, the total dissolved Fe^{2+} is expected to be $\sim 1\,\mu M$. Thus, the relationship between $[Fe^{2+}]$, $[SO_4^{2-}]$, and total $[H_2S]$ is explained by precipitation of FeS_m until the supply of dissolved Fe^{2+} from dissimilatory Fe reduction is exhausted. Then, dissolved $[H_2S]$ begins to increase. In many cases, Fe(II) and S(−II) do not remain sequestered in mackinawite. Mackinawite is metastable, and is subject to oxidation, either to Fe(III) and S(VI) or to the stable mineral, pyrite (FeS_2). In 'sulfide-dominated' systems such as that shown in **Figure 4**, pyrite (FeS_2) can form either through

oxidation of S in FeS by H_2S:

$$FeS + H_2S \Rightarrow FeS_2 + H_2 \qquad [10]$$

or by the reaction of aqueous FeS 'clusters' with polysulfide:

$$FeS_{aq} + S_n^{2-} \Rightarrow FeS_2 + S_{n-1}^{2-} \qquad [11]$$

Thus, mackinawite is best viewed as an unstable intermediate product of iron and sulfate reduction. The fate of Fe and S present in mackinawite may be recycling to Fe(III) and S(VI), to be used again as electron acceptors for organic matter oxidation, or sequestration in pyrite.

As noted above, not all sediments – particularly those in low-salinity environments – have elevated

H$_2$S levels in pore waters. In these locations, Fe phosphate minerals may control Fe solubility. The combination of laboratory equilibration studies and pore water solute concentration measurements led Martens *et al.*, for example, to conclude that deep pore waters were in equilibrium with vivianite – Fe$_3$PO$_4 \cdot$8H$_2$O – in a coastal sediment. Hyacinthe *et al.* found that iron was sequestered as an Fe(III) phosphate in low-salinity, estuarine sediments. This ferric phosphate may have been formed in surface sediments (see below) or in the water column.

Figure 4 also shows the removal of Mn^{2+} from pore waters below the zone where it is added by reductive dissolution of Mn oxides. Equilibrium calculations have been used to infer that the mineral controlling Mn solubility in coastal and estuarine sediments is most likely to be a mixed Mn, Ca carbonate,

$$x\text{Mn}^{2+} + (1-x)\text{Ca}^{2+} + \text{CO}_3{}^{2-}$$
$$\Rightarrow \text{Mn}_x\text{Ca}_{(1-x)}\text{CO}_3 \qquad [12]$$

with a solubility product between those of calcite (CaCO$_3$) and rhodochrosite (MnCO$_3$):

$$\log K_{\text{Mn}_x\text{Ca}_{(1-x)}\text{CO}_3} = x \log K_{\text{MnCO}_3}$$
$$+ (1-x)\log K_{\text{CaCO}_3} \qquad [13]$$

Oxidized Phases

The removal of reduced Fe, Mn, and S to solid phases is readily apparent from the pore water solute profiles in **Figure 4**. Close inspection shows that dissolved Fe^{2+} is also removed from solution above its pore water maximum, as dissolved [Fe^{2+}] drops to near zero before Fe^{2+} reaches the sediment–water interface by diffusion. In addition, there is a small maximum in dissolved [SO$_4{}^{2-}$] just below the sediment–water interface, indicating its addition by oxidation of S($-$II). Finally, although the pore water Mn^{2+} profile does not clearly show its removal,

direct measurement of the Mn^{2+} flux across the sediment–water interface at this site (using *in situ* benthic flux chambers) showed that the actual flux is significantly smaller than is calculated by diffusion driven by the pore water concentration gradient: Mn^{2+} must be removed in a thin layer at the sediment–water interface.

These Fe and Mn removal processes are shown more clearly by pore water profiles at a site with a lower organic matter oxidation rate than the **Figure 4** site. **Figure 5** shows the seasonality in oxygen penetration depth that is common in shallow-water sediments. In March, the depth of removal of Fe^{2+} from pore waters coincides closely with the O$_2$ penetration depth, suggesting that there is rapid oxidation of Fe^{2+} by O$_2$ to insoluble oxides. In August, O$_2$ penetration is much shallower and Fe^{2+} production during organic matter oxidation is much more rapid, but the dissolved Fe^{2+} profile still shows evidence of removal (the upward curvature of the profile is indicative of removal from solution). In this case, the removal coincides with both the zone of Mn^{2+} release into the pore waters and with O$_2$ penetration; oxidation by both O$_2$ and Mn oxides may be occurring. The figure shows removal of Fe from solution by upward diffusion of dissolved Fe^{2+} and its oxidation. As at the **Figure 4** site, directly measured fluxes of dissolved Mn^{2+} across the sediment–water interface show that removal of dissolved Mn in a thin layer at the sediment–water interface impedes its transport from sedimentary pore waters to bottom water.

The removal of Fe(II), S($-$II), and Mn(II) when they are transported toward the sediment–water interface is predicted by the pe scale shown in **Figure 2**. The oxidation of Fe^{2+} (the reverse of the reaction in **Table 2**) could be coupled to the reduction of MnO$_2$,

$$\text{Fe}^{2+} + \tfrac{1}{2}\text{MnO}_2 + \text{H}_2\text{O} \Rightarrow \text{FeOOH} + \tfrac{1}{2}\text{Mn}^{2+} + \text{H}^+$$
$$\Delta G_{7.5} = -2.303RT(8.69 + 0.52) = -53\,\frac{\text{kJ}}{\text{mol Fe}}$$
$$[14]$$

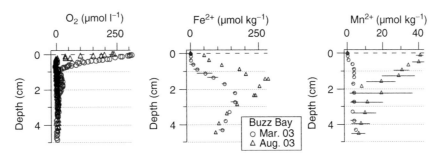

Figure 5 Pore water concentration vs. depth profiles, measured at a site in Buzzards Bay, Massachusetts, in Mar. and Aug., 2003.

or to the reduction of O_2,

$$Fe^{2+} + \tfrac{1}{4}O_2 + \tfrac{3}{2}H_2O \Rightarrow FeOOH + 2H^+$$

$$\Delta G = -2.303RT(13.1 + 0.52) = -78\frac{kJ}{mol\ Fe} \quad [15]$$

These reactions release energy, and can occur both abiotically or by microbial catalysis. Similar calculations using the data in **Table 2** and **Figure 2** show that $S(-\text{II})$ can be oxidized by Mn oxides or O_2, and $Mn(\text{II})$ can be oxidized by O_2. The rapid oxidation of Fe and S by O_2 is well known and occurs abiotically. Oxidation by O_2 was demonstrated by Aller using sediments from Long Island Sound. Aller and Rude used incubation experiments to demonstrate the occurrence of microbially catalyzed oxidation of solid-phase $Fe(\text{II})$ and $S(-\text{II})$ by Mn oxides.

Figure 5 illustrates these oxidative removal processes through loss of dissolved components of pore waters. Equally important is the upward mixing of solid phase Fe sulfides to zones of O_2 or Mn oxide reduction, followed by their oxidation. Particle mixing by bioturbation is an important mechanism for transport of reduced phases toward the sediment–water interface, particularly in warm-weather (and high productivity) months.

Measurement

The above discussion highlights the key role that authigenic solid phases play in sedimentary processes in estuarine sediments. Because rapid particle transport due to the activities of benthic fauna is present in virtually all of these sediments, solid phases are important participants in cycling between reduced and oxidized forms of Fe, S, and Mn. Unfortunately, the quantification of the solid phases in these sediments is difficult. Because the solids are predominantly made up of nonreactive, terrigenous minerals, and because the authigenic minerals tend to be very fine-grained and poorly crystalline, they cannot be measured directly. Instead, their presence is inferred by removal of dissolved constituents from pore waters and calculation of solubilities based on laboratory studies. The only means currently available to quantify their concentrations are selective chemical leaches of sediments. Several studies have evaluated the use of reducing agents to determine reactive Fe and Mn oxides in sediments. Hyacinthe *et al.* showed recently that ligand-enhanced reductive dissolution using a buffered ascorbate–citrate solution dissolves approximately the same amount of Fe as microbial processes, but the correspondence is still quite uncertain. The quantification of reduced phases is still more difficult. Cornwell and Morse compared

several procedures for quantifying reduced sulfur in sediments, concluding that dissolution in 6 N or 12 N HCl released S associated with FeS (and other relatively reactive reduced S species), but that the procedures also release Fe associated with oxides and some silicates. The procedure achieved reasonable separation of FeS from pyrite, which was dissolved by a more intense chemical leach. Rickard and Morse have emphasized the difficulty in interpreting the results of these procedures. Clearly, given the importance of cycling of solid phases in estuarine and coastal sediments, the quantification of their concentrations remains an important problem.

Elemental Cycling within Estuarine and Coastal Sediments

It is clear that chemical processes do not involve just alteration of particles that fall to the sediment–water interface. Rather, these systems are dominated by cycling between oxidized and reduced chemical forms within the sediment column. Heterotrophic respiration leads to the oxidation of organic matter, releasing C, N, and P into solution, and the reduction of O_2, Fe, Mn, and S. The latter three elements are then subject to transport, both in dissolved and solid forms, back toward the sediment–water interface, where they may be reoxidized either abiotically or by lithotrophic bacteria. The burial of reduced Fe, Mn, and S is a slow leak from these rapid internal cycles (**Figure 6**).

The upward transport and oxidation of reduced phases introduces complications into the determination of the relative importance of different organic matter oxidation pathways. For instance, O_2 is consumed not only for direct organic matter oxidation, but also for oxidation of reduced products of organic matter oxidation by Mn and Fe oxides and SO_4^{2-} ($Mn(\text{II})$, $Fe(\text{II})$, $S(-\text{II})$). Similarly, Mn oxides are consumed by oxidation of $Fe(\text{II})$ and $S(-\text{II})$ as well as by oxidation of organic matter. Thus, in considering the role played by each electron acceptor in organic matter decomposition, allowance must be made for oxidation of chemical constituents other than organic matter. Jorgensen used measurements of dissolved fluxes across the sediment–water interface to infer that oxidation of organic matter by O_2 and SO_4^{2-} were of roughly equal importance in shallow-water marine sediments. More recently, Thamdrup used sediment incubation techniques to show that those results overestimated the role of O_2 but underestimated that of Fe. The uncertainties in earlier estimates were both due to the failure to account for oxidation of species other than organic matter and due to the failure to account for the

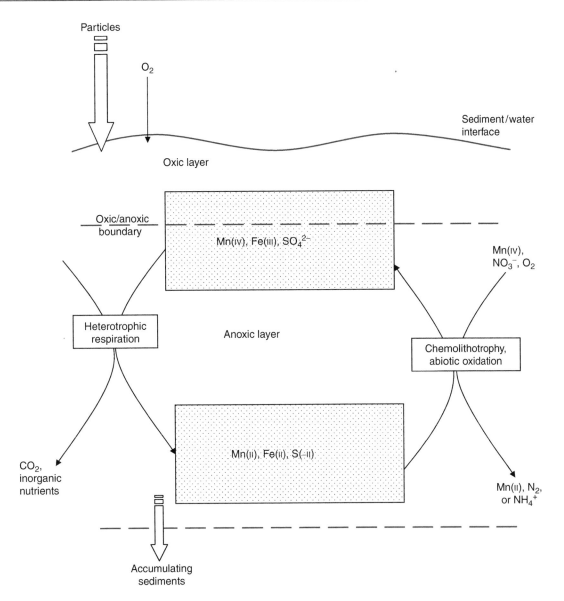

Figure 6 The cycling of Fe, Mn, and S within the sediment column. Transport of reduced Fe, Mn, and S occurs both in the dissolved and solid phases.

formation of solid-phase FeS. O_2 is still likely to be of great importance, however. For instance, while the $Fe(III)/Fe(II)$ pathway may be used directly for organic matter oxidation, it is the internal cycling between $Fe(III)$ and $Fe(II)$ that allows this pathway to be important: the ultimate sink for electrons is likely to be O_2. This linking of oxidation/reduction reactions in series is illustrated in **Figure 7** using a diagram of the sort introduced by Aller.

The Effect of Sedimentary Redox Cycling on Nutrient Cycles

Sedimentary respiration results in the decomposition of the majority of the organic matter that falls from the water column to the sediments. However, sedimentary processes can result in less efficient cycling of P and N than C, with potentially significant implications for coastal productivity.

Phosphate has a strong tendency to adsorb to Fe oxides. Therefore, the precipitation of Fe oxides near the interface of coastal and estuarine sediments can impede the return of phosphate, released into solution by the decomposition of organic matter, to the water column. **Figure 4** shows that dissolved reactive phosphate ('TRP') can reach very high levels in sediments to which there is a large flux of organic matter. In fact, the TRP level in **Figure 4** is significantly greater than would be predicted by the decomposition of organic matter alone. This phenomenon is further illustrated in **Figure 8**.

Figure 8(a) shows that P is highly enriched in the solid phase in the upper centimeter of the sediment column. This enrichment cannot be explained simply by larger organic matter concentrations in surface seciments. **Figure 8(b)** illustrates that internal P

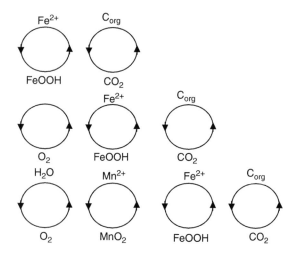

Figure 7 The oxidation of organic matter by FeOOH. In each panel, the oxidation of organic carbon ('C_{org}') to CO_2 is coupled to the reduction of FeOOH to Fe^{2+}. In the top panel, the source of the FeOOH is the particle flux to the sediments. The lower two panels show that internal cycling within the sediments can also supply Fe(III) for organic matter oxidation. In the middle panel, Fe(II) that was produced by organic matter oxidation is oxidized by O_2 to Fe(III), to be reused ot oxidize organic matter. In the lower panel, Fe(II) is oxidized to Fe(III) by MnO_2. The MnO_2 for this reaction, in turn, may be supplied either by the particle flux to the sediments or by oxidation of Mn(II) to Mn(IV) by O_2. The figure illustrates that the coupling of organic matter oxidation to Fe(III) reduction can be quantitatively important even if the supply of reactive Fe(III) through the fall or particles to the seafloor is small. When this occurs, the ultimate sink for the electrons for organic matter oxidation by the Fe(III)/Fe(II) pathway is O_2.

cycling with Fe establishes the enrichment. The dashed line in the figure shows the slope of a pore water TPO_4 versus TCO_2 plot if TPO_4 is simply released by organic matter oxidation. If the only process affecting dissolved TPO_4 were decomposition of organic matter, then the pore water concentration data would plot along a line of that slope from the origin. The pore water data deviate from that simple relationship in two ways: TPO_4 is essentially completely removed from pore waters near the sediment–water interface by upward diffusion and removal with freshly formed Fe oxides. When Fe(III) is reduced to oxidize organic matter, the rate of P release is much greater than can be accounted for by the organic matter alone: the Fe oxide-associated P is released. The result is a strong internal cycle of P. A fraction of the P that arrives at the sediment surface with organic matter is sequestered within the sediments by precipitation with Fe oxides. **Figure 8** illustrates this process at a marine sediment site. Hyacinthe *et al.* showed that P can also be sequestered in estuarine sediments when P that is removed with Fe oxides remains in the solid phase after incomplete reduction of the oxide phase.

The sediments are also important sites for ther removal of fixed nitrogen from coastal waters. Because there is relatively little NO_3^- in the shallow waters overlying coastal and estuarine sediments, diffusion of NO_3^- from bottom water is not a major source of N for denitrification. However, rapid organic matter oxidation results in the release of NH_4^+ to the pore waters. NH_4^+ can be converted to N_2 by a nitrification/denitrification cycle or by NH_4^+ oxidation coupled to the reduction of NO_3^-:

$$5NH_4^+ + 3NO_3^- \Rightarrow 4N_2 + 9H_2O + 2H^+ \quad [16]$$

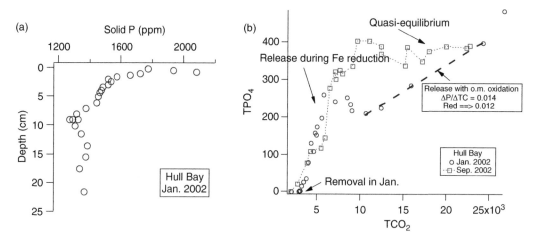

Figure 8 (a) Solid-phase P, measured at a site in Hingham Bay, Boston Harbor. (b) The relationship between the pore water concentrations of total reactive phosphate (TPO_4) and TCO_2 in pore waters at that site. Concentrations of both species are in $\mu mol\ l^{-1}$.

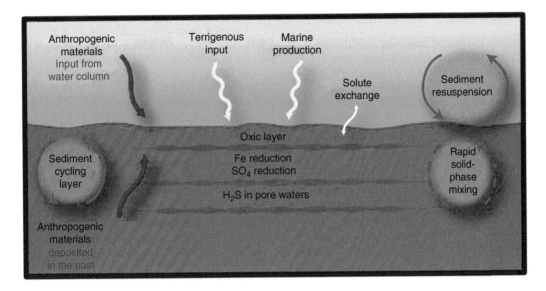

Figure 9 The structure of the chemical environment in shallow-water marine sediments.

Thamdrup and Dalsgaard illustrated the occurrence of this N_2 formation mechanism. The result of these processes is that 15–75% of the flux of N to the surface of coastal and estuarine sediments may be removed to N_2.

Contaminant Cycling: Anthropogenic Metals

The 'layered' structure of the chemical environment in estuarine and coastal sediments has important implications for the cycling of contaminants. Anthropogenic materials can both be present in the sediments from deposition in the past and can arrive with current deposits. They encounter an environment similar to that depicted in **Figure 9**. The upper centimeters of the sediment can be seen as an active, 'sediment-cycling layer', in which particles are rapidly mixed by bioturbation. This layer consists of an oxic 'cap' just below the sediment–water interface, where Fe and Mn oxides precipitate. Below that is a region of Fe and S reduction, but without buildup of dissolved H_2S. Below this layer – if the supply of organic matter supports extensive sulfate reduction – is a layer of elevated dissolved H_2S. An extensive discussion of contaminant cycling in estuarine and coastal sediments is beyond the scope of this article. Two examples are cited to show the importance of the redox structure of these environments to contaminant cycling.

Many contaminant metals – Cu, Ag, and Pb are examples – both form insoluble sulfides and tend to sorb onto precipitating Fe oxides. Thus, they are released to pore waters when Fe oxide reduction occurs, but tend to be immobilized by precipitation as sulfides. Dissolved fluxes to the overlying water are determined by a balance between their rates of diffusion from the site of Fe reduction toward the sediment/water interface and of removal by precipitation with Fe oxides. Because both the thickness of the oxic layer and the rate of Fe reduction vary seasonally and spatially, these metals have seasonally and spatially variable rates of release from sediments to the water column. Their precipitation with Fe oxides can lead to enrichments in fine-grained solids at the sediment–water interface. Thus, sediment resuspension and subsequent reaction and transport in the water column can be an important mechanism for redistribution of contaminant metals. Because the FeS that is formed by precipitation of the products of Fe(III) and SO_4^{2-} reduction is metastable, the formation of pyrite may be an important step in the immobilization of these metals. One example of a study of these processes in coastal systems is that of Kalnejais.

A different type of metal is Hg. The toxic, bioavailable form of Hg (methylmercury) may be formed in sediments when sulfate-reducing bacteria incorporate HgS^0 and methylate the Hg. This process appears to occur where sulfate reduction is important, but dissolved sulfide levels are less than $\sim 10\,\mu M$. Under these conditions, sediments can be an important source of methylmercury to the coastal water column.

See also

Phosphorus Cycle.

Further Reading

Aller RC (1980) Diagenetic processes near the sediment–water interface of Long Island Sound. Part II: Fe and Mn. In: Saltzman B (ed.) *Estuarine Physics and Chemistry: Studies in Long Island Sound*, vol. 22, pp. 351–415. New York: Academic Press.

Aller RC and Rude PD (1988) Complete oxidation of solid phase sulfides by manganese and bacteria in anoxic marine sediments. *Geochimica et Cosmochimica Acta* 52: 751–765.

Benoit JM, Gilmour CC, and Mason RP (2001) The influence of sulfide on solid-phase mercury bio-availability for methylation by pure cultures of *Desulfobulbus propionicus*. *Environmental Science and Technology* 35: 127–132.

Cornwell JC and Morse JW (1987) The characterization of iron sulfide minerals in anoxic marine sediments. *Marine Chemistry* 22: 193–206.

Elderfield H, Luedtke, McCaffrey RJ, and Bender M (1981) Benthic studies in Narragansett Bay. *American Journal of Science* 281: 768–787.

Froelich PN, Klinkhammer GP, Luedtke NA, *et al.* (1979) Early oxidation of organic matter in pelagic sediments of the eastern equatorial Atlantic: Suboxic diagenesis. *Geochimica et Cosmochimica Acta* 43: 1075–1090.

Hammerschmidt CR, Fitzgerald WC, Lamborg CH, Balcom PH, and Visscher PT (2004) Biogeochemistry of methylmercury in sediments of Long Island Sound. *Marine Chemistry* 90: 31–52.

Huerta-Diaz MA and Morse JW (1992) Pyritization of trace metals in anoxic marine sediments. *Geochimica et Cosmochimica Acta* 56: 2681–2702.

Hyacinthe C, Bonneville S, and Van Cappellen P (2006) Reactive iron(III) in sediments: Chemical versus microbial extractions. *Geochimica et Cosmochimica Acta* 70: 4166–4180.

Hyacinthe C and Van Cappellen P (2004) An authigenic iron phosphate phase in estuarine sediments: Composition, formation and chemical reactivity. *Marine Chemistry* 91: 227–251.

Jorgensen BB (1982) Mineralization of organic matter in the sea bed – the role of sulphate reduction. *Nature* 296: 643–645.

Kalnejais LH (2005) *Mechanisms of Metal Release from Contaminated Coastal Sediments*, p. 238. Woods Hole, MA: Massachusetts Institute of Technology, Woods Hole Oceanographic Institution.

Kostka JE and Luther GW (1994) Partitioning and speciation of solid phase iron in saltmarsh sediments. *Geochimica et Cosmochimica Acta* 58: 1701–1710.

Martens CS, Berner RA, and Rosenfeld JK (1978) Interstitial water chemistry of anoxic Long Island Sound sediments. Part 2: Nutrient regeneration and phosphate removal. *Limnology and Oceanography* 23(4): 605–617.

Morford JL, Martin WR, Kalnejais LH, François R, Bothner M, and Karle I-M (2007) Insights on geochemical cycling of U, Re, and Mo from seasonal sampling in Boston Harbor, Massachusetts, USA. *Geochimica et Cosmochimica Acta* 71: 895–917.

Rickard D (2006) The solubility of FeS. *Geochimica et Cosmochimica Acta* 70: 5779–5789.

Rickard D and Morse JW (2005) Acid volatile sulfide (AVS). *Marine Chemistry* 97: 141–197.

Seitzinger SP (1988) Denitrification in freshwater and coastal marine ecosystems: Ecological and geochemical significance. *Limnology and Oceanography* 33: 702–724.

Stumm W and Morgan JJ (1996) *Aquatic Chemistry: Chemical Equilibria and Rates in Natural Waters*, 3rd edn. New York: Wiley.

Thamdrup B (2000) Bacterial manganese and iron reduction in aquatic sediments. *Advances in Microbial Ecology* 16: 41–84.

Thamdrup B and Dalsgaard T (2002) Production of N_2 through anaerobic ammonium oxidation coouled to nitrate reduction in marine sediments. *Applied and Environmental Microbiology* 68: 1312–1318.

FIORD CIRCULATION

A. Stigebrandt, University of Gothenburg, Gothenburg, Sweden

Introduction

Fiords are glacially carved oceanic intrusions into land. They are often deep and narrow with a sill in the mouth. Waters from neighboring seas and locally supplied fresh water fill up the fiords, often leading to strong stratification. During transport into and stay in the fiord, mixing processes modify the properties of imported water masses. From the top downward, the fiord water is appropriately partitioned into surface water, intermediary water, and, beneath the sill level, basin water. Fiord circulation is forced both externally and internally. External forcing is provided by temporal variations of both sea level (e.g., tides) and density of the water column outside the fiord mouth. Internal forcing is provided by freshwater supply, winds, and tides in the fiord. The response of circulation and mixing in the different water masses to a certain forcing depends very much on characteristics of the fiord topography. The circulation of basin water is critically dependent on diapycnal mixing. Hence we focus on fiord circulation from a hydrodynamic point of view. Major

hydrodynamic processes and simple quantitative models of the main types of circulation are presented.

Basic Concepts in Fiord Descriptions

Some key elements of the hydrography and dynamics of fiords are shown in **Figure 1**. The surface water may have reduced salinity due to freshwater supply. It is kept locally well mixed by the wind and the thickness is typically of the order of a few meters. The thickness of the intermediary layer, reaching down to the sill level, depends strongly on the sill depth. This layer may be thin and even missing in fiords with shallow sills. Surface and intermediary waters have free connection with the coastal area through the fiord's mouth. Basin water, the densest water in the fiord, is trapped behind the sill. It is vertically stratified but the density varies less than in the layers above. A typical vertical distribution of density, $\rho(z)$, in a strongly stratified fiord is shown in **Figure 1**. The area (water) outside the fiord is denoted here as coastal area (water).

In most fiords, temporal variations of the density of the coastal water are crucial for the water exchange, both above and below the sill level. Surface and intermediary waters in short fiords with relatively wide mouths may be exchanged quickly (i.e., in days). The vertical stratification in the intermediary layer in such a fiord is usually quite similar to the

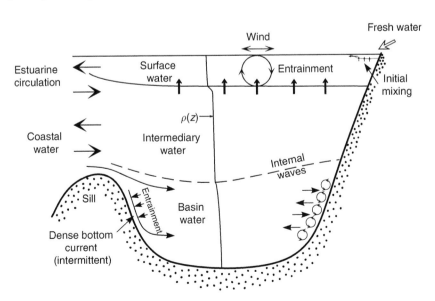

Figure 1 Basic features of fiord hydrography and circulation. Turbulent mixing in the basin water occurs mainly close to the bottom boundary, with increased intensity close to topographic features where internal tides are generated.

stratification outside the fiord although there is some phase lag, with the coastal stratification leading before that in the fiord. Short residence time for water above sill level means that the pelagic ecology may change rapidly due to advection. Denser coastal water, occasionally appearing above sill level, intrudes into the fiord and sinks down along the seabed. It then forms a turbulent dense bottom current that entrains ambient water, decreasing the density and increasing the volume flow of the current (**Figure 1**). The intruding water replaces residing basin water. During so-called stagnation periods, when the density of the coastal water above sill level is less than the density of the basin water, a pycnocline develops at or just below sill depth in the fiord. In particular, during extended stagnation periods, oxygen deficit and even anoxia may develop. The density of basin water decreases slowly due to turbulent mixing, transporting less-dense water from above into this layer. Tides are usually the main energy source for deep-water turbulence in fiords. Baroclinic wave drag, acting on barotropic tidal flow across sills separating stratified basins, is the process controlling this energy transfer.

Major Hydrodynamic Processes in Fiord Circulation

Understanding fiord circulation requires knowledge of some basic physical oceanographic processes such as diapycnal mixing and a variety of strait flows, regulating the flow through the mouth. Natural modes of oscillation (seiches) as well as sill-induced processes like baroclinic wave drag, tidal jets, and hydraulic jumps may constitute part of the fiord response to time-dependent forcing.

Flow through fiord mouths is driven mainly by barotropic and baroclinic longitudinal pressure gradients. The barotropic pressure gradient, constant from sea surface to seabed, is due to differing sea levels in the fiord and the coastal area. Baroclinic pressure gradients arise from differing vertical density distributions in the fiord and coastal area. The baroclinic pressure gradient varies with depth. Different types of flow resistance and mechanisms of hydraulic control may modify flow through a mouth. Barotropic forcing usually dominates in shallow fiord mouths while baroclinic forcing may dominate in deeper mouths.

Three main mechanisms cause resistance to barotropic flow in straits. First is the friction against the seabed. Second is the large-scale form drag, due to large-scale longitudinal variations of the vertical cross-sectional area of the strait causing contraction followed by expansion of the flow. And third is the baroclinic wave drag. This is due to generation of baroclinic (internal) waves in the adjacent stratified basins.

Barotropic flow Q through a straight, rectangular, narrow, and shallow strait of width B, depth D, and length L, connecting a wide fiord and the wide coastal area may be computed from

$$Q^2 = \frac{2g\Delta\eta B^2 D^2}{1 + 2C_D(L/D)} \qquad [1]$$

The sign of Q equals that of $\Delta\eta = h_o - h_i$, h_o (h_i) is the sea level in the coastal area (fiord). This equation contains both large-scale topographic drag and drag due to bottom friction (drag coefficient C_D). Large-scale topographic drag is the greater of the two in short straits (i.e., where $L/D < 2/C_D$).

To get first estimates of baroclinic flows and processes, it is often quite relevant to approximate a continuous stratification by a two-layer stratification, with two homogeneous layers of specified thickness and density difference $\Delta\rho$ on top of each other. The reduced gravity of the less-dense water is $g' = g\Delta\rho/\rho_0$, where g is the acceleration of gravity and ρ_0 a reference density.

In stratified waters, fluctuating barotropic flows (e.g., tides) over sills are subject to baroclinic wave drag. This is important in fiords because much of the power transferred to baroclinic motions apparently ends up in deep-water turbulence. Assuming a two-layer approximation of the fiord stratification, with the pycnocline at sill depth, the barotropic to baroclinic energy transfer E_j from the jth tidal component is given by

$$E_j = \frac{\rho_0}{2}\omega_j^2 a_j^2 \frac{A_f^2}{A_m}\frac{H_b}{H_b + H_t}c_i \qquad [2]$$

Here ω_j is the frequency and a_j the amplitude of the tidal component. A_f is the horizontal surface area of the fiord, A_m the vertical cross-sectional area of the mouth, H_t (H_b) sill depth (mean depth of the basin water), and

$$c_i = \sqrt{g'\frac{H_t H_b}{H_t + H_b}}$$

the speed of long internal waves in the fiord. Baroclinic wave drag occurs if the speed of the barotropic flow in the mouth is less than c_i. If the barotropic speed is higher, a tidal jet develops on the lee side of the sill together with a number of flow phenomena like internal hydraulic jumps and associated internal waves of supertidal frequencies.

Baroclinic flows in straits may be influenced by stationary internal waves imposing a baroclinic hydraulic control. For a two-layer approximation of the stratification in the mouth the flow is hydraulically controlled when the following condition, formulated by Stommel and Farmer in 1953, is fulfilled:

$$\frac{u_{1m}^2}{g'H_{1m}} + \frac{u_{2m}^2}{g'H_{2m}} = 1 \qquad [3]$$

Here u_{1m} (u_{2m}) and H_{1m} (H_{2m}) are speed and thickness, respectively, of the upper (lower) layer in the mouth. Equation [3] may serve as a dynamic boundary condition for fiord circulation as it does in the model of the surface layer presented later in this article. It has also been applied to very large fiords like those in the Bothnian Bay and the Black Sea. Experiments show that superposed barotropic currents just modulate the flow. However, if the barotropic speed is greater than the speed of internal waves in the mouth these are swept away and the baroclinic control cannot be established and the transport capacity of the strait with respect to the two water masses increases. In wide fiords, the rotation of the Earth may limit the width of baroclinic currents to the order of the internal Rossby radius $\sqrt{g'H_1}/f$. Here H_1 is the thickness of the upper layer in the fiord and f the Coriolis parameter. The outflow from the surface layer is then essentially geostrophically balanced and the transport is $Q_1 = g'H_1^2/2f$. This expression has been used as boundary condition for the outflow of surface water, for example, from Kattegat and the Arctic Ocean through Fram Strait.

Diapycnal (vertical) mixing processes may modify the water masses in fiords. In the surface layer, the wind creates turbulence that homogenizes the surface layer vertically and entrains seawater from below. The rate of entrainment may be described by a vertical velocity, w_e, defined by

$$w_e = \frac{m_0 u_*^3}{g'H_1} \qquad [4]$$

Here u_* is the friction velocity in the surface layer, linearly related to the wind speed, and m_0 (~ 0.8) is essentially an efficiency factor, well known from seasonal pycnocline models. H_1 is the thickness of the surface layer in the fiord and g' the buoyancy of surface water relative to the underlying water. Buoyancy fluxes through the sea surface modify the entrainment velocity.

Due to their sporadic and ephemeral character, there are only few direct observations of dense bottom currents in fiords. However, from both observations and modeling of dense bottom currents in the Baltic it appears that entrainment velocity may be described by eqn [4]. For this application, u_* is proportional to the current speed, H_1 is the current thickness, and g' the buoyancy of ambient water relative to the dense bottom current.

Observational evidence strongly supports the idea that diapycnal mixing in the basin water of most fiords is driven essentially by tidal energy, released by baroclinic wave drag at sills. The details of the energy cascade, from baroclinic wave drag to small-scale turbulence, are still not properly understood. Figure 1 leaves the impression that energy transfer to small-scale turbulence and diapycnal mixing takes place in the inner reaches of a fiord. Here internal waves, for example, waves generated by baroclinic wave drag at the sill, are supposed to break against sloping bottoms. A tracer experiment in the Oslo Fiord suggests that mixing essentially occurs along the rim of the basin. Recently, the temporal and spatial distributions of turbulent mixing in the basin waters of a few fiords have been mapped. Measurements in the Gullmar Fiord suggest that much of the mixing takes place close to the sill where most of the barotropic to baroclinic energy transfer takes place.

In a column of the basin water the mean rate of work against the buoyancy forces is given by

$$W = W_0 + \frac{Rf \sum_{j=1}^{n} E_j}{A_t} \qquad [5]$$

Here W_0 is the nontidal energy supply, n the number of tidal components, and E_j may be obtained from eqn [2]. A_t is the horizontal surface area of the fiord at sill level and Rf the flux Richardson number, the efficiency of turbulence with respect to diapycnal mixing. Estimates from numerous fiords show that $Rf \sim 0.06$. Experimental evidence shows that in fiords with a tidal jet at the mouth, most of the released energy dissipates above sill level and only a small fraction contributes to mixing in the deep water.

Simple Quantitative Models of Fiord Circulation

The Surface Layer

The upper layers in fiords may be exchanged due to so-called estuarine circulation, caused by the combination of freshwater supply and vertical mixing. This is essentially a baroclinic circulation, driven by density differences between the upper layers in the fiord and coastal area, respectively. However, if the sill is very shallow, the water exchange tends to be performed by barotropic flow with alternating direction, forced by the fluctuating sea level outside the fiord.

The following equations describe the steady-state volume and salt conservation of the surface layer:

$$Q_1 = Q_2 + Q_f \qquad [6]$$

$$Q_1 S_1 = Q_2 S_2 \qquad [7]$$

Here Q_f is the freshwater supply, Q_1 (Q_2) the outflow (inflow) of surface (sea-) water, and S_1 (S_2) the salinity of the surface water (seawater). Equations [6] and [7] give:

$$Q_1 = Q_f \frac{S_2 - S_1}{S_2} \qquad [8]$$

This equation has been used throughout the past century for diagnostic estimates of the magnitude of estuarine circulation from measurements of S_1, S_2, and Q_f.

Density varies with both salinity and temperature. In brackish waters, density (ρ) variations are often dominated by salinity (S) variations. For simplified analytical models, one may take advantage of this and use the equation of state for brackish water:

$$\rho = \rho_f (1 + \beta S) \qquad [9]$$

Here ρ_f is the density of fresh water and the so-called salt contraction coefficient β equals $0.0008\,S^{-1}$. The density difference $\Delta\rho$ between two homogeneous layers, with salinity difference ΔS, then equals $\rho_f \beta \Delta S$, and the buoyancy $g' = g\Delta\rho/\rho$ equals $g\beta\Delta S$. A continuous stratification in a salt-stratified system may be replaced by a dynamically equivalent two-layer stratification. This requires that the two-layer and observed stratification (1) contain the same amount of fresh water and (2) have the same potential energy.

Stationary estuarine circulation in fiords with deep sills To investigate how salinity S_1 and thickness H_1 of the surface layer depend on wind and freshwater supply, the following simple model may be illustrative. It is assumed that the fiord mouth is deep and even narrow compared to the fiord. Then the so-called compensation current into the fiord is deep and slow compared with the current of outflowing surface water. The hydraulic control condition in the mouth, eqn [3], is then simplified to $u_{1m}^2 = g' H_{1m}$. The thickness of the surface layer H_1 in the fiord is related to that in the mouth by $H_1 = \varphi H_{1m}$. Entrainment of seawater of salinity S_2 into the surface layer is described by eqn [4]. Under

these assumptions, expressions for the thickness H_1 and salinity S_1 of the surface layer in the fiord may be derived:

$$H_1 = \frac{G}{2Q_f g'} + \varphi \left(\frac{Q_f^2}{g' B_m^2} \right)^{1/3} \qquad [10]$$

$$S_1 = \frac{S_2 G}{G + 2\varphi \left[Q_f^5 \left(\frac{g'}{B_m} \right)^2 \right]^{1/3}} \qquad [11]$$

Here B_m is the width of the control section in the mouth, $G = CW_s^3 A_f$, $C = 2.5 \times 10^{-9}$ is an empirical constant containing, among others, the drag coefficient for air flow over the sea surface, W_s the wind speed and $g' = g\beta S_2$. Theoretically, the value of φ is expected to be in the range 1.5–1.7 for fiords where $B_m/B_f \leq 1/4$ where B_f is the width of the fiord inside the mouth. The value of φ should be smaller for wider mouths. Observations in fiords with narrow mouths give φ values in the range 1.5–2.5.

The left term in the expression for H_1 is the so-called Monin–Obukhov length, known from the theory of geophysical turbulent boundary layers with vertical buoyancy fluxes, and the right term is the freshwater thickness H_{1f}, hydraulically controlled by the mouth. The salinity of the surface layer S_1 increases with increasing wind speed and decreasing freshwater supply. For a given freshwater supply, strong winds may apparently multiply the outflow as compared with Q_f (cf. eqn [8]).

The freshwater volume in the fiord is $V_f = H_{1f} A_f$. The residence time of fresh water in the fiord, $\tau_f = V_f/Q_f$ is given by

$$\tau_f = \varphi A_f \left(\frac{1}{g' B_m^2} \right)^{1/3} Q_f^{-1/3} \qquad [12]$$

The residence time thus decreases with the freshwater supply. It should be noted that H_{1f}, τ_f, and V_f are independent of the rate of wind mixing.

Water exchange through very shallow and narrow mouths In very shallow and narrow fiord mouths, barotropic flow usually dominates. The instantaneous flow which can be estimated using eqn [1] is typically unidirectional and the direction depends on the sign of $\Delta\eta$, the sea level difference across the mouth. If the mouth is extremely shallow and narrow, tides and other sea level fluctuations in the fiord will have smaller amplitude than in the coastal area due to the choking effect of the mouth. A number of choked fiords and other semi-enclosed water bodies have been described in the literature, such as Framvaren and Nordaasvannet (Norway), Sechelt Inlet (Canada), the Baltic Sea, the Black Sea, and tropical lagoons.

The Intermediary Layer

Density variations in the coastal water above sill level give rise to water exchange across the mouth, termed intermediary water exchange. The stratification in the fiord strives toward that in the coastal water. Intermediary circulation increases in importance with increasing sill depth. It has been found that baroclinic intermediary circulation is the dominating circulation component in a majority of Scandinavian and Baltic fiords and bays. This is probably true also in other regions, although there have been few investigations quantifying intermediary circulation.

Despite its often-dominating contribution to water exchange in fiords, the intermediary circulation has remained astonishingly anonymous and in many studies of inshore waters even completely overlooked. One obvious reason for this is that a simple formula to quantify the mean rate of intermediary water exchange, Q_i, has been available only during the last decade (eqn [13]).

$$Q_i = \gamma \sqrt{B_m H_t A_f \frac{g \Delta M}{\rho}} \qquad [13]$$

Here the dimensionless empirical constant γ equals 17×10^{-4}, as estimated for Scandinavian conditions, and ΔM the standard deviation of the weight of the water column down to sill level (kg m^{-2}) in the coastal water. The latter should be a hydrodynamically reasonable measure of the mean strength of the baroclinic forcing. Statistics of scattered historic hydrographic measurements may be used to compute ΔM. Equation [13] should be regarded as a precursor to a formula, which is yet to be developed, accommodating for the frequency dependence of ΔM.

A conservative estimate of the mean residence time for water above the sill is $\tau_i = V_i/Q_i$, where V_i is the volume of the fiord above sill level. The residence time may be shorter if other types of circulation contribute to the water exchange.

The Basin Water

The time between two consecutive exchanges of basin water, often called the residence time, can be partitioned into the stagnation time (when there is no inflow into the basin) and the filling time (the time it takes to fill the basin with new deep water). The filling time is determined by the flow rate of new deep water through the mouth and the fiord volume

below sill level. In most fiords the filling time is much shorter than the stagnation time. The spectral distribution of the density variability in the coastal water determines the recurrence time of water of density higher than a certain value. Knowing the spectral distribution of the variability of the coastal density and the rate of density decrease of the basin water due to diapycnal mixing, one may estimate the recurrence time of water exchange. In fiords with short filling time, this should be inversely proportional to the rate of diapycnal mixing. A very long stagnation period may occur only if the rate of vertical mixing is very low and the density in the coastal water has a long-period component of appreciable amplitude.

A rough estimate of the mean rate of diapycnal mixing in the basin water may be obtained from eqn [14]:

$$\frac{d\rho}{dt} = -\frac{CW}{gH_b^2} \qquad [14]$$

Here the empirical constant C equals 2.0 and W may be obtained from eqn [5]. The vertical diffusivity κ at the level z in the basin water may be computed from the empirical expression in eqn [15]:

$$\kappa(z) = \frac{W/H_b}{\rho \bar{N}^2} c_\kappa \left(\frac{N(z)}{\bar{N}} \right)^{-1.5} \qquad [15]$$

Here $\bar{N}$ is the volume-weighted vertical average of the buoyancy frequency $N(z)$ and c_κ (~ 1) is an empirical constant.

If the filling time of the fiord basin is very long, the basin will be filled not only with the densest but also with less dense coastal water. The basin water in such a fiord will thus have lower density than the basin water in a neighboring fiord with similar conditions except for a much shorter filling time. If the filling time is sufficiently long, this will determine the residence time, which will not change if the rate of vertical mixing changes. The fiord basin is then said to be overmixed. The Baltic and the Black Seas are two overmixed systems for which the transport capacities of the mouths determine the residence time.

Conclusions

This article on fiord circulation demonstrates that several oceanographic processes of general occurrence are involved in fiord circulation. Being sheltered from winds and waves, fiords are excellent large-scale laboratories for studying these processes. The mechanics of water exchange of surface and intermediary layers, as described here, should apply equally well to narrow bays lacking sills.

Nomenclature

a_j	amplitude of the jth tidal component
A_f	horizontal surface area of the fiord
A_m	vertical cross-sectional area of the mouth
A_t	horizontal surface area of the fiord at sill depth
B	width of the mouth channel
B_f	width of the fiord inside the mouth
B_m	width of the mouth at the control section
c_κ	empirical constant (≈ 1)
C	empirical constant ($= 2.5 \times 10^{-9}$)
C_D	drag coefficient for flow over the seabed
D	depth of the mouth channel
E_j	barotropic to baroclinic energy transfer from the jth tidal component
g	acceleration of gravity
g'	($= g\Delta\rho/\rho_0$) buoyancy
G	wind factor
h_i	sea level in the fiord
h_o	sea level in the coastal area
$H_{1m}(H_{2m})$	thickness of upper (lower) layer in the mouth
H_b	mean depth of the basin water
H_t	sill depth
L	length of mouth channel
m_0	efficiency factor ($= 0.8$)
$N(z)$	buoyancy frequency at depth z
$\bar{N}$	volume-weighted buoyancy frequency in the deepwater
Q_1	flow out of the surface layer
Q_2	flow into the fiord beneath the surface layer
Q_f	freshwater supply
Rf	Richardson flux number, efficiency factor
$S_1 (S_2)$	salinity of upper (lower) layer
t	time
$u_{1m} (u_{2m})$	speed of upper (lower) layer in the mouth
u_*	friction velocity
V_f	volume of fresh water in the upper layer
W	rate of work against buoyancy forces in a column of basin water
Ws	wind speed
z	vertical coordinate
β	salt contraction of water ($= 0.0008$ S^{-1})
γ	empirical constant ($= 17 \times 10^{-4}$)
ΔM	standard deviation of weight of water column between sea surface and sill level
$\Delta\eta$	sea level difference across the mouth ($h_o - h_i$)
$\Delta\rho$	density difference
ρ_0	reference density
ρ_f	density of fresh water
$\rho(z)$	density as function of depth z
τ_f	residence time for fresh water in the upper layer
φ	contraction factor ($= 1.5$ for fiords with narrow mouths)
ω_j	frequency of the jth tidal component

Further Reading

Arneborg L, Janzen C, Liljebladh B, Rippeth TP, Simpson JH, and Stigebrandt A (2004) Spatial variability of diapycnal mixing and turbulent dissipation rates in a stagnant fiord basin. *Journal of Physical Oceanography* 34: 1679–1691.

Aure J, Molvær J, and Stigebrandt A (1997) Observations of inshore water exchange forced by fluctuating offshore density field. *Marine Pollution Bulletin* 33: 112–119.

Farmer DM and Freeland HJ (1983) The physical oceanography of fiords. *Progress in Oceanography* 12: 147–220.

Freeland HJ, Farmer DM and Levings DC (eds.) (1980) *Fiord Oceanography*, 715pp. New York: Plenum.

Stigebrandt A (1999) Resistance to barotropic tidal flow by baroclinic wave drag. *Journal of Physical Oceanography* 29: 191–197.

Stigebrandt A and Aure J (1989) Vertical mixing in the basin waters of fiords. *Journal of Physical Oceanography* 19: 917–926.

FIORDIC ECOSYSTEMS

K. S. Tande, Norwegian College of Fishery Science,
Tromsø, Norway

Introduction

Fiords and semienclosed marine systems are characterized by distinct vertical gradients in environmental
factors such as salinity, temperature, nutrients, and
oxygen; sampling of biological variables is considerably less adversely influenced by currents and weather conditions than in offshore systems. Therefore,
the fiords are particularly well-suited for detailed
studies of the pelagic habitat and have traditionally
attracted marine scientists, mainly as a site for curi-
osity-driven research. They are easily accessible and
have therefore been used as experimental laboratories
for testing new methodologies, exploring trophic relations, and building new theories.

Compared with the fisheries on the shelf outside or
in open oceanic waters, the catch from the fiords does
not seem proportionately great. Nevertheless, fiords
play an important role in many local communities in
productive coastal areas in Norway, Scotland, British
Columbia, and Greenland. There are fisheries locally
renowned for local herring stocks – Loch Fyne for
example – and for early year classes of Atlantic–
Scandian herring. The salmon fisheries of British
Columbia, Scotland, and Norway, highly seasonal
affairs catching the fish as they return from the open
sea to spawn in fresh water, are also a prominent
resource. In more recent time, fiords have become
refuse pits of cities and highly populated areas. A
growing awareness of the importance of these sites
for the bordering societies has led to a more cau-
tionary use of the fiords. This is in particular true for
the development of salmon and mariculture production facilities, where future demand for area is
being met in coastal developmental plans developed
from our knowledge of the physical and the biological setting of the particular fiord habitats.

The trophic relationships play an essential role in
shaping the pattern of species interactions. De-
lineating their existence in the form of food chains
and webs provides important knowledge of ecosystem functioning and dynamics. This basic knowledge is essential for the understanding that has to
form the basis for ecosystem and management

models for the marine biota; this article presents
some of this generic knowledge for fiordic
ecosystems.

General Features

Globally, most of the fiords are found in high latitudes:
in the Northern Hemisphere north of 50°N in Canada,
the United States, Greenland, Scandinavia, Scotland
and Svalbard; and south of 40°S in Peru, Chile, and
New Zealand. They are subjected to temperate, boreal, or arctic climates, and have many oceanographic
processes that occur over a wide range of space and
timescales. At the high-frequency end of the spectrum,
tidal flows can generate turbulence, internal waves,
and eddies, while processes such as renewal of oxygen
in deep enclosed basins may in extreme cases have
timescales of tens of years. Another property of fiords
is the presence of strong gradients in a number of
abiotic factors, which may affect organisms. The gradients may exist in the horizontal or vertical plane and
may be permanent or variable.

Topography and Estuarine Circulation

The fiord systems of the world are all quite young in
evolutionary time, yet they offer opportunities that
are quite unique among coastal systems. Their
common origin as glaciated coastal U-shaped valleys
and the common hydrographic conditions give them
a number of characteristic features. They are generally long, narrow, and often deeper than the continental shelf outside. A deep basin connects directly
or indirectly with the open sea at one end over a
relatively shallow sill, typically one-half to one-tenth
the basin depth (**Figure 1**). The sill, formed by mo-
raine material as it accumulated at the ice edge,
limits the exchange of water between the fiord and
the coastal region outside. Most fiords have large
rivers that often enter at the head of the fiord, but a
substantial amount of fresh water is drained into the
fiord from the bordering mainland.

The water masses can be separated into three vertical layers: brackish surface, intermediate, and basin
water. The brackish surface layer varies in time and
space, dictated by the fresh water runoff, wind, and
tide. At the river outlet, a local elevation of the surface
generates a density-driven fresh water current out of
the fiord, which is gradually being mixed with saline
water from below. This vertical mixing process is
called entrainment and leads to an overall increase in

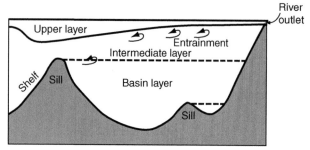

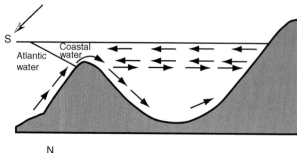

Figure 1 Basic topography and water layers of a fiord with estuarine circulation.

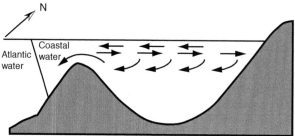

Figure 2 General pattern of wind-induced water exchanges between coastal and fiords on the western European continental margins, with prevailing northerly wind (upper panel) with exchange of basin water, and southerly wind (lower panel) with exchange of water in the intermediate layer.

salinity in the surface brackish water. Entrainment leads to a compensation current below that is going in the opposite direction. This pattern of water transport is called estuarine circulation and is a prominent feature in most fiords. Owing to the Coriolis force there is a tendency for water currents to be deflected to the right (in the Northern Hemisphere), which will generate a larger outflow of surface water on the right side of the fiords compared to the left side.

The tides vary locally, but will in general add only little to the net water transport. Therefore the wind-generated and estuarine circulations, are important for the exchange of water between fiords and the outside shelf. The wind-driven circulation is dictated by the wind stress on the surface and the vertical density gradient. The effect of the wind decreases vertically with increasing density gradient. Since the wind blows in or out the fiord, the wind-driven circulation sets up the estuarine circulation, as either an inward or an outward flowing surface layer depending on the magnitude, duration, and direction of the wind.

The main reason for water exchange in the intermediate layer is the existence of density gradients of water between the fiord and the shelf outside, which follow wind driven upwelling and downwelling at the shelf (**Figure 2**). Characteristically, this exchange occurs mainly as a two-layer circulation with inflow in the surface layer and a compensating outflow at the lower intermediate layer. This happens when the prevailing wind is northward along the western side of the continent (downwelling), and an oppositely directed circulation occurs when the wind is coming from the north (upwelling).

The sill creates a natural barrier that prevents water exchange between the basin and the corresponding depth outside. Partial or total renewal of the basin water takes place when water of higher density than the deep basin water is found above the sill. The major periods of deep-water renewals are found in late winter during upwelling conditions on the shelf outside. Total renewal of the deep water in the fiords happens occasionally, with a frequency of 5–10

years. The basin water in most fiords is being renewed at a rate that prevents anoxia. Nevertheless, in some fiords with very shallow sills, the renewal of the basin water is low and takes place through vertical diffusion. Here one can find anoxic conditions with high levels of hydrogen sulfide, due to the organic drain-off from land and sedimentation from the surface plankton production.

The above outline of the basic physical functioning of fiords underlines the open-ended nature of fiord communities, which nevertheless seem to maintain individual and group characteristics in terms of their biology. There may be a useful distinction between fiords *sensu stricto* and smaller, more shallow enclosed systems (i.e., 'polls'). The basic distinction between these two systems is that fiords have a sill depth greater than the depth of the pycnocline, while shallow enclosed systems normally have a sill depth less than the depth of the pycnocline. In biological terms, the latter are in general an ultraplankton–microzooplankton community, whereas the fiord community is chiefly a net phytoplankton–macro/megazooplankton community.

Fresh Water Runoff and Nutrient Cycling

The fresh water runoff plays a major modifying role in the dynamics of the lower trophic organisms in a fiord. The magnitude of this fresh water runoff varies depending on the extent of the surrounding landmass. Fiords can be classified as high-runoff and low-runoff, arbitrarily separated at an annual mean river runoff of $150 \, m^3 \, s^{-1}$. On the other hand, peak values of a few weeks' duration are found in the range from 500 to $20\,000 \, m^3 \, s^{-1}$. There is a strong seasonal variation in fresh water runoff, with annual maximum around April in southern fiords and around June in subarctic regions. Low runoff in the beginning of the season will enhance the stabilization of the water masses, facilitating the onset of the spring bloom. Later, increased runoff enhances the washout of the surface water layer, and modifies the phytoplankton biomass and species composition.

Fresh water runoff plays a minor direct role in the nutrient cycle of fiords, where the nitrate supplied has been found to account for a maximum of 10–20% of the potential uptake by phytoplankton in June in western Norwegian fiords. Indirect factors influencing the transport of new nutrients into the photic zone from below are therefore of major importance for the new production within the fiord. Although most of the nitrate loss out of fiords takes place below the euphotic zone, such losses may affect the future vertical transport of nitrate from the nonphotic to the photic zone. Hence, advective nutrient loss in the layer close to the photic zone may lead to reduced new production within the fiord. Since loss rates of nutrient appears to exceed loss rates from phytoplankton, an advective nutrient exchange, even below the euphotic zone, may therefore be of greater importance to the primary production than the exchange of the phytoplankton itself. The above picture prevails under specific wind conditions on the coast, and in case of reversed winds a nutrient loss may be turned into net nutrient supply to the fiord.

Seasonality in Energy Input to the System

Primary Production

Year-round measurements of primary production exist for a number of fiords and demonstrate that they are in the range of the general level of production in coastal waters of $100–150 \, g \, C \, m^{-2} \, y^{-1}$ (Table 1). Some of the data indicate that remarkably high levels of primary production can exist where boundary conditions or land runoff enhance the nutrient load to the fiord. Vertical stability can also generate off-seasonal blooms, even in November and December in fiords at lower latitudes, but it is unlikely that these blooms effectively stimulate secondary production.

A significant proportion of the annual phytoplankton production occurs before the macrozooplankton grazing population becomes established. The estimated annual flux of phytoplankton based

Table 1 Estimates of primary production in representative fiords on the Northern Hemisphere

Region	Year	Primary production ($g \, Cm^{-2} \, y^{-1}$)
Norway		
Balsfiorden	1977	110
Lindspolls	1976	90–100
Sweden		
Kungsbackafjord	1970	100
Greenland		
Godthaabfjord	1953–56	98
Canada		
How Sound		
entrance	1973	300
entrance	1974	516
inner stations	1973	118
inner stations	1974	163
Port Moody Arm	1975–76	532
Indian Arm and the Narrows regions	1975–76	260
Strait of Georgia	1965–68	120
River Plume	1975–77	149
USA		
Puget Sound	1966–67	465
Port Valdez	1971–72	150
Valdez Arm	1971–72	200

carbon to the bottom of fiords is $\sim 10\%$ of the total particulate material sedimentation. This means the majority of the particulate material reaching the bottoms of the fiord is inorganic. Although there is a strong component of copepods in fiord communities, krill, when present, are the major pellets producers contributing to the recognizable organic matter reaching the fiord bottom. Surface sediments show negligible seasonal variation in total organic matter, organic carbon and nitrogen, amino acids, and lipids. This may be due to a rapid conversion of sediment material into a pool of sediment microorganisms. The macrobenthos in the deep basin can be dominated by specialized deposit-eaters, such as the echinoderm (*Ctenodiscus crispatus*), which accumulates fatty acids indicative of an extensive microbial input.

Variability in Time and Space

Scaling of Exchange Processes

The pelagic community in fiords is often found to have a higher variability in time and space than that in oceanic water. This is mostly related to differences in advection between the habitats, where the strongest flushing is usually found in fiords.

As the physical scale of fiords varies, the balance between internal and external forcing is also likely to vary. The cross-sectional area above the sill is therefore an important boundary property, and the ratio between the cross-sectional area and the total fiord volume may indicate the impact of the sill boundary conditions on the fiord. This ratio varies considerably from one fiord to another (**Table 2**). In a sill fiord, the cross-sectional area of the fiord mouth (A) is believed to constrain the water exchange over the sill. The advective impact has been found to be larger in fiords with a high ratio of cross-sectional area to total fiord volume (see A/V in **Table 2**).

The extent to which the planktonic part of a fiord system is controlled by internal biological processes rather than advective processes depends on the physical scale of the fiord versus the timescale of these processes. As a general simplification we may write eqns [1] and [2].

$$\frac{\delta B}{\delta t} = rB + 0.5vR(B_B - B) \qquad [1]$$

$$R = \frac{A}{V} \qquad [2]$$

Here B = biomass concentration within the system (mg m^{-3}); t = time (s); r = local instantaneous growth rage of B (s^{-1}); v = mean absolute current above the sill (m s^{-1}); B_B = biomass concentration in incoming current (mg m^{-1}); A = cross-sectional area above the sill (m^2); and V = fiord volume (m^3).

As B approaches B_B, the net advective effect becomes zero. This does not mean, however, that the advective effect has ceased, since the biomass renewal within the system may still be dominated by advection rather than local growth. The growth rate (r) and the advective rate ($\beta = 0.5\,vR$) have the same dimension (s^{-1}) and the ratio $r/\beta > 1$ decides which of the two processes dominates the biomass formation within the system. If $r/\beta > 1$, growth is the dominating process, while $r/\beta < 1$ indicates advective dominance.

The importance of advection relative to the growth of phytoplankton and zooplankton is given in **Figure 3**. The much lower growth rate of zooplankton compared with that of phytoplankton implies that the transport influences primarily the zooplankton biomass. Phytoplankton is, however, constrained to the upper, photic, zone where transport processes are most prominent. Zooplankton may utilize the entire water column and the advective influence may thereby be diminished. The zooplankton confined to the advective layer (dotted line) in **Figure 3** is influenced three times more strongly by advection than is zooplankton distributed in the entire fiord volume (solid line). Similarly, vertical migration (diel and seasonal) may also reduce the influence of advection for populations depending on the food availability in the advective layer.

From eqns [1] and [2] we see that the value of R (ratio between the cross-sectional area above the sill and the fiord volume) gives the order of magnitude of the advective influence in a particular system. This ratio varies considerably from one fiord to another (**Table 2**), and it may serve as an index indicating the potential advective influence on a system.

Impact of Tidal Currents

The tidal amplitude varies by approximately a factor of 2 within the regions where fiords are found. Tidal currents result in no net water exchange when averaged over a tidal period (12.42 h). On the other hand, there is a significant vertical difference in tidal currents, where the highest flow rates are found in surface layer, with concurrent flow in the opposite direction close to the bottom. Therefore, total exchange of organisms is not necessarily proportional to the net exchange of water, and the vertical position of the organisms plays a major role for the advection rate.

Table 2 Examples of ratios of cross-sectional area (A) to total fiord volume (V) in Norwegian fiords

Fiord	A/V	Advective influence
Lindåspolls	10^{-7}	Advective influence of *Calanus* population < internal processes
Ryfylke fjord	2×10^{-6}	Advective exchange of zooplankton < internal production
Masfjord	8×10^{-6}	Advective exchange of zooplankton = internal production
Malangen	4×10^{-5}	Advective exchange of zooplankton biomass > internal production
Korsfjorden	10^{-4}	*Calanus* heavily influenced by advective processes

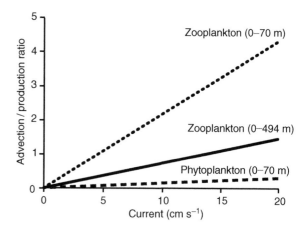

Figure 3 The relation between the ratio advection/production and current velocity for phytoplankton and zooplankton in a fiord. Modified from Aksnes *et al.* (1989).

The renewal rate of water and zooplankton can be calculated as a percentage of the total water and zooplankton biomass volume in a fiord with moderate to high tidal amplitudes (**Table 3**). Comparing contrasting periods in spring and autumn, the total renewal rate of the water is ~6% and 14% per day in spring and autumn, respectively. The numbers for the two given size categories of zooplankton are 6% in spring, 3.5% for zooplankton >500 μm and 12% for zooplankton between 180 and 500 μm in size. The importance of the tidal current for water and zooplankton transport can be estimated by comparing the residual current (i.e., the current velocity after the removal of the tidal component) and the total current. For example, the tidal water is responsible for 23% and 66% of the renewal of zooplankton >500 μm during spring and autumn, respectively. Equivalent numbers for the smallest size fraction of zooplankton are 19% and 39%. This demonstrates that the tide can play an important role in the transport of zooplankton in fiords.

Impact of Vertical Behavior

The zooplankton community structure in fiords appears to be determined primarily by the species–environment relationship, rather than species–species relationships as in the central oceanic gyres. Different spatial distribution patterns are often observed along the length of the fiord, where populations decrease as they are moved away from their population centers. This along-fiord difference in abundance usually evolves from homogenous low stocks of overwintering populations early the same year.

Neritic species prevail at the inner regions, with a downstream decline in the abundance. Most oceanic species have been unable to establish viable populations even in the deepest part of the fiords. However, some, mesopelagic oceanic species have succeeded and the species-specific along-fiord differences in abundance vary and are linked to their differences in vertical behavior.

Table 3 Advective daily transport of water and zooplankton biomass in Malangen, expressed as percentage of the entire volume inside the fjord

	Spring (%)		Autumn (%)	
	Total	Residual	Total	Residual
Water	6.1	3.6	14.6	8.1
Zooplankton >500 μm	6.6	5.1	3.5	1.2
Zooplankton 180–500 μm	6.4	5.2	12.0	7.4

Most zooplankton migrates on different timescales, and by diel vertical migration (DVM) the population tends to move from deep waters during daytime to surface during night. The residence time during day and night is normally much longer than the transition time between upper and deep waters. Combining the residence time of the animals with the total current in the same depth strata of the water column, a simple prediction of the displacement over time can be estimated. For a species with wide and regular vertical migration, for instance, *Chiridius armatus*, there is a strong tendency to reside in the same geographical region over time (**Figure 4**). This species does best at deeper and more oceanic sites, and has been found to reduce its population size by a factor of 4 over a distance of 3–4 km, with an overall decline in abundance toward the head of the fiord.

Other species have very low migration amplitude, where ontogenetic migration tends to keep the recruits in surface water for a long time, with an induction of a slow downward migration at older life stages. This is the case for the oceanic species *Calanus finmarchicus*, which has its population center in the North Atlantic and the Nordic Seas but is a very widely distributed and quantitative species in the fiords of the Northern Hemisphere. *C. finmarchicus* spends as much as 8–9 months in the deep waters

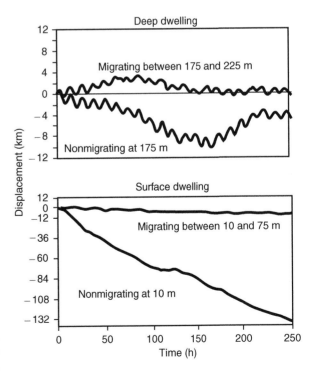

Figure 4 The travel distance for copepods migrating between 175 and 225 m and residing at 175 m during a 24 h cycle (upper panel), and for copepods migrating between 10 and 75 m and residing at 10 m during a 24 h cycle (lower panel).

>500 m while reproduction and growth takes place at the surface from April to July.

C. finmarchicus is a very prominent species on the shelves and in the fiords during the productive season but descends during summer and early autumn, being drained off from the shallower areas toward overwintering depths in deep basins in the fiords and the continental shelves. In some fiords the biomass build-up can reach 10–20% per day, and the high biomass, in particular in the basins during the autumn, clearly indicates that physical/biological aggregation mechanisms overrule the local production. The mechanisms can be explained by the same simple model given above (**Figure 4**), where a declining flush rate with depth tends to retain older stages as they initiate the downward migration during the summer. The biomass aggregation will thus continue for several months, draining older stages from the upper and intermediate water layers in the fiord.

Food Web Structure and Functioning

The Zooplankton Community

The community structure varies extensively between fiords but reflects mostly the shelf habitats found at similar latitudes (**Figure 5**). In subarctic waters, the zooplankton is composed of few species, but with high biomass. Small copepods may be abundant, especially during summer and autumn, and are not major pathways to the juvenile and adult fish. The larger copepods forms (i.e., *Calanus finmarchicus* and *Metridia longa*), the chaetognath *Sagitta elegans*, and the two krill species (*Thysanoessa* spp.) form easily identifiable trophic links in the transfer of materials to higher trophic levels. They all spawn during spring, matching the spring bloom to variable degrees, and each has a restricted growth period within the time-window from April to October. During the long overwintering period a marked decrease in organic lipid-based reserves takes place in both copepods and krill, accounting for 40–70% of that present at the end of the primary production season. Copepods and krill are often found as sound-scattering layers (SSLs) in the basin water of the fiord, and are heavily preyed upon both by demersal and pelagic fish.

The Higher Trophic Animals

In fiords there may be around 30 species that have a commercial potential, although only half of these are exploited regularly by man. Some pelagic and demersal fishes are separated into ocean and coastal stocks, where the former spends their time mostly in

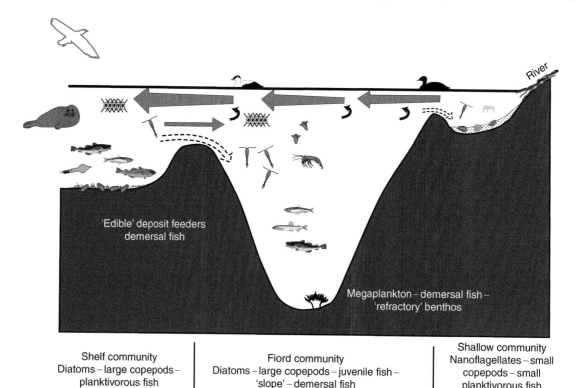

'Edible' deposit feeders
demersal fish

Megaplankton – demersal fish –
'refractory' benthos

Shelf community
Diatoms – large copepods –
planktivorous fish

Fiord community
Diatoms – large copepods – juvenile fish –
'slope' – demersal fish

Shallow community
Nanoflagellates – small
copepods – small
planktivorous fish
or medusae

Figure 5 A generalized structure of the biological community from the shelf to the inner part of the fiord. Modified from Matthews and Heimdal (1980).

the ocean and visit the coast during spawning or overwintering. The latter are more confined to the fiords and coastal zones during their entire lifetime.

The communities in fiords may have from one to several apex predators. In some fiords cod plays this role, and as many as 70 taxa have been found in cod stomachs from Balsfiorden, northern Norway. Although as many as 15 fish species were identified, krill, capelin, and herring are considered its dominant prey. This emphasizes that cod is weakly linked with the benethic community. Prawns, also, may be more trophically dependent on the pelagial, a feature contrary to the general tendency in the literature to classify them as part of the benthic food chain. The cod stocks in fiords are mostly coastal cod and undertake very little migration. Tagging experiments have documented that individuals are confined to an area from 5 to 8 km in extent during their entire lifetime. Artificially released cod tend to migrate over longer distance and are more easily trapped by fishing gears and other predators, and are thereby subjected to a higher risk of predation. Cod stocks in fiords may have lower growth rate, length, and age at maturity than oceanic stocks, which indicates that they may be subjected to different management strategies in future.

Consumption of cod by mammals other than man is substantial, although it varies extensively between fiords. Species such as sea otter, harbor seals, and porpoises do visit fiords for short time-windows, having a more limited affect as predators on the local fish fauna. Cormorants have a less varied diet and are known to roost and feed in fiords from late September to early April, within 8 km of their night roost. Although their toll on local cod stocks may be low, around 1/20 in terms of numbers, compared to cannibalism within the cod population, their local impact is still important since they prefer juvenile cod in the length range 4–50 cm.

Other Structural Forces of the Pelagic Community in Fiordic Ecosystems

Light is an important limiting factor for the visual foraging process in fishes, and the light regime may potentially affect the competition between visual and tactile predators. Food demand and risk of mortality are regulated by balance between catching and avoidance between predator and prey, which ultimately may be regulated by visibility in the water column. The seasonal variation in the light may therefore be an important structural force for vertical distributions of important components in the community in fiords. Zooplankton size and density increase with depth; the most visible forms are found only in deep waters. At night, macroplankton and mesopelagic

fishes are dispersed in the water column, with a tendency for dispersion of the SSL. All components of the SSL respond to changes in light intensity during day, in order to balance vision versus visibility. The pelagic juvenile fishes in the upper part of the SSL migrate to the surface to maximize their feeding period. The more visible fishes stay in the deeper part of the SSL. At dawn, euphausiids (*Meganyctiphanes norvegica*) descend from surface waters to midwater depths, and the larger mesopelagic fishes (*Benthosema glaciale*) and the pelagic prawns (*Sergestes arcticus* and *Pasiphaea multidentata*) migrate to even darker water. Large pelagic fishes are found in the entire water column, feeding with the highest densities mainly below and at the deepest end of the SSL.

A strong faunal difference is often found between adjacent fiords. This has been linked to the differences in the light climate, fiords with low visibility tending to have a higher component of jellyfish, with the jellyfish being replaced by fish in fiords with higher visibility. This has prompted the hypothesis that the visibility regime may affect the distribution of tactile and visual predators such as jellyfish and fish. The implication is that light is a forcing factor on the marine ecosystem dynamics through the visual feeding process, with potential secondary links to eutrophication. The mechanistic relationship between differences in sea water absorbance and the biological components is not established, but fiords with different visibility regimes will still play an important basis for research needed for ecologically based future management of these important marine biotopes.

See also

Beaches, Physical Processes Affecting. Coastal Circulation Models. Macrobenthos.

Further Reading

Aksnes DL, Aure J, Kaartved S, Magnesen T, and Richard J (1989) Significance of advection for the carrying capacities of fiord populations. *Marine Ecology Progress Series* 50: 263–274.

Falkenhaug T (1997) *Studies of Spatio-temporal Variations in a Zooplankton Community. Interactions between Vertical Behaviour and Physical Processes.* Dr Scient thesis, University of Troms; ISBN 82-91086-12-5.

Howell B, Moksness E, and Svsand T (eds.) (1999) *Stock Enhancement and Sea Ranching.* Oxford: Fishing News Books.

Matthews JBL and Heimdal BR (1980) Pelagic productivity and food chains in fiord systems. In: Freeland HJ, Farmer DM, and Levings CD (eds.) *Fiord Oceanography.* New York: Plenum.

SALT MARSHES, LAGOONS, BEACHES AND ROCKY SHORES

SALT MARSHES AND MUD FLATS

J. M. Teal, Woods Hole Oceanographic Institution, Rochester, MA, USA

Structure

Salt marshes are vegetated mud flats. They are above mean sea level in the intertidal area where higher plants (angiosperms) grow. Sea grasses are an exception to the generalization about higher plants because they live below low tide levels. Mud flats are vegetated by algae.

Geomorphology

Salt marshes and mud flats are made of soft sediments deposited along the coast in areas protected from ocean surf or strong currents. These are long-term depositional areas intermittently subject to erosion and export of particles. Salt marsh sediments are held in place by plant roots and rhizomes (underground stems). Consequently, marshes are resistant to erosion by all but the strongest storms. Algal mats and animal burrows bind mud flat sediments, although, even when protected along tidal creeks within a salt marsh, mud flats are more easily eroded than the adjacent salt marsh plain.

Salinity in a marsh or mud flat, reported in parts per thousand (ppt), can range from about 40 ppt down to 5 ppt. The interaction of the tides and weather, the salinity of the coastal ocean, and the elevation of the marsh plain control salinity on a marsh or mud flat. Parts of the marsh with strong, regular tides (1 m or more) are flooded twice a day, and salinity is close to that of the coastal ocean. Heavy rain at low tide can temporarily make the surface of the sediment almost fresh. Salinity may vary seasonally if a marsh is located in an estuary where the river volume changes over the year. Salinity varies within a marsh with subtle changes in surface elevation. Higher marshes at sites with regular tides have variation between spring and neap tides that result in some areas being flooded every day while other, higher, areas are flooded less frequently. At higher elevations flooding may occur on only a few days each spring tide, while at the highest elevations flooding may occur only a few times a year.

Some marshes, on coasts with little elevation change, have their highest parts flooded only seasonally by the equinoctial tides. Other marshes occur in areas with small lunar tides where flooding is predominantly wind-driven, such as the marshes in the lagoons along the Texas coast of the United States. They are flooded irregularly and, between flooding, the salinity is greatly raised by evaporation in the hot, dry climate. The salinity in some of the higher areas becomes so high that no rooted plants survive. These are salt flats, high enough in the tidal regime for higher plants to grow, but so salty that only salt-resistant algae can grow there. The weather further affects salinity within marshes and mud flats. Weather that changes the temperature of coastal waters or varying atmospheric pressure can change sea level by 10 cm over periods of weeks to months, and therefore affect the areas of the marsh that are subjected to tidal inundation.

Sea level changes gradually. It has been rising since the retreat of the continental glaciers. The rate of rise may be increasing with global warming. For the last 10 000 years or so, marshes have been able to keep up with sea level rise by accumulating sediment, both through deposition of mud and sand and through accumulation of peat. The peat comes from the underground parts of marsh plants that decay slowly in the anoxic marsh sediments. The result of these processes is illustrated in **Figure 1**, in which the basement sediment is overlain by the accumulated marsh sediment. Keeping up with sea level rise creates a marsh plain that is relatively flat; the elevation determined by water level rather than by the geological processes that determined the original, basement sediment surface on which the marsh developed. Tidal creeks, which carry the tidal waters on and off the marsh, dissect the flat marsh plain.

Organisms

The duration of flooding and the salinities of the sediments and tidal waters control the mix of higher vegetation. Competitive interactions between plants and interactions between plants and animals further determine plant distributions. Duration of flooding duration controls how saturated the sediments will be, which in turn controls how oxygenated or reduced the sediments are. The roots of higher plants must have oxygen to survive, although many can survive short periods of anoxia. Air penetrates into the creekbank sediments as they drain at low tide. Evapotranspiration from plants at low tide also removes water from the sediments and facilitates entry

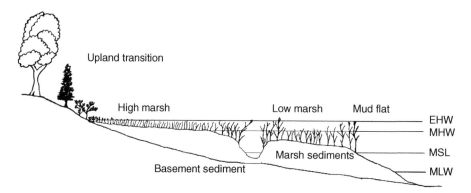

Figure 1 Cartoon of a typical salt marsh of eastern North America. The plants shown are mostly grasses and may differ in other parts of the world. MLW, mean low water; MSL, mean sea level; MHW, mean high water; EHW, extreme high water. The mud flat is shown as a part of the marsh but mud flats also exist independently of marshes.

of air. Most salt marsh higher plants have aerenchyma (internal air passages) through which oxygen reaches the roots and rhizomes by diffusion or active transport from the above-ground parts. However, they also benefit from availability of oxygen outside the roots.

The species of higher plants that dominate salt marshes vary with latitude, salinity, region of the world, and tidal amplitude. They are composed of relatively few species of plants that have invested in the ability to supply oxygen to roots and rhizomes in reduced sediments and to deal with various levels of salt. Grasses are important, with *Spartina alterniflora* the dominant species from mid-tide to high-tide levels in temperate Eastern North America. *Puccinellia* is a dominant grass in boreal and arctic marshes. The less regularly flooded marshes of East Anglia (UK) support a more diverse vegetation community in which grasses are not dominant. The salty marshes of the Texas coast are covered by salt-tolerant *Salicornia* species. Adjacent to the upper, landward edge of the marsh lie areas flooded only at times when storms drive ocean waters to unusual heights. Some land plants can survive occasional salt baths, but most cannot. An extreme high-water even usually results in the death of plants at the marsh border.

Algae on the marsh and mud flat are less specialized. Depending upon the turbulence of the tidal water, macroalgae (seaweeds) may be present, but a diverse microalgal community is common. Algae live on or near the surface of the sediments and obtain oxygen directly from the air or water and from the oxygen produced by photosynthesis. Their presence on surface sediment is controlled by light. In highly turbid waters they are almost entirely limited to the intertidal flats. In clearer waters, they can grow below low-tide levels. Algae growing on the vegetated marsh plain and on the stems of marsh plants

get less light as the plants mature. Production of these algae is greatest in early spring, before the developing vegetation intercepts the light.

Photosynthetic bacteria also contribute to marsh and mud flat production. Blue-green bacteria can be abundant enough to forms mats. Photosynthetic sulfur bacteria occupy a thin stratum in the sediment where they get light from above and sulfide from deeper reduced levels for their hydrogen source but are below the level of oxygen penetration that would kill them. These strange organisms are relicts from the primitive earth before the atmosphere contained oxygen.

Salt marsh animals are from terrestrial and marine sources; mud flat inhabitants are limited to marine sources. Insects, spiders, and mites live in marsh sediments and on marsh plants. Crabs, amphipods, isopods and shrimps, polychaete and oligochaete annelids, snails, and bivalves live in and on the sediments. Most of these marine animals have planktonic larval stages that facilitate movement between marshes and mud flats. Although burrowing animals, such as crabs that live at the water edge of the marsh, may be fairly large (2–15 cm), in general burrowers in marshes are smaller than those in mud flats, presumably because the root mats of the higher plants interfere with burrowing.

Fish are important faunal elements in regularly flooded salt marshes and mud flats. They can be characterized as permanent marsh residents; seasonal residents (species that come into the marsh at the beginning of summer as new post-larvae and live in the marsh until cold weather sets in); species that are primarily residents of coastal waters but enter the marshes at high tide; and predatory fish that come into marshes on the ebb tide to feed on the smaller fishes forced off the marsh plain and out of the smaller creeks by falling water levels.

A few mammals live in the marshes, including those that flee only the highest tides by retreating to land, such as voles, or those that make temporary refuges in tall marsh plants, such as raccoons. The North American muskrat builds permanent houses on the marsh from the marsh plants, although muskrats are typically found only in the less-saline marshes. Grazing mammals feed on marsh plants at low tide. In Brittany, lambs raised on salt marshes are specially valued for the flavor of their meat.

Many species of birds use salt marshes and mud flats. Shore bird species live in the marshes and/or use associated mud flats for feeding during migration. Northern harriers nest on higher portions of salt marshes and feed on their resident voles. Several species of rails dwell in marshes as do bitterns, ducks, and some wrens and sparrows. The nesting species must keep their eggs and young from drowning, which they achieve by building their nests in high vegetation, by building floating nests, and by nesting and raising their young between periods of highest tides.

Functions

Marshes, and to some extent mud flats, produce animals and plants, provide nursery areas for marine fishes, modify nutrient cycles, degrade organic chemicals, immobilize elements within their sediments, and modify wave action on adjacent uplands.

Production and Nursery

Plant production from salt marshes is as high as or higher than that of most other systems because of the ability of muddy sediments to serve as nutrient reservoirs, because of their exposure to full sun, and because of nutrients supplied by sea water. Although the plant production is food for insects, mites and voles, large mammalian herbivores that venture onto the marsh, a few crustacea, and other marine animals, most of the higher plant production is not eaten directly but enters the food web as detritus. As the plants die, they are attacked by fungi and bacteria that reduce them to small particles on the surface of the marsh. Since the labile organic matter in the plants is quickly used as food by the bacteria and fungi, most of the nutrient value of the detritus reaches the next link in the food chain through these microorganisms. These are digested from the plant particles by detritivores, but the cellulose and lignin from the original plants passes through them and is deposited as feces that are recolonized by bacteria and fungi. Besides serving as a food source in the marsh itself, a portion of the detritus–algae mixture is exported by tides to serve as a food source in the marsh creeks and associated estuary.

The primary plant production supports production of animals. Fish production in marshes is high. Resident fishes such as North American *Fundulus heteroclitus* live on the marsh plain during their first summer, survive low tide in tiny pools or in wet mud, feed on the tiny animals living on the detritus–algae–microorganism mix, and grow to migrate into small marsh creeks. At high tide they continue to feed on the marsh plain, where they are joined by the young-of-the-year of those species that use the marsh principally as a nursery area. The warm, shallow waters promote rapid growth and are refuge areas where they are protected from predatory fishes, but not from fish-eating herons and egrets.

The fishes are the most valuable export from the marshes to estuaries and coastal oceans. Some of the fishes are exported in the bodies of predatory fishes that enter the marsh on the ebb tide to feed. Many young fishes, raised in marshes, migrate offshore in the autumn after having spent the summer growing in the marsh.

Nutrient and Element Cycling

Nitrogen is the critical nutrient controlling plant productivity in marshes. Phosphorus is readily available in muddy salt marsh sediments and potassium is sufficiently abundant in sea water. Micronutrients, such as silica or iron, that may be limiting for primary production in deeper waters are abundant in marsh sediments. Thus nitrogen is the nutrient of interest for marsh production and nutrient cycling.

In marshes where nitrogen is in short supply, blue-green bacteria serve as nitrogen fixers, building nitrogen gas from the air into their organic matter. Nitrogen-fixing bacteria associated with the roots of higher plants serve the same function. Nitrogen fixation is an energy-demanding process that is absent where the supply is sufficient to support plant growth.

Two other stages of the nitrogen cycle occur in marshes. Organic nitrogen released by decomposition is in the form of ammonium ion. This can be oxidized to nitrate by certain bacteria that derive energy from the process if oxygen is present. Both nitrate and ammonium can satisfy the nitrogen needs of plants, but nitrate can also serve in place of oxygen for the respiration of another group of bacteria that release nitrogen gas as a by-product in the process called denitrification. Denitrification in salt marshes and mud flats is significant in reducing eutrophication in estuarine and coastal waters.

Phosphorus, present as phosphate, is the other plant nutrient that can be limiting in marshes, especially in regions where nitrogen is in abundant supply. It can also contribute to eutrophication of coastal waters, but phosphate is readily bound to sediments and so tends to be retained in marshes and mud flats rather than released to the estuary.

Sulfur cycling in salt marshes, while of minor importance as a nutrient, contributes to completing the production cycle. Sulfate is the second most abundant anion in sea water. In anoxic sediments, a specialized group of decomposing organisms living on the dead, underground portions of marsh plants can use sulfate as an electron acceptor – an oxygen substitute – in their respiration. The by-product is hydrogen sulfide rather than water. Sulfate reduction yields much less energy than respiration with oxygen or nitrate reduction, so these latter processes occur within the sediment surface, leaving sulfate reduction as the remaining process in deeper parts of the sediments. The sulfide carries much of the free energy not captured by the bacteria in sulfate reduction. As it diffuses to surface layers, most of the sulfide is oxidized by bacteria that grow using it as an energy source. A small amount is used by the photosynthetic sulfur bacteria mentioned above.

Pollution

Marshes, like the estuaries with which they tend to be associated, are depositional areas. They tend to accumulate whatever pollutants are dumped into coastal waters, especially those bound to particles. Much of the pollution load enters the coast transported by rivers and may originate far from the affected marshes. For example, much of the nitrogen and pesticide loading of marshes and coastal waters of the Mississippi Delta region of the United States comes from farming regions hundreds or thousands of kilometers upstream.

Many pollutants, both organic and inorganic, bind to sediments and are retained by salt marshes and mud flats. Organic compounds are often degraded in these biologically active systems, especially since many of them are only metabolized when the microorganisms responsible are actively growing on other, more easily degraded compounds. There are, unfortunately, some organics, the structures of which are protected by constituents such as chlorine, that are highly resistant to microbial attack. Some polychlorinated biphenyls (PCBs) have such structures, with the result that a PCB mixture will gradually lose the degradable compounds while the resistant components will become relatively more concentrated.

Metals are also bound to sediments and so may be removed and retained by marshes and mud flats.

Mercury is sequestered in the sediment, while cadmium forms soluble complexes with chloride in sea water and is, at most, temporarily retained.

Since marshes and mud flats tend, in the long term, to be depositional systems, they remove pollutants and bury them as long as the sediments are not remobilized by erosion. Since mud flats are more easily eroded than marsh sediments held in place by plant roots and rhizomes, they are less secure long-term storage sites.

Storm Damage Prevention

While marshes and mud flats exist only in relatively protected situations, they are still subject to storm damage as are the uplands behind them. During storms, the shallow waters and the vegetation on the marshes offer resistance to water flow, making them places where wave forces are dissipated, reducing the water and wave damage to the adjoining upland.

Human Modifications

Direct Effects

Many marshes and mud flats in urban areas have been highly altered or destroyed by filling or by dredging for harbor, channel or marina development. Less intrusive actions can have large impacts. Since salt marshes and mud flats typically lie in indentations along the coast, the openings where tides enter and leave them are often sites of human modification for roads and railroads. Both culverts and bridges restrict flow if they are not large enough. Flow is especially restricted at high water unless the bridge spans the entire marsh opening, a rare situation because it is expensive. The result of restriction is a reduction in the amount of water that floods the marsh. The plants are submerged for a shorter period and to a lesser depth, and the floodwaters do not extend as far onto the marsh surface. The ebb flow is also restricted and the marsh may not drain as efficiently as in the unimpeded case. Poor drainage could freshen the marsh after a heavy rain and runoff. Less commonly, it could increase salinity after an exceptionally prolonged storm-driven high tide.

The result of the disturbance will be a change in the oxygen and salinity relations between roots and sediments. Plants may become oxygen-stressed and drown. Tidal restrictions in moist temperate regions usually result in a freshening of the sediment salinity. This favors species that have not invested in salt control mechanisms of the typical salt marsh plants. A widespread result in North America has been the spread of the common reed, *Phragmites australis*, a brackish-water and freshwater species. Common reed is a tall (3 m) and vigorous plant that can spread horizontally by rhizomes at 10 meters per year. Its robust stem decomposes more slowly than that of the salt marsh cord grass, *Spartina alterniflora* and as a result, it takes over a marsh freshened by tidal restrictions. Since its stems accumulate above ground and rhizomes below ground, it tends to raise the marsh level, fill in the small drainage channels, and reduce the value of the marsh for fish and wildlife. Although *Phragmites australis* is a valuable plant for many purposes (it is the preferred plant for thatching roofs in Europe), its takeover of salt marshes is considered undesirable.

The ultimate modification of tidal flow is restriction by diking. Some temperate marshes have been diked to allow the harvest of salt hay, valued as mulch because it lacks weed seeds. Since some diked marshes are periodically flooded in an attempt to maintain the desired vegetation, they are not completely changed and can be restored. Other marshes, such as those in Holland, have been diked and removed from tidal flow so that the land may be used for upland agriculture. Many marshes and mud flats have been modified to create salt pans for production of sea salt and for aquaculture. The latter is a greater problem in the tropics, where the impact is on mangroves rather than on the salt marshes of more temperate regions.

Indirect Effects

Upland diking The upper borders of coastal marshes were often diked to prevent upland flooding. People built close to the marshes to take advantage of the view. With experience they found that storms could raise the sea level enough to flood upland. The natural response was to construct a barrier to prevent flooding. Roads and railroads along the landward edges of marshes are also barriers that restrict upland flooding. They are built high enough to protect the roadbed from most flooding and usually have only enough drainage to allow rain runoff to pass to the sea. In both cases, the result is a barrier to landward migration of the marsh. As the relative sea level rises, sandy barriers that protect coastal marshes are flooded and, during storms, the sand is washed onto the marsh. As long as the marsh can also move back by occupying the adjacent upland it may be able to persist without loss in area, but if a barrier prevents landward transgression the marsh will be squeezed between the barrier and the rising sea and will eventually disappear. During this process, the drainage structures under the barrier gradually become flow restrictors. The sea will flood the land behind the barrier through the culverts, but these are inevitably too small to permit unrestricted marsh development. When flooding begins, the culverts are typically fitted with tide gates to prevent whatever flooding and marsh development could be accommodated by the capacity of the culverts.

Changes in sediment loading Increases in sediment supplies can allow the marshes to spread as the shallow waters bordering them are filled in. The plant stems further impede water movement and enhance spread of the marsh. This assumes that storms do not carry the additional sediment onto the marsh plain and raise it above normal tidal level, which would damage or destroy rather than extend the marsh.

Reduced sediment supply can destroy a marsh. In a river delta where sediments gradually de-water and consolidate, sinking continually, a continuous supply of new sediment combined with vegetation remains, accumulating as peat, and maintains the marsh level. When sediment supplies are cut off, the peat accumulation may be insufficient to maintain the marsh at sea level. Dams, such as the Aswan Dam on the Nile, can trap sediments. Sediments can be channeled by levées so that they flow into deep water at the mouth of the river rather than spreading over the delta marshes, as is happening in the Mississippi River delta. In the latter case, the coastline of Louisiana is retreating by kilometers a year as a result of the loss of delta marshlands.

Introduction of foreign species Dramatic changes in the marshes and flats of England and Europe occurred after *Spartina alterniflora* was introduced from the east coast of North America and probably hybridized with the native *S. maritima* to produce *S. anglica*. The new species was more tolerant of submergence than the native forms and turned many mud flats into salt marshes. This change reduced populations of mud flat animals, many with commercial value, and reduced the foraging area for shore birds that feed on mud flats. A similar situation has developed in the last decades on the north-west coast of the United States, where introduced *Spartina alterniflora* is invading mud flats and reducing the available area for shellfish.

Marsh restoration Salt marshes and mud flats may be the most readily restored of all wetlands. The source and level of water is known. The vascular plants that will thrive are known and can be planted if a local seed source is not available. Many of the marsh animals have planktonic larvae that can invade the restored marsh on their own. Although many of the properties of a mature salt marsh take time to develop, such as the nutrient-retaining capacity of the sediments, these will develop if the marsh is allowed to survive.

See also

Salt Marsh Vegetation. Tides.

Further Reading

Adam P (1990) *Salt Marsh Ecology*. Cambridge: Cambridge University Press.

Mitsch WJ and Gosselink JG (1993) *Wetlands*. New York: Wiley.

Peterson CH and Peterson NM (1972) *The Ecology of Intertidal Flats of North Carolina: A Community Profile*. Washington, DC: US Fish and Wildlife Service, Office of Biological Services, FWS/OBS-79/39.

Streever W (1999) *An International Perspective on Wetland Rehabilitation*. Dordrecht: Kluwer.

Teal JM and Teal M (1969) *Life and Death of the Salt Marsh*. New York: Ballentine.

Weinstein MP and Kreeger DA (2001) *Concepts and Controversies in Tidal Marsh Ecology*. Dordrecht: Kluwer.

Whitlatch RB (1982) *The Ecology of New England Tidal Flats: A Community Profile*. Washington, DC: US Fish and Wildlife Service, Biological Services Program, FES/OBS-81/01.

SALT MARSH VEGETATION

C. T. Roman, University of Rhode Island, Narragansett, RI, USA

Introduction

Coastal salt marshes are intertidal features that occur as narrow fringes bordering the upland or as extensive meadows, often several kilometers wide. They occur throughout the world's middle and high latitudes, and in tropical/subtropical areas they are mostly, but not entirely, replaced by mangrove ecosystems. Salt marshes develop along the shallow, protected shores of estuaries, lagoons, and behind barrier spits. Here, low energy intertidal mud and sand flats are colonized by halophytes, plants that are tolerant of saline conditions. The initial colonizers serve to enhance sediment accumulation and over time the marsh expands vertically and spreads horizontally, encroaching the upland or growing seaward. As salt marshes mature they become geomorphically and floristically more complex with establishment of creeks, pools, and distinct patterns or zones of vegetation.

Several interacting factors influence salt marsh vegetation patterns, including frequency and duration of tidal flooding, salinity, substrate, surface elevation, oxygen and nutrient availability, disturbance by wrack deposition, and competition among plant species. Moreover, the ability of individual flowering plant species to adapt to an environment with saline and waterlogged soils plays an important role in defining salt marsh vegetation patterns. Morphological and physiological adaptations that halophytes may possess to manage salt stress include a succulent growth form, salt-excreting glands, mechanisms to reduce water loss, such as few stomates and low surface area, and a C4 photosynthetic pathway to promote high water use efficiency. To deal with anaerobic soil conditions, many salt marsh plants have well-developed aerenchymal tissue that delivers oxygen to below-ground roots.

Salt Marsh Vegetation Patterns

Vegetation patterns often reflect the stage of maturation of a salt marsh. Early in the development, halophytes, such as *Spartina alterniflora* along the east coast of the United States, colonize intertidal flats. These initial colonizers are tolerant of frequent flooding. Once established, the plants spread vegetatively by rhizome growth, the plants trap sediments, and the marsh begins to grow vertically. As noted from A. C. Redfield's classic study of salt marsh development along a coast with a rising sea level, salt marshes often extend seaward over tidal flats, while also accreting vertically and encroaching the upland or freshwater tidal wetlands. With this vertical growth and maturity of the marsh, discrete patterns of vegetation develop: frequently flooded low marsh vegetation borders the seaward portion of the marsh and along creekbanks, while high marsh areas support less flood-tolerant species, such as *Spartina patens*. There is some concern that vertical growth or accretion of salt marshes may not be able to keep pace with accelerated rates of sea level rise, resulting in submergence or drowning of marshes. This has been observed in some areas of the world.

Plant species and patterns of vegetation that dominate the salt marsh vary from region to region of the world and it is beyond the scope of this article to detail this variability; however, the general pattern of low marsh and high marsh remains throughout. There is often variation in vegetation patterns from marshes within a region and even between marshes within a single estuary, but in general, the low marsh is dominated by a limited number of species, often just one. On the Atlantic coast of North America, *Spartina alterniflora* is the early colonizer and almost exclusively dominates the low marsh. In European marshes, *Puccinellia maritima* often dominates the intertidal low marsh. *Spartina anglica*, a hybrid of *Spartina alterniflora* and *Spartina maritima*, has invaded many of the muddier low marsh sites of European marshes over the past century. With increasing elevation of the high marsh, species richness tends to increase. *Spartina patens*, *Distichlis spicata*, *Juncus gerardi*, *Juncus roemerianus*, and short-form *Spartina alterniflora* occupy the US east coast high marsh, with each species dominating in patches or zones to form a mosaic vegetation pattern. In European marshes, the low marsh may give way to a diverse high marsh of *Halimione portulacoides*, *Limonium* sp., *Suaeda maritima*, and *Festuca* sp., among others.

Factors Controlling Vegetation Patterns

The frequency and duration of tidal flooding are mostly responsible for the low and high marsh delineation, but many other factors contribute to the wide variation found in salt marsh vegetation patterns (**Figure 1**). Soil salinity is relatively constant within the low marsh because of frequent tidal flooding, but extremes in soil salinity can occur on the less-flooded high marsh contributing to the vegetation mosaic. Concentrations in excess of 100 parts per thousand can occur, resulting in hypersaline pannes that remain unvegetated or are colonized by only the most salt-tolerant halophytes (e.g., *Salicornia*). Salt marshes of southern temperate and subtropical/tropical latitudes tend to have higher soil salinity because of more intense solar radiation and higher evaporation rates. At the other extreme, soil salinity of the high marsh can be dramatically depressed by rainfall or by discharge of groundwater near the marsh upland border. On salt marshes of the New England coast (USA), *Juncus gerardi* and the shrub, *Iva fructescens*, are less tolerant of salt, and thus, grow at higher elevations where tidal flooding is only occasional or near upland freshwater sources.

For successful growth in environments of high soil salinity, halophytes must be able to maintain a flow of water from the soil into the roots. Osmotic pressure in the roots must be higher than the surrounding soil for water uptake. To maintain this osmotic difference, halophytes have high concentrations of solutes in their tissues (i.e., high osmotic pressure), concentrations greater than the surrounding environment. This osmotic adjustment can be achieved by an accumulation of sodium and chloride ions or organic solutes. To effectively tolerate high internal salt concentrations, many halophytes have a succulent growth form (e.g., *Salicornia*, *Suaeda*). This high tissue water content serves to dilute potentially toxic salt concentrations. Other halophytes, such as *Spartina alterniflora*, have salt glands to actively excrete salt from leaves.

Waterlogged or anaerobic soil conditions also strongly influence the pattern of salt marsh vegetation. Plants growing in waterlogged soils must deal

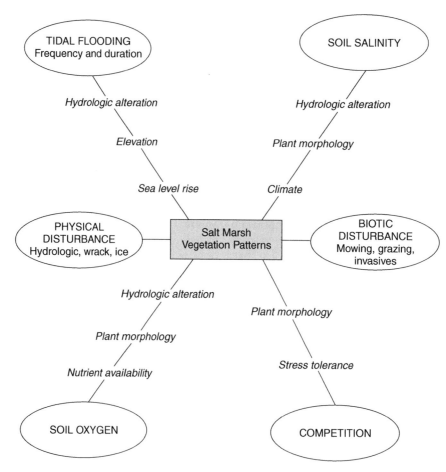

Figure 1 Conceptual model of factors controlling vegetation patterns in salt marshes. Ovals denote major factors and the key interacting parameters are shown in italics. For example, soil salinity responds to hydrologic alterations, climate (e.g., temperature/evaporation and precipitation) influences soil salinity, and a species response to soil salinity is dependent on morphological considerations (e.g., succulence, salt glands, etc.).

with a lack of oxygen at the rhizosphere and the accumulation of toxins resulting from biogeochemical soil processes (i.e., sulfate reduction). High concentrations of hydrogen sulfide are toxic to root metabolism, inhibiting nitrogen uptake and resulting in decreased plant growth. Many salt marsh plants are able to survive these soil conditions because they have well-developed aerenchyma tissue, or a network of intercellular spaces that serve to deliver oxygen from above-ground plant parts to below-ground roots. Extensive research has been conducted on the relationships between sediment oxygen levels and growth of *Spartina alterniflora*. Aerenchyma transports oxygen to roots to support aerobic respiration, facilitating nitrogen uptake, and the aerated rhizosphere serves to oxidize reduced soil compounds, such as sulfide, thereby promoting growth. Under severely reducing soil conditions, the aerenchyma may not supply sufficient oxygen for aerobic respiration. *Spartina alterniflora* then has the ability to respire anaerobically, but growth is reduced. It is clear that the degree of soil aeration can dramatically influence salt marsh vegetation patterns. Plants that have the ability to form aerenchyma (e.g., *Spartina alterniflora*, *Juncus roemerianus*, *Puccinellia maritima*) or develop anaerobic respiration will be able to tolerate moderately reduced or waterlogged soils, whereas other species will be limited to better-drained areas of the marsh.

Coupled with the physical factors of salinity stress and waterlogging, competition is another process that controls salt marsh vegetation patterns. It has been suggested that competition is an important factor governing the patterning of vegetation near the upland boundaries of the high marsh. Plants that dominate the low marsh, like *Spartina alterniflora*, or forbs in salt pannes (e.g., *Salicornia*), can tolerate harsh environmental conditions and exist with few competitors. However, plant assemblages of the high marsh, and especially near the upland, may occur there because of their exceptional competitive abilities. For example, the high marsh plant, *Spartina patens*, has a dense growth form and is able to outcompete species such as *Distichlis spicata* and *Spartina alterniflora*, each with a diffuse or clumped morphology.

Natural and human-induced disturbances also influence salt marsh vegetation patterns. Vegetation can be killed and bare spots created following deposition of wrack on the marsh surface. The resulting bare areas may then become vegetated by early colonizers (e.g., *Salicornia*). In northern regions, ice scouring can dramatically alter the creek or bayfront of marshes. Also, large blocks of ice, laden with sediment, are often deposited on the high marsh

creating a new microrelief habitat, or these blocks may transport plant rhizomes to mud or sand flats and initiate the process of salt marsh development.

Regarding human-induced disturbances, hydrologic alterations have dramatic effects on salt marsh vegetation patterns throughout the world. Some salt marshes have been diked and drained for agriculture. In others, extensive ditching has drained salt marshes for mosquito-control purposes. In yet another hydrologic alteration, water has been retained or impounded within salt marshes to alter wildlife habitat functions or to control mosquitoes. Impoundments generally restrict tidal inflow and retain fresh water, resulting in conversion from salt-tolerant vegetation to brackish or freshwater vegetation. Practices that drain the marsh, such as ditching, tend to lower the water-table level and aerate the soil, and the resulting vegetation may shift toward that typical of a high marsh. Another type of hydrologic alteration is the restriction of tidal flow by bridges, culverts, roads, and causeways. This is particularly common along urbanized shorelines, where soil salinity can be reduced and water-table levels altered resulting in vegetation changes, such as the conversion from *Spartina*-dominated salt marsh to *Phragmites australis*, as is most evident throughout the north-eastern US. The role of hydrologic alterations in controlling vegetation patterns is clearly identified in **Figure 1** as a key physical disturbance, and also as a variable that influences tidal flooding, soil salinity, and soil oxygen levels.

Grazing by domestic animals, mostly sheep and cattle, and mowing for hay are two practices that have been ongoing for many centuries on salt marshes worldwide, although they seem to be declining in some regions. Studies in Europe have demonstrated that certain plant species are favored by intensive grazing (e.g., *Puccinellia*, *Festuca rubra*, *Agrostis stolonifera*), whereas others become dominant on ungrazed salt marshes (e.g., *Halimione portulacoides*, *Limonium*, *Suaeda*).

Conclusions

Vegetation zones of salt marshes have been described as belts of plant communities from creekbank or bayfront margins to the upland border, or most appropriately, as a mosaic of communities along this elevation gradient. All salt marsh sites display some zonation but because of the complex of interacting factors that influence marsh vegetation patterns, there is extraordinary variability in zonation among individual marshes. Moreover, salt marsh vegetation patterns are constantly changing on seasonal to decadal timescales. Experimental research and long-term monitoring efforts are needed to further evaluate vegetation pattern responses to the myriad of interacting environmental factors that influence salt marsh vegetation. An ultimate goal is to model and predict the response of marsh vegetation as a result of natural or human-induced disturbance, accelerated rates of sea level rise, or marsh-restoration strategies.

See also

Coastal Circulation Models. Intertidal Fishes. Salt Marshes and Mud Flats. Sea Level Change.

Further Reading

Adam P (1990) *Saltmarsh Ecology.* Cambridge: Cambridge University Press.

Bertness MD (1999) *The Ecology of Atlantic Shorelines.* Sunderland, MA: Sinauer.

Chapman VJ (1960) *Salt Marshes and Salt Deserts of the World.* New York: Interscience.

Daiber FC (1986) *Conservation of Tidal Marshes.* New York: Van Nostrand Reinhold.

Flowers TJ, Hajibagheri MA, and Clipson NJW (1986) Halophytes. *Quarterly Review of Biology* 61: 313–337.

Niering WA and Warren RS (1980) Vegetation patterns and processes in New England salt marshes. *BioScience* 30: 301–307.

Redfield AC (1972) Development of a New England salt marsh. *Ecological Monographs* 42: 201–237.

Reimold RJ and Queen WH (eds.) (1974) *Ecology of Halophytes.* New York: Academic Press.

LAGOONS

R. S. K. Barnes, University of Cambridge, Cambridge, UK

Introduction

A 'lagoon' is any shallow body of water that is semi-isolated from a larger one by some form of natural linear barrier. In ocean science, it is used to denote two rather different types of environment: 'coastal lagoons' (the main subject of this article) and lagoons impounded by coral reefs. Coastal lagoons are a product of rising sea levels and hence are geologically transient. Whilst they exist, they are highly productive environments that support abundant crustaceans and mollusks, and their fish and bird predators. Harvests from lagoonal aquaculture of up to 2500 kg ha^{-1}y^{-1} of penaeid shrimp, of 400 tonnes ha^{-1}y^{-1} of bivalve molluscs, and in several cases of >200 kg ha^{-1} y^{-1} of fish are taken.

What is a Lagoon?

Coastal lagoons are bodies of water that are partially isolated from an adjacent sea by a sedimentary barrier, but which nevertheless receive an influx of water from that sea. Some 13% of the world's coastline is faced by sedimentary barriers, with only Canada, the western coast of South America, the China Sea coast from Korea to south-east Asia, and the Scandinavian peninsula lacking them and therefore also being without significant lagoons (**Table 1, Figure 1**). The lagoons behind such barrier coastlines range in size from small ponds of <1 ha through to large bays exceeding 10 000 km². The median size has been suggested to be about 8000 ha. Although several do bear the word 'lagoon' in their name, many do not. Usage of the same titles as for fresh water habitats is widespread (e.g. *Étang* de Vaccarès, France; *Swan-Pool*, UK; Oyster *Pond*, USA; *Lake* Menzalah, Egypt; Benacre *Broad*, UK; *Ozero* Sasyk, Ukraine; Kiziltashskiy *Liman*, Russia, etc.), as is those for coastal marine regions (Peel-Harvey *Estuary*, Australia; Great South *Bay*, USA; Ringkbing *Fjord*, Denmark; *Zaliv* Chayvo, Russia; Pamlico *Sound*, USA; Charlotte *Harbor*, USA; and even Gniloye-*More*, Ukraine; *Mer* des Bibans, Tunisia; *Mar* Menor, Spain, etc.), whilst lagoons liable to hypersalinity are often termed Sebkhas in the Arabic-speaking world

(e.g. Sebkha el Melah, Tunisia). Conversely, a few systems with 'lagoon' in their name fall out with the definition; the Knysna Lagoon in South Africa, for example, is an estuarine mouth dilated behind rocky headlands between which only a narrow channel occurs.

Coastal lagoons are most characteristic of regions with a tidal range of <2 m, since large tidal ranges (those >4 m) generate powerful water movements usually capable of breaching if not destroying incipient sedimentary barriers. Furthermore, the usual meaning of 'lagoon' requires the permanent presence of at least some water and large tidal ranges are likely to result in ebb of water from open systems during periods of low tide in the adjacent sea. Thus in Europe, for example, lagoons are abundant only around the shores of the microtidal Baltic, Mediterranean, and Black Seas. However, they are also present along some macrotidal coasts – such as the Atlantic north of about 47°N (in the east) and 40°N (in the west) (**Figure 2**) – where off shore deposits of pebbles or cobbles ('shingle') are to be found as a result of past glacial action. Here, shingle can replace the more characteristic sand of microtidal seas as the barrier material, as indeed it partially does in the lagoon-rich microtidal East Siberian, Chukchi, and Beaufort Seas in the Arctic, because it is less easily redistributed by tidal water movements. Nevertheless, sandy sedimentary barriers have developed (and persisted) in a few relatively macrotidal areas (e.g. in southern Iceland and Portugal), although the regions impounded to landwards retain water only during high tide for the reasons outlined above. Those environments that are true lagoons only during high tide are often referred to as 'tidal-flat lagoons'. They are therefore the relatively rare macrotidal coast equivalent of the typical lagoons of microtidal seas.

In many cases, the salinity of a lagoonal water mass is exactly the same as that of the adjacent sea, although where fresh water discharges into a lagoon its water may be brackish and a (usually relatively stable) salinity gradient can occur between river-mouth and lagoonal entrance channel. In regions where evaporation exceeds precipitation for all or part of a year, lagoons are often hypersaline.

The second environment to which the word lagoon is applied is associated with coral reefs; coral here replaces the unconsolidated sedimentary barrier of the coastal lagoon. Circular atoll reefs enclose the 'lagoon' within their perimeter, whilst barrier reefs are separated from the mainland by an equivalent although less isolated body of water. Indeed barrier reef lagoons are virtually sheltered stretches of coastal sea. Atoll lagoons, however, are distinctive in being floored by coral sand which supports submerged beds of seagrasses and fringing mangrove swamps justas do the coastal lagoons of similar latitudes. The similarity between the two types of lagoon is nevertheless purely physiographic since the atoll lagoon fauna is that typical of coral reefs in general and not related in any way to those of coastal lagoons. Coral lagoons are covered in greater detail in the article Coral Reefs.

Table 1 The contribution of different continents to the world total barrier/lagoonal coastline of 3200 km[a]

Continent	Percent of coastline barrier/lagoonal	Percent of world's lagoonal resource
N. America	17.6	33.6
Asia	13.8	22.2
Africa	17.9	18.7
S. America	12.2	10.3
Europe	5.3	8.4
Australia	11.4	6.8

[a](Reproduced with permission from estimates by Cromwell JE (1971) *Barrier Coast Distribution: A World-wide Survey*, p. 50. Abstracts Volume, 2nd National Coastal Shallow Water Research Conference, Baton Rouge, Louisiana.)

The Formation of Lagoons

Lagoons have existence only by virtue of the barriers that enclose them. They are therefore characteristic only of periods of rising sea level, when wave action can move sediment on and along shore, and – for a limited period of time – of constant sea level. At times of marine regression, they either drain or their basins may fill with fresh water. Many of the enclosing barriers have today been greatly augmented by wind-blown sand; for example, most of the larger systems along the South African coast (Wilderness, St Lucia, Kosi, etc.), and such enclosed lagoons have a particularly lake-like appearance (**Figure 3**).

The precise physiographic nature of a lagoon then depends on the relationship of the barrier to the adjacent coastline. Starting at a point at which the barriers are some distance offshore, we can erect a sequence of situations in which the barriers are moved ever shorewards. This starting point can be exemplified by Pamlico and Albemarle Sounds in North Carolina, USA (**Figure 4**). There a chain of long narrow barrier islands located some 30 km off the coast enclosesan area of shallow bay of around 5000 km².

Further landwards movement of the barriers, often to such an extent that some of the larger barriers become attached to the mainland at one end to produce spits, leads to the situation currently characterizing most lagoonal coastlines and to the typical

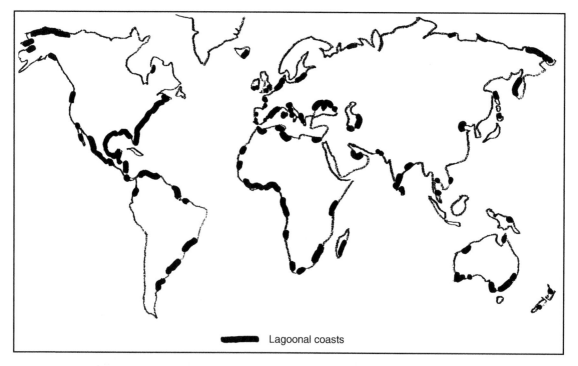

Lagoonal coasts

Figure 1   Major barrier/lagoonal coastlines.

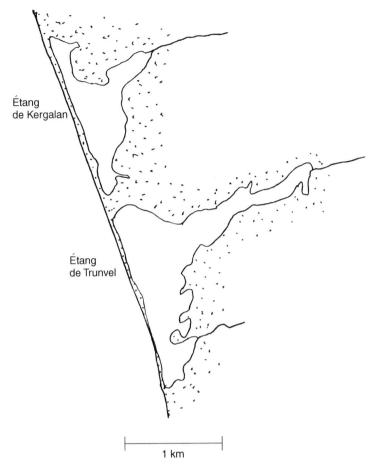

Figure 2 Shallow lagoons along the coast of the Baie d'Audierne (Brittany, France) formed behind a shingle barrier beach. Sea water enters these systems only via overtopping of the barrier; lagoonal water leaves by percolation through the barrier.

coastal lagoon. Sea water then enters and leaves these lagoons through the channels between the islands or around the end of the spit. More rarely, the lagoons are enclosed by multiple tombolos connecting an offshore island to the mainland, or even tombolos joining a number of islands (as, for example, in Te Whanga, New Zealand). With the exception of tombolo lagoons, the long axis of typical lagoons – and many are extremely elongate – lies parallel to the coastline. Typical coastal lagoons are especially characteristic of microtidal seas. They range in size up to the 10 000 km² Lagoa dos Patos, Brazil.

The abundant lagoons that occur within river deltas form a special case of this category. Several deltas (e.g. those of the Danube and Nile) have developed within – and have now largely or completely obliterated – former lagoons enclosed by spits and intervening barrier island chains. The filling of lagoons with sediment when they receive river discharge is an inevitable process and many surviving examples are but small remnants of their past extents (**Figure 5**) – islands of water in a sea of marshland.

Some 5000 years BP, the Lake St Lucia lagoon in South Africa, for example, had a length in excess of 110 km and a surface area of nearly 1200 km²; infilling with sediment has reduced the body of free water to a length of 40 km and an area of 310 km² (see **Figure 3**). Allowing for the shallowing that has also taken place, this is equivalent to an average input of 623 000 m³ of sediment per year. Infill, coupled with barrier transgression via wash-over fans, is probably the ultimate fate of virtually all of the world's lagoons.

The next stage in the hypothetical sequence sees the barrier very close to the mainland, if not plastered onto it. This has several consequences, of which the first is impingement on emerging river systems. 'Estuarine lagoons', which as their name implies merge in to estuaries, have formed where barriers have partially blocked existing drowned river valleys. For this reason they usually have their long axis perpendicular to the coastline. Even if blockage of the river is complete, lagoonal status may remain if sea water can enter by overtopping of the barrier during high water of spring tides.

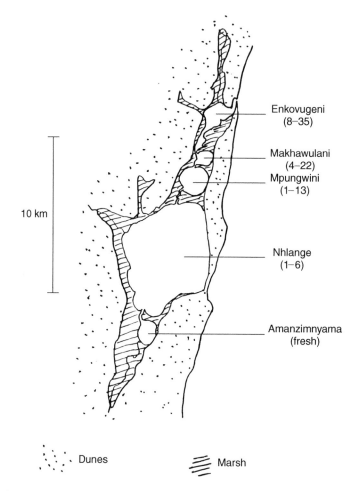

Figure 3 The Kosi Lakes (KwaZulu/Natal, South Africa), with approximate salinity ranges (in practical salinity units) in parentheses. They are enclosed within Pleistocene sand dune fields, and form a classic case of 'segmentation' of an original linear lagoon into aseries of rounded basins. (Salinity data reproduced with permission from Begg G (1978) *The Estuaries of Natal.* Natal Town and RegionalPlanning Report.).

Nevertheless, many former estuarine lagoons are now completely fresh water habitats with seawater entry being prevented by the barrier. In many tropical and subtropical regions, closure of the estuarine lagoons is seasonal. During the wet season, river flow is sufficient to maintain breaches through the barriers. In the dry season, however, flow is reduced and drift can seal the barrier. If fresh water still flows, the whole isolated basin then becomes fresh for several months. The mouths of many natural lagoonal barriers are periodically breached by man for a variety of reasons ranging from ensuring the entry of juveniles of commercially important mollusks, crustaceans, and fish to temporarily lowering water levels to avoid damage to property.

Longshore movement of barrier sediment may also divert the mouths of discharging estuaries several kilometers along the coast, as for example the 30 km deflection of that of the Senegal River by the Langue de Barbarie in West Africa, and the creation of the Indian River Lagoon in Florida, USA. Not infrequently a new mouth is broken through the barrier, naturally or more usually as a result of human intervention, and the diverted stretch can become a dead-end backwater lagoon fed by backflow.

Other consequences of on shore barrier migration also involve human intervention. Many small scale lagoons have (unwittingly) been created by land reclamation schemes. Regions to landwards of longshore ridges that were once, for example, low-lying salt marsh but which are now reclaimed usually receive an influx of water from out of the barrier water-table and this collects in depressions that were once part of the creeks draining the marshes. Equivalently, gravel pits or borrow pits in coastal shingle masses, from which building materials have been extracted, similarly receive an influx of water from out of the shingle. In both cases, the shingle water-table is derived from sea water soaking into the barrier during high tide in the adjacent sea

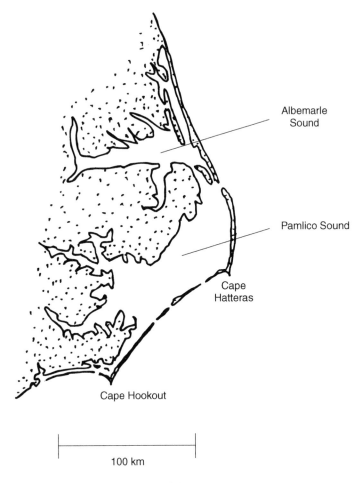

Figure 4 Pamlico and Albemarle Sounds (North Carolina, USA).

together with such rainfall as has also soaked in. Some of the natural limans of the Black Seacoast also receive their salt input via equivalent seepage.

Finally in this category, barriers may come onshore in such a fashion as to straddle the mouth of a pre-existing bay. 'Bahira lagoons' (from the Arabic for 'littlesea') are pre-existing partially land-locked coastal embayments, drowned by the post glacial rise in sea level, that have later had their mouths almost completely blocked by the development of sedimentary barriers. Also included here are systems in which the sea has broken through a pre-existing sedimentary barrier to flood part of the hinterland, but inwhich the entrance/exit channel remains narrow. Bahira lagoons clearly merge into semi-isolated marine bays.

Although lagoons may come and go during geological time, some of the larger lagoon systems are not solely features of the present interglacial period. Some, including the Lagoa dos Patos, the Gippsland Lakes of Victoria, Australia, and the Lake St Lucia lagoon, may incorporate elements of barriers and basins formed during previous marine transgressions.

The history of the Gippsland lakes over the last 70 000 years has been reconstructed in some detail (**Figure 6**). Fossil assemblages of foraminiferans suggest that lagoons may have been a feature of the Gulf Coast of the USA and Mexico at intervals ever since the Jurassic.

Lagoonal Environments

It can be argued that the main force structuring lagoons is the extent to which they are connected to the adjacent sea, and they have been divided into three or four types on this basis(**Figure 7**) which largely reflect points along the hypothetical evolutionary sequence described above.

'Leaky lagoons' function virtually as sheltered marine bays (see also **Figure 4**). As a result of large tidal ranges in the adjacent sea or the existence of many and/or wide connecting channels, interactions with the ocean are dominant and the lagoons have tidally fluctuating water levels, short flushing times and a salinity equal to that ofthe local sea water.

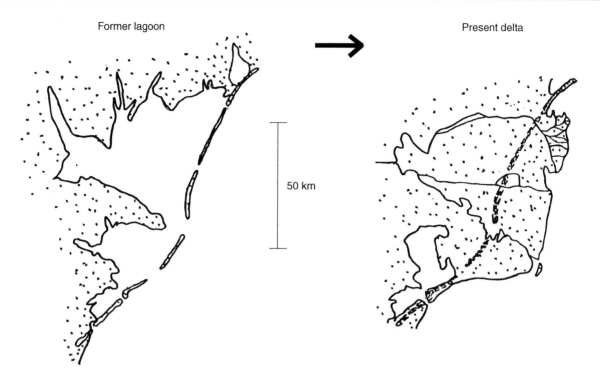

Figure 5 Infill of a former lagoon (established around 6000 years ago) on the Romanian/Ukrainian coast of the Black Sea with sediment carried by the River Danube. The current Danube delta occupies virtually all the former lagoon.

'Restricted lagoons' have asmall number of narrow entrance channels through elements of a barrier is land chain or enclosing spits and are therefore more isolated from the ocean(see also **Figure 7**). Characteristics include: longer flushing times, a vertically well-mixed water column, both wind and tidal water movements as forcing agents, and brackish to oceanic salinity.

'Choked lagoons' are connected to the ocean by a single, often long and narrow, entrance channel that serves as a filter largely eliminating tidal currents and fluctuations in water level(see also **Figure 4**). The $1000 \, km^2$ Chilka Lake lagoon, for example, communicates with the Bay of Bengal via a channel 8 km long and 130 m across at its widest. Choked lagoons are typical of coasts with high wave energy and significant longshore drift. Characteristics include: very long flushing times, intermittent stratification of the water column by thermoclines and/or haloclines, wind action as thed ominant forcing agent, purely freshwater zones near inflowing rivers, and otherwise brackish, and in some climatic zones periodic hypersaline, waters. Wet season rainfall in the $44 \, km^2$ Laguna Unare, Venezuela (at the end of the dry season), for example, can change the salinity from 60–92 to 18–25, at the same time increasing lagoonal depth by 1 m, area by $20 \, km^2$ and volume by $76 \times 10^6 m^3$.

The 'closed lagoons' shown in **Figure 2** form the limiting condition of effective isolation from direct inputs from the ocean except via occasional overtopping or after percolation through the barrier system. Most are not only 'former lagoons', but also current coastal freshwater lakes; for example, the much studied Slapton Ley in England.

Characteristically restricted and choked lagoons are very shallow (usually about 1 m and almost always $<5 \, m$ deep), floored by soft sediments, fringed by reedbeds(*Phragmites* and/or *Scirpus*), mangroves (especially *Rhizophora*) or salt marsh vegetation, and support dense beds of submerged macrophytes such as seagrasses (e.g. *Zostera*), pondweeds (*Potamogeton*) or *Ruppia*, together with green algae such as *Chaetomorpha*. The action of the dense beds of submerged vegetation may be to raise levels of pH and to contribute considerable quantities of organic matter. In stratified choked lagoons, decomposition of this organic load below the thermocline or halocline can then lead to anoxic conditions, not withstanding surface waters super saturated with oxygen (**Figure 8**).

Lagoonal Biotas and Ecology

Insofar as is known, although the species may be different, the ecology of coastal lagoons shows the

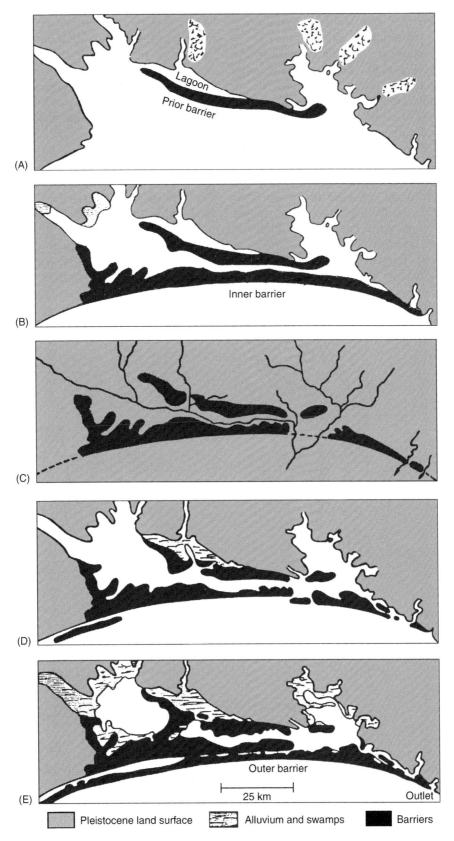

Figure 6 Evolution of the Gippsland Lagoon system (Victoria, Australia). (Reproduced with permission from Barnes 1980; after Bird ECF (1966) The evolution of sandy barrier formations on the East Gippsland coast. *Proceedings of the Royal Society of Victoria* 79:75–88.) The prior barrier (A) can be dated to about 70 000 years BP; the inner barrier (B) was formed during the late Pleistocene; the situation in (C) represents a low sea level phase of the late glacial, and (D) and (E) further evolution during the current marine transgression.

same general pattern as seen in estuaries and other regions of coastal soft sediment, although the relative importance of submerged macrophytes may be greater in lagoons (**Figure 9**). The action of predators as important forces structuring the communities in lagoons as in similar areas, for example, is reflected by the young stages of many species occurring within the dense beds of submerged macrophytes, as the hunting success of predatory species is lower there.

The primary productivities of lagoons appear to vary with their general nature, their latitude, and their depth. Choked lagoons tend to be the most productive, with recorded maxima approaching $2000 \, \text{g C m}^{-2} \, \text{y}^{-1}$, and with productivities (allowing for the effect of latitude) some 50% more than inrestricted lagoons. Temperate zone lagoons achieve only about 50–70% of the productivity of low latitude lagoons of comparable nature; and whilst shallow phytoplankton-dominated lagoons are more productive than deeper ones (depth $> 2 \, \text{m}$), the converse may be true of macrophyte-dominated systems. An order of magnitude calculation (assuming that the total area of the world's lagoons is around $320\,000 \, \text{km}^2$ and that average lagoonal productivity is $300 \, \text{g C m}^{-2} \, \text{y}^{-1}$) yields a total annual lagoonal production of 10^{11} kg fixed C. The contribution of lagoons to total oceanic carbon fixation is therefore minor, although they are as productive per unit area as are estuaries, and are only less productive than some regions of upwelling, coral reefs, and kelp forests.

What happens to this primary production? Data are still scarce. There has been much argument as to whether estuaries are sources or sinks for fixed

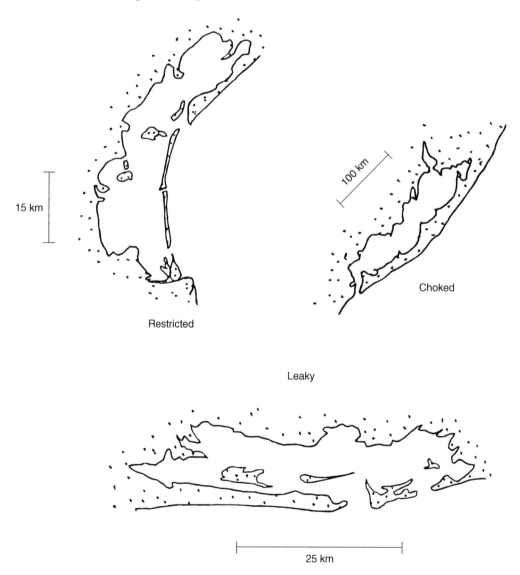

Figure 7 'Choked' (Lagoa dos Patos, Brazil), 'restricted' (Laguna di Venezia, Italy) and 'leaky' (Laguna Jiquilisco, El Salvador) lagoons.

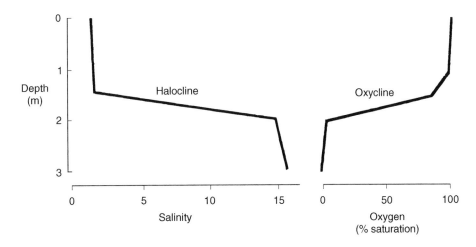

Figure 8 A halocline and consequent oxycline in the choked Swanpool Lagoon, England, after a prolonged period of high freshwater input (salinity inpractical salinity units). (Reproduced with permission from data in Dorey AE *et al.* (1973) An ecological study of the Swanpool, Falmouth II. Hydrography and its relation to animal distributions. *Estuarine Coastla and Marine Science* 1: 153–176.)

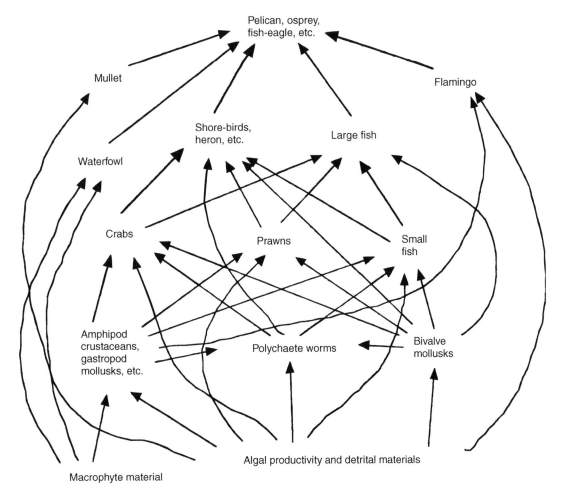

Figure 9 Simplified lagoonal food web.

carbon. In so far as information is available it appears that lagoons are more nearly in balance, as is appropriate for their more isolated nature, although all those so far studied do function as slight sinks.

Even though relatively isolated, however, no lagoon is self-contained, not least in that many elements in the faunas migrate between lagoons and other habitats.

Coastal lagoons are heavily used by wetland and shore birds as feeding areas, most noticeably by grebes, pelicans, ibis, egrets, herons, spoonbills, avocets, stilts, storks, flamingoes, cormorants, kingfishers, and fish eagles. Such birds prey on the abundant lagoonal invertebrate and fish faunas. These comprise mixtures of three different elements: both (a) essentially freshwater and (b) essentially marine and/or estuarine species capable of with standing adegree of brackishness, as well as (c) specialist lagoonal or 'paralic' species that are not in fact restricted to lagoons but also (or closely related species) occur in habitats like the Eurasian inland seas (e.g. the Caspian and Aral), as well as in (the usually man-made) tideless brackish ponds and drainage ditches that are abundant in reclaimed coastal regions. The main feature of the lagoonal species category is that they are species of marine ancestry that seem to be restricted to shallow, relatively tideless maritime habitats; over their ranges as a whole they are also clearly capable of inhabiting a wide range of salinity, including full-strength and concentrated sea water, but not fresh water. Their relatedness to marine species is evidenced by the fact that they often form species-pairs with marine or estuarine species. The fresh-water component of lagoonal faunas is often greater than would be expected in comparable salinities in estuaries, with the common occurrence of adult water beetles and hemipteran bugs besides the omnipresent dipteran larvae. The reasons for this are not understood, but may relate to decreased water movements.

Because oceanic and freshwater inputs into alagoon occur at different points around its perimeter, these three elements of the fauna tend to be broadly zoned in relation to these inputs. Indeed, an influential French school, the 'Group d'Etude du Domaine Paralique', identifies six characteristic biotic zones in lagoons based not upon salinity but upon 'a complex and abstract value which cannot be measured in the present state of knowledge' which the group terms 'confinement'. This is a function, at least, of the extent to which the water mass at any given point is isolated from oceanic influence. Beginning with the Mediterranean Sea coast, disciples of this school have now published maps of the location of the 'six degrees of confinement' in many lagoons throughout the world.

Lagoons also form the nursery grounds and adult feeding areas for a large number of commercially important fish and crustaceans that migrate between this habitat and the sea, most of which spawn outside the lagoons and only later move in actively or in some cases probably passively. In respect of the fish,

temperate regions are amongst others used byeels (Anguillidae), sea bass (Moronidae), drum (Sciaenidae), sea bream (Sparidae), grey mullet (Mugilidae), flounder (Pleuronectidae), and various clupeids, and in warmer waters these are joined by many others, including cichlids, milkfish (Chanidae), silver gars (Belonidae), grouper (Serranidae), puffer fish (Tetraodontidae), grunts(Pomadasyidae), rabbit fish (Siganidae), various flatfish (Soleidae, Cynoglossidae), and rays (Dasyatidae).

Therefore, throughout the world lagoons support (often artisanal) fisheries, as well as being the location of bivalve shellfish culture (mussels, oysters, and clams). For obvious reasons, most data on secondary productivity have been collected in relation to these fisheries. The fish yield from lagoons is greater than from all other similar aquatic systems, averaging (n $= 107$) 113 kg ha^{-1} y^{-1} and with a median value of 51 kg (**Table 2**) and with 13% exceeding 200 kg ha^{-1} y^{-1}. Yields vary with the intensity of human involvement in the catching and stocking processes. In the Mediterranean Sea, the region for which most data are available, the yield without intensive intervention is some 82 kg ha^{-1}y^{-1}; with the installation of permanent fish traps this increases to 185 kg; and with the addition of artificial stocking with juvenile fish it increases to 377 kg. For a 10 year period, the fishermen's cooperative based on the Stagno di Santa Giusta in Sardinia harvested nearly 700 kg ha^{-1} y^{-1}. West African lagoons with 'fish parks' – areas that attract fish because of the provision of artificial refuges – are considerably more productive, with average fish yields of 775 kg ha^{-1} y^{-1}.

Amongst the invertebrates, penaeid shrimp, mudcrabs (Scylla), oysters, mussels, and the arcid 'cockle' Anadara are the major harvested organisms. Yields of Penaeus may attain 2500 kg ha^{-1} y^{-1}in lagoonal systems of aquaculture, whilst those of the oyster Crassostrea and the mussel Perna can both be 400 tonnesha^{-1}y^{-1}. In the Indian subcontinent wholecommunities can be economically entirely

Table 2 Mean fisheries yield from coastal lagoons in relation to those from other aquatic habitats[a]

Habitat	Fish yield (kg ha^{-1}y^{-1})	Percent of sites exceeding yield of 200 kg ha^{-1}y^{-1}
Coastal lagoons	113	13
Continental shelf	59	5
Coral reef	49	0
Fresh waters	34	2

[a](Data reproduced with permission from Chauvet, 1988.)

Table 3 Discharges in the early 1990s into two lagoons on the southern (Polish/Russian/Lithuanian) coast of the Baltic Sea(inputs in tonnes yr^{-1})[a]

Parameter	Curonian Lagoon	Vistula Lagoon
Surface area	1584 km^2	838 km^2
Mean depth	3.8 m	2.6 m
Maximum depth	5.8 m	5.2 m
Volume	6.0 km^3	2.3 km^3
Annual water exchange	27.3 km^3	?
BOD input	150 000	47 000
Inorganic N	20 000	25 000
Inorganic P	5000	2100
Chlorinated substances	1	'High concentrations'
Cu	30	?
Zn	50	?
Pb	600	?

[a](Reproduced with permission from data in *Coastal Lagoons and Wetlands in the Baltic*. WWF Baltic Bulletin 1994, no 1.)

dependent on these shrimp, crab, and fish harvests; for example, the 60 000 people living around Chilka Lake, India.

However, many of the world's lagoons are threatened habitats, suffering deterioration as a result of pollution and destruction, through reclamation and from the natural losses resulting from succession to freshwater and swamp habitats and from landwards barrier migration. The discharge of materials into lagoons and its consequences are essentially similar to those in any other semi-enclosed coastal embayment, but there is one specifically lagoonal feature. As they are often used for aquaculture and yields are generally related to primary productivity, lagoonal waters are frequently deliberately enriched to boost catches. Such enrichment varies from domestic organic wastes from the surrounding communities to commercial processed fish foods. Probably in the majority of such cases, however, the result of this nutrient injection has been eutrophication, loss of macrophytes, deoxygenation, and in several areas a change in the primary producers in the direction of a bacteria-dominated plankton and, across wide areas, benthos as well. In Mediterranean France, this all too frequent state of affairs is known as 'malaïgue'. Culture of mussels in the Thau Lagoon in France produces an input to the benthos of some 45 000 tonnes (dry weight) of pseudofecal material. Not surprisingly, at times of minimum throughput of water, malaïgues can cause mass mortality of the cultured animals and degradation of the whole habitat. Thus in Europe, malaïgues in the south, pollution in the Baltic lagoons(**Table 3**), and reclamation ofthose on the Atlantic seaboard have rendered the habitat especially threatened even at a continental level. For this reason they are now a 'priority habitat' under the European Union's Habitats Directive. Intensiv elagoonal aquaculture also injects not only nutrients, but antibiotics, hormones, vitamins, and a variety of other compounds, and the wider effects of these are giving cause for concern.

See also

Geomorphology. Macrobenthos. Molluskan Fisheries. Phytobenthos. Salt Marshes and Mud Flats.

Further Reading

Ayala Castañares and Phleger F.B. (eds.) (1969) *Lagunas Costeras. Un Simposio.* Universidad Nacional Autónoma de México.

Barnes RSK (1980) *Coastal Lagoons.* Cambridge: Cambridge University Press.

Bird ECF (1984) *Coasts*, 3rd edn. London: Blackwell.

Chauvet C (1988) *Manuel sur l'Aménagement des Pêches dans les Lagunes Côtières: La Bordigue Méditerranéenne.* FAO Document Technique sur les Pches No. 290. FAO:

Cooper JAG (1994) Lagoons and microtidal coasts. In: Carter RWG and Woodroffe CD (eds.) *Coastal Evolution*, pp. 219–265. Cambridge: Cambridge University Press.

Emery KO (1969) *A Coastal Pond Studied by Oceanographic Methods.* Elsevier.

Guélorget O, Frisoni GF, and Perthuisot J-P (1983) La zonation biologique des milieux lagunaires: définition d'une échelle de confinement dans le domaine paralique mediterranéen. *Journal de Recherche Océanographique, Paris* 8>: 15–36.

Kapetsky JM and Lasserre G (eds.) (1984) *Management of Coastal Lagoon Fisheries*, vol. 2 vols.. Rome: FAO.

Kjerfve B (ed.) (1994) *Coastal Lagoon Processes*. Elsevier.

Lasserre P and Postma H (eds) (1982) Les Lagunes Ctières. *Oceanologica Acta* (Volume Spécial.)

Rosecchi E and Charpentier B (1995) *Aquaculture in Lagoon and Marine Environments*. Tour du Valat.

Sorensen J, Gable F, and Bandarin F (1993) *The Management of Coastal Lagoons and Enclosed Bays*. American Society of Civil Engineers.

Special issue (1992) *Vie et Milieu* 42(2): 59–251.

Yáñez-Arancibia A (ed.) (1985) *Fish Community Ecology in Estuaries and Coastal Lagoons*. Universidad Nacional Autónoma de México.

BEACHES, PHYSICAL PROCESSES AFFECTING

A. D. Short, University of Sydney, Sydney, Australia

Introduction

Beach systems occur on shorelines throughout the world and are a product of three basic factors – wave, sediment, and substrate. They are, however, also influenced by tides and impacted by a range of geological, biotic, chemical, and temperature factors. Beach systems therefore occupy a considerable range of environments, from the picturesque tropical beaches, to those exposed to the full forces of North Atlantic storms, to the usually frozen shores of the Arctic Ocean. In addition formative waves can range from low to extreme, and sediment from fine sand to boulders. This article looks at how the two main processes – waves and tides – interact with sediment to form three types of beach systems: wave-dominated, tide-modified, and tide-dominated, the three representing all the world's beach types. It then considers some of the major factors that modify beach systems on a global and regional scale, including geology, temperature, biota, and chemical processes.

Beaches

Beaches are a wave-deposited accumulation of sediment lying between wave base and the limit of wave uprush or swash. The wave base is the seaward limit or depth from which average waves can entrain and transport sediment shoreward. The essential requirements for a beach to form are therefore an underlying substrate, sediment, and waves. In two-dimensions as waves shoal across the nearshore zone, between wave base and the breaker zone or shoreline, they generate a concave upward profile. As the waves transform at breaking to swash, the swash builds a steep beach face or swash zone, which caps the beach system.

In reality beaches are usually more complex. Waves can range in size from a few centimeters to several meters, and period from 1 to 20 s, and additional processes, particularly tides influence both the waves and the resulting beach morphology. Beaches can also be composed of sediments ranging from fine sand to boulders. In addition, the underlying substrate, known as the geological inheritance provides both the two-dimensional foundations, as well as potential two- and three-dimensional irregularities such as reefs and headlands, all of which influence wave transformation, sediment transport, and two- and three-dimensional beach morphology. Finally when waves break across a surf zone, they generate three-dimensional flows that produce a more complex three-dimensional surf zone and beach morphology, that lies between the nearshore and swash zones. Beaches, therefore, all contain a nearshore zone dominated by wave shoaling processes; they may contain a surf zone dominated by wave breaking and surf zone currents, and finally a swash zone dominated by wave uprush and backwash (**Figure 1**).

Beaches also range from relatively simple low-energy, two-dimensional, tideless systems with waves surging against a uniform shoreline, to more complex higher energy systems, incorporating surf zones with three-dimensional circulation, such as rip currents, associated with highly rhythmic bars and shorelines. In addition geological inheritance can introduce a wide spectrum of bedrock and sedimentology inputs that cause additional wave transformation through variable wave refraction and attenuation, as well as influence sediment type and mobility.

This article will cover the full range of physical beach systems from the low to high wave energy wave-dominated beaches. It will then address the influence of increasing tide range in the tide-modified and tide-dominated beaches. Finally it will briefly review the role of geology (e.g., headlands and reefs) and other physical, chemical, and biotic factors in modifying beach systems.

Wave-dominated Beaches

Wave-dominated beaches occur in environments where waves are high relative to the tide range. This can be defined quantitatively by the relative tide range (RTR)

$$RTR = TR/Hb \qquad [1]$$

where TR is the spring tide range and Hb the average breaker wave height. When $RTR < 3$ beaches are wave dominated, when $3 < RTR < 15$ they are tide-modified and when the $RTR > 15$ they become tide-dominated.

Within any beach environment, wave height and period and sediment size can all vary substantially. In

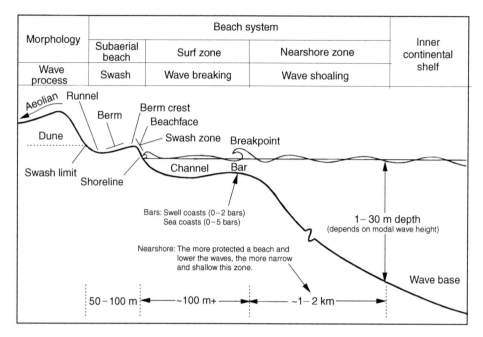

Figure 1 Definition sketch of a high-energy wave-dominated beach system. The beach contains a subaerial beach dominated by swash, a surf zone dominated by breaking waves and the nearshore zone across which waves shoal. (Reproduced from Short, 1999.)

order to accommodate the potentially wide range of wave height–period–sediment combinations **Figure 2** presents a sensitivity plot which defines the three wave-dominated beach types, reflective, intermediate and dissipative, using the dimensionless fall velocity

$$\Omega = \text{Hb}/T\text{Ws} \qquad [2]$$

where T is wave period (s) and W_s is sediment fall velocity (ms^{-1}).

Reflective Beaches

Reflective beaches occur when $\Omega < 1$, which requires a combination of lower waves, longer periods and particularly coarser sands. They occur on sandy open swell coasts when waves average less than 0.5 m, and on all coasts when beach sediments are coarse sand or coarser, including all gravel through boulder beaches. On gravel–boulder reflective beaches waves may exceed a few meters. They are, however, all characterized by a nearshore zone of wave shoaling that extends to the shoreline. Waves then break by plunging and/or surging across the base of the beach face. The ensuing strong swash rushes up the beach, combining with the coarse sediments to build a steep beach face, commonly capped by well-developed beach cusps and/or a berm (**Figures 3** lower and **4**). When the sediment consists of a range of grain sizes, the coarser grains accumulate as a coarser steep step below the zone of wave breaking, at the base of the beach face.

Reflective beaches therefore consist of a concave upward nearshore zone composed of wave ripples and relatively fine sand across which waves shoal. As the waves break at the base of the beach there may be a coarser step up to a few decimeters high, then a relatively steep (4–10°) beach face, possibly capped with cusps. The cusps are a product of cellular circulation on the high tide beach resulting from sub-harmonic edge waves produced from the interaction of the incoming and reflected backwash. The high degree of incident wave reflection off the beach face is responsible for naming of this beach type, i.e., reflective. Apart from the cosmetic beach cusps and swash circulation, if present, these are essentially two-dimensional beaches with no alongshore variation in either processes or morphology. On sand beaches they also represent the lower energy end of the beach spectrum and as such are relatively stable systems, only experiencing beach change during periods of higher waves which tend to erode the berm and deposit it as an attached low tide bar or terrace (**Figure 3** lower). During following lower waves the terrace quickly migrates back up onto the beach as a new berm or cusps.

Intermediate Beaches

Intermediate beaches are called such as they represent a suite of beach types between the lower energy reflective and higher energy dissipative. They are the beaches that form under moderate (Hb > 0.5 m) to high waves, on swell and seacoasts, in fine to

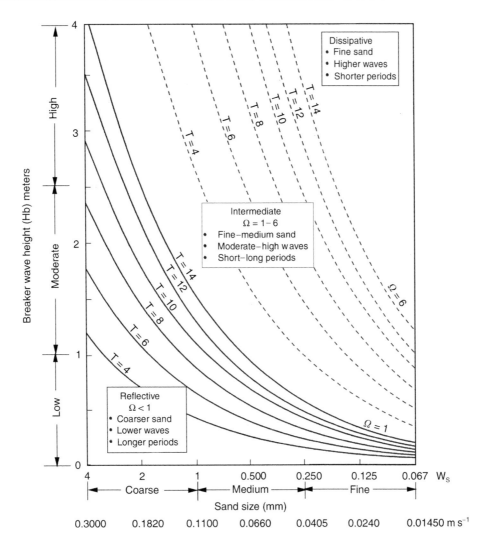

Figure 2 Sensitivity plot of the contribution of wave height, sediment size and wave period to Ω and beach type. To use the chart determine the breaker height, period (T) and grain size/fall velocity (phi or m s^{-1}). Read off the wave height and grain size, then use the period to determine where the boundary of reflective/intermediate, or intermediate/dissipative beaches lie. $\Omega = 1$ along solid T lines, and $\Omega = 6$ along dashed T lines. Below the solid lines $\Omega < 1$ the beach is reflective, above the dashed lines $\Omega > 6$ the beach is dissipative, between the solid and dashed lines Ω is between 1 and 6 and the beach is intermediate. (Reproduced from Short, 1999.)

medium sand (**Figure 2**). The two most distinguishing characteristics of intermediate beaches is first a surf zone, and second cellular rip circulation, usually associated with rhythmic beach topography. Since intermediate beaches can occur across a wide range of wave conditions, they consist of four beach states ranging from the lower energy low tide terrace to the high energy longshore bar and trough (**Figure 3**).

Intermediate beaches are controlled by processes related to wave dissipation across the surf zone which transfers energy from incident waves with periods of 2–20 s, to longer infragravity waves with periods > 30 s. As a consequence, incoming long waves associated with wave groupiness, increase in energy and amplitude across the surf zone and are manifest at the shoreline as wave set up (crest) and

set down (trough). They then reflect off the beach leading to an interaction between the incoming and outgoing waves to produce a standing wave across the surf zone. It is believed that standing edge waves trapped in the surf zone are responsible for the cellular circulation that develops into rip current circulation, that in turn is responsible for the high degree of spatial and temporal variability in intermediate beach morphodynamics.

Low tide terrace Low tide terrace (or ridge and runnel) beaches are characterized by a continuous attached bar or terrace located at low tide (**Figure 5**). Usually smaller waves (0.5–1 m) break across the bar at low tide, whereas at high tide they may remain unbroken and surge up the reflective

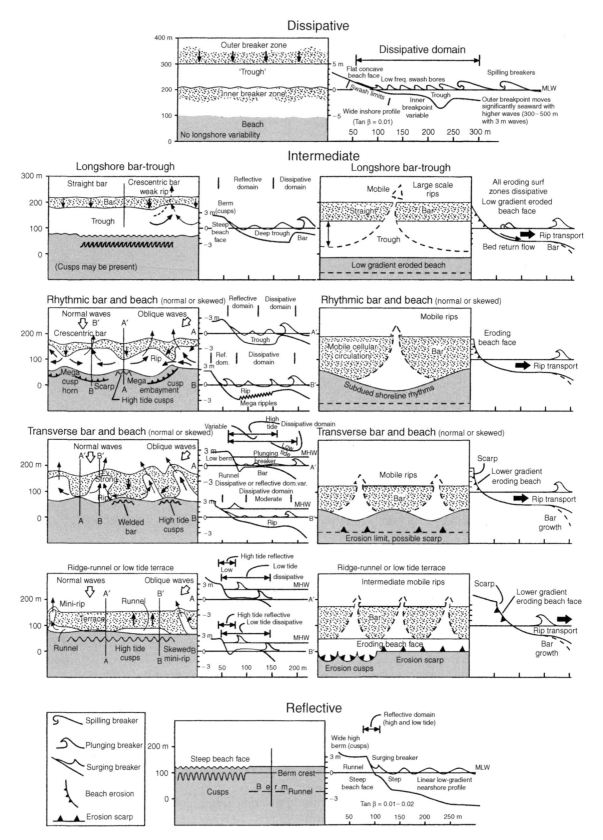

Figure 3 Three-dimensional sequence of wave-dominated beach changes for accretionary (left) and erosional (right) wave conditions. The sequence ranges from dissipative (top), through intermediate, to reflective (lower). (Reproduced from Short, 1999.)

Figure 4 Pearl Beach near Sydney, Australia, is a modally reflective beach composed of medium to coarse sand and receiving usually lower swell waves. Note the very narrow zone of wave breaking, well-developed beach cusps and cusp-controlled segregated swash circulation (A.D. Short).

Figure 6 Well-developed transverse bar and rip beach with high rhythmic shoreline (megacusp horns and embayments), attached bars and well-developed rip channels, Johana Beach, Victoria (A.D. Short).

Figure 5 A wide, well-developed low tide terrace at Crowdy Head, New South Wales. Note the transition from transverse bar and rips (foreground) to mini-rips, then continuous low tide terrace (A.D. Short).

high tide beach face. At mid to low tide cellular circulation in the surf zone usually produces weak rip current flows, which may be transitory across a planar bar, or may reside in shallow rip channels (mini-rips) (**Figure 3**).

Transverse bar and rip Transverse bar and rip beaches form under moderate waves (1–1.5 m) on swell coasts. They consist of well-developed rip channels, which are separated by shallow bars, which are attached and perpendicular or transverse to the beach. Variable wave breaking and refraction across the shallow bars and deeper rip channels leads to a longshore variation in swash height and approach, which reworks the beach to form prominent megacusp horns, in lee of bars, and embayments, in

lee of channels (**Figures 3** and **6**). Water tends to flow shoreward over the bars, then into the rip feeder channels. The flow moves close to the shoreline and converges laterally in the rip embayment. It then moves seaward in the rip channel as a relatively narrow (a few meters), strong flow $(0.5–1\,\mathrm{m\ s^{-1}})$, called a rip current. The current usually moves through the surf zone and beyond the breakers as a rip head. This beach state has extreme spatial–longshore variation in wave breaking, surf zone and swash circulation, and beach and surf zone topography, leading to a high unstable and variable beach system.

Rhythmic bar and beach The rhythmic bar and beach state forms during periods of moderate to high waves (1.5–2 m) on swell coasts. The high waves lead to greater surf zone discharges that require deeper and wider rip feeder channels and rip channels to accommodate the flows. Rips flow in well-developed rip channels, separated by transverse bars, but the bars are detached from the beach by the wider feeder channels (**Figure 7**). The bar is highly rhythmic moving seaward where rips exit the surf, and shoreward over the bars, whereas the shoreline is also highly rhythmic for the reason stated above. It is not uncommon for the more energetic swash in the rip embayments to scarp the beach face, while the adjoining horns are accreting (**Figure 3**).

Longshore bar and trough The longshore bar and trough systems are a product of periods of higher waves which excavate a continuous longshore trough between the bar and the beach. Waves break heavily on the outer bar, reform in the trough and then break again at the shoreline, often producing a steep reflective beach face (coarser sand, **Figure 3**)

Figure 7 Rhythmic bar and beach at Egmond, The Netherlands. Note the shoreline rhythms and detached crescentic bar (A.D. Short).

Figure 8 Aracaju Beach in north-east Brazil is a well-developed double bar dissipative beach produced by persistent trade wind waves breaking across a fine sand beach. Note the straight beach scarp, a characteristic of dissipative beach erosion. Photograph by AD Short.

or a low tide terrace (finer sand). On swell coasts waves must exceed 2 m to produce this intermediate beach state. The high level of surf zone discharge excavates the deep trough to accommodate the flow. Surf zone circulation is both cellular rip flows, as well as increasingly shore normal bed return flows (see below).

Dissipative Beaches

Dissipative beaches represent the high-energy end of the beach spectrum. They occur in areas of high waves, prefer short wave periods, and must have fine sand. They are relatively common in exposed sea environments where occasional periods of high, short storm waves produce multibarred dissipative beach systems, as in the North and Baltic seas. They also occur on high-energy midlatitude swell and storm wave coasts as in northwest USA, southern Africa, Australia, and New Zealand. On swell coasts waves must exceed 2–3 m for long periods to generate fully dissipative beaches. They are characterized by a wide long gradient beach face and surf zone, with two and more shore parallel bars forming across the surf zone (**Figures 3** upper and 8). The low

gradient is a product of both the fine sand, and the dominance of lower frequency infragravity swash and surf zone circulation, which act to plane down the beach. Their name comes from the fact that waves dissipate their energy across the many bars and wide surf zone. The dissipation of the incident wave energy leads to a growth in the longer infragravity energy, which becomes manifest as a strong set up and set down at the shoreline. The standing wave generated by the interaction of the incoming and outgoing waves may have two or more nodes across the surf zone. It is believed that the bar crests from under the standing wave nodes and troughs under the antinodes. Surf zone circulation is vertically segregated. Wave bores move water shoreward toward the surface of the water column. This water builds up against the shoreline as wave set up. As it sets down the return flows tends to concentrate toward the bed, which propels a current across the bed of the surf zone (below the wave bores) called bed return flow.

Like the reflective end of the beach spectrum dissipative beaches are remarkably stable systems. They are designed to accommodate high waves, and can accommodate still higher waves by simply widening their surf zone and increasing the amplitude of the standing waves, while periods of lower waves are often too short to permit substantial onshore sediment migration.

Bar Number

All dissipative beaches have at least two bars and in sea environments commonly have three or more, with up to 10 recorded in the Baltic Sea. In addition, whereas the reflective and intermediate beaches described above assume one bar, all may occur in the lee of an outer bar or bars. In energetic waves the number of bars can be empirically predicted by the bar parameter

$$B^* = \chi_s/g \tan \beta T^2 \qquad [3]$$

where χ_s is the distance to where nearshore gradient approaches zero, g is the gravitational constant and β the beach gradient. Bar number increases as the gradient and/or period decrease, with no bar occurring when $B^* < 20$, one bar between 20–50, and two or more bars > 50.

In two or more bar environments sequential wave breaking lowers the height across the surf zone. As a consequence, Ω (eqn [2]) decreases shoreward and the bar type adjusts to both the lower wave height and Ω. A common multibar sequence is a dissipative outer (storm) bar, with more rhythmic (intermediate) inner bar/s, and a lower energy beach (reflective/low tide terrace).

Modes of Beach Change and Erosion

Figure 3 illustrates an idealized sequence of wave-dominated beaches from dissipative to reflective. It also suggests that beaches can change from one type or state to another, either though beach accretion (movement downward, on left side) or beach erosion (upward, on right side). Beach change, whether erosion or accretion is simply a manifestation of a beach adapting to changing wave conditions by attempting to become more reflective and move sediment shoreward (lower waves) or more dissipative and move sediment seaward (higher waves). Shoreward sediment movement normally requires onshore bar migration, leading in **Figure 3**, from a longshore bar and trough, to a rhythmic bar and beach, the bars then attaching to the beach in the transverse bar and rip, the rips infilling at the low tide terrace, and finally migrating up onto the reflective beach. The rip channels tend to remain stationary during beach accretion, as they are topographically fixed by the bars. Conversely, higher waves and greater surf zone discharges require a wider and deeper surf zone to accommodate the flows. Sediment is eroded from the beach faces and inner surf zone (see scarps, **Figure 2**) and moved seaward to build a wider bar/s. Large erosional transitory rips and increasingly bed return flow transport the sediment seaward.

In nature, movement through this entire sequence is rare, requiring several weeks of low waves for accretion, and many days of high waves for the full erosion. In reality waves rarely stay so low or so high long enough, and most beaches shift between one of two states as they adjust to ever-changing wave conditions, particularly in swell environments. In sea environments, calm periods between storms lead to a stagnation of often high-energy beach topography, which will await the next storm to be reactivated.

Tide-modified Beaches

Most of the world's beaches are affected by tide. On most open coasts where tides are low (< 2 m) waves dominate and tidal impacts are minimized, as in the wave-dominated systems. However, as tide range increases and/or wave height decreases tidal influences become increasingly important. To accommodate these influences beaches, still by definition wave-formed, can be divided into three tide-modified and three tide-dominated types, as defined by eqn [1].

The major impact of increasing tide range is to shift the location of the shoreline between high and low tide, depending on the shoreline gradient, by tens to hundreds of meters. This shift not only moves the shoreline and accompanying swash zone, but also the surf zone, if present, and the nearshore zone. Whereas wave-dominated beaches have a 'fixed' swash–surf–nearshore zone, on tide-modified beaches they are more mobile. The net result is a smearing of the three dominant wave processes of shoaling (nearshore zone), breaking (surf zone) and swash (swash zone), as a section of intertidal beach can be exposed to all three processes at different states of the tide. Because all three zones are mobile, except for a brief period at high and low tide, there is a reduction in the time any one process can fully imprint its dominance on a particular part of the beach. As a consequence there is a tendency for swash processes to dominate only the spring high tide beach, for surf zone processes to dominate only the beach morphology around low tide, during the turn of the low tide, while the shoaling wave processes become increasingly dominant overall, producing a lower gradient more concave beach cross-section.

If we take the three wave-dominated beach types and increase tide range to an RTR of between 3 and 15 the following three beach types result.

Reflective Plus Low Tide Terrace

When $\Omega < 2$ and RTR > 3 the high tide beach remains reflective, much like its wave-dominated counterpart. However, the higher tide range leads to an exposure of the inner nearshore zone, which when swept by surf and swash processes becomes an attached low tide terrace (**Figure 9**). It has a sharp break in slope between the steep beach face and

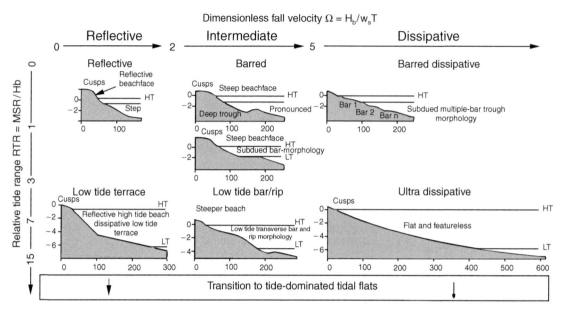

Figure 9 A simplified classification of beaches based on the dimensionless fall velocity and relative tide range. Wave-dominated beaches (top) have an RTR<3, whereas tide-modified beach types have RTR 3–15 (lower). (Reproduced from Short, 1999.)

Figure 10 Reflective high tide beach and wide low tide terrace, Embleton Bay, Northumberland (A.D. Short).

morphology, which may include transverse and rhythmic bars and rips. The surf zone processes, including the rips currents, are however only active at low tide. In this system the higher waves maintain a surf right across the systems as the tide rises and falls, but only during the low tide turn-around is there sufficient time for the cellular surf zone processes to imprint themselves on the morphology, producing surf zone topography essentially identical to the wave-dominated intermediate counterparts. However, there is no associated shoreline rhythmic forms, as the tide-modified surf zone is detached from the shoreline by as much as a few hundred meters (**Figure 11**). At mid to high tide the low tide morphology is dominated by shoaling waves as the surf zone moves rapidly shoreward.

flatter terrace at the point where the water table exits from the beach (**Figure 10**). The low tide drainage from the beach is another feature of tide-modified beaches. At low tide waves dissipate across the flat terrace, whereas at high tide they surge across the steep reflective high tide beach.

Reflective Plus Low Tide Bar and Rips

In areas of moderate waves ($\Omega = 2$–5) and tide range (RTR $= 3$–7) the tide-modified beaches consist of a high tide reflective beach, a usually wider intertidal zone, which may contain transitory swash ridges, and a low tide bar dominated by surf zone

Ultradissipative

Higher energy tide-modified beaches composed of fine sand are characterized by a wide, low gradient concave upward, flat, and featureless, beach and intertidal system (**Figures 9** and **12**). These beaches are swept by both high incident waves and their associated long waves (set up and set down), as well as shoaling waves at high tide. Even at low tide the waves cannot imprint their dissipative bar morphology on the topography. The only surface features usually present on the low tide beach are wave and swash ripples.

Figure 11 Nine Mile Beach, Queensland, showing a dye-highlighted low tide rip current, the 100 m wide intertidal zone with a low swash ridge, and the linear high tide reflective beach (right) (A.D. Short).

Figure 12 Rhossili Beach in Wales is an exposed high-energy tide-modified beach shown here at low tide. Note the wide, flat, featureless intertidal zone, and wide dissipative surf zone, which moves across the beach every six hours (A.D. Short).

Tide-dominated Beach

When the RTR exceeds 15, the tide range is more than 15 times the wave height. As the maximum global tide range is about 12 m, and usually much less, this means that most tide-dominated beaches also receive low (<1 m) to very low waves, and are commonly dominated by locally generated wind waves. Tide-dominated beaches in fact represent a transition from beach to tidal flats, though they must still maintain a high tide beach. Tide-dominated beaches are characterized by a usually steep high tide beach, and a wide, low gradient (<1°) sandy intertidal zone, which in temperate to tropical locations is usually bordered by subtidal seagrass meadows.

Beach and Sand Ridges

The beach and sand ridge type consists of a steep, reflective low-energy high tide beach, which is only active during spring high tides, and in seagrass locations is usually littered with seagrass debris. It is fronted by a low gradient sand flats which contain multiple, sinuous, shore-parallel, equally spaced, low amplitude sand ridges. On the Queensland coast the number of ridges range from one to 19 and averages seven. What causes the ridges is still not understood, and they are very different in size and spacing from the standing wave-generated multiple dissipative bars. They do, however, occur on the more exposed sand flat locations, compared to the next beach type.

Beach and Sand Flats

A beach and sand flats is just that; a very low energy periodically active high tide reflective beach, fronted by a wide flat, featureless intertidal sand flat. The only features on the sand flats are usually associated ground-water drainage from the high tide beach, and bioturbation particularly crab holes and balls.

Tidal Sand Flat

The final 'beach' type is the tidal sand flat, which is a beach in so far as it consists of sand and is exposed to sufficient waves to winnow out any finer sediments. It is flat and featureless; it may have irregular, discontinuous, shelly high tide beach deposits, interspersed occasionally with mangroves in tropical locations, or salt marsh in high latitudes. It is transitional between the beaches and the often muddy, tidal flats with associated inter- and supra-tidal flora.

Beach Modification

The above beach systems assume beaches are solely a product of the wave-tides and sediment regimes. Most beaches are, however, influenced to some degree by geological inheritance, and, depending on latitude, by temperature-controlled processes.

Geological Inheritance

The dominant physical factors modifying beaches are related to the role of what is generally termed geological inheritance, that is, pre-existing bedrock (or artificial structures) and sedimentological controls. Bedrock influences beaches as headlands, which form boundaries and determine beach length. In addition headlands and associated subaqueous substrate (reefs, rocky seabed) induce wave refraction and attenuation, which in turn influences beach shape and breaker wave height. Wave refraction around

Figure 13   A small embayed, headland-controlled beach, with a topographically controlled rip exiting against the headland, Point Peter, South Australia (A.D. Short).

Figure 14   Frozen tundra (dark patches left), beach covered by snow and ice (center) and shore-parallel ice ridges (right); Point Lay, north Alaska (A.D. Short).

headlands and reefs will tend to produce swash aligned beaches in their lee, as well as lower energy beaches. Wave attenuation over rock substrates will reduce wave energy and through breaker height influence beach type. Eroding basement rocks and cliffs can also supply sediment to the beach budget.

In energetic embayed (headland-controlled) beaches, beach systems can be classified as normal, where the headland exerts no influences, as in areas with low waves or on long beaches. On shorter beaches and during moderate waves the beaches become transitional, with normal surf zone circulation in the center, and headland-induced rips to either end. Finally in high energy situations (Hb > 3 m) and on shorter beaches (< few kilometers) the entire surf zone circulation becomes cellular, meaning it is controlled by the bedrock topography, leading to large-scale megarips draining the surf zone, and often aligned against a headland (**Figure 13**).

Temperature

Frozen beaches   Ocean temperature is responsible for two dramatic physical impacts on beach systems. In the high latitudes when air temperature falls below zero the beach surface freezes and when ocean temperature falls below − 4°C the water surface freezes. In places like the Arctic Ocean beach processes cease for several months of the year, as a frozen ocean surface and snow-covered beach negate any physical activity (**Figure 14**) other than tides. When the ice and snow finally melts, near normal beach processes return and control the long-term beach morphology.

Biotic influences   In the tropics, warm water (> 26°C) permits the prolific growth of coral and algal reefs in suitable shallow water environments.

Figure 15   Three bands of beachrock at Shoal Cape, Western Australia. Note the wave breaking over the successive reefs, leading to a low energy inner beach, on an exposed high-energy coast (A.D. Short).

Where these reefs lie seaward of beach systems, they lower the wave height through wave breaking and therefore lead to solely reflective beaches in their lee. The reefs also supply carbonate detritus to the beaches leading to coarse carbonate rich beach systems.

In temperate semiarid to arid shelf regions such as across southern Australia and South Africa, the prolific growth of shelf carbonate supplies the bulk of the beach sediments, leading to extensive mid-latitude carbonate coasts and beach systems.

Chemical influences   Water chemistry modifies tropical beaches through the formation of beachrock, the cementation of the intertidal sand grains by carbonate cement. This commonly occurs in all tropical beach locations and during beach erosion leads to exposure of the beachrock parallel to the shoreline.

On arid to temperate carbonate-rich coasts sub-aerial pedogenesis led to the formation of calcarenite in beach and dune systems, particularly during Pleistocene low sea levels. When these systems are subsequently drowned by rising sea level, they remain as dune or beach calcarenite, and are manifest as reefs, islands, and rocky bluffs and cliffs, all of which exert a geological influence on the shoreline (**Figure 15**).

See also

Coastal Circulation Models. Geomorphology. Sandy Beaches, Biology of. Storm Surges. Tides. Waves on Beaches.

Further Reading

Carter RWG (1988) *Coastal Environments. An Introduction to the Physical, Ecological and Cultural Systems of Coastlines.* London: Academic Press.

Hardisty J (1990) *Beaches: Form and Process.* London: Unwin and Hyman.

Horikawa K (ed.) (1988) *Nearshore Dynamics and Coastal Processes.* Tokyo: University of Tokyo Press.

King CAM (1972) *Beaches and Coasts.* London: Edward Arnold.

Komar PD (1998) *Beach Processes and Sedimentation.* 2nd edn. Englewood Cliffs, NJ: Prentice-Hall.

Pethick J (1984) *An Introduction to Coastal Geomorphology.* London: Edward Arnold.

Schwartz ML and Bird ECF (eds) (1990) Artificial Beaches. *Journal of Coastal Research Special Issue 6.*

Short AD (ed.) (1993) Beach and surf zone morphodynamics. *Journal of Coastal Research Special Issue 15.*

Short AD (ed.) (1999) *Handbook of Beach and Shoreface Morphodynamics.* Chichester: John Wiley.

SANDY BEACHES, BIOLOGY OF

A. C. Brown, University of Cape Town, Cape Town, Republic of South Africa

Introduction

Some 75% of the world's ice-free shores consists of sand. Nevertheless, biological studies on these beaches lagged behind those of other coastal marine habitats for many decades. This was probably because there are very few organisms to be seen on a beach at low tide during the day, when biological activity is at a minimum. A very different impression may be gained at night, but the beach fauna is then much more difficult to study. Even at night few species may be observed but these can be present in vast numbers. Most sandy-beach animals are not seen at all, as they are very tiny, virtually invisible to the naked eye, and they live between the sand grains (i.e., they are interstitial). A large number of these species, comprising the meiofauna, may be present. A further striking characteristic of ocean beaches is the total absence of attached plants intertidally and in the shallow subtidal. The animals are thus deprived of a resident primary food source, such as exists in most other habitats. We may, therefore, suspect that ecosystem functioning differs considerably from that found elsewhere and research has shown that this is, indeed, the case.

The Dominating Factor

In harsh, unpredictably varying environments, physical factors are far more important in shaping the ecosystem than are biological interactions; ocean beaches are no exception. Here the 'superparameter' controlling community structure, species diversity, and the modes of life of the organisms is water movement – waves, tides, and currents. Water movements determine the type of shore present and then interact with the sand particles so that a highly dynamic, unstable environment results. Instability reaches a peak during storms, when many tons of sand may be washed out to sea and physical conditions on the beach become chaotic. To exist in the face of such instability, beach animals must be very mobile and all must be able to burrow. Indeed the ability to burrow before being swept away by the next swash is critical for those animals that habitually emerge from the sand onto the beach face. The more exposed the beach to wave action, the more rapid does burrowing have to be. Furthermore, on most ocean beaches the sand is too unstable not only to support plant growth but also to allow permanent burrows to exist intertidally. Consequently animals such as the bloodworm, *Arenicola*, and the burrowing prawn, *Callianassa*, are found only on sheltered beaches. On exposed beaches, there is virtually nothing the aquatic fauna can do to modify their environment; they are entirely at the mercy of whatever the physical regime may dictate. During storms, behavior patterns must often change if the animals are to survive. Some animals simply dig themselves deeper into the sand, others may move offshore, and some semiterrestrial species (e.g., sand hoppers) move up into the dunes until the storm has passed. Notwithstanding these behavior patterns, mortality due to physical stresses often outweighs mortality due to predation.

Tidal Migrations and Zonation

Despite this extremely harsh environment, the essential mobility of the beach fauna permits exploitation of tidal rise and fall to an extent not available to more-sedentary animals. The meiofauna migrate up and down through the sand column in time with the tides, achieving the most optimal conditions available, and maximizing food resources and, in some cases, promoting photosynthesis. Many of the larger species (i.e., macrofauna) emerge from the sand and follow the tide up and down the slope of the beach face. The aquatic species tend to keep within the swash zone as it rises and falls; this is the area in which their food is likely to be most plentiful. As a bonus, this behavior also reduces predation, as the swash is too shallow for predatory fish to get to them, and most shore birds do not enter the water. Many semiterrestrial crustaceans, above the waterline, follow the falling tide down the slope, feed intertidally, and then migrate back to their positions above high water-mark as the tide rises. They do so only at night, remaining buried during the day, so that predation by birds is at a minimum.

If the macrofauna is sampled at low tide, a zonation of species is often apparent (**Figure 1**) and attempts have been made to relate these zones to those found on rocky shores or with changing physical conditions up the beach. However, the number of zones differs on different beaches and the zones tend to be blurred, with considerable overlap. Moreover,

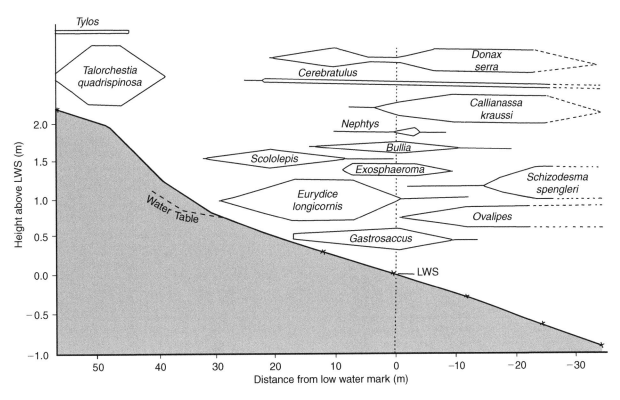

Figure 1 Low-tide distribution of the macrofauna up the beach slope at Muizenberg, near Cape Town. Some zonation is apparent. *Nephtys* and *Scololepis* are annelid worms; *Schizodesma* is a bivalve mollusk, larger than *Donax*; *Cerebratulus* is a nemertean worm; *Eurydice* and *Exosphaeroma* are isopods (pill bugs), *Ovalipes* is a swimming crab. The remaining genera are mentioned in the text. LWS, the level of Low Water of Spring tide.

due to tidal migrations, the zones change as the tide rises. It has, therefore, been suggested that only two zones are apparent on all beaches at all states of the tide – a zone of aquatic animals and, higher up the shore, a zone of semiterrestrial air breathers; these are known as Brown's Zones.

Longshore Distribution

Not only is the distribution of the macrofauna up the beach slope not uniform but their longshore distribution is also typically discontinuous, or patchy. There is no one reason for this patchy distribution in all species. Variations in penetrability of the sand account for discontinuity in some species, breeding behavior or food maximization in others. In the semiterrestrial pill bug, *Tylos*, aggregations are apparently an incidental effect of their tendency to use existing burrows.

Diversity and Abundance: Meiofauna

The meiofauna consists of interstitial animals that will pass through a 1.0 mm sieve. They tend to be slender and worm-shaped, a necessary adaptation for gliding or wriggling between the grains. Nematoda (round worms) and harpacticoid copepods (small Crustacea, **Figure 2**) are usually dominant. However, most animal phyla are represented, including Platyhelminthes (flat worms), Nemertea (ribbon worms), Rotifera (wheel animalcules), Gastrotricha (**Figure 2**), Kinorhyncha, Annelida (segmented worms) and various Arthropoda such as mites and Collembola (spring tails). Rarer meiofaunal animals include an occasional cnidarian (hydroid), a few species of nudibranchs (sea slugs), tardigrades, a bryozoan, and even a primitive chordate. One whole phylum of animals, the Loricifera, is found only in beach meiofauna, as is the crustacean subclass Mystacocarida (**Figure 2**).

Both the diversity and abundance of the meiofauna are commonly highest on beaches intermediate between those with very gentle slopes (i.e., dissipative) and those with abrupt slopes (reflective). This is largely because these intermediate beaches also present an intermediate particle size range; the occurrence of most meiofaunal species is governed not by particle size itself but by the sizes of the pores between the grains. Pore size (porosity) also governs the depth to which adequate oxygen tensions penetrate, this depth being least in very fine sand. This

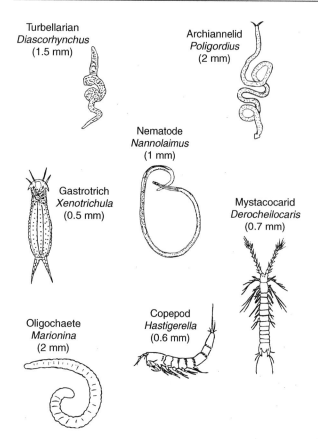

Figure 2 Some characteristic meiofaunal genera found in sandy beaches. (Reproduced from Brown and McLachlan, 1990 with permission from Elsevier Science.)

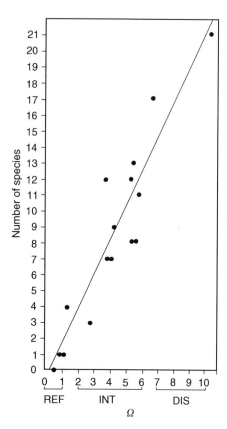

Figure 3 Relationship betwseen macrofaunal diversity and beach morphodynamic state (Dean's parameter, Ω) for 23 beaches in Australia, South Africa, and the USA. REF, reflective; INT, intermediate; DIS, dissipative. (Reproduced from Brown and McLachlan, 1990 with permission from Elsevier Science.)

limits the vertical distribution of aerobic members of the meiofauna. Groups such as the Gastrotricha are completely absent from fine sands and harpacticoid copepods are relatively uncommon. As with the macrofauna, a zonation of the meiofauna is often apparent; maximum diversity and abundance are usually attained in the sand of the mid to upper part of the intertidal beach. Meiofaunal numbers typically average $10^6 \, m^{-2}$ but they may be as low as $0.05 \times 10^6 \, m^{-2}$ or as high as $3 \times 10^6 \, m^{-2}$. One of the few beaches whose meiofauna has been studied intensively is a beach on the island of Sylt, in the North Sea. No less than 652 meiofaunal species have been identified here.

Diversity and Abundance: Macrofauna

As with the meiofauna, the diversity and abundance of the macrofauna vary greatly with beach morphodynamics. Maximum numbers, species, and biomass are found on gently sloping beaches with wide surf zones (i.e., dissipative beaches). Important criteria here, apart from slope, are the relatively fine sand, the extensive beach width and the long,

essentially laminar flow of the swash, which results in minimum disturbance of sand, facilitates the transport of some species (see below) and renders burrowing easier. At the other extreme are abruptly sloping, reflective beaches, with waves breaking on the beach face itself, leading to turbulent swash, considerable sand movement and large particle size. Such beaches are inhospitable to virtually all species of aquatic macrofauna. There is, in fact, some correlation between diversity/abundance/biomass and beach slope, breaker height or the particle size of the sand, as these are all related. The mean individual size of the macrofaunal animals correlates best with particle size, whereas biomass correlates best with breaker height. Diversity and abundance are usually best correlated with Dean's Parameter (Ω) (**Figure 3**), which gives a good indication of the overall morphodynamic state of the beach:

$$\Omega = \frac{Hb}{WsT}$$

where Hb is average breaker height, Ws the mean fall velocity of the sand and T the wave period.

Beaches with extensive surf zones and gentle slopes may harbour over 20 intertidal species of aquatic macrofauna but this number decreases with increasing slope until fully reflective beaches may have no aquatic macrofauna at all.

Biomass varies greatly but on dissipative beaches averages about 7000 g dry mass per meter stretch of beach. One beach in Peru was found to have a biomass of no less than $25\,700\,\mathrm{g\,m}^{-1}$. Although diversity and abundance generally decrease with increasing wave exposure, some very exposed beaches display a surprisingly high biomass due to the larger individual sizes of the macrofauna. Species diversity and biomass tend to decrease from low to high tide marks but often increase again immediately above the high water level, especially if algal debris (e.g., kelp or wrack) is present.

On oceanic beaches, there is a gap in size between the meiofaunal animals and the macrofauna. This is probably because, although the meiofauna move between the grains, macrofaunal animals burrow by displacing the sand. They therefore have to be relatively robust and much larger than the sand grains. Common aquatic macrofauna include filter-feeding bivalve mollusks (clams) of several genera, of which the most wide-spread is *Donax*. Scavenging gastropod mollusks (whelks) may also occur; most is known about the plough-shell, *Bullia* (**Figure 4**). On relatively sheltered shores, a variety of annelid worms occurs, but these decrease in numbers and diversity with increasing wave action, small crustaceans becoming more dominant. On tropical and subtropical beaches, the mole-crabs *Hippa* and *Emerita* (**Figure 5**) may achieve dense populations, whereas on temperate beaches mysid shrimps (**Figure 6**) and aquatic isopods (pill bugs) may be much in evidence.

Above the water-line, semiterrestrial isopods, such as *Tylos* (**Figure 7**), are often abundant, and sand hoppers (or 'beach fleas') (*Talitrus*, **Figure 8**, *Orchestia* and *Talorchestia*) may occur in their millions on temperate beaches. On tropical and subtropical beaches, semiterrestrial crabs, such as the ghost crab, *Ocypode* (**Figure 9**), are important. In addition, insects may be present in some numbers; these include kelp flies and their relatives, staphylinids (rove beetles) and other beetles, and mole crickets.

Although the above forms are those most commonly encountered, some beaches have a quite different fauna. One beach has been found to be dominated by small Tanaidacea (related to pill bugs), another by picno-mole crickets (tiny insects). Seasonal visitors may also make considerable impact; people who know the beaches of the Atlantic seaboard of the United States or the beaches of Japan, Korea, or Malaysia will be familiar with the horseshoe crab, *Limulus* (actually not a crab at all), which may come ashore in vast numbers, and on some tropical beaches female turtles invade the beach to breed, their eggs and hatchlings providing a valuable food supply for a variety of marauding animals.

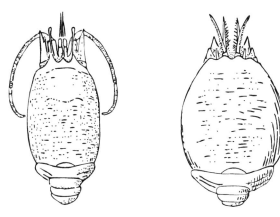

Figure 5 The mole-crabs *Emerita* and *Hippa* in dorsal view. (Reproduced from Brown and McLachlan, 1990 with permission from Elsevier Science.)

Figure 4 The 'plough shell' *Bullia*, about to surf.

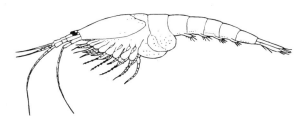

Figure 6 The mysid shrimp, *Gastrosaccus*.

Figure 7 The pill bug, *Tylos granulatus*, showing postcopulatory grooming. The male is mounted back-to-front on the female and is stroking her with his antennae. (Photo: Claudio Velasquez.) (Reproduced with permission from Brown and Odendaal, 1994.)

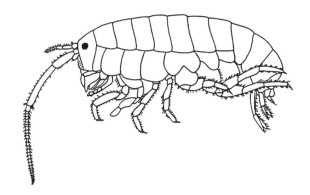

Figure 8 The semiterrestrial sand hopper, *Talitrus saltator*.

Figure 9 The ghost crab, *Ocypode*. (Reproduced from Brown and McLachlan, 1990 with permission from Elsevier Science.)

Food Relationships

As the instability of the substratum precludes the growth of attached plants in the intertidal area, the sandy-beach community is almost entirely dependent on imported food. A few diatoms may be present in the surface layers of sheltered beaches, but they are in general too sparse to be of much nutritional significance. Some potential food may be blown in from the land (plant debris, insects, etc.) and birds may defecate on the beach or die there; however, food washed up by the sea is of overwhelming importance. Three types of economy based on water-transported food may be distinguished, although they are not mutually exclusive.

One of these is based on detached algae (such as kelp), washed up onto the beach. In most cases, the rising tide pushes this material towards the top of the intertidal slope, where the semiterrestrial fauna make short work of it. However, not only are these animals messy feeders but their assimilation of the ingested material is poor, so that their feces are nutrient-rich. Algal fragments and feces find their way into the sand, where they form a food supply for bacteria and the meiofauna. From this community, mineralized nutrients leach back to sea, where they may support further algal growth – and so the cycle is completed. Some nutrients also pass to the land, because shore birds, and even birds such as passerines (e.g., swallows), feed on the semiterrestrial crustaceans and insects, as do some mammals and reptiles. On reflective beaches with a negligible aquatic macrofauna, this kelp-based economy may be the only nutrient cycle of significance.

A second type of food web depends on carrion, such as stranded jellyfish and animals detached from nearby rocky shores. Even dead seals, penguins, sea snakes, and fish may add to the food supply. Macrofaunal scavengers (swimming crabs, ghost crabs, and whelks such as *Bullia*) come into their own where carrion is plentiful. They will eat any animal matter and their assimilation is good, so they pass very little on to the bacteria and meiofauna. The aquatic scavengers are preyed on by the fishes and crabs of the surf zone, as the tide rises, and ghost crabs and other semiterrestrial crabs are taken by birds during the day and more particularly by small mammals and snakes which invade the beach at night.

The third type of economy is found on beaches with extensive surf zones displaying circulating cells of water. A number of phytoplankton (diatom) species are adapted to live in such cells (**Figure 10**), where they may achieve vast numbers, giving primary production rates of up to $10 \, \text{g} \, \text{C} \, \text{m}^{-3} \, \text{h}^{-1}$. Some of this production is inevitably exported to sea but the remaining dissolved and particulate organic material supports three types of community (**Figure 11**). Firstly, it drives a 'microbial loop' in the surf, consisting of bacteria, which are eaten by flagellated protozoans, which in turn are consumed by microzooplankton of various kinds. Secondly, the production supports the interstitial meiofauna of both the surf zone and the intertidal beach, again largely through bacteria. Thirdly, the surf phytoplankton is eaten by a number of species of zooplankton (notably swimming prawns), which in turn are eaten by fish; the phytoplankton also supports the filter-feeders (*Donax*, *Emerita*, etc.) of the beach and these too are consumed by fish as the tide rises. The surf zone of such a beach is thus highly productive, displays considerable diversity and biomass, and forms an important nursery area for fish.

Adaptations of the Macrofauna

Locomotion

It has already been stressed that beach animals must be highly mobile and able to burrow into the sand. Two distinct types of burrowing are in evidence. The Crustacea, having hard exoskeletons, typically use their jointed walking legs as spades to dig themselves in, often with amazing rapidity; *Emerita* takes less than 1.5 s to completely bury itself. Soft-bodied invertebrates (annelid worms and mollusks) have to employ totally different methods (**Figure 12**). The sand surface is probed repeatedly, to liquefy it and so increase its permeability, and the head (of a worm) or the foot (of a clam or whelk) is inserted deep enough

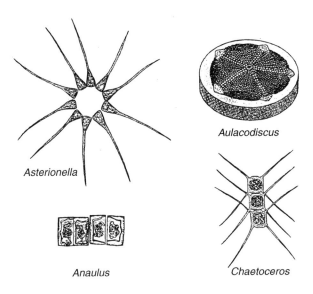

Asterionella

Aulacodiscus

Anaulus

Chaetoceros

Figure 10 Common surf-zone diatoms (Reproduced from Brown and McLachlan, 1990 with permission from Elsevier Science.)

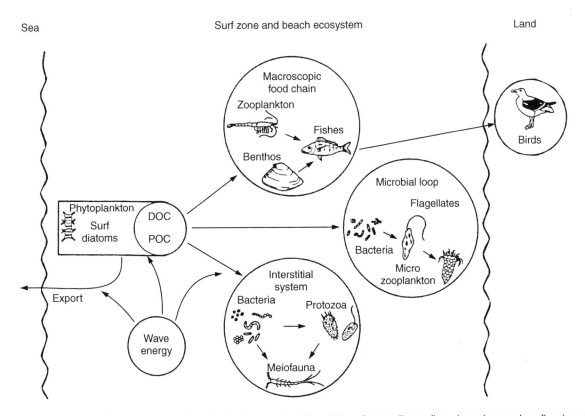

Figure 11 Food relationships on a typical dissipative beach with a rich surf-zone diatom flora, based on carbon flux (greatly simplified). DOC, dissolved organic carbon; POC, particulate organic carbon, including the diatoms themselves. (Reproduced from Brown and McLachlan, 1990 with permission from Elsevier Science.)

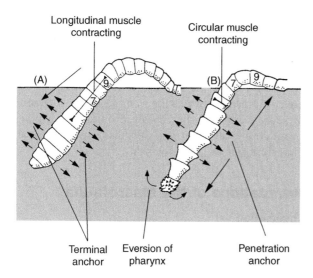

Figure 13 The swimming crab, *Ovalipes*, breaking into the shell of a living *Bullia*. (Photo: George Branch.)

Figure 12 Principal stages of burrowing in the annelid worm *Arenicola*. (A) Anterior segments dilated to form a terminal anchor. (B) Penetration anchor formed by flanging. (Based on Trueman (1966). The mechanism of burrowing in the polychaete worm *Arenicola marina* (L.). *Biological Bulletin* 131: 369–377.

Nutrition and Energy Conservation

Many of the special adaptations of beach animals are concerned with making the best use of a highly erratic food supply **Figure 13**. Thus many members of the macrofauna have more than one method of feeding. The mysid shrimp *Gastrosaccus* filter-feeds while swimming, scoops up detritus deposited on the sand, and will also tear off pieces of carrion. The ghost crab, *Ocypode*, is both a scavenger and a voracious predator **Figure 13** and, when all else fails, it sucks sand for any organic matter it may contain. *Bullia* will also turn predator in the absence of carrion; in addition, it grows an algal garden on its shell and uses its long proboscis to harvest the algae in a manner similar to mowing a lawn.

Not only must the fauna be opportunistic feeders and able to maximize the resource, they must also be able to conserve energy when no food at all is available. Some species remain inactive for long periods and allow their metabolic rates to decline to very low levels. *Tylos*, in its burrows during the day, has very modest rates of oxygen consumption, which increase dramatically during its short period of activity (**Figure 14**). *Bullia digitalis* also allows its metabolic rate to decline to low levels when inactive; in addition, its oxygen consumption is independent of temperature at all levels of activity, thus saving energy as the temperature rises.

Sensory Adaptations

The mobility of the macrofauna and the tidal migrations of many species up and down the shore, present considerable potential danger to them. If the animal moves too far down the slope it may be washed out to sea, and too far up the slope it will die of exposure (heat and desiccation). Sensory adaptations to ensure

for it to swell to form a terminal anchor. With such an anchor in place, the rest of the body can be drawn down towards it. The swelling of the terminal anchor now subsides, while a new anchor appears, by swelling or flanging, closer to the surface, forming a penetration anchor. This allows the front part of the worm or the foot of the mollusk to penetrate deeper into the substratum; the cycle is repeated and so burrowing continues until the animal is completely buried.

Burrowing, crawling and swimming are common forms of locomotion, and a few semiterrestrial crustaceans (e.g., ghost crabs) run. However, some members of the macrofauna exploit not only the tides but also the waves in their quest for food. These are the surfers or swash-riders. Mollusks such as *Donax* and *Bullia* surf by maximally extending their feet and allowing the swash to transport them, whereas some crustaceans, such as juvenile *Tylos*, roll up into a ball when caught by the surf and get carried up the beach. Surfing has been studied in detail in *Bullia digitalis*. The whelk emerges from the sand in response to olfactory stimulation from stranded carrion or in response to increased water currents, spreads its foot to form a turgid underwater sail and then tacks at an angle to the direction of flow of the swash up the slope. As the whelk detects a decrease in olfactory stimulus (volatile amines), it flips over onto its other side and tacks in the opposite direction. It does this repeatedly and so reaches the carrion quickly and efficiently.

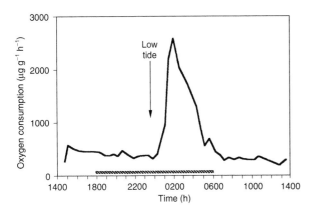

Figure 14 Respiratory rhythm of an individual of *Tylos granulatus*, showing a very low rate of oxygen consumption during the day but with a marked peak immediately after the time of low water at night. The horizontal bar denotes hours of darkness.

maintenance of position in the optimal area are thus essential. Some of these adaptations are relatively simple. For example, *Gastrosaccus* always swims against the current, facing the sea on an incoming swash and turning to face the land as the swash retreats; and the faster the current, the faster it swims. *Donax* and *Bullia* have sensory cells in the foot, which detect the degree of saturation of the sand; as the animal migrates up the beach, it eventually finds itself in unsaturated sand and migrates no further upshore but allows itself to be carried down the slope.

In contrast, the sensory adaptations of semi-terrestrial crustaceans are extremely complex and sophisticated. Most is known about the sand hopper, *Talitrus*. This little amphipod, with a 'brain' that can only be seen under the microscope, nevertheless has a whole suite of responses in its repertoire. It can sense where the shoreline of its particular beach lies relative to the sun and the moon, an orientation learnt while it was carried in the maternal brood pouch. Moreover, it has an internal clock which constantly adjusts this orientation as the sun or moon moves across the sky. It also senses beach slope and will move up the slope if it finds itself on wet sand (i.e., near the sea), or down the slope if the sand is very dry. It will react to broken horizons (e.g., dunes), which reinforces its sense of direction. Some have a magnetic compass and the animal may also orientate to polarized light patterns if the sun or moon is obscured. It also has a second internal clock geared to tidal rise and fall, which tells it when to emerge from the sand (on a falling tide) and when to retreat up the slope (on a rising tide). Some hoppers can detect an approaching storm (how is not known) and then reverse their normal response to slope, migrating as far away from the intertidal zone as possible. It is not surprising that these animals are so successful.

Plasticity of Behavior

In relatively constant environments, animals can often survive with a set of routine behavioral responses but in harsh, unpredictably changing habitats a much greater degree of flexibility (or plasticity) of behavior is called for. The sandy-beach macrofauna provides an outstanding opportunity to study this plasticity among invertebrates. Variations in behavior may in part be determined by genetic selection but the ability to learn plays a most important role in most cases. Both genetically determined plasticity and learning ability are necessary for populations of a species inhabiting different beaches, as no two beaches present exactly the same environment.

For example, *Tylos* typically moves down the intertidal slope to feed but on Mediterranean beaches it commonly moves up the slope after emergence, as food is more plentiful there. In the Eastern Cape Province of South Africa, some populations of *Tylos* have moved permanently into the dunes and have abandoned their tidal rhythm of emergence and reburial; this is also food-related. Another example, *Bullia digitalis* is typically an intertidal, surfing whelk, but in some localities where the beach is inhospitable to it, it occurs offshore, although it may return to the beach and surf if conditions become more favorable.

Perhaps the most remarkable example of the ability to adapt to circumstances concerns observations of *Tylos* on a beach in Japan. Detecting an approaching storm, the animals moved *en masse* towards the sea (i.e., towards the coming danger) and onto an artificial jetty, to which they clung until the storm had passed. This phenomenon has also been observed on a Mediterranean beach. Such behavioral flexibility plays a major role in the survival of the macrofauna and must have been rigorously selected for during the course of evolution.

Conclusion

All of the above phenomena can be linked directly or indirectly to the movements of the water. Waves, tides, and currents determine the type of shore, patterns of erosion, and deposition, the particle size distribution of the substratum, and the slope of the beach. The interaction of waves and sand results in an instability, which precludes attached plants, leading to a unique series of food webs and a series of faunal adaptations to deal with an erratic and unpredictable supply of imported food. Tides and waves are exploited to take maximum advantage of this food. Instability requires that the fauna be mobile and able to burrow, but the three-dimensional nature of the substratum also allows the development of a diverse meiofauna living between the grains. Being of necessity extremely mobile, the macrofauna require sophisticated sensory responses to maintain an optimum position on the beach and the ability to react appropriately to the conditions and circumstances they encounter on their particular beach. Finally, energy conservation is essential, so that metabolic, biochemical adaptations are ultimately dictated by an erratic food supply, again dependent on water movements.

See also

Beaches, Physical Processes Affecting. Storm Surges. Tides.

Further Reading

Brown AC (1996) Behavioural plasticity as a key factor in the survival and evolution of the macrofauna on exposed sandy beaches. *Revista Chilena de Historia Natural* 69: 469–474.

Brown AC and McLachlan A (1990) *Ecology of Sandy Shores*. Amsterdam: Elsevier.

Brown AC and Odendaal FJ (1994) The biology of oniscid Isopoda of the genus. *Tylos Advances in Marine Biology* 30: 89–153.

Brown AC, Stenton-Dozey JME, and Trueman ER (1989) Sandy-beach bivalves and gastropods: a comparison between *Donax serra* and *Bullia digitalis*. *Advances in Marine Biology* 25: 179–247.

Campbell EE (1996) The global distribution of surf diatom accumulations. *Revista Chilena de Historia Natural* 69: 495–501.

Little C (2000) *The Biology of Soft Shores and Estuaries*. Oxford: Oxford University Press.

McLachlan A and Erasmus T (eds.) (1983) *Sandy Beaches as Ecosystems*. The Hague: W. Junk.

McLachlan A, De Ruyck A, and Hacking N (1996) Community structure on sandy beaches: patterns of richness and zonation in relation to tide range and latitude. *Revista Chilena de Historia Natural* 69: 451–467.

ROCKY SHORES

G. M. Branch, University of Cape Town, Cape Town, Republic of South Africa

Introduction

Intertidal rocky shores have been described as 'superb natural laboratories' and a 'cauldron of scientific ferment' because a rich array of concepts has arisen from their study. Because intertidal shores form a narrow band fringing the coast, the gradient between marine and terrestrial conditions is sharp, with abrupt changes in physical conditions. This intensifies patterns of distribution and abundance, making them readily observable. Most of the organisms are easily visible, occur at high densities, and are relatively small and sessile or sedentary. Because of these features, experimental tests of concepts have become a feature of rocky-shore studies, and the critical approach encouraged by scientists such as Tony Underwood has fostered rigor in marine research as a whole.

Rocky shores are a strong contrast with sandy beaches. On sandy shores, the substrate is shifting and unstable. Organisms can burrow to escape physical stresses and predation, but experience continual turnover of the substrate by waves. Most of the fauna relies on imported food because macroalgae cannot attach in the shifting sands, and primary production is low. Physical conditions are relatively uniform because waves shape the substrate. On rocky shores, by contrast, the physical substrate is by definition hard and stable. Escape by burrowing may not be impossible, but is limited to a small suite of creatures capable of drilling into rock. Macroalgae are prominent and *in situ* primary production is high. Rocks alter the impacts of wave action, leading to small-scale variability in physical conditions.

Research on rocky shores began with a phase describing patterns of distribution and abundance. Later work attempted to explain these patterns – initially focusing on physical factors before shifting to biological interactions. Integration of these focuses is relatively recent, and has concentrated on three issues: the relative roles of larval recruitment versus adult survival; the impact of productivity; and the effects of stress or disturbance on the structure and function of rocky shores.

Zonation

The most obvious pattern on rocky shores is an upshore change in plant and animal life. This often creates distinctive bands of organisms. The species making up these bands vary, but the high-shore zone is frequently dominated by littorinid gastropods, the upper midshore prevalently occupied by barnacles, and the lower section by a mix of limpets, barnacles, and seaweeds. The low-shore zone commonly supports mussel beds or mats of algae. Such patterns of zonation were of central interest to Jack Lewis in Britain, and to Stephenson, who pioneered descriptive research on zonation, first in South Africa and then worldwide.

In general, physical stresses ameliorate progressively down the shore. In parallel, biomass and species richness increase downshore. Three factors powerfully influence zonation: the initial settlement of larvae and spores; the effects of physical factors on the survival or movement of subsequent stages; and biotic interactions between species.

Settlement of Larvae or Spores

Many rocky-shore species have adults that are sessile, including barnacles, zoanthids, tubicolous polychaete worms, ascidians, and macroalgae. Many others, such as starfish, anemones, mussels, and territorial limpets, are extremely sedentary, moving less than a few meters as adults. For such species, settlement of the dispersive stages in their life cycles sets initial limits to their zonation (often further restricted by later physical stresses or biological interactions).

Some larvae selectively settle where adults are already present. Barnacles are a classic example. This gregarious behavior, which concentrates individuals in particular zones, has several possible advantages. The presence of adults must indicate a habitat suitable for survival. Furthermore, individuals of sessile species that practice internal fertilization (e.g., barnacles) are obliged to live in close proximity. Even species that broadcast their eggs and sperm will enhance fertilization if they are closely spaced, because sperm becomes diluted away from the point of release. Finally, adults may themselves shelter new recruits. As examples, larvae of the sabellariid reef-worm *Gunnarea capensis* that settle on adult colonies suffer less desiccation, and sporelings of kelps that settle among the holdfasts of adults experience less intense grazing than those that are isolated.

Cues used by larvae to select settlement sites are diverse. Barnacle larvae differ in their preferences for light intensity, water movement, substrate texture, and water depth. All of these responses influence the type of habitat or zone in which the larvae will settle. Most barnacle larvae are attracted to species-specific chemicals in the exoskeletons of their own adults, which persists on the substratum even after adults are eliminated. (Incidentally, this behavior is not just of academic interest: gregarious settlement of barnacles on the hulls of ships costs billions of dollars each year due to increased fuel costs caused by the additional drag of 'fouling' organisms.)

Cues influencing larval settlement can also be negative. Rick Grosberg elegantly demonstrated that the larvae of a wide range of sessile species that are vulnerable to overgrowth avoid settling in the presence of *Botryllus*, a compound ascidian known to be an aggressive competitor for space.

A different aspect of 'supply-side ecology' is the rate at which dispersive larvae or spores settle. The relative importance of recruit supply versus subsequent survival is a topic of intense research. In situations of low recruitment, rate of supply critically influences population and community dynamics. At high levels of recruitment, however, supply rates become less important than subsequent biological interactions such as competition.

Control of Zonation by Physical Factors

Early research on the causes of zonation focused strongly on physical factors such as desiccation, temperature, and salinity, all of which increase in severity upshore. Measurements showed a correlation between the zonation of species and their tolerance of extremes of these factors. In some cases – particularly for sessile species living at the top of the shore – physical conditions become so severe that they kill sections of the population, thus imposing an upper zonation limit by mortality.

For those species blessed with mobility, zonation is more often set by behavior than death, and most individuals live within the 'zone of comfort' that they can tolerate. One example illustrates the point. The trochid gastropod *Oxystele variegata* increases in size from the low- to the high-shore. This gradient is maintained by active migration, and animals transplanted to the 'wrong' zone re-establish themselves in their original zones within 24 hours. The underlying causes of this size gradient seem to be twofold:

desiccation is too high in the upper shore for small individuals to survive there; and predation on adults is greatest in the lower shore.

Adaptations to minimize the effects of physical stresses are varied. Physiological adaptation and tolerance are one avenue of escape. Avoidance by concealment in microhabitats is another. Morphological adaptations are a third route. For instance, desiccation and heat stress can be reduced by large size (reducing the ratio of surface area to size), and by differences in shape, color, and texture (**Figure 1A**).

Physical factors can clearly limit the upper zonation of species. It is often difficult, however, to imagine them setting lower limits. For this, we turn to biological interactions.

Biological Interactions

Interactions between species – particularly competition and predation – only began to influence the thinking of intertidal rocky-shore ecologists in the mid 1950s. Extremely influential was Connell's work, exploring whether the zonation of barnacles in Scotland is influenced by competition. He noted that a high-shore species, *Chthamalus stellatus*, seldom penetrates down into the midshore, where another species, *Semibalanus balanoides*, prevails. Was competition from *Semibalanus* excluding *Chthamalus* from the midshore? In field experiments in which *Semibalanus* was eliminated, *Chthamalus* not only occupied the midshore, but survived and grew there better than in its normal high-shore zone. *Chthamalus* is more tolerant of physical stresses than *Semibalanus*, and can therefore survive in the high-shore, where it has a 'spatial refuge' beyond the limits of *Semibalanus*. In the midshore, however, *Semibalanus* thrives and competitively excludes *Chthamalus* by undercutting or overgrowing it.

Other forms of competition have since been discovered. For instance, territorial limpets defend patches of algae by aggressively pushing against other grazers. In areas where they occur densely, they profoundly influence the zonation of other species and the nature of their associated communities.

The role of predators came to the fore following the work of Bob Paine who showed that experimental removal of the starfish *Pisaster ochraceus* from open-coast shores in Washington State led to encroachment of the low-shore by mussels, which are usually restricted to the midshore. Thus, predation sets lower limits to the zonation of the mussels. More importantly, it was shown that the downshore advance of mussel beds reduced the number of space-occupying species there. In other

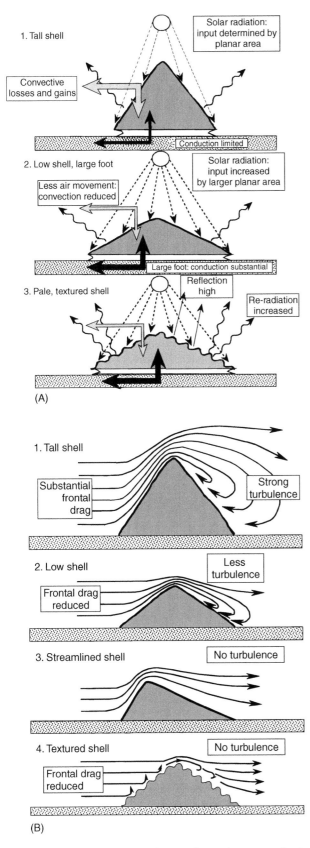

1. Tall shell

Solar radiation: input determined by planar area

Convective losses and gains

Conduction limited

2. Low shell, large foot

Solar radiation: input increased by larger planar area

Less air movement: convection reduced

Large foot: conduction substantial

3. Pale, textured shell

Reflection high

Re-radiation increased

(A)

1. Tall shell

Substantial frontal drag

Strong turbulence

2. Low shell

Frontal drag reduced

Less turbulence

3. Streamlined shell

No turbulence

4. Textured shell

No turbulence

Frontal drag reduced

(B)

Figure 1 (A) Effects of shell shape, color, and texture on heat uptake and loss. (B) Influence of shell proportions and texture on hydrodynamic drag.

words, predation normally prevents the competitively superior mussels from ousting subordinate species, thus maintaining a higher diversity of species. This concept – the 'predation hypothesis' – has since been broadened to include all forms of biological or physical disturbance. If disturbance is too great, few species survive and diversity is low. On the other hand, if it is absent or has little effect, a few species may competitively monopolize the system, reducing diversity. At intermediate levels of disturbance, diversity is highest – the 'intermediate disturbance hypothesis'. (Incidentally, the idea is not new. Darwin gives an accurate description of this effect in sheep-grazed meadows.)

Paine's work led to the idea of 'keystone predators': those that have a strong effect on community structure and function. Unfortunately the term has been debased by general application to any species that an author feels is somehow 'important'. Consequently, not all scientists are enamoured of the concept. More recently, it has been redefined to mean those species whose effects are disproportionately large relative to their abundance. This more usefully allows recognition of species that can be regarded as 'strong interactors', and which powerfully influence community dynamics.

Many researchers have shown that grazers (particularly limpets) profoundly influence algal zonation, excluding them from much of the shore by eliminating sporelings before they develop. Hawkins and Hartnoll suggested that interactions between grazers and algae depend on the upshore gradient in physical stress, and argued that low on the shore, algae are likely to be sufficiently productive to escape grazing and proliferate, forming large, adult growths that are relatively immune to grazing. In the mid- to high-shore, grazers become dominant and algae seldom develop beyond the sporeling stage. A more subtle reverse effect, that of algae-limiting grazers, has also been described. Low on the shore where productivity is high, algae form dense mats. This not only deprives grazers of a firm substrate for attachment, but also of their primary food source, namely microalgae. Grazers experimentally transplanted into low-shore algal beds starve in the midst of apparent plentitude.

The influences of grazers and predators extend beyond their direct impacts on prey. A bird consuming limpets has positive effects on algae that are grazed by the limpets. Such indirect effects occupy ecologists because their consequences are often difficult to predict. For example, experimental removal of a large, grazing chiton, *Katharina tunicata*, from the shores of Washington State might logically have been expected to improve the lot of intertidal limpets, on the grounds that elimination of a competitor must be good for the remaining grazers. In fact, elimination of the chiton led to starvation of the limpets because macroalgae proliferated, excluding microalgae on which the limpets depend. Indirectly, *Katharina* facilitates microalgae, thus benefiting limpets.

The mix of physical and biological controls affecting zonation was investigated by Bruce Menge in 1976 by using cages to exclude predators from plots in the high-, mid-, and low-shore. In the high-shore, only barnacles became established, irrespective of whether predators were present or absent. In the midshore, mussels became dominant and outcompeted barnacles, again independently of the presence or absence of predators. Competition ruled. In the low-shore, however, mussels only dominated where cages excluded the predators. Elsewhere, predators eliminated mussels, thus allowing barnacles to persist. These results elucidated the interplay between physical stress and biological control and were instrumental in formalizing 'environmental stress models'. These suggest that predation will only be important (exerting a 'top-down' control on community structure) when physical conditions are mild. As stress rises, first predation, and then competition, diminish in importance.

Wave Action

It has been shown that wave action is probably the most important factor affecting distribution patterns along the shore. In a negative sense, waves physically remove organisms, damage them by throwing up logs and boulders, reduce their foraging excursions, and increase the amount of energy devoted to clinging on. One manifestation is a reduction of grazer biomass (**Figure 2F**). Adaptations can, however, counter these adverse effects. Tenacity can be increased by cementing the shell to the rock face (e.g., oysters), developing temporary attachments (e.g., the byssus threads of mussels), or employing adhesion (e.g., the feet of limpets and chitons). Shape

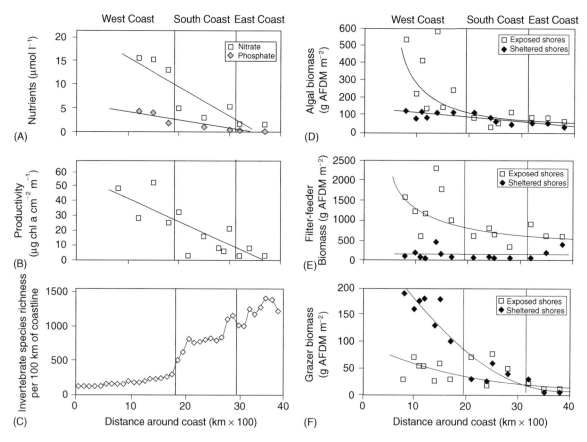

Figure 2 Moving west to east around the coastline of southern Africa (from northern Namibia to southern Mozambique), (A) nutrient input and (B) intertidal primary productivity decrease, (C) invertebrate species richness rises, and there are declines in the biomasses (ash-free dry mass) of (D) algae, (E) filter-feeders, and (F) grazers. Note that biomass is strongly influenced by wave action, either positively (algae and filter-feeders) or negatively (grazers). AFDM.

can be modified to reduce drag, turbulence, and lift (**Figure 1B**). Each organism is, however, a compromise between conflicting stresses. For instance, desiccation is reduced if a limpet has a small aperture; but this implies a smaller foot and thus less ability to cling to the rocks.

Wave action also brings benefits. It enhances nutrient supply, reduces predation and grazing, and increases food supply for filter-feeders. The biomass on South African rocky shores has been shown to rise steeply as wave action increases, mostly due to increases in filter-feeders (**Figure 2E**). Possible explanations include enhanced larval supply and reduced predation, but measurements of food abundance and turnover showed that wave action vitally enhances particulate food.

Wave action varies over short distances. Headlands and bays influence the magnitude of waves at a scale of kilometers, but even a single large boulder will alter wave impacts at a scale of centimeters to meters. As a result, community structure can be extremely patchy on rocky shores.

Productivity

'Nutrient/productivity models' (NPMs) attempt to explain community structure in terms of nutrient input and, thus, productivity. This 'bottom-up' approach concerns how the influences of primary production filter up to higher levels. Nutrient/productivity models were developed with terrestrial systems in mind, but their application to rocky shores has been enlightening. Terrestrial systems depend largely on the productivity of plants, which is usually limited by availability of nutrients and water. On rocky shores, neither constraint is necessarily relevant. The shore is washed by the rise and fall of the tide, which also imports particulate material that fuels filter-feeders independently of the productivity on the shore itself. Even so, local productivity does influence rocky-shore community structure. In one theory, the lower the productivity, the fewer the steps that can be supported in the trophic web. If production is very low, plant life cannot sustain grazers. As productivity rises, grazers may be supported, and begin to control the standing stocks of plants. Further increases may lead to three trophic levels, with predators controlling grazers, which thus lose their capacity to regulate plants.

Nowhere else in the world is the coastline better configured to test ideas about NPMs than in South Africa. The cold, nutrient-rich, upwelled Benguela Current bathes the west coast. The east coast receives fast-moving, warm, nutrient-poor waters from the southward-flowing Agulhas Current. Between the two, the Agulhas swings away from the south coast, creating conditions that are intermediate. From west to east, the coast has a strong gradient of nutrients and productivity (**Figure 2A, B**). As productivity drops, so do the average biomasses of algae, filter-feeders, and grazers (**Figure 2D–F**), and the total biomass. On the other hand, species richness rises (**Figure 2C**). At a more local scale, guano input on islands achieves the same effects (**Figure 3**).

Productivity also has more subtle effects on the functioning of rocky shores. For instance, the frequency of territoriality in limpets is inversely correlated with productivity. It seems that the need to defend patches of food diminishes as the ratio of productivity : consumption rises. Indirectly, this has profound effects on community dynamics, because these territorial algal species reduce species richness and biomass but greatly increase local productivity.

Increased productivity is, however, not an unmixed blessing. The upwelling that fuels coastal productivity also results in a net offshore movement of water. This may export the recruits of species with dispersive stages. The scarcity of barnacles on the west coast of South Africa may be one consequence. In California it has been shown that barnacle settlement is inversely related to an index of upwelling. Furthermore, nutrient input can lead to heavy blooms of phytoplankton (often called red or brown tides), that subsequently decay, causing anoxia or even development of hydrogen sulfide. Either eventuality is lethal, and mass mortalities ensue. Records of thousands of tons of rock lobsters spectacularly stranding themselves on the shore in a futile attempt to escape anoxic waters testify to the ecological and economic consequences.

Energy Flow

Flows of energy (or of any material such as carbon or nitrogen) through an ecosystem can be used to quantify rates of turnover and passages between elements of the food web. Developing a complete energy-flow model for a rocky shore is a formidable task. For at least the major species, energy uptake must be measured directly, or estimated by summing the requirements for growth and reproduction and the losses associated with respiration, excretion, and secretion. Critics of ecosystem energy-flow models emphasize the huge investment of research time needed to complete the task, and argue that rocky shores differ so much from place to place that a model describing one shore will often be inapplicable elsewhere.

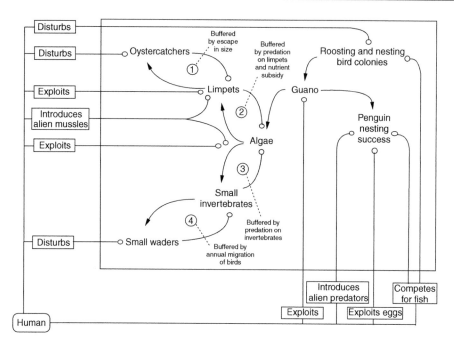

Figure 3 Interactions between organisms on nearshore islands on the west coast of South Africa. Natural interactions, and the processes buffering them (numbered 1–4), are shown inside the box; human impacts influencing them lie outside the box. Lines terminating in arrows and circles indicate positive and negative effects, respectively.

Nevertheless, energy-flow models of rocky shores have revealed features difficult to grasp by other approaches, as the following examples reveal. First, measurements show that most macroalgal productivity is not consumed by grazers, but adds to a detrital pool that fuels filter-feeders. On sheltered shores, production exceeds demands, and they are net exporters of material. On wave-beaten shores, however, the needs of filter-feeders far exceed intertidal production, and these shores depend on a subsidy of materials from the subtidal zone or from offshore. Tides, waves and currents play a vital role in turning over this supply of particulate materials. Intertidal standing stocks of particulate matter are in the order of $0.25 \, \mathrm{g \, C \, m^{-3}}$, far below the annual requirements of filter-feeders (about $500 \, \mathrm{g \, C \, m^{-2} \, y^{-1}}$). But if a hypothetical flow of $20 \, \mathrm{m^3}$ passes over each square meter of shore per day, it will supply $1825 \, \mathrm{g \, C \, m^{-2} \, y^{-1}}$. This stresses the need for small-scale hydrographic research to predict flows that are meaningful at the level of individual filter-feeders.

Second, some elements of the ecosystem have consistently been overlooked because of their small contribution to biomass. An obvious case is the almost invisible 'skin' of microbiota (sporelings, diatoms, bacteria, and fungi) coating rocks. On most shores, their share of the biomass is an apparently insignificant 0.1%. However, in terms of productivity, they contribute 12%, and most grazers depend on this food source.

Finally, there has always been a tacit assumption that sedentary intertidal grazers must depend on *in situ* algal productivity. However, on the west coast of South Africa, grazers reach extraordinarily high biomasses (up to $1000 \, \mathrm{g}$ wet flesh $\mathrm{m^{-2}}$). Modeling shows that their needs greatly exceed *in situ* primary production. Instead, they survive by trapping drift and attached kelp. This subtidal material effectively subsidizes the intertidal system. Thus sustained, dense beds of limpets dominate sections of the shore, eliminating virtually all macroalgae and most other grazers. Both 'bottom-up' and 'top-down' processes are at play.

Energy-flow models are seldom absolute measures of how a system operates. Changes in time and space preclude this. Rather, their power lies in identifying bottlenecks, limitations, and overlooked processes, which can then be investigated by complementary approaches.

Integration of Approaches

Ecology as a whole, and that addressing rocky shores in particular, has suffered from polarized viewpoints. Classic examples are arguments over competition versus predation, or the merits of 'top-down' versus 'bottom-up' approaches. In reality, all of these are valid. What is needed is an integration that identifies the circumstances under which one or other model has greatest predictive power. A single example demonstrates the multiplicity of factors operating on intertidal rocky shores (**Figure 3**).

Islands off the west coast of South Africa support dense colonies of seabirds. These have two important effects on the dynamics of rocky shores. First, their guano fuels primary production. Second, oystercatchers that aggregate on the islands prey on limpets. Indirect effects complicate this picture. Reduction of limpets adds to the capacity of seaweeds to escape grazing, leading to luxurious algal mats. These sustain small invertebrates, in turn a source of food for waders. The limpets benefit indirectly from the guano because their growth rates and maximum sizes are increased by the high algal productivity. This allows some to reach a size where they are immune to predation by oystercatchers, and also increases their individual reproductive output. But this 'interaction web' embraces less obvious connections. Dense bird colonies only exist on the islands for two reasons: absence of predators, and food in the form of abundant fish, sustained ultimately by upwelling. Comparison with adjacent mainland rocky shores reveals a contrast: roosting and nesting birds are scarce or absent, oystercatchers are much less numerous, limpets are abundant but small, and algal beds are absent.

This case emphasizes the complexity of driving forces and the difficulty of making predictions about their consequences. Top-down and bottom-up effects, direct and indirect impacts, productivity, grazing, predation, competition, and physical stresses all play their role.

Human Impacts

In one sense intertidal rocky-shore communities are vulnerable to human effects because they are accessible and many of the species have no refuge. In another sense, they are relatively resilient to change. One reason is that humans seldom change the structure and physical factors influencing rocky shores. Tidal excursions and wave action, the two most important determinants of rocky-shore community structure, are seldom altered. The physical rock itself is also rarely modified by human actions. These circumstances are an important contrast with systems such as mangroves, estuaries, and coral reefs, where the structure is determined by the biota. Remove or damage mangrove forests, salt marshes, or corals, and the fundamental nature of these systems is changed and slow to recover. Estuaries are especially vulnerable because their two most important physical attributes – input of riverine water and tidal

exchange – can be revolutionized by human actions. Even after massive abuse such as oil pollution, rocky shores recover relatively quickly once the agent of change is brought under control; and there is good evidence that the kinds of communities appearing after recovery resemble those originally present.

Humans impact rocky-shore communities in many ways, including trampling, harvesting, pollution, introduction of alien species and by altering global climate. Harvesting is of specific interest because it has taught much about the functioning of rocky shores.

Almost without exception, harvesting reduces mean size, density, and total reproductive output of the target species, although compensatory increases of growth rate and reproductive capacity of surviving individuals are not unusual. Of greater interest, however, are the consequences for community structure and dynamics. Clear demonstrations of this come from Chile, where intense artisanal harvesting occurs on rocky shores. In particular, a lucrative trade has developed for giant keyhole limpets, *Fissurella* spp., and for a predatory muricid gastropod, *Concholepas concholepas*, colloquially known as 'loco.' Decimated along most of the coast, the natural roles of these species are only evident inside marine protected areas. There, locos consume a small mussel, *Perumytilus purpuratus*, which otherwise outcompetes barnacles. Macroalgae and barnacles compete for space, but only on a seasonal basis. Keyhole limpets control macroalgae and outcompete smaller acmaeid limpets. *Perumytilus* acts as a settlement site for conspecifics and for recruits of keyhole limpets and locos. With a combination of predation, grazing, competition and facilitation, and both direct and indirect effects, the consequences of harvesting locos or keyhole limpets would have been impossible to resolve without the existence of marine protected areas. Even then, careful manipulative experiments inside and outside these areas were required to disentangle these interacting effects.

A second issue of general interest is whether human impacts are qualitatively different from those of other species. The short answer is 'yes', and is best illustrated by a return to an earlier example – interactions between species on rocky shores of islands on the west coast of South Africa (**Figure 3**). In its undisturbed state, each of the key interactions in this ecosystem is buffered in some way. Limpets are consumed by oystercatchers, but some escape by growing too large to be eaten, aided by the high primary production. Limpets and other invertebrates graze on algae, but their effects are muted by predation on them and by the enhancement of algal growth by guano. Waders eat small seaweed-associated invertebrates, but emigrate in winter.

For several reasons, human impacts are not constrained in these subtle ways. First, human populations do not depend on rocky shores in any manner limiting their own numbers. They can harvest these resources to extinction with impunity. Second, modern human effects are too recent a phenomenon for the impacted species to have evolved defenses. Thirdly, humans are supreme generalists. Simultaneously, they can act as predators, competitors, amensal disturbers of the environment, and 'commensal' introducers of alien species. Fourthly, money, not returns of energy, determines profitability. Fifthly, long-range transport means that local needs no longer limit supply and demand. Sixthly, technology denies resources any refuge.

Thus, humans supersede the ecological and evolutionary rules under which natural systems operate; and only human-imposed rules and constraints can replace them in meeting our self-proclaimed goals of sustainable use and maintenance of biodiversity.

See also

Beaches, Physical Processes Affecting. Coastal Circulation Models. Coastal Trapped Waves. Exotic Species, Introduction of. Intertidal Fishes. Macrobenthos. Tides. Waves on Beaches.

Further Reading

Branch GM and Griffiths CL (1988) The Benguela ecosystem Part V. The coastal zone. *Oceanography and Marine Biology Annual Review* 26: 395–486.

Castilla JC (1999) Coastal marine communities: trends and perspectives from human-exclusion experiments. *Trends in Ecology and Evolution* 14: 280–283.

Connell JH (1975) Some mechanisms producing structure in natural communities: a model and evidence from field experiments. In: Cody ML and Diamond JM (eds.) *Ecology and Evolution of Communities*, pp. 460–490. Cambridge, MA: Belknap Press.

Denny MW (1988) *Biology and the Mechanics of the Wave-swept Environment*. Princeton, NJ: Princeton University Press.

Hawkins SJ and Hartnoll RG (1983) Grazing of intertidal algae by marine invertebrates. *Oceanography and Marine Biology Annual Review* 21: 195–282.

Menge BA and Branch GM (2001) Rocky intertidal communities. In: Bertness MD, Gaines SL, and Hay ME (eds.) *Marine Community Ecology*. Sunderland: Sinauer Associates.

Moore PG and Seed R (1985) *The Ecology of Rocky Coasts*. London: Hodder and Stoughton.

Newell RC (1979) *Biology of Intertidal Animals*. Faversham, Kent: Marine Ecological Surveys.

Paine RT (1994) *Marine Rocky Shores and Community Ecology: An Experimentalist's Perspective*. Oldendorf: Ecology Institute.

Siegfried WR (ed.) (1994) *Rocky Shores: Exploitation in Chile and South Africa*. Berlin: Springer-Verlag.

Underwood AJ (1997) *Experiments in Ecology: their Logical Design and Interpretation Using Analysis of Variance*. Cambridge: Cambridge University Press.

CORALS AND REEFS

CORAL REEFS

J. W. McManus, University of Miami, Miami, FL, USA

Introduction

Coral reefs are highly diverse ecosystems that provide food, income, and coastal protection for hundreds of millions of coastal dwellers. They are found in a diverse range of geomorphologies, from small coral communities of little or no relief, to calcareous structures hundreds of kilometers across. The most diverse coral reefs occur in the waters around southeast Asia. This is primarily because of extinctions that have occurred in other regions due to the gradual restriction of global oceanic circulation associated with continental drift. However, rates of speciation of coral reef organisms in this region may also have been high within the last 50 million years.

Human activities have caused the degradation of coral reefs to varying degrees in all areas of the world. A major focus of present research is on the resilience of coral reefs to disturbances such as storms, diseases of reef species, bleaching (the expulsion of endosymbiotic photosynthetic zooxanthellae) and harvesting to local extinction. Along many coastlines, a combination of increased eutrophication due to coastal runoff and the extraction of herbivorous fish and invertebrates appears to favor the replacement of corals with macroalgae following disturbances. Another important aspect of resilience is the degree to which a depleted population of a given species on one reef can be replenished from other reefs. Most coral reef organisms undergo periods of free-swimming (pelagic) life ranging from a few hours to a few months. Genetic studies show evidence of broad dispersal of the progeny of some species among reefs, but most of the replenishment of a given population from year to year is believed to be from the same or nearby coral reefs.

Coral reefs are complex biophysical systems that are generally linked to similarly complex socioeconomic systems. Their proper management calls for system-level approaches such as integrated coastal management.

General

Types of Coral Reefs

The term 'coral reef' commonly refers to a marine ecosystem in which a prominent ecological functional role is played by scleractinian corals. A 'structural coral reef' differs from a 'nonstructural coral community' in being associated with a geomorphologically significant calcium carbonate (limestone) structure of meters to hundreds of meters height above the surrounding substrate, deposited by components of a coral reef ecosystem. The term 'coral reef' is often applied to both structural and nonstructural coral ecosystems or their fossil remains, although many scientists, especially geomorphologists, reserve the term for structural coral reefs and their underlying limestone. Both types of ecosystem occur within a wide range of tropical and subtropical marine environments, although structural development tends to be greater in waters of lower silt or mud concentration and oceanic salinity. Many reefs survive well amid open ocean waters with low nutrients, aided by efficient 'combing' of waters for plankton, high levels of nitrogen fixation and fast and thorough nutrient cycling. However, extensive coral reefs also occur in coastal waters of much higher nutrient concentrations.

Scleractinian (stony) corals grow as colonies or solitary polyps on a wide variety of substrates, including fallen trees, metal wreckage, rubber tires and rocks. Rates of settlement are often enhanced by the presence of calcareous encrusting algae. Soft sand, silt and mud tend to inhibit the settlement of stony corals, and so few coral ecosystems occur in modern or ancient deltaic deposits. However, coral can grow very near the mouths of small rivers and steams, and fresh or brackish groundwater often percolates through reef structures or emerges periodically through tunnels and caves. Vertical caves are often called 'blue holes'.

Nonstructural coral communities are common on rocky outcrops in shallow seas in many tropical and subtropical regions. They can range from a few clumps of coral to very substantial communities covering many square kilometers of wave-cut shelves near deeper areas.

Structural coral reefs come in many shapes and sizes, from less than a kilometer to many tens of kilometers in linear dimension. It is helpful to differentiate individual coral reefs from systems of coral reefs, such as the misnamed 'Great Barrier Reef' of

Australia, which actually consists of thousands of densely packed coral reefs. Common types of structural coral reefs include fringing reefs, barrier reefs, knoll reefs, pinnacle reefs, platform reefs, ribbon reefs, crescent reefs, and atolls. The term 'patch reef' may refer either to a patch of coral and limestone a few meters across within a structural coral reef (typically in a lagoon or on a reef flat), or to a platform or knoll reef. Thus, the term is best avoided. There is a wide range of structures intermediate between crescent reefs, platform reefs, and atolls in areas such as the Great Barrier Reef System.

A fringing reef is, by definition, always found adjacent to a land mass. Most fringing reefs include a wave-breaking reef crest, one or more meters above the rest of the reef, forming a thin strip offshore (**Figure 1**). Between the crest and the land, there is usually a relatively level area, broken by channels, called a 'reef flat' (**Figure 2**). Fringing reefs differ from barrier reefs, in that the latter are separated

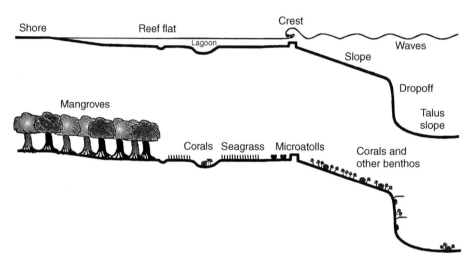

Figure 1   Profiles of a hypothetical fringing reef showing geomorphological and ecological zonation relative to wave action.

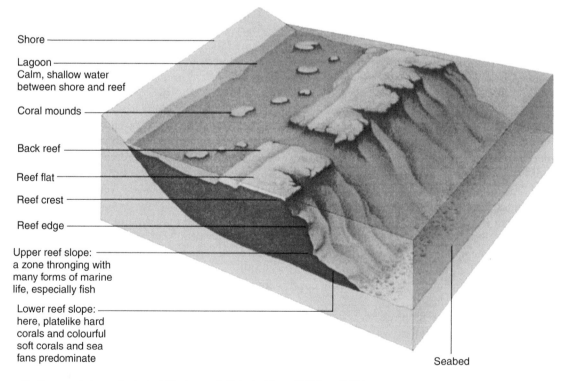

Figure 2   Geomorphology of a typical fringing reef. (Adapted from Holliday, 1989.)

from land by a 'navigable' water body (lagoon). It is useful to differentiate a lagoon from a reef flat in terms of depth; a lagoon is at least 2 m deep at mean tide. Fringing reefs often include lagoons, but the separation of a reef crest and slope from land is more complete in a barrier reef (**Figure 3**).

Structural reefs such as knoll, pinnacle, platform, ribbon, or crescent reefs, are arbitrarily labeled based on their shape. Atolls are donut-shaped structures which, although often supporting islands along the outer rim, do not have an island in the central portion (**Figure 4**). The large Apo Reef, east of the Philippine island of Mindoro, is a double atoll, with two lagoons within adjacent triangular rims, the whole reef being roughly diamond-shaped. Atolls can be quite large, such as the North Male Atoll, which houses the capital of the Maldives (**Figure 5**).

Although one commonly thinks of structural coral reefs as reaching to the sea surface, most of the coral reefs of the world do not. For example, there is a system of atolls and other reefs to approximately 50 km off the north-west of Palawan Island (Philippines) that closely resembles portions of the Australian Great Barrier Reef. However, very few reefs of the Palawan 'barrier system' come within 10 m of the sea surface. Some estimates of coral reef area are based on reefs at or near the sea surface, partly because the larger examples of such reefs tend to be well-charted, whereas most other coral reefs are poorly known in terms of location and characteristics.

Importance

Coral reefs support the highest known biodiversity of marine life, and constitute the largest biologically generated structures on Earth. Coral reefs are of substantial social, cultural, and economic importance. Coral reef systems in Florida, Hawaii, the Philippines, and Australia each account for more than $1 billion in tourist-related income each year. Coral reefs provide food and livelihoods for several tens of millions of fishers and their families, most of whom live in developing countries on low incomes, and have limited occupational mobility. Coral reefs protect coastal developments and farm lands from erosion. In many countries, particularly among Pacific islands, coral reefs are culturally very important, as they are involved in social structuring and interaction, and in religion. Despite poor taxonomic understanding and increasingly strict controls on bioprospecting, coral reef species are yielding substantial numbers of important drugs and other products.

Zonation

The ecological zonation of most coral reefs depends on physical factors, including depth, exposure to

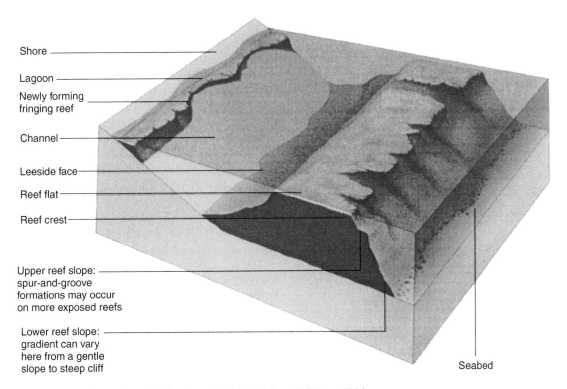

Figure 3 Geomorphology of a typical barrier reef. (Adapted from Holliday, 1989.)

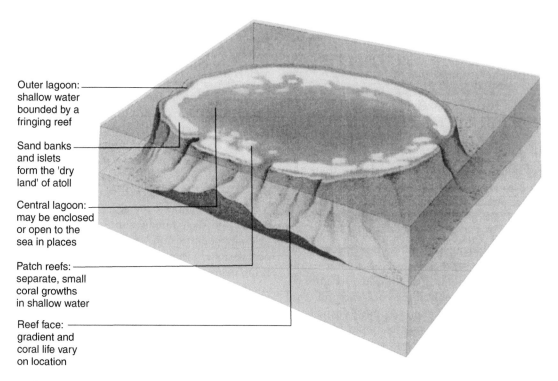

Outer lagoon: shallow water bounded by a fringing reef

Sand banks and islets form the 'dry land' of atoll

Central lagoon: may be enclosed or open to the sea in places

Patch reefs: separate, small coral growths in shallow water

Reef face: gradient and coral life vary on location

Figure 4 Geomorphology of a typical atoll. (Adapted from Holliday, 1989.)

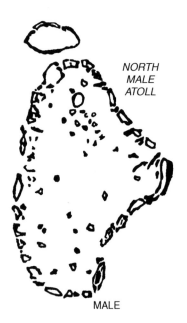

NORTH MALE ATOLL

MALE

Figure 5 North Male Atoll of the Maldives. Large atolls often support substantial human population that will be threatened as sealevels rise. (Adapted from Holliday, 1989.)

waves and currents, and oxygen limitation. A fringing reef is basically a large slab of limestone jutting out from land. The landward margin may support extensive mangrove forests. A sandy channel may separate these from extensive seagrass beds on a reef flat. Other channels, depressions, and basins may be predominated by sandy bottoms, studded with clumps of coral. Clumps in deeper waters, rising 2 m or more from the bottom and consisting of several species of coral are known as 'bommies.' Some lagoons hold large numbers of 'pillars,' tall shafts of limestone, similarly supporting corals. Throughout the reef flat, there are, typically, low clumps of coral. Macroalgae and seagrass are often kept away from these clumps by the feeding action of the herbivorous fish and sea urchins, particularly active at night and resting during the day in the clumps of coral. Low massive coral colonies may form 'microatolls,' in which central portions of the colony are dead, while the raised outer edge and sides continue to flourish. On some reef flats, branching corals form patches which extend over hectares. Other reef flats and lagoons may be packed with high densities of diverse coral colonies, and have little or no seagrass.

The seaward edge of the reef flat often leads into low branching or massive corals, including microatolls, which become increasingly dense to seaward. A thin band of macroalgae, such as *Turbinaria*, may be present, just before the clumps of coral and hard substrate rise to form a reef crest or 'pavement.' In some areas, the crest may be made of living coral. In areas protected from heavy wave action, the crest may consist of fingerlike *Montipora* or *Porites* forming a band several meters wide, broken periodically by whorl-shaped colonies of leafy corals, all apparently formed to efficiently comb breaking waves for zooplankton. On higher-energy coasts, the corals may be primarily dense growths of wave-adapted *Pocillopora*. Other crests may be covered with various species and growth forms of the brown alga *Sargassum*. The edible and commercially important alga *Caulerpa* may also be abundant. Other reef crests may be densely packed with small clumps of articulated calcareous algae such as *Halimeda* and *Amphiroa*. Still others consist of a pavement of calcareous material or sheared-off ancient corals, and are relatively devoid of all macrobenthos except tiny clumps of algal turf or encrusting algae, sometimes forming a white or pink algal crest. The height of a reef crest above the reef flat is often determined by the height of local tides.

Most reef crests are broken by channels of varying width and depth – exit routes for water piled up behind the reef crest by the breaking waves. These may be studded with corals or smoothed by scouring sand and rock carried along with the exiting water. They are particularly important as breeding grounds for a variety of reef fish.

Beyond the reef crest, on the upper reaches of the 'reef slope,' one often finds small thick clumps of branching *Pocillopora* coral or various species of *Acropora* colonies in similar, wave-resistant growth forms. Encrusting and low lump-like (submassive) colonies may be common. More of the brown algae *Turbinaria* and *Sargassum* may be present. Along a rounded or gentle upper reef slope in the Indo-Pacific, there may be table-shaped *Acropora* colonies of gradually larger size as one proceeds to deeper waters. On many reefs, the reef slope is the most active area of coral growth, because of the oxygen, nutrients, and plankton brought in on the waves and currents. Coral cover may exceed 100%, as colonies overgrow colonies, all competing for light. Soft corals dominate some reef slopes, *Sargassum* others, and on some, the profusion of corals, sponges, algae, and other benthic organisms may prevent the identification of a dominant group. The mean and median stony coral cover (the percentage of the substrate covered with coral) on a reef slope globally are both approximately 40%, with a broad variance.

On most reefs, small channels on the upper slope consolidate into increasingly wider and deeper channels along lower reaches of the slope. Some may be steepened or converted into tunnels by coral growth. The channels may occur fairly regularly at distances of tens of meters, resulting in a 'ridge and rift' or 'spur (or buttress) and groove' structure resembling the toes of giant feet. The bottoms are generally filled with a mix of sand and debris from the reefs, which makes its way downwards, particularly during storms. Sand may drop continuously from escarpments, in flows resembling waterfalls, and spectacular columns of limestone cut away from the reef proper may border deeper rifts.

Many reef slopes lead to a steep 'wall' or 'drop-off,' often beginning about 10–20 m depth. On shelf areas, the drop-off may end at 20–30 m depth, followed by a 'talus' slope of deposited reef materials, often 30°–60°, leading into the more gradual shelf slope. Typically, this shelf will be interrupted by outcrops of limestone and bommies for considerable distances from the reef itself. On reefs jutting into deep waters, the drop-off may extend downwards for hundreds of thousands of meters. Corals and a myriad of other organisms generally cover the slopes, leaving little or no bare substrate.

On some reefs, such as some small fringing reefs along the Sinai, the upper portion consists almost entirely of live massive, platy or occasionally thick branching corals extending from land or from a small reef flat. The corals are joined together tangentially, leaving large spaces of water between. There may be no identifiable reef crest, and the mass simply juts out over a steep dropoff to hundreds of meters depth.

The zonation of atolls and other surface reefs is generally similar. However, as one proceeds across a lagoon basin from the windward to the leeward side of an atoll, one encounters a 'backreef' area (note that the term is also applied by some to the coral-dominated area behind a reef crest). The leeward side of the atoll is subject to less energetic waves and currents. It tends to have less well-defined zonation, less of a distinct crest, and often broader, more gradual slopes.

Storms are very important in determining features of the reef. It is common to find large chunks of limestone on upper slopes, crests and reef flats, representing masses of coral and substrate pulled up from the lower slope and deposited during a storm. Pieces of this jetsam may weight several tonnes each. Smaller pieces of coral and substrate, and masses of sand may pile up during storms, forming islands. Processes of calcification in intertidal areas may glue together pieces of coral and reef substrate thrown up by storms, forming 'beach rock,' which helps to prevent erosion at the edge of low islands. Other islands may be formed on portions of a reef uplifted by tectonic processes. Islands developed by storm action or uplift often have very little relief, but may support human communities. Entire nations, as with the Maldives or some Pacific island nations, may consist of such low islands on atoll reefs.

Geology

Calcification

Carbon dioxide and water combine to form carbonic acid, which can then dissociate into hydrogen ions (H^+) and (HCO_3^-)bicarbonate, or carbonate (CO_3^{2-}) ions as follows:

$$CO_2 + H_2O \Leftrightarrow H^+ + HO_3^- \Leftrightarrow 2H^+ + CO_3^{2-}$$

Colder water can hold more CO_2 in solution than warmer water. Calcium carbonate reacts with CO_2 and water as follows:

$$CO_2 + H_2O + CaCO_3 \Leftrightarrow Ca^{2+} + 2HCO_3^-$$

The dissolution of $CaCO_3$ depends directly on the amount of dissolved CO_2 present in the water. Thus, in colder waters the higher levels of CO_2 inhibit calcification. Structural coral reefs are primarily limited to a circumglobal band stretching from the Line Islands in Hawaii to Perth, Australia, with nonstructural coral communities being increasingly dominant along the fringes. The distribution band varies, such that cold waters (e.g., Peru) narrow the range of coral reefs closer to the equator, and warm currents facilitate higher latitude development (e.g., Bermuda).

Calcification in stony corals is greatly enhanced by the presence of zooxanthellae in the tissues. Zooxanthellae are nonmotile stages of a dinoflagellate algae. They are contained within specialized coral cells, growing on excess nutrients and trace metals from the carnivorous corals, producing sugars for utilization by the host, and using up excess CO_2 to enhance calcification by the coral. Zooxanthellae are found in other reef organisms, including giant clams, and the tiny foraminifera, amoeba-like organisms (Order Sarcodina) with calcareous skeletons that create substantial amounts of the sand on coral reefs and nearby white-sand beaches. Much of the biology and ecology of zooxanthellae is poorly known.

Paleoecology

Structural coral reefs are forms of 'biogenic reefs,' distinct geomorphological structures constructed by living organisms. Biogenic reefs have existed in various forms since approximately 3.5 billion years ago, at which time cyanobacteria began building stromatolites (**Figure 6**). Bioherms, biogenic reefs constructed from limestone produced by shelled animals, became prominent by 570 million years ago. During the mid-Triassic, the Jurassic and early Cretaceous periods (roughly 200–100 million years ago), scleractinian corals had become significant

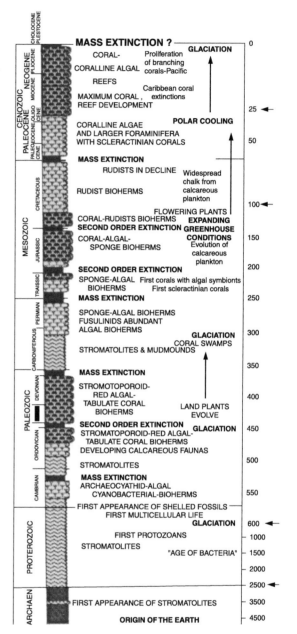

Figure 6 Geological time-scale (in thousand years) highlighting biogenic reef development. Arrows denote changes in scale. (Adapted from Hallock, 1989.)

components of coral–algal–sponge bioherms. Some of these corals may have included zooxanthellae. However, by the mid-Cretaceous, rudist bivalves dominated shallow-water reefs, and scleractinian corals were mainly restricted to deeper shelf-slope environments. This may have been due to seawater chemistry or competition with rudists or both. The massive extinction at the end of the Cretaceous, which ended the reign of dinosaurs on land, also led to the extinction of rudists. Many scleractinian corals survived in their deeper habitats, and evolved into

EARLY PLIOCENE

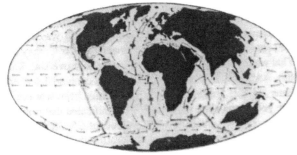

MIDDLE OLIGOCENE

MIDDLE PALEOCENE

Figure 7 Major tectonic movements and inferred patterns of major currents since the Paleocene. Gradually, the circumglobal waterway of the Paleocene, the Tethys Sea, was closed off, resulting in regional extinction of coral reef organisms. This process explains many of the differences in biodiversity now seen among the world's coral reefs. (Adapted from Hallock, 1997.)

most of the modern genera by the Eocene. However, CO_2 concentrations in the atmosphere were high at the time, apparently inhibiting reef development by aragonite-producing corals. Instead, major limestone deposits of the time were formed by calcite-producing organisms such as larger foraminifera (including the limestones from which the Egyptian pyramids were later to be built) and red coralline algae. Calcite is less stable over time than aragonite, but can form in waters of higher CO_2 concentration. Major reef-building by stony corals and their associates did not occur until the middle to late Oligocene. By then, CO_2 concentrations were falling and sea water was warming in tropical areas, whereas at high-latitudes it became cooler. By the late Oligocene, Caribbean coral reefs achieved their greatest development, and by the early Miocene, coral reefs globally had extended to beyond 10° north and south.

During the late Eocene and early Oligocene, tropical oceans were openly connected, in a system of equatorial waterways known as the Tethys Sea. During this period, most scleractinian corals were cosmopolitan. The upward movements of Africa and India restricted circulation, particularly with the closure of the Qatar arch in the Middle East around the time of the Oligocene–Miocene boundary (**Figure 7**). The Central American passageway became restricted, but did not close until the middle Pliocene. However, nearly half of the Caribbean coral genera became extinct at the Oligocene–Miocene boundary, and many more disappeared during the Miocene. Larger foraminifera suffered similar extinctions. Far less dramatic extinctions occurred in the Indo-Pacific.

Reef Geomorphology

Charles Darwin proposed that atolls were formed by a process involving the sinking of islands (**Figure 8**). He noted that most high islands in the Pacific were

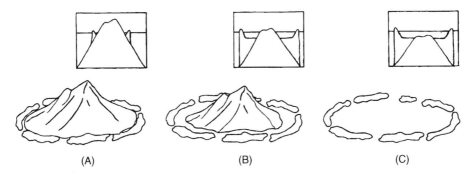

(A) (B) (C)

Figure 8 Stages in the development of an atoll as envisioned by Darwin. Fringing reefs surrounding a volcano (A) become barrier reefs as the volcano sinks (B) and finally form a donut-shaped atoll (C) Sinking islands and/or rising sea levels do help to explain the structures of many reefs, but other factors have been more important in the case of other reefs. (Adapted from Slafford-Deitsch, 1991.)

the tops of volcanic cones. A rocky, volcanic island would tend to form fringing reefs around its shores. Should the island begin to sink, the highly oxygenated outer edges would be able to maintain themselves at the sea surface, while the less actively growing reef flats would tend to sink, forming lagoons defining barrier reefs. Eventually, a sinking cone would disappear altogether, leaving a lagoon in the midst of an atoll. The feasibility of this explanation was confirmed by the mid-twentieth century when drilling on Enewetak and Bikini reefs yielded signs that the atolls were perched over islands that had gradually subsided over thousands of years.

Modern researchers have identified a variety of explanations for the morphology of various reefs. Variations in relative sea level have had profound influences on most structural coral reefs. Few coral reefs are believed to have exhibited continuous growth prior to 8000 years ago, and most are considerably younger. Coral reefs located on extensive continental shelf areas, such as the Florida reef tract and the vast shelf areas off the Yucatan and peninsular south-east Asia, grow on substrates that were generally far inland during the previous ice age.

The three-dimensional shapes of many coral reefs have been highly influenced by underlying topography, often an ancient reef. Wind, storms, waves, and currents have helped to shape many reefs, resulting in tear-drop shapes with broad, raised, steep edges facing predominant winds and currents. The shapes of some reefs are believed to have been particularly dependent on the erosion of much larger blocks of reef limestone during low stands of sea level. It has been demonstrated in the laboratory that small blocks of limestone subjected to rain-like erosion can form many features now identifiable on coral reefs, including lagoons, rifts and ridges, and channels. Evidence has also been found that the ridge and rift structures and some other features of coral reefs can result from collective ecological growth processes as the biota accommodates effects of waves, channeled water, and scouring by sand and other reef matter as calcifying benthic organisms compete for exposure to sunlight. It has been suggested that the double lagoon structure of the Atoll reef off Mindoro, Philippines, resulted from coral reef growth around the subsurface rim of a volcanic crater. Drilling on some reefs has demonstrated continuous growth for 30 m or more. On other reefs, little or no substantial calcification has occurred, the extreme case being nonstructural coral communities.

The Hawaiian and Line Islands are part of a vast chain of islands formed as a tectonic plate has drifted north and west across the Pacific. Volcanic lava has tended to erupt in a fixed 'hot-spot', forming volcanic islands one after the other. As the islands have drifted north, they have subsided gradually into deeper waters until coral growth has ceased. This process has had a strong influence over the structure of coral reefs along these islands and seamounts. However, at smaller scales, coral reefs along the coasts of the Hawaiian islands also show the effects of heavy storm action, exhibiting mound-like features at the sea surface often dominated by calcareous algae, and sometimes broad platforms which produce the famous surfing waves of the islands.

Role in Oil Exploration

Coral reefs are generally highly porous, filled with holes, tunnels, and caves of all sizes. The most famous 'blue-holes' of the Bahamas are the vertical seaward entrances of huge cave and tunnel systems formed by erosion at low stands of sea level, through which sea water flows beneath the islands as water through a sponge. Much of the porosity of coral reefs results from incomplete filling of spaces between coral heads during the active growth of the reef.

Much of the world's crude oil comes from ancient coral reefs that have been subsequently overlain with terrigenous (land-based) sediments. Dense, horizontally layered sediments have often trapped oil within the mounds of porous coral reef limestone. In some areas, subsequent tectonic activity has produced faults, and much of the oil has been lost. In other areas, however, the ancient reefs have yielded vast amounts of oil.

In some parts of the world, modern coral reefs have grown in areas where ancient reefs once occurred and were subsequently covered by dense sediments that trapped oil. Thus, it is common to find oil drilling platforms on modern reefs. This has frequently raised environmental concerns, centering on the damage done to modern corals by the construction and associated activities, the occasional spillages of drilling muds of various compositions and the leakage of oil.

Biology

Biogeography

The highest species diversity on coral reefs occurs in the seas of south-east Asia, often referred to as the Central Indo-West Pacific region. In the Caribbean or Hawaii, there are 70–80 species of reef-building corals. In south-east Asia, there are more than 70 genera and 400 or more species of reef-building corals (**Figure 9**). The diversities of most other groups of reef-associated species, including fish, tend to be higher in south-east Asia than in other regions.

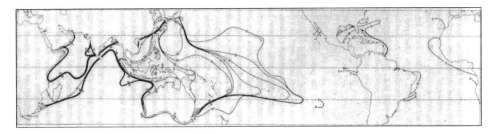

Figure 9 Patterns of generic diversity is scleractinian corals. Contours were estimated from generic patterns and existing data on species distributions. (From Veron 1993; reprinted with permission of the author.)

Diversity begins to drop off gradually to the south below Indonesia, and north above the Philippines. The decline eastward is very dramatic, such that only a few species of corals live in Tahiti and the Galapagos. Westward, the decline is only slight; at least 50 genera of corals live along the coasts of East Africa and the Red Sea. There are only a few species of reef-building corals off the coast of West Africa. Species in south-east Asia tend to have very broad ranges, and endemism is higher in peripheral areas such as the central Pacific.

Much of the global pattern of diversity, particularly at the level of genera and families, can be explained in terms of selective regional extinctions following the gradual breakup of the circumglobal Tethys Sea during the Cenozoic. This is well supported in the fossil records of corals, seagrasses and mangrove trees. Although fish fossils are uncommon, there is a highly diverse assemblage of fossil coral reef fish in Italy, a remnant from the Tethys Sea.

However, there is some reason to believe that rates of speciation have been higher in south-east Asia than elsewhere. The evidence is best known for mollusks, and include unusually high ratios of species to genera and particularly highly evolved armament (such as the spines on many murex shells) in this region. There is a myriad of explanations for this potentially higher speciation rate. Two popular explanations, the Pacific island vicariance hypothesis and the basin isolation hypothesis, both involve changes in sea level. In the former hypothesis, high levels of the sea tend to isolate groups of Pacific islands, facilitating speciation. At low sea levels, the species spread to the heterogeneous refugia habitat of south-east Asia and are gradually lost on the less-heterogeneous Pacific islands due to competition from other species. In the latter hypothesis, low stands of sea level, and, more importantly, periods of mountain building, have isolated particular marine basins within south-east Asia in the past, possibly permitting rapid speciation among whole biotas.

Further understanding of these processes must await increased efforts in taxonomy and systematics.

The vast majority of coral reef organisms, particularly in the Indo-West Pacific region, have never been formally identified. The ranges of known organisms are poorly known. Furthermore, recent decades have seen a rapid decline in interest in and support for taxonomy and systematics. It is likely that human impacts will result in vast changes to the distributions and abundances of coral reef species long before existing biogeography and ecology have been well understood.

Coral Reef Ecosystem Health

Resilience and phase shifts   Most coral reefs are subjected to periodic disturbances, such as storms. The ability of an ecosystem to maintain constancy in terms of ecological functions and in the abundances and distributions of organisms is ecological resistance. The capacity of an ecosystem to revert to a previous state (or near to it) in terms of these characteristics is ecological resilience. Terms such as ecosystem health, integrity, and stability usually infer degrees of resistance and resilience. Because profound changes often occur to a coral reef following a perturbation such as a storm, there is increasing focus on the resilience of a coral reef.

Many coral reefs have been known to undergo losses of coral cover from greater than 50% to less than 10%, and then to recover to the former level within 4–10 years. Naturally, in reefs in which colonies may have been decades or centuries old, the age structure of the corals present have often been disrupted. This, in turn, may affect resilience to future perturbations, as certain corals are believed to exhibit higher fecundities in older colonies.

Some coral reefs do not seem to return to high levels of coral cover after a perturbation, especially if they have been subsequently overgrown by fleshy macroalgae. Human intervention, in the form of increased eutrophication and the removal of herbivorous fish and invertebrates, are suspected to favor the growth of macroalgae following perturbations.

Reefs around Jamaica have shown little recovery in more than a decade following coral damage by a hurricane, and both forms of intervention are suspected of being causes of this loss of resilience. Widespread losses of resilience are a concern throughout most coral reef areas.

Reef interconnectivity   Most coral reef organisms undergo pelagic life stages before settling into a reef community. Most corals live as planulae for a few hours to a few days before setting, although longer periods have been recorded. The average benthic invertebrate spends roughly two weeks in waterborne stages, but some survive for months. The average coral reef fish appears to require nearly a month before settling, and many require two months.

The pelagic stages are by no means passive, and although the sizes are very small, the organisms may be adapted to swim into currents and eddies that facilitate their retention or return to particular reefs or groups of reefs. Their success in doing so is believed to depend on factors such as reef geomorphology and the nature and predictability of local oceanography. Analyses indicate that some fish populations regularly exchange genetic material over thousands of kilometers. Although most coral-reef populations are believed to be replenished each year from local progeny, a severely depleted reef may be restocked to some degree from other reefs. This process is crucial to the problem of resilience, and is an active area of controversy and research. The results of this work may have profound implications for the design of marine protected areas, for international agreements on the coordination of management schemes, and for the regulation of harvesting on coral reefs. Furthermore, climate change is likely to alter local oceanographic processes, and has important implications for reef management as such disruptions occur.

Coral Reef Management

Disturbances

Various types of perturbations have affected coral reefs with increasing frequency within recent decades. Some of these are clearly directly related to human activities, whereas others are suspected to be the indirect results of human interventions.

The majority of coral reefs are located in developing countries. In many of these, crowded low-income human populations increasingly overfish, often reducing herbivorous fish and invertebrates, thereby decreasing reef resilience. In a process known as Malthusian overfishing, social norms break down and fishers turn to destructive fishing methods such as the use of poisons and explosives to capture fish.

Organic pollution from coastal habitations and agricultural fertilizing activities are common along coastlines. Runoff from deforested hillsides, mining operations, and coastal construction often contains materials that favor macroalgal growth, and silt and mud that restrict light to zooxanthellae, abrade reef benthos or bury portions of coral reefs entirely.

Increasingly common outbreaks of the coral-eating crown-of-thorns starfish, *Acanthaster planci*, may be related to reductions in predators, such as lethrinid fishes. During 1997–98, a worldwide epidemic of coral bleaching occurred, in which the zooxanthellae of many colonies were expelled and high rates of coral mortality resulted. The cause was unusually warm patches of sea water associated with a strong El Niño event, which some believe to be related to increasing levels of atmospheric CO_2 and global warming. A more controversial suggestion is that the increasing CO_2 levels will cause acidification of the oceans sufficient to result in the net erosion of some coral reefs, especially at high latitudes. Although there have recently been major epidemics of diseases killing corals and associated organisms in western Atlantic reefs, only certain coral diseases appear to be directly linked to stress from human activities.

Assessments

Efforts are underway to gather empirical information on coral reefs via the quantification of benthic organisms and fish by divers. These efforts are being supplemented by the use of remotely sensed data from satellites, space shuttles, aircraft, ships, manned and unmanned underwater vehicles. The usefulness of these approaches ranges from identifying or mapping reefs, to quantifying bleaching and disease. A global database, ReefBase, has been developed by the International Center for Living Aquatic Resources Management (ICLARM) to gather together existing information about the world's reefs and to make it widely available via CD-ROM and the Internet.

Integrated Coastal Management

Management is a process of modifying human behavior. Biophysical scientists can provide advice and predictions concerning factors such as levels of fishing pressure, siltation, and pollution. However, the management decisions must account for social, cultural, political, and economic considerations. Furthermore, almost all management interventions will have both positive and negative effects on various aspects of the ecosystem and the societies that impact it and depend upon it. For example, diverting fishers into forestry may lead to increased deforestation, siltation, and further reef degradation.

It is increasingly recognized that effective management is achieved only through approaches that integrate biophysical considerations with socio-economic and related factors. Balanced stakeholder involvement is generally a prerequisite for compliance with management decisions. The field of integrated coastal management is rapidly evolving, as is the set of scientific paradigms on which it is based. To a large degree, the future of the world's coral reefs is directly linked to this evolution.

See Also

Beaches, Physical Processes Affecting. Coral Reef Fishes. Geomorphology. Lagoons. Macrobenthos. Rocky Shores. Sandy Beaches, Biology of. Sea Level Change. Storm Surges.

Further Reading

Birkeland C (ed.) (1997) *Life and Death of Coral Reefs.* New York: Chapman and Hall.

Bryant D, Burke L, McManus J, and Spalding M (1998) *Reefs at Risk: a Map-based Indicator of the Threats to the World's Coral Reefs.* Washington, DC: World Resources Institute.

Davidson OG (1998) *The Enchanted Braid: Coming to Terms with Nature on the Coral Reef.* New York: John Wiley.

Holliday L (1989) *Coral Reefs: a Global View by a Diver and Aquarist.* London: Salamander Press.

McManus JW, Ablan MCA, Vergara SG, *et al.* (1997) *ReefBase Aquanaut Survey Manual.* Manila, Philippines: International Center for Living Aquatic Resources Management.

McManus JW and Vergara SG (eds.) (1998) *ReefBase: a Global Database of Coral Reefs and Their Resources. Version 3.0 (Manual and CD-ROM).* Manila, Philippines: International Center for Living Aquatic Resource Management.

Polunin NVC and Roberts C (eds.) (1996) *Reef Fisheries.* New York: Chapman and Hall.

Sale PF (ed.) (1991) *The Ecology of Fishes on Coral Reefs.* New York: Academic Press.

Stafford-Deitsch J (1993) *Reef: a Safari Through the Coral World.* San Francisco: Sierra Club Books.

Veron JEN (1993) *A Biogeographic Database of Hermatypic Corals. Species of the Central Indo-Pacific, Genera of the World. Monograph Series 10.* Townsville, Australia: Australian Institute of Marine Science.

CORALS AND HUMAN DISTURBANCE

N. J. Pilcher, Universiti Malaysia Sarawak, Sarawak, Malaysia

Introduction

Coral reefs are the centers of marine biodiversity on the planet. Reefs are constructed by a host of hermatypic (reef-building) coral species, but also are home to ahermatypic (non-calcium-carbonate depositing) corals, such as soft corals, black corals and gorgonians. The major structural components of reefs are the scleractinian corals. Much like their terrestrial rivals the tropical rainforests, reefs combine a host of microhabitats and a diverse array of life forms that is still being discovered and described. Coral reefs are mostly distributed throughout the tropical belt, and a large fraction are located in developing countries.

To understand how human activities affect coral reefs, it is necessary to briefly review their basic life history. Coral reefs are mostly made up of numerous smaller coral colonies; these colonies are in turn made up of thousands of minute polyps, which secrete a calcium carbonate skeleton. The deposition rate for individual coral species varies, but generally ranges between 0.1 mm and 10.0 cm per year. The accumulation of these skeletons over a long period of time results in massive, three-dimensional geological structures. The actual living tissue, however, is only the thin layer of living coral polyps on the surface. Corals are particularly susceptible to contaminants in sea water because the layer of tissue covering the coral skeleton is thin ($\sim 100\,\mu$m) and rich in lipids, facilitating direct uptake of chemicals. Coral polyps feed by filtering plankton using nematocyst (stinging cell)-tipped tentacles, and also receive organic matter through their symbiotic relationship with minute dinoflagellates called zooxanthellae. Zooxanthellae live within the gastrodermal tissues, and chemical communication (exchange) occurs via the translocation of metabolites. These small algal cells use sunlight to photosynthesize carbonates and water into organic matter and oxygen, both of which are used by the polyp.

Coral reefs support complex food and energy webs that are interlinked with nutrient inputs from outside sources (such as those brought with ocean currents and runoff from nearby rivers) and from the reef itself (where natural predation and die-off recirculate organic matter). These complex webs mean that any effect on one group of individuals will ultimately impact another, and single disturbances can have multiple effects on reef inhabitants. For example, the complete eradication of the giant Triton *Charonia trinis* through overfishing can result in outbreaks of the crown-of-thorns starfish *Acanthaster planci*. This can lead to massive coral mortality as the starfish reproduce and feed on the coral polyps. The mortality in turn may reduce habitats and food sources for reef fishes, which again, in turn, could lead to declines of larger predatory fishes. Similarly, the introduction of an invasive species either by accident or through ignorance (e.g., dumping of a personal aquarium contents into local habitats) might disrupt feeding processes and kill resident fishes. The death of key organisms on the reef (which then shifts from an autotrophic to a heterotrophic, suspension/detritus-feeding community) changes the dominant ecological process from calcium carbonate deposition to erosion, and ultimate loss of coral reef. Reef ecosystems may respond to environmental change by altering their physical and ecological structure, and through changes in rates of accretion and biogeochemical cycling. However, the potential for adaptation in reef organisms may be overwhelmed by today's anthropogenic stresses. The following sections provide a review of human disturbances and their general effects on coral reefs.

Collection of Corals

Corals have been mined for construction purposes in numerous Pacific Ocean islands and in South-East Asia. Usually the large massive life forms such as *Porites*, *Platygyra*, *Favia*, and *Favites* are collected and broken into manageable sizes or crushed for cement and lime manufacture. Similarly, the shells of the giant clam *Tridacna* and the conch shell *Strombus* are collected. Coral blocks and shells are used for construction of houses, roads, and numerous other projects. Corals are also collected for use in the ornamental trade, either as curios and souvenirs or as jewelry. Entire, small colonies of branching species such as *Acropora* and *Seriatophora* are used in the souvenir trade and for decoration, while black corals *Antipathes* and blue coral *Heliopora* are used for jewelry. The aquarium industry is also responsible for coral collection either for direct sale as live colonies or through the process of fish collecting. In many cases entire colonies full of fish are brought to the surface and are then smashed and discarded.

The removal of coral colonies decreases the shelter and niche areas available to numerous other reef inhabitants. Juvenile stages of fish that seek shelter among the branching species of corals, and worms and ascidians that take up residence on massive life forms, are deprived of protection and may become prey to other reef organisms. Removal of adult colonies also results in a reduction of overall reproductive output, as the corals no longer serve as a source of replenishment larvae. Further, removal of entire colonies reduces the overall structural stability of the reef, and increases rates of erosion through wave and surge damage.

Destructive Fishing

Destructive fishing pressures are taking their toll on coral reefs, particularly in developing countries in South-East Asia. The use of military explosives and dynamite was common shortly after the Second World War, but today this has shifted to the use of home-made explosives of fertilizer, fuel, and fuse caps inserted into empty beer bottles. Bombs weigh approximately 1 kg and have a destructive diameter of 4–5 m. Blast fishers hunt for schooling fish such as sweetlips and fusiliers, which aggregate in groups in the open or hide under large coral heads. Parrot fish and surgeon fish schools grazing on the reef crest are also actively sought. The bombs are usually set on five-second fuses and are dropped into the center of an area judged to have many fish. After the bomb has exploded, the fishers use dip nets, either from the boats or from underwater, to collect the stunned and dying fish. Many larger boats collect the fish using 'hookah' compressors and long air hoses to support divers working underwater.

The pressure wave from the explosion kills or stuns fish, but also damages corals. Natural disturbances may also fragment stony reef corals, and there are few quantitative data on the impacts of skeletal fragmentation on the biology of these corals. Lightly bombed reefs are usually pockmarked with blast craters, while many reefs in developing countries comprise a continuous band of coral rubble instead of a reef crest and upper reef slope. The lower reef slope is a mix of rubble, sand, and overturned coral heads. Typically at the base of the reef slope is a mound of coral boulders that have been dislodged by a blast and then rolled down the slope in an underwater avalanche. The reef slopes are mostly dead coral, loose sand, rubble, or rock and occasionally have overturned clams or coral heads with small patches of living tissue protruding from the rubble. The blasts also change the three-dimensional structure of reefs, and blasted areas no longer provide food or shelter to reef inhabitants. Further, once the reef structure has been weakened or destroyed by blast fishing, it is much more susceptible to wave action and the reef is unable to maintain its role in coastline protection. Larvae do not settle on rubble and thus replenishment and rehabilitation is minimal. Additionally, the destruction of adult colonies also results in a reduction of overall reproductive output, and reefs no longer serve as a source of replenishment larvae. Experimental findings, for instance, indicate that fragmentation reduces sexual reproductive output in the reef-building coral *Pocillopora damicornis*. The recovery of such areas has been measured in decades, and only then with complete protection and cessation of fishery pressure of any kind.

Another type of destructive fishing is 'Muro Ami', in which a large semicircular net is placed around a reef. Fish are driven into the net by a long line of fishermen armed with weighted lines. The weights are repeatedly crashed onto the reef to scare fish in the direction of the net, reducing coral colonies to rubble. The resulting effects are similar to the effects of blast fishing, spread over a larger area.

Cyanide fishing is also among the most destructive fishing methods, in which an aqueous solution of sodium cyanide is squirted at fish to stun them, after which they are collected and sold to the live-fish trade. Other chemicals are also used, including rotenone, plant extracts, fertilizers, and quinaldine. These chemicals all narcotize fish, rendering them inactive enough for collection. The fish are then held in clean water for a short period to allow them to recover, before being hauled aboard boats with live fish holds. In the process of stunning fish, the cyanide affects corals and small fish and invertebrates. The narcotizing solution for large fish is often lethal to smaller ones. Cyanide has been shown to limit coral growth and cause diseases and bleaching, and ultimately death in many coral species.

Among other destructive aspects of fishing are lost fishing gear, and normal trawl and purse fishing operations, when these take place near and over reefs. Trawlers operate close to reefs to take advantage of the higher levels of fish aggregated around them, only to have the trawls caught on the reefs. Many of these have to be cut away and discarded, becoming further entangled on the reefs, breaking corals and smothering others. Similarly, fishing with fine-mesh nets that get entangled in coral structures also results in coral breakage and loss. In South-East Asia, fishing companies have been reported to pull a chain across the bottom using two boats to clear off corals, making it accessible to trawlers.

Spearfishing also damages corals as fishermen trample and break coral to get at fish that disappear into crevices, and crowbars are frequently used to break coral. The collection of reef invertebrates along the reef crest results in breakage of corals that have particular erosion control functions, reducing the reef's potential to act as a coastal barrier.

Discharges

Mankind also effects corals through the uncontrolled and often unregulated discharge of a number of industrial and domestic effluents. Many of these are 'point-source' discharges that affect local reef areas, rather than causing broad-scale reef mortality. Sources of chemical contamination include terrestrial runoff from rivers and streams, urban and agricultural areas, sewage outfalls near coral reefs, desalination plants, and chemical inputs from recreational uses and industries (boat manufacturing, boating, fueling, etc.). Landfills can also leach directly or indirectly into shallow water tables. Industrial inputs from coastal mining and smelting operations are sources of heavy metals. Untreated and partially treated sewage is discharged over reefs in areas where fringing reefs are located close to shore, such as the reefs that fringe the entire length of the Red Sea. Raw sewage can result in tumors on fish, and erosion of fins as a result of high concentrations of bacteria. The resulting smothering sludge produces anaerobic conditions under which all benthic organisms perish, including corals. In enclosed and semi-enclosed areas the sewage causes eutrophication of the coral habitats. For instance, in Kaneohe bay in Hawaii, which had luxuriant reefs, sewage was dumped straight into the bay and green algae grew in plague quantities, smothering and killing the reefs. Evidence indicates that branching species might be more susceptible to some chemical contaminants than are massive corals.

Abattoir refuse is another localized source of excessive nutrients and other wastes that can lead to large grease mats smothering the seabed, local eutrophication, red tides, jellyfish outbreaks, an increase in biological oxygen demand (BOD), and algal blooms. Similarly, pumping/dumping of organic compounds such as sugar cane wastes also results in oxygen depletion.

The oil industry is a major source of polluting discharges. Petroleum hydrocarbons and their derivatives and associated compounds have caused widespread damage to coastal ecosystems, many of which include coral reefs. The effects of these discharges are often more noticeable onshore than offshore where reefs are generally located, but nonetheless have resulted in the loss of reef areas, particularly near major exploration and drilling areas, and along major shipping routes. Although buoyant eggs and developing larvae are sometimes affected, reef flats are more vulnerable to direct contamination by oil. Oiling can lead to the increased incidence of mortality of coral colonies.

In the narrow Red Sea, where many millions of tonnes per annum pass through the region, there have been more than 20 oil spills along the Egyptian coast since 1982, which have smothered and poisoned corals and other organisms. Medium spills from ballast and bilgewater discharges, and leakages from terminals, cause localized damage and smothering of intertidal habitats. Oil leakage is a regular occurrence from the oil terminal and tankers in Port Sudan harbor. Seismic blasts during oil exploration are also a threat to coral reefs. Refineries discharge oil and petroleum-related compounds, resulting in an increase in diatoms and a decrease in marine fauna closer to the refineries. Throughout many parts of the world there is inadequate control and monitoring of procedures, equipment, and training of personnel at refineries and shipping operations.

Drilling activities frequently take place near reef areas, such as the Saudi Arabian shoreline in the Arabian Gulf. Drilling muds smother reefs and contain compounds that disrupt growth and cause diseases in coral colonies. Field assessment of a reef several years after drilling indicated a 70–90% reduction in abundance of foliose, branching, and platelike corals within 85–115 m of a drilling site. Research indicates that exposure to ferrochrome lignosulfate (FCLS) can decrease growth rates in *Montastrea annularis*, and growth rates and extension of calices (skeleton supporting the polyps) decrease in response to exposure to $100 \, \text{mg} \, \text{l}^{-1}$ of drilling mud.

Oil spills affect coral reefs through smothering, resulting in a lack of further colonization, such as occurred in the Gulf of Aqaba in 1970 when the coral *Stylophora pistillata* did not recolonize oil-contaminated areas after a large spill. Effects of oil on individual coral colonies range from tissue death to impaired reproduction to loss of symbiotic algae (bleaching). Larvae of many broadcast spawners pass through sensitive early stages of development at the sea surface, where they can be exposed to contaminants and surface slicks. Oiling affects not only coral growth and tissue maintenance but also reproduction. Other effects from oil pollution include degeneration of tissues, impairment of growth and reproduction (there can be impaired gonadal development in both brooding and broadcasting species, decreased egg size and decreased fecundity), and decreased photosynthetic rates in zooxanthellae.

In developing countries, virtually no ports have reception facilities to collect these wastes and the problem will continue mostly through a lack of enforcement of existing regulations. The potential exists for large oil spills and disasters from oil tank ruptures and collisions at sea, and there are no mechanisms to contain and clean such spills. The levels of oil and its derivatives (persistent carcinogens) were correlated with coral disease in the Red Sea, where there were significant levels of diseases, especially Black Band Disease. In addition to the impacts of oils themselves are the impacts of dispersants used to combat spills. These chemicals are also toxic and promote the breakup of heavier molecules, allowing toxic fractions of the oil to reach the benthos. They also promote erosion through limiting adhesion among sand particles. The full effect of oil on corals is not fully understood or studied, and much more work is needed to understand the full impact. Although natural degradation by bacteria occurs, it is slow and, by the time bacteria consume the heavy, sinking components, these have already smothered coral colonies.

Industrial effluents, from a variety of sources, also impact coral reefs and their associated fauna and habitats. Heavy metal discharges lead to elevated levels of lead, mercury, and copper in bivalves and fish, and to elevated levels of cadmium, vanadium, and zinc in sediments. Larval stages of crustaceans and fish are particularly affected, and effluents often inhibit growth in phytoplankton, resulting in a lack of zooplankton, a major food source for corals. Industrial discharges can increase the susceptibility of fish to diseases, and many coral colonies end up with swollen tissues, excessive production of mucus, or areas without tissue. Reproduction and feeding in surviving polyps is affected, and such coral colonies rarely contribute to recolonization of reef areas.

Organisms in low-nutrient tropical waters are particularly sensitive to pollutants that can be metabolically substituted for essential elements (such as manganese). Metals enter coral tissues or skeleton by several pathways. Exposed skeletal spines (in response to environmental stress), can take up metals directly from the surrounding sea water. In Thailand, massive species such as *Porites* tended to be smaller in areas exposed to copper, zinc, and tin, there was a reduced growth rate in branching corals, and calcium carbonate accretion was significantly reduced.

Symbiotic algae have been shown to accumulate higher concentrations of metals than do host tissues in corals. Such sequestering in the algae might diminish possible toxic effects to the host. In addition, the symbiotic algae of corals can influence the skeletal concentrations of metals through enhancement of calcification rates. There is evidence, however, that corals might be able to regulate the concentrations of metals in their own tissues. For example, elevated iron in Thai waters resulted in loss of symbiotic algae in corals from pristine areas, but this response was lower in corals that has been exposed to daily runoff from an enriched iron effluent, suggesting that the corals could develop a tolerance to the metal.

Cooling brine is another industrial effluent that affects shoreline-fringing reefs, often originating from industrial installations or as the outflow from desalination plants. These effluents are typically up to 5–10°C higher in temperature and up to 3–10 ppt higher in salinity. Discharges into the marine environment from desalination plants in Jeddah include chlorine and antiscalant chemicals and 1.73 billion $m^3 d^{-1}$ of brine at a salinity of 51 ppt and 41°C. The higher temperatures decrease the water's ability to dissolve oxygen, slowing reef processes. Increases in temperature are particularly threatening to coral reefs distributed throughout the tropics, where reef-building species generally survive just below their natural thermal thresholds. Higher-temperature effluents usually result in localized bleaching of coral colonies. The higher-salinity discharges increase coral mucus production and result in the expulsion of zooxanthellae and eventual bleaching and algal overgrowth in coral colonies. Often these waters are chlorinated to limit growth of fouling organisms, which increases the effects of the effluents on reef areas. The chlorinated effluents contain compounds that are not biodegradable and can circulate in the environment for years, bringing about a reduction in photosynthesis, with blooms of blue/green and red algae. Chlorinated hydrocarbon compounds include aldrin, lindrane, dieldrin, and even the banned DDT. These oxidating compounds are absorbed by phytoplankton and in turn by filter-feeding corals. Through the complex reef food webs these compounds concentrate in carnivorous fishes, which are often poisonous to humans.

Many airborne particles are also deposited over coral reefs, such as fertilizer dust, dust from construction activities and cement dust. At Ras Baridi, on the Red Sea coast of Saudi Arabia, a cement plant that operates without filtered chimneys discharges over 100 t d^{-1} of partially processed cement over the nearby coral reefs, which are now smothered by over 10 cm of fine silt.

Solid Waste Dumping

The widespread dumping of waste into the seas has continued for decades, if not centuries. Plastics, metal, wood, rubber, and glass can all be found littering coral reefs. These wastes are often not biodegradable, and those that are can persist over long periods. Damage to reefs through solid waste dumping is primarily physical. Solid wastes damage coral colonies at the time of dumping, and thereafter through natural tidal and surge action. Sometimes the well-intentioned practice of developing artificial reefs backfires and the artificial materials are thrown around by violent storms, wrecking nearby reefs in the process.

Construction

Construction activities have had a major effect on reef habitats. Such activity includes coastal reclamation works, port development, dredging, and urban and industrial development. A causeway across Abu Ali bay in the northern Arabian Gulf was developed right over coral reefs, which today no longer exist. Commercial and residential property developments in Jeddah, on the Red Sea, have filled in reef lagoon areas out to reef crest and bulldozed rocks over reef crest for protection against erosion and wave action. Activities of this type result in increased levels of sedimentation as soils are nearly always dumped without the benefit of screens or silt barriers.

Siltation is invariably the consequence of poorly planned and poorly implemented construction and coastal development, which can result in removal of shoreline vegetation and sedimentation. Coral polyps, although able to withstand moderate sediment loading, cannot displace the heavier loads and perish through suffocation. Partial smothering also limits photosynthesis by zooxanthellae in corals, reducing feeding, growth, and reproductive rates.

The development of ports and marinas involves dredging deep channels through reef areas for safe navigation and berthing. Damage to reefs comes through the direct removal of coral colonies, sediment fallout, churning of water by dredger propellers, which increases sediment loads, and disruption of normal current patterns on which reefs depend for nutrients.

Landfilling is one of the most disruptive activities for coastal and marine resources, and has caused severe and permanent destruction of coastal habitats and changed sedimentation patterns that damage adjacent coral resources. Changes in water circulation caused by landfilling can alter the distribution of coral communities through redistribution of nutrients or increased sediment loads.

Recreation

The recreation industry can cause significant damage to coral reefs. Flipper damage by scuba divers is widespread. Some will argue that today's divers are more environmentally conscious and avoid damaging reefs, but certain activities, such as irresponsible underwater photography finds divers breaking corals to get at subjects and trampling reef habitats in order to get the 'perfect shot'. In areas where divers walk over a reef lagoon and crest to reach the deeper waters, there is a degree of reef trampling, heightened in cases where entry and exit points are limited.

Anchor damage from boats is a common problem at tourist destinations. In South-East Asia many diving operations are switching to nonanchored boat operations, but many others continue the practice unabated. Large tracts of reef can be found in Malaysia that have been scoured by dragging anchors, breaking corals and reducing reef crests to rubble. Experiments have shown that repeated break-age of corals, such as is caused by intensive diving tourism, may lead to substantially reduced sexual reproduction in corals, and eventually to lower rates of recolonization. In the northern Red Sea, another popular diving destination, and in the Caribbean, efforts are underway to install permanent moorings to minimize the damage to reefs from anchors.

Shipping and Port Activities

Congested and high-use maritime areas such as narrow straits, ports, and anchorage zones often lead to physical damage and/or pollution of coral reef areas. Ship groundings and collisions with reefs occur in areas where major shipping routes traverse coral reef areas, such as the Spratley Island complex in the South China Sea, the Red Sea, the Straits of Bab al Mandab and Hormuz, and the Gulf of Suez, to name only a few. Major groundings have occurred off the coast of Florida in the United States, such as the one off Key Largo in State park waters in the 1980s, causing extensive damage to coral reefs. Often these physical blows are severe and destroy decades, if not centuries, of growth. Fish and other invertebrates lose their refuges and foraging habitats, while settlement of new colonies is restricted by the broken-up nature of the substrate. Seismic cables towed during seabed surveys and exploration activities may damage the seabed. Cable damage from towing of vessels (e.g., a tug and barge) has been reported snagging on shallow reefs in the Gulf of Mexico, causing acute damage to sensitive reefs.

Discharges from vessels include untreated sewage, solid wastes, oily bilge, and ballast water. On the high seas these do not have a major noticeable effect on marine ecosystems, but close to shore, particularly at anchorages and near ports, the effects become more obvious. At low tides, oily residues may coat exposed coral colonies, and sewage may cause localized eutrophication and algal blooms. Algal blooms in turn deplete dissolved oxygen levels and prevent penetration of sunlight.

Port activities can have adverse effects on nearby reefs through spills of bulk cargoes and petrochemicals. Fertilizers, phosphates, manganese, and bauxite, for instance, are often shipped in bulk, granular form. These are loaded and offloaded using massive mechanical grabs that spill a little of their contents on each haul. In Jordan, the death of corals was up to four times higher near a port that suffered frequent phosphate spills when compared to control sites. The input of these nutrients often reduces light penetration, inhibits calcification, and increases sedimentation, resulting in slower feeding and growth rates, and limited settlement of new larva.

War-related Activities

The effects of war-related activities on coral reef health and development are often overlooked. Nuclear testing by the United States in Bimini in the early 1960s obliterated complete atolls, which only in recent years have returned to anything like their original form. This redevelopment is nothing like the original geologic structure that had been built by the reefs over millenia. The effects of the nuclear fallout at such sites is poorly understood, and possibly has long-term effects that are not appreciable on a human timescale. The slow growth rate of coral reefs means that those blast areas are still on the path to recovery.

Target practice is another destructive impact on reefs, such as occurred in 1999 in Puerto Rico, where reefs were threatened by aerial bombing practice operations. In Saudi Arabia, offshore islands were used for target practice prior to the Gulf war in 1991. The bombs do not always impact reefs, but those that do cause acute damage that takes long periods to recover.

In the Spratley islands, the development of military structures to support and defend overlapping claims to reefs and islands has brought about the destruction of large tracts of coral reefs. Man-made islands, aircraft landing strips, military bases, and housing units have all used landfilling to one extent or another, smothering complete reefs and resulting in high sediment loads over nearby reefs. Dredging to create channels into reef atolls has also wiped out extensive reef areas.

Indirect Effects

Most anthropogenic effects and disturbances to coral reefs are easily identifiable. Blast fishing debris and discarded fishing nets can be seen. Pollutant levels and sediment loads can be measured. However, many other man-made changes can have indirect impacts on coral reefs that are more difficult to link directly to coral mortality. Global warming is generally accepted as an ongoing phenomenon, resulting from the greenhouse effect and the buildup of carbon dioxide in the atmosphere. Temperatures generally have risen by 1–2°C across the planet, bringing about secondary effects that have had noticeable consequences for coral reefs. The extensive coral beaching event that took place in 1998, which was particularly severe in the Indian Ocean region, is accepted as having been the result of surface sea temperature rise. Bleaching of coral colonies occurs through the expulsion of zooxanthellae, or reductions in chlorophyll content of the zooxanthellae, as coral polyps become stressed by adverse thermal gradients. Some corals are able to survive the bleaching event if nutrients are still available, or if the period of warm water is short.

Coupled with global warming is change of sea level, which is predicted to rise by 25 cm by the year 2050. This sea level rise, if not matched by coral growth, will mean corals will be submerged deeper and will not receive the levels of sunlight required for zooxanthellae photosynthesis. Additionally, the present control of erosion by coral reefs will be lost if waves are able to wash over submerged reefs.

Coral reef calcification depends on the saturation state of carbonate minerals in surface waters, and this rate of calcification may decrease significantly in the future as a result of the decrease in the saturation level due to anthropogenic release of CO_2 into the atmosphere. The concentration of CO_2 in the atmosphere is projected to reach twice the pre-industrial level by the middle of the twenty-first century, which will reduce the calcium carbonate saturation state of the surface ocean by 30%. Carbonate saturation, through changes in calcium concentration, has a highly significant short-term effect on coral calcification. Coral reef organisms do not seem to be able to acclimate to the changing saturation state, and, as calcification rates drop, coral reefs will be less able to cope with rising sea level and other anthropogenic stresses.

The Future

Mankind has contributed to the widespread destruction of corals, reef areas, and their associated fauna through a number of acute and chronic pollutant discharges, through destructive processes, and through uncontrolled and unregulated development. These effects are more noticeable in developing countries, where social and traditional practices have changed without development of infrastructure, finances, and educational resources. Destructive fishing pressures are destroying large tracts of reefs in South-East Asia, while the development of industry affects reefs throughout their range. If mankind is to be the keeper of coral reefs into the coming millennium, there is going to have to be a shift in fishing practices, and adherence to development and shipping guidelines and regulations, along with integrated coastal management programs that take into account the socioeconomic status of people, the environment, and developmental needs.

Glossary

Ahermatypic Non-reef-building corals that do not secrete a calcium carbonate skeletal structure.

DDT Dichlorodiphenyltrichloroethane.

Dinoflagellates One of the most important groups of unicellular plankton organisms, characterized by the possession of two unequal flagella and a set of brownish photosynthetic pigments.

Eutrophication Pollution by excessive nutrient enrichment.

Gastrodemal The epithelial (skin) lining of the gastric cavity.

Hermatypic Reef-building corals that secrete a calcium carbonate skeletal structure.

Quinaldine A registered trademark fish narcotizing agent.

Scleractinians Anthozoa that secrete a calcareous skeleton and are true or stony corals (Order Scleractinia).

Zooxanthellae Symbiotic algae living within coral polyps.

See also

Gas Exchange in Estuaries. Oil Pollution.

Further Reading

Birkland C (1997) *Life and Death of Coral Reefs*. New York: Chapman and Hall.

Connel DW and Hawker DW (1991) *Pollution in Tropical Aquatic Systems*. Boca Raton, FL: CRC Press.

Ginsburg RN (ed.) (1994) *Global Aspects of Coral Reefs: Health, Hazards and History*, 7–11 June 1993, p. 420. Miami: University of Miami.

Hatziolos ME, Hooten AJ, and Fodor F (1998) Coral reefs: challenges and opportunities for sustainable management. In: *Proceedings of an Associated Event of the Fifth Annual World Bank Conference on Environmentally and Socially Sustainable Development*. Washington, DC: World Bank.

Peters EC, Glassman NJ, Firman JC, Richmonds RH, and Power EA (1997) Ecotoxicology of tropical marine ecosystems. *Environmental Toxicology and Chemistry* 16(1): 12–40.

Salvat B (ed.) (1987) *Human Impacts on Coral Reefs: Facts and Recommendations*, p. 253. French Polynesia: Antenne Museum E.P.H.E.

Wachenfeld D, Oliver J, and Morrisey JI (1998) *State of the Great Barrier Reef World Heritage Area 1998*. Townsville: Great Barrier Reef Marine Park Authority.

Wilkinson CR (1993) Coral reefs of the world are facing widespread devastation: can we prevent this through sustainable management practices. In: *Proceedings of the Seventh International Coral Reef Symposium* Guam, Micronesia. Mangilao: University of Guam Marine Laboratory.

Wilkinson CR and Buddemeier RW (1994) *Global Climate Change and Coral Reefs: Implications for People and Reefs*. Report of the UNEO-IOC-ASPEI-IUCN Task Team on Coral Reefs. Gland: IUCN.

Wilkinson CR, Sudara S, and Chou LM (1994) Living coastal resources of Southeast Asia: Status and review. In: *Proceedings of the Third ASEAN-Australia Symposium on Living Coastal Resources*, vol. 1. Townsville: ASEAN-Australia Marine Science Project, Living Coastal Resources.

CORAL REEF FISHES

M. A. Hixon, Oregon State University, Corvallis, OR, USA

Introduction: Diversity, Distribution, and Conservation

Coral reef fishes comprise the most speciose assemblages of vertebrates on the Earth. The variety of shapes, sizes, colors, behavior, and ecology exhibited by reef fishes is amazing. Adult body sizes range from gobies (Gobiidae) less than 1 cm in length to tiger sharks (Carcharhinidae) reportedly over 9 m long. It has been estimated that about 30% of the some 15 000 described species of marine fishes inhabit coral reefs worldwide, and hundreds of species can coexist on the same reef. Taxonomically, reef fishes are dominated by about 30 families, mostly the perciform chaetodontoids (butterflyfish and angelfish families), labroids (damselfish, wrasse, and parrotfish families), gobioids (gobies and related families), and acanthuroids (surgeonfishes and related families).

The latitudinal distribution of reef fishes follows that of reef-building corals, which are usually limited to shallow tropical waters bounded by the 20 °C isotherms (roughly between the latitudes of 30° N and S). The longitudinal center of diversity is the Indo-Australasian archipelago of the Indo-Pacific region. Local patterns of diversity are correlated with those of corals, which provide shelter and harbor prey. There is a high degree of endemism in reef fishes, especially on more isolated reefs, and many species (about 9%) have highly restricted geographical ranges.

The major human activities that threaten reef fishes include overfishing (especially by destructive fishing practices and live collections for restaurants and aquariums), and habitat destruction, which includes both local effects near human population centers and the ongoing worldwide decline of reefs due to coral bleaching and ocean acidification caused by anthropogenic carbon emissions and global warming. Worldwide, about 31% of coral reef fishes are now considered critically endangered and 24% threatened. The major solution for local conservation is fully protected marine reserves, which have proven effective in replenishing depleted populations.

Fisheries

Where unexploited by humans, coral reef fishes typically exhibit high standing stocks, the maximum being about $240\,t\,km^{-2}$ (about $24\,t\,C\,km^{-2}$). High standing crops reflect the high primary productivity of coral reefs, often exceeding $10^3\,g\,C\,m^{-2}\,yr^{-1}$, much of which is consumed directly or indirectly by fishes. Correspondingly, reported fishery yields have reached $44\,t\,km^{-2}\,yr^{-1}$, with an estimated global potential of $6\,Mt\,yr^{-1}$. These fisheries provide food, bait, and live fish for the restaurant and aquarium trades. However, the estimated maximum sustainable yield from shallow areas of actively growing coral reefs is around $20–30\,t\,km^{-2}\,yr^{-1}$, so many reefs are clearly overexploited. Indeed, overfishing of coral reefs occurs worldwide, due primarily to unregulated multispecies exploitation in developing nations. Few and inadequate stock assessments or other quantitative fishery analyses, susceptibility of fish at spawning aggregations (see below), and destructive fishing practices (including the use of dynamite, cyanide, and bleach) are contributing factors. In the Pacific, some 200–300 reef fish species are taken by fisheries, about 20 of which comprise some 75% of the catch by weight. As fishing intensifies in a given locality, large fishes, especially piscivores (see below), are typically depleted first, followed by less-preferred, smaller, and more-productive planktivores and benthivores. (Note that fishing of some piscivores is naturally inhibited in some regions by ciguatera fish poisoning, caused by dinoflagellate toxins concentrated in the tissues of some species.) The indirect effects of overfishing include the demise of piscivores, perhaps enhancing local populations of prey species or causing a trophic cascade. Overfishing of urchin-eating species (such as triggerfishes, Balistidae) and various herbivorous fishes may provide sea urchins predatory and competitive release, respectively. Although urchin grazing may help maintain benthic dominance by corals, overabundant urchins may overgraze and bioerode reefs.

Morphology

A typical perciform reef fish (virtually an oxymoron) is laterally compressed, with a closed swimbladder and fins positioned in a way that facilitate highly maneuverable slow-speed swimming. Compared to more generalized relatives, reef fishes have a greater proportion of musculature devoted to both locomotion and feeding. Their jaws and pharyngeal apparatus are complex and typically well developed for suction feeding of smaller invertebrate prey, with tremendous variation reflecting a wide variety of diets. For example, most butterflyfish (Chaetodontidae) have forceps-like jaws that extract individual polyps from corals, many damselfish (Pomacentridae: e.g., genus *Chromis*) have highly protrusible jaws that facilitate pipette-like suction feeding of zooplankton, and parrotfishes (Scaridae) have fused beak-like jaw teeth and molar-like pharyngeal teeth enabling some species to excavate algae from dead reef surfaces. (This excavation and subsequent defecation of coral sand can bioerode up to 9 kg of calcium carbonate per square meter annually.) Tetraodontiform reef fishes typically swim relatively slowly with their dorsal and anal fins, and consequently are morphologically well defended from predation by large dorsal–ventral spines (triggerfishes, Balistidae), toxins (puffers, Tetraodontidae), or quill-like scales (porucupinefishes, Diodontidae). The latter two families have fused dentition which is well adapted for consuming hard-shelled invertebrates.

Diurnal reef fishes are primarily visual predators. Visual acuity is high and retinal structure indicates color vision. At least some planktivorous damselfishes have ultraviolet-sensitive cones, which may assist in detecting zooplankton by enhancing contrast against background light. Coloration is highly variable (including ultraviolet reflectance), ranging from cryptic to dazzling. Bright 'poster' colors are hypothesized to serve as visual signals in aggression, courtship, and other social interactions. Sexually dimorphic coloration is associated with haremic social systems (see below). Nocturnal reef fishes are either visually oriented, having relatively large eyes (e.g., squirrelfishes, Holocentridae), or rely on olfaction (e.g., moray eels, Muraenidae).

Behavior

Overt behavioral interactions between coral reef fishes include mutualism (when both species benefit), interference competition (often manifested as territoriality), and predator–prey relationships.

Mutualism

Three of the best-documented cases of marine mutualism occur in reef fishes. 'Cleaning symbiosis' occurs when small microcarnivorous fish consume ectoparasites or necrotic tissue off larger host fish, which often allow cleaners to feed within their mouths and gill cavities. The major cleaners are various gobies (Gobiidae) and wrasses (Labridae). Some of the cleaner wrasses are specialists that maintain fixed cleaning stations regularly visited by hosts, which assume solicitous postures. The interaction is not always mutualistic in that cleaners occasionally bite their hosts, and some saber-tooth blennies (Blenniidae) mimic cleaner wrasse and thereby parasitize host fish. Anemonefishes (Pomacentridae, especially the genus *Amphiprion*) live in a mutualistic association with several genera of large anemones. By circumventing discharge of the cnidarian's nematocysts, the fish gain protection from predators by hiding in the stinging tentacles of the anemone. In turn, the fish defend their host from butterflyfishes and other predators that attack anemones. However, some host anemones survive well without anemonefish, in which case the relationship is commensal rather than mutualistic. Finally, some gobies cohabit the burrows of digging shrimp. The shrimp provides shared shelter and the goby alerts the shrimp to the presence of predators.

Territoriality

The most overt form of competition involves territoriality or defense of all or part of an individual's home range. Many reef fishes behave aggressively toward members of both their own and other species, but the most obviously territorial species are benthic-feeding damselfishes (Pomacentridae: e.g., genus *Stegastes*). By pugnaciously defending areas about a meter square from herbivorous fishes, damselfish prevent overgrazing and can thus maintain dense patches of seaweeds. These algal mats serve as a food source for the damselfish as well as habitat for small juvenile fish of various species that manage to avoid eviction. At a local spatial scale, the algal mats can both smother corals as well as maintain high species diversity of seaweeds. By forming dense schools, nonterritorial herbivores (parrotfishes and surgeonfishes) can successfully invade damselfish territories.

Piscivory and Defense

Predation is a major factor affecting the behavior and ecology of reef fishes. There are three main modes of piscivory. Open-water pursuers, such as reef sharks (Carcharhinidae) and jacks (Carangidae), simply overtake their prey with bursts of speed. Bottom-oriented stalkers, such as grouper (Serranidae) and trumpetfishes (Aulostomidae), slowly approach their prey before a sudden attack. Bottom-sitting ambushers, such as lizardfishes (Synodontidae) and anglerfishes (Antennariidae), sit and wait cryptically for prey to approach them. The vision of piscivores is often suited for crepuscular twilight, when the vision of their prey is least acute (being adapted for either diurnal or nocturnal foraging). Hence, many prey species are inactive during dawn and dusk, resulting in crepuscular 'quiet periods' when both diurnal and nocturnal species shelter in the reef framework. (Parrotfishes may further secrete mucous cocoons around themselves at night, and small wrasses may bury in the sand.) Otherwise, prey defensive behavior when foraging or resting typically involves remaining warily near structural shelter and shoaling either within or among species. Associated with day–night shifts in activity are daily migrations between safe resting areas and relatively exposed feeding areas. Caribbean grunts (Haemulidae) spend the day schooling inactively on reefs, and after dusk migrate to nearby seagrass beds and feed. Reproducing reef fishes may avoid predation by spawning (in some combination) offshore, in midwater, or at night. Spawning during ebbing spring tides that carry eggs offshore or guarding broods of demersal eggs further defends propagules from reef-based predators. Subsequent settlement of larvae back to the reef, which occurs mostly at night, is also an apparent antipredatory adaptation.

Reproduction

Social Systems and Sex Reversal

The best-studied examples of highly structured social systems in reef fishes are the harems of wrasses and parrotfishes. Typically, these fish are born as females that defend individual territories or occupy a shared home range. A larger male defends a group of females from other males, thereby sequestering matings. When the male dies, the dominant (typically largest) female changes sex (protogyny) and becomes the new harem master. At high population sizes, some fish may be born as males, develop huge testes, resemble females, infiltrate harems, and sneak spawnings with the resident females. Spatially isolated at their home anemones, anemonefishes have monogamous social systems in which the largest individual is female, the second largest is male, and the remaining fish are immature. Upon the death of the female, the male changes sex (protandry) and the behaviorally dominant juvenile fish matures into a male. Simultaneous hermaphroditism occurs among a few sparids and serranine sea basses. These fish have elaborate courtship behaviors during which individuals switch male and female roles between successive pair spawnings. Regardless of the broad variety of mating systems found in reef fishes, each individual behaves in a way that tends to maximize lifetime reproductive success.

Life Cycle

The typical bony reef fish has a bipartite life cycle: a pelagic egg and larval stage followed by a demersal (seafloor-oriented) juvenile and adult stage. Most bony reef fishes broadcast spawn, releasing gametes directly into the water column where they are swept to the open ocean. Smaller species spawn at their home reefs and some larger species, such as some grouper (Serranidae) and snapper (Lutjanidae), migrate to traditional sites and form massive spawning aggregations. Gametes are released during a paired or group 'spawning rush' followed by rapid return to the seafloor. Exceptions to broadcast spawning include demersal spawners that brood eggs until they hatch, either externally (e.g., egg masses defended by damselfishes) or internally (e.g., mouthbrooding cardinalfishes, Apogonidae), and a few ovoviviparous or viviparous species that give birth to well-developed juveniles (including reef sharks and rays). Annual fecundity of broadcast spawners ranges from about 10 000 to over a million eggs per female. Spawning is weakly seasonal compared to temperate species, typically peaking during summer months but not strongly related to any particular environmental variable. Lunar and semilunar spawning cycles are common. These are presumably adaptations that transport larvae offshore away from reef-based predation, maximize the number of settlement-stage larvae returning during favorable conditions that vary on lunar cycles, and/or benefit spawning adults in some way.

Little is known about the behavior and ecology of reef-fish larvae. Duration of the pelagic larval stage ranges from about 9 to well over 100 days, averaging about a month. Larval prey include a variety of small zooplankters. Comparisons of fecundity at spawning to subsequent larval settlement back to the reef suggest that larval mortality is both extremely high and extremely variable, apparently due mostly to predation. Patterns of endemism, settlement to isolated islands, and limited data tracking larvae directly suggest that there is considerable larval retention at the scale of large islands, yet substantial larval dispersal nonetheless. Later-stage larvae are active swimmers and may control their dispersal by selecting currents among depths. The overall reproductive strategy is apparently to disperse the larvae offshore from reef-based predators, but then to retain offspring close enough to shore for subsequent settlement in suitable habitat.

Settlement, the transition from pelagic larva to life on the reef (or nearby nursery habitat), occurs at a total length of c. 8 to c. 200 mm. Larger larvae are either morphologically distinct (e.g., the acronurus of surgeonfishes) or essentially pelagic juveniles (e.g., squirrelfishes and porcupinefishes). Choice of settlement habitat is apparent in some species, and both seagrass beds and mangroves can serve as nursery habitats. Some wrasse larvae bury in the sand for several days before emerging as new juveniles. There is typically weak metamorphosis during settlement involving the growth of scales and onset of pigmentation. Estimates of settlement are generally called 'recruitment' and are based on counts of the smallest juveniles that can be found by divers some time after settlement. Once settled, most reef fish are thought to live less than a decade, although some small damselfish live at least 15 years.

Ecology

Coral reef fishes are superb model systems for studying population dynamics and community structure of demersal marine fishes because they are eminently observable and experimentally manipulable *in situ*. These characteristics make studies of reef fishes conceptually relevant to demersal fisheries and ecology in general.

Population Dynamics

Because reefs are patchy at all spatial scales and reef fish are largely sedentary, coral reef fishes form metapopulations: groups of local populations linked by larval dispersal. Many local populations are demographically open, such that reproductive output drifts away and is thus unrelated to subsequent larval settlement originating from elsewhere. Ultimately, the degree of openness depends on the spatial scale examined. For example, anemonefish populations are completely open at the scale of each anemone, may be partially closed at the scale of an oceanic island, and mostly closed at the scale of an archipelago. It is clear that variability in population size is driven by variation in recruitment due to larval mortality (and perhaps spawning success). Input to local populations via recruitment varies considerably at virtually every spatial and temporal scale examined. Increasing evidence indicates that two mechanisms predominate in regulating reef fish populations. First, given that density-dependent growth is common and that there is a general exponential relationship between body size and egg production in fish, density-dependent fecundity is likely. Second, early postsettlement mortality is often density-dependent, and has been demonstrated experimentally to be caused by predation in a variety of species.

Community Structure

Due to high local species diversity, reef fish communities are complex. There are about five major feeding guilds, each containing dozens of species

locally (with approximate percentage of total fish biomass): zooplanktivores (up to 70%), herbivores (up to 25%), and piscivores (up to 55%), with the remainder being benthic invertebrate eaters or detritivores. The benthivores can be further subdivided based on prey taxa (e.g., corallivores) or other categories (e.g., consumers of hard-shelled invertebrates). Grunts that migrate from reefs at night and feed in surrounding seagrass beds subsequently return nutrients to the reef as feces. There is also considerable consumption of fish feces (coprophagy) by other fish on the reef. Fishes thus contribute substantially to nutrient trapping (via planktivory and nocturnal migration) and recycling (via coprophagy and detritivory) on coral reefs. Within each feeding guild, there is typically resource partitioning: each species consumes a particular subset of the available prey or forages in a distinct microhabitat. Communities are also structured temporally, with a diurnal assemblage being replaced by a nocturnal assemblage (the resting assemblage sheltering in the reef framework). The diurnal assemblage is dominated by perciform and tetraodontiform fishes, whereas the nocturnal assemblage is dominated by beryciform fishes (evolutionary relics apparently relegated to the night by more recently evolved fishes).

Maintenance of Species Diversity

Four major hypotheses have been proposed to explain how many species of ecologically similar coral reef fishes can coexist locally. There are data that both corroborate and falsify each hypothesis in various systems, suggesting that no universal generalization is possible. The first two hypotheses are based on the assumption that local populations are not only saturated with settlement-stage larvae, but also regularly reach densities where resources become limiting. First, the 'competition hypothesis', borrowed from terrestrial vertebrate ecology, suggests that coexistence is maintained despite ongoing interspecific competition by fine-scale resource partitioning (or niche diversification) among species. Second, the 'lottery hypothesis', derived to explain coexistence among similar territorial damselfishes that did not appear to partition resources, is based on the assumptions that, in the long run, competing species are approximately equal in larval supply, settlement rates, habitat and other resource requirements, and competitive ability. Thus, settling larvae are likened to lottery tickets, and it becomes unpredictable which species will replace which following the random appearance of open space due to the death of a territory holder or the creation of new habitat. The relatively restrictive assumptions of this hypothesis can be relaxed if one considers the 'storage effect', which is based on the multiyear life span of reef fishes and the fact that settlement varies through time. Even though a species is at times an inferior competitor, as long as adults can persist until the next substantial settlement event, that species can persist in the community indefinitely. The third hypothesis, 'recruitment limitation', assumes that larval supply is so low that populations seldom reach levels where competition for limiting resources occurs, so that postsettlement mortality is density-independent and coexistence among species is guaranteed (assuming a storage effect). Finally, the 'predation hypothesis' predicts that early postsettlement predation, rather than limited larval supply, keeps populations from reaching levels where competition occurs, thereby ensuring coexistence.

See also

Coral Reef and Other Tropical Fisheries. Coral Reefs.

Further Reading

Böhlke JE and Chaplin CCG (1993) *Fishes of the Bahamas and Adjacent Tropical Waters*. Austin, TX: University of Texas Press.

Caley MJ (ed.) (1998) Recruitment and population dynamics of coral-reef fishes: An international workshop. *Australian Journal of Ecology* 23(3).

Lieske E and Myers R (1996) *Coral Reef Fishes: Indo-Pacific and Caribbean*. Princeton, NJ: Princeton University Press.

Randall J (1998) *Shore Fishes of Hawai'i*. Honolulu, HI: University of Hawai'i Press.

Randall JE, Allen GR, and Steene RC (1997) *Fishes of the Great Barrier Reef and Coral Sea*, 2nd edn. Honolulu, HI: University of Hawai'i Press.

Sale PF (ed.) (1991) *The Ecology of Fishes on Coral Reefs*. San Diego, CA: Academic Press.

Sale PF (ed.) (2002) *Coral Reef Fishes: Dynamics and Diversity in a Complex Ecosystem*. San Diego, CA: Academic Press.

CORAL REEF AND OTHER TROPICAL FISHERIES

V. Christensen, University of British Columbia, Vancouver, BC, Canada
D. Pauly, University of British Columbia, Vancouver, BC, Canada

Introduction

Until recently studying and reporting on tropical fisheries tended to be done in the context of development aid projects initiated with the perception that the transfer of technology, management approaches, and scientific models to tropical countries would assist them in a way that would help to raise the standard of living and the food supply. Many of these aims have been achieved, although not necessarily through such North/South transfer but rather through local growth of the preexisting fisheries, and through access rights granted to distant water fleets of developed countries to operate in the waters of developing countries. These developments have turned tropical and reef fisheries from the marginal activities they were in the 1960s and 1970s to key players in international fisheries. Catches in tropical and subtropical fisheries presently exceed those in developed countries, as do exports from developing to developed countries – which dwarf exports in the opposite direction. In the process tropical fisheries have become globalized to a much further extent than other food commodities such as rice, for example, which is overwhelmingly consumed locally.

The experience gained in developing, managing, and studying developing country fisheries have in the process become part of the mainstream of fisheries research. Some of these findings are presented below to characterize trends in tropical and coral reef fisheries, which have paralleled or in some case preceded developments in higher latitude fisheries.

Coral Reef and Tropical Fisheries

Fisheries Development

Through the first half of the last century the tropical seas were only lightly exploited by humans, even if coastal areas were harvested on a small scale using sustainable methods and often with locally based regulation methods in place. However, during the inter-war period many colonial states tried to introduce more industrial fishing methods to increase productivity and food supply. In general, the early attempts were not particularly successful, an exception being the introduction of trawling in the South China Sea area by the Japanese in the late 1920s.

After the Second World War surplus engines and crafts allowed for an expansion of effort in many areas and the following decades saw the introduction of trawling in most of the coastal shelf areas of the tropics. Stagnating catches and signs of overfishing soon followed the expansion of trawling; an early example of this came from the Manila Bay in the Philippines where trawling was introduced shortly after the war, and where overfishing was apparent by the late 1950s.

The classical, well-documented case of how fisheries 'development' with the introduction of trawling quickly leads to stagnating catches in spite of continuing build-up of effort comes from the Gulf of Thailand. Here a development project in the early part of the 1960s initiated a demersal trawl fishery in hitherto unexploited parts of the Gulf. This led to a rapid build-up of commercial fisheries in the mid-1960s, and has often been hailed as a leading example of a successful development project. Fortunately, a stratified research survey series has been in place in the Gulf continuously since 1966 documenting the ecological changes brought about by the fisheries. Initially, the catches increased steadily with effort, but stagnated after a decade and since then they have remained at about the same level in spite of increasing effort. At first glance this may be indicative of a sustainable fishery – after all the catch level has been maintained for several decades. However, a closer look shows a different and unfortunately very typical picture.

As the fishing pressure increased, the longer-lived species rapidly declined, many by a factor of 10 or more within a decade. Sharks and the larger rays were among the first to disappear because of their low fecundity and long life spans, while the dominant group (by weight) in the ecosystem changed from ponyfishes to squids within a decade from the onset of the trawl fisheries. The average trophic level of the catches, as well as of the ecosystem resources in the Gulf, has decreased steadily as fishing pressure mounted. Hence, catch level may be maintained but what is being extracted is smaller and generally lower-value catches; the average fish is finger-sized. The groupers, snappers, and most other high-value species are gone, and the catches include some 40%

of 'trashfish', i.e. small species and juveniles of larger, commercial species. The trashfish are being used to produce fish oil and meal mainly to supply a growing aquaculture industry.

The degradation of the Gulf of Thailand ecosystem as described is typical of what is happening to the fisheries of the world because of overexploitation, and is indicative of a process now generally being called 'fishing down the food web'. This term recognizes how we change ecosystems by systematically eradicating the upper trophic levels, how we follow up by removing the more intermediate levels to end up catching squids, shrimps, and other organisms low in the food web. This may be considered an economically interesting alternative to the unexploited state given the high value of squids and shrimps, but a closer examination shows that it is a dangerous path, beset with increased risk of unwanted structural changes in the underlying ecosystems. The scientific and practical challenge lies in balancing the harvesting so as to maintain healthy ecosystems. Also, the path does not recognize the growing public concern for marine ecosystems – they may be largely out of sight but they are no longer out of mind.

The Ecosystem Perspective

Fisheries in temperate areas are often described in the form of their target, e.g. herring or cod fisheries. In tropical fisheries the taxonomic diversity, until recently not fully mastered by fish taxonomists (but see FishBase at //www.fishbase.org), along with the unselective nature of the fishing gear used (e.g. lift-nets or trawls in shallow waters), result in a widely diverse catch, often comprising hundreds of species in a single haul. This precludes the notion of using single-species management procedures, which so far have dominated the management of temperate fisheries. From the onset, management in the tropics had to be based on the yields of aggregate of species of target groups (e.g. groupers), implicitly or explicitly taking account of their biological interactions. In contrast, in temperate waters managers have only recently fully realized the consequences of the nonselective nature of the gear used, and thus the ecological consequences of their impact through the food webs.

The trend in fisheries research in recent years has been toward incorporating an ecosystem perspective into the assessment and management of living aquatic resources. This is done in recognition both of the fact that there is biological interaction among the resources ('fish eat fish') and acceptance that exploitation has consequences not just for the exploited resource but for their predators, competitors, and prey as well. Of especial significance here is what may

be termed 'charismatic' organisms, notably marine mammals, sea birds, and turtles. As the environmental movement has gained strength over the past decades it is becoming increasingly clear that there are more players to be recognized than the hunters and gatherers. Examples are the 'dolphin-free' tuna stamp now required to export tuna to the USA, and the turtle-excluding devices required for shrimp fisheries in order to maintain export markets.

Incorporating the ecosystem perspective into management has been an arduous task that has kept the fisheries research community challenged for decades. The pioneers in the field were E. Ursin working in the North Sea and T. Laevastu in the north-east Pacific back in the 1970s. Since then the major steps in the northern temperate areas have been focused on 'multispecies virtual population analysis' (MSVPA), based on single species assessment methodologies but incorporating biological interaction. Building directly on traditional assessment methodologies, MSVPA has had a wide support base in the fisheries assessment circles of the North, but it has not been widely used, partly because of a heavy price tag caused by its extensive data requirements, partly because it was designed to include only the commercial fish species, and partly because its ability to address ecosystem-level questions is limited.

In the tropics the development has taken a different route. A US fisheries scientist, J. Polovina, who had the task of developing an ecosystem model (Ecopath) for a Hawaiian reef system, was inspired by Laevastu's approach and developed a simple version of it incorporating all important functional groups of the ecosystem studied. The model required comparatively little information for parameterization and its ecological book-keeping system ensured that gross errors were likely to be recognized, hence providing a feedback to the biological sampling system. In the nearly two decades since it was first described, the Ecopath approach has been developed considerably by a growing team of scientists, and it has now reached a level where it is being used for ecosystem-based fisheries management in many countries. The approach (and the software that makes it operational, freely available from http://www.ecopath.org) is easy to adapt to new areas, as illustrated by the fact that it is being used in more than 100 countries in almost equal measures in temperate and tropical countries.

Fisheries Affect the Physical Environment

Many of traditional small-scale fisheries previously prevailing in the developing countries of the inter-tropical belt were benign, i.e. they did not modify the habitat in which the resources they used were embedded. The introduction of industrialized fishing methods and the modernization of local techniques both impacted local ecosystems very severely. Thus trawling in the narrow coastal shelf destroys structured habitats with sponges and soft coral to an extent that was not visible in the fisheries of the higher latitudes (where benthic habitats are also destroyed). Furthermore, coral reefs, which are important in the inter-tropical belt, are now exploited by extremely destructive methods such as dynamite and cyanide fishing; the latter usually to catch live large fish for restaurants or for the ornamental trade, both of which support valuable export sectors. Here again it is the visibility of the impacts that have subsequently led in the North to reassessment of the physical impact of fisheries on habitat. An example of this is the growing concern about how trawling impacts ecosystems; for instance, how heavy beam trawls have eradicated cold-water coral reef structures.

Globalization and Poverty Jointly Destroy Fisheries Resources

The bulk of the countries of the inter-tropical belt do not have the administrative structure that would make it possible to limit entry into fisheries. The result is that millions of landless farmers and other rural poor have become small-scale fishers in recent decades – the fisheries being the last resort for livelihood. These fishers have no tradition of restrained and community control of effort as they start to compete with more established fishers. Their lack of access to capital forces them to use cheap and often, unfortunately, destructive methods – most notably explosives and poisons (cyanide or insecticides).

Growing populations with a need for cheap fish provide an outlet for the low quality products that emerge from these activities, while a largely insatiable world market (notably Western Europe, USA, Japan, and China) absorbs the high quality products, especially expensive invertebrates such as shrimps and sea urchins. This massive uncontrolled inflow of effort leads to what has been called 'Malthusian overfishing', wherein extreme poverty reduces the cost of labor (usually a limiting factor in fisheries) to virtually nothing; thus making fishing 'profitable' even when productivity is very low. Adult fishers are often seen operating over an entire day to land only one pound of small fish per person. In such cases poverty and other nonfishers in the family subsidize the fishery.

Other subsidies impacting small-scale fisheries and leading to gross overexploitation are provided by development banks that continue to encourage

increasing fishing effort, and the access rights that are negotiated by powerful developed countries for operating distant water fleets in the tropics, e.g. for trawling off west Africa by EU vessels or tuna fishing in the Indian and Pacific Oceans. Such developments, which have contributed to an enormous excess capacity in the developing world, have also contributed to making the subsidy issue more visible in developing countries, where attempts to address these problems have so far been stymied by shipbuilding and other powerful lobbies. There is, however, a growing realization of the need to reduce the overcapitalization, as perhaps best exemplified by the growing reluctance of development agencies to fund 'fisheries development' projects due to the recognized high failure rate of such projects.

Marine Protected Areas are Part of the Solution

While the foregoing presented a number of problems, which appeared or became visible first in the inter-tropical belt and only later were seen as global problems, this section presents elements of solutions to the global ills, which also have appeared first in the developing world. One of these is marine protected areas (MPAs), first demonstrated in the Philippines to have the potential both to allow resources to regenerate/rebuild themselves and to increase the yield for fishers. Such marine protection implies permanent closure of a significant part of an area traditionally exploited by fisheries, 40% in the case of the island first studied by the pioneering team of Dr Angel Alcala in the Central Philippines. These closures allow the growth of animals, in this case especially groupers and snappers, which would otherwise be caught as juveniles. Their increased biomass and the high percentage of adults within the MPA lead to migration from the MPAs to the surrounding waters, and hence to increased catches in the surrounding areas, and as has been demonstrated, to larger overall catch.

As widely noted, MPAs are not a panacea; notably they require that fishing outside the MPAs must be regulated. Even more importantly the communities whose fishing grounds have been impacted by an MPA must accept the gamble that the MPA represents, and must indeed be the advocates and protectors of the MPA. This implies that they have accepted, through fruitful interaction with scientists or other communities that are patrons of an MPA, that their fishing impacts the resource, something still hotly denied by the less enlightened part of the fishing community in many countries. Further, the communities must be self-organized so that they can prevent defection, i.e. fishing within the MPAs for the resources that rapidly build up as the fishing pressure is released. Third, for the MPA to be set in place, the central government for the country in question must have relinquished enough power to the communities for them to be in a position to take pertinent decisions and see them implemented. Finally, for the MPA approach to work fishing communities must have accepted the notion that the immediate gains from exploiting a resource must be put in a longer-term context. Essentially the community must take a conservation-oriented standpoint, i.e. it must overcome the huge chasm which until now has separated fisheries from conservation.

Demanding as these requirements may seem, they have been met in many parts of the Philippines, the Caribbean, and other parts of the world – including New Zealand, a developed country. Given the difficulties in implementing other forms of effort reduction measures and the very destructive trends in the areas where MPAs are not implemented it is likely that the trend toward using MPAs as an integral part of sustainable fisheries management will continue. Indicative of this trend is the fact that in his last year of office US President Clinton signed an executive order calling for 'protection of existing MPAs and the establishment of new MPAs, as appropriate.'

Further Reading

Alcala AC (1988) Effects of marine reserves on coral fish abundance and yields of Philippine coral reefs. *Ambio* XVII: 194–199.

Christensen V (1996) Managing fisheries involving top predator and prey species components. *Reviews in Fish Biology and Fisheries* 6: 417–442.

Christensen V and Pauly D (eds.) (1993) *Trophic Models of Aquatic Ecosystems. ICLARM Conference Proceedings 26.* Manila: ICLARM.

Hilborn R and Walters CJ (1992) *Quantitative Fisheries Stock Assessment, Choices, Dynamics and Uncertainty.* New York: Chapman and Hall.

Longhurst A and Pauly D (1987) *Ecology of Tropical Oceans.* San Diego: Academic Press.

Munro JL (ed.) (1983) *Caribbean Coral Reef Fishery Resources, ICLARM Study Review 7.* Manila: ICLARM.

Munro JL and Munro PE (eds.) (1994) *The Management of Coral Reef Resource Systems. ICLARM Conference Proceedings 44.* Manila: ICLARM Studies and Reviews.

Pauly D (1979) *Theory and Management of Tropical Multispecies Stocks. ICLARM Study Review 1.* Manila: ICLARM Studies and Reviews.

Pauly D, Christensen V, Dalsgaard A, Froese R, and Torres J (1998) Fishing down marine food webs. *Science* 279: 860–863.

Polunin NVC and Roberts CM (eds.) (1996) *Reef Fisheries.* London: Chapman and Hall.

Safina C (1998) *Song for the Blue Ocean.* New York: H Holt & Co.

ARTIFICIAL REEFS

W. Seaman and W. J. Lindberg, University of Florida, Gainesville, FL, USA

Introduction

Artificial reefs are intentionally placed benthic structures built of natural or man-made materials, which are designed to protect, enhance, or restore components of marine ecosystems. Their ecological structure and function, vertical relief, and irregular surfaces vary according to location, construction, and degree to which they mimic natural habitats, such as coral reefs. For ages, humans have taken advantage of the behavior of some organisms to seek shelter at submerged objects by introducing structures into shallow waters, where biological communities could form and fishes could be harvested. Contemporary purposes for artificial reefs include increasing the efficiency of artisanal, commercial, and recreational fisheries, producing new biomass in fisheries and aquaculture, boosting underwater recreation and ecotourism opportunities, preserving and renewing coastal habitats and biodiversity, and advancing research. This article reviews the evolution, spread, and increased scale of such practices globally, discusses their scientific basis, and describes trends concerning artificial reef planning, evaluation, and their appropriate application in natural resource management.

Utilization of Reefs in the World's Oceans

The near-shore ocean ecosystems of all inhabited continents contain artificial reefs. Their scales and purposes vary widely, ranging, for example, from placement in local waters of small structures by individuals in artisanal fishing communities, to deploying complex systems of heavy modules in distant areas by organizations concerned with commercial seafood production, to pilot studies using experimentally based structures for restoration of seagrass, kelp, and coral ecosystems. The origins of artificial reef technology are traced to places as diverse as Japan and Greece. Modern deployment of reefs has the longest history and has been most widespread in Eastern Asia, Australia, Southern Europe, and North America. In recent years, the technology has been introduced or more extensively established in scores of nations worldwide, including in Central and South America and the Indo-Pacific basin. No organization maintains a database for artificial reef development globally, so that assessments of reef-related activity must be derived from the records of economic development, fishery, and environment organizations, as well as from scientific journal articles which serve as a proxy for gauging national efforts.

For centuries structures of natural materials have been used to support artisanal and subsistence fishing in coastal communities, particularly in tropical areas (**Figure 1**). In India, traditionally tree branches have been weighted and sunken. Brush parks made of

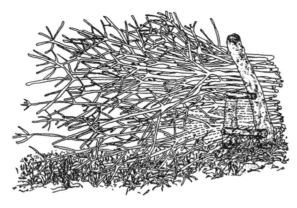

Figure 1 In some areas of long-standing artisanal fisheries, structures such as the *pesquero* (left), a bundle of mangrove tree branches, are used for benthic structure, as in Cuba. More recently, small fabricated modules, such as this example from India (right), are used to complement or replace natural materials.

branches stuck into the substrates have been used in estuaries of several African countries, Sri Lanka, and Mexico. In the Caribbean basin, casitas of wood logs provide benthic shelters from which spiny lobsters (*Panulirus argus*) are harvested. Indigenous knowledge of local fisherfolk is important in sustainably managing these reef systems, even when in their initial stages they may function principally to aggregate larger mobile fishes, such as around bundles of mangrove branches, *pesqueros*, in Cuba.

In recent decades, some artisanal fishing communities have deployed newer designs of artificial reefs, and at larger scales. In part, this has been in response to damage of habitat and fisheries by coastal land-use practices and more intense fishing practices, such as trawling. In India, for example, a national effort has deployed reef modules of steel plates, to increase the time fishing cooperatives actually devote to harvest as a means of promoting social and economic well-being. The coastal waters of Thailand received concrete modules as part of a plan to balance small- and large-scale fisheries, whereby areas of 14–22 km^2 received between 2400 and 3300 units.

The most extensive deployment of artificial reefs, for any purpose, is in Japan. There, fishing enhancement is the goal of a national plan begun in 1952 and since greatly expanded. By 1989, 9% of the coastal shelf (<200 m depth) had been affected by reef development. Early emphasis was on engineering and construction of reefs to withstand rigorous high seas environmental conditions. Industrial manufacturers have used materials such as steel, fiberglass, and concrete in structures that are among the largest units in the world, attaining heights of 9 m, widths of 27 m, and volumes of 3600 m^3 (**Figure 2**). Research at sea has been augmented by laboratory studies in which scale models of reefs have been analyzed for effects including deflection of

bottom water currents to create nutrient upwellings for enhancing primary production. Observations of fish behavior led to a classification of the affinity of fishes to associate with physical structure. The Japanese aims have been to develop nursery, reproduction, and fishing grounds to supply seafood including abalone, clams, sea urchins, and finfishes. This partially was in response to the closure of distant waters to fishing by other nations.

In contrast to Japan (and many other nations more recently involved), the United States, while also an early entrant in this field, built thousands of artificial reefs to enhance recreational fishing through numerous, independent small-scale efforts organized by local interest groups or governmental organizations at the state level. Until recently, materials of opportunity predominated, including heavier and more durable materials such as concrete rubble from bridge and building demolition sites. Designed structures are becoming more common, especially as reefs gain acceptance in habitat restoration. One state, Florida, has about half of the nation's total number of reefs, and in one four-county area annual expenditures by nonresidents and visitors on fishing and diving at dozens of artificial reefs were US $1.7 billion (2001 values). Recreational fishing has also been the focus of widespread reef building in Australia, and lesser efforts in areas such as Northern Ireland and Brazil.

Other regions more recently deploying artificial reefs to enhance or simply maintain fisheries to supply human foodstuffs include southern Europe and Southeast Asia. From inception, their approach has been to design, experimentally test, and document reefs. As early as the 1970s, 8 m^3, 13 t concrete blocks (and other designs) were deployed in the Adriatic Sea of Italy, forming reefs of volumes up to 13 000 m^3 and coverage up to 2.4 ha. They serve multiple, fishery-related objectives that are sought in many nations: shelter for juvenile and adult organisms; reproduction sites; capture nutrients in shellfish biomass; protect fish spawning and nursery areas from illegal trawling; and protect artisanal set fishing gear from illegal trawling damage. Higher catches and profits for small-scale fisheries were the result. This is especially true in Spain, where heavy concrete and steel rod reef designs have been used effectively to protect seagrass beds from illegal trawling (**Figure 3**).

Artificial reefs are being used as part of a larger strategy to implement marine ranching programs. In Korea, for example, the decline of distant water fisheries led to a transition from capture-based to culture-based fisheries. From 1971 to 1999, artificial reef placements have affected 143 000 ha of submerged

Figure 2 The largest reef modules have been fabricated and deployed in Japan.

Figure 3 Concrete reef modules with projecting steel beams such as railroad tracks are deployed in protected marine areas of Spain to discourage illegal trawling, so as to promote restoration of seagrass habitats and fish populations. Photograph Courtesy of Juan Goutayer.

Figure 4 This larval enclosure tent is for settlement of the coral, *Montastraea faveolata*, at the site of a ship grounding on Molasses Reef, Florida Keys National Marine Sanctuary. Coral are mass-spawned, larvae cultured through the planktonic phase, and then introduced into the tent. In the laboratory, larvae may settle onto rubble pieces which subsequently are cemented to the reef. Photograph by D. Paul Brown.

coastal areas, out of the total 307 000 ha planned. At certain reef sites, hatchery-reared juvenile invertebrates and fishes have been released, and enhanced fishery yield reported, as part of long-term experiments. Artificial reefs are also used exclusively for aquaculture, such as in settings where excess nutrients stimulate primary production that is transferred into biomass of harvestable species, such as mussels (*Mytilus galloprovincialis*) and oysters (*Ostrea edulis* and *Crassostrea gigas*) in Italy.

Restoration of marine ecosystems using artificial reefs has focused on plant and coral communities. On the Pacific coast of the United States, for example, a project to mitigate for electric powerplant impacts is developing a 61 ha artificial reef for kelp (*Macrocystis pyrifera*) colonization, survival, and growth. Along the coast of the Northwest Mediterranean Sea, artificial reefs are deployed in protected areas to preserve and promote colonization of seagrasses (*Posidonia*). In numerous tropical areas, artificial reefs have been used in repairs to damaged coral reefs (**Figure 4**), or to hasten replacement of dead or removed coral. In the Philippines, for example, hollow concrete cubes are deployed as sites for coral colonization, sometimes being located in marine reserves.

Increasing popularity of artificial reefs to promote tourism is seen in the development of diving opportunities, both for people in submersible vehicles and for scuba divers. In places such as Mexico, the Bahamas, Monaco, and Hawaii, submarine operators provide trips to artificial reefs. On a larger scale, obsolete ships have been sunk to create recreational dive destinations in areas as diverse as British Columbia, Canada, Oman, New Zealand, Mauritius, and Israel. Purposes include generation of economic revenues in local communities and diversion of diver pressure from natural reefs.

As strictly works of engineering, structures such as rock jetties for shore protection also serve as *de facto* reefs with biological communities. Submerged breakwaters have been used to create waves for recreational surfboarding.

Progress toward Scientific Understanding

Advances in research on artificial reefs include methodologies, subjects and rigor of approaches. Methodological advances include improved underwater biological census techniques, and the application of remote sensing. Investigations have become more ecological process-oriented in explaining biological phenomena such as sheltering, reproduction, recruitment, and feeding, better able to explain physical and hydrodynamic behavior of reefs, and have led to specifying reef designs consistent with the ecology of organisms. Newer approaches include an expansion of hypothesis-driven and experimental research, scaling up of pilot designs into larger reefs, ecological modeling and forecasting, and interdisciplinary studies with various combinations of physical, social, life, and mathematical sciences. Comparative studies of natural and artificial habitats have advanced understanding of both systems. The earlier history of research on artificial reefs may be observed in the subjects, methods, and findings reported at eight international conferences held from

1974 to 2005. Subsequent to a peak number of peer-reviewed articles contained in a proceedings from the 1991 conference (84), fewer resulted from the latest conferences (1999 (56) and 2005 (20)). Scientists studying artificial reefs increasingly are reporting in journals devoted to fish biology, fisheries, and marine ecology.

Colonization of artificial reefs has the oldest and most extensive record of scientific inquiry. The appearance of large fishes at some reef sites almost immediately after placement of structure is the most visible form of colonization (and also a factor in one of the most controversial artificial reef issues, discussed below). However, microbes, plants, and invertebrates also colonize reefs, and patterns of succession occur that converge toward naturally occurring assemblages of species (**Figure 5**). Augmenting earlier studies describing diversity and abundance of artificial reef flora and fauna, later research addressed the influence of environmental variables upon reef ecology, interactions of individual species with the artificial reef structure and biota, and long-term life history studies of selected species. Comparison of patch reefs of stacked concrete blocks with translocated coral reefs in the Bahamas (**Figure 6**) determined that the reefs had similar fish species composition, while the natural reefs supported more individuals and species, owing to their greater structural complexity (i.e., variety of hole sizes) and associated forage base. Meanwhile, long-term studies of the spiny lobster experimentally established that artificial reef structures, with dark recessed crevices approximately one body-length deep and multiple entrances, indeed augmented recruitment in a situation of habitat limitation in the Florida Keys, USA. While these situations allowed observation by scuba diving, in deeper waters hydroacoustic and video-recording instruments have been used to document fish species, such as at petroleum production platforms in the North Sea and Gulf of Mexico.

Prominent among the scientific issues regarding the efficacy of artificial reefs for fisheries production and management is the 'attraction-production question': Do artificial reefs merely attract fishes thereby improving fishing efficiency, or do they contribute to biological production so as to enhance fisheries stocks? While this popularized dichotomy is an oversimplification, and an expectation of one all-encompassing answer is unrealistic, this question has had great heuristic value for the advance of reef science and responsible reef development. The origins of this issue derive from early observations and assumptions about artificial reefs. High catch rates and densities of fish at artificial reefs were popularly

Figure 5 Aspects of attachment surfaces for marine plants and shelter and feeding sites for invertebrates and fishes provided by artificial reefs. Drawing courtesy of S. Riggio.

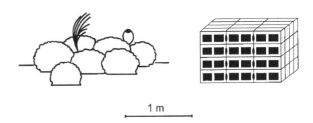

1 m

Figure 6 Experimental modules such as the one pictured on the right from the Bahamas exemplify a global trend toward greater manipulation of artificial reefs as tools for testing of hypotheses concerning ecological processes in the ocean, and comparison with systems such as coral reefs, pictured left. Drawing courtesy of American Fisheries Society.

taken as proof that artificial reefs benefited fisheries stocks, based on assumptions that hard substrata were limiting and reef fish standing stocks were dependent on food webs that have hard substrata as their foundations. However, a few fisheries scientists challenged this reasoning in the 1980s. They noted that before heavy exploitation the existing natural habitat supported abundant reef fishes, presumably at or near carrying capacity. Fishing then reduced populations to some lower level, yet the amount of natural habitat remained the same, still capable of supporting higher population numbers. The fisheries scientists reasoned that with fish stocks substantially below carrying capacity, the amount of hard-bottom habitat could not be the factor limiting population size, so the addition of artificial reefs would not benefit fish stocks. Thus, observed high densities of fish and high catch rates at artificial reefs were considered an artifact of fish behavioral preferences, which simply concentrated them at reef sites and intensified fishing mortality, what conservationists today call an ecological trap.

The fisheries scientists' formal argument was more thorough and complex than the popularized version, as they recognized a continuum from solely attraction, with little if any direct biological benefit, to potentially high levels of added biological production from each unit of additional reef structure. Species fit along that continuum according to their life history characteristics and use of reef habitat ecologically. Thus small, highly site-attached fishes (e.g., blennies and gobies) that derive all resource needs (i.e., food, shelter, and mates) directly from the occupied reef structure, and that complete all but the planktonic phase of their life cycle in one place, would have the greatest net production potential from artificial reef development. Conversely, large, transient reef-visiting fishes (e.g., jacks and mackerels) were expected to have the least production potential from artificial reef development. In between these extremes were the majority of exploited reef fishes for which one would expect some combination of attraction and production affected by reef ecological setting and overall stock abundance. (Implicit in the argument by the fisheries scientists was the assumption that fishing pressure would be intensely focused at the artificial reefs, which has more to do with the management of human activities than the intrinsic ecological characteristics of the reefs themselves.)

Differing professional positions concerning attraction-production still remain, and the debate continues to stimulate research questions and artificial reef management issues. In part, this is because the complexity of reef ecological functions and the effects of spatially explicit fishing mortality are beyond simple answers to the seemingly dichotomous question that has been popularized among lay audiences. And in part, this is because researchers necessarily address this issue from their own disciplinary perspectives and study systems, generally not integrating the multiple levels of biological organization or spatial-temporal domains over which this question can be legitimately addressed. Pertinent studies can focus on life history stages of populations, communities, or ecosystems and encompass individual reefs, broad landscapes, or geographic ranges of fisheries stocks, while covering seasonal, interannual, or long-term time frames. Practical solutions to the fisheries management implications will require more specific attention to the desired species and assemblages, and the processes that sustain them.

Now a pivotal question is how the manipulation of habitat affects demographic rates and exploitation rates across spatial and temporal scales. An example of a way to integrate large sets of ecological data to analyze different scenarios of artificial reef development is through ecosystem simulations such as for Hong Kong where fishing levels in and out of protected areas have been predicted as part of a fishery recovery program.

Advances in Planning, Design, and Construction

The increase of artificial reefs in the world's coastal seas, as measured by the growing number, cost, size, intricacy, and footprint of structures deployed, has prompted an emphasis on their planning. This is to promote more efficient and cost-effective structures, enable work at larger scales and with more precision, satisfy regulatory requirements, reduce conflicts with other natural and human aspects of ecosystems, and minimize the prospects of unintended consequences from improperly constructed reefs. The scientific basis for reefs has been strengthened by research efforts in dozens of countries, coupled with the practical experiences of numerous individuals and groups that have worked independently to build, manage, and, increasingly, evaluate reefs.

Independently in different geographic areas, common approaches and some published guidelines for developing artificial reefs have emerged. The earliest handbooks focused on physical and oceanographic conditions as first encountered by the Coastal Fishing Ground Enhancement and Development Project of Japan. (Translations of certain documents into English in the 1980s by the United States National Marine Fisheries Service disseminated information.) The United States produced a national plan in 1984 (in revision in 2006) which spurred promulgation of plans among the coastal states. In some nations, planning is done through a coordinated reef program run by a federal or provincial marine or fishery resources organization, as in Italy or China. Finally, where reef deployment is directed to selected localities or areas, but not nationwide, master plans or other guidelines for siting, materials, and other aspects have been produced, such as for the Aegean Sea of Turkey and northern Taiwan.

With a legitimate intent assured, the initial component of reef planning is to frame measurable objectives, define expectations for success, and forecast interactions of that reef with the ecosystem. Subsequently, a valid design and site plan for the reef and procedures for its construction must be developed and meet regulatory requirements. The context for management of the reef must be defined. For example, an artificial reef program in Hong Kong was based on an extensive consultation with community interests, leading to stakeholder support for regulatory practices in the local fisheries and protection of marine reserve areas containing some of the reefs. Finally, protocols for evaluation of reef performance and management of assessment data must be established, with communication of results to all interested parties.

Construction is a three-phase process: fabrication, transportation, and placement. Representative costs of larger reefs include US $16 000 000 for a 56 ha reef in California, USA, and between US $38 and $57 million annually during 1995–99 for a national program in Korea. The ecological and economic impacts of poorly designed artificial reefs are exemplified by an effort in the United States – projected to cost US $5 000 000 – to remove 2 000 000 automobile tires that had been deployed in the 1970s, but which were drifting onto adjacent live coral reefs and beaches.

The design phase of reef development has changed dramatically due to growing emphasis on making the structural attributes of intentional, fabricated reef materials conform and contribute to the biological life history requirements of organisms that are particularly desired as part of the reef assemblage. Reef design should be dictated by: (1) a set of measurable and justified objectives for the reef, (2) the ecology of species of concern associated with the reef, and (3) predicted and understandable environmental and socioeconomic consequences of introduction of the reef into the aquatic ecosystem. This requires a multidisciplinary approach using expertise of biologists, engineers, economists, planners, sociologists, and others.

From a physical standpoint, key aspects of design and construction include stability of a reef site, durability of the reef configuration, and potential adverse impacts in the environment. Also, the use of physical processes to enhance reef performance is a desirable aspect. To build a quantitative understanding of the environment into which a reef is to be placed, large-scale oceanographic processes and local conditions must be determined by site surveys early in reef planning, including water circulation driven by tides, wind and baroclinic/density fields, locally generated wind waves, swells propagated from distance, sediment/substrate composition, distribution and transport, and depth. These factors are then coupled with the attributes of the reef material, such as weight, density, dimensions, and strength, in order to forecast reef physical performance.

From an ecological standpoint, abiotic and biotic influences considered in design include geographic location, type and quality of substrate surrounding the reef site, isolation, depth, currents, seasonality, temperature regime, salinity, turbidity, nutrients, and productivity. Substrate attributes affecting reef ecology include its composition and surface texture, shape, height, profile, surface area, volume and hole size, which taken together contribute to the structural complexity of the reef. Spacing of individual and groups of reef structure is important.

A plethora of designs for artificial reefs exist (**Figure 7**). For a mitigation reef aiming to create new kelp beds off California, USA, biologists and engineers concluded that the most effective design should place boulders and concrete rubble in low-relief piles (<1 m height), at depths of 12–14.5 m on a sand layer of 30–50 cm overlaying hard substrate; success criteria include (1) support of four adult plants per $100 \, m^2$ and (2) invertebrate and fish populations similar to natural reefs. As part of a system of artificial reefs in Portugal, experiments quantified production of sessile invertebrates (upon which fishes could feed) according to location on different facets of cubic settlement structures placed on reefs. For fishes, meanwhile, one of five designs used in Korea is the box reef, a $3 \times 3 \times 3$ m concrete cube (**Figure 8**), targeted to two species: small, dark spaces in the lower two-thirds of the reef are provided for rockfish (*Sebastes schlegeli*), while the upper third is more open to satisfy behavioral preferences of porgy (*Pagrus major*). Repeatedly, authorities cite complexity of structure as a primary factor in design.

As large individual and sets of reefs are planned, more organizations are using pilot projects to determine physical, biological, economic, and even

Figure 7 Artificial reefs of concrete modules are used worldwide, with designs intended to meet ecological requirements of designated species and habitats.

Figure 8 The box reef design used in Korea resulted from a collaboration of engineers and biologists. Modules have been placed in research plots within Tongyong Marine Reserve for use in experiments with stocking juvenile fishes. Photograph by Kim Chang-gil.

political feasibility. The 42 000 t Loch Linne Artificial Reef in Scotland was started in 2001 as a platform for scientific investigation of the performance of different structures, ultimately to establish fisheries for target species, specifically lobster (*Homarus gammarus*). This effort also is representative of a trend toward increased predeployment research – in this case a 4-year study of seabed, water-column, and biological parameters – intended to enable better forecasting of reef impacts and measurement of results.

A special case of working with reef materials concerns the deployment of obsolete ships and petroleum/gas production platforms, due to the need to handle, prepare, and place them in environmentally compatible ways. In Canada, for example, decommissioned naval vessels require extensive removal of electrical wiring and other components to eliminate release of pollutants into the sea, in conformance with strict federal rules. In the northern Gulf of Mexico (where over 4000 platforms provide a considerable area of hard surface for sessile organism attachment), eastern Pacific, North Sea, and Adriatic Sea offshore platforms either act as *de facto* reefs or in a limited number of cases are being toppled in place or transported to new locations to serve as dedicated reefs, which costs less than removal to land.

Integration of Reefs in Ecological and Human Systems

Increasingly, artificial reef technology is being applied globally in fisheries and ecosystem management. Early disappointments and healthy skepticism

have led to more realistic expectations bolstered by two decades of scientific advances. National plans (e.g., Japan, Korea, and United States), regional programs (e.g., Hong Kong, Singapore, and Turkey), and large-scale pilot projects (e.g., Loch Linne, Scotland; San Onofre, USA) concerning artificial reefs are better defining their role in ecosystem and fishery management. International scientific bodies such as the North Pacific Marine Sciences Organization (PICES) have addressed the relevance of artificial reefs to core fisheries issues, including stock enhancement, fishing regulations, and conservation. Integrated coastal management in the Philippines, India, Spain, and elsewhere now includes artificial reefs in multifaceted responses to issues of habitat destruction, fishery decline, and socioeconomic development. Finally, private consultants and businesses have found markets for their services and products, with one patented design being deployed in over 40 countries, and ecological engineers have recognized reefs as constructed ecosystems for use in restoration ecology.

The evaluation of reef performance is fostering an increased acceptance and utilization of artificial reef technology by resource management organizations. Quantitative evaluation is increasingly driven by agency concerns for demonstrating positive returns on their investments, and the increasing scale of artificial reef projects. Evaluations vary in intensity and complexity, ranging from descriptive studies of short duration (e.g., pre- and postdeployment) to extensive studies of ecological processes and dynamics, to the synthesis of complex databases through quantitative modeling. One response of the scientific community was through formation of the European Artificial Reef Research Network, which promulgated priorities and protocols for research across its membership.

A significant trend is for artificial reef projects to be planned and evaluated in an adaptive management framework, in which expectations are more explicitly stated, the projects implemented and rigorously evaluated, and then adjustments made to the management practices based on findings from the evaluations. When applied consistently, this cycle will continue to evolve the application of artificial reef technologies toward an ever-increasing standard of practical effectiveness.

See also

Coastal Zone Management. Coral Reef and Other Tropical Fisheries. Coral Reef Fishes. Coral Reefs. Fiordic Ecosystems. Mariculture Diseases and

Health. Mariculture, Economic and Social Impacts. Mariculture Overview.

Further Reading

American Fisheries Society (1997) *Special Issue on Artificial Reef Management. Fisheries* 22: 4–36.

International Council for the Exploration of the Sea (2002) *ICES Journal of Marine Science* 59(supplement S1-5363).

Jensen AC, Collins KJ, and Lockwood APM (eds.) (2000) *Artificial Reefs in European Seas.* Dordrecht: Kluwer.

Lindberg WJ, Frazer TK, Portier KM, *et al.* (2006) Density-dependent habitat selection and performance by a large mobile reef fish. *Ecological Applications* 16(2): 731–746.

Love MS, Schroeder DM, and Nishimoto MM (2003) The ecological role of oil and gas production platforms and natural outcrops on fishes in southern and central California: A synthesis of information. OCS Study MMS 2003-032. Seattle. WA: US Department of the Interior, US Geological Survey, Biological Resources Division.

Seaman W (ed.) (2000) *Artificial Reef Evaluation.* Boca Raton, FL: CRC Press.

Svane I and Petersen JK (2001) On the problems of epibioses, fouling and artificial reefs: A review. *Marine Ecology* 22: 169–188.

GROUNDWATER SEEPAGE

GROUNDWATER FLOW TO THE COASTAL OCEAN

A. E. Mulligan and M. A. Charette, Woods Hole
Oceanographic Institution, Woods Hole, MA, USA

Introduction

Water flowing through the terrestrial landscape ultimately delivers fresh water and dissolved solutes to the coastal ocean. Because surface water inputs (e.g., rivers and streams) are easily seen and are typically large point sources to the coast, they have been well studied and their contributions to ocean geochemical budgets are fairly well known. Similarly, the hydrodynamics and geochemical importance of surface estuaries are well known. Only recently has significant attention turned toward the role of groundwater inputs to the ocean. Historically, such inputs were considered insignificant because groundwater flow is so much slower than riverine flow. Recent work however has shown that groundwater flow through coastal sediments and subsequent discharge to the coastal ocean can have a significant impact on geochemical cycling and it is therefore a process that must be better understood.

Groundwater discharge into the coastal ocean generally occurs as a slow diffuse flow but can be found as large point sources in certain terrain, such as karst. In addition to typically low flow rates, groundwater discharge is temporally and spatially variable, complicating efforts to characterize site-specific flow regimes. Nonetheless, the importance of submarine groundwater discharge (SGD) as a source of dissolved solids to coastal waters has become increasingly recognized, with recent studies suggesting that SGD-derived chemical loading may rival surface water inputs in many coastal areas. So while the volume of water discharged as SGD may be small relative to surface discharge, the input of dissolved solids from SGD can surpass that of surface water inputs. For example, SGD often represents a major source of nutrients in estuaries and embayments. Excess nitrogen loading can result in eutrophication and its associated secondary effects including decreased oxygen content, fish kills, and shifts in the dominant flora.

First, let us define 'groundwater' in a coastal context. We use the term to refer to any water that resides in the pore spaces of sediments at the land–ocean boundary. Hence, such water can be fresh terrestrially derived water that originates as rainwater infiltrating through the subsurface or it can represent saline oceanic water that flows through the sediments (**Figure 1**). Therefore, groundwater discharging to coastal waters can have salinity that spans a large range, being some mixture of the two end members. We therefore use the terms fresh SGD and saline SGD to distinguish these sources of fluid and brackish SGD to mean a mixture of the fresh and saline end members.

Basics of Groundwater Flow

Groundwater flow in the subsurface is driven by differences in energy – water flows from high energy areas to low energy. The mechanical energy of a unit volume of water is determined by the sum of gravitational potential energy, pressure energy, and kinetic energy:

$$\text{Energy per unit volume} = \rho g z + P + \frac{\rho V^2}{2} \quad [1]$$

where ρ is fluid density, g is gravitational acceleration, z is elevation of the measuring point relative to a datum, P is fluid pressure at the measurement point, and V is fluid velocity. Because groundwater flows very slowly (on the order of $1\,\text{m day}^{-1}$ or less), its kinetic energy is very small relative to its gravitational potential and pressure energies and the kinetic energy term is therefore ignored. By removing the kinetic energy term and rearranging eqn [1] to express energy in terms of mechanical energy per unit weight, the concept of hydraulic head is developed:

$$\text{Energy per unit weight} = \text{hydraulic head} = z + \frac{P}{\rho g}$$
$$[2]$$

Groundwater therefore flows from regions of high hydraulic head to areas of low hydraulic head.

Because groundwater flows through a porous media, the rate of flow depends on soil properties such as the degree to which pore spaces are interconnected. The property of interest in groundwater flow is the permeability, **k**, which is a measure of the ease with which a fluid flows through the soil matrix. Groundwater flow rate can then be calculated using Darcy's law, which says that the flow rate is linearly proportional to the hydraulic gradient:

$$\mathbf{q} = -\frac{\rho \mathbf{gk}}{\mu}(\nabla h) \quad [3]$$

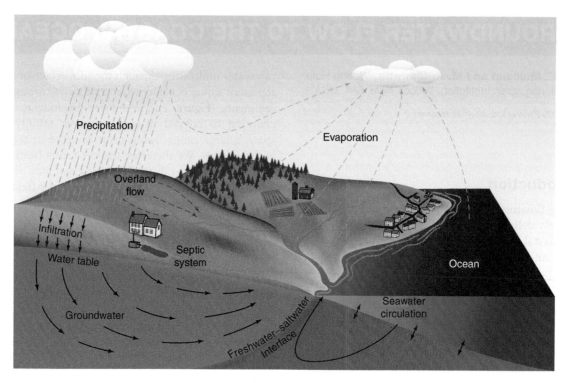

Figure 1 Simplified water cycle at the coastal margin. Modified from Heath R (1998) Basic ground-water hydrology. *US Geological Survey Water Supply Paper 2220*. Washington, DC: USGS.

where **q** is the Darcy flux, or flow rate per unit surface area, and μ is fluid viscosity. A more general expression of Darcy's law is:

$$\mathbf{q} = -\frac{\mathbf{k}}{\mu}(\nabla P + \rho \mathbf{g} \nabla z) \qquad [4]$$

In inland aquifers, the density of groundwater is constant and eqn [4] is reduced to the simpler form of Darcy's law (eqn [3]). In coastal aquifers, however, the presence of saline water along the coast means that the assumption of constant density is not valid and so the more inclusive form of Darcy's law, eqn [4], is required.

Groundwater Flow at the Coast

Several forces drive groundwater flow through coastal aquifers, leading to a complex flow regime with significant variability in space and time (**Figure 2**). The primary driving force of fresh SGD is the hydraulic gradient from the upland region of a watershed to the surface water discharge location at the coast. Freshwater flux is also influenced by several other forces at the coastal boundary that also drive seawater through the sediments. For example, seawater circulates through a coastal aquifer under the force of gravity, from oceanic forces such as

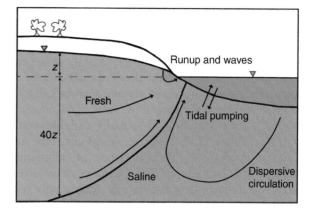

Figure 2 Simplification of an unconfined coastal groundwater system. Water flow is driven by the inland hydraulic gradient, tides, beach runup and waves, and dispersive circulation. Other forcing mechanisms can drive fluid through coastal sediments, including seasonal changes in recharge to the inland groundwater system and tidal differences across islands.

waves and tides, as a result of dispersive circulation along the freshwater–saltwater boundary within the aquifer, and from changes in upland recharge. Several other forcing mechanisms exist, but they are generally only present in specific settings. For example, tidal height differences across many islands can drive flow through the subsurface. All of these forcing mechanisms affect the rate of fluid flow for both fresh and saline groundwater and are ultimately important in controlling the submarine discharge of both fluids.

The analysis of coastal groundwater flow must account for the presence of both fresh- and saline water components. When appropriate, such as in regional-scale analysis or for coarse estimation purposes, an assumption can be made that there is a sharp transition, or interface, between the fresh and saline groundwater. While this assumption is not strictly true, it is often appropriate, and invoking it results in simplifying the analysis. For example, we can estimate the position of the freshwater–saltwater interface by assuming a sharp interface, no flow within the saltwater region, and only horizontal flow within the fresh groundwater. Invoking these assumptions means that the pressures at adjacent points along the interface on both the freshwater and saltwater sides are equal. Equating these pressures and rearranging, the depth to the interface can be calculated as follows:

$$\text{Interface depth} = \frac{\rho_1}{\rho_2 - \rho_1}z = 40z \qquad [5]$$

where ρ_1 is the density of fresh water ($1000 \, \text{kg m}^{-3}$) and ρ_2 is the density of seawater ($1025 \, \text{kg m}^{-3}$). This equation states that the depth of the interface is 40 times the elevation of the water table relative to mean sea level. While eqn [5] is only an approximation of the interface location, it is very helpful in thinking about freshwater and saltwater movement in response to changes in fresh groundwater levels. As recharge to an aquifer increases, water levels increase and the interface is pushed downward. This is also equivalent to pushing the interface seaward and the net effect is to force saltwater out of the subsurface and to replace it with fresh water. The opposite flows occur during times of little to no recharge or if groundwater pumping becomes excessive.

While the sharp-interface approach is useful for conceptualizing flow at the coast, particularly in large-scale problems, the reality is more complex. Not only does the saline groundwater flow but also a zone of intermediate salinity extends between the fresh and saline end members, establishing what many refer to as a 'subterranean estuary'. Like their surface water counterparts, these zones are hotbeds of chemical reactions. Because the water in the interface zone ultimately discharges into coastal waters, the flow and chemical dynamics within the zone are critically important to understand. Research into these issues has only just begun.

Detecting and Quantifying Submarine Groundwater Discharge

As a first step in quantifying chemical loads to coastal waters, the amount of water flowing out of the subsurface must be determined. This is particularly challenging because groundwater flow is spatially and temporally variable. A number of qualitative and quantitative techniques have been developed to sample SGD, with each method sampling a particular spatial and temporal scale. Because of limitations with each sampling method, several techniques should be used at any particular site.

Physical Approaches

Infrared thermography Infrared imaging has been used to identify the location and spatial variability of SGD by exploiting the temperature difference between surface water and groundwater at certain times of the year. While this technique is quite useful for identifying spatial discharge patterns, it has not yet been applied to estimating flow rates.

An example of thermal infrared imagery is shown in **Figure 3**, an image of the head of Waquoit Bay, a small semi-enclosed estuary on Cape Cod, Massachusetts, USA. In late summer, the groundwater temperature is approximately 13 °C, whereas surface water is about 7–10 °C warmer. Locations of SGD can be seen in the infrared image as locations along the beach face with cooler temperature than the surrounding surface water. The image clearly shows spatial variability in SGD along the beach face, information that is extremely valuable in designing an appropriate field sampling campaign.

Hydrologic approaches There are two hydrologic approaches to estimating SGD: the mass balance method and Darcy's law calculation. Both methods are typically applied to estimating fresh groundwater discharge, although Darcy's law can also be used to estimate saline flow into and out of the seafloor.

To apply Darcy's law, one must measure the soil permeability and hydraulic head at several locations (at least two) at the field site. Data must also be gathered to determine the cross-sectional flow area. The field data are then used with Darcy's law to calculate a groundwater flow rate into the coastal ocean. The main disadvantages of this approach include the fact that permeability is highly heterogeneous, often ranging over several orders of magnitude, and so an 'average' value to use with Darcy's law is seldom, if ever, well known. Furthermore, hydraulic head measurements require invasive, typically expensive, well installations. Finally, hydraulic head

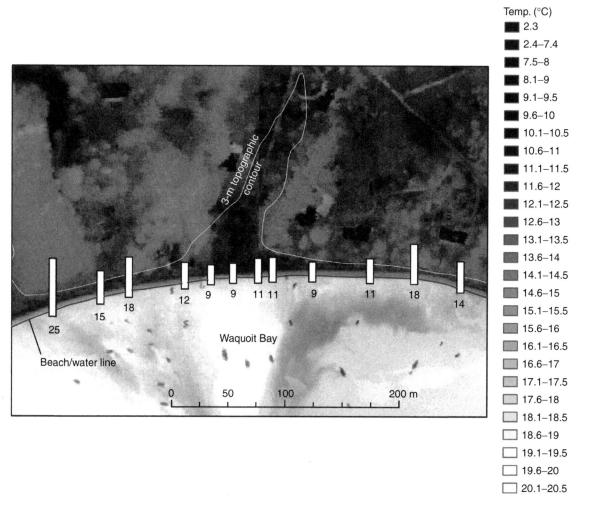

Temp. (°C)
- 2.3
- 2.4–7.4
- 7.5–8
- 8.1–9
- 9.1–9.5
- 9.6–10
- 10.1–10.5
- 10.6–11
- 11.1–11.5
- 11.6–12
- 12.1–12.5
- 12.6–13
- 13.1–13.5
- 13.6–14
- 14.1–14.5
- 14.6–15
- 15.1–15.5
- 15.6–16
- 16.1–16.5
- 16.6–17
- 17.1–17.5
- 17.6–18
- 18.1–18.5
- 18.6–19
- 19.1–19.5
- 19.6–20
- 20.1–20.5

Figure 3 Thermal infrared image of the head of Waquoit Bay, Massachusetts. The beach/water line is shown as a red curve. Light grays imply higher temperatures and darker shades show lower temperatures. Lower temperatures within the bay indicate regions influenced by SGD. The bars show average groundwater seepage rates as measured by manual seep meters from high tide to low tide. Numbers below the bars are average rates in cm d^{-1}. Modified from Mulligan AE and Charette MA (2006) Intercomparison of submarine ground water discharge estimates from a sandy unconfined aquifer. *Journal of Hydrology* 327: 411–425.

is a point measurement and capturing the spatial variability therefore requires installing many wells. The primary advantage of this approach is that it is well established and easy to implement: head measurements are easy to collect once wells are installed and the flux calculations are simple.

The mass balance approach to estimating SGD requires ascertaining all inputs and outputs of water flow, except SGD, through the groundwatershed. Assuming a steady-state condition over a specified time frame, the groundwater discharge rate is calculated as the difference between all inputs and all outputs. Implementing this approach can be quite simple or can result in complex field campaigns, but the quality of the data obviously affect the level of uncertainty. Even with extensive field sampling, water budgets are seldom known with certainty and

so should be used with that in mind. Furthermore, if the spatial and temporal variability of SGD is needed for a particular study, the mass balance approach is not appropriate.

Direct measurements: Seepage meters   SGD can be measured directly with seepage meters. Manual seepage meters (**Figure 4**) are constructed using the tops of 55 gallon drums, where one end is open and placed into the sediment. The top of the seepage meter has a valve through which water can flow; a plastic bag prefilled with a known volume of water is attached to the valve, so that inflow to or outflow from the sediments can be determined. After a set length of time, the bag is removed and the volume of water in the bag is measured. The change in water volume over the sampling period is then used to determine the average flow rate of fluid across the water–sediment boundary over the length of the sampling period. These meters are very simple to operate, but they are manually intensive and are sensitive to wave disturbance and currents. Furthermore, they only sample a small flow area and so many meters are needed to characterize the spatial variability seen at most sites (**Figure 3**).

Recently, several other technologies have been applied toward developing automated seep meters. Technologies include the heat-pulse method, continuous

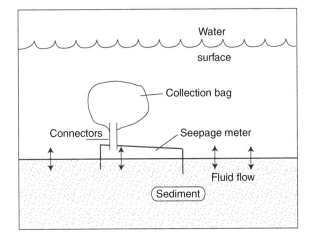

Figure 4 Graphic of a manual seepage meter deployed at the sediment–water interface. These meters can be constructed using the top of a 55 gallon drum. The collection bag serves as a fluid reservoir so that both inflow to and outflow from the sediments can be measured. Modified from Lee DR (1977) A device for measuring seepage in lakes and estuaries. *Limnology and Oceanography* 22: 140–147.

heat, ultrasonic, and dye dilution. These meters can be left in place for days and often weeks and will measure seepage without the manual intervention needed using traditional seepage meters. The trade-off with these meters is that they are expensive and therefore only a limited number are typically employed at any given time. An example of seepage measurements made using the dye dilution meter is shown in **Figure 5**. Note that the seepage rates are inversely proportional to tidal height. As the tide rises, the hydraulic gradient from land to sea is reduced, seepage slows, and the flow reverses, indicating that seawater is flowing into the aquifer at this location during high tide. Conversely, at low tide, the hydraulic gradient is at its steepest and SGD increases.

Chemical Tracers

The chemical tracer approach to quantifying SGD has an advantage over seepage meters in that it provides an integrated flux over a wide range of spatial scales from estuaries to continental shelves. The principle of using a chemical tracer is simple: find an element or isotope that is highly enriched (or depleted) in groundwater relative to other sources of water, like rivers or rainfall, to the system under study. If SGD is occurring, then the flux of this element via groundwater will lead to enrichment in the coastal zone that is well above background levels in the open ocean (**Figure 6**). A simple mass balance/box model for the system under study can be performed, where all sources of the tracer other than groundwater are subtracted from the total inventory of the chemical. The residual inventory, or 'excess', is then

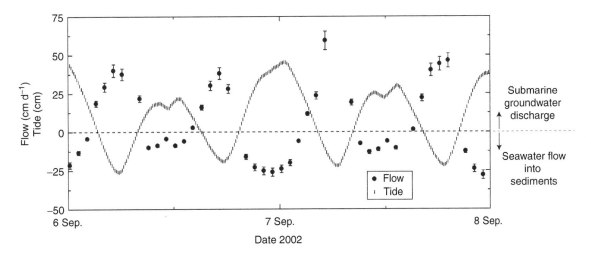

Figure 5 Seepage rates at Waquoit Bay, Massachusetts, USA, measured using a dye dilution automated seepage meter. Modified from Sholkovitz ER, Herbold C, and Charette MA (2003) An automated dye-dilution based seepage meter for the time-series measurement of submarine groundwater discharge. *Limnology and Oceanography: Methods* 1: 17–29.

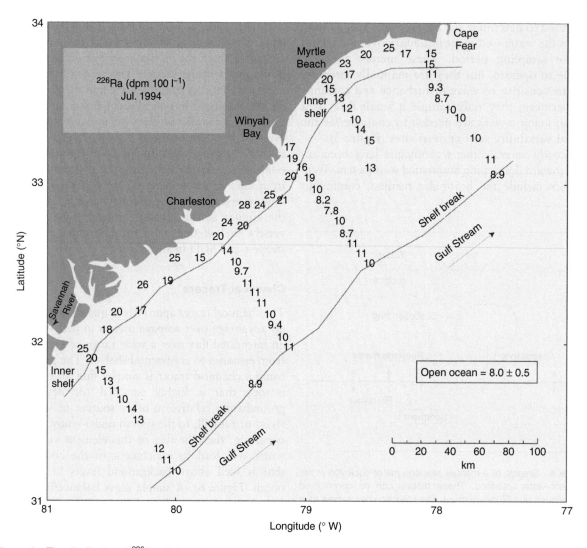

Figure 6 The distribution of ^{226}Ra offshore South Carolina reveals high activities on the inner shelf that decrease offshore. Moore (1996) used the excess ^{226}Ra to estimate regional SGD fluxes.

divided by the concentration of the tracer in the discharging groundwater to calculate the groundwater flow rate.

Naturally occurring radionuclides such as radium isotopes and radon-222 have gained popularity as tracers of SGD due to their enrichment in groundwater relative to other sources and their built-in radioactive 'clocks'. The enrichment of these tracers is owed to the fact that the water–sediment ratio in aquifers is usually quite small and that aquifer sediments (and sediments in general) are enriched in many U and Th series isotopes; while many of these isotopes are particle reactive and remain bound to the sediments, some like Ra can easily partition into the aqueous phase. Radon-222 ($t_{1/2} = 3.82$ days) is the daughter product of ^{226}Ra ($t_{1/2} = 1600$ years) and a noble gas; therefore, it is even more enriched in groundwater than radium.

A key issue when comparing different techniques for measuring SGD is the need to define the fluid composition that each method is measuring (i.e., fresh, saline, or brackish SGD). For example, whereas hydrogeological techniques are estimates of fresh SGD, the radium and radon methods include a component of recirculated seawater. Therefore, it is often not possible to directly compare the utility of these techniques. Instead, they should be regarded as complementary.

One of the seminal studies that showed how SGD can impact chemical budgets on the scale of an entire coastline was conducted off North and South Carolina, USA. Literature estimates of water residence time, riverine discharge/suspended sediment load, and the activity of desorbable ^{226}Ra on riverine particles were used to determined that only $\sim 50\%$ of the ^{226}Ra inventory on the inner continental shelf

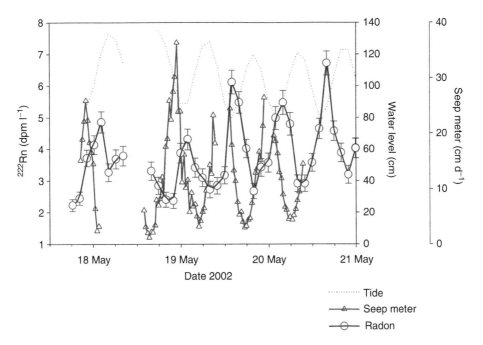

Figure 7 Plot comparing variations in SGD at Shelter Island, NY, between an automated seepage meter and the continuous radon method.

off North and South Carolina could be explained by surface inputs (**Figure 6**). The remaining inventory enters the system via SGD. Using an estimate of groundwater ^{226}Ra, it is calculated that the groundwater flux to this region of the coastline is on the order of 40% of the river water flux.

The approach for quantifying SGD using ^{222}Rn is similar to that for radium (^{226}Ra), except for a few key differences: (1) ^{222}Rn loss to the atmosphere must be accounted for in many situations, (2) there is no significant source from particles in rivers, and (3) decay must be accounted for owing to its relatively short half-life. The first example of ^{222}Rn use to quantify SGD to the coastal zone occurred in a study in the northeastern Gulf of Mexico. The strong pycnocline that develops in the summer time means that fluid that flows from the sediments into the bottom boundary layer is isolated from the atmosphere and therefore no correction for the air–sea loss of ^{222}Rn is needed. Using this approach, diffuse SGD in a 620 km^2 area of the inner shelf was estimated to be equivalent to ~20 first-magnitude springs (a first-magnitude spring has a flow equal to or greater than 245 000 m^3 water per day, which is equivalent to ~60% of the daily water supply for the entire state of Rhode Island, USA).

Since 2000, a number of SGD estimation technique intercomparison experiments have been conducted through a project sponsored by the Scientific Committee on Oceanic Research (SCOR) and the Land–Ocean Interaction in the Coastal Zone (LOICZ) Project. During these intercomparisons, several methods (chemical tracers, different types of seep meters, hydrogeologic approaches, etc.) were run side by side to evaluate their relative strengths and weaknesses. **Figure 7** displays a comparison from the Shelter Island, NY, experiment of calculated radon fluxes (based on measurements from a continuous radon monitor), with seepage rates measured directly via the dye dilution seepage meter. During the period (17–20 May) when both devices were operating at the same time, there is a clear and reproducible pattern of higher radon and water fluxes during the low tides. There is also a suggestion that the seepage spikes slightly led the radon fluxes, which is consistent with the notion that the groundwater seepage is the source of the radon (the radon monitor was located offshore of the seepage meter). The excellent agreement in patterns and overlapping calculated advection rates (seepage meter: 2–37 cm day^{-1}, average = 12 ± 8 cm day^{-1}; radon model: 0–34 cm day^{-1}, average = 11 ± 7 cm day^{-1}) by these two completely independent assessment tools is reassuring.

Geochemistry of the Subterranean Estuary

The magnitude of chemical fluxes carried by SGD is influenced by biogeochemical processes occurring in the subterranean estuary, defined as the mixing zone between groundwater and seawater in a coastal

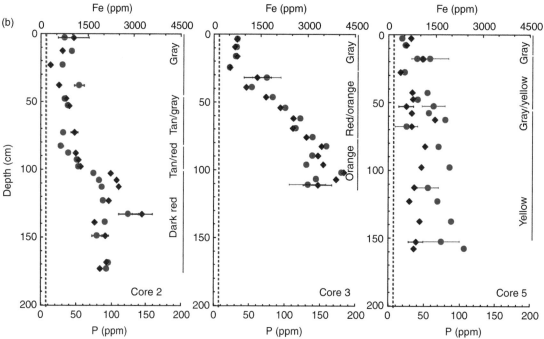

Figure 8 (a) Scientists extruding a sediment core taken through the subterranean estuary of Waquoit Bay, MA. Note the presence of iron oxides within the sediments at the bottom of the core (orange-stained sediments in foreground). (b) Changes in iron and phosphate concentration with depth in three sediment cores similar to the one shown in (a). The red circles indicate Fe concentration (ppm:μg Fe/g dry sediment) while the blue diamonds represent P (ppm:μg P/g dry sediment). Error bars indicate the standard deviation for triplicate leaches performed on a selected number of samples. The dashed lines represent the concentration of Fe and P in 'off-site' quartz sand. Also shown is the approximate color stratigraphy for each core. The R^2 value for Fe vs. P in cores 2, 3, and 5 is 0.80, 0.91, and 0.16, respectively.

aquifer. The biogeochemical reactions in such underground estuaries are presumed to be similar to the surface estuary (river water/seawater) counterpart, though few comprehensive studies of chemical cycling in subterranean estuaries have been undertaken. Drivers of biogeochemical reactions in these environments include oxidation–reduction gradients, desorption–sorption processes, and microbially driven diagenesis.

In a study of the Waquoit Bay subterranean estuary, a large accumulation of iron (hydr)oxide-coated sediments within the fresh–saline interface was encountered. These iron-oxide-rich sands could act as a geochemical barrier by retaining and accumulating certain dissolved chemical species carried to the subterranean estuary by groundwater and/or coastal seawater. Significant accumulation of phosphorus in the iron oxide zones of the Waquoit cores exemplifies this process (**Figure 8**).

Phosphorous is not the only nutrient that can be retained/removed via reactions in the subterranean estuary. The microbial reduction of nitrate to inert dinitrogen gas, a process known as denitrification, is known to occur in the redox gradients associated with fresh and saline groundwater mixing. Conversely, ammonium, which is more soluble in saline environments, may be released within the subterranean estuary's mixing zone. While the overall importance of SGD on the 'global cycle' of certain chemical species remains to be seen, there is little doubt that SGD is important at the local scale both within the United States and throughout the world.

Summary

Groundwater discharge to the coastal ocean can be an important source of fresh water and dissolved solutes. Although a significant amount of research into the role of SGD in solute budgets has only occurred in the past decade or so, there is increasing evidence that solute loading from groundwater can be significant enough to affect solute budgets and even ecosystem health. Proper estimation of solute loads from groundwater requires confident estimates of both the groundwater discharge rate and average solute concentrations in the discharging fluid, neither of which are easily determined.

Estimating groundwater discharge rates is complicated by the spatial and temporal variability of groundwater flow. A multitude of time-varying driving mechanisms complicate analysis, as does geologic heterogeneity. Nonetheless, a suite of tools from hydrogeology, geophysics, and geochemistry have been developed for sampling and measuring SGD.

The nature of geochemical reactions within nearshore sediments is not well understood, yet recent studies have shown that important transformations occur over small spatial scales. This is an exciting and critically important area of research that will reveal important process in the near future.

See also

Chemical Processes in Estuarine Sediments. Estuarine Circulation. Gas Exchange in Estuaries. Submarine Groundwater Discharge.

Further Reading

Burnett WC, Aggarwal PK, Bokuniewicz H, *et al.* (2006) Quantifying submarine groundwater discharge in the coastal zone via multiple methods. *Science of the Total Environment* 367: 498–543.

Burnett WC, Bokuniewicz H, Huettel M, Moore WS, and Taniguchi M (2003) Groundwater and pore water inputs to the coastal zone. *Biogeochemistry* 66: 3–33.

Cable JE, Burnett WC, Chanton JP, and Weatherly GL (1996) Estimating groundwater discharge into the northeastern Gulf of Mexico using radon-222. *Earth and Planetary Science Letters* 144: 591–604.

Charette MA and Sholkovitz ER (2002) Oxidative precipitation of groundwater-derived ferrous iron in the subterranean estuary of a coastal bay. *Geophysical Research Letters* 29: 1444.

Charette MA and Sholkovitz ER (2006) Trace element cycling in a subterranean estuary. Part 2: Geochemistry of the pore water. *Geochimica et Cosmochimica Acta* 70: 811–826.

Heath R (1998) Basic ground-water hydrology. *US Geological Survey Water Supply Paper* 2220. Washington, DC: USGS.

Kohout F (1960) Cyclic flow of salt water in the Biscayne aquifer of southeastern Florida. *Journal of Geophysical Research* 65: 2133–2141.

Lee DR (1977) A device for measuring seepage in lakes and estuaries. *Limnology and Oceanography* 22: 140–147.

Li L, Barry DA, Stagnitti F, and Parlange J-Y (1999) Submarine groundwater discharge and associated chemical input to a coastal sea. *Water Resources Research* 35: 3253–3259.

Michael HA, Mulligan AE, and Harvey CF (2005) Seasonal water exchange between aquifers and the coastal ocean. *Nature* 436: 1145–1148 (doi:10.1038/nature03935).

Miller DC and Ullman WJ (2004) Ecological consequences of ground water discharge to Delaware Bay, United States. *Ground Water* 42: 959–970.

Moore WS (1996) Large groundwater inputs to coastal waters revealed by Ra-226 enrichments. *Nature* 380: 612–614.

Moore WS (1999) The subterranean estuary: A reaction zone of ground water and sea water. *Marine Chemistry* 65: 111–125.

Mulligan AE and Charette MA (2006) Intercomparison of submarine ground water discharge estimates from a sandy unconfined aquifer. *Journal of Hydrology* 327: 411–425.

Paulsen RJ, Smith CF, O'Rourke D, and Wong T (2001) Development and evaluation of an ultrasonic ground water seepage meter. *Ground Water* 39: 904–911.

Portnoy JW, Nowicki BL, Roman CT, and Urish DW (1998) The discharge of nitrate-contaminated groundwater from developed shoreline to marsh-fringed estuary. *Water Resources Research* 34: 3095–3104.

Shinn EA, Reich CD, and Hickey TD (2002) Seepage meters and Bernoulli's revenge. *Estuaries* 25: 126–132.

Sholkovitz ER, Herbold C, and Charette MA (2003) An automated dye-dilution based seepage meter for the time-series measurement of submarine groundwater discharge. *Limnology and Oceanography: Methods* 1: 17–29.

Slomp CP and Van Cappellen P (2004) Nutrient inputs to the coastal ocean through submarine groundwater discharge: Controls and potential impact. *Journal of Hydrology* 295: 64–86.

Taniguchi M, Burnett WC, Cable JE, and Turner JV (2002) Investigation of submarine groundwater discharge. *Hydrological Processes* 16: 2115–2129.

Valiela I, Foreman K, LaMontagne M, *et al.* (1992) Couplings of watersheds and coastal waters: Sources and consequences of nutrient enrichment in Waquoit Bay, Massachusetts. *Estuaries* 15: 443–457.

SUBMARINE GROUNDWATER DISCHARGE

W. S. Moore, University of South Carolina, Columbia, SC, USA

Introduction

The most evident interface between terrestrial and ocean waters is the river mouth or estuary. Less obvious is the subsurface interface between terrestrial groundwater and ocean water. The direct flow of groundwater into the ocean and the recycling of seawater through coastal aquifers are processes that have been largely ignored in estimating material fluxes between the land and the sea. Submarine groundwater discharge (SGD) includes any and all flow of water on continental margins from the seabed to the coastal ocean, regardless of fluid composition or driving force. SGD is becoming recognized as an important process along many coasts that leads to fluxes of nutrients, metals, and carbon that may exceed riverine and atmospheric inputs.

Where sediments are saturated, as expected in submerged materials, groundwater is synonymous with pore water; thus, advective pore water exchange falls under this definition of SGD. The definition of SGD does not include such processes as deep-sea hydrothermal circulation, deep fluid expulsion at convergent margins, and density-driven cold seeps on continental slopes.

The direct flow of groundwater into the ocean has been reported since ancient times and likely recognized much earlier. The Roman geographer Strabo (63 BC–AD 21) noted a submarine spring 4.5 km offshore Syria in the Mediterranean. Water from this spring was collected from a boat, utilizing a lead funnel and leather tube, and transported to the city as a source of fresh water.

Although submarine springs constitute obvious and in some cases spectacular manifestations of SGD, visible springs are thought to contribute only a small fraction of the water that enters the ocean via SGD. Diffuse flow through permeable coastal sediments is probably a much more important component.

Because the direct flow of groundwater into the ocean was thought to be a relatively minor component of the hydrological cycle, this phenomenon has been slow to receive attention from hydrologists and geochemists. Some early workers recognized that groundwater discharge may be an important source of nutrients for coral reefs, but there was little appreciation of the process in other settings or of its global importance.

There are two issues that must be considered: (1) What is the flux of fresh water due to SGD? (2) What are the material fluxes due to chemical reactions of seawater and meteoric water with aquifer solids? Hydrologists are primarily concerned with the first question as it relates directly to the freshwater reserve in coastal aquifers and salinization of these aquifers through seawater intrusion. Oceanographers are also interested in this question because there may be buoyancy effects on the coastal ocean associated with direct input of fresh water from the bottom. Chemical, biological, and geological oceanographers are more concerned with the second question as it relates directly to alteration of coastal aquifers and nutrient, metal, and carbon inputs to the ocean (and in some cases removal from the ocean).

We now recognize that SGD is an important component of the hydrological cycle due to the transfer of nutrients, metals, and carbon to the coastal ocean. New studies have found that SGD provides a major source of nutrients to salt marshes, estuaries, and other communities on the continental shelf. For example, it is estimated that the fluxes of nitrogen and phosphorus to the South Carolina and Georgia, USA, shelf from submarine groundwater discharge exceed fluxes from local rivers. Similar results have been obtained in South Korea and Thailand. Because nutrient concentrations in coastal groundwater may be several orders of magnitude greater than surface waters, groundwater input may be a significant factor in the eutrophication of coastal waters. Some have linked the nutrients supplied by SGD to harmful algae blooms.

Composition of Submarine Groundwater Discharge

As freshwater flows through an aquifer driven by an inland hydraulic head, it can entrain seawater diffusing and dispersing upward from the salty aquifer that underlies it (**Figure 1**). Superimposed upon this terrestrially driven circulation are a variety of marine-induced forces that result in flow into and out of the seabed even in the absence of a hydraulic head. Most SGD is a mixture of fresh groundwater and seawater. Although the salinity of SGD emerging from

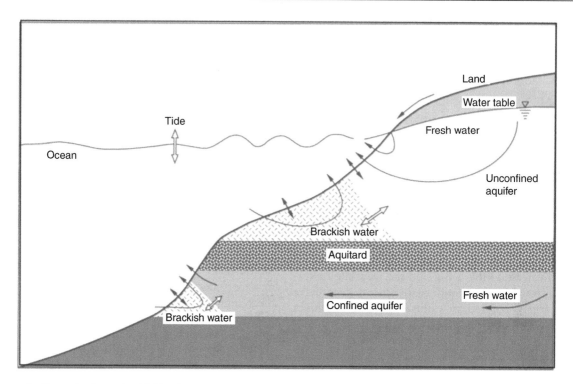

Figure 1 General schematic of SGD. Groundwater in the unconfined aquifer may discharge directly or mix with seawater and discharge as brackish water. Both terrestrial and ocean forces determine the discharge, which is expected to occur close to shore. Below the aquitard, groundwater may discharge further seaward. Here again the discharge depends on terrestrial and ocean forces. Biogeochemical reactions within each aquifer modify the composition of the fluids. Reproduced from Burnett WC, Bokuniewicz H, Huettel M, Moore WS, and Taniguchi M (2003) Groundwater and pore water inputs to the coastal zone. *Biogeochemistry* 66: 3–33, **Figure 1**, with permission from Elsevier.

coastal aquifers may be used to determine the fraction of fresh groundwater that has diluted recycled seawater, the fluids that comprise SGD are often chemically distinct from these end members. To emphasize the importance of mixing and chemical reaction in coastal aquifers, these systems have been called 'subterranean estuaries' because they are characterized by biogeochemical reactions that influence the transfer of nutrients, metals, and carbon to the coastal zone in a manner similar to that of surface estuaries. Several studies estimate that on some coastlines the flux of nutrients into salt marshes and the coastal ocean through subterranean estuaries rivals the nutrient flux to the coast by rivers or the atmosphere.

Geochemists studying coastal aquifers have long recognized the importance of chemical reactions between aquifer solids and a mixture of seawater and fresh groundwater. For example, mixing of seawater supersaturated with respect to calcite with fresh groundwater saturated with respect to calcite can result in solutions that are either supersaturated or undersaturated with respect to calcite. The undersaturated solutions result primarily from the nonlinear dependence of activity coefficients on ionic strength and on changes in the distribution of inorganic carbon species as a result of mixing. This

mechanism has been used to explain the massive subsurface dissolution of limestone along the northern Yucatan Peninsula.

Calcite dissolution may also be driven by addition of CO_2 to fluids in the subterranean estuary. For example, salt water penetrating the Floridan aquifer near Savannah, Georgia, is enriched in inorganic carbon and calcium, as well as ammonia and phosphate, relative to seawater and fresh groundwater end members. These enrichments may be due to oxidation of organic carbon within the aquifer or CO_2 infiltration from shallower aquifers.

Seawater–groundwater mixing has been invoked to explain the formation of dolomite in coastal limestone. Surface seawater is supersaturated with respect to calcite and dolomite, yet the inorganic precipitation of these minerals from seawater is rarely observed. A mixture of seawater and groundwater could be undersaturated with respect to calcite, yet remain supersaturated with respect to dolomite, leading to the replacement of calcite by dolomite. The SGD from such an aquifer should be enriched in calcium with respect to magnesium.

Because of organic matter oxidation, some subterranean estuaries are suboxic or anoxic. In these cases, metals such as iron and manganese may be released from their oxides to solution. Other metals that were adsorbed to these oxide coatings would also be released. SGD from such subterranean estuaries will transfer these dissolved metals to the coastal ocean. Unlike dissolved metals that enter surface estuaries from rivers, metals that enter the coastal water from subterranean estuaries may bypass the estuary filter, that is, the particle-rich and high-productivity zones in many surface estuaries, where dissolved metals are sequestered. Important additions of iron to the ocean are thought to occur along the southern coast of Brazil.

Seawater that recharges subterranean estuaries may transport dissolved metals that precipitate in reducing environments. Uranium may be removed from seawater by this mechanism.

The natural supply of nutrients and carbon by SGD may be necessary to sustain biological productivity in some environments. There is also an awareness that in some cases SGD may be involved in harmful algal blooms and other negative effects. As coastal development continues, changes in the flux and composition of SGD are expected to occur. To evaluate the effects of SGD on coastal waters, it is necessary to know the current flux of SGD and its composition. Studies are underway along many coastlines to evaluate SGD and its effect on the environment.

Mechanisms that Drive Submarine Groundwater Discharge

Submarine springs are obviously caused by artesian flow of groundwater in response to pressure gradients in aquifers and breaches in confining layers. However, just as these springs constitute only a small component of SGD, additional mechanisms are required to explain the range of SGD.

In the simplest case, fresh groundwater flows through an aquifer driven by an inland hydraulic head to the sea. As it nears the sea it encounters salty groundwater that has infiltrated from the ocean (**Figure 1**). Because the density of fresh groundwater is lower than salt water, it tends to flow above the salt water. In an unconfined homogenous aquifer with no other forces, the region of discharge of the SGD can be predicted by the density difference between the fluids. The classic Ghyben–Herzberg relation predicts that the discharge of such mixtures will be restricted to a few hundred meters from shore in such homogenous unconfined aquifers. However, coastal sediments often comprise a complex assemblage of confined, semiconfined, and unconfined aquifers. Simple models do not consider the anisotropic nature of the coastal sediments, dynamic processes of dispersion, tidal

pumping, wave setup, thermal gradients, inverted density stratification, and sea level change, or the nonsteady state nature of coastal aquifers due to groundwater mining. Flow through these systems often represents both terrestrial and marine forces and components. Because the timescales of these forcing functions cover a wide range, SGD may appear to be out of phase with expected forces.

In the real world, fresh groundwater flowing toward the sea encounters an irregular interface where mixing of the fluids is driven by diffusion and dispersion augmented by tides and waves. This pattern of two-layer circulation and mixing is similar to that observed in many surface estuaries. Dispersion along the interface may be enhanced by tidal forces operating in an anisotropic medium. During rising tide, seawater enters the aquifer through permeable outcrops and breaches in confining units; during falling tide, it flows back to the sea. Through each tidal cycle, the net movement of the freshwater–saltwater interface may be small, but, during rising tide, preferential flow of salt water along channels of high hydraulic conductivity may be caused by anisotropic permeability. This leads to irregular penetration of salt water into freshwater zones. Diffusion mixes the salt into the adjacent fresh water and chemical reactions may ensue. During falling tide, the reverse occurs as fresher water penetrates along these same paths and gains salt by diffusion. The permeability and preferential flow paths may be changed by chemical reactions within the aquifer. Precipitation of solids can restrict or seal some paths, while dissolution will enlarge existing paths or open new ones. The fluid expelled to the sea may be quite different chemically from the fresh water or seawater that entered.

Cyclic flow of seawater through coastal aquifers may also be driven by geothermal heating in what is known as the Kohout cycle. The Floridan aquifer in western Florida is in hydraulic contact with seawater at depths of 500–1000 m in the Gulf of Mexico. Cool seawater (5–10 °C) entering the deep aquifer is heated as it penetrates to the interior of the platform. The warm water rises through fractures or solution channels, mixes with fresh groundwater, and emerges as warm springs along the Florida coast. Temperature profiles in deep wells on the platform reveal a negative horizontal temperature gradient from the interior of the Florida platform to the margin, where cool seawater penetration is hypothesized. Wells near the margin display negative vertical temperature gradients at depths of 500–1000 m. These observations are consistent with the proposed large-scale cycling of seawater–groundwater mixtures through the aquifer. This process may operate along many continental shelves as well as ocean islands (**Figure 2**).

The upward flow of fluids from deep, pressurized aquifers may also give rise to SGD. The continental shelf off the southeastern US coast consists of a thick sequence of Tertiary limestone overlain by sand and some impermeable clay. Drilling has revealed that substantial fresh water lies trapped in the deep sediments. The Atlantic Marine Coring (AMCOR) project drilled 19 cores on the continental shelf and noted that fresh or slightly brackish waters underlie much of the Atlantic continental shelf. JOIDES test hole J-1B drilled 40 km offshore from Jacksonville, Florida, was set in the Eocene limestone (equivalent to the Floridan aquifer onshore) 250 m below sea level. This well tapped an aquifer that produced a flow of fresh water to an height of 10 m above sea level.

Another mechanism that may be important for localized cycling is evaporation of seawater in restricted coastal systems. This reflux mechanism is driven by the increased density of isolated seawater after evaporation. Reflux cycling can occur in response to daily, monthly, or longer sea level changes.

In sandy sediments where bottom currents are sufficiently strong to create ripples, wave-generated pressure gradients may drive pore water exchange, a type of SGD. In the ripple troughs water penetrates into the sediment and flows on a curved path toward the ripple crests, where the pore water is released (Figure 3). The resulting pore water exchange carries organic matter and oxygen into the sediment, creates horizontal concentration gradients that can be as strong as the vertical gradients, and increases the flux of pore water constituents across the sediment–water interface. The depth of the sediment layer that is affected by this advective exchange is related to the size and spacing of the ripples and, in homogenous sediment, reaches down to approximately 2 times the ripple wavelength. It has been estimated that worldwide waves filter a volume of $97 \times 10^3 \, \text{km}^3 \, \text{yr}^{-1}$ through the permeable shelf sediments. Additionally, the intertidal pump, driven by swash and tidal water level changes, moves another $1.2 \times 10^3 \, \text{km}^3 \, \text{yr}^{-1}$ through the sandy beaches of the world. These estimates suggest that wave action alone can filter the

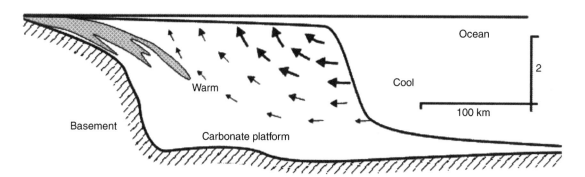

Figure 2 Effect of geothermal heating on movement of water through a submarine carbonate platform on the continental shelf.

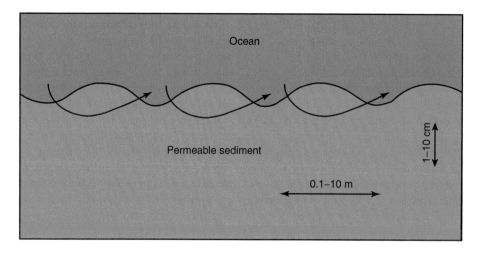

Figure 3 Pore water exchange driven by differential pressure gradients on bottom topography.

total ocean volume through permeable sediments within *c.* 14 ky.

Changes in the density of coastal waters overlying permeable sediments may also drive advective pore water exchange. If upwelling of deeper-ocean water replaces warm coastal water with colder water, the warm pore waters may become buoyant. In this case convective overturn of the pore fluids may occur suddenly. This process has been documented by sudden temperature changes in sandy sediments 1.5 m below the sea bed (**Figure 4**).

Like surface estuaries, subterranean estuaries are affected strongly by sea level changes. The occurrence of both systems today is due to the drowning of coastlines by the last transgression of the sea. During low glacial sea stands, modern surface and subterranean estuaries did not exist in their present locations. Twenty thousand years ago, SGD was probably restricted to the edge of the continental shelf. At this time river channel cutting may have caused breaches in overlying confining units, resulting in coastal springs. If these channels were filled with coarse clastics during sea level rise, the channel fill may currently serve as an effective conduit for shallow groundwater flow across the shelf and the breaches may provide communication between the surficial and deeper confined aquifers. As the rising sea approached its current level, both surface and subterranean estuaries developed as seawater infiltrated into river basins and coastal aquifers.

Demand for fresh water by coastal communities has intensified infiltration of seawater into subterranean estuaries by decreasing the hydraulic pressure in coastal aquifers. In many cases, seasonal groundwater usage causes considerable fluctuation in hydraulic pressure and salt infiltration in these systems. Changes in groundwater usage are also reflected in the position of the freshwater–saltwater interface in semiconfined coastal aquifers. Overpumping of coastal aquifers has caused some water to become nonpotable due to salt encroachment. When such aquifers are abandoned, they may recharge with fresh water. In this case, salty fluids in the aquifer may be discharged to the sea as SGD.

Measuring Submarine Groundwater Discharge

Attempts to quantify SGD have focused on three approaches: (1) using seepage meters to directly measure discharge; (2) using tracer techniques to integrate SGD signals on a regional scale; and (3) groundwater modeling.

Direct Measurement

Seepage meters are constructed from the top of a steel drum driven into the sediment. In the simplest application, a small plastic bag attached to a port on the top of the meter collects the SGD. More advanced seepage meters are based on heat pulse and acoustic Doppler technologies or dye dilution.

In calm conditions, seepage meters provide a direct measure of SGD. Discharge rates from numerous meters are required to evaluate the pattern of SGD. These patterns often include higher seepage at low tide and a decrease in seepage with increasing distance from the shoreline. Caution is required in using seepage meters during rough conditions when they may be dislodged or biased by currents and waves.

Tracer Techniques

Regional estimation of SGD is complicated due to the fact that direct measurement over large temporal and spatial scales is not possible by conventional means. Measurement of a range of tracers at the aquifer–marine interface and in the coastal ocean provides a method to produce integrated flux estimates of discharge not possible by other means. Tracer techniques utilize chemicals (often naturally occurring radionuclides in the uranium and thorium decay series) that have high concentrations in groundwater relative to coastal waters and low reactivity in the coastal ocean. These techniques require an assessment of the

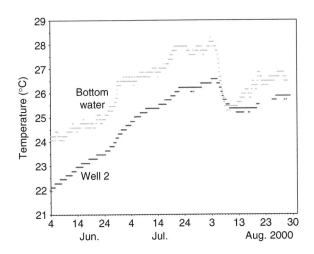

Figure 4 Temperature record from the seabed and 1.5 m below the seabed in a monitoring well 15 km off the coast of North Carolina, Jun.–Aug. 2000. Note the abrupt drop in the well temperature as the bottom water temperature suddenly decreased. Reproduced from Moore WS and Wilson AM (2005) Advective flow through the upper continental shelf driven by storms, buoyancy, and submarine groundwater discharge. *Earth and Planetary Science Letters* 235: 564–576, Figure 7, with permission from Elsevier.

tracer concentration in the groundwater, evaluation of other sources of the tracer, and a measure of the residence time of the coastal water. With this information, an inventory of the tracer in coastal waters is converted to an offshore flux of the tracer. This tracer flux must be replaced by new inputs of the tracer from SGD.

The discovery of high activities of ^{226}Ra (half-life = 1600 years) in the coastal ocean (**Figure 5**) that could not be explained by input from rivers or sediments coupled with measured high ^{226}Ra in salty coastal aquifers led to the hypothesis that SGD was responsible for the elevated activities. These elevated ^{226}Ra activities, which have been measured along many coasts, provide the primary evidence for large SGD fluxes to the coastal ocean.

There are good reasons why radium is closely linked to SGD studies. Because radium is highly enriched in salty coastal groundwater relative to the ocean, small inputs of SGD can be recognized as a strong radium signal. In many cases, this signal can be separated into different SGD components using ^{228}Ra (half-life = 5.7 years) as well as ^{226}Ra, the two long-lived Ra isotopes. The short-lived Ra isotopes, ^{223}Ra (half-life = 11 days) and ^{224}Ra (half-life = 3.66 days) are useful in evaluating the residence time or mixing rate of estuaries and coastal waters. Such estimates coupled with distributions of the long-lived Ra isotopes enable a direct estimate of radium fluxes. At steady state, these fluxes must be balanced by fresh inputs of Ra to the system. If other sources of Ra can be evaluated, the Ra flux that must be sustained by SGD can be evaluated. By measuring the Ra content of water in the coastal aquifers, the amount of SGD necessary to supply the Ra is determined. SGD fluxes of other components are

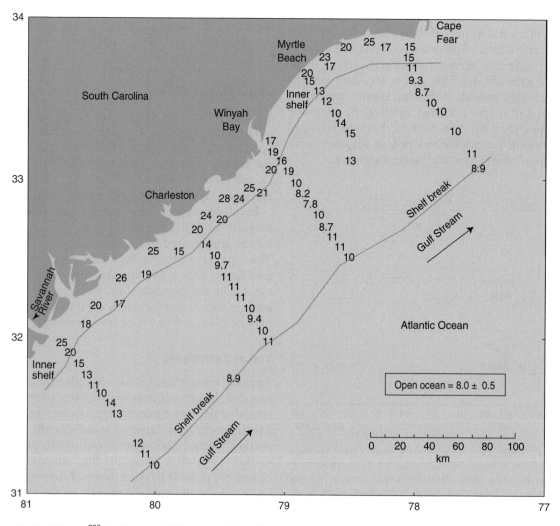

Figure 5 Activities of ^{226}Ra (dpm per 100 l) measured in surface waters of the South Atlantic Bight from 8–11 Jul. 1994. These high near-shore activities are explained by SGD containing high concentrations of radium. Adapted from Moore WS (1996) Large groundwater inputs to coastal waters revealed by ^{226}Ra enrichments. *Nature* 380: 612–614, **Figure 1**, with permission.

determined by measuring their concentrations in the coastal aquifer or by determining relationships between the component and Ra and multiplying the component/Ra ratio with the Ra flux.

Other tracers also provide evidence of SGD and means to estimate the fluxes. The immediate daughter of ^{226}Ra, ^{222}Rn (half-life $= 3.8$ days), is highly enriched in fresh and salty groundwater. High ^{222}Rn activities reveal regions of high SGD to coastal waters. Because ^{222}Rn can be measured continuously *in situ*, patterns of SGD through tidal cycles may be evaluated and compared to results from seepage meters. To estimate SGD fluxes via ^{222}Rn, losses to the atmosphere must also be considered.

Hydrologic Models

Simple water balance calculations have been shown to be useful in some situations as an estimate of the fresh groundwater discharge. Hydrogeologic, dual-density, groundwater modeling can also be conducted as simple steady-state (annual average flux) or non-steady-state (requires real-time boundary conditions) methods. Unfortunately, at present, neither approach generally compares well with seepage meter and tracer measurements, often because of differences in scaling in both time and space. Apparent inconsistencies between modeling and direct measurement approaches often arise because different components of SGD (fresh and salt water) are being evaluated or the models do not include transient terrestrial (e.g., recharge cycles) or marine processes (tidal pumping, wave setup, storms, thermal gradients, etc.) that drive part or all of the SGD. Geochemical tracers and seepage meters measure total flow, very often a combination of fresh groundwater and seawater. Water balance calculations and most models evaluate just the fresh groundwater flow.

Worldwide Importance

Many researchers have applied a variety of methods to estimate SGD. A large spread of global estimates for SGD fluxes (both fresh and saline) illustrates the high degree of variability and uncertainty of present estimates. It should be noted that estimates based on water balance considerations and some models usually include only the freshwater fraction of the total hydrologic flow. Recirculated seawater and saline groundwater fluxes are often volumetrically important and may increase these flows substantially. In regions where SGD nutrient fluxes have been quantified, they rival river and atmospheric nutrient fluxes to the study areas.

Glossary

Aquifer Geological formation that is water bearing.

Aquitard Low-permeability geologic layer that retards groundwater flow.

Brackish water Slightly salty water, having salinity between that of fresh water and seawater.

Confined aquifer Aquifer between two aquitards.

DPM Decay per minute (1 dpm $= 1/60$ Bq) of a radioactive substance.

Estuary A semi-enclosed coastal body of water which has a free connection with the open sea and within which seawater is measurably diluted with fresh water derived from land drainage.

Groundwater Water in the saturated zone of geologic material.

Meteoric water Water that is derived from the atmosphere.

See also

Chemical Processes in Estuarine Sediments. Estuarine Circulation. Gas Exchange in Estuaries. Groundwater Flow to the Coastal Ocean.

Further Reading

Burnett WC, Bokuniewicz H, Huettel M, Moore WS, and Taniguchi M (2003) Groundwater and pore water inputs to the coastal zone. *Biogeochemistry* 66: 3–33.

Burnett WC, Chanton JP and Kontar E (eds.) (2003) *Special Issue: Submarine Groundwater Discharge. Biogeochemistry.* 66: 1–202.

Charette MA and Sholkovitz ER (2006) Trace element cycling in a subterranean estuary. Part 2: Geochemistry of the pore water. *Geochimica et Cosmochimica Acta* 70: 811–826.

Huettel M, Ziebis W, Forster S, and Luther G (1998) Advective transport affecting metal and nutrient distribution and interfacial fluxes in permeable sediments. *Geochimica et Cosmochimica Acta* 62: 613–631.

Kim G and Swarzenski PW (2008) Submarine groundwater discharge (SGD) and associated nutrient fluxes to the coastal ocean. In: Liu K-K, Atkinson L, Quinones R and Talaue-McManus L (eds.) *IGBP Global Change Book Series: Carbon and Nutrient Fluxes in Continental Margin.* New York: Springer.

Kohout FA, Meisler H, Meyer FW, *et al.* (1988) Hydrogeology of the Atlantic continental margin. In: Sheridan RE and Grow JA (eds.) *The Geology of North America*, vol. I–2. pp. 463–480. Boulder, CO: Geological Society of America.

Michael HA, Mulligan AE, and Harvey CF (2005) Seasonal oscillations in water exchange between aquifers and the coastal ocean. *Nature* 436: 1145–1148.

Moore WS (1996) Large groundwater inputs to coastal waters revealed by ^{226}Ra enrichments. *Nature* 380: 612–614.

Moore WS (1999) The subterranean estuary: A reaction zone of ground water and sea water. *Marine Chemistry* 65: 111–126.

Moore WS and Wilson AM (2005) Advective flow through the upper continental shelf driven by storms, buoyancy, and submarine groundwater discharge. *Earth and Planetary Science Letters* 235: 564–576.

Plummer LN (1975) Mixing of sea water with calcium carbonate ground water. In: Whitten EHT (ed.) *Geological Society of America Memoir: Quantitative Studies in the Geological Sciences*, pp. 219–238. Boulder, CO: Geological Society of America.

Povince PP and Aggarwal PK (eds.) (2006) *Special Issue: Submarine Groundwater Discharge Studies Offshore*

South-Eastern Sicily. Continental Shelf Research. 26: 825–884.

Simmons GM, Jr. (1992) Importance of submarine groundwater discharge (SGWD) and seawater cycling to material flux across sediment/water interfaces in marine environments. *Marine Ecology Progress Series* 84: 173–184.

Valiela I and D'Elia C (eds.) (1990) *Special Issue: Groundwater Inputs to Coastal Waters. Biogeochemistry.* 10: 1–328.

Windom HL, Moore WS, Niencheski LFH, and Jahnke RA (2006) Submarine groundwater discharge: A large, previously unrecognized source of dissolved iron to the South Atlantic Ocean. *Marine Chemistry* 102: 252–266.

Zektser IS (2000) *Groundwater and the Environment: Applications for the Global Community*, 175pp. Boca Raton, FL: Lewis Publishers.

ICE AND PERMAFROST

ICE-INDUCED GOUGING OF THE SEAFLOOR

W. F. Weeks, Portland, OR, USA

Introduction

Inuit hunters have long known that both sea ice and icebergs could interact with the underlying sea floor, in that sea floor sediments could occasionally be seen attached to these icy objects. As early as 1855, no less a scientific luminary than Charles Darwin speculated that gouging icebergs could traverse isobaths, concluding that an iceberg could be driven over great inequalities of [a seafloor] surface easier than could a glacier. Only in 1924 did similar inferences of the possibility of sea ice and icebergs affecting the seafloor again begin to reappear in the scientific literature. However, direct observations of the nature of these interactions were lacking until the early 1970s. The reason for the increased interest in this rather exotic phenomena was initially the discovery of the supergiant oil field on the edge of the Beaufort Sea at Prudhoe Bay, Alaska. Because the initial successful well was located on the coast, there was immediate interest in the possibility of developing offshore fields to the north of Alaska and Canada. It was also apparent that if major oil resources occurred to the north of Alaska and Canada, similar resources might be found on the world's largest continental shelf located to the north of the Russian mainland. Offshore wells are typically tied together by subsea pipelines which take the oil from the individual wells either to a central collection point where tanker pickup is possible or to the coast where the oil can be fed into a pipeline transportation system. If sea ice processes could result in major distrubances of the seafloor, this would clearly become a major consideration in pipeline design. Some time after the Prudhoe Bay find, another large offshore oil discovery was made to the east of Newfoundland. Here the ice-induced gouging problem was caused not by sea ice but by icebergs. In both cases the initial step in treating these perceived problems was to investigate the nature of these ice–seafloor interactions. One needs to know the frequency of gouging events in time and space as well as the widths and depths of the gouges, the water depth range in which this phenomenon occurs, and the effective lifetime of a gouge after its initial formation. Also of importance are the type of subsea soil and the nature and extent of the soil movements below the gouges, as this information is essential in calculating safe burial depths for subsea structures. The following attempts to summarize our current knowledge of this type of naturally occurring phenomenon.

As extensive studies of ice-induced disruptions of the seafloor have been undertaken only during the last 30 years, there is as yet no commonly agreed upon terminology for this phenomenon. In the literature the process has been described as scouring, scoring, plowing and gouging. Here gouging is used because at least in the case of sea ice, it is felt that it more accurately describes the process than do the other terms.

Observational Techniques

A variety of techniques have been used to study the gouging phenomenon. Typically, a fathometer is used to resolve the seafloor relief directly beneath the ship with a precision of better than 10 cm (**Figure 1**) while, at the same time, a sidescan sonar system provides a sonar map of the seafloor on either side of the ship (**Figure 2**). Total sonar swath widths have been typically 200 to 250 m not including a narrow area directly beneath the ship that was not imaged. The simultaneous use of these two different types of records allows one to both measure the depth and width of the gouge (fathometer) at the point where it is crossed by the ship track and also to observe the general orientation and geometry of the gouge track (sonar). Along the Alaskan coast the ships used have been small, allowing them to operate in shallow water. In some cases divers have been used to examine the gouges and, at a few deeper water sites off the Canadian east coast, manned submersibles have been used to gather direct observations of the gouging process.

Results

Arctic Shelves

The most common features gouging the shelves of the Arctic Ocean are the keels of pressure ridges that are made of deformed sea ice. As the ice pack moves under the forces exerted on it by the wind and currents, pressure ridges typically form at floe boundaries as the result of differential movements between the floes. These features can be very large. Pressure ridge keels with drafts up to 50 m have been

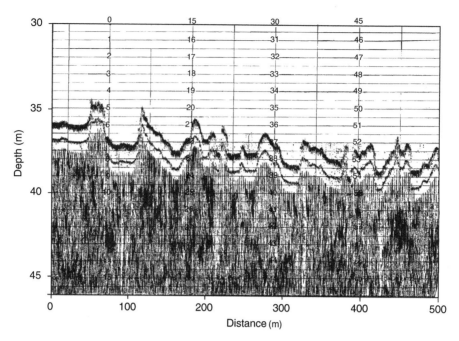

Figure 1 Fathogram of an ice-gouged seafloor. Water depth is 36 m. Record taken 25 km NE of Cape Halkett, Beaufort Sea, offshore Alaska. The multiple reflections from the upper layer of the seafloor are the result of the presence of a thawed active layer in the subsea permafrost. (From Weeks *et al.*, 1983.)

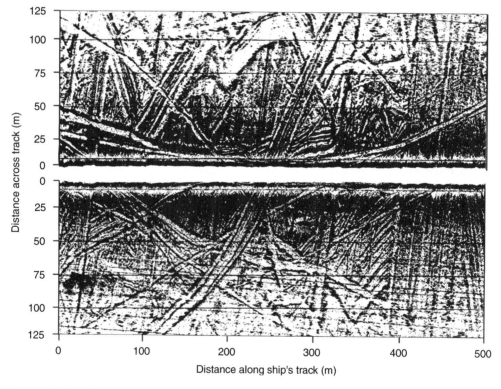

Figure 2 Sonograph of an ice-gouged seafloor. Water depth is 20 m. Record taken 20 km NE of Cape Halkett, Beaufort Sea, offshore Alaska. (From Weeks *et al.*, 1983.)

observed in the upward-looking sonar records collected by submarines passing beneath the ice. The distribution of keel depths of these deformed features is approximately a negative exponential, in that there are many shallow ridge keels while deep keels are rare. Deformation can be particularly intense in nearshore regions where the moving ice pack contacts an immovable coast with the formation of large shear ridges that can extend for many kilometers. Such ridges are frequently anchored at shoal areas where large accumulations of highly deformed sea ice can build up. Such grounded ice features, which are referred to using the Russian term stamuki, can have freeboards of as much as 10 m and lateral extents of 10–15 km. As the offshore ice field moves against and along the coast, it exerts force on the sides of any such grounded features, causing them to scrape and plough their way along the seafloor. Considering that surficial sediments along many areas of the Beaufort Sea Coast of Canada and Alaska are fine-grained silts, it is hardly surprising that over a period of time such a process can cause extensive gouging of the seafloor sediments. **Figure 3** is a photograph showing active gouging. The relatively undeformed first-year sea ice in the foreground is moving away from the viewer and pushing against a piece of grounded multiyear sea ice (indicated by its rounded upper surface formed during the previous summer's melt period), pushing it along the coast. The interaction between the first-year and the multiyear ice has resulted in a pileup of broken first-year ice, which when the photograph was taken was higher than the upper surface of the multiyear ice. Also evident is the track cut in the first-year ice as it moves past the multiyear ice. The fact that both the multiyear ice

and the pileup of first year ice are interacting with the seafloor is indicated by the presence of bottom sediment in the deformed first-year ice and on the far side of the multiyear ice.

The maximum water depth in which contemporary sea ice gouging is believed to occur is roughly 50–60 m, corresponding approximately to the draft of the largest pressure ridges. Although gouges occur in water up to 80 m deep, it is generally believed that these deep-water gouges are relicts that formed during periods when sea level was lower than at present.

Not surprisingly the depths of gouges in the seafloor mirrors the keel depth distributions observed in pressure ridge keels in that they are also well approximated by a negative exponential with the character of the falloff as well as the number of gouges varying with water depth. As with ridge keels, shallow gouges are common and deep gouges are rare. As might be expected, there are fewer deep gouges in shallow water as the large ice masses required to produce them have already grounded farther out to sea. For instance, along the Beaufort Coast in water 5 m deep, a 1 m gouge has an exceedance probability of approximately 10^{-4}; that is, 1 gouge in 10 000 will on the average be expected to have a depth equal to or greater than 1 m. In water 30 m deep, a 3.4 m gouge has the same probability of occurrence. Gouges in excess of 3 m deep are not rare and 8 m gouges have been reported in the vicinity of the Mackenzie Delta.

As might be expected, the nature of gouging varies with changes in seafloor sediment types. Along the Beaufort coast where there are two distinct soil types, the gouges in stiff, sandy, clayey silts are typically more frequent and slightly deeper than those found in more sandy sediments. Presumably the gouges in the more sandy material are more easily obliterated by wave and current action. It is also reasonable to assume that the slightly deeper gouges in the silts provide a better picture of the original incision depths. An extreme example of the effect of bottom sediment type on gouging can be found between Sakhalin Island and the eastern Russian mainland. Here the seafloor is very sandy and the currents are very strong. As a result, even though winter observations have conclusively shown that seafloor gouging is common, summer observations reveal that the gouges have been completely erased by infilling.

Gouge tracks in the Beaufort Sea north of the Alaskan and western Canadian coast generally run roughly parallel to the coast, with some regions having an excess of 200 gouges per kilometer. Histograms of the frequency of occurrence of

Figure 3 Photograph of active ice gouging occurring along the coast of the Beaufort Sea. The grounded multiyear ice floe that is being pushed by the first-year ice has a freeboard of ~2 m. The thickness of the first-year ice is ~0.3 m. (Photograph by GFN Cox.)

distances between gouges are also well described by a negative exponential, a result suggesting that the spatial occurrence of gouges may be described as a Poisson process. Gouge occurrence is hardly uniform, however, with significantly higher concentrations of gouges occurring on the seaward sides of shoal areas and barrier islands and fewer gouges on their more protected lee sides. This is clearly shown in **Figure 4** in that the degree of exposure to the moving pack ice decreases in the sequence Jones Islands–Lonely–Harrison Bay–Lagoons. **Figure 5** is a sketch based on divers' observations showing gouging along the Alaskan coast of the Beaufort Sea.

The second features producing gouges on the surface of the outer continental shelf of the Arctic Ocean are ice islands. These features are, in fact, an unusual type of tabular iceberg formed by the gradual breakup of the ice shelves located along the north coast of Ellesmere Island, the northernmost of the Canadian Arctic Islands. Once formed, an ice island can circulate in the Arctic Ocean for many years. For instance, the ice island T-3 drifted in the Arctic

Ocean between 1952 and 1979, ultimately completing three circuits of the Beaufort Gyre (the large clockwise oceanic circulation centered in the offshore Beaufort Sea) before exiting the Arctic Ocean through Fram Strait between Greenland and Svalbard (Spitzbergen). The lateral dimensions of ice islands can vary considerably from a few tens of meters to over ten kilometers. Thicknesses, although variable, are typically in the 40–50 m range; ice islands possess freeboards in the same range as exhibited by larger sea ice pressure ridges. In the study of fathometer and sonar data on gouge distributions and patterns, no attempt is usually made to separate gouges made by pressure ridges from gouges made by ice islands, as the ice features that made the gouges are commonly no longer present. However, it is reasonable to assume that many of the very wide, uniform gouges are the result of ice islands interacting with the seafloor.

It is relatively easy to characterize the state of the gouging existing on the seafloor at any given time. It is another matter to answer the question of how deep one must bury a pipeline, a cable, or some other type of fixed structure beneath the seafloor to reduce the chances of it being impacted by moving ice to some acceptable level. To answer this question one needs to know the rates of occurrence of new gouges; these values are in many locations still poorly known, as they require replicate measurements over a period of many years so that new gouges can be counted and the rate of infilling of existing gouges can be estimated. The best available data on gouging occurrence along the Beaufort Coast has been collected by the Canadian and US Geological Surveys and indicates that gouging occurs in rather large scale regional events related to severe storms that drive the pack ice inshore – about once every 4–5 years.

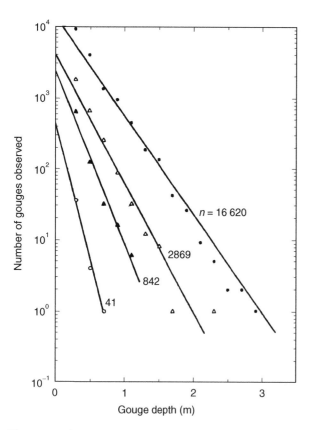

Figure 4 Semilogarithmic plot of the number of gouges observed versus gouge depth for four regions along the Alaskan coast of the Beaufort Sea. ●, Jones Island and east; △, Lonely;, △ Harrison Bay; ○, Lagoons. (From Weeks *et al.*, 1983.)

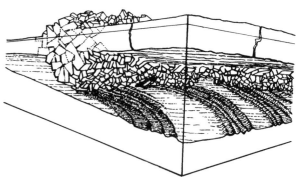

Figure 5 Diver's sketch of active gouging occurring off the coast of the Alaskan Beaufort Sea. Some sense of the scale of this drawing can be gained from the observation that commonly the thickness of the ice blocks in such ridges ranges between 0.3 and 1.0 m. (Drawing by TR Alpha, US Geological Survey.)

Because there is no absolute technique for dating the age of existing gouges, there is no current method for determining whether a particular gouge that one observes on the seafloor formed during the last year or sometime during the last 6000 years (after the Alaskan shelf was submerged as the result of rising sea levels at the end of the Pleistocene). It is currently believed that gouges in shallow water formed quite recently. For instance, during the late summer of 1977 when the Beaufort Coast was comparatively ice free, strong wave action during late summer storms obliterated gouges in water less than 13 m deep and caused pronounced infilling of gouges in somewhat deeper water. It is generally estimated that such storms have an average recurrence interval of approximately 25 years. On the other hand, gouges in water deeper than ∼60 m are believed to be relatively old in that they are presumed to be below the depth of currently active gouging. The lengths of time represented by the gouges observed in water depths between 20 and 50 m are less well known. Unfortunately, this is the water depth range in which the largest gouges are found. Stochastic models of gouging occurrence, which incorporate approximate simulations of subsea sediment transport, suggest that if seafloor currents are sufficiently strong to exceed the threshhold for sediment movement, gouge infilling will occur comparatively rapidly (within a few years).

To date, two different procedures have been used to estimate the depth of gouges with specified recurrence intervals. In the first case, the available rates and geometric characteristics of new gouges at a particular site are used. If a multiyear high-quality dataset of new gouge data is available for the region of interest, this would clearly appear to be the favored approach. In the other case, information on the draft distribution of pressure ridge keels and pack ice drift rates are combined to calculate the rates of gouging. The problem with using pack ice drift rates and pressure ridge keel depths is that drift rates in the near shore where grounding is occurring are undoubtedly less than in the offshore where the ice can move relatively unimpeded. In addition, it is doubtful that offshore keel depths observed in water sufficiently deep to allow submarine operations provide accurate estimates of keel depths in the near-shore.

One might suggest that offshore engineers should simply bury pipelines and cables at depths below those of known gouges and forget about all these statistical considerations. Although this sounds attractive, it is not a realistic resolution of the problem. In the first place, for a pipeline to avoid deformation, buckling, and failure it must, depending on the nature of the seafloor sediments, be buried at depths up to two times that of the design gouging event. Considering that gouges in excess of 3 m are not rare at many locations along the Beaufort Coast of Alaska and Canada, this means burial depths >5 m. Burial at such depths is extremely costly and time consuming considering the very short operating season in the offshore Arctic. It is also near the edge of existing subsea trenching technology. Another factor to be considered here is that the deeper the pipeline is buried, the more likely it is that its presence will result in the thawing of subsea permafrost that is known to exist in many areas of the Beaufort and Chukchi Shelves. Such thawing could result in settlement in the vicinity of the pipe, which could threaten the integrity of the pipeline. There are engineering remedies to this problem but, as one might expect, they are expensive.

Canadian East Coast

The primary problem to the east of Newfoundland and off the coast of Labrador is not sea ice but icebergs. Although the majority of these originate from the Greenland Ice Sheet calving to the west into Baffin Bay, some icebergs do originate from the East Greenland coast and from the Canadian Arctic. Considering their great size and deep draft, icebergs are formidable adversaries. Iceberg gouges with lengths >60 km are known and many have lengths >20 km. Gouge depths can exceed 10 m and recent gouging is known to occur at depths of at least 230 m along the Baffin and Labrador Shelves. Icebergs also appear to be able to traverse vertical ranges of bathymetry up to at least 45 m. As with the sea-ice-induced gouges of the Arctic shelves, iceberg gouge depths are exponentially distributed, with small gouges being common and deep gouges being rare. Iceberg gouges can be straight or curved. Pits are also common, which occur when the iceberg draft is suddenly increased through splitting and rolling. The iceberg can then remain fixed to the seafloor while it rocks and twists as a result of the wave and current forces that impinge upon it. At such times pits can be produced that are deeper than the maximum gouge depth otherwise associated with the iceberg.

Although considerable information is available on the statistics of iceberg gouge occurrences, in studies of iceberg groundings more attention has been paid to the energetics of the ice–sediment interactions than to the statistics. Also, more emphasis has been placed on iceberg 'management' in order either to reduce or to remove the threat to a specific offshore operation. For instance, as icebergs are discrete objects as compared to pressure ridges, which are giant piles of ice blocks that are frequently poorly cemented together, icebergs can be deflected away from a production site via the use of tugboats in combination with trajectory modeling. Such schemes can also utilize aerial, satellite, and ship reconnaissance techniques to provide operators with an adequate warning of a potential threat. Surface-based over-the-horizon radar technology has also been developed as an iceberg reconnaissance tool, but at the time of writing it has not been used operationally.

Other Locations

Although the discussion here has primarily focused on the Beaufort Coast of the Arctic Ocean and the offshore regions of Newfoundland and Labrador, this is only because the interest in offshore oil and gas reserves at these locations has resulted in the collection of observational information on the local nature of the ice-induced gouging phenomenon. However, regions where ice-induced gouging is presently active are very large and include the complete continental shelf of the Arctic Ocean, the continental shelves of Greenland, of eastern Canada, and of Svalbard, as well as the continental shelf of the Antarctic continent where typical shelf icebergs are known to have drafts of ∼200 m with lateral dimensions as large as 100 km. In addition, relict iceberg gouges exist and as a result affect seafloor topography in regions where icebergs are no longer common or no longer occur, such as the Norwegian Shelf and even the northern slope of Little Bahama Bank and the Straits of Florida. Finally, although 230 m is a reasonable estimate for the maximum depth of active iceberg gouging, relict gouges presumed to have occurred during the Pleistocene have been discovered at depths from 450 m to at least 850 m on the Yermak Plateau located to the northwest of Svalbard in the Arctic Ocean proper.

Conclusions

There is clearly much more that we need to know concerning processes acting on the poorly explored continental shelves of the polar and subpolar regions. The ice-induced gouging phenomenon is clearly one of the more important of these, in that an understanding of this process is essential to both the engineering and the scientific communities. At first glance, icebergs and stamuki zones might appear to be exotic entities that are out of sight somewhere way to the north or south and that can be safely put out of mind. However, as oil production moves to ever more difficult frontier areas such as the offshore Arctic, a quantitative understanding of processes such as ice gouging becomes essential for safe development. Lack of such understanding leads to poor, inefficient designs and the increased possibility of failures. The last thing that either the environmental community or the petroleum industry needs is an offshore oil spill in the high Arctic.

See also

Rigs and Offshore Structures. Sub-Sea Permafrost.

Further Reading

Barnes PW, Schell DM, and Reimnitz E (eds.) (1984) *The Alaskan Beaufort Sea: Ecosystems and Environments.* Orlando: Academic Press.

Colony R and Thorndike AS (1984) An estimate of the mean field of arctic sea ice motion. *Journal of Geophysical Research* 89(C6): 10623–10629.

Darwin CR (1855) On the power of icebergs to make rectilinear, uniformly-directed grooves across a submarine undulatory surface. *London, Edinburgh, and Dublin Philosophical Magazine and Journal of Science* 10: 96–98.

Goodwin CR, Finley JC and Howard LM (1985) *Ice Scour Bibliography.* Environmental Studies Revolving Funds Rept. No. 010, Ottawa.

Lewis CFM (1977) The frequency and magnitude of drift ice groundings from ice-scour tracks in the Canadian Beaufort Sea. *Proceedings of 4th International Conference on Port Ocean Engineering Under Arctic Conditions.* Newfoundland: Memorial University. vol. 1, 567–576.

Palmer AC, Konuk I, Comfort G, and Been K (1990) Ice gouging and the safety of marine pipelines. *Offshore Technology Conference*, Paper 6371, 235–244.

Reed JC and Sater JE (eds.) (1974) *The Coast and Shelf of the Beaufort Sea.* Arlington, VA: Arctic Institute of North America.

Reimnitz E, Barnes PW, and Alpha TR (1973) *Bottom features and processes related to drifting ice, US Geological Survey Miscellaneous Field Studies*, Map MF-532. Washington, DC: US Geological Survey.

Vogt PR, Crane K, and Sundvor E (1994) Deep Pleistocene iceberg plowmarks on the Yermak Plareau: sidescan and 3.5 kHz evidence for thick calving ice fronts and a possible marine ice sheet in the Arctic Ocean. *Geology* 22: 403–406.

Wadhams P (1988) Sea ice morphology. In: Leppäranta M (ed.) *Physics of Ice-Covered Seas*, pp. 231–287. Helsinki: Helsinki University Printing House.

Weeks WF, Barnes PW, Rearic DM, and Reimnitz E (1983) *Statistical aspects of ice gouging on the Alaskan Shelf of the Beaufort Sea. Cold Regions Research and Engineering Laboratory Report.* New Hampshire, USA: Hanover. 83–21.

Weeks WF, Tucker WB III, and Niedoroda AW (1985) A numerical simulation of ice gouge formation and infilling on the shelf of the Beaufort Sea. *Proceedings, International Conference on Port Ocean Engineering Under Arctic Conditions, Narssarssuaq, Greenland* 1: 393–407.

Woodworth-Lynas CMT, Simms A, and Rendell CM (1985) Iceberg grounding and scouring on the Labrador continental shelf. *Cold Regions Science and Technology* 10(2): 163–186.

ICE SHELF STABILITY

C. S. M. Doake, British Antarctic Survey, Cambridge, UK

Introduction

Ice shelves are floating ice sheets and are found mainly around the Antarctic continent (**Figure 1**). They can range in size up to $500\,000\,km^2$ and in thickness up to 2000 m. Most are fed by ice streams and outlet glaciers, but some are formed by icebergs welded together by sea ice and surface accumulation. They exist in embayments where the shape of the bay and the presence of ice rises, or pinning points, plays an important role in their stability. Ice shelves lose mass by basal melting, which modifies the properties of the underlying water mass and eventually influences the circulation of the global ocean, and by calving. Icebergs, sometimes more than 100 km in length, break off intermittently from the continent and drift to lower latitudes, usually breaking up and melting by the time they reach the Antarctic Convergence. The lowest latitude that ice shelves can exist is determined by the mean annual air temperature. In the Antarctic Peninsula a critical isotherm of about $-5°C$ seems to represent the limit of viability. In the last 40 years or so, several ice shelves have disintegrated in response to a measured atmospheric warming trend in the western and northern part of the Antarctic Peninsula. However, this warming trend is not expected to affect the stability of the larger ice shelves further south such as Filchner–Ronne or Ross, in the near future.

What are Ice Shelves?

Physical and Geographical Setting

An ice shelf is a floating ice sheet, attached to land where ice is grounded along the coastline. Nourished mainly by glaciers and ice streams flowing off the land, ice shelves are distinct from sea ice, which is formed by freezing of sea water. Most ice shelves are found in Antarctica where the largest can cover areas of 500 $000\,km^2$ (e.g. Ross Ice Shelf and Filchner–Ronne Ice Shelf). Thicknesses vary from nearly 2000 m, around for example parts of the grounding line of Ronne Ice Shelf, to about 100 m at the seaward edge known as the ice front.

Typically, ice shelves exist in embayments, constrained by side walls until they diverge too much for the ice to remain in contact. Thus the geometry of the coastline is important for determining both where an ice shelf will exist and the position of the ice front. There are often localized grounding points on sea bed shoals, forming ice rises and ice rumples, both in the interior of the ice shelf and along the ice front. These pinning points provide restraint and cause the ice shelf to be thicker than if it were not pinned. Ice flows from the land to the ice front. Input velocities range from near zero at a shear margin (e.g., with a land boundary), to several hundred meters per year at the grounding line where ice streams and outlet glaciers enter. At the ice front velocities can reach up to several kilometers per year. A characteristic of ice shelves is that the (horizontal) velocity is almost the same at all depths, whereas in glaciers and grounded ice sheets the velocity decreases with depth (**Figure 2**).

In the Antarctic, most ice shelves have net surface accumulation although there may be intensive summer melt which floods the surface. The basal regime is controlled by the subice circulation. Basal melting is often high near both the grounding line and the ice front. Marine ice can accumulate in the intermediate areas, where water at its *in situ* freezing point upwells and produces frazil ice crystals. Surface temperatures will be near the mean annual air temperature whereas the basal temperature will be at the freezing point of the water. Therefore the coldest ice is normally in the upper layers.

Ice shelves normally form where ice flows smoothly off the land as ice streams or outlet glaciers. In some areas, however, ice breaks off at the coastline and reforms as icebergs welded together by frozen sea ice and surface accumulation. The processes of formation and decay are likely to operate under very different conditions. An ice shelf is unlikely to reform under the same climatic conditions which caused it to decay. Complete collapse is a catastrophic process and rebuilding requires a major change in the controlling parameters. However, there is evidence of cyclicity on periods of a few hundred years in some areas.

Collapse of the northernmost section of Larsen Ice Shelf within a few days in January 1995 indicates that, after retreat beyond a critical limit, ice shelves can disintegrate rapidly. The breakup history of two northern sections of Larsen Ice Shelf (Larsen A and Larsen B) between 1986 and 1997 has been used to determine a stability criterion for ice shelves.

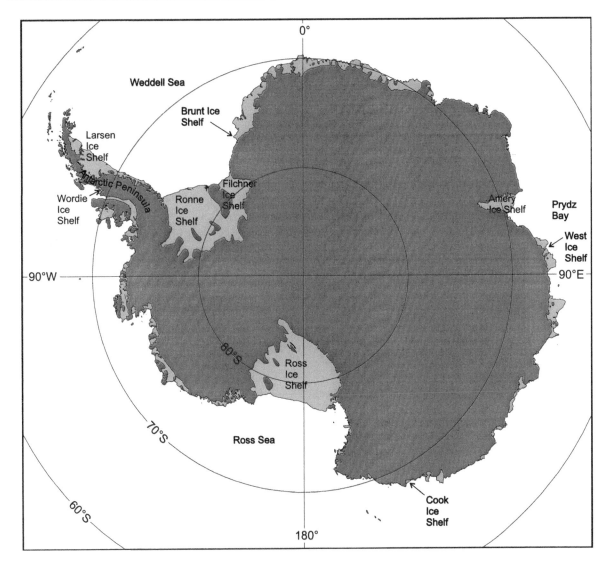

Figure 1 Distribution of ice shelves around the coast of Antarctica.

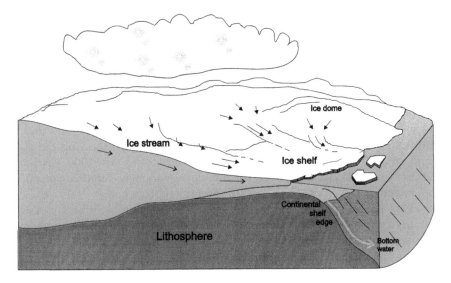

Figure 2 Cartoon of ice shelf in bay with ice streams.

Analysis of various ice-shelf configurations reveals characteristic patterns in the strain-rates near the ice front which have been used to describe the stability of the ice shelf.

Why are Ice Shelves Important?

Ice shelves are one of the most active parts of the ice sheet system. They interact with the inland ice sheet, glaciers and ice streams which flow into them, and with the sea into which they eventually melt, either directly or as icebergs. A typical Antarctic ice shelf will steadily advance until its ice front undergoes periodic calving, generating icebergs. Calving can occur over a wide range of time and spatial scales. Most ice fronts will experience a quasi-continuous 'nibbling' away of the order of tens to hundreds of meters per year, where ice cliffs collapse to form bergy bits and brash ice. The largest icebergs may be more than 100 km in size, but will calve only at intervals of 50 years or more. Thus when charting the size and behavior of an ice shelf it can be difficult to separate and identify stable changes from unstable ones.

Mass balance calculations show that calving of icebergs is the largest factor in the attrition of the Antarctic Ice Sheet. Estimates based on both ship and satellite data suggest that iceberg calving is only slightly less than the total annual accumulation. Ice shelf melting at the base is the other principal element in the attrition of the ice sheet, with approximately 80% of all ice shelf melting occurring at distances greater than 100 km from the ice front. Although most of the mass lost in Antarctica is through ice shelves, there is still uncertainty about the role ice shelves play in regulating flow off the land, which is the important component for sea-level changes. Because ice shelves are already floating, they do not affect sea level when they breakup.

Ice shelves are sensitive indicators of climate change. Those around the Antarctic Peninsula have shown a pattern of gradual retreat since about 1950, associated with a regional atmospheric warming and increased summer surface melt. The effects of climate change on the mass balance of the Antarctic Ice Sheet and hence global sea level are unclear.

There have been no noticeable changes in the inland ice sheet from the collapse of the Antarctic Peninsula ice shelves. This is because most of the margins with the inland ice sheet form a sharp transition zone, making the ice sheet dynamics independent of the state of the ice shelf. This is not the case with a grounding line on an ice stream where there is a smooth transition zone, but generally the importance of the local stresses there diminishes rapidly upstream.

An important role of ice shelves in the climate system is that subice-shelf freezing and melting processes influence the formation of Antarctic Bottom Water, which in part helps to drive the oceanic thermohaline circulation.

How have Ice Shelves Changed in the Past?

As the polar ice sheets have waxed and waned during ice age cycles, ice shelves have also grown and retreated. During the last 20–25 million years, the Antarctic Ice Sheet has probably been grounded out to the edge of the continental shelf many times. It is unlikely that any substantial ice shelves could have existed then, because of unsuitable coastline geometry and lack of pinning points, although the extent of the sea ice may have been double the present day area. At the last glacial maximum, about 20 000 years ago, grounded ice extended out to the edge of the continental shelf in places, for example in Prydz Bay, but not in others such as the Ross Sea. When the ice sheets began to retreat, some of the large ice shelves seen today probably formed by the thinning and eventual flotation of the formerly grounded ice sheets.

Large ice shelves also existed in the Arctic during the Pleistocene. Abundant geologic evidence shows that marine Northern Hemisphere ice sheets disappeared catastrophically during the climatic transition to the current interglacial (warm period). Only a few small ice shelves and tidewater glaciers exist there today, in places like Greenland, Svalbard, Ellesmere Island and Alaska.

There have been periodic changes in Antarctic ice-shelf grounding lines and ice fronts during the Holocene. The main deglaciation on the western side of the Antarctic Peninsula which started more than 11 000 years ago was initially very rapid across a wide continental shelf. By 6000 years ago, the ice sheet had cleared the inner shelf, whereas in the Ross Sea retreat of ice shelves ceased about the same time. Retreat after the last glacial maximum was followed by a readvance of the grounding line during the climate warming between 7000 and 4000 years ago. Open marine deposition on the continental shelf is restricted to the last 4000 years or so. Short-term cycles (every few hundred years) and longer-term events (approximately 2500 year cycles) have been detected in marine sediment cores that are likely related to global climate fluctuations.

More recently, ice-shelf disintegration around the Antarctic Peninsula has been associated with a regional atmospheric warming which has been occurring since about 1950. Ice front retreat of marginal ice shelves elsewhere in Antarctica (e.g., Cook Ice Shelf, West Ice Shelf) is probably also related to atmospheric warming. There is little sign of any significant impact on the grounded ice.

Where are Ice Shelves Disintegrating Now? Two Case Studies

The detailed history of the breakup of two ice shelves, Wordie and Larsen, illustrate some of the critical features of ice-shelf decay.

Wordie Ice Shelf

Retreat of Wordie Ice Shelf on the west coast of the Antarctic Peninsula has been documented using high resolution visible satellite images taken since 1974 and the position of the ice front in 1966 mapped from aerial photography. First seen in 1936, the ice front has fluctuated in position but with a sustained retreat starting around 1966 (**Figure 3**). The ice shelf area has decreased from about 2000 km² in 1966 to about 700 km² in 1989. However, defining the position of the ice front can be very uncertain, with large blocks calving off to form icebergs being difficult to classify as being either attached to, or separated from, the ice shelf.

Ice front retreat until about 1979 occurred mainly by transverse rifting along the ice front, creating icebergs up to 10 km by 1 km. The western area was the first to be lost, between 1966 and 1974. Results from airborne radio-echo sounding between 1966 and 1970 suggested that the ice shelf could be divided into a crevassed eastern part and a rifted western part where brine could well-up and infiltrate the ice at sea level. By 1974 there were many transverse rifts south of Napier Ice Rise and by 1979 longitudinal rifting was predominant. Many longitudinal rifts had formed upstream of several ice rises by 1979, some following preexisting flowline features. A critical factor in the break-up was the decoupling of Buffer Ice Rise; by 1986 ice was streaming past it apparently unhindered, in contrast to the compressive upstream folding seen in earlier images. Between 1988 and 1989 the central part of the ice shelf, consisting of broken ice in the lee of Mount Balfour, was lost, exposing the coastline and effectively dividing the ice shelf in two. By 1989, longitudinal rifting has penetrated to the grounding line north of Mount Balfour. Further north, a growing shear zone marked the boundary between fast-flowing ice from the north Forster Ice Piedmont and slower moving ice from Hariot Glacier. There has been no discernible change in the grounding line position.

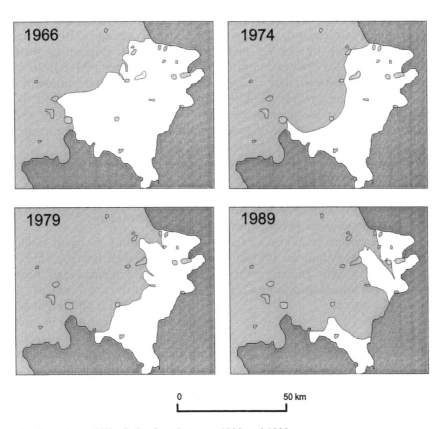

0   50 km

Figure 3 Cartoon showing retreat of Wordie ice front between 1966 and 1989.

The major ice rises have played several roles in controlling ice-shelf behavior. When embedded in the ice shelf, they created broken wakes downstream, and zones of compression upstream which helped to stabilize the ice shelf. During ice front retreat, they temporarily pinned the local ice front position and also acted as nucleating points for rifting which quickly stretched upstream, suggesting that a critical fracture criterion had been exceeded. At this stage, an ice rise, instead of protecting the ice shelf against decay, aided its destruction by acting as an indenting wedge.

Breakup was probably triggered by a climatic warming which increased ablation and the amount of melt water. Laboratory experiments show that the fracture toughness of ice is reduced at higher temperatures and possibly by the presence of water. Instead of refreezing in the upper layers of firn, free water could percolate down into crevasses and, by increasing the pressure at the bottom, allow them to grow into rifts or possibly to join up with basal crevasses. Processes like these would increase the production rate of blocks above that required for a 'steady state' ice front position. The blocks will drift away as icebergs if conditions, such as bay geometry and lack of sea ice, are favorable. Thus, ice front retreat would be, *inter alia*, a sensitive function of mean annual air temperature.

Some ice shelves, such as Brunt, are formed from blocks that break off at the coast line and, unable to float away, are 'glued' together by sea ice and snowfall. These heterogeneous ice shelves contrast with those, such as Ronne, where glaciers or ice streams flow unbroken across the grounding line to form a more homogeneous type of ice shelf. Before 1989, Wordie Ice Shelf consisted of a mixture of both types, the main tongues being derived from glacier inputs, while the central portion was formed from blocks breaking off at the grounding line and at the sides of the main tongues. The western rifted area that broke away between 1966 and 1974 was described in early 1967 as 'snowed-under icebergs' and it was the heterogeneous central part that broke back to the coastline around Mount Balfour in 1988/89. Increased ablation would not only enhance rifting along lines of weakness but would also loosen the 'glue' that held the blocks together.

Larsen Ice Shelf

The most northerly ice shelf in the Antarctic, Larsen Ice Shelf, extends in a ribbon down the east coast of the Antarctic Peninsula from James Ross Island to the Ronne Ice Shelf. It consists of several distinct ice shelves, separated by headlands. The ice shelf in Prince Gustav Channel, between James Ross Island and the mainland, separated from the main part of Larsen Ice Shelf in the late 1950s and finally disappeared by 1995.

The section between Sobral Peninsula and Robertson Island, known as Larsen A, underwent a catastrophic collapse at the end of January 1995 when it disintegrated into a tongue of small icebergs, bergy bits and brash ice. Previously the ice front had been retreating for a number of years, but in a more controlled fashion by iceberg calving. Before the final breakup, the surface that had once been flat and smooth had become undulating, suggesting that rifting completely through the ice had occurred. Collapse during a period of intense north-westerly winds and high temperatures was probably aided by a lack of sea ice, allowing ocean swell to penetrate and add its power to increasing the disintegration processes. The speed of the collapse and the small size of the fragments of ice (the largest icebergs were less than 1 km in size) implicate fracture as the dominant process in the disintegration (**Figure 4**).

The ice front of the ice shelf known as Larsen B, between Robertson Island and Jason Peninsula, steadily advanced for about 6 km from 1975 until 1992. Small icebergs broke away from the heavily rifted zone south of Robertson Island after July 1992 and a major rift about 25 km in length had opened up by then. This rift formed the calving front when an iceberg covering 1720 km^2 and smaller pieces corresponding to a former ice shelf area of 550 km^2 broke away between 25 and 30 January 1995, coincident with the disintegration of Larsen A. The ice front has retreated continuously by a few kilometers per year since then (to 1999) and the ice shelf is considered to be under threat of disappearing completely.

Why do Ice Shelves Break-up?

Calving and Fracture

Calving is one of the most obvious processes involved in ice shelf breakup. Although widespread, occurring on grounded and floating glaciers, including ice shelves, as well as glaciers ending in freshwater lakes, it is not well understood. The basic physics, tensile propagation of fractures, may be the same in all cases but the predictability of calving may be quite different, depending on the stress field, the basal boundary conditions, the amount of surface water, etc. Fracture of ice has been studied mainly in laboratories and applying these results to the conditions experienced in naturally occurring ice masses requires extrapolations which may not be valid.

Figure 4 Photo of the Larsen breakup. Picture taken approximately 10 km north of Robertson Island, looking west north-west. In the far background the mountains of the Antarctic Peninsula can be seen.

Questions such as how is crevasse propagation influenced by the ice shelf geometry and by the physical properties of the ice such as inhomogeneity, crystal anisotropy, temperature and presence of water, need to be answered before realistic models of the calving process can be developed. The engineering concepts of fracture mechanics and fracture toughness promise a way forward.

Fracture of ice is a critical process in ice shelf dynamics. Mathematical models of ice shelf behavior based on continuum mechanics, which treat ice as a nonlinear viscous fluid will have to incorporate fracture mechanics to describe iceberg calving and how ice rises can initiate fracture both upstream and downstream. High stresses generated at shear margins around ice rises, or at the 'corner points' of the ice shelf where the two ends of the ice front meet land, can exceed a critical stress and initiate crevassing. These crevasses will propagate under a favorable stress regime, and if there is sufficient surface water may rift through the complete thickness. Transverse crevasses parallel to the ice front act as sites for the initiation of rifting and form lines where calving may eventually occur.

Calving is a very efficient method of getting rid of ice from an ice shelf – once an iceberg has formed, it can drift away within a few days. Sometimes, however, an iceberg will ground on a sea bed shoal, perhaps for several years. Once an Antarctic iceberg has drifted north of the Antarctic Convergence, it usually breaks-up very quickly in the warmer waters.

Climate Warming

Circumstantial evidence links the retreat of ice shelves around the Antarctic Peninsula to a warming trend in atmospheric temperatures. Observations at meteorological stations in the region show a rise of about 2.5°C in 50 years. Ice shelves appear to exist up to a climatic limit, taken to be the mean annual −5°C isotherm, which represents the thermal limit of ice-shelf viability. The steady southward migration of this isotherm has coincided with the pattern of ice-shelf disintegration.

There are too few sea temperature measurements to show if the observed atmospheric warming has resulted in warmer waters, and thus increased basal melting. Although basal melting may have increased, the indications are that it is surface processes that have played the dominant role in causing breakup. The mechanism for connecting climate warming with ice-shelf retreat is not fully understood, but increased summer melting producing substantial amounts of water obviously plays a part in enhancing fracture processes, leading to calving.

The retreat of ice shelves around the Antarctic Peninsula that has occurred in the last half of the twentieth century has raised questions about whether or not this reflects global warming or whether it is only a regional phenomenon. The warming trend seems to be localized to the Antarctic Peninsula region and there is no significant correlation with temperature changes in the rest of Antarctica.

Stress Patterns

Numerical models of ice shelves have been used to examine their behavior and stability criteria. The strain-rate field can be specified by the principal values ($\dot{\varepsilon}_1$ and $\dot{\varepsilon}_2$ where $\dot{\varepsilon}_1 > \dot{\varepsilon}_2$) and the direction of the principal axes. A simple representation is given by the trajectories, which are a set of orthogonal curves whose directions at any point are the directions of the principal axes. Analyses of the strain-rate trajectories for Filchner Ronne Ice Shelf and for different ice shelf configurations of Larsen Ice Shelf show characteristic patterns of a 'compressive arch' and of isotropic points (**Figure 5**). The 'compressive arch' is seen in the pattern formed by the smallest principal component ($\dot{\varepsilon}_2$) of the strain-rate trajectories. Seaward of the arch both principal strain-rate components are extensive, whereas inland of the arch the $\dot{\varepsilon}_2$ component is compressive. The arch extends from the two ends of the ice front across the whole width of the ice shelf. It is a generic feature of ice shelves studied so far and structurally stable to small perturbations. It is probably related to the geometry of the ice shelf bay. A critical arch, consisting entirely of compressive trajectories, appears to correspond to a criterion for stability. If the ice front breaks back through the arch then an irreversible retreat occurs, possibly catastrophically, to another stable configuration. The exact location of the critical arch cannot yet be determined *a priori*, but it is probably close to the compressive arch delineated by the transition from extension to compression for $\dot{\varepsilon}_2$.

Another pattern seen in the strain-rate trajectories is that of isotropic points. They are indicators of generic features in the surface flow field which are stable to small perturbations in the flow. This means that their existence should not be sensitive either to reasonable errors in data used in the model or to simplifications in the model itself. They act as (permanent) markers in a complicated (tensor) strain-rate field and are often located close to points where the two principal strain-rates are equal or where the velocity field is stationary (usually either a maximum or a saddle point). Isotropic points are classified by a number of properties, but only two categories can be reliably distinguished observationally, by the way the trajectory of either of the principal strain-rates varies

around a path enclosing the isotropic point. In one case the trajectory varies in a prograde sense and the isotropic point is called a 'monstar', whereas in the other case the trajectory varies in a retrograde sense and the isotropic point is called a 'star' (**Figure 5**).

The two categories of isotropic points can be identified in the model strain-rate trajectories, 'stars' occurring near input glaciers and 'monstars' near ice fronts if they are in a stable configuration. Icebergs calving off Filchner Ice Shelf in 1986 moved the position of the 'monstar' up to the newly formed ice front, whereas on Ronne Ice Shelf the 'monstar' is about 50 km inland of the ice front. This suggests that the existence of a monstar can be used as a 'weak' indicator of a stable ice front. Calving can remove a monstar even if the subsequent ice front position is not necessarily an unstable one and may readvance.

How Vulnerable are the Large Ice Shelves and how Stable are Grounding Lines?

The two largest ice shelves (Ross and Filchner–Ronne) are too far south to be attacked by the atmospheric warming that is predicted for the twenty-first century. Basal melting will increase if warmer water intrudes onto the continental shelf, but the effects of a warming trend could be counterintuitive. If the warming was sufficient to reduce the rate of sea ice formation in the Weddell Sea, then the production of High Salinity Shelf Water (HSSW) would reduce as well. This would affect the subice shelf circulation, replacing the relatively warm HSSW under the Filchner–Ronne Ice Shelf with a colder Ice Shelf Water which would reduce the melting and thus thicken the ice shelf. Further climate warming would restore warmer waters and eventually thin the ice shelf by increasing the melting rates, possibly to values of around 15 m year^{-1} as seen under Pine Island Glacier.

The supplementary question is what effect, if any, would the collapse of the ice shelves have on the ice sheet, especially the West Antarctic Ice Sheet (WAIS). It has been a tenet of the latter part of the twentieth century that the WAIS, known as a marine ice sheet because much of its bed is below sea level, is potentially unstable and may undergo disintegration if the fringing ice shelves were to disappear. However, the theoretical foundations of this belief are shaky and more careful consideration of the relevant dynamics suggests that the WAIS is no more vulnerable than any other part of the ice sheet.

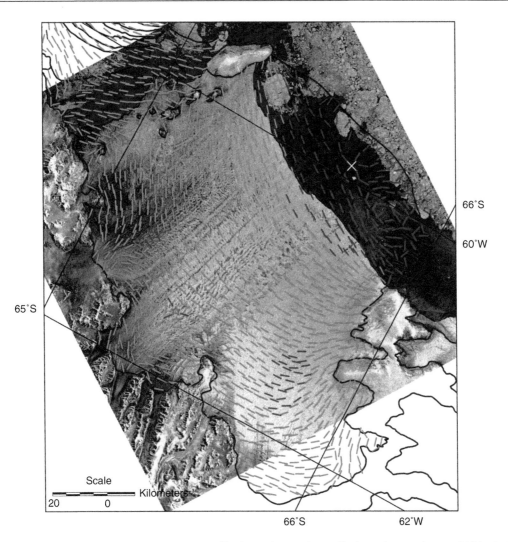

Figure 5 Modeled strain-rate trajectories on Larsen Ice Shelf superimposed on a Radarsat image taken on 21 March 1997. The trajectory of the smallest principal strain-rate ($\dot{\varepsilon}_2$) is negative (compressing flow, blue) over nearly all the Larsen Ice Shelf in its post-iceberg calving configuration (January 1995) but is positive (red) in the region where the iceberg calved. The pattern shows isotropic points ('stars,' green) near where glaciers enter the ice shelf and, for the 'iceberg' area a 'monstar' (yellow cross) near the ice front. 'Stars' are seen for all the different ice front configurations, indicating that no fundamental change is occurring near the grounding line due to retreat of the ice front. A 'monstar' is only seen for Larsen B before the iceberg calved in 1995. The disappearance of the isotropic point is perhaps a 'weak' indication that the iceberg calving event was greater than expected for normal calving from a stable ice shelf and suggests further retreat may be expected.

The problem lies in how to model the transition zone between ice sheet and ice shelf. If the transition is sharp, where the basal traction varies over lengths of the order of the ice thickness, then it can be considered as a passive boundary layer and does not affect the mechanics of the sheet or shelf to first order. Mass conservation must be respected but there is no need to impose further constraints. Requiring the ice sheet to have the same thickness as the ice shelf at the grounding line permits only two stable grounding line positions for a bed geometry which deepens inland. Depending on the depth of the bed below sea level, the ice sheet will either shrink in size until it disappears (or, if the bed shallows again, the

grounding line is fixed on a bed near sea level), or grow to the edge of the continental shelf. The lack of intermediate stable grounding line positions in this kind of model has been used to support the idea of instability of marine ice sheets. However, permitting a jump in ice thickness at the transition zone means the ice sheet system can be in neutral equilibrium, with an infinite number of steady-state profiles.

Another kind of transition zone is a smooth one, where the basal traction varies gradually, for example, along an ice stream. The equilibrium dynamics have not been worked out for a full three-dimensional flow, but it seems that the presence of ice streams destroys the neutral equilibrium and helps to stabilize marine ice sheets. This emphasizes the importance of understanding the dynamics of ice streams and their role in the marine ice sheet system.

Attempts to include moving grounding lines in whole ice sheet models suffer from incomplete specification of the problem. Assumptions built into the models predispose the results to be either too stable or too unstable. Thus there are no reliable models that can analyze the glaciological history of the Antarctic Ice Sheet. Predictive models rely on linearizations to provide acceptable accuracy for the near future but become progressively less accurate the longer the timescale.

Further Reading

Doake CSM and Vaughan DG (1991) Rapid disintegration of Wordie Ice Shelf in response to atmospheric warming. *Nature* 350: 328–330.

Doake CSM, Corr HFJ, Rott H, Skvarca P, and Young N (1998) Breakup and conditions for stability of the northern Larsen Ice Shelf, Antarctica. *Nature* 391: 778–780.

Hindmarsh RCA (1993) Qualitative dynamics of marine ice sheets. In: Peltier WR (ed.) *Ice in the Climate System*, NATO ASI Series, vol. 12, pp. 67–99. Berlin: Springer-Verlag.

Kellogg TB and Kellogg DE (1987) Recent glacial history and rapid ice stream retreat in the Amundsen Sea. *Journal of Geophysical Research* 92: 8859–8864.

Robin G and de Q (1979) Formation, flow and disintegration of ice shelves. *Journal of Glaciology* 24(90): 259–271.

Scambos TA, Hulbe C, Fahnestock M, and Bohlander J (2000) The link between climate warming and break-up of ice shelves in the Antarctic Peninsula. *Journal of Glaciology* 46(154): 516–530.

van der Veen CJ (ed.) (1997) *Calving Glaciers: Report of a Workshop 28 February–2 March 1997*. BPRC Report No. 15. Columbus, Ohio: Byrd Polar Research Center, The Ohio State University.

van der Veen CJ (1999) *Fundamentals of Glacier Dynamics*. Rotterdam: A.A. Balkema.

Vaughan DG and Doake CSM (1996) Recent atmospheric warming and retreat of ice shelves on the Antarctic Peninsula. *Nature* 379: 328–331.

SUB-SEA PERMAFROST

T. E. Osterkamp, University of Alaska,
Alaska, AK, USA

Introduction

Sub-sea permafrost, alternatively known as submarine permafrost and offshore permafrost, is defined as permafrost occurring beneath the seabed. It exists in continental shelves in the polar regions (**Figure 1**). When sea levels are low, permafrost aggrades in the exposed shelves under cold subaerial conditions. When sea levels are high, permafrost degrades in the submerged shelves under relatively warm and salty boundary conditions. Sub-sea permafrost differs from other permafrost in that it is relic, warm, and generally degrading. Methods used to investigate it include probing, drilling, sampling, drill hole log analyses, temperature and salt measurements, geological and geophysical methods (primarily seismic and electrical), and geological and geophysical models. Field studies are conducted from boats or, when the ocean surface is frozen, from the ice cover. The focus of this article is to review our understanding of sub-sea permafrost, of processes ocurring within it, and of its occurrence, distribution, and characteristics.

Sub-sea permafrost derives its economic importance from current interests in the development of offshore petroleum and other natural resources in the continental shelves of polar regions. The presence and characteristics of sub-sea permafrost must be considered in the design, construction, and operation of coastal facilities, structures founded on the seabed, artificial islands, sub-sea pipelines, and wells drilled for exploration and production.

Scientific problems related to sub-sea permafrost include the need to understand the factors that control its occurrence and distribution, properties of warm permafrost containing salt, and movement of heat and salt in degrading permafrost. Gas hydrates that can occur within and under the permafrost are a potential abundant source of energy. As the sub-sea permafrost warms and thaws, the hydrates destabilize, producing gases that may be a significant source of global carbon.

Nomenclature

'Permafrost' is ground that remains below 0°C for at least two years. It may or may not contain ice. 'Ice-bearing' describes permafrost or seasonally frozen soil that contains ice. 'Ice-bonded' describes ice-bearing material in which the soil particles are mechanically cemented by ice. Ice-bearing and ice-bonded material may contain unfrozen pore fluid in addition to the ice. 'Frozen' implies ice-bearing or ice-bonded or both, and 'thawed' implies non-ice-bearing. The 'active layer' is the surface layer of sediments subject to annual freezing and thawing in areas underlain by permafrost. Where seabed temperatures are negative, a thawed layer ('talik') exists near the seabed. This talik is permafrost but does not contain ice because soil particle effects, pressure, and the presence of salts in the pore fluid can depress the freezing point 2°C or more. The boundary between a thawed region and ice-bearing permafrost is a phase

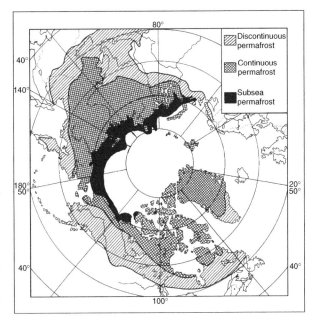

Figure 1 Map showing the approximate distribution of sub-sea permafrost in the continental shelves of the Arctic Ocean. The scarcity of direct data (probing, drilling, sampling, temperature measurements) makes the map highly speculative, with most of the distribution inferred from indirect measurements, primarily water temperature, salinity, and depth (100 m depth contour). Sub-sea permafrost also exists near the eroding coasts of arctic islands, mainlands, and where seabed temperatures remain negative. (Adapted from Pewe TL (1983). *Arctic and Alpine Research* 15(2):145–156 with the permission of the Regents of the University of Coloroado.)

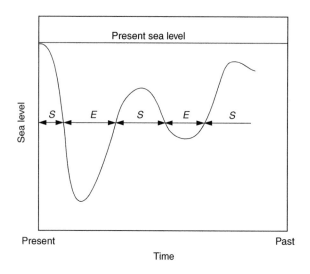

Figure 2 Schematic sea level curve during the last glaciation. The history of emergence (E) and submergence (S) can be combined with paleotemperature data on sub-aerial and sub-sea conditions to construct an approximate thermal boundary condition for sub-sea permafrost at any water depth.

boundary. 'Ice-rich' permafrost contains ice in excess of the soil pore spaces and is subject to settling on thawing.

Formation and Thawing

Repeated glaciations over the last million years or so have caused sea level changes of 100 m or more (**Figure 2**). When sea levels were low, the shallow continental shelves in polar regions that were not covered by ice sheets were exposed to low mean annual air temperatures (typically −10 to −25°C). Permafrost aggraded in these shelves from the exposed ground surface downwards. A simple conduction model yields the approximate depth (X) to the bottom of ice-bonded permafrost at time t, (eqn [1]).

$$X(t) = \sqrt{\frac{2K(T_e - T_g)t}{h}} \qquad [1]$$

K is the thermal conductivity of the ice-bonded permafrost, T_e is the phase boundary temperature at the bottom of the ice-bonded permafrost, T_g is the long-term mean ground surface temperature during emergence, and h is the volumetric latent heat of the sediments, which depends on the ice content. In eqn [1], K, h, and T_e depend on sediment properties. A rough estimate of T_g can be obtained from information on paleoclimate and an approximate value for t can be obtained from the sea level history (**Figure 2**). Eqn [1] overestimates X because it neglects geothermal heat flow except when a layer of ice-bearing permafrost from the previous transgression remains at depth.

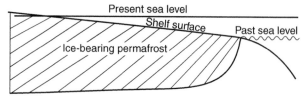

Figure 3 A schematic illustration of ice-bearing sub-sea permafrost in a continental shelf near the time of minimum sea level. Typical thicknesses at the position of the present shoreline would have been about 400–1000 m with shelf widths that are now typically 100–600 km.

Timescales for permafrost growth are such that hundreds of meters of permafrost could have aggraded in the shelves while they were emergent (**Figure 3**).

Cold onshore permafrost, upon submergence during a transgression, absorbs heat from the seabed above and from the geothermal heat flux rising from below. It gradually warms (**Figures 4** and **5**), becoming nearly isothermal over timescales up to a few millennia (**Figure 4**, time t_3). Substantially longer times are required when unfrozen pore fluids are present in equilibrium with ice because some ice must thaw throughout the permafrost thickness for it to warm.

A thawed layer develops below the seabed and thawing can proceed from the seabed downward, even in the presence of negative mean seabed temperatures, by the influx of salt and heat associated with the new boundary conditions. Ignoring seabed erosion and sedimentation processes, the thawing rate at the top of ice-bonded permafrost during submergence is given by eqn [2].

$$\dot{X}_{top} = \frac{J_t}{h} - \frac{J_f}{h} \qquad [2]$$

J_t is the heat flux into the phase boundary from above and J_t is the heat flux from the phase boundary into the ice-bonded permafrost below. J_t depends on the difference between the long-term mean temperature at the seabed, T_s, and phase boundary temperature, T_p, at the top of the ice-bonded permafrost. For $J_t = J_t$, the phase boundary is stable. For $J_t < J_t$, refreezing of the thawed layer can occur from the phase boundary upward. For thawing to occur, T_s must be sufficiently warmer than T_p to make $J_t > J_t$. T_s is determined by oceanographic conditions (currents, ice cover, water salinity, bathymetry, and presence of nearby rivers). T_p is determined by hydrostatic pressure, soil particle effects, and salt concentration at the phase boundary (the combined effect of *in situ* pore fluid salinity, salt transport from the seabed through the thawed layer, and changes in concentration as a result of freezing or thawing).

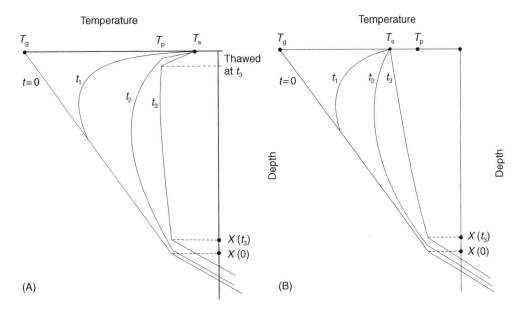

Figure 4 Schematic sub-sea permafrost temperature profiles showing the thermal evolution at successive times (t_1, t_2, t_3) after submergence when thawing occurs at the seabed (A) and when it does not (B). T_g and T_s are the long-term mean surface temperatures of the ground during emergence and of the seabed after submergence. T_p is the phase boundary temperature at the top of the ice-bonded permafrost. $X(0)$ and $X(t_3)$ are the depths to the bottom of ice-bounded permafrost at times $t = 0$ and t_3. (Adapted with permission from Lachenbruch AH and Marshall BV (1977) Open File Report 77-395. US Geological Survey, Menlo Park, CA.)

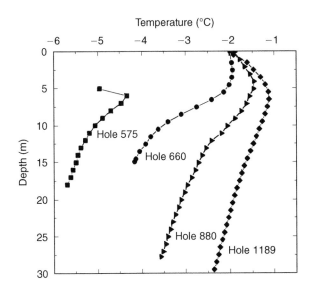

Figure 5 Temperature profiles obtained during the month of May in sub-sea permafrost near Barrow, Alaska, showing the thermal evolution with distance (equivalently time) offshore. Hole designation is the distance (m) offshore and the shoreline erosion rate is about 2.4 m y^{-1}. Sea ice freezes to the seabed within 600 m of shore. (Adapted from Osterkamp TE and Harrison WD (1985) Report UAGR-301. Fairbanks, AK: Geophysical Institute, University of Alaska.)

Nearshore at Prudhoe Bay, $\dot{X}_{top}$ varies typically from centimeters to tens of centimeters per year while farther offshore it appears to be on the order of millimeters per year. The thickness of the thawed layer at the seabed is typically 10 m to 100 m, although values of less than a meter have been observed. At some sites, the thawed layer is thicker in shallow water and thinner in deeper water.

Sub-sea permafrost also thaws from the bottom by geothermal heat flow once the thermal disturbance of the transgression penetrates there. The approximate thawing rate at the bottom of the ice-bonded permafrost is given by eqn [3].

$$\dot{X}_{bot} \cong \frac{J_g}{h} - \frac{J_f'}{h} \qquad [3]$$

J_g is the geothermal heat flow entering the phase boundary from below and J_f' is the heat flow from the phase boundary into the ice-bonded permafrost above. J_f' becomes small within a few millennia except when the permafrost contains unfrozen pore fluids. $\dot{X}_{bot}$ is typically on the order of centimeters per year. Timescales for thawing at the permafrost table and base are such that several tens of thousands of years may be required to completely thaw a few hundred meters of sub-sea permafrost.

Modeling results and field data indicate that impermeable sediments near the seabed, low T_s, high ice contents, and low J_g favor the survival of ice-bearing sub-sea permafrost during a transgression. Where conditions are favorable, substantial thicknesses of ice-bearing sub-sea permafrost may have survived previous transgressions.

Characteristics

The chemical composition of sediment pore fluids is similar to that of sea water, although there are detectable differences. Salt concentration profiles in thawed coarse-grained sediments at Prudhoe Bay (**Figure 6**) appear to be controlled by processes occurring during the initial phases of submergence. There is evidence for highly saline layers within ice-bonded permafrost near the base of gravels overlying a fine-grained sequence both onshore and offshore. In the Mackenzie Delta region, fluvial sand units deposited during regressions have low salt concentrations (**Figure 7**) except when thawed or when lying under saline sub-sea mud. Fine-grained mud sequences from transgressions have higher salt concentrations. Salts increase the amount of unfrozen pore fluids and decrease the phase equilibrium temperature, ice content, and ice bonding. Thus, the sediment layering observed in the Mackenzie Delta region can lead to unbonded material (clay) between layers of bonded material (fluvial sand).

Thawed sub-sea permafrost is often separated from ice-bonded permafrost by a transition layer of ice-bearing permafrost. The thickness of the ice-bearing layer can be small, leading to a relatively sharp (centimeters scale) phase boundary, or large, leading to a diffuse boundary (meters scale). In general, it appears that coarse-grained soils and low salinities produce a sharper phase boundary and fine-grained soils and higher salinities produce a more diffuse phase boundary.

Sub-sea permafrost consists of a mixture of sediments, ice, and unfrozen pore fluids. Its physical and mechanical properties are determined by the individual properties and relative proportions. Since ice and unfrozen pore fluid are strongly temperature dependent, so also are most of the physical and mechanical properties.

Ice-rich sub-sea permafrost has been found in the Alaskan and Canadian portions of the Beaufort Sea and in the Russian shelf. Thawing of this permafrost can result in differential settlement of the seafloor that poses serious problems for development.

Processes

Submergence

Onshore permafrost becomes sub-sea permafrost upon submergence, and details of this process play a major role in determining its future evolution. The rate at which the sea transgresses over land is determined by rising sea levels, shelf topography, tectonic setting, and the processes of shoreline erosion,

thaw settlement, thaw strain of the permafrost, seabed erosion, and sedimentation. Sea levels on the polar continental shelves have increased more than 100 m in the last 20 000 years or so. With shelf widths of 100–600 km, the average shoreline retreat rates would have been about 5 to 30 m y^{-1}, although maximum rates could have been much larger. These average rates are comparable to areas with very rapid shoreline retreat rates observed today on the Siberian and North American shelves. Typical values are 1–6 m y^{-1}.

It is convenient to think of the transition from sub-aerial to sub-sea conditions as occurring in five regions (**Figure 8**) with each region representing different thermal and chemical surface boundary conditions. These boundary conditions are successively applied to the underlying sub-sea permafrost during a transgression or regression. Region 1 is the onshore permafrost that forms the initial condition for sub-sea permafrost. Permafrost surface temperatures range down to about $-15\,°C$ under current sub-aerial conditions and may have been 8–10°C colder during glacial times. Ground water is generally fresh, although salty lithological units may exist within the permafrost as noted above.

Region 2 is the beach, where waves, high tides, and resulting vertical and lateral infiltration of sea water produce significant salt concentrations in the active layer and near-surface permafrost. The active layer and temperature regime on the beach differ from those on land. Coastal banks and bluffs are a trap for wind-blown snow that often accumulates in insulating drifts over the beach and adjacent ice cover.

Region 3 is the area where ice freezes to the seabed seasonally, generally where the water depth is less than about 1.5–2 m. This setting creates unique thermal boundary conditions because, when the ice freezes into the seabed, the seabed becomes conductively coupled to the atmosphere and thus very cold. During summer, the seabed is covered with shallow, relatively warm sea water. Salt concentrations at the seabed are high during winter because of salt rejection from the growing sea ice and restricted circulation under the ice, which eventually freezes into the seabed. These conditions create highly saline brines that infiltrate the sediments at the seabed.

Region 4 includes the areas where restricted under-ice circulation causes higher-than-normal sea water salinities and lower temperatures over the sediments. The ice does not freeze to the seabed or only freezes to it sporadically. The existence of this region depends on the ice thickness, on water depth, and on flushing processes under the ice. Strong currents or steep bottom slopes may reduce its extent or eliminate it.

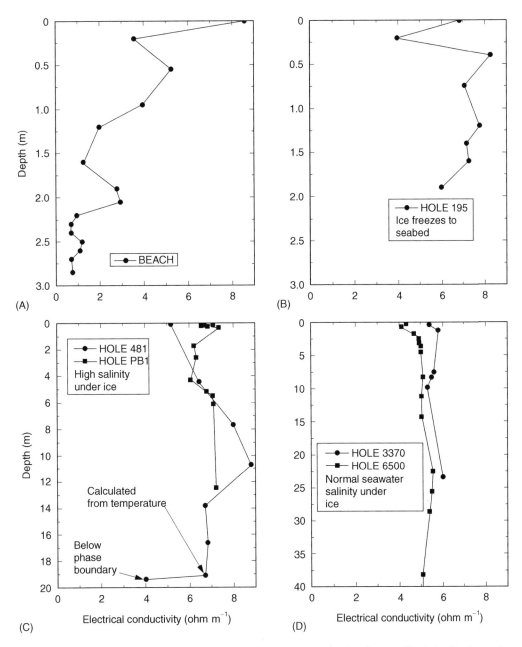

Figure 6 Electrical conductivity profiles in thawed, relatively uniform coarse-grained sediments. The holes lie along a line extending offshore near the West Dock at Prudhoe Bay, Alaska except for PB1, which is in the central portion of Prudhoe Bay. Hole designation is the distance (m) offshore. On the beach (A), concentrations decrease by a factor of 5 at a few meters depth. There are large variations with depth and concentrations may be double that of normal seawater where ice freezes to the seabed (B) and where there is restricted circulation under the ice (C). Farther offshore (D), the profiles tend to be relatively constant with depth with concentrations about the same or slightly greater than the overlying seawater. (Adapted from Iskandar IK *et al.* (1978) *Proceedings, Third International Conference on Permafrost*, Edmonton, Alberta, Canada, pp. 92–98. Ottawa, Ontario: National Research Council.)

The setting for region 5 consists of normal sea water over the seabed throughout the year. This results in relatively constant chemical and thermal boundary conditions.

There is a seasonal active layer at the seabed that freezes and thaws annually in both regions 3 and 4. The active layer begins to freeze simultaneously with the formation of sea ice in shallow water. Brine drainage from the growing sea ice increases the water salinity and decreases the temperature of the water at the seabed because of the requirement for phase equilibrium. This causes partial freezing of the less saline pore fluids in the sediments. Thus, it is not necessary for the ice to contact the seabed for the seabed to freeze. Seasonal changes in the pore fluid salinity show that the partially frozen active layer

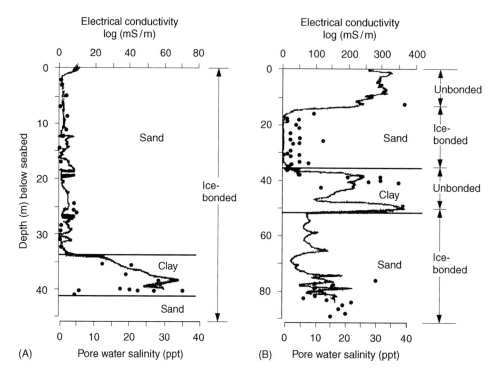

Figure 7 Onshore (A) and offshore (B) (water depth 10 m) electrical conductivity log (lines) and salinity (dots) profiles in the Mackenzie Delta region. The onshore sand–clay–sand sequence can be traced to the offshore site. Sand units appear to have been deposited under sub-aerial conditions during regressions and the clay unit under marine conditions during a transgression. At the offshore site, the upper sand unit is thawed to the 11 m depth. (Adapted with permission from Dallimore SR and Taylor AE (1994). *Proceedings, Sixth International Conference on Permafrost*, Beijing, vol. 1, pp.125–130. Wushan, Guangzhou, China: South China University Press.)

redistributes salts during freezing and thawing, is infiltrated by the concentrated brines, and influences the timing of brine drainage to lower depths in the sediments. These brines, derived from the growth of sea ice, provide a portion or all of the salts required for thawing the underlying sub-sea permafrost in the presence of negative sediment temperatures.

Depth to the ice-bonded permafrost increases slowly with distance offshore in region 3 to a few meters where the active layer no longer freezes to it (**Figure 8**) and the ice-bonded permafrost no longer couples conductively to the atmosphere. This allows the permafrost to thaw continuously throughout the year and depth to the ice-bonded permafrost increases rapidly with distance offshore (**Figure 8**).

The time an offshore site remains in regions 3 and 4 determines the number of years the seabed is subjected to freezing and thawing events. It is also the time required to make the transition from sub-aerial to relatively constant sub-sea boundary conditions. This time appears to be about 30 years near Lonely, Alaska (about 135 km southeast of Barrow) and about 500–1000 years near Prudhoe Bay, Alaska. **Figure 9** shows variations in the mean seabed temperatures with distance offshore near Prudhoe Bay where the shoreline retreat rate is about $1 \, \mathrm{m \, y}^{-1}$.

The above discussion of the physical setting does not incorporate the effects of geology, hydrology, tidal range, erosion and sedimentation processes, thaw settlement, and thaw strain. Regions 3 and 4 are extremely important in the evolution of sub-sea permafrost because the major portion of salt infiltration into the sediments occurs in these regions. The salt plays a strong role in determining T_p and, thus, whether or not thawing will occur.

Heat and Salt Transport

The transport of heat in sub-sea permafrost is thought to be primarily conductive because the observed temperature profiles are nearly linear below the depth of seasonal variations. However, even when heat transport is conductive, there is a coupling with salt transport processes because salt concentration controls T_p. Our lack of understanding of salt transport processes hampers the application of thermal models.

Thawing in the presence of negative seabed temperatures requires that T_p be significantly lower than T_s, so that generally salt must be present for thawing to occur. This salt must exist in the permafrost on submergence and/or be transported from the seabed

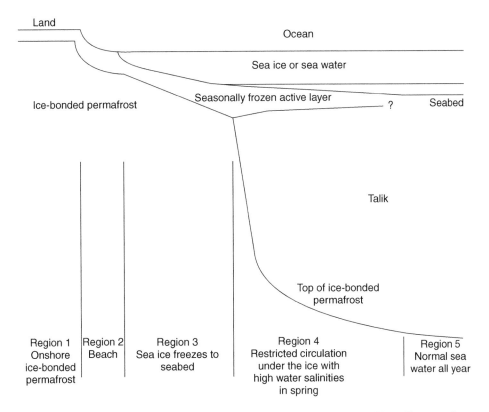

Figure 8 Schematic illustration of the transition of permafrost from sub-aerial to sub-sea conditions. There are five potential regions with differing thermal and chemical seabed boundary conditions. Hole 575 in **Figure 5** is in region 3 and the rest of the holes are in region 4. (Adapted from Osterkamp TE (1975) Report UAGR-234. Fairbanks, AK: Geophysical Institute, University of Alaska.)

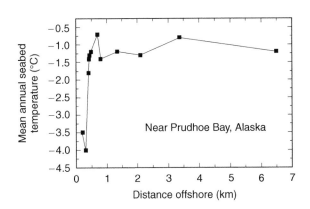

Figure 9 Variation of mean annual seabed temperatures with distance offshore. Along this same transect, about 6 km offshore from Reindeer Island in 17 m of water, the mean seabed temperature was near − 1.7°C. Data on mean annual seabed salinities in regions 3 and 4 do not appear to exist. (Adapted from Osterkamp TE and Harrison WD (1985) Report UAGR-301. Fairbanks, AK: Geophysical Institute, University of Alaska.)

to the phase boundary. The efficiency of salt transport through the thawed layer at the seabed appears to be sensitive to soil type. In clays, the salt transport process is thought to be diffusion, a slow process; and in coarse-grained sands and gravels, pore fluid convection, which (involving motion of fluid) can be rapid.

Diffusive transport of salt has been reported in dense overconsolidated clays north of Reindeer Island offshore from Prudhoe Bay. Evidence for convective transport of salt exists in the thawed coarse-grained sediments near Prudhoe Bay and in the layered sands in the Mackenzie Delta region. This includes rapid vertical mixing as indicated by large seasonal variations in salinity in the upper 2 m of sediments in regions 3 and 4 and by salt concentration profiles that are nearly constant with depth and decrease in value with distance offshore in region 5 (**Figures 5–7**). Measured pore fluid pressure profiles (**Figure 10**) indicate downward fluid motion. Laboratory measurements of downward brine drainage velocities in coarse-grained sediments indicate that these velocities may be on the order of $100 \, \mathrm{m \, y^{-1}}$.

The most likely salt transport mechanism in coarse-grained sediments appears to be gravity-driven convection as a result of highly saline and dense brines at the seabed in regions 3 and 4. These brines infiltrate the seabed, even when it is partially frozen, and move rapidly downward. The release of relatively fresh and buoyant water by thawing ice at the phase boundary may also contribute to pore fluid motion.

Occurrence and Distribution

The occurrence, characteristics, and distribution of sub-sea permafrost are strongly influenced by regional and local conditions and processes including the following.

1. Geological (heat flow, shelf topography, sediment or rock types, tectonic setting)
2. Meteorological (sub-aerial ground surface temperatures as determined by air temperatures, snow cover and vegetation)
3. Oceanographic (seabed temperatures and salinities as influenced by currents, ice conditions, water depths, rivers and polynyas; coastal erosion and sedimentation; tidal range)
4. Hydrological (presence of lakes, rivers and salinity of the ground water)
5. Cryological (thickness, temperature, ice content, physical and mechanical properties of the onshore permafrost; presence of sub-sea permafrost that has survived previous transgressions; presence of ice sheets on the shelves)

Lack of information on these conditions and processes over the long timescales required for permafrost to aggrade and degrade, and inadequacies in the theoretical models, make it difficult to formulate reliable predictions regarding sub-sea permafrost. Field studies are required, but field data are sparse and investigations are still producing surprising results indicating that our understanding of sub-sea permafrost is incomplete.

Pechora and Kara Seas

Ice-bonded sub-sea permafrost has been found in boreholes with the top typically up to tens of meters below the seabed. In one case, pure freshwater ice was found 0.3 m below the seabed, extending to at least 25 m. These discoveries have led to difficult design conditions for an undersea pipeline that will cross Baydaratskaya Bay transporting gas from the Yamal Peninsula fields to European markets.

Laptev Sea

Sea water bottom temperatures typically range from $-0.5°C$ to $-1.8°C$, with some values colder than $-2°C$. A 300–850 m thick seismic sequence has been found that does not correlate well with regional tectonic structure and is interpreted to be ice-bonded permafrost. The extent of ice-bonded permafrost appears to be continuous to the 70 m isobath and widespread discontinuous to the 100 m isobath. Depth to the ice-bonded permafrost ranges from 2 to 10 m in water depths from 45 m to the shelf edge. Deep taliks may exist inshore of the 20 m isobath. A shallow sediment core with ice-bonded material at its base was recovered from a water depth of 120 m. Bodies of ice-rich permafrost occur under shallow water at the locations of recently eroded islands and along retreating coastlines.

Bering Sea

Sub-sea permafrost is not present in the northern portion except possibly in near-shore areas or where shoreline retreat is rapid.

Chukchi Sea

Seabed temperatures are generally slightly negative and thermal gradients are negative, indicating ice-bearing permafrost at depth within 1 km of shore near Barrow, Rabbit Creek and Kotzebue.

Alaskan Beaufort Sea

To the east of Point Barrow, bottom waters are typically $-0.5°C$ to $-1.7°C$ away from shore, shoreline erosion rates are rapid ($1-10\ m\ y^{-1}$) and sediments are thick. Sub-sea permafrost appears to be thicker in the Prudhoe Bay region and thinner west of Harrison Bay to Point Barrow. Ice layers up

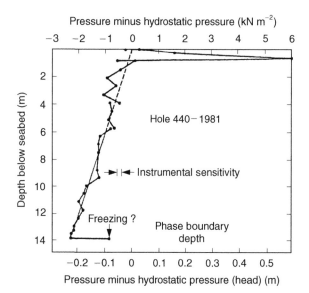

Figure 10 Measured *in situ* pore fluid pressure minus calculated hydrostatic pressure through the thawed layer in coarse-grained sediments at a hole 440 m offshore in May 1981 near Prudhoe Bay, Alaska. High pressure at the 0.54 m depth is probably related to seasonal freezing and the solid line is a least-squares fit to the data below 5.72 m. The negative pressure head gradient (-0.016) indicates a downward component of pore fluid velocity. (Adapted from Swift DW *et al.* (1983) *Proceedings, Fourth International Conference on Permafrost*, Fairbanks, Alaska, vol. 1, pp. 1221–1226. Washington DC: National Academy Press.)

to 0.6 m in thickness have been found off the Sagavanirktok River Delta.

Surface geophysical studies (seismic and electrical) have indicated the presence of layered ice-bonded sub-sea permafrost. A profile of sub-sea permafrost near Prudhoe Bay (**Figure 11**) shows substantial differences in depth to ice-bonded permafrost between coarse-grained sediments inshore of Reindeer Island and fine-grained sediments farther offshore. Offshore from Lonely, where surface sediments are fine-grained, ice-bearing permafrost exists within 6–8 m of the seabed out to 8 km offshore (water depth 8 m). Ice-bonded material is deeper (~15 m). In Elson Lagoon (near Barrow) where the sediments are fine-grained, a thawed layer at the seabed of generally increasing thickness can be traced offshore.

Mackenzie River Delta Region

The layered sediments found in this region are typically fluvial sand and sub-sea mud corresponding to regressive/transgressive cycles (see **Figure 12**). Mean seabed temperatures in the shallow coastal areas are generally positive as a result of warm river water, and negative farther offshore. The thickness of ice-bearing permafrost varies substantially as a result of a complex history of transgressions and regressions, discharge from the Mackenzie River, and possible effects of a late glacial ice cover. Ice-bearing permafrost in the eastern and central Beaufort Shelf exceeds 600 m. It is thin or absent beneath Mackenzie Bay and may be only a few hundred meters thick toward the Alaskan coast. The upper surface of

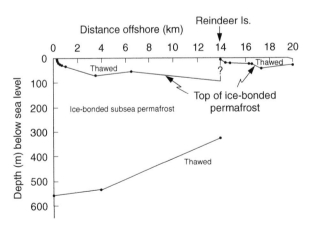

Figure 11 Sub-sea permafrost profile near Prudhoe Bay, Alaska, determined from drilling and well log data. The sediments are coarse-grained with deep thawing inshore of Reindeer Island and fine-grained with shallow thawing farther offshore. Maximum water depths are about 8 m inshore of Reindeer Island and 17 m about 6 km north. (Adapted from Osterkamp TE *et al.* (1985) *Cold Regions Science and Technology* 11: 99–105, 1985, with permission from Elsevier Science.)

ice-bearing permafrost is typically 5–100 m below the seabed and appears to be under control of seabed temperatures and stratigraphy.

The eastern Arctic, Arctic Archipelago and Hudson Bay regions were largely covered during the last glaciation by ice that would have inhibited permafrost growth. These regions are experiencing isostatic uplift with permafrost aggrading in emerging shorelines.

Antarctica

Negative sediment temperatures and positive temperature gradients to a depth of 56 m below the seabed exist in McMurdo Sound where water depth is 122 m and the mean seabed temperature is −1.9°C. This sub-sea permafrost did not appear to contain any ice.

Models

Modeling the occurrence, distribution and characteristics of sub-sea permafrost is a difficult task. Statistical, geological, analytical, and numerical models are available. Statistical models attempt to combine geological, oceanographic, and other information into algorithms that make predictions about sub-sea permafrost. These statistical models have not been very successful, although new GIS methods could potentially improve them.

Geological models consider how geological processes influence the formation and development of sub-sea permafrost. It is useful to consider these models since some sub-sea permafrost could potentially be a million years old. A geological model for the Mackenzie Delta region (**Figure 12**) has been developed that provides insight into the nature and complex layering of the sediments that comprise the sub-sea permafrost there.

Analytical models for investigating the thermal regime of sub-sea permafrost include both one- and two-dimensional models. All of the available analytical models have simplifying assumptions that limit their usefulness. These include assumptions of one-dimensional heat flow, stable shorelines or shorelines that undergo sudden and permanent shifts in position, constant air and seabed temperatures that neglect spatial and temporal variations over geological timescales, and constant thermal properties in layered sub-sea permafrost that is likely to contain unfrozen pore fluids. Neglect of topographical differences between the land and seabed, geothermal heat flow, phase change at the top and bottom of the sub-sea permafrost, and salt effects also limits their application. An analytical model exists that addresses the coupling between heat and

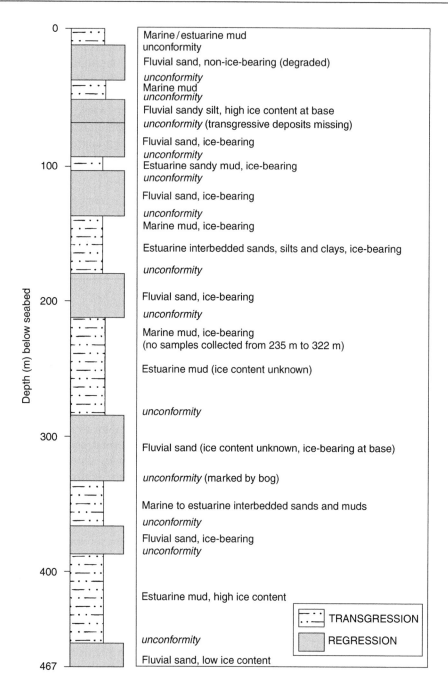

Figure 12 Canadian Beaufort Shelf stratigraphy in 32 m of water near the Mackenzie River Delta. Eight regressive/transgressive fluvial sand/marine mud cycles are shown. It is thought that, except for thawing near the seabed, the ice-bearing sequence has been preserved through time to the present. (Adapted with permission from Blasco S (1995) GSC Open File Report 3058. Ottawa, Canada: Geological Survey of Canada.)

salt transport but only for the case of diffusive transport with simplifying assumptions. Nevertheless, these analytical models appear to be applicable in certain special situations and have shaped much of the current thinking about sub-sea permafrost.

Two-dimensional numerical thermal models have addressed most of the concerns related to the assumptions in analytical models except for salt

transport. Models have been developed that address salt transport via the buoyancy of fresh water generated by thawing ice at the phase boundary. Models for the infiltration of dense sea water brines derived from the growth of sea ice into the sediments do not appear to exist.

Successful application of all models is limited because of the lack of information over geological timescales on initial conditions, boundary conditions, material properties, salt transport and the coupling of heat and salt transport processes. There is also a lack of areas with sufficient information and measurements to fully test model predictions.

Nomenclature

h	Volumetric latent heat of the sediments (1 to 2×10^8 J m^{-3})
J_g	Geothermal heat flow entering the bottom phase boundary from below
J_f	Heat flow from the bottom phase boundary into the ice-bonded permafrost above
J_t	Heat flux into the top phase boundary from above
J_f	Heat flux from the top phase boundary into the ice-bonded permafrost below
K	Thermal conductivity of the ice-bonded permafrost (1 to 5 W m^{-1} K^{-1})
t	Time
T_s	Long-term mean temperature at the seabed
T_p	Phase boundary temperature at the top of the ice-bonded permafrost (0 to $-2\,^\circ$C)
T_e	Phase boundary temperature at the bottom of the ice-bonded permafrost (0 to $-2\,^\circ$C)
T_g	Long-term mean ground surface temperature during emergence
$X(t)$	Depth to the bottom of ice-bonded permafrost at time, t
$\dot{X}_{top}$	Thawing rate at top of ice-bonded permafrost during submergence
$\dot{X}_{bot}$	Thawing rate at the bottom of ice-bonded permafrost during submergence

See also

Coastal Circulation Models. Glacial Crustal Rebound, Sea Levels, and Shorelines. River Inputs. Sea Level Change.

Further Reading

Dallimore SR and Taylor AE (1994) Permafrost conditions along an onshore–offshore transect of the Canadian Beaufort Shelf. *Proceedings, Sixth International Conference on Permafrost, Beijing, vol. 1, pp 125–130.* South China University Press: Wushan, Guangzhou, China.

Hunter JA, Judge AS, MacAuley HA, *et al.* (1976) *The Occurrence of Permafrost and Frozen Sub-sea Bottom Materials in the Southern Beaufort Sea. Beaufort Sea Project, Technical Report 22.* Ottawa: Geological Survey Canada.

Lachenbruch AH, Sass JH, Marshall BV, and Moses TH Jr (1982) Permafrost, heat flow, and the geothermal regime at Prudhoe Bay, Alaska. *Journal of Geophysical Research* 87(B11): 9301–9316.

Mackay JR (1972) Offshore permafrost and ground ice, Southern Beaufort Sea, Canada. *Canadian Journal of Earth Science* 9(11): 1550–1561.

Osterkamp TE, Baker GC, Harrison WD, and Matava T (1989) Characteristics of the active layer and shallow sub-sea permafrost. *Journal of Geophysical Research* 94(C11): 16227–16236.

Romanovsky NN, Gavrilov AV, Kholodov AL, *et al.* (1998) *Map of predicted offshore permafrost distribution on the Laptev Sea Shelf. Proceedings, Seventh International Conference on Permafrost, Yellowknife, Canada, pp. 967–972.* University of Laval, Quebec: Center for Northern Studies.

Sellmann PV and Hopkins DM (1983) *Sub-sea permafrost distribution on the Alaskan Shelf. Final Proceedings, Fourth International Conference on Permafrost, Fairbanks, Alaska, pp. 75–82.* Washington, DC: National Academy Press.

WAVES, TIDES, SURGES, TSUNAMI AND SEICHES

WAVES ON BEACHES

R. A. Holman, Oregon State University, Corvallis, OR, USA

Introduction

Wave motions are one of the most familiar of oceanographic phenomena. The waves that we see on beaches were originally generated by ocean winds and storms, sometimes at long distances from their final destination. In fact, groups of waves, generated by large storms, have been tracked from the Southern Ocean near Australia all the way to Alaska.

Open ocean waves can be thought of as simple sinusoids that are superimposed to yield a realistic sea. Waves entering the nearshore, called incident waves, can have wave periods (T, the time between consecutive passages of wave crests) ranging from 2 to 20 s, with 10 s a typical value. Wave heights (H, the vertical distance from the trough to peak of a wave) can exceed 10 m, but are typically 1 m, representing an energy density, of $1250 \, \mathrm{J \, m^{-2}}$ (ρ is the density of sea water, g is the acceleration of gravity) and a flux of power impinging on the coast of about 10 kW per meter of coastline. Although this is a substantial amount of power, it is not enough to make broad commercial exploitation of wave power economical at the time of writing.

Of interest in this section are the dynamics of waves once they progress into the shallow beach environment such that the ocean bottom begins to restrict the water motions. Most people are familiar with refraction (the turning of waves toward the beach), wave breaking, and swash (the back and forth motion of the water's edge), but are less familiar with the other types of fluid motion that are generated near the beach.

Figure 1 illustrates schematically the evolution of ocean wave energy as it moves from deep water (top of the figure) through progressively shallower water toward the beach (bottom of the figure). Offshore, most energy lies in waves of roughly 10 s period (middle of the figure). However, the processes of shoaling distribute that energy to both higher frequencies (right half of the figure) and lower frequencies (left half) including mean flows. In general, these processes can be distinguished as those that occur offshore of the surf zone (where waves become

overly steep and break) and those that occur within the surf zone.

The axes of **Figure 1**, cross-shore position and frequency, are two of several variables that can be used to structure a discussion of near-shore fluid dynamics. Other important distinctions that will be made include whether the incident waves are monochromatic (single frequency) versus random (including a range of frequencies), depth-averaged versus depth-dependent, longshore uniform (requiring consideration of only one horizontal dimension, 1HD) versus long shore variable (2HD), and linear versus nonlinear.

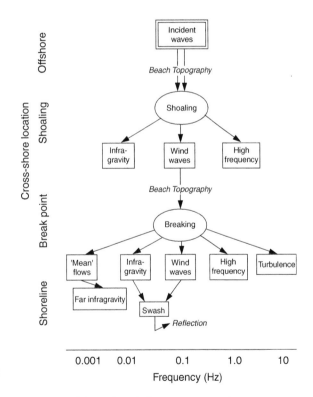

Figure 1 Schematic of important near-shore processes showing how the incident wave energy that drives the system evolves as the waves progress from offshore to the shoreline (top to bottom of figure). Wave evolution is grouped into processes occurring seaward of the breakpoint (labeled 'shoaling') and those within the surf zone (denoted 'breaking'). In both cases, energy is spread to lower (left) and higher (right) frequencies. The beach topography provides the bottom boundary condition for flow, so is important to wave processes. In turn, the waves move sediment, slowly changing the topography. Wind and tides may be important in some settings, but are not shown here.

The Dynamics of Incident Waves

Much of the early progress in understanding near-shore waves was based on the examination of a monochromatic wave train, propagating onto a long shore uniform beach (1HD). Many observable properties can be explained in terms of a few principles including conservation of wave crests, of momentum, and of energy. Most dynamics are depth-averaged.

Table 1 lists a number of properties of monochromatic ocean waves, in the linear limit of infinitesimal wave amplitude (known as linear, or Airy wave theory). The general expressions (center column) contain complicated forms that can be substantially simplified for both shallow (depths less than 1/20 of the deep water wavelength, L_0) and deep (depths greater than $1/2\ L_0$) water limits.

The speed of wave propagation is known as the celerity or phase speed, c, to distinguish it from the velocity of the actual water particles. In deep water, c depends only on the wave period (independent of depth). However, as the wave enters shallower water, the wavelength decreases and the phase speed becomes slower (contrary to common belief, this is not a result of bottom friction). An interesting consequence of the slowing is wave refraction, the turning of waves toward the coast. For a wave approaching the coast at any angle, the end in shallower water will always progress more slowly than the deeper end. By propagating faster, the deeper end will begin to catch up to the shallow end, effectively turning the wave toward the beach (refraction). In shallow water, the general expression for celerity, $c = (gh)^{1/2}$, depends only on depth so that waves of all periods propagate at the same speed.

The energy density of Airy waves (energy per unit area) is the sum of kinetic and potential energy components and depends only on the square of the wave height (**Table 1**). Perhaps of more interest is the rate at which this energy is propagated by the wave train, known as the wave power, P, or wave energy flux. In deep water, wave energy progresses at half the speed of wave phase (individual wave crests will out-run the energy packet), whereas in shallow water energy travels at the same speed as wave phase and the flux depends only on $H^2 h^{1/2}$. Offshore of the surf zone, wave energy is conserved since there is no breaking dissipation and energy loss through bottom friction has been shown to be negligible except over very wide flat seas. Thus, as the depth, h, decreases, the wave height, H, must increase to conserve $H^2 h^{1/2}$. This is a phenomenon familiar to beach-goers as the looming up of a wave just before breaking.

The combined result of shoaling is reduced wavelength and increased wave height, hence waves that become increasingly steep and may become unstable and break. One criterion for breaking is that

Table 1 Kinematic relationships of linear waves[a]

Parameter	Shallow water[b]	General expression	Deep water[c]
Surface elevation	$\eta = \dfrac{H}{2}\cos(kX - \sigma t)$	$\eta = \dfrac{H}{2}\cos(kX - \sigma t)$	$\eta = \dfrac{H}{2}\cos(kX - \sigma t)$
Energy density	$E = \dfrac{1}{8}\rho g H^2$	$E = \dfrac{1}{8}\rho g H^2$	$E = \dfrac{1}{8}\rho g H^2$
Wave length	$L = T\sqrt{(gh)}$	$L = \dfrac{gT^2}{2\pi}\tanh(kh)$	$L_0 = \dfrac{gT^2}{2\pi}$
Phase velocity	$c = \sqrt{(gh)}$	$c = \dfrac{gT}{2\pi}\tanh(kh)$	$c_0 = \dfrac{gT}{2\pi}$
Group velocity	$c_g = c = \sqrt{(gh)}$	$c_g = cn = c\left[\dfrac{1}{2} + \dfrac{kh}{\sinh(2kh)}\right]$	$c_g = \dfrac{c}{2} = \dfrac{gT}{4\pi}$
Wave power, $P = Ec_g$	$P = \dfrac{1}{8}\rho g H^2 \sqrt{(gh)}$	$\dfrac{\rho g H^2}{8}\dfrac{gT\tanh(kh)}{2\pi}\left[\dfrac{1}{2} + \dfrac{kh}{\sinh(2kh)}\right]$	$P = \dfrac{1}{8}\rho g H^2\dfrac{gT}{4\pi}$
Wave momentum flux	$S_{xx} = \dfrac{3}{2}E$	$S_{xx} = E\left[\dfrac{2kh}{\sinh(2kh)} + \dfrac{1}{2}\right]$	$S_{xx} = \dfrac{1}{2}E$
Horizontal velocity magnitude	$u = \dfrac{H}{2}\sqrt{\left(\dfrac{g}{h}\right)}$	$u = \dfrac{\pi H}{T}\dfrac{\cosh(k(z + h))}{\sinh(kh)}$	$u = \dfrac{\pi H}{T}e^{kz}$

[a]H is the peak to trough wave height and is twice the amplitude, a; k is the wave number $= 2\pi/L$; X is distance in the direction of propagation; σ is the frequency $= 2\pi/T$ where T is the wave period; ρ is the density of water; g is the acceleration of gravity; z is the depth below the surface within the water column of total depth h. Units of each quantity depend on units used in each equation.
[b]$h < L_0/20$. h is the water depth; L_0 is the deep water wavelength.
[c]$h < L_0/2$.

the increasing water particle velocities (u in **Table 1**) exceed the decreasing wave phase speed such that the water leaps ahead of the wave in a curling or plunging breaker. From the relationships in **Table 1**, it can be found that this occurs. when γ, the wave height to depth ratio ($\gamma = H/b$) exceeds a value of 2. Of course, for waves that have steepened to the point of breaking, the approximations of infinitesimal waves, inherent in Airy wave theory, are badly violated. However, observations show that γ does reach a limiting value of approximately 1 for monochromatic waves and 0.4 for a random wave field. As waves continue to break across the surf zone, the wave height decreases in a way that γ is approximately maintained, and the wave field is said to be saturated (cannot get any larger).

The above-saturation condition implies that wave heights will be zero at the shoreline and there will be no swash, in contradiction to common observation. Instead, it can be shown that very small amplitude waves, incident on a sloping beach, will not break unless their shoreline amplitude, a_s, exceeds a value determined by

$$\frac{\sigma^2 a_s}{g\beta^2} \leq \kappa \qquad [1]$$

where κ is an O(1) constant. For larger amplitudes, the wave amplitude at any cross-shore position is the sum of a standing wave contribution of this maximum value plus a dissipative residual that obeys the saturation relationship.

The ratio of terms on the left-hand side of eqn.[1] is important to a wide range of nearshore phenomena and is often re-written as the Iribarren number,

$$\xi_0 = \frac{\beta}{(H_s/L_0)^{1/2}} \qquad [2]$$

where the measure of wave amplitude is replaced by the offshore significant wave height,[1] H_s. This form clarifies the importance of beach steepness, made dynamically important by comparing it to the wave steepness, H_s/L_0. For very large values of ξ_0, the beach acts as a wall and is reflective to incident waves (the non-breaking case from eqn. [1]). For smaller values, the presence of the sloping beach takes on increasing importance as the waves begin to break as the plunging breakers that surfers like, where water is thrown ahead of the wave and the advancing crest resembles a tube. Still smaller beach steepnesses (and ξ_0) are associated with spilling breakers in which a volume of frothy turbulence is pushed along with the advancing wave front.

Radiation Stress: the Forcing of Mean Flows and Set-up

The above discussion is based on the assumption of linearity, strictly true only for waves of infinitesimal amplitude. Because the dynamics are linear, energy in a wave of some particular period, say 10 s, will always be at that same period. In fact, once wave amplitude is no longer negligible, there are a number of nonlinear interactions that may transfer energy to other frequencies, for example to drive currents (zero frequency).

Nonlinear terms describe the action of a wave motion on itself and arise in the momentum equation from the advective terms, $u \cdot \nabla u$, and from the integrated effect of the pressure term. For waves, the time-averaged effect of these terms can easily be calculated and expressed in terms of the radiation stress, S, defined as the excess momentum flux due to the presence of waves. Since a rate of change of momentum is the equivalent of a force by Newton's second law, radiation stress allows us to understand the time-averaged force exerted by waves on the water column through which they propagate. A spatial gradient in radiation stress, for instance a larger flux of momentum entering a particular location than exiting, would then force a current.

Radiation stress is a tensor such that S_{ij} is the flux of i-directed momentum in the j-direction. For waves in shallow water, approaching the coast at an angle θ, the components of the radiation stress tensor are

$$S = \begin{bmatrix} S_{xx} & S_{xy} \\ S_{yx} & S_{yy} \end{bmatrix} = E \begin{bmatrix} (\cos^2\theta + 1/2) & \cos\theta\sin\theta \\ \cos\theta\sin\theta & (\sin^2\theta + 1/2) \end{bmatrix} \qquad [3]$$

where x is the cross-shore distance measured positive to seaward from the shoreline and y is the long-shore distance measured in a right-hand coordinate system with z positive upward from the still water level.

For a wave propagating straight toward the beach ($\theta = 0$) $S_{xx} = 3/2E$ is the shoreward flux of shoreward-directed momentum. The increase of wave height (hence energy, E) associated with shoaling outside the surf zone must be accompanied by an

---

[1]Although the peak to trough vertical distance for monochromatic waves is a unique and hence sensible measure of wave height, for random waves this scale is a statistical quantity, representing a distribution. A single measure, often chosen to represent the random wave Reld, is the significant wave height, Hs, defined as the average height of the largest one-third of the waves. This statistic was chosen historically as best representing the value that would be visually estimated by a semitrained observer. It is usually calculated as four times the standard deviation of the sea surface times series.

increasing shoreward flux of momentum (radiation stress). This gradient, in turn, provides an offshore thrust, pushing water away from the break point and yielding a lowering of mean sea level called set-down. As the waves start to break and decrease in height through the surf zone, the decreasing radiation stress pushes water against the shore until an opposing pressure gradient balances the radiation stress gradient. The resulting set-up at the shoreline, $\bar{\eta}$, a contributor to coastal erosion and flooding, is found to depend on the offshore significant wave height, H_s, as

$$\bar{\eta}_{\max} = KH_s\xi_0 \qquad [4]$$

where K is found empirically to be 0.45.

If waves approach the beach at an angle, they also carry with them a shoreward flux of longshore-directed momentum, S_{yx}. Cross-shore gradients in this quantity, due to the breaking of waves in the surf zone, provide a net long shore force that accelerates a long shore current, $\bar{V}$ along the beach until the forcing just balances bottom friction. If the cross-shore structure of $\bar{V}$ is solved for, a discontinuity is evident at the seaward limit of the surf zone, where the radiation stress forcing jumps from zero (seaward of the break point) to a large value (where the wave just begin to break).

This discontinuity is an artifact of the fact that every wave breaks at exactly the same location for an assumed monochromatic wave forcing, and must be artificially smoothed by an assumed horizontal mixing for this case. However, a natural random wave field consists of an ensemble of waves with (for linear waves) a Rayleigh distribution of heights. Depth-limited breaking of such a wave field will be spread over a region from offshore, where a few largest waves break, to onshore where the smallest waves finally begin to dissipate. The spatially distributed nature of these contributions to the average radiation stress provides a natural smoothing, often obviating the need for additional horizontal smoothing.

Nonlinear Incident Waves

The above discussion dwelt on the nonlinear transfer of energy from incident waves to mean flows. Nonlinearities will also transfer energy to higher frequencies, yielding a transformation of incident wave shape from sinusoidal to peaky and skewed forms. The Ursell number, $(H/L)(L/b)^3$, measures the strength of the nonlinearity. For monochromatic incident waves, this evolution was often modeled in terms of an ordered Stokes expansion of the wave form to produce a series of harmonics (multiples of the incident

wave frequency) that are locked to the incident wave. For waves with Ursell number of $O(1)$, propagating in depths that are not large compared to the wave height, higher order theories must be used to model the finite amplitude dynamics. For a random sea under such theories, the total evolution of the spectrum must be found by summing the spectral evolution equations for all possible Fourier pairs (in other words, all frequencies in the sea can and will interact with all other frequencies). Such approaches are very successful in predicting the evolving shape and nonlinear statistics (important for driving sediment transport) for natural random wave fields outside the surf zone.

Vertically Dependent Processes

Depth-independent models are successful in reproducing many nearshore fluid processes but cannot explain several important phenomena, for example undertow, offshore-directed currents that exist in the lower part of the water column under breaking waves. The primary cause of depth dependence arises from wave-breaking processes. When waves break, the organized orbital motions break down, either through the plunge of a curling jet of water thrown ahead of the advancing wave or as a turbulent foamy mass (called a roller) carried on the advancing crest. Both processes originate at the surface but drive turbulence and bubbles into the upper part of the water column.

The transfer of momentum from wave motions to mean currents described by radiation stress gradients above does not account for the existence of an intermediate repository, the active turbulence of the roller, that decays slowly as it is carried with the progressing wave. This time delay causes a shift of the forcing of longshore currents, such that a current jet will occur landward of the location expected from study of the breaking locations of incident waves.

The other consequence of the vertical dependence of the momentum transfer is that the shoreward thrust provided by wave breaking is concentrated near the surface. Set-up, the upward slope of sea level against the shore, will balance the depth averaged wave forcing. However, due to the vertical structure of the forcing, shoreward flows are driven near the top of the water column and a balancing return flow, the undertow, occurs in the lower water column. Undertow strengths can reach $1\,\mathrm{m\,s^{-1}}$.

2HD Flows – Circulation

All of the previous discussion was based on the assumption that all processes were long-shore uniform

(1 HD) so that no long-shore gradients existed. It is rare in nature to have perfect long shore uniformity. Most commonly, some variability (often strong) exists in the underlying bathymetry. This can lead to refractive focusing (the concentration of wave energy by refraction of waves onto a shallower area) and the forcing of long-shore gradients in wave height, hence of setup. Since setup is simply a pressure head, long-shore gradients will drive long-shore currents toward low points where the converging water will turn seaward in a jet called a rip current.

It is possible to develop long-shore gradients in wave height (hence rips) in the absence of long-shore variations in bathymetry. Interactions between two elements of the wave field (either two incident wave trains from different directions or an incident wave and an edge wave, defined below) can force rip currents if the interacting trains always occur with a fixed relative phase.

Infragravity Waves and Edge Waves

There is a further, very important consequence of the fact that natural wave fields are not monochromatic, but instead are random. For random waves, wave height is no longer constant but varies from wave to wave. Usually these variations are in the form of groups of five to eight waves, with heights gradually increasing then decreasing again. This observation is known as surf beat and is particularly familiar to surfers. A consequence of these slow variations is that the radiation stress of the waves is no longer constant, but also fluctuates with wave group time-scales and forces flows (and waves) in the near-shore with corresponding wave periods. These waves have periods of 30–300 s and are called infragravity waves, in analogy to infrared light being lower frequency than its visible light counterpart.

The direct forcing of infragravity motions described above can be thought of as a time-varying setup, with the largest waves in a group forcing shoreward flows that pile up in setup, followed by seaward flow as the setup gradients dominate over the weaker forcing of the small waves. If the modulations of the incident wave group are long shore uniform, this result of this setup disturbance will simply be a free (but low frequency) wave motion that propagates out to sea. However, in the normal case of wave groups with longshore (as well as time) variability, we can think of the response by tracing rays as the setup disturbance tries to propagate away. Rays that travel offshore at an angle to the beach will refract away from the beach normal (essentially the opposite of incident wave refraction, discussed

earlier). For rays starting at a sufficiently steep angle to the normal, refraction can completely turn the rays such that they re-approach and reflect from the shore in a repeating way and the energy is trapped within the shallow region of the beach. These trapped motions are called edge waves because the wave motions are trapped in the near-shore wave guide. (Any region wherein wave celerity is a minimum can similarly trap energy by refraction and is known as a wave guide. The deep ocean sound channel is a well-known example and allows propagation of trapped acoustic energy across entire ocean basins.) Motions that do not completely refract and thus are lost to the wave guide are called leaky modes.

The requirement that wave rays start at a sufficiently steep angle to be trapped by refraction can be expressed in terms of the long-shore component of wavenumber, k_y. For large k_y (waves with a large angle to the normal), rays will be trapped in edge waves whereas small k_y motions will be leaky modes. The cutoff between these is σ^2/g.

In the same sense that waves that slosh in a bathtub occur as a discrete set of modes that exactly fit into the tub, edge waves occur in a set of modes that exactly fit between reflection at the shoreline and an exponentially decaying tail offshore. The detailed form of the waves depends on the details of the bathymetry causing the refraction. However, for the example of a plane beach of slope $\beta(h = xtan\beta)$, the cross-shore forms of the lowest four modes (mode numbers, $n = 0, 1, 2, 3$), are shown in **Figure 2** and are given in **Table 2**.

The existence of edge waves as a resonant mode of wave energy transmission in the near-shore has several impacts. First, the dispersion relation provides a selection for particular scales. For example, **Figure 3** shows a spectrum of infragravity wave energy collected at Duck, North Carolina, USA. The concentration of energy into very clear, preferred scales is striking and has led to suggestions that edge waves may be responsible for the generation of sand bars with corresponding scales. Second, because edge-wave energy is trapped in the near-shore, it can build to substantial levels even in the presence of weak, incremental forcing. Moreover, because edge-wave energy is large at the shoreline where the incident waves have decayed to their minimum due to breaking, edge waves may feasibly be the dominant fluid-forcing pattern on near-shore sediments in these regions.

The magnitudes of infragravity energy (including edge waves and leaky modes) have been found to depend on the relative beach steepness as expressed by the Iribarren number (eqn. [2]). For steep beaches (high ξ_0), very little infragravity energy can be

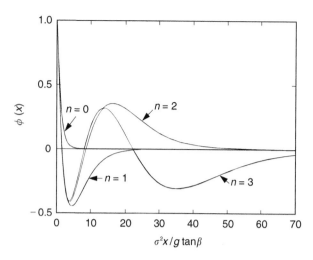

Figure 2 Cross-shore structure of edge waves. Only the lowest four mode numbers of the larger set are shown. The mode number, n, describes the number of zero crossings of the modes. So, for example, a mode 1 edge wave always has a low offshore, opposite a shoreline high, and visa versa. Edge waves propagate along the beach.

Table 2 The average abundance of the refractory elements in the Earth's crust, and their degree of enrichment, relative to aluminum, in the oceans

$k_y < \dfrac{\sigma^2}{g}$, leaky modes

$k_y > \dfrac{\sigma^2}{g}$, edge waves

Velocity potential	$\Phi = \dfrac{ag}{\sigma}\phi(x)\cos(ky - \sigma t)$
Sea surface elevation	$\eta = -a\phi(x)\sin(ky - \sigma t)$
Cross-shore velocity	$u = \dfrac{ag}{\sigma}\dfrac{\partial\phi(x)}{\partial x}\cos(ky - \sigma t)$
Long-shore velocity	$v = -\dfrac{a\sigma}{\sin[(2n + 1)\beta]}\phi(x)\sin(ky - \sigma t)$
Dispersion relationship	$L = \dfrac{gT^2}{2\pi}\sin[(2n + 1)\beta]$

Cross-shore shape functions	n	$\phi(x)$
	0	$1 \cdot e^{-kx}$
	1	$(1 - 2kx) \cdot e^{-kx}$
	2	$(1 - 4kx + 2k^2x^2) \cdot e^{-kx}$
	3	$(1 - 6kx + 6k^2x^2 - 4/3k^3x^3) \cdot e^{-kx}$

generated and the beaches are termed reflective due to the high reflection coefficient for the incident waves. However, for low-sloping beaches (small ξ_0), infragravity energy can be dominant, especially compared to the highly dissipated incident waves. On the Oregon Coast of the USA, for example, swash spectra have been analyzed in which 99% of the variation is at infragravity timescales (making beachcombing an energetic activity). **Table 3** lists five

Table 3 Magnitude of infragravity waves on different beach types

Location	ξ_0	m^a
New South Wales, Australia	0.40	0.185 ± 0.157
Duck, NC, USA	0.43	0.237 ± 0.057
Martinique Beach, Canada	0.57	0.323 ± 0.073
Torrey Pines, CA, USA	0.60	0.554 ± 0.100
Santa Barbara, CA, USA	0.83	0.459 ± 0.089

[a]Least squares regression slope between the significant swash height (computed from the infragravity band energy only) and the offshore significant wave height.
Reproduced from Howd PA, Oltman-Shay J, and Holman RA (1991) Wave variance partitioning in the trough of a barred beach. Journal of Geophysical Research 96 (C7), 12781} 12795.)

representative beach locations, with mean values of ξ_0 and of m, the linear regression slope between the measured significant swash magnitude,[2]R_s, in the infragravity band, and the offshore significant wave height, H_s.

Shear Waves

Up until the mid-1980s long shore currents were viewed as mean flows whose dynamics were readily described as in the above sections. However, field data from the Field Research Facility in Duck, North Carolina, provided surprising evidence that as longshore currents accelerated on a beach with a well-developed sand bar, the resulting current was not steady but instead developed slow fluctuations in strength and a meandering pattern in space. Typical wave periods of these wave are hundreds of seconds and long-shore wavelengths are just hundreds of meters (**Figure 3**). These very low frequencies are called far infragravity waves, in analogy to the relationship of far infrared to infrared optical frequencies. However, the wavelengths are several orders of magnitude shorter than that which would be expected for gravity waves (e.g., edge waves or leaky modes) of similar periods.

These meanders have been named shear waves and arise due to an instability of strong currents, similar to the instability of a rising column of smoke. The name comes from the dependence of the dynamics (described briefly below) on the shear of the longshore current (the cross-shore gradient of the longshore current). A jet-like current with large shear, such as might develop on a barred beach where the wave forcing is concentrated over and near the bar,

---

[2] Swash oscillations are commonly expressed in terms of their vertical component.

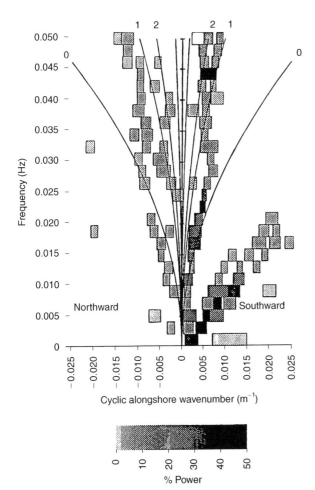

Figure 3 The spectrum of low frequency wave motions, as measured by current meters sampling the long shore component of velocity, from Duck, North Carolina, USA. The vertical axis corresponds to frequency and the horizontal axis to along shore component of wavenumber. Positive wavenumbers describe wave motions propagating along the beach to the south. Surprisingly, wave energy is not spread broadly in frequency–wavenumber space, but concentrates on specific ridges. Black lines, indicating the theoretical dispersion lines for edge waves (mode numbers marked at figure top), provide a good match to much of the data at low frequencies with some offset at higher frequency associated with Doppler shifting by the long-shore current. The concentration of low-frequency energy angling up to the right corresponds to shear waves. (After Oltman-Shay J and Guza RT (1987) Infragravity edge wave observations on two California beaches. *Journal of Physical Oceanography* 17(5): 644–663.)

have been shown to develop similar instabilities although with very larges scales. Similarly, under wave motions, the bottom boundary layer, matching the moving wave oscillations of the water column interior with zero velocity at the fixed boundary, is also unstable.

It can be shown that a necessary condition for such an instability is the presence of an inflection point in the velocity profile (the spatial curvature of the current field changes sign), a requirement satisfied by long shore currents on a beach. In that case, cross-shore perturbations will extract energy from the mean long-shore current at a rate that depends on the strength of the current shear. Thus, these perturbations will grow to become first wave-like meanders, then if friction is not strong, to become a field of turbulent eddies. The extent of this evolution (wave-like versus eddy-like) is not yet known for natural beach environments.

Shear waves can clearly form an important component of the near-shore current field.) Root mean square (RMS) velocity fluctuations can reach 35 cm s^{-1} in both cross-shore and long-shore components of flow. This corresponds to an RMS swing of the current of 70 cm s^{-1}, with many oscillations much larger.

Conclusions

As ocean waves propagate into the shoaling waters of the nearshore, they undergo a wide range of changes. Most people are familiar with the refraction, shoaling and eventual breaking of waves in a near-shore surf zone. However, this same energy can drive strong secondary flows. Wave breaking pushes water shoreward, yielding a super-elevation at the shoreline that can accentuate flooding and erosion. Waves arriving at an angle to the beach will drive strong currents along the beach that can transport large amounts of sediment. Often these currents form circulation cells, with strong rip currents spaced along the beach.

Natural waves occur in groups, with heights that vary. The breaking of these fluctuating groups drives waves and currents at the same modulation timescale, called infragravity waves. These can be trapped in the nearshore by refraction as edge waves. Even long shore currents can develop instabilities called shear waves that drive meter-per-second fluctuations in the current strength with timescales of several minutes.

The apparent physics that dominates different beaches around the world often appears to vary. For example, on low-sloping energetic beaches,

can develop strong shear waves. In contrast, for a broad, featureless planar beach, the shear of any generated long-shore current will be weak so that shear wave energy may be undetectable. This explains why shear waves were not discovered until field experiments were carried out on barred beaches.

The instability by which shear waves are generated has a number of other analogs in nature. Large-scale coastal currents, flowing along a continental shelf, infragravity energy often dominates the surf zone, whereas shear waves can be very important on barred beaches. In fact, the physics is unchanging in these environments, with only the observable manifestations of that physics changing. The unification of these diverse observations through parameters such as the Iribarren number is an important goal for future research.

Symbols Used

E	wave energy density
H	wave height
H_s	significant wave height
L	wavelength
L_0	deep water wave length
P	wave power or energy flux
R_s	significant swash height
RMS	root mean square statistic
S	radiation stress (wave momentum flux)
T	wave period
$\bar{V}$	mean longshore current
X	distance coordinate in the direction of wave propagation
a	wave amplitude
a_s	wave amplitude at the shoreline
c	wave celerity, or phase velocity
c_g	wave group velocity
g	acceleration of gravity
h	water depth
k	wavenumber (inverse of wavelength)
n	ratio of group velocity to celerity
m	ratio of infragravity swash height to offshore wave height
n	edge wave mode number
u	water particle velocity under waves
v	long-shore component of wave particle velocity
x	cross-shore position coordinate
y	long-shore position coordinate
z	vertical coordinate
β	beach slope
γ	ratio of wave height to local depth for breaking waves
η	sea surface elevation
$theta$	angle of incidence of waves relative to normal
ξ_0	Iribarren number
ρ	density of water
σ	*radial frequency* $2\pi/T$
φ	cross-shore structure function for edge waves
$\bar{\eta}_{max}$	mean set-up at the shoreline
∇	gradient operator

See also

Beaches, Physical Processes Affecting. Coastal Trapped Waves. Sea Level Change.

SEICHES

D. C. Chapman, Woods Hole Oceanographic
Institution, Woods Hole, MA, USA
G. S. Giese, Woods Hole Oceanographic
Institution, Woods Hole, MA, USA

Introduction

Seiches are resonant oscillations, or 'normal modes', of lakes and coastal waters; that is, they are standing waves with unique frequencies, 'eigenfrequencies', imposed by the dimensions of the basins in which they occur. For example, the basic behavior of a seiche in a rectangular basin is depicted in **Figure 1**. Each panel shows a snapshot of sea level and currents every quarter-period through one seiche cycle. Water moves back and forth across the basin in a periodic oscillation, alternately raising and lowering sea level at the basin sides. Sea level pivots about a 'node' in the middle of the basin at which the sea level never changes. Currents are maximum at the center (beneath the node) when the sea level is horizontal, and they vanish when the sea level is at its extremes.

Seiches can be excited by many diverse environmental phenomena such as seismic disturbances, internal and surface gravity waves (including other normal modes of adjoining basins), winds, and atmospheric pressure disturbances. Once excited, seiches are noticeable under ordinary conditions because of the periodic changes in water level or currents associated with them (**Figure 1**). At some locations and times, such sea-level oscillations and currents produce hazardous or even destructive conditions. Notable examples are the catastrophic seiches of Nagasaki Harbor in Japan that are locally known as 'abiki', and those of Ciutadella Harbor on Menorca Island in Spain, called 'rissaga'. At both locations extreme seiche-produced sea-level oscillations greater than 3 m have been reported. Although seiches in most harbors do not reach such heights, the currents associated with them can still be dangerous, and for this reason the study of coastal and harbor seiches and their causes is of practical significance to harbor management and design. In this article we place emphasis on marine seiches, especially those in coastal and harbor waters.

History

In 1781, J. L. Lagrange found that the propagation velocity of a 'long' water wave (one whose wavelength is long compared to the water depth h) is given by $(gh)^{1/2}$, where g is gravitational acceleration. Merian showed in 1828 that such a wave, reflecting back and forth from the ends of a closed rectangular basin of length L, produces a standing wave with a period T, given by

$$T = \frac{2L}{n(gh)^{1/2}} \qquad [1]$$

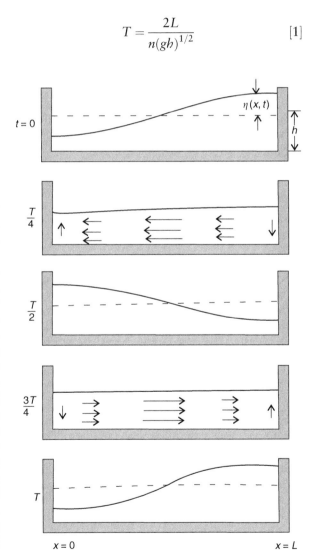

Figure 1 Diagram of a mode-one seiche oscillation in a closed basin through one period T. Panels show the sea surface and currents (arrows) each quarter-period. The basin length is L. The undisturbed water depth is h, and the deviation from this depth is denoted by η. Note the node at $x = L/2$ where the sea surface never moves (i.e. $\eta = 0$).

where $n = 1, 2, 3, \ldots$ is the number of nodes of the wave (**Figure 1**) and designates the 'harmonic mode' of the oscillation. Eqn [1] is known as Merian's formula.

F. A. Forel, between 1869 and 1895, applied Merian's formula with much success to Swiss lakes, in particular Lake Geneva, the oscillations of which had long been recognized by local inhabitants who referred to them as 'seiches', apparently from the Latin word 'siccus' meaning 'dry'. Forel's seiche studies were of great interest to scientists around the world and by the turn of the century many were contributing to descriptive and theoretical aspects of the phenomenon. Perhaps most noteworthy was G. Chrystal who, in 1904 and 1905, developed a comprehensive analytical theory of free oscillations in closed basins of complex form.

By the end of the nineteenth century, it was widely recognized that seiches also occurred in open basins, such as harbors and coastal bays, either as lateral oscillations reflecting from side-to-side across the basin, or more frequently, as longitudinal oscillations between the basin head and mouth. Longitudinal harbor oscillations are dynamically equivalent to lake seiches with a node at the open mouth (**Figure 2**), and a modified version of Merian's formula for such open basins gives the period as

$$T = \frac{4L}{(2n-1)(gh)^{1/2}} \quad [2]$$

where $n = 1, 2, 3, \ldots$. The dynamics leading to both eqns [1] and [2] are discussed below.

Interest in coastal seiches was fanned by the development of highly accurate mechanical tide recorders (*see* Tides) which frequently revealed surprisingly regular higher-frequency or 'secondary' oscillations in addition to ordinary tides. Even greater motivation was provided by F. Omori's observation, reported in 1900, that the periods of destructive sea waves (*see* Tsunami) in harbors were often the same as those of the ordinary 'secondary waves' in those same harbors. This led directly to a major field, laboratory, and theoretical study that was carried out in Japan from 1903 through 1906 by K. Honda, T. Terada, Y. Yoshida, and D. Isitani, who concluded that coastal bays can be likened to a series of resonators, all excited by the same sea with its many frequencies of motion, but each oscillating at its own particular frequencies – the seiche frequencies.

Most twentieth century seiche research can be traced back to problems or processes recognized in the important work of Honda and his colleagues. In particular, it became widely accepted that most coastal and harbor seiching is forced by open sea

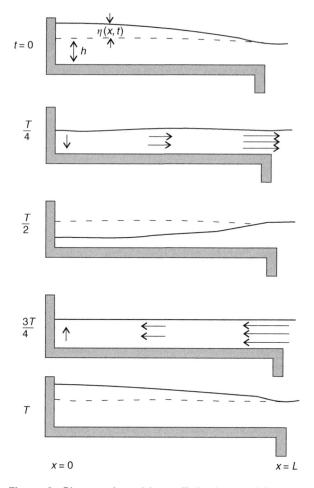

Figure 2 Diagram of a seiche oscillation in a partially open basin through one period T. Panels show the sea surface and currents (arrows) each quarter-period. The basin length is L with the open end at $x = L$. The undisturbed water depth is h, and the deviation from this depth is denoted by η. Note that the node (where $\eta = 0$) is located at the open end.

processes. A. Defant developed numerical modeling techniques which modern computers have made very efficient, and B. W. Wilson applied the theory to ocean engineering problems. Many others have made major contributions during the twentieth century.

Dynamics

The dynamics of seiches are easiest to understand by considering several idealized situations with simplified physics. More complex geometries and physics have been considered in seiche studies, but the basic features developed here apply qualitatively to those studies.

In a basin in which the water depth is much smaller than the basin length, fluid motions may be described by the depth-averaged velocity u and the deviation of the sea surface from its resting position

η. Changes in these quantities are related through momentum and mass conservation equations:

$$\frac{\partial u}{\partial t} = -g\frac{\partial \eta}{\partial x} - ru \qquad [3]$$

$$\frac{\partial \eta}{\partial t} + h\frac{\partial u}{\partial x} = 0 \qquad [4]$$

in which h is the fluid depth at rest, r is a coefficient of frictional damping, g is gravitational acceleration, x is the horizontal distance and t is time (see **Figure 1**). Nonlinear and rotation effects have been neglected, and only motions in the x direction are considered. Eqn [3] states that the fluid velocity changes in response to the pressure gradient introduced by the tilting of the sea surface, and is retarded by frictional processes. Eqn [4] states that the sea-surface changes in response to convergences and divergences in the horizontal velocity field; that is, where fluid accumulates $\partial u/\partial x < 0$ the sea surface must rise, and vice versa. Eqns [3] and [4] can be combined to form a single equation for either η or u, each having the same form. For example,

$$\frac{\partial^2 u}{\partial t^2} + r\frac{\partial u}{\partial t} - gh\frac{\partial^2 u}{\partial x^2} = 0 \qquad [5]$$

Closed Basins

The simplest seiche occurs in a closed basin with no connection to a larger body of water, such as a lake or even a soup bowl. **Figure 1** shows a closed basin with constant depth h and vertical sidewalls. At the sides of the basin, the velocity must vanish because fluid cannot flow through the walls, so $u = 0$ at $x = 0$ and L. Solutions of eqn [5] that satisfy these conditions and oscillate in time with frequency ω are

$$u = u_0 e^{-rt/2}\cos(\omega t)\sin\left(\frac{n\pi}{L}x\right) \qquad [6]$$

where u_0 is the maximum current, $n = 1, 2, 3, \ldots$, and $\omega = [gh(n\pi/L)^2 - r^2/4]^{1/2}$. The corresponding sea-surface elevation is

$$\eta = -\frac{u_0 \omega L}{gn\pi} e^{-rt/2}\left[\sin(\omega t) - \frac{r}{2\omega}\cos(\omega t)\right]\cos\left(\frac{n\pi}{L}x\right) \qquad [7]$$

Eqns [6] and [7] represent the normal modes or seiches of the basin. The integer n defines the harmonic mode of the seiche and corresponds to the number of velocity maxima and sea-level nodes (locations where sea level does not change) which occur where $cos(n\pi x/L) = 0$.

The spatial structure of the lowest or fundamental mode seiche ($n = 1$) is shown schematically in **Figure 1** through one period and was described above. Sea level rises and falls at each sidewall, pivoting about the node at $x = L/2$. The velocity vanishes at the sidewalls and reaches a maximum at the node. Sea level and velocity are almost 90° out of phase; the velocity is zero everywhere when the sea level has its maximum displacement, whereas the velocity is maximum when the sea level is horizontal.

The effect of friction is to cause a gradual exponential decay or damping of the oscillations and a slight decrease in seiche frequency with a shift in phase between u and η. If friction is weak (small r), the seiche may oscillate through many periods before fully dissipating. In this case, the frequency is close to the undamped value, $(gh)^{1/2}n\pi/L$ with period ($T = 2\pi/\omega$) given by Merian's formula, eqn [1]. If friction is sufficiently strong (very large r), the seiche may fully dissipate without oscillating at all. This occurs when $r > 2(gh)^{1/2}n\pi/L$, for which the frequency ω becomes imaginary.

The speed of a surface gravity wave in this basin is $(gh)^{1/2}$, so the period of the fundamental seiche ($n = 1$) is equivalent to the time it takes a surface gravity wave to travel across the basin and back. Thus, the seiche may be thought of as a surface gravity wave that repeatedly travels back and forth across the basin, perfectly reflecting off the sidewalls and creating a standing wave pattern.

Partially Open Basins

Seiches may also occur in basins that are connected to larger bodies of water at some part of the basin boundary (**Figure 2**). For example, harbors and inlets are open to the continental shelf at their mouths. The continental shelf itself can also be considered a partially open basin in that the shallow shelf is connected to the deep ocean at the shelf edge. The effect of the opening can be understood by considering the seiche in terms of surface gravity waves. A gravity wave propagates from the opening to the solid boundary where it reflects perfectly and travels back toward the opening. However, on reaching the opening it is not totally reflected. Some of the wave energy escapes from the basin into the larger body of water, thereby reducing the amplitude of the reflected wave. The reflected portion of the wave again propagates toward the closed sidewall and reflects back toward the open side. Each reflection from the open side reduces the energy in the oscillation, essentially acting like the frictional effects described above. This loss of energy due to the radiation of waves into the deep basin is called 'radiation damping.' Its effect is to produce a decaying response in the partially open basin, similar to frictional decay.

In general, a wider basin mouth produces greater radiation damping, and hence a weaker resonant seiche response to any forcing. Conversely, a narrower mouth reduces radiation damping and hence increases the amplification of fundamental mode seiches, theoretically becoming infinite as the basin mouth vanishes. However, seiches are typically forced through the basin mouth (see below), so a narrow mouth is expected to limit the forcing and yield a decreased seiche response. J. Miles and W. Munk pointed out this apparent contradiction in 1961 and referred to it as the 'harbor paradox.' Later reports raised a number of questions concerning the validity of the harbor paradox, among them the fact that frictional damping, which would increase with a narrowing of the mouth, was not included in its formulation.

The seiche modes of an idealized partially open basin (**Figure 2**) can be found by solving eqn [5] subject to a prescribed periodic sea-level oscillation at the open side; $\eta = \eta_0 \cos(\sigma t)$ at $x = L$ where σ is the frequency of oscillation. For simplicity, friction is neglected by setting $r = 0$. The response in the basin is

$$\eta = \eta_0 \cos(\sigma t) \frac{\cos(kx)}{\cos(kL)} \qquad [8]$$

$$u = \eta_0 (g/h)^{1/2} \sin(\sigma t) \frac{\sin(kx)}{\cos(kL)} \qquad [9]$$

where the wavenumber k is related to the frequency by $\sigma = (gh)^{1/2}k$. The response is similar to that in the closed basin, with the velocity and sea level again $90°$ out of phase. The spatial structure (k) is now determined by the forcing frequency. Notice that both the velocity and sea-level amplitudes are inversely proportional to $\cos(kL)$, which implies that the response will approach infinity (resonance) when $\cos(kL) = 0$. This occurs when $k = (n\pi - \pi/2)/L$, or equivalently when $\sigma = (gh)^{1/2}(n\pi - \pi/2)/L$ where $n = 1, 2, 3 \ldots$ is any integer.

These resonances correspond to the fundamental seiche modes for the partially open basin. They are sometimes called 'quarter-wave resonances' because their spatial structure consists of odd multiples of quarter wavelengths with a node at the open side of the basin ($x = L$). The first mode ($n = 1$) contains one-quarter wavelength inside the basin (as in **Figure 2**), so its total wavelength is equal to four times the basin width L, and its period is $T = 2\pi/\sigma = 4L/(gh)^{1/2}$. Other modes have periods given by eqn [2].

Despite the fact that these modes decay in time owing to radiation damping, they are expected to be the dominant motions in the basin because their amplitudes are potentially so large. That is, if the forcing consists of many frequencies simultaneously, those closest to the seiche frequencies will cause the largest response and will remain after the response at other frequencies has decayed. Furthermore, higher modes ($n \geq 2$) have shorter length scales and higher frequencies, so they are more likely to be dissipated by frictional forces, leaving the first mode to dominate the response. This is much like the ringing of a bell. A single strike of the hammer excites vibrations at many frequencies, yet the fundamental resonant frequency is the one that is heard. Finally, the enormous amplification of the resonant response means that a small-amplitude forcing of the basin can excite a much larger response in the basin.

Observations in the laboratory as well as nature reveal that seiches have somewhat longer periods than those calculated for the equivalent idealized open basins discussed above. This increase is similar to that which would be produced by an extension in the basin length, L, and it results from the fact that the water at the basin mouth has inertia and therefore is disturbed by, and participates in, the oscillation. This 'mouth correction', which was described by Lord Rayleigh in 1878 with respect to air vibrations, increases with the ratio of mouth width to basin length. In the case of a fully open square harbor, the actual period is approximately one-third greater than in the idealized case.

In nature, forcing often consists of multiple frequencies within a narrow range or 'band'. In this case, the response depends on the relative strength of the forcing in the narrow band and the resonant response at the seiche frequency closest to the dominant band. If the response at the dominant forcing frequency is stronger than the response at the resonant frequency, then oscillations will occur primarily at the forcing frequency. For example, the forcing frequency σ in eqn any seiche frequency, and the energy in the forcing at the resonant seiche frequency may be so small that it is not amplified enough to overwhelm the response at the dominant forcing frequency. In this case, the observed oscillations, sometimes referred to as 'forced seiches', will have frequencies different from the 'free seiche' frequencies discussed above.

Generating Mechanisms and Observations

Seiches in harbors and coastal regions may be directly generated by a variety of forces, some of which are depicted in **Figure 3**: (1) atmospheric

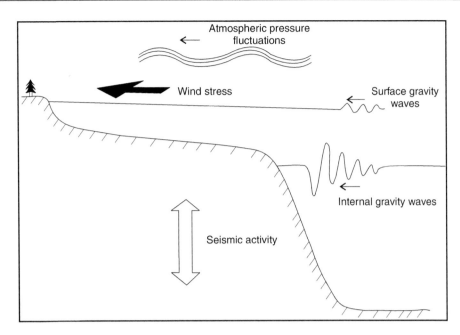

Figure 3 Sketch of various forcing mechanisms that are known to excite harbor and coastal seiches. Arrows for atmospheric pressure fluctuations, surface gravity waves, and internal gravity waves indicate propagation. Arrows for wind stress and seismic activity indicate direction of forced motions.

pressure fluctuations; (2) surface wind stress (*see* Storm Surges); (3) surface gravity waves caused by seismic activity (*see* Tsunami); (4) surface gravity waves formed by wind; and (5) internal gravity waves. It should be kept in mind that each of these forcing mechanisms can also generate or enhance other forcing mechanisms, thereby indirectly causing seiches. Thus, precise identification of the cause of seiching at any particular harbor or coast can be difficult.

To be effective the forcing must cause a change in the volume of water in the basin, and hence the sea level, which is usually accomplished by a change in the inflow or outflow at the open side of the basin. The amplitude of the seiche response depends on both the form and the time dependence of the forcing. The first mode seiche typically has a period somewhere between a few minutes and an hour or so, so the forcing must have some energy near this period to generate large seiches.

As an example, a sudden increase in atmospheric pressure over a harbor could force an outflow of water, thus lowering the harbor sea level. When the atmospheric pressure returns to normal, the harbor rapidly refills, initiating harbor seiching. However, although atmospheric pressure may change rapidly enough to match seiche frequencies, the magnitude of such high frequency fluctuations typically produces sea-level changes of only a few centimeters, so direct forcing is unlikely though not unknown. Several examples of direct forcing of fundamental

and higher-mode free oscillations of shelves, bays, and harbors by atmospheric pressure fluctuations have been described. In the case of the observations at Table Bay Harbor, in Cape Town, South Africa, it was noted that 'a necessary ingredient ...was found to be that the pressure waves approach ... from the direction of the open sea.'

In lakes, seiches are frequently generated by relaxation of direct wind stress, and since wind stress acting on a harbor can easily produce an outflow of water, this might seem to be a significant generation mechanism in coastal waters as well. However, strong winds rarely change rapidly enough to initiate harbor seiching directly. That is, the typical timescales for changes in strong winds are too long to match the seiche mode periods. Nevertheless, wind relaxation seiches have been observed in fiords and long bays such as Buzzards Bay in Massachusetts, USA.

Tsunamis are rare, but they consist of large surface gravity waves that can generate enormous inflows into coastal regions, causing strong seiches, especially in large harbors. The resulting seiches, which may be a mix of free and forced oscillations, were a major motivation for early harbor seiche research as noted earlier (*see* Tsunami). Direct generation of seiches by local seismic disturbances (as distinct from forcing by seismically generated tsunamis) is well established but very unusual. For example, the great Alaskan earthquake of 1964 produced remarkable seiches in the bayous along the Gulf of Mexico coast of Louisiana, USA. Similar phenomena are the sometimes very destructive oscillations excited by sudden slides of earth and glacier debris into high-latitude fiords and bays.

Most wind-generated surface gravity waves tend to occur at higher frequencies than seiches, so they are not effective as direct forcing mechanisms. However, in some exposed coastal locations, wind-generated swells combine to form oscillations, called 'infragravity' waves, with periods of minutes. These low-frequency surface waves are a well-known agent for excitation of seiches in small basins with periods less than about 10 minutes. Noteworthy are observations of 2–6 min seiches in Duncan Basin of Table Bay Harbor, Cape Town, South Africa, that occur at times of stormy weather. In 1993, similar short period seiches were reported in Barbers Point Harbor at Oahu, Hawaii, and their relationship to local swell and infragravity waves was demonstrated.

Perhaps the most effective way of directly exciting harbor and coastal seiches is by internal gravity waves. These internal waves can have large amplitudes and their frequency content often includes seiche frequencies. Furthermore, internal gravity waves are capable of traveling long distances in the ocean before delivering their energy to a harbor or coastline. In recent years this mechanism has been suggested as an explanation for the frequently reported and sometimes hazardous harbor seiches with periods in the range of 10–100 min. There is little or no evidence of a seismic origin for these seiches, and their frequency does not match that of ordinary ocean wind-generated surface waves. Their forcing has often been ascribed to meteorologically produced long surface waves. For example, it has been suggested that the 'abiki' of Nagasaki Harbor and the 'Marrobbio' in the Strait of Sicily may be forced by the passage of large low-pressure atmospheric fronts. In 1996, evidence was found that the hazardous 10-minute 'rissaga' of Ciutadella Harbor, Spain, and offshore normal modes are similarly excited by surface waves generated by atmospheric pressure oscillations, and it was proposed that the term 'meteorological tsunamis' be applied to all such seiche events.

However, attributing the cause of remotely generated harbor seiches to meteorologically forced surface waves does not account for observations that such seiches are frequently associated with ocean tides. In 1908 it was noted that in many cases harbor seiche activity occurs at specific tidal phases. In the 1980s, a clear association was found between tidal

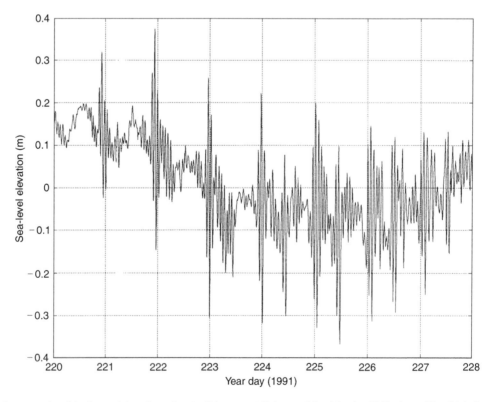

Figure 4 An example of harbor seiches from Puerto Princesa at Palawan Island in the Philippines. The tidal signal has been removed from this sea-level record to accentuate the bursts of 75-min harbor seiches. The seiches are excited by the arrival at the harbor mouth of internal wave packets produced by strong tidal current flow across a shallow sill some 450 km away.

and seiche amplitudes and it was suggested that tide-generated internal waves could be a significant agent for excitation of coastal and harbor seiches. In 1990, a study of the fundamental theoretical questions concerning transfer of momentum from internal waves to seiche modes and the wide frequency gap between tides and harbor seiches indicated that the high-frequency energy content of tide-generated internal solitary waves is sufficient to account for the energy of the recorded seiches, and a dynamical model for the generating process was published.

Observations at Palawan Island in the Philippines have demonstrated that harbor seiches can be forced by tide-generated internal waves and, as might be expected, there was also a strong dependency between seiche activity and water column density stratification. Periods of maximum seiche activity are associated with periods of strong tides, an example of which is given in **Figure 4**, which shows an 8-day sea-level record from Puerto Princesa at Palawan Island with the tidal signal removed. Bursts of 75-min harbor seiches are excited by the arrival at the harbor mouth of internal wave packets produced by strong tidal current flow across a shallow sill some 450 km away. The internal wave packets require 2.5 days to reach the harbor, producing a similar delay between tidal and seiche patterns. As an illustration, note the change in seiche activity from a diurnal to a semidiurnal pattern that is evident in **Figure 4**. A similar shift in tidal current patterns occurred at the internal wave generation site several days earlier.

More recent observations at Ciutadella Harbor in Spain point to a second process producing internal wave-generated seiches. Often the largest seiche events at that harbor occur under a specific set of conditions – seasonal warming of the sea surface and extremely small tides – which combine to produce very stable conditions in the upper water column. It has been suggested that under those conditions, meteorological processes can produce internal waves by inducing flow over shallow topography and that these meteorologically produced internal waves are responsible for the observed seiche activity.

See also

Storm Surges. Tides. Tsunami.

Further Reading

Chapman DC and Giese GS (1990) A model for the generation of coastal seiches by deep-sea internal waves. *Journal of Physical Oceanography* 20: 1459–1467.

Chrystal G (1905) On the hydrodynamic theory of seiches. *Transactions of the Royal Society of Edinburgh* 41: 599–649.

Defant A (1961) *Physical Oceanography*, vol 2. New York: Pergamon Press.

Forel FA (1892) *Le Leman (Collected Papers)*, 2 vols. Lausanne, Switzerland: Rouge.

Giese GS, Chapman DC, Collins MG, Encarnacion R, and Jacinto G (1998) The coupling between harbor seiches at Palawan Island and Sulu Sea internal solitons. *Journal of Physical Oceanography* 28: 2418–2426.

Honda K, Terada T, Yoshida Y, and Isitani D (1908) Secondary undulations of oceanic tides. *Journal of the College of Science, Imperial University, Tokyo* 24: 1–113.

Korgen BJ (1995) *Seiches. American Scientist.* 83: 330–341.

Miles JW (1974) Harbor seiching. *Annual Review of Fluid Mechanics* 6: 17–35.

Okihiro M, Guza RT, and Seymour RT (1993) Excitation of seiche observed in a small harbor. *Journal of Geophysical Research* 98: 18 201–18 211.

Rabinovich AB and Monserrat S (1996) Meteorological tsunamis near the Balearic and Kuril islands: descriptive and statistical analysis. *Natural Hazards* 13: 55–90.

Wilson BW (1972) Seiches. *Advances in Hydroscience* 8: 1–94.

COASTAL TRAPPED WAVES

J. M. Huthnance, CCMS Proudman Oceanographic
Laboratory, Wirral, UK

Introduction

Many shelf seas are dominated by shelf-wide motions that vary from day to day. Oceanic tides contribute large coastal sea-level variations and (on broad shelves) large currents. Atmospheric pressure and (especially) winds generate storm surges; strong currents and large changes of sea level. Other phenomena on these scales are wind-forced upwelling, along-slope currents and poleward undercurrents common on the eastern sides of oceans, responses to oceanic eddies, and alongshore pressure gradients.

All these responses depend on natural waves that travel along or across the continental shelf and slope. These waves, which have scales of about one to several days and tens to hundreds of kilometers according to the width of the continental shelf and slope, are the subject of this article. Also included are 'Kelvin' waves, also coastally trapped, that travel cyclonically around ocean basins but with typical scales of thousands of kilometers both alongshore and for offshore decrease of properties.

The waves have been widely observed through their association with the above phenomena. In fact they have been identified along coastlines of various orientations and all continents in both the Northern and Southern Hemispheres. Typically, the identification involves separating forced motion from the accompanying free waves. The 'lowest' mode with simplest structure (see below) has been most often identified; its peak coastal elevation is relatively easily measured. More complex forms need additional offshore measurements (usually of currents) for identification. This has been done (for example) off Oregon, the Middle Atlantic Bight and New South Wales (Australia). Observations substantiate many of the features described in the following sections.

Formulation

Analysis is based on Boussinesq momentum and continuity equations for an incompressible sea of near-uniform density between a gently-sloping seafloor $z = -h(\mathbf{x})$ and a free surface $z = \eta(\mathbf{x}, t)$ where the surface elevation $\eta = 0$ for the sea at rest. Cartesian coordinates $\mathbf{x} \equiv (x, y), z$ (vertically up) rotate with a vertical component $f/2$. The motion, velocity components (u, v, w), is assumed to be nearly horizontal and in hydrostatic balance. (These assumptions are almost always made for analysis on these scales; they are probably not necessary but certainly simplify the analysis.) At the surface, pressure and stress match atmospheric forcing (for free waves). There is no component of flow into the seabed (generalizing to zero onshore transport uh at the coast); $u \to 0$ far from the coast (the trapping condition) or is specified by forcing.

Straight Unstratified Shelf

This is the simplest context. Taking x offshore (and y alongshore; **Figure 1**) the depth is $h(x)$. Uniformity along shelf suggests wave solutions $\{u(x), v(x), \eta(x)\} \exp(iky + i\sigma t)$. For positive wave frequency, σ, $k > 0$ corresponds to propagation in $-y$, with the coast on the right ('forward' in the Northern Hemisphere). Then the momentum equations give u, v in terms of η satisfying

$$(h\eta')' + K\eta = 0 \qquad [1]$$

where $K(\alpha) \equiv kfh'/\sigma + (\sigma^2 - f^2)/g - k^2h$ (uniform f); primes (') denote cross-shelf differentiation $\partial/\partial x$. The boundary conditions become

$$h(\sigma\eta' + fk\eta) \to 0 (x \to 0); \quad \eta \to 0 (x \to \infty) \qquad [2]$$

Free wave modes are represented by eigensolutions of eqns. [1] and [2]. Successive modes with more offshore nodes correspond to large positive K and arise in two ways.

The term $(\sigma^2 - f^2)/g - k^2h$ in K represents the gravity wave mechanism, modified by rotation; it increases with frequency σ. For $kf > 0$ it gives rise to the 'Kelvin wave' – the mode with simplest offshore form; decay but no zeros of elevation. Other forms depending on this term ($kf < 0$ and/or nodes of elevation offshore) are termed edge waves and discussed elsewhere. If there is no slope or boundary, K equals this term alone and plane inertiogravity waves are solutions of eqn. [1].

The term kfh'/σ in K increases with decreasing σ if everywhere h increases offshore and $kf > 0$. It represents the following potential vorticity (angular momentum) restoring mechanism. If fluid is

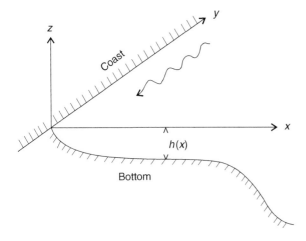

Figure 1 The wavy arrow denotes the sense of wave propagation for $k > 0$. This is 'forwards' if $f > 0$ (Northern Hemisphere).

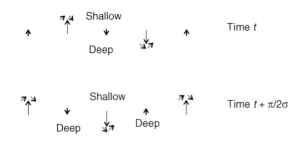

Figure 2 Topographic wave mechanism. ↑ displacement, velocity, ↗↘ relative vorticity (Northern Hemisphere).

displaced into shallower water, it spreads laterally to conserve volume and therefore spins more slowly in total; taking account of the earth's rotation, it acquires anticyclonic relative vorticity. 'Forwards' along the slope, the resulting up-slope velocity implies an up-slope displacement in time. Hence the up-slope displacement propagates 'forwards' along the slope. (Behind it, the anticyclonic relative vorticity implies down-slope flow restoring the fluid location from its previous up-slope displacement.) This sequence is depicted in **Figure 2**. These modes are referred to as continental shelf waves.

The mode forms and frequencies are known for several analytic models, e.g., level, uniformly sloping, exponential concave and convex shelves bordering an ocean of uniform depth. Numerical solutions for the waveforms and dispersion relations $\sigma(k)$ are easily found for any depth profile $h(x)$.

For any monotonic profile $h(x)$ the following have been proved. Phase propagates 'forwards' for all modes with $\sigma < |f|$. Waves forms with 1, 2,... nodes offshore have frequencies $|f| > \sigma_1 > \sigma_2 > \dots$ defined for all k (subject to $kf > 0$). The Kelvin wave frequency $\sigma_0(k) > \sigma_1(k)$ is likewise defined for all

k ($kf > 0$) and passes smoothly through $|f|$ to $\sigma_0 > |f|$ for large enough k. Edge waves with 0 (if $kf < 0$), 1, 2,... nodes in the offshore form have increasing frequencies $\sigma > |f|$; however, low wavenumbers (and frequencies) are excluded where the dispersion curves break the trapping criterion $0 < K(\infty) \equiv (\sigma^2 - f^2)/g - k^2 h(\infty)$ (**Figure 3**).

Besides these properties, the following features are typical. Bounded h'/h ensures a maximum σ_M in $\sigma(k)$; near σ_{1M}, mode 1 velocity tends to be maximal near the shelf edge, and polarized anticyclonically; the nearest approach to inertial motion in this topographic context; here the group velocity $\partial\sigma/\partial k$ of energy propagation (in $-y$) reverses through zero. If $k \to \infty$, then $\sigma_n \to f/(2n+1)$ for the n-node continental shelf wave which becomes concentrated over the 'beach' at the coast. As σ, $k \to 0$, the shelf wave and Kelvin wave speeds σ/k approach constant (maximum) values and $u/\sigma, v, \eta$ approach constant forms so that cross-slope velocities tend to zero. Variables v, η and η' are in phase or antiphase, v and η' being near the geostrophic balance $fv = g\eta'$; u is 90° out of phase. Typically, Kelvin-wave currents in shallow shelf waters are polarized cyclonically but first-mode continental shelf wave currents are anticyclonic.

Continental shelf wave-forms depend on the shape rather than the horizontal scale L of the depth profile. Phase speeds scale as fL and (u, v, η) scale as $(\sigma U/f, U, fUL/g)$ where the velocity scale U may be typically 0.1 m s^{-1}. Kelvin wave forms depend more on the depth; usually the phase speed is just less than $[gh(\infty)]^{1/2}$ and (u, v, η) scale as $(\sigma ZL/h, (g/h)^{1/2}Z, Z)$ where Z is typically 0.1 to 1 m. Quantitative results depend on the strength of forcing and accurate profile modeling; numerical calculations should be used for real shelves.

The typical maximum in $\sigma(k)$ and associated reversal of group velocity $\partial\sigma/\partial k$ appears to be of practical significance. Shelf waves with frequency near the maximum (for some mode) appear in several observations, e.g., North Carolina sea levels, Scottish and Vancouver Island diurnal tides, wind-driven flow north of Scotland and over Rockall Bank. There may be a bias in seeking motion correlated with local forcing, i.e., responses with nonpropagating energy. **Figure 4** shows modeled rotary currents over the shelf edge, continental shelf waves near the maximum frequency with slow energy propagation, as a response to impulsive wind forcing.

Other Geometry

Continental shelf waves exist in more general contexts than a straight shelf, as identification in nature testifies. Analyses also verify their possibility in rectangular and circular basins. Perfect trapping around islands is only possible if $\sigma < |f|$; then results are qualitatively as for a straight shelf except that wavelength around the island (and hence frequency) is quantized.

For a broad shelf (distant coast), with the continental slope regarded as a scarp, again only waves in $\sigma < |f|$ are trapped and results are qualitatively as for a straight shelf. An exception is the lowest mode, a 'double' Kelvin wave decaying to both sides. A seamount is again similar but introduces the same quantization as an island.

A ridge comprises two scarps back-to-back. Each has its set of waves propagating 'forwards' (relative to the local slope) in $\sigma < |f|$; a double Kelvin wave is associated with any net depth difference. Edge waves also propagate in both senses for $\sigma/|f|$ large enough to make $K > 0$ (see eqn. [1]). Similarly, a trench has sets of waves appropriate to each side.

All cases $h(x)$ and radial geometry $h(r)$ can be treated numerically in the same way as a straight monotone profile.

Stratification

We consider the simplest context, a straight shelf with rest state density $\rho_0(z)$; let $N^2 \equiv -g/\rho_0 d\rho_0/dz$. A wave form $\{u(x,z), v(x,z), w(x,z), \rho(x,z), p(x,z)\}$ $\exp(iky + i\sigma t)$ is posed. Hydrostatic balance, density, and momentum equations give ρ, w, u and v, respectively, in terms of p; then

$$\partial^2 p/\partial x^2 + (f^2 - \sigma^2)\partial/\partial z(N^{-2}\partial p\partial z) - k^2 p = 0 \quad [3]$$

by continuity. The boundary conditions for no flow through the bottom, zero pressure at the surface and trapping become

$$\partial h/\partial x(\partial p/\partial x + fkp/\sigma) + (f^2 - \sigma^2)N^{-2}\partial p/\partial z = 0$$
$$(z = -h) \quad [4]$$

$$\partial p/\partial z + N^2 p/g = 0 \quad (z = 0) \; p \to 0(x \to \infty) \quad [5]$$

A flat bottom (uniform h) with uniform $N^2 \ll g/h$ admits the simplest Kelvin wave and (for $n \geq 0$) internal Kelvin wave solutions

$$\cos[N(z + h)/c_n] \exp[i\sigma t + i\sigma y/c_n - fx/c_n] \quad [6]$$

where $c_n = (gh)^{1/2}(n = 0), c_n = Nh/n\pi (n = 1, 2, \ldots)$. Similar solutions, with vertical structure distributed roughly as N, exist for any $N(z) > 0$. Propagation is 'forwards' (cyclonic around the deep sea; anticyclonic around a cylindrical island) but depends only on density stratification.

For a sloping bottom $h(x)$ with offshore scale L (shelf width) and depth scale H, the parameter $S \equiv N^2H^2/f^2L^2$ indicates the importance of stratification. For small S, the solutions in $\sigma < |f|$ are depth-independent Kelvin and continental shelf waves. As S increases, wave speeds σ/k increase and nodes of u, v, p in the (x, z) cross-section tilt outwards from the vertical towards horizontal. Correspondingly, seasonal changes have been observed in the vertical structure and offshore scale of currents on the Oregon shelf (for example) and off Vancouver Island, where stratification increases the offshore decay scale of upper-level currents. Quite moderate S may imply σ monotonic increasing in k, contrasting with the

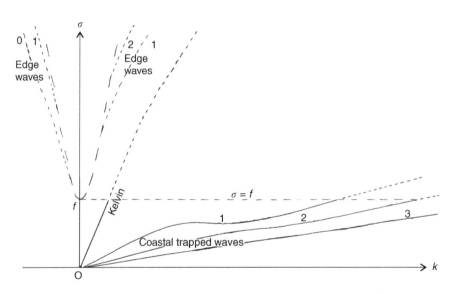

Figure 3 Qualitative dispersion diagram. —— trapped waves, - - - - nearly trapped waves, - - - $\sigma^2 = f^2 + k^2 gh(\infty)$.

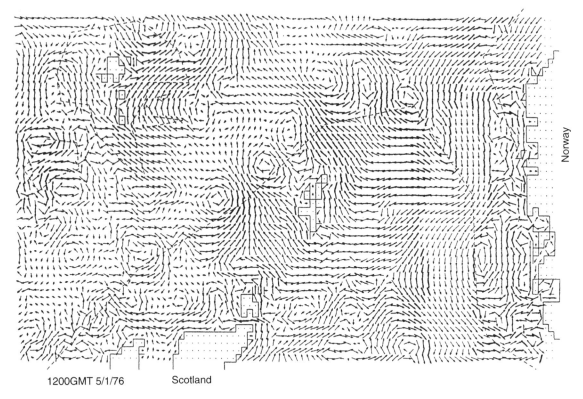

1200GMT 5/1/76   Scotland

Figure 4 Modeled continental shelf waves around Scotland at 1200Z, 5 January 1976 after impulsive wind forcing near Shetland on 3 January 1976. Dashed line shows 200 m depth contour. (Reproduced with permission from Huthnance JM (1995) Circulation, exchange and water masses at the ocean margin: the role of physical processes at the shelf edge. *Progress in Oceanography* 35: 353–431.)

maximum in $\sigma(k)$ common among unstratified modes. For large S, the modes in $\sigma < |f|$ become internal Kelvin-like waves with x replaced by $x - h^{-1}(z)$; $S \to \infty$ corresponds to a shelf width L much less than the internal deformation scale NH/f; the slope is 'seen' only as a coastal wall. Internal Kelvin waves have been observed in the Great Lakes and around Bermuda, where the bottom slope is steep. Similarly, records off Peru show an offshore scale ~ 70 km, greater than the shelf width, because c_n/f is large near the equator.

For $\sigma > |f|$ and nonzero S, trapping is imperfect. However, there are frequencies $\sigma(k)$ at which waves are almost trapped, radiate energy only slowly or respond with maximal amplitude to sustained forcing. These $\sigma(k)$ appear to correspond to dispersion curves in $\sigma < |f|$ but there is still some uncertainty about the role of these waves; they may need some oceanic forcing.

Bottom-trapped waves are an idealized form with motion everywhere parallel to a plane sloping seafloor (in uniform N^2); they decay away from the seafloor. They may propagate for $\sigma < |f|$ or $\sigma > |f|$ and up or down the slope, but always with a component 'forwards' along the slope. If this phase propagation direction is ϕ relative to the along-slope direction, then the velocity being transverse is at angle ϕ to the slope and the frequency is $\sigma = N \partial h / \partial x \cos \phi$. In $\sigma < |f|$ the general coastal trapped wave form tends for large k to bottom-trapped waves confined near the seafloor maximum of $N \partial h / \partial x < |f|$; then $\sigma = N \partial h / \partial x$. If the maximum $N \partial h / \partial x < |f|$, then σ increases to $|f|$ as k increases, a qualitative difference from unstratified behavior; formally there is a smooth transition at $\sigma = |f|$. Bottom-trapped wave identification may be difficult (despite the apparent prevalence of near-bottom currents), requiring knowledge of the local slope and stratification. There is evidence from continental slopes off the eastern USA and NW Africa.

Friction

Friction causes cross-shelf phase shifts and significant damping of coastal trapped waves. The depth-integrated alongshore momentum balance for idealized uniform conditions $(\partial p / \partial y = 0, \int u \, dz = 0)$ is $\partial v / \partial t + rv/h = \tau / \rho h$ suggesting that the flow v lags the forcing stress τ less for low frequency, shallow water, and large friction r. For example, nearshore currents lag the wind less than currents in deeper water offshore. Damping rates may be estimated as $r/h = O(0.003\,Uh^{-1})$, i.e., a decay time less than 4 days for a typical current $U = 0.1\,\mathrm{m\ s^{-1}}$ and depth $h = 100\,m$. In this estimate of friction, U should represent all currents present, e.g., tidal currents can provide strong damping. This decay time converts in to a decay distance $c_g h/r$ for a wave with energy propagation speed c_g. Such decay distances are largest (hundreds to a thousand kilometers or more) for long waves with 'forward' energy propagation; much less for (short) waves with 'backward' energy propagation.

Mean Flows

Mean currents are significant in many places where continental shelf waves have been observed, e.g., adjacent to the Florida current. Waves with phase speeds of a few meters per second or less may be significantly affected by boundary currents of comparable speed, or by vorticity of order f. By linear theory, advection in a uniform mean current V is essentially trivial, but could reverse the propagation of slower waves (higher modes or short wavelengths). Shear $V' \equiv \mathrm{d}V/\mathrm{d}x$ modifies the background potential vorticity to $P(x) \equiv (f + V')/h$; then the gradient of P (rather than f/h) underlies continental shelf wave propagation.

'Barotropic' instability is possible if V is strong enough; necessary conditions are $P'(x_s) = 0$ (some x_s) and $P'[V(x_s) - V] > 0$ for some x; the growth rate is bounded by max $|V'/2|$. Gulf Stream meanders have been interpreted as barotropically unstable shelf waves from Blake Plateau. In a stratified context, these effects of mean flow V may still apply. Additionally, vertical shear $\partial V/\partial z$ is associated with horizontal density gradients:

$$f\partial V/\partial z = -g\rho_0^{-1}\partial \rho/\partial x \qquad [7]$$

and associated 'baroclinic' instability extracting gravitational potential energy. Two-layer models represent $\partial \rho/\partial x$ by a sloping interface; the varying layer depths are another source of gradients $\partial P/\partial x$ in each layer. Thus baroclinic instability may occur even if the current and total depth are uniform. More generally, slow flows over a gently sloping bottom are unstable only if $\partial P/\partial x$ has both signs in the system. Such a two-layer channel model predicts instability at peak-energy frequencies in Shelikov Strait, Alaska (for example) and a corresponding wavelength roughly matching that observed. However, we caution that two-layer models may unduly segregate internal Kelvin and continental shelf wave

types, say, exaggerating the multiplicity of wave forms and scope for instability.

In continuous stratification, the equivalent potential vorticity gradient $\partial f/\partial x + \partial^2 V/\partial x^2 + f^2(N^{-2}\partial V/\partial z)/\partial z$ may support waves in the interior. Density contours rising coastward in association with a shelf edge surface jet modify the fastest continental shelf wave to an inshore 'frontal-trapped' form. The bottom slope also supports waves. Over a uniform bottom slope, additional bottom features can couple and destabilize the interior and bottom modes, even if $V(x,z)$ is otherwise stable. However, the bottom slope stabilizes bottom-intensified waves under an intermediate uniformly stratified layer.

Non-linear Effects

These mirror typical nonlinear effects for waves. For example, each part of the nonlinear Kelvin wave form moves with the local speed $v + [g(h + \eta)]^{1/2}$; crests (η and v positive) gain on troughs (η and v negative) and wave fronts steepen. Similarly, for internal Kelvin waves between an upper layer, depth h, and a deep lower layer (density difference $\Delta\rho$) the local speed is $(gh\Delta\rho/\rho)^{1/2}$ everywhere; troughs gain on crests. Dispersion limits this nonlinear steepening; the associated shorter wavelengths propagate more slowly; and the steepening is left behind by the wave. For small amplitudes and long waves, steepening and dispersion are small and can balance in permanent-form $\mathrm{sech}^2(ky + \sigma t)$ solutions. Steepening may be important for internal Kelvin waves, but for typical continental shelf wave amplitudes (near linear) it takes weeks or months, longer than likely frictional decay times.

Mean currents may be forced via frictional contributions to the time-averaged nonlinear convective derivatives in the momentum equations. The mean flows, scale $h^{-1}h'f^{-1}\hat{u}^2$, are typically confined close to the coast or the shelf break ($\hat{u}$ denotes on–offshore excursion in the waves). Another mechanism is wave-induced form drag over an irregular bottom, giving a biased response to variable forcing; a 'forward' flow along the shelf. For example, low-frequency sinusoidal wind forcing may give a mean current up to a maximum fraction $(2\pi)^{-1}$ of the value under a steady wind, i.e., some centimeters per second, principally in shallower shelf waters.

Three-way interactions between coastal-trapped waves can occur if $\sigma_3 = \sigma_1 \pm \sigma_2$, $k_3 = k_1 \pm k_2$, possible for particular combinations according to the shape of the dispersion curve. Typical timescales for energy exchange are many days; effects may be masked by frictional decay. Near a group velocity of zero, there may be more response to a range of energy inputs.

Alongshore Variations

If changes in the stratification and continental shelf form are small in one wavelength, then individual wave modes conserve a longshore energy flux; local wave forms are as for a uniform shelf. Thus Kelvin wave amplitudes increase as $f^{1/2}$ and are confined closer to the coast at higher latitudes. Energy flux conservation implies a large amplitude increase if waves of frequency σ approach a shelf region where the maximum (σ_M) for their particular mode is near σ, as for Scottish and Vancouver Island diurnal tides; the energy tends to 'pile up.' If shelf variations cause σ_M to fall well below σ, then the waves are totally reflected, with large amplitudes near where $\sigma_M = \sigma$. Poleward-propagating waves experience changing conditions. Near the equator, f is small, S (effective stratification) is large and internal Kelvin-like waves are expected, as off Peru; as f increases poleward, waves evolve to less stratified forms, more like continental shelf waves. (Variations of f are special in supporting offshore energy leakage to Rossby waves in the ocean.)

Small irregularities in the shelf (lateral, vertical scales $\varepsilon(gH)^{1/2}/f, \varepsilon H$) generally cause $O(\varepsilon^2)$ effects, but $O(\varepsilon)$ nearby and in phase shifts after depth changes. Scattering occurs, preferentially to adjacent wave modes and (if unstratified) the highest mode at the incident frequency (having near-zero group velocity). However, long waves, L_W, on long topographic variations (as above; L_T) adopt the appropriate local form; scattering is slow unless $L_W \sim L_T$.

If the depth profile has a self-similar form $h[(x - c(y))/L(y)]$ then long $(\gg L)$ continental shelf waves propagate with changes of amplitude but no scattering or change of form, provided that c and L also vary slowly $(c/c', L/L' \gg L)$. Likewise, there is no scattering if the depth is $h(\xi)$ where $\nabla^2 \xi(x, y) = 0$, representing approximately uniform topographic convexity. In these cases, stronger currents and shorter wavelengths are implied on narrow sections of shelf.

Abrupt features are apt to give the strongest scattering, substantial local changes or eddies on the flow. Scattering is the means of slope–current adjustment to a changed depth profile. However, all energy must remain trapped in $\sigma < |f|$; even in $\sigma > |f|$ special interior angles $\pi/(2n + 1)$ can give perfect Kelvin wave energy transmission (for example). Successive reflections in a finite shelf (embayment) may synthesize near-resonant waves with small energy leakage. A complete barrier across the shelf implies reflection into (short, slow) waves of opposite group velocity. There is a considerable literature of particular calculations. However, it is difficult to generalize, because of the several non-scattering cases interspersed among those with strong scattering.

Generation and Role of Coastal-trapped Waves

Oceanic motion may impinge on the continental shelf. Notably at the equator, waves travel eastward to the coast and divide to travel north and south. In general, oceanic motions accommodate to the presence of the coast and shelf; at a wall, by internal Kelvin waves (vertical structure modes); for more realistic shelf profiles, by coastal trapped waves. Oceanic signals tend to be seen at the coast if alongshelf scale > wave-decay distance or if the feature is shallower than the shelf-water depth.

Natural modes of the ocean are significantly affected by continental shelves. Modes depending on f/h gradients have increased frequencies and forms concentrated over topography. Numerical models have shown 13 modes with periods between 30 and 80 hours, each mode being localized over one shelf area.

Atmospheric pressure forcing the sea surface can be effective in driving Kelvin and edge waves, especially if there is some match of speed and scale, more likely in shallower (shelf) seas.

Longshore wind stress τ is believed to be the most effective means of generating coastal trapped waves. Within the forcing region, the flow tends to match the wind field; when or where the forcing ceases, the wave travels onwards ('forwards') and is then most recognizable. A simple view of this forcing is that τ accelerates the alongshore transport hu. A more sophisticated view is that τ induces a cross-shelf surface transport $|\tau|/\rho f$; coastal blocking induces a compensating return flow beneath, which is acted upon by the Coriolis force to give the same accelerating alongshore transport. A typical stress $0.1\,\mathrm{N\,m}^{-2}$ for $10^5\,\mathrm{s}$ (~ 1 day) accelerates $100\,\mathrm{m}$ water to $0.1\,\mathrm{m\,s}^{-1}$. Winds blowing across depth contours may be comparably effective if the coast is distant.

Other generation mechanisms include scattering (of alongshore flow, especially) by shelf irregularities as above, variable river runoff, and a co-oscillating sea.

Shelf-sea motion is often dominated by tides and responses to wind forcing. On a narrow shelf, oceanic elevation signals penetrate more readily to the coast as the higher-mode decay distances are short; the tide is represented primarily by a Kelvin wave spanning the ocean and shelf. Model fits to semidiurnal measurements, showing a dominant

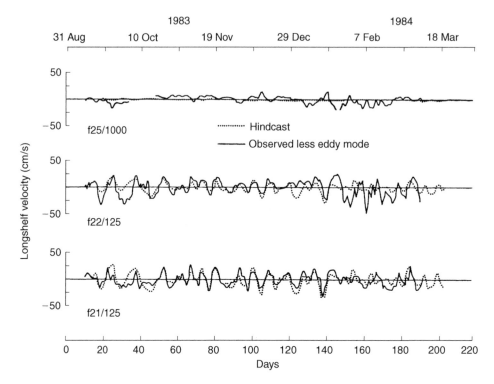

Figure 5 Comparison of hindcast along-shelf currents using three coastal-trapped wave modes with measured currents in a cross-shelf section after band-pass filtering and removing an eddy mode. (Reproduced with permission from Church JA, White NJ, Clarke AJ, Freeland HJ and Smith RL (1986) Coastal-trapped waves on the East Australian continental shelf. Part II: model verification. *Journal of Physical Oceanography* 16: 1945–1957.)

Kelvin wave, have been made off California, Scotland and north-west Africa, for example. Several areas at higher latitudes show dominant continental shelf wave contributions to diurnal tidal currents, e.g., west of Scotland, Vancouver Island, Yermak Plateau. On a wide shelf, there is correspondingly greater scope for wind-driven elevations and currents. The extensive forcing scale, typically greater than the shelf width, induces flow with minimal structure (typically no reversals across the shelf) corresponding to the lowest-mode continental shelf wave with maximum elevation signal at the coast. In stratified conditions, the upwelling or downwelling response also corresponds to a wave (or waves).

As a wave travels, its amplitude is continually incremented by local forcing. A model based on this approach was used to make the hindcast of measured currents on the south-east Australian shelf shown in **Figure 5**. At any fixed position, the motion results from local forcing and from arriving waves, bringing the influence of forcing (e.g., upwelling) 'forwards' from the 'backward' direction. In the Peruvian upwelling regime, for example, variable currents are not well correlated with local winds but include internal Kelvin-like features coming from nearer the equator. This is hardly compatible with (common) simplifications of a zero alongshore pressure

gradient. Moreover, the waves carry the influence of assumed 'backward' boundary conditions far into a model. The same applies for steady flow. Friction introduces a 'forward' decay distance for a coastal-trapped wave; this distance has a definite low-frequency limit. Currents decay over these distances according to their structure as a wave combination. Thus alongshore evolution or adjustment of flow (however forced) is affected by coastal-trapped waves whose properties should guide model design.

Summary

This article considers waves extending across the continental shelf and/or slope and having periods of the order of one day or longer. Their phase propagation is generally cyclonic, with the coast to the right in the Northern Hemisphere, a sense denoted 'forward'; cross-slope displacements change water-column depth and relative vorticity, causing cross-slope movement of adjacent water columns. At short-scales, energy propagation can be in the opposite 'backward' sense. Strict trapping occurs only for periods longer than half a pendulum day; shorter-period waves leak energy to the deep ocean, albeit only slowly for some forms. The waves travel faster in stratified seas and on broad shelf-slope profiles;

speeds can be affected, even reversed, by along-shelf flows and reverses of bottom slope. Large amplitudes and abrupt alongshore changes in topography cause distortion and transfers between wave modes. The waves form a basis for the behavior (response to forcing, propagation) of shelf and slope motion on scales of days and the shelf width. Hence, they are important in shelf and slope–sea responses to forcing by tides, winds (e.g., upwelling), density gradients, and oceanic features. Their propagation (distance before decay) implies nonlocal response (over a comparable distance), especially in the 'forward' direction.

See also

Coastal Circulation Models. Regional and Shelf Sea Models. Storm Surges. Tides.

Further Reading

Brink KH (1991) Coastal trapped waves and wind-driven currents over the continental shelf. *Annual Review of Fluid Mechanics* 23: 389–412.

Brink KH and Chapman DC (1985) Programs for computing properties of coastal-trapped waves and wind-driven motions over the continental shelf and slope. Woods Hole Oceanographic Institution, Technical Report 85–17, 2nd edn, 87–24.

Dale AC and Sherwin TJ (1996) The extension of baroclinic coastal-trapped wave theory to superinertial frequencies. *Journal of Physical Oceanography* 26: 2305–2315.

Huthnance JM, Mysak LA and Wang D-P (1986) Coastal trapped waves. In: Mooers CNK (ed.) *Baroclinic Processes on Continental Shelves*. Coastal and Estuarine Sciences, 3, pp. 1–18. Washington DC. American Geophysical Union.

LeBlond PH and Mysak LA (1978) *Waves in the Ocean*. New York: Elsevier.

STORM SURGES

R. A. Flather, Proudman Oceanographic Laboratory, Bidston Hill, Prenton, UK

Introduction and Definitions

Storm surges are changes in water level generated by atmospheric forcing; specifically by the drag of the wind on the sea surface and by variations in the surface atmospheric pressure associated with storms. They last for periods ranging from a few hours to 2 or 3 days and have large spatial scales compared with the water depth. They can raise or lower the water level in extreme cases by several meters; a raising of level being referred to as a 'positive' surge, and a lowering as a 'negative' surge. Storm surges are superimposed on the normal astronomical tides generated by variations in the gravitational attraction of the moon and sun. The storm surge component can be derived from a time-series of sea levels recorded by a tide gauge using:

$$\text{surge residual} = (\text{observed sea level}) - (\text{predicted tide level}) \quad [1]$$

producing a time-series of surge elevations. **Figure 1** shows an example.

Sometimes, the term 'storm surge' is used for the sea level (including the tidal component) during a storm event. It is important to be clear about the usage of the term and its significance to avoid confusion. Storms also generate surface wind waves that have periods of order seconds and wavelengths, away from the coast, comparable with or less than the water depth.

Positive storm surges combined with high tides and wind waves can cause coastal floods, which, in terms of the loss of life and damage, are probably the most destructive natural hazards of geophysical origin. Where the tidal range is large, the timing of the surge relative to high water is critical and a large surge at low tide may go unnoticed. Negative surges reduce water depth and can be a threat to navigation. Associated storm surge currents, superimposed on tidal and wave-generated flows, can also contribute to extremes of current and bed stress responsible for coastal erosion. A proper understanding of storm surges, the ability to predict them and measures to mitigate their destructive effects are therefore of vital concern.

Storm Surge Equations

Most storm surge theory and modeling is based on depth-averaged hydrodynamic equations applicable to both tides and storm surges and including non-linear terms responsible for their interaction. In vector form, these can be written:

$$\frac{\partial \zeta}{\partial t} + \nabla \cdot (D\mathbf{q}) = 0 \quad [2]$$

$$\frac{\partial \mathbf{q}}{\partial t} + \mathbf{q} \cdot \nabla \mathbf{q} - f\mathbf{k} \times \mathbf{q} = -g\nabla(\zeta - \bar{\zeta}) - \frac{1}{\rho}\nabla p_a \\ + \frac{1}{\rho D}(\tau_s - \tau_b) + A\nabla^2 \mathbf{q} \quad [3]$$

where t is time; ζ the sea surface elevation; $\bar{\zeta}$ the equilibrium tide; $\mathbf{q}$ the depth-mean current; τ_s the wind stress on the sea surface; τ_b the bottom stress; p_a atmospheric pressure on the sea surface; D the total water depth ($D = h + \zeta$, where h is the undisturbed depth); ρ the density of sea water, assumed to be uniform; g the acceleration due to gravity; f the Coriolis parameter ($= 2\omega \sin\varphi$, where ω is the angular speed of rotation of the Earth and φ is the latitude); $\mathbf{k}$ a unit vector in the vertical; and A the coefficient of horizontal viscosity. Eqn [2] is the continuity equation expressing conservation of volume. Eqn [3] equates the accelerations (left-hand side) to the force per unit mass (right-hand side).

In this formulation, bottom stress, τ_b is related to the current, $\mathbf{q}$, using a quadratic law:

$$\tau_b = k\rho\mathbf{q}|\mathbf{q}| \quad [4]$$

where k is a friction parameter (~ 0.002). Similarly, the wind stress, τ_s, is related to $\mathbf{W}$, the wind velocity at a height of $10\,\text{m}$ above the surface, also using a quadratic law:

$$\tau_s = c_D\rho_a\mathbf{W}|\mathbf{W}| \quad [5]$$

where ρ_a is the density of air and c_D a drag coefficient. Measurements in the atmospheric boundary layer suggest that c_D increases with wind speed, W, accounting for changes in surface roughness associated with wind waves. A typical form due to J.

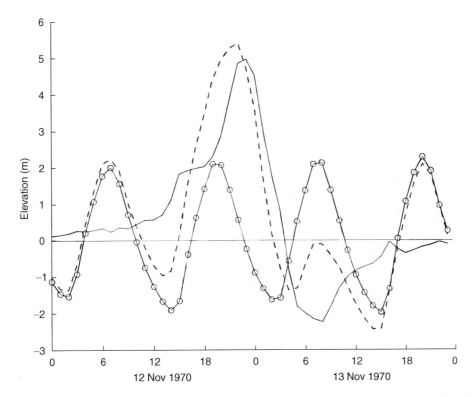

Figure 1 Water level (dashed line), predicted tide (line with ○), and the surge residual (continuous line) at Sandwip Island, Bangladesh, during the catastrophic storm surge of 12–13 November 1970 (times are GMT).

Wu is:

$$10^3 c_D = 0.8 + 0.065\,W \qquad [6]$$

Alternatively, from dimensional analysis, H. Charnock obtained $gz_0/u_*^2 = \alpha$, where z_0 is the aerodynamic roughness length associated with the surface wavefield, u_* is the friction velocity ($u_*^2 = \tau_s/\rho_a$), and α is the Charnock constant. So, the roughness varies linearly with surface wind stress. Assuming a logarithmic variation of wind speed with height z above the surface, $W(z) = (u_*/\kappa)\,\ln(z/z_0)$, where κ is von Kármán's constant. It follows that for $z = 10$ m:

$$c_D = \left[(1/\kappa)\ln\{gz/(\alpha c_D W^2)\}\right]^{-2} \qquad [7]$$

Estimates of α range from 0.012 to 0.035.

Generation and Dynamics of Storm Surges

The forcing terms in eqn [3] which give rise to storm surges are those representing wind stress and the horizontal gradient of surface atmospheric pressure. Very simple solutions describe the basic mechanisms. The sea responds to atmospheric pressure variations by adjusting sea level such that, at depth, pressure in the water is uniform, the hydrostatic approximation.

Assuming in eqn [3] that $\mathbf{q} = 0$ and $\tau_s = 0$, then $g\nabla\zeta + (1/\rho)\nabla p_a = 0$, so $\rho g\zeta + p_a =$ constant. This gives the 'inverse barometer effect' whereby a decrease in atmospheric pressure of 1 hPa produces an increase in sea level of approximately 1 cm. Wind stress produces water level variations on the scale of the storm. Eqns [5] and [6] imply that the strongest winds are most important since effectively $\tau_s \propto W^3$. Both pressure and wind effects are present in all storm surges, but their relative importance varies with location. Since wind stress is divided by D whereas ∇p_a is not, it follows that wind forcing increases in importance in shallower water. Consequently, pressure forcing dominates in the deep ocean whereas wind forcing dominates in shallow coastal seas. Major destructive storm surges occur when extreme storm winds act over extensive areas of shallow water.

As well as the obvious wind set-up, with the component of wind stress directed towards the coast balanced by a surface elevation gradient, winds parallel to shore can also generate surges at higher latitudes. Wind stress parallel to the coast with the coast on its right will drive a longshore current, limited by bottom friction. Geostrophic balance gives (in the Northern Hemisphere) a surface gradient raising levels at the coast (see Tides).

Amplification of surges may be caused by the funneling effect of a converging coastline or estuary and by a resonant response; e.g. if the wind forcing travels at the same velocity as the storm wave, or matches the natural period of oscillation of a gulf, producing a seiche.

As a storm moves away, surges generated in one area may propagate as free waves, contributing as externally generated components to surges in another area. Generally, away from the forcing center, the response of the ocean consists of longshore propagating coastally trapped waves. Examples include external surges in the North Sea (see **Figure 2**), which are generated west and north of Scotland and propagate anticlockwise round the basin like the diurnal tide; approximately as a Kelvin wave. Low mode continental shelf waves have been identified in surges on the west coast of Norway, in the Middle Atlantic Bight of the US, in the East China Sea, and on the north-west shelf of Australia. Currents associated with low mode continental shelf waves generated by a tropical cyclone crossing the north-west shelf of Australia have been observed and explained by numerical modeling. Edge waves can also be generated by cyclones travelling parallel to the coast in the opposite direction to shelf waves.

From eqns [3] and [4], bottom stress (which dissipates surges) also depends on water depth and is non-linear; the current including contributions from tide and surge, $q = q_T + q_s$. Consequently, dissipation of surges is stronger in shallow water and where tidal currents are also strong. In deeper water and where tides are weak, free motions can persist for long times or propagate long distances. For example, the Adriatic has relatively small tides and seiches excited by storms can persist for many days.

In areas with substantial tides and shallow water, non-linear dynamical processes are important, resulting in interactions between the tide and storm surge such that both components are modified. The main contribution arises from bottom stress, but time-dependent water depth, $D = h + \zeta(t)$, can also be significant (e.g. $\tau_s/(\rho D)$ will be smaller at high tide than at low tide). An important consequence is that the linear superposition of surge and tide without accounting for their interaction gives substantial errors in estimating water level. For example, for surges propagating southwards in the North Sea into the Thames Estuary, surge maxima tend to occur on the rising tide rather than at high water (**Figure 3**).

Interactions also occur between the tide–surge motion and surface wind waves (see below).

Areas Affected by Storm Surges

Major storm surges are created by mid-latitude storms and by tropical cyclones (also called hurricanes and typhoons) which generally occur in geographically separated areas and differ in their scale. Mid-latitude storms are relatively large and evolve slowly enough to allow accurate predictions of their wind and pressure fields from atmospheric forecast models. In tropical cyclones, the strongest winds occur within a few tens of kilometers of the storm center and so are poorly resolved by routine weather prediction models. Their evolution is also rapid and much more difficult to predict. Consequently prediction and mitigation of the effects of storm surges is further advanced for mid-latitude storms than for tropical cyclones.

Tropical cyclones derive energy from the warm surface waters of the ocean and develop only where the sea surface temperature (SST) exceeds 26.5°C. Since their generation is dependent on the effect of the local vertical component of the Earth's rotation, they do not develop within 5° of the equator. **Figure 4** shows the main cyclone tracks. Areas affected include: the continental shelf surrounding the Gulf of Mexico and on the east coast of the US (by hurricanes); much of east Asia including Vietnam, China, the Philippines and Japan (by typhoons); the Bay of Bengal, in particular its shallow north-east corner, and northern coasts of Australia (by tropical cyclones). Areas affected by mid-latitude storms include the North Sea, the Adriatic, and the Patagonian Shelf. Inland seas and large lakes, including the Great Lakes, Lake Okeechobee (Florida), and Lake Biwa (Japan) also experience surges.

The greatest loss of life due to storm surges has occurred in the northern Bay of Bengal and Meghna Estuary of Bangladesh. A wide and shallow continental shelf bounded by extensive areas of low-lying poorly protected land is impacted by tropical cyclones. Cyclone-generated storm surges on 12–13 November 1970 and 29–30 April 1991 (**Figure 5**) killed approximately 250 000 and 140 000 people, respectively, in Bangladesh.

A severe storm in the North Sea on 31 January–1 February 1953 generated a large storm surge, which coincided with a spring tide to cause catastrophic floods in the Netherlands (**Figure 6**) and south-east England, killing approximately 2000 people. Subsequent government enquiries resulted in the 'Delta Plan' to improve coastal defences in Holland, led to the setting up of coastal flood warning authorities, and accelerated research into storm surge dynamics.

The city of Venice, in Italy, suffers frequent 'acqua alta' which flood the city, disrupting its life and accelerating the disintegration of the unique historic buildings.

In 1969 Hurricane Camille created a surge in the Gulf of Mexico which rose to 7 m above mean sea

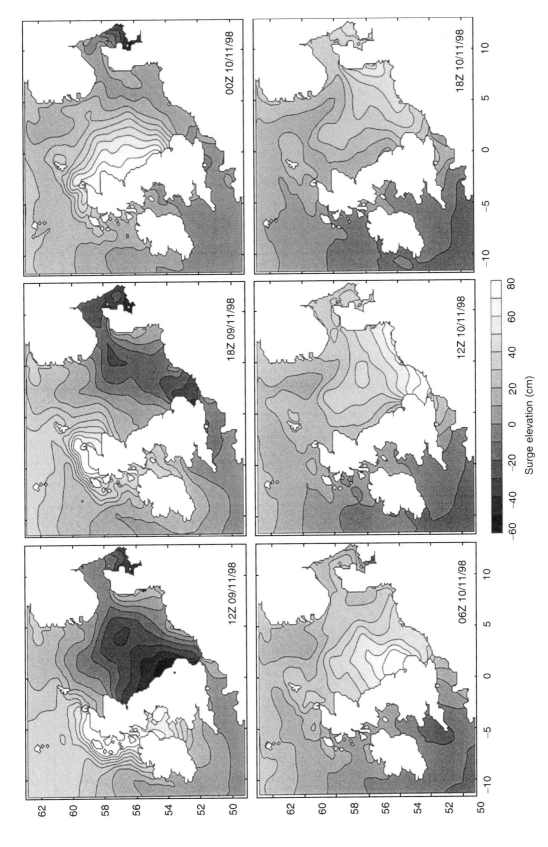

Figure 2 Propagation of an external surge in the North Sea from a numerical model simulation.

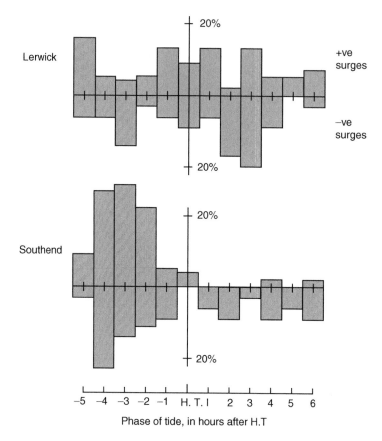

Figure 3 Frequency distribution relative to the time of tidal high water of positive and negative surges at Lerwick (northern North Sea) and Southend (Thames Estuary). The phase distribution at Lerwick is random, whereas due to tide–surge interaction most surge peaks at Southend occur on the rising tide (re-plotted from Prandle D and Wolf J, 1978, The interaction of surge and tide in the North Sea and River Thames. *Geophys. J.R. Astr. Soc.*, 55: 203–216, by permission of the Royal Astronomical Society).

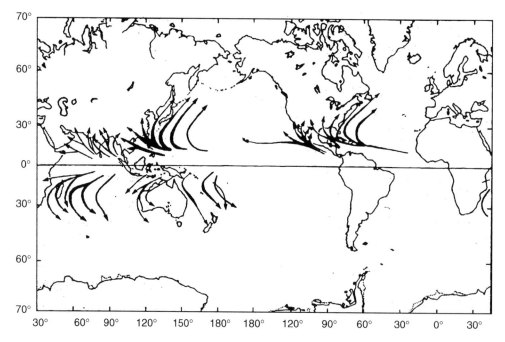

Figure 4 Tropical cyclone tracks (from Murty, 1984, reproduced by permission of the Department of Fisheries and Oceans, Canada).

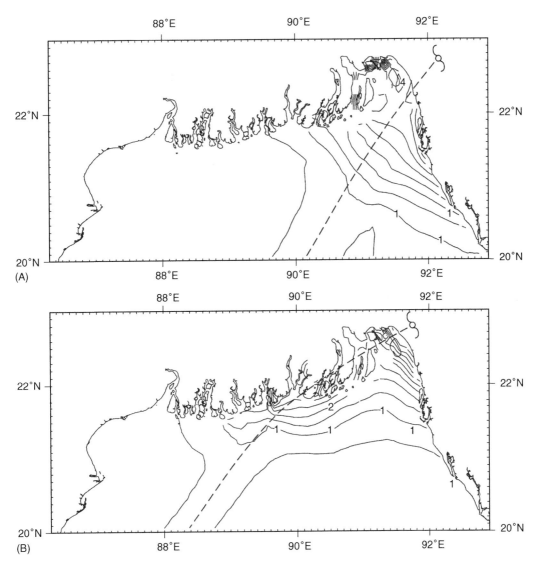

Figure 5 Cyclone tracks (dashed lines) and maximum computed surge elevation measured in meters in the northern Bay of Bengal during the cyclones of (A) 1970 and (B) 1991 from numerical model simulations (contour interval 0.5 m). (Re-plotted from Flather, 1994, by permission of the American Meteorological Society.)

level, causing more than 100 deaths and about 1 billion dollars worth of damage. An earlier cyclone in the region, in 1900, flooded the island of Galveston, Texas, with the loss of 6000 lives.

Storm Surge Prediction

Early research on storm surges was based on analysis of observations and solution of simplified – usually linearized – equations for surges in idealized channels, rectangular gulfs, and basins with uniform depth. The first self-recording tide gauge was installed in 1832 at Sheerness in the Thames Estuary, England, so datasets for analysis were available from an early stage. Interest was stimulated by events such

as the 1953 storm surge in the North Sea, which highlighted the need for forecasts.

First prediction methods were based on empirical formulae derived by correlating storm surge elevation with atmospheric pressure, wind speed and direction and, where appropriate, observed storm surges from a location 'upstream'. Long time-series of observations are required to establish reliable correlations. Where such observations existed, e.g. in the North Sea, the methods were quite successful.

From the 1960s, developments in computing and numerical techniques made it possible to simulate and predict storm surges by solving discrete approximations to the governing equations (eqns [1] and [2]). The earliest and simplest methods, pioneered in Europe by W. Hansen and N.S. Heaps and in the USA by

Figure 6 Breached dyke in the Netherlands after the 1953 North Sea storm surge. (Reproduced by permission of RIKZ, Ministry of Public Works, The Netherlands.)

R.O. Reid and C.P. Jelesnianski, used a time-stepping approach based on finite difference approximations on a regular grid. Surge–tide interaction could be accounted for by solving the non-linear equations and including tide. Effects of inundation could also be included by allowing for moving boundaries; water levels computed with a fixed coast can be O(10%) higher than those with flooding of the land allowed.

Recent developments have revolutionized surge modeling and prediction. Among these, coordinate transformations, curvilinear coordinates and grid nesting allow better fitting of coastal boundaries and enhanced resolution in critical areas. A simple example is the use of polar coordinates in the SLOSH (Sea, Lake and Overland Surges from Hurricanes) model focusing on vulnerable sections of the US east coast. Finite element methods with even greater flexibility in resolution (e.g. **Figure 7**) have also been used in surge computations in recent years.

There has also been increasing use of three-dimensional (3-D) models in storm surge studies. Their main advantage is that they provide information on the vertical structure of currents and, in particular, allow the bottom stress to be related to flow near the seabed. This means that in a 3-D formulation the bottom stress need not oppose the direction of the depth mean flow and hence of the water transport. Higher surge estimates result in some cases.

In the last decades many countries have established and now operate model-based flood warning systems. Although finite element methods and 3-D models have been developed and are used extensively for research, most operational models are still based on depth-averaged finite-difference formulations.

A key requirement for accurate surge forecasts is accurate specification of the surface wind stress. Surface wind and pressure fields from numerical weather prediction (NWP) models are generally used for mid-latitude storms. Even here, resolution of small atmospheric features can be important, so preferably NWP data at a resolution comparable with that of the surge model should be used. For tropical cyclones, the position of maximum winds at landfall is critical, but prediction of track and evolution (change in intensity, etc.) is problematic. Presently, simple models are often used based on basic parameters: p_c, the central pressure; W_m, the maximum sustained 10 m wind speed; R, the radius to maximum winds; and the velocity, V, of movement of the cyclone's eye. Assuming a pressure profile, e.g. that due to G.J. Holland:

$$p_a(r) = p_c + \Delta p \exp\left[-(R/r)^B\right] \qquad [8]$$

where r is the radial distance from the cyclone center, Δp the pressure deficit (difference between the ambient and central pressures), and B is a 'peakedness' factor typically $1.0 < B < 2.5$. Wind fields can then be estimated using further assumptions and approximations. First, the gradient or cyclostrophic wind can be calculated as a function of r. An empirical factor (~ 0.8) reduces this to W, the 10 m wind. A contribution from the motion of the storm (maybe 50% of V) can be added, introducing asymmetry to the wind field, and finally to account for frictional effects in the atmospheric boundary layer, wind vectors may be turned inwards by a cross-isobar angle of $10°$–$25°$. Such procedures are rather crude, so that cyclone surges computed using the resulting winds

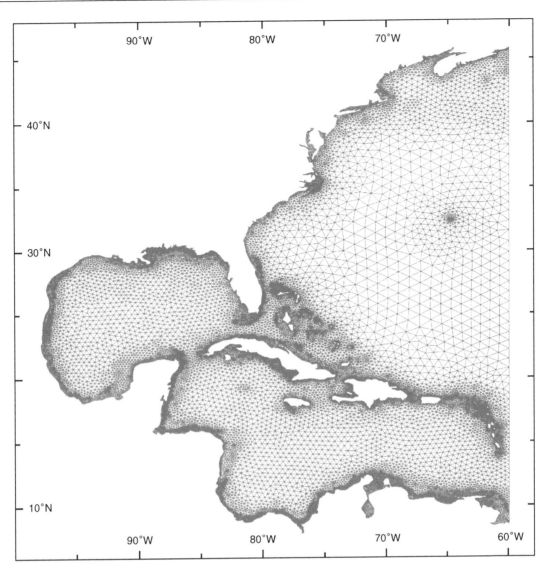

Figure 7 A finite element grid for storm surge calculations on the east coast of the USA, the Gulf of Mexico, and Caribbean (from Blain CA, Westerink JJ and Leuttich RL (1994), The influence of domain size on the response characteristics of a hurricane storm surge model. *J. Geophys. Res.*, 99(C9) 18467–18479. Reproduced by permission of the American Geophysical Union.).

are unlikely to be very accurate. Simple vertically integrated models of the atmospheric boundary layer have been used to compute winds from a pressure distribution such as eqn [8], providing a more consistent approach. In reality, cyclones interact with the ocean. They generate wind waves, which modify the sea surface roughness, z_0, and hence the wind stress generating the surge. Wind- and wave-generated turbulence mixes the surface water changing its temperature and so modifies the flux of heat from which the cyclone derives its energy. Progress requires improved understanding of air–sea exchanges at extreme wind speeds and high resolution coupled atmosphere–ocean models.

Interactions with Wind Waves

As mentioned above, observations suggested that Charnock's α was not constant but depended on water depth and 'wave age', a measure of the state of development of waves. Young waves are steeper and propagate more slowly, relative to the wind speed, than fully developed waves and so are aerodynamically rougher enhancing the surface stress. These effects can be incorporated in a drag coefficient which is a function of wave age, wave height, and water depth and agrees well with published datasets over the whole range of wave ages. Further research, considering the effects of waves on airflow in the atmospheric boundary layer, led P.A.E.M. Janssen to propose a wave-induced stress enhancing the effective roughness. Application of this theory requires the dynamical coupling of surge and wave models such that friction velocity and roughness determine and are determined by the waves. Mastenbroek *et al.* obtained improved agreement with observed surges on the Dutch coast by including the wave-induced stress in a model experiment (**Figure 8**). However, they also found that the same improvement could be obtained by a small increase in the standard drag coefficient.

In shallow water, wave orbital velocities also reach the seabed. The bottom stress acting on surge and tide is therefore affected by turbulence introduced at the seabed in the wave boundary layer. With simplifying assumptions, models describing these effects have been developed and can be used in storm surge modeling. Experiments using both 2-D and 3-D surge models have been carried out. 3-D modeling of surges in the Irish Sea using representative waves shows significant effects on surge peaks and improved agreement with observations. Bed stresses are much enhanced in shallow water. Because the processes depend on the nature of the bed, a more complete treatment should take account of details of bed types.

Non-linear interactions give rise to a wave-induced mean flow and a change in mean water depth (wave set-up and set-down). The former has contributions from a mean momentum density produced by a non-zero mean flow in the surface layer (above the trough level of the waves), and from wave breaking. Set-up and set-down arise from the 'radiation stress', which is defined as the excess momentum flux due to the waves (*see* Waves on Beaches). Mastenbroek *et al.* showed that the radiation stress has a relatively small influence on the calculated water levels in the North Sea but cannot be neglected in all cases. It is important where depth-induced changes in the waves, as shoaling or breaking, dominate over propagation and generation, i.e. in coastal areas. The effects should be included in the momentum equations of the surge model.

Although ultimately coupled models with a consistent treatment of exchanges between atmosphere and ocean and at the seabed remain a goal, it appears that with the present state of understanding the benefits may be small compared with other inherent uncertainties. In particular, accurate definition of the wind field itself and details of bed types (rippled or smooth, etc.) are not readily available.

Data Assimilation

Data assimilation plays an increasing role, making optimum use of real-time observations to improve the accuracy of initial data in forecast models. Bode

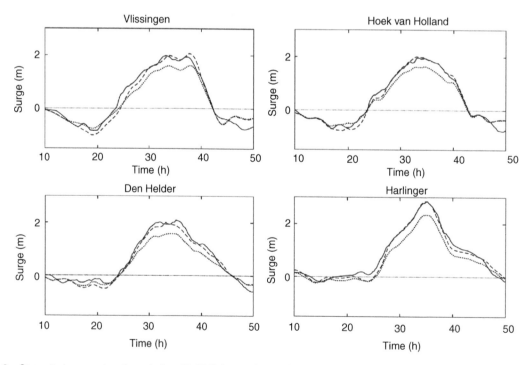

Figure 8 Computed surge elevations during 13–16 February 1989 with (dashed line) and without (dotted) wave stress compared with observations (continuous line). (Re-plotted from Mastenbroek *et al.*, 1994 by permission of the American Geophysical Union.)

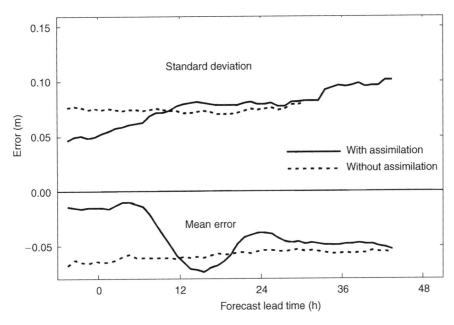

Figure 9 Variation of forecast errors in surge elevation with forecast lead-time from the Dutch operational model with (continuous line) and without (dashed line) assimilation of tide gauge data. (Replotted from Flather, 2000, by permission of Elsevier Science BV.)

and Hardy (see Further Reading section) reviewed two approaches, involving solution of adjoint equations and Kalman filtering. The Dutch operational system has used Kalman filtering since 1992, incorporating real-time tide gauge data from the east coast of Britain. Accuracy of predictions (**Figure 9**) is improved for the first 10–12 h of the forecast.

Related Issues

Extremes

Statistical analysis of storm surges to derive estimates of extremes is important for the design of coastal defenses and safety of offshore structures. This requires long time- series of surge elevation derived from observations of sea level where available or, increasingly, from model hindcasts covering O(50 years) forced by meteorological analyses.

Climate Change Effects

Climate change will result in a rise in sea level and possible changes in storm tracks, storm intensity and frequency, collectively referred to as 'storminess'. Changes in water depth with rising mean sea level (MSL) will modify the dynamics of tides and surges, increasing wavelengths and modifying the generation, propagation, and dissipation of storm surges. Increased water depth implies a small reduction in the effective wind stress forcing, suggesting smaller surges. However, effects of increased storminess may offset this. It has been suggested, for example, that increased temperatures in some regions could raise sea surface temperatures resulting in more intense and more frequent tropical cyclones. Regions susceptible to tropical cyclones could also be extended. Research is in progress to assess and quantify some of these effects, e.g. with tide–surge models forced by outputs from climate GCMs. An important issue is that of distinguishing climate-induced change from the natural inter-annual and decadal variability in storminess and hence surge extremes.

See also

Tides. Waves on Beaches.

Further Reading

Bode L and Hardy TA (1997) Progress and recent developments in storm surge modelling. *Journal of Hydraulic Engineering* 123(4): 315–331.

Flather RA (1994) A storm surge prediction model for the northern Bay of Bengal with application to the cyclone disaster in April 1991. *Journal of Physical Oceanography* 24: 172–190.

Flather RA (2000) Existing operational oceanography. *Coastal Engineering* 41: 13–40.

Heaps NS (1967) Storm surges. In: Barnes H (ed.) *Oceanography and Marine Biology Annual Review 5*, pp. 11–47. London: Allen and Unwin.

Mastenbroek C, Burgers G, and Janssen PAEM (1993) The dynamical coupling of a wave model and a storm surge model through the atmospheric boundary layer. *Journal of Physical Oceanography* 23: 1856–1866.

Murty TS (1984) Storm surges – meteorological ocean tides. *Canadian Bulletin of Fisheries and Aquatic Sciences* 212: 1–897.

Murty TS, Flather RA, and Henry RF (1986) The storm surge problem in the Bay of Bengal. *Progress in Oceanography* 16: 195–233.

Pugh DT (1987) *Tides, Surges, and Mean Sea Level*. Chichester: John Wiley Sons.

World Meteorological Organisation (1978) *Present Techniques of Tropical Storm Surge Prediction*. Report Report 13. Marine science affairs, WMO No. 500. Geneva, Switzerland.

TSUNAMI

P. L.-F. Liu, Cornell University, Ithaca, NY, USA

Introduction

Tsunami is a Japanese word that is made of two characters: *tsu* and *nami*. The character *tsu* means harbor, while the character *nami* means wave. Therefore, the original word *tsunami* describes large wave oscillations inside a harbor during a 'tsunami' event. In the past, tsunami is often referred to as 'tidal wave', which is a misnomer. Tides, featuring the rising and falling of water level in the ocean in a daily, monthly, and yearly cycle, are caused by gravitational influences of the moon, sun, and planets. Tsunamis are not generated by this kind of gravitational forces and are unrelated to the tides, although the tidal level does influence a tsunami striking a coastal area.

The phenomenon we call a tsunami is a series of water waves of extremely long wavelength and long period, generated in an ocean by a geophysical disturbance that displaces the water within a short period of time. Waves are formed as the displaced water mass, which acts under the influence of gravity, attempts to regain its equilibrium. Tsunamis are primarily associated with submarine earthquakes in oceanic and coastal regions. However, landslides, volcanic eruptions, and even impacts of objects from outer space (such as meteorites, asteroids, and comets) can also trigger tsunamis.

Tsunamis are usually characterized as shallow-water waves or long waves, which are different from wind-generated waves, the waves many of us have observed on a beach. Wind waves of 5–20-s period ($T =$ time interval between two successive wave crests or troughs) have wavelengths ($\lambda = T^2(g/2\pi)$ distance between two successive wave crests or troughs) of c. 40–620 m. On the other hand, a tsunami can have a wave period in the range of 10 min to 1 h and a wavelength in excess of 200 km in a deep ocean basin. A wave is characterized as a shallow-water wave when the water depth is less than 5% of the wavelength. The forward and backward water motion under the shallow-water wave is felted throughout the entire water column. The shallow water wave is also sensitive to the change of water depth. For instance, the speed (celerity) of a shallow-water wave is equal to the square root of the product

of the gravitational acceleration ($9.81\,\mathrm{m\,s^{-2}}$) and the water depth. Since the average water depth in the Pacific Ocean is 5 km, a tsunami can travel at a speed of about $800\,\mathrm{km\,h^{-1}}$ ($500\,\mathrm{mi\,h^{-1}}$), which is almost the same as the speed of a jet airplane. A tsunami can move from the West Coast of South America to the East Coast of Japan in less than 1 day.

The initial amplitude of a tsunami in the vicinity of a source region is usually quite small, typically only a meter or less, in comparison with the wavelength. In general, as the tsunami propagates into the open ocean, the amplitude of tsunami will decrease for the wave energy is spread over a much larger area. In the open ocean, it is very difficult to detect a tsunami from aboard a ship because the water level will rise only slightly over a period of 10 min to hours. Since the rate at which a wave loses its energy is inversely proportional to its wavelength, a tsunami will lose little energy as it propagates. Hence in the open ocean, a tsunami will travel at high speeds and over great transoceanic distances with little energy loss.

As a tsunami propagates into shallower waters near the coast, it undergoes a rapid transformation. Because the energy loss remains insignificant, the total energy flux of the tsunami, which is proportional to the product of the square of the wave amplitude and the speed of the tsunami, remains constant. Therefore, the speed of the tsunami decreases as it enters shallower water and the height of the tsunami grows. Because of this 'shoaling' effect, a tsunami that was imperceptible in the open ocean may grow to be several meters or more in height.

When a tsunami finally reaches the shore, it may appear as a rapid rising or falling water, a series of breaking waves, or even a bore. Reefs, bays, entrances to rivers, undersea features, including vegetations, and the slope of the beach all play a role modifying the tsunami as it approaches the shore. Tsunamis rarely become great, towering breaking waves. Sometimes the tsunami may break far offshore. Or it may form into a bore, which is a step-like wave with a steep breaking front, as the tsunami moves into a shallow bay or river. **Figure 1** shows the incoming 1946 tsunami at Hilo, Hawaii.

The water level on shore can rise by several meters. In extreme cases, water level can rise to more than 20 m for tsunamis of distant origin and over 30 m for tsunami close to the earthquake's epicenter. The first wave may not always be the largest in the series of waves. In some cases, the water level will fall significantly first, exposing the bottom of a bay

Figure 1   1946 tsunami at Hilo, Hawaii (Pacific Tsunami Museum). Wave height may be judged from the height of the trees.

or a beach, and then a large positive wave follows. The destructive pattern of a tsunami is also difficult to predict. One coastal area may see no damaging wave activity, while in a neighboring area destructive waves can be large and violent. The flooding of an area can extend inland by 500 m or more, covering large expanses of land with water and debris. Tsunamis may reach a maximum vertical height onshore above sea level, called a runup height, of 30 m.

Since scientists still cannot predict accurately when earthquakes, landslides, or volcano eruptions will occur, they cannot determine exactly when a tsunami will be generated. But, with the aid of historical records of tsunamis and numerical models, scientists can get an idea as to where they are most likely to be generated. Past tsunami height measurements and computer modeling can also help to forecast future tsunami impact and flooding limits at specific coastal areas.

Historical and Recent Tsunamis

Tsunamis have been observed and recorded since ancient times, especially in Japan and the Mediterranean areas. The earliest recorded tsunami occurred in 2000 BC off the coast of Syria. The oldest reference of tsunami record can be traced back to the sixteenth century in the United States.

During the last century, more than 100 tsunamis have been observed in the United States alone. Among them, the 1946 Alaskan tsunami, the 1960 Chilean tsunami, and the 1964 Alaskan tsunami were the three most destructive tsunamis in the US history. The 1946 Aleutian earthquake $(M_w = 7.3)$ generated catastrophic tsunamis that attacked the Hawaiian Islands after traveling about 5 h and killed 159 people.

(The magnitude of an earthquake is defined by the seismic moment, M_0 (dyn cm), which is determined from the seismic data recorded worldwide. Converting the seismic moment into a logarithmic scale, we define $M_w = (1/1.5)\log_{10} M_0 - 10.7$.) The reported property damage reached \$26 million. The 1960 Chilean tsunami waves struck the Hawaiian Islands after 14 h, traveling across the Pacific Ocean from the Chilean coast. They caused devastating damage not only along the Chilean coast (more than 1000 people were killed and the total property damage from the combined effects of the earthquake and tsunami was estimated as \$417 million) but also at Hilo, Hawaii, where 61 deaths and \$23.5 million in property damage occurred (see **Figure 2**). The 1964 Alaskan tsunami triggered by the Prince William Sound earthquake $(M_w = 8.4)$, which was recorded as one of the largest earthquakes in the North American continent, caused the most destructive damage in Alaska's history. The tsunami killed 106 people and the total damage amounted to \$84 million in Alaska.

Within less than a year between September 1992 and July 1993, three large undersea earthquakes strike the Pacific Ocean area, causing devastating tsunamis. On 2 September 1992, an earthquake of magnitude 7.0 occurred *c.* 100 km off the Nicaraguan coast. The maximum runup height was recorded as 10 m and 168 people died in this event. A few months later, another strong earthquake $(M_w = 7.5)$ attacked the Flores Island and surrounding area in Indonesia on 12 December 1992. It was reported that more than 1000 people were killed in the town of Maumere alone and two-thirds of the population of Babi Island were swept away by the tsunami. The maximum runup was estimated as 26 m. The final toll of this Flores earthquake stood at 1712 deaths and more than 2000 injures. Exactly

Figure 2   The tsunami of 1960 killed 61 people in Hilo, destroyed 537 buildings, and damages totaled over $23 million.

7 months later, on 12 July 1993, the third strong earthquake ($M_w = 7.8$) occurred near the Hokkaido Island in Japan (Hokkaido Tsunami Survey Group 1993). Within 3–5 min, a large tsunami engulfed the Okushiri coastline and the central west of Hokkaido, impinging extensive property damages, especially on the southern tip of Okushiri Island in the town of Aonae. The runup heights on the Okushiri Island were thoroughly surveyed and they varied between 15 and 30 m over a 20-km stretch of the southern part of the island, with several 10-m spots on the northern part of the island. It was also reported that although the runup heights on the west coast of Hokkaido are not large (less than 10 m), damage was extensive in several towns. The epicenters of these three earthquakes were all located near residential coastal areas. Therefore, the damage caused by subsequent tsunamis was unusually large.

On 17 July 1998, an earthquake occurred in the Sandaun Province of northwestern Papua New Guinea, about 65 km northwest of the port city of Aitape. The earthquake magnitude was estimated as $M_w = 7.0$. About 20 min after the first shock, Warapo and Arop villages were completely destroyed by tsunamis. The death toll was at over 2000 and many of them drowned in the Sissano Lagoon behind the Arop villages. The surveyed maximum runup height was 15 m, which is much higher than the predicted value based on the seismic information. It has been suggested that the Papua New Guinea tsunami could be caused by a submarine landslide.

The most devastating tsunamis in recent history occurred in the Indian Ocean on 26 December 2004.

An earthquake of $M_w = 9.0$ occurred off the west coast of northern Sumatra. Large tsunamis were generated, severely damaging coastal communities in countries around the Indian Ocean, including Indonesia, Thailand, Sri Lanka, and India. The estimated tsunami death toll ranged from 156 000 to 178 000 across 11 nations, with additional 26 500–142 000 missing, most of them presumed dead.

Tsunami Generation Mechanisms

Tsunamigentic Earthquakes

Most tsunamis are the results of submarine earthquakes. The majority of earthquakes can be explained in terms of plate tectonics. The basic concept is that the outermost part the Earth consists of several large and fairly stable slabs of solid and relatively rigid rock, called plates (see **Figure** 3). These plates are constantly moving (very slowly), and rub against one another along the plate boundaries, which are also called faults. Consequently, stress and strain build up along these faults, and eventually they become too great to bear and the plates move abruptly so as to release the stress and strain, creating an earthquake. Most of tsunamigentic earthquakes occur in subduction zones around the Pacific Ocean rim, where the dense crust of the ocean floor dives beneath the edge of the lighter continental crust and sinks down into Earth's mantle. These subduction zones include the west coasts of North and South America, the coasts of East Asia (especially Japan), and many Pacific island chains (**Figure** 3). There are different types of faults along subduction margins. The interplate fault usually accommodates a large relative motion between two tectonic plates and the overlying plate is typically pushed upward. This upward push is impulsive; it occurs very quickly, in a few seconds. The ocean water surface responds immediately to the upward movement of the seafloor and the ocean surface profile usually mimics the seafloor displacement (see **Figure 4**). The interplate fault in a subduction zone has been responsible for

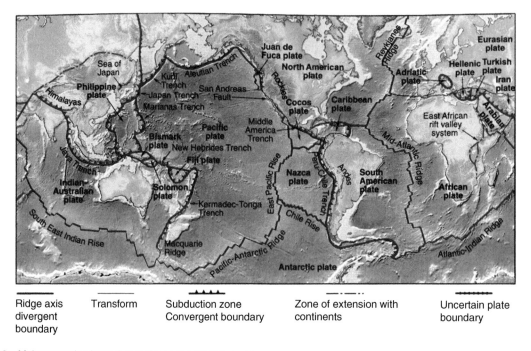

Ridge axis   Transform   Subduction zone   Zone of extension with   Uncertain plate
divergent   Convergent boundary   continents   boundary
boundary

Figure 3 Major tectonic plates that make up the Earth's crust.

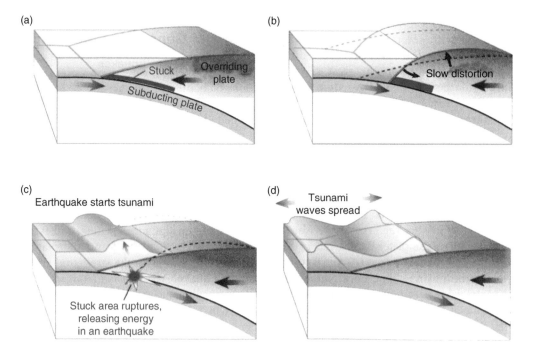

Figure 4 Sketches of the tsunami generation mechanism caused by a submarine earthquake. An oceanic plate subducts under an overriding plate (a). The overriding plate deforms due to the relative motion and the friction between two tectonic plates (b). The stuck area ruptures, releasing energy in an earthquake (c). Tsunami waves are generated due to the vertical seafloor displacement (d).

most of the largest tsunamis in the twentieth century. For example, the 1952 Kamchatka, 1957 Aleutian, 1960 Chile, 1964 Alaska, and 2004 Sumatra earthquakes all generated damaging tsunamis not only in the region near the earthquake epicenter, but also on faraway shores.

For most of the interplate fault ruptures, the resulting seafloor displacement can be estimated based on the dislocation theory. Using the linear elastic theory, analytical solutions can be derived from the mean dislocation field on the fault. Several parameters defining the geometry and strength of the fault rupture need to be specified. First of all, the mean fault slip, D, is calculated from the seismic moment M_0 as follows:

$$M_0 = \mu DS \qquad [1]$$

where S is the rupture area and μ is the rigidity of the Earth at the source, which has a range of 6–$7 \times 10^{11}\,\mathrm{dyn\,cm^{-2}}$ for interplate earthquakes. The seismic moment, M_0, is determined from the seismic data recorded worldwide and is usually reported as the Harvard Centroid-Moment-Tensor (CMT) solution within a few minutes of the first earthquake tremor. The rupture area is usually estimated from the aftershock data. However, for a rough

estimation, the fault plane can be approximated as a rectangle with length L and width W. The aspect ratio L/W could vary from 2 to 8. To find the static displacement of the seafloor, we need to assign the focal depth d, measuring the depth of the upper rim of the fault plane, the dip angle δ, and the slip angle λ of the dislocation on the fault plane measured from the horizontal axis (see **Figure 5**). For an oblique slip on a dipping fault, the slip vector can be decomposed into dip-slip and strike-slip components. In general, the magnitude of the vertical displacement is less for the strike-slip component than for the dip-slip component. The closed form expressions for vertical seafloor displacement caused by a slip along a rectangular fault are given by Mansinha and Smylie.

For more realistic fault models, nonuniform stress-strength fields (i.e., faults with various kinds of barriers, asperities, etc.) are expected, so that the actual seafloor displacement may be very complicated compared with the smooth seafloor displacement computed from the mean dislocation field on the fault. As an example, the vertical seafloor displacement caused by the 1964 Alaska earthquake is sketched in **Figure 6**. Although several numerical models have considered geometrically complex faults, complex slip distributions, and elastic layers of variable thickness, they are not yet disseminated in

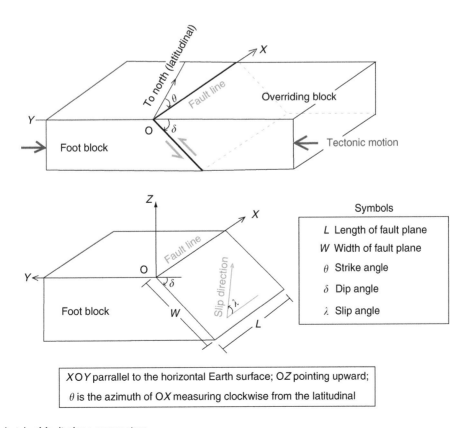

Figure 5 A sketch of fault plane parameters.

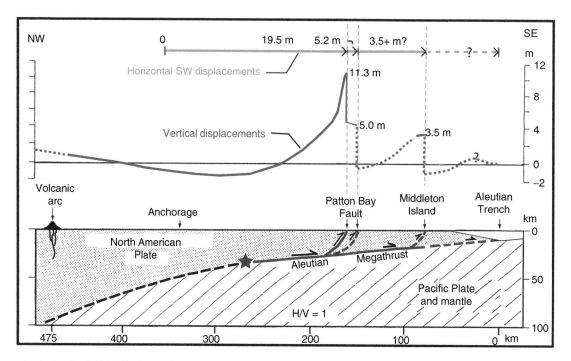

Figure 6   A sketch of 1964 Alaska earthquake generated vertical seafloor displacement (G. Plafker, 2006).

tsunami research. One of the reasons is that our knowledge in source parameters, inhomogeneity, and nonuniform slip distribution is too incomplete to justify using such a complex model.

Certain earthquakes referred to as tsunami earthquakes have slow faulting motion and very long rupture duration (more than several minutes). These earthquakes occur along the shallow part of the interplate thrust or décollement near the trench (the wedge portion of the thin crust above the interface of the continental crust and the ocean plate). The wedge portion consisting of thick deformable sediments with low rigidity, and the steepening of rupture surface in shallow depth all favor the large displacement of the crust and possibility of generating a large tsunami. Because of the extreme heterogeneity, accurate modeling is difficult, resulting in large uncertainty in estimated seafloor displacement.

Landslides and Other Generation Mechanisms

There are occasions when the secondary effects of earthquakes, such as landslide and submarine slump, may be responsible for the generation of tsunamis. These tsunamis are sometimes disastrous and have gained increasing attentions in recent years. Landslides are generated when slopes or sediment deposits become too steep and they fail to remain in equilibrium and motionless. Once the unstable conditions are present, slope failure can be triggered by storm, earthquakes, rains, or merely continued

deposition of materials on the slope. Alternative mechanisms of sediment instability range between soft sediment deformations in turbidities, to rotational slumps in cohesive sediments. Certain environments are particularly susceptible to the production of landslides. River delta and steep underwater slopes above submarine canyons are likely sites for landslide-generated tsunamis. At the time of the 1964 Alaska earthquake, numerous locally landslide-generated tsunamis with devastating effects were observed. On 29 November 1975, a landslide was triggered by a 7.2 magnitude earthquake along the southeast coast of Hawaii. A 60-km stretch of Kilauea's south coast subsided 3.5 m and moved seaward 8 m. This landslide generated a local tsunami with a maximum runup height of 16 m at Keauhou. Historically, there have been several tsunamis whose magnitudes were simply too large to be attributed to the coseismic seafloor movement and landslides have been suggested as an alternative cause. The 1946 Aleutian tsunami and the 1998 Papua New Guinea tsunami are two significant examples.

In terms of tsunami generation mechanisms, two significant differences exist between submarine landslide and coseismic seafloor deformation. First, the duration of a landslide is much longer and is in the order of magnitude of several minutes or longer. Hence the time history of the seafloor movement will affect the characteristics of the generated wave and needs to be included in the model. Second, the effective size of the landslide region is usually much smaller than the coseismic seafloor deformation zone. Consequently, the typical wavelength of the tsunamis generated by a submarine landslide is also shorter, that is, *c.* 1–10 km. Therefore, in some cases, the shallow-water (long-wave) assumption might not be valid for landslide-generated tsunamis.

Although they are rare, the violent geological activities associated with volcanic eruptions can also generate tsunamis. There are three types of tsunami-generation mechanism associated with a volcanic eruption. First, the pyroclastic flows, which are mixtures of gas, rocks, and lava, can move rapidly off an island and into an ocean, their impact displacing seawater and producing a tsunami. The second mechanism is the submarine volcanic explosion, which occurs when cool seawater encounters hot volcanic magma. The third mechanism is due to the collapse of a submarine volcanic caldera. The collapse may happen when the magma beneath a volcano is withdrawn back deeper into the Earth, and the sudden subsidence of the volcanic edifice displaces water and produces a tsunami. Furthermore, the large masses of rock that accumulate on the sides of volcanoes may suddenly slide down the slope into the sea, producing tsunamis. For example, in 1792, a large mass of the mountain slided into Ariake Bay in Shimabara on Kyushu Island, Japan, and generated tsunamis that reached a height of 10 m in some places, killing a large number of people.

In the following sections, our discussions will focus on submarine earthquake-generated tsunamis and their coastal effects.

Modeling of Tsunami Generation, Propagation, and Coastal Inundation

To mitigate tsunami hazards, the highest priority is to identify the high-tsunami-risk zone and to educate the citizen, living in and near the risk zone, about the proper behaviors in the event of an earthquake and tsunami attack. For a distant tsunami, a reliable warning system, which predicts the arrival time as well as the inundation area accurately, can save many lives. On the other hand, in the event of a near-field tsunami, the emergency evacuation plan must be activated as soon as the earth shaking is felt. This is only possible, if a predetermined evacuation/inundation map is available. These maps should be produced based on the historical tsunami events and the estimated 'worst scenarios' or the 'design tsunamis'. To produce realistic and reliable inundation maps, it is essential to use a numerical model that calculates accurately tsunami propagation from

a source region to the coastal areas of concern and the subsequent tsunami runup and inundation.

Numerical simulations of tsunami have made great progress in the last 50 years. This progress is made possible by the advancement of seismology and by the development of the high-speed computer. Several tsunami models are being used in the National Tsunami Hazard Mitigation Program, sponsored by the National Oceanic and Atmospheric Administration (NOAA), in partnership with the US Geological Survey (USGS), the Federal Emergency Management Agency (FEMA), to produce tsunami inundation and evacuation maps for the states of Alaska, California, Hawaii, Oregon, and Washington.

Tsunami Generation and Propagation in an Open Ocean

The rupture speed of fault plane during earthquake is usually much faster than that of the tsunami. For instance, the fault line of the 2004 Sumatra earthquake was estimated as 1200-km long and the rupture process lasted for about 10 min. Therefore, the rupture speed was *c.* 2–3 km s^{-1}, which is considered as a relatively slow rupture speed and is still about 1 order of magnitude faster than the speed of tsunami (0.17 km s^{-1} in a typical water depth of 3 km). Since the compressibility of water is negligible, the initial free surface response to the seafloor deformation due to fault plane rupture is instantaneous. In other words, in terms of the tsunami propagation timescale, the initial free surface profile can be approximated as having the same shape as the seafloor deformation at the end of rupture, which can be obtained by the methods described in the previous section. As illustrated in **Figure 6**, the typical cross-sectional free surface profile, perpendicular to the fault line, has an N shape with a depression on the landward side and an elevation on the ocean side. If the fault plane is elongated, that is, $L \gg W$, the free surface profile is almost uniform in the longitudinal (fault line) direction and the generated tsunamis will propagate primarily in the direction perpendicular to the fault line. The wavelength is generally characterized by the width of the fault plane, W.

The measure of tsunami wave dispersion is represented by the depth-to-wavelength ratio, that is, $\mu^2 = h/\lambda$, while the nonlinearity is characterized by the amplitude-to-depth ratio, that is, $\varepsilon = A/h$. A tsunami generated in an open ocean or on a continental shelf could have an initial wavelength of several tens to hundreds of kilometers. The initial tsunami wave height may be on the order of magnitude of several meters. For example, the 2004 Indian Ocean tsunami had a typical wavelength of 200 km in the Indian Ocean basin with an amplitude of 1 m. The water depth varies from several hundreds of meters on the continental shelf to several kilometers in the open ocean. It is quite obvious that during the early stage of tsunami propagation both the nonlinear and frequency dispersion effects are small and can be ignored. This is particularly true for the 2004 Indian tsunami. The bottom frictional force and Coriolis force have even smaller effects and can be also neglected in the generation area. Therefore, the linear shallow water (LSW) equations are adequate equations describing the initial stage of tsunami generation and propagation.

As a tsunami propagates over an open ocean, wave energy is spread out into a larger area. In general, the tsunami wave height decreases and the nonlinearity remains weak. However, the importance of the frequency dispersion begins to accumulate as the tsunami travels a long distance. Theoretically, one can estimate that the frequency dispersion becomes important when a tsunami propagates for a long time:

$$t \gg t_d = \sqrt{\frac{h}{g}} \left(\frac{\lambda}{h}\right)^3 \qquad [2]$$

or over a long distance:

$$x \gg x_d = t_d \sqrt{gh} = \frac{\lambda^3}{h^2} \qquad [3]$$

In the case of the 2004 Indian Ocean tsunami, $t_d \approx 700$ h and $x_d \approx 5 \times 10^5$ km. In other words, the frequency dispersion effect will only become important when tsunamis have gone around the Earth several times. Obviously, for a tsunami with much shorter wavelength, for example, $\lambda \approx 20$ km, this distance becomes relatively short, that is, $x_d \approx 5 \times 10^2$ km, and can be reached quite easily. Therefore, in modeling transoceanic tsunami propagation, frequency dispersion might need to be considered if the initial wavelength is short. However, nonlinearity is seldom a factor in the deep ocean and only becomes significant when the tsunami enters coastal region.

The LSW equations can be written in terms of a spherical coordinate system as:

$$\frac{\partial \zeta}{\partial t} + \frac{1}{R\cos\varphi}\left[\frac{\partial P}{\partial \psi} + \frac{\partial}{\partial \varphi}(\cos\varphi Q)\right] = -\frac{\partial h}{\partial t} \qquad [4]$$

$$\frac{\partial P}{\partial t} + \frac{gh}{R\cos\varphi}\frac{\partial \zeta}{\partial \psi} = 0 \qquad [5]$$

$$\frac{\partial Q}{\partial t} + \frac{gh}{R}\frac{\partial \zeta}{\partial \varphi} = 0 \qquad [6]$$

where (ψ, φ) denote the longitude and latitude of the Earth, R is the Earth's radius, ζ is free surface elevation, P and Q the volume fluxes ($P = hu$ and $Q = hv$, with u and v being the depth-averaged velocities in longitude and latitude direction, respectively), and h the water depth. Equation [4] represents the depth-integrated continuity equation, and the time rate of change of water depth has been included. When the fault plane rupture is approximated as an instantaneous process and the initial free surface profile is prescribed, the water depth remains time-invariant during tsunami propagation and the right-hand side becomes zero in eqn [4].

The 2004 Indian Ocean tsunami provided an opportunity to verify the validity of LSW equations for modeling tsunami propagation in an open ocean. For the first time in history, satellite altimetry measurements of sea surface elevation captured the Indian Ocean tsunami. About 2 h after the earthquake occurred, two NASA/French Space Agency joint mission satellites, *Jason-1* and *TOPEX/Poseidon*, passed over the Indian Ocean from southwest to northeast (*Jason-1* passed the equator at 02:55:24UTC on 26 December 2004 and *TOPEX/Poseidon* passed the equator at 03:01:57UTC on 26 December 2004) (see **Figure 7**). These two altimetry satellites measured sea surface elevation with accuracy better than 4.2 cm.

Using the numerical model COMCOT (Cornell Multi-grid Coupled Tsunami Model), numerical simulations of tsunami propagation over the Indian Ocean with various fault plane models, including a transient seafloor movement model, have been carried out. The LSW equation model predicts accurately the arrival time of the leading wave and is insensitive of the fault plane models used. However, to predict the trailing waves, the spatial variation of seafloor deformation needs to be taken into consideration. In **Figure 8**, comparisons between LSW results with an optimized fault plane model and *Jason-1/TOPEX* measurements are shown. The excellent agreement between the numerical results and satellite data provides a direct evidence for the validity of the LSW modeling of tsunami propagation in deep ocean.

Coastal Effects – Inundation and Tsunami Forces

Nonlinearity and bottom friction become significant as a tsunami enters the coastal zone, especially during the runup phase. The nonlinear shallow water (NLSW) equations can be used to model certain aspects of coastal effects of a tsunami attack. Using the same notations as those in eqns [4]–[6], the NLSW

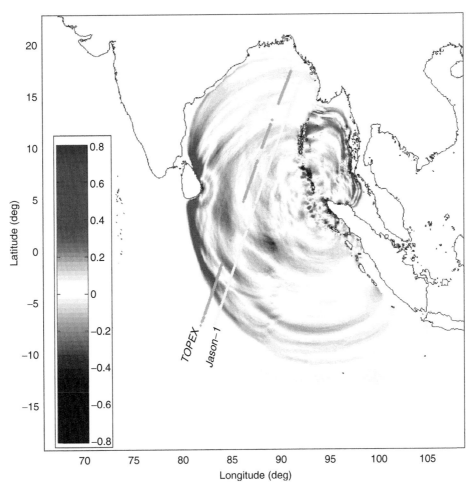

Figure 7 Satellite tracks for *TOPEX* and *Jason-1*. The colors indicate the numerically simulated free surface elevation in meter at 2 h after the earthquake struck.

equations in the Cartesian coordinates are

$$\frac{\partial \zeta}{\partial t} + \frac{\partial P}{\partial x} + \frac{\partial Q}{\partial y} = 0 \qquad [7]$$

$$\frac{\partial P}{\partial t} + \frac{\partial}{\partial x}\left(\frac{P^2}{H}\right) + \frac{\partial}{\partial y}\left(\frac{PQ}{H}\right) + gH\frac{\partial \zeta}{\partial x} + \tau_x H = 0 \qquad [8]$$

$$\frac{\partial Q}{\partial t} + \frac{\partial}{\partial x}\left(\frac{PQ}{H}\right) + \frac{\partial}{\partial y}\left(\frac{Q^2}{H}\right) + gH\frac{\partial \zeta}{\partial y} + \tau_y H = 0 \qquad [9]$$

The bottom frictional stresses are expressed as

$$\tau_x = \frac{gn^2}{H^{10/3}}P(P^2 + Q^2)^{1/2} \qquad [10]$$

$$\tau_y = \frac{gn^2}{H^{10/3}}Q(P^2 + Q^2)^{1/2} \qquad [11]$$

where n is the Manning's relative roughness coefficient. For flows over a sandy beach, the typical value for the Manning's n is 0.02.

Using a modified leapfrog finite difference scheme in a nested grid system, COMCOT is capable of solving both LSW and NLSW equations simultaneously in different regions. For the nested grid system, the inner (finer) grid adopts a smaller grid size and time step compared to its adjacent outer (larger) grid. At the beginning of a time step, along the interface of two different grids, the volume flux, P and Q, which is product of water depth and depth-averaged velocity, is interpolated from the outer (larger) grids into its inner (finer) grids. And at the end of this time step, the calculated water surface elevations, ζ, at the inner finer grids are averaged to update those values of the larger grids overlapping the finer grids, which are used to compute the volume fluxes at next time step in the outer grids. With this procedure, COMCOT can capture near-shore

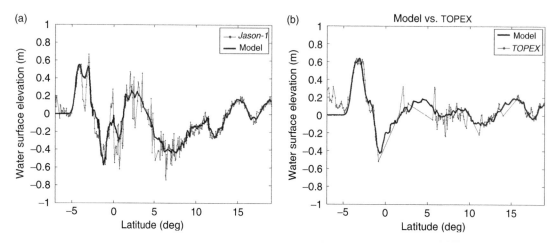

Figure 8 Comparisons between optimized fault model results and *Jason-1* measurements (a)/*TOPEX* measurements (b).

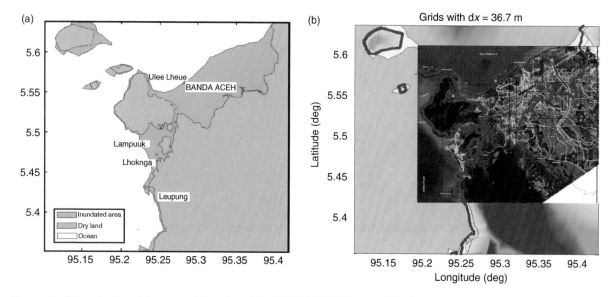

Figure 9 Calculated inundation areas (a) and overlaid with *QUICKBIRD* image (b) in Banda Aceh, Indonesia.

features of a tsunami with a higher spatial and temporal resolution and at the same time can still keep a high computational efficiency.

To estimate the inundation area caused by a tsunami, COMCOT adopts a simple moving boundary scheme. The shoreline is defined as the interface between a wet grid and its adjacent dry grids. Along the shoreline, the volume flux is assigned to be zero. Once the water surface elevation at the wet grid is higher than the land elevation in its adjacent dry grid, the shoreline is moved by one grid toward the dry grid and the volume flux is no longer zero and need to be calculated by the governing equations.

COMCOT, coupled up to three levels of grids, has been used to calculate the runup and inundation areas at Trincomalee Bay (Sri Lanka) and Banda Aceh (Indonesia). Some of the numerical results for Banda Aceh are shown here.

The calculated inundation area in Banda Aceh is shown in **Figure 9**. The flooded area is marked in blue, the dry land region is rendered in green, and the white area is ocean region. The calculated inundation area is also overlaid with a satellite image taken by *QUICKBIRD* in **Figure 9(b)**.

In the overlaid image, the thick red line indicates the inundation line based on the numerical simulation. In the satellite image, the dark green color (vegetation) indicates areas not affected by the tsunami and the area shaded by semitransparent red color shows flooded regions by this tsunami. Obviously, the calculated inundation area matches reasonably well with the satellite image in the neighborhood of Lhoknga and the western part of Banda Aceh. However, in the region of eastern Banda Aceh, the simulations significantly underestimate the inundation area. However, in general, the agreement between the numerical simulation and the satellite observation is surprisingly good.

In **Figure 10**, the tsunami wave heights in Banda Aceh are also compared with the field measurements by two Japan survey teams. On the coast between Lhoknga and Leupung, where the maximum height is measured more than 30 m, the numerical results match very well with the field measurements. However, beyond Lhoknga to the north, the numerical results, in general, are only half of the measurements, except in middle regions between Lhoknga and Lampuuk.

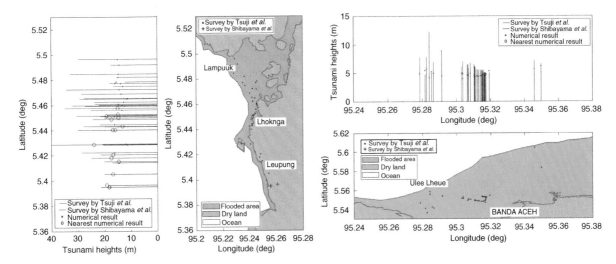

Figure 10 Tsunami heights on eastern and northern coast of Banda Aceh, Indonesia. The field survey measurements are from Tsuji *et al.* (2005) and Shibayama *et al.* (2005).

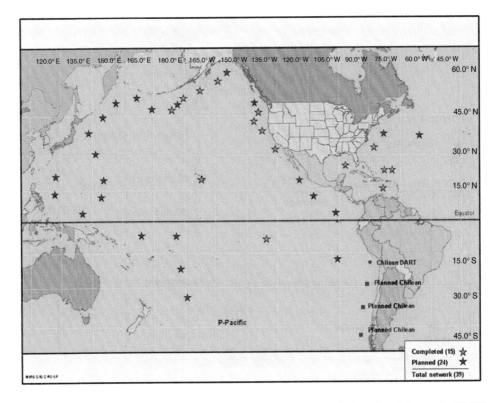

Figure 11 The locations of the existing and planned Deep-Ocean Assessment and Reporting of Tsunamis (DART) system in the Pacific Ocean (NOAA magazine, 17 Apr. 2006).

Tsunami Hazard Mitigation

The ultimate goal of the tsunami hazard mitigation effort is to minimize casualties and property damages. This goal can be met, only if an effective tsunami early warning system is established and a proper coastal management policy is practiced.

Tsunami Early Warning System

The great historical tsunamis, such as the 1960 Chilean tsunami and the 1964 tsunami generated near Prince William Sound in Alaska, prompted the US government to develop an early warning system in the Pacific Ocean. The Japanese government has also developed a tsunami early warning system for the entire coastal community around Japan. The essential information needed for an effective early warning system is the accurate prediction of arrival time and wave height of a forecasted tsunami at a specific location. Obviously, the accuracy of these predictions relies on the information of the initial water surface displacement near the source region, which is primarily determined by the seismic data. In many historical events, including the 2004 Indian Ocean tsunami, evidences have shown that accurate seismic data could not be verified until those events were over. To delineate the source region problem, in the United States, several federal agencies and states

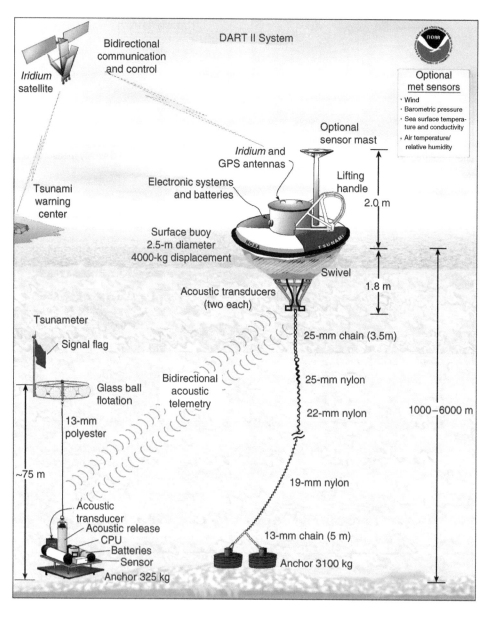

Figure 12 A sketch of the second-generation DART (II) system.

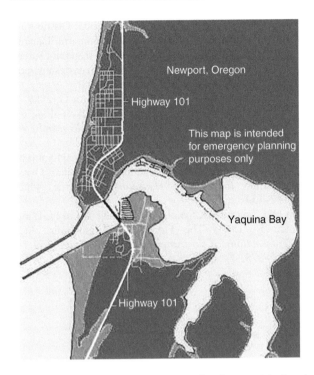

Figure 13 Tsunami inundation map for the coastal city of Newport, Oregon.

have joined together to create a warning system that involves the use of deep-ocean tsunami sensors to detect the presence of a tsunami.

These deep-ocean sensors have been deployed at different locations in the Pacific Ocean before the 2004 Indian Ocean tsunami. After the 2004 Indian Ocean tsunami, several additional sensors have been installed and many more are being planned (see **Figure 11**). The sensor system includes a pressure gauge that records and transmits the surface wave signals instantaneously to the surface buoy, which sends the information to a warning center via *Iridium* satellite (**Figure 12**). In the event of a tsunami, the information obtained by the pressure gauge array can be used as input data for modeling the propagation and evolution of a tsunami.

Although there have been no large Pacific-wide tsunamis since the inception of the warning system, warnings have been issued for smaller tsunamis, a few of which were hardly noticeable. This tends to give citizens a lazy attitude toward a tsunami warning, which would be fatal if the wave was large. Therefore, it is very important to keep people in a danger areas educated of tsunami hazards.

Coastal Inundation Map

Using numerical modeling, hazards in areas vulnerable to tsunamis can be assessed, without the area ever having experienced a devastating tsunami.

These models can simulate a 'design tsunami' approaching a coastline, and they can predict which areas are most at risk to being flooded. The tsunami inundation maps are an integral part of the overall strategy to reduce future loss of life and property. Emergency managers and local governments of the threatened communities use these and similar maps to guide evacuation planning.

As an example, the tsunami inundation map (**Figure 13**) for the coastal city of Newport (Oregon) was created using the results from a numerical simulation using a design tsunami. The areas shown in orange are locations that were flooded in the numerical simulation

Acknowledgment

The work reported here has been supported by National Science Foundation with grants to Cornell University.

See also

Glacial Crustal Rebound, Sea Levels, and Shorelines. Land–Sea Global Transfers. Waves on Beaches.

Further Reading

Geist EL (1998) Local tsunami and earthquake source parameters. *Advances in Geophysics* 39: 117–209.

Hokkaido Tsunami Survey Group (1993) Tsunami devastates Japanese coastal region. *EOS Transactions of the American Geophysical Union* 74: 417–432.

Kajiura K (1981) Tsunami energy in relation to parameters of the earthquake fault model. *Bulletin of the Earthquake Research Institute, University of Tokyo* 56: 415–440.

Kajiura K and Shuto N (1990) Tsunamis. In: Le Méhauté B and Hanes DM (eds.) *The Sea: Ocean Engineering Science*, pp. 395–420. New York: Wiley.

Kanamori H (1972) Mechanism of tsunami earthquakes. *Physics and Earth Planetary Interactions* 6: 346–359.

Kawata Y, Benson BC, Borrero J, *et al.* (1999) Tsunami in Papua New Guinea was as intense as first thought. *EOS Transactions of the American Geophysical Union* 80: 101, 104–105.

Keating BH and Mcguire WJ (2000) Island edifice failures and associated hazards. *Special Issue: Landslides and Tsunamis. Pure and Applied Geophysics* 157: 899–955.

Liu PL-F, Lynett P, Fernando H, *et al.* (2005) Observations by the International Tsunami Survey Team in Sri Lanka. *Science* 308: 1595.

Lynett PJ, Borrero J, Liu PL-F, and Synolakis CE (2003) Field survey and numerical simulations: A review of the

1998 Papua New Guinea tsunami. *Pure and Applied Geophysics* 160: 2119–2146.

Mansinha L and Smylie DE (1971) The displacement fields of inclined faults. *Bulletin of Seismological Society of America* 61: 1433–1440.

Satake K, Bourgeois J, Abe K, *et al.* (1993) Tsunami field survey of the 1992 Nicaragua earthquake. *EOS Transactions of the American Geophysical Union* 74: 156–157.

Shibayama T, Okayasu A, Sasaki J, *et al.* (2005) The December 26, 2004 Sumatra Earthquake Tsunami, Tsunami Field Survey in Banda Aceh of Indonesia. http://www.drs.dpri.kyoto-u.ac.jp/sumatra/indonesia-ynu/indonesia_survey_ynu_e.html (accessed Feb. 2008).

Synolakis CE, Bardet J-P, Borrero JC, *et al.* (2002) The slump origin of the 1998 Papua New Guinea tsunami. *Proceedings of Royal Society of London, Series A* 458: 763–789.

Tsuji Y, Matsutomi H, Tanioka Y, *et al.* (2005) Distribution of the Tsunami Heights of the 2004 Sumatra Tsunami in Banda Aceh measured by the Tsunami Survey Team. http://www.eri.u-tokyo.ac.jp/namegaya/sumatera/surveylog/eindex.htm (accessed Feb. 2008).

von Huene R, Bourgois J, Miller J, and Pautot G (1989) A large tsunamigetic landslide and debris flow along the Peru trench. *Journal of Geophysical Research* 94: 1703–1714.

Wang X and Liu PL-F (2006) An analysis of 2004 Sumatra earthquake fault plane mechanisms and Indian Ocean tsunami. *Journal of Hydraulics Research* 44(2): 147–154.

Yeh HH, Imamura F, Synolakis CE, Tsuji Y, Liu PL-F, and Shi S (1993) The Flores Island tsunamis. *EOS Transactions of the American Geophysical Union* 74: 369–373.

WAVE ENERGY

M. E. McCormick and D. R. B. Kraemer, The Johns
Hopkins University, Baltimore, MD, USA

Introduction

In the last half of the twentieth century, humankind
finally realized that fossil fuel resources are finite and
that use of those fuels has environmental consequences. These realizations have prompted the
search for other energy resources that are both renewable and environmentally 'friendly'. One such
resource is the ocean wind wave. This is a form of
solar energy in that the sun is partly responsible for
the winds that generate water waves.

The exploitation of water waves has been a goal
for thousands of years. Until recent times, however,
only sporadic efforts were made, and these were
generally directed at a specific function. In the 1960s,
Yoshio Masuda, the 'renaissance man' of wave energy conversion, came up with a scheme to convert
the energy of water waves into electricity by using a
floating pneumatic device. Originally, the Masuda
system was used to power remote navigation aides,
such as buoys. One such buoy system was purchased
by the US Coast Guard which, in turn, requested an
analysis of the performance of the system. The results of that analysis were reported by McCormick
(1974). This was the first of a long list of theoretical
and experimental studies of the pneumatic and other
wave energy conversion systems. (For summaries of
some of the works, see the Further Reading section.)
The most recent collective type of publication is
that edited by Nicholls (1999), written under the
joint sponsorship of the Engineering Committee on
Oceanic Resources (ECOR) and the Japan Marine
Science and Technology Center (JAMSTEC). In the
late 1970s and early 1980s, JAMSTEC co-sponsored
a full-scale trial of a floating, offshore pneumatic
system called the Kaimei. The 80 m long, 10 m wide
Kaimei (**Figures 1** and **2**) was designed to produce
approximately 1.25 MW of electricity while operating in the Sea of Japan. This power was to be
produced by 10 pneumatic turbo-generators. Eight
of these (produced in Japan) utilized a unidirectional
turbine. The other two utilized bi-directional turbines designed by Wells in the UK and McCormick in
the USA. Unfortunately, the designed electrical
power production was never attained by the system.
The Wells turbine was found to be the most effective
for wave energy conversion, and is now being used to
power fixed pneumatic systems in the Azores, in
India, and on the island of Islay off of the coast of
Scotland.

Most of the published works resulting from research, development, and demonstration efforts are
directed at the production of electrical energy.
However, Hicks *et al.* (1988) described a wave energy conversion technique that could be used to
produce potable water from ocean salt water. This
technique had been developed earlier by Pleass and
Hicks. Their work resulted in a commercial system
called the Del Buoy, and inspired the efforts of others

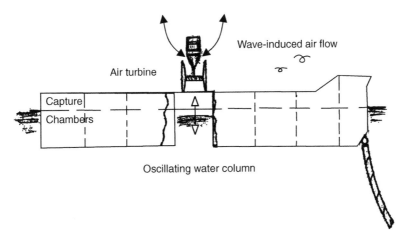

Figure 1 Schematic diagram of the Kaimei wave energy conversion system, consisting of 10 capture chambers and 10 pneumatic-electric generating systems.

Figure 2 The Kaimei deployed in the Sea of Japan.

Figure 4 The McCabe wave pump located 500 m off the coast of Kilbaha, County Clare, Ireland.

to apply the McCabe wave pump (**Figures 3** and **4**) to the production of potable water. The high-pressure pumps located between the barges pump sea water through a reverse osmosis (RO) desalination system located on the shore. The first deployment of the McCabe wave pump occurred in 1996 in the Shannon River, western Ireland. A second deployment of the system is expected in the spring of the year 2000 at the same location.

Wave Power: Resource and Exploitation

A mathematical expression for the power of water waves is obtained from the linear wave theory. Simply put, the expression is based on the waves having a sinusoidal profile, as sketched in **Figure 5**. The wave power (energy flux) expression is:

$$P = \frac{1}{8}\rho g H^2 b c_G \qquad [1]$$

where ρ is the mass density of salt water (approximately 1030 kg m^{-3}), g is the gravitational acceleration (9.81 m s^{-2}), H is the wave height in meters, b is the

wave crest width of interest in meters, and the vector c_G is called the group velocity. In deep water, defined as water depth (h) greater than half of the wave length (λ), the group velocity (in m s^{-1}) is approximately:

$$|c_G| \simeq \frac{c}{2} \simeq \frac{gT}{4\pi} \qquad [2]$$

where c is the wave celerity (the actual speed of the wave), and T is the period of the wave in seconds. In shallow water, defined as where $h \leq \lambda/20$, the group velocity is approximated by:

$$|c_G| \simeq c \simeq \sqrt{gh} \qquad [3]$$

Consider an average wave approaching the central Atlantic states of the contiguous United States. The average wave height and period of waves in deep water are approximately 1 m and 7 s, respectively. For this wave, the wave power per crest width is:

$$\frac{|P|}{b} = \frac{1}{32\pi}\rho g^2 H^2 T \simeq 6.90 (kW/m) \qquad [4]$$

from eqn[1] combined with the expression in eqn [2].

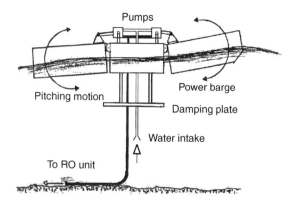

Figure 3 Sketch of the McCabe wave pump.

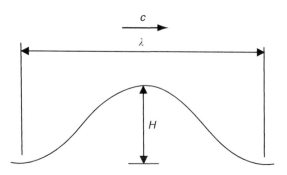

Figure 5 Wave notation.

The percentage of this power that can be captured depends on both the width of the capturing system and the frequency (or period) characteristics of the system. Each wave power system has one fundamental frequency (f_n). If the inverse of that frequency ($1/f_n = T_n$) is the same as the average wave period, then the system is in resonance with the average wave, and the maximum amount of wave power will be extracted by the wave power system. This, then, is the design goal, i.e. to design the system to resonate with the design wave. When resonance is achieved, then another phenomenon occurs which is of benefit to the system: i.e. resonant focusing, where diffraction draws energy toward the system. For a single degree of freedom wave power system, such as the heaving buoy sketched in **Figure 6**, the wave power absorbed by the system comes from a crest width equal to the width of the system (B) plus an additional width equal to the wavelength divided by 2π.

Hence, in deep water, the total power available to the single degree of freedom system operating in the average wave is:

$$|P| = \frac{1}{32\pi}\rho g^2 H^2 T \left(B + \frac{\lambda}{\pi} \right) \qquad [5]$$

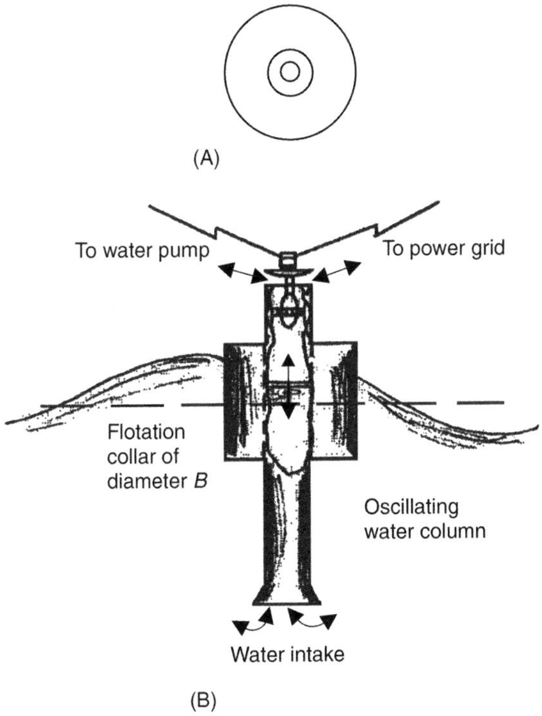

(A)

(B)

Figure 6 Floating oscillating water column wave energy converter, designed to produce electrical power for either potable water production or electricity. (A) Plan view; (B) elevation.

where the wavelength in deep water is approximately:

$$\lambda \simeq \frac{gT^2}{2\pi} \qquad [6]$$

Thus, for the aforementioned average wave, the deep-water wavelength is about 76.5 m.

Consider an ideal 1 m diameter heaving system operating in the 1 m, 7 s average wave. For this wave, the wave power captured by the system is $6.90(1 + 76.5/2\pi)$ kW, or approximately 91 kW. If bus-bar conversion efficiency is 50%, then about 45.5 kW will be supplied to the power grid for consumption. In the contiguous United States, each citizen requires about 1 kW, on average, at any time. Hence, this system would supply 87.5 citizens.

To use the same system coupled with a RO desalinator to supply potable water, the value of the osmotic pressure of the desalinator's membranes is required. This value is 23 atmospheres or approximately 23 bars. From fluid mechanics, the power is equal to the volume rate of flow in the system multiplied by the back pressure. Hence, the 1 m diameter system would supply approximately 5 (US) gallons of salt water per second to the RO unit. Half of this flow would become product (potable) water, while the other half would be brine waste. This ideal system would supply 2.5 gallons per second (about 225 000 gallons per day) of potable water. Each US citizen residing in the contiguous United States uses about 60 gallons per day, on average. So, the wave-powered desalination system would satisfy the daily potable water needs of approximately 3700 citizens.

The numbers presented in the last two paragraphs illustrate the potential of wave energy conversion. The electrical and water producing systems described are ideal. In actuality, the waves in the sea are random in nature. The system, then, must be tuned to some design wave, such as that having an average wave period.

Economics of Wave Power Conversion

The economics of ocean wave energy conversion vary, depending on both the product (electricity or potable water) and the location. To illustrate this, consider the following two cases. First, on Lord Howe Island in the South Pacific, the cost of electrical energy is about 45 (US) cents per kilo-Watt hour (kWh). Electricity produced by wave energy conversion would cost about 15 cents/kWh. Hence, for such an application, wave energy conversion would be extremely cost-effective. On the other hand, in San Diego, California, the energy cost is about 13 cents/kWh, making wave energy conversion cost-ineffective. For the production of potable water, the McCabe wave pump, coupled with a RO system, will produce potable water at approximately US $1.10 per cubic meter (265 US gallons). On some remote islands, the cost of potable water is approximately $4.00 per-gallon. On the coast of Saudi Arabia, on the Arabian Sea, the cost of potable water is $3.10 per cubic meter. Therefore, these locations, wave-powered desalination systems are very cost-effective.

Concluding remarks

The reader is encouraged to consult the Further Reading section for more information on wave energy conversion. There are many activities presently underway in this area of technology. These can be found on the Internet by searching the world wide web for wave energy conversion.

See also

Coastal Trapped Waves. Seiches. Storm Surges. Tides. Tsunami. Waves on Beaches.

Further Reading

Count B (ed.) (1980) *Power from Sea Waves*. London: Academic Press.

Hicks D, Pleass CM, and Mitcheson G (1988) *Delbuoy Wave-Powered Seawater Desalination System*. Proceedings of OCEANS'88, US Department of Energy, pp. 1049–1055.

McCormick ME (1981) *Ocean Wave Energy Conversion*. New York: Wiley-Interscience.

McCormick ME (1974) An analysis of power generating buoys. *Journal of Hydronautics* 8: 77–82.

McCormick ME, McCabe RP, and Kraemer DRB (1999) Utilization of a hinged-barge wave energy conversion system. *International Journal of Power Energy Systems* 19: 11–16.

McCormick ME and Murtagh JF (1992) *Large-Scale Experimental Study of the McCabe Wave Pump*. US Naval Academy Report EW-3-92, January.

McCormick M, Murtagh J, and McCabe P (1998) *Large-Scale Experimental Study of the McCabe Wave Pump*. European Wave Energy Conference (European Union), Patras, Greece, Paper H1.

Nicholls HB (1999) *Workshop Group on Wave Energy Conversion*. Engineering Committee on Oceanic Resources (ECOR), St Johns, Newfoundland, Canada.

Ross D (1998) *Power from the Waves*. Oxford: Oxford University Press.

Shaw R (1982) *Wave Energy. A Design Challenge*. Chichester, UK: Ellis Horwood.

TIDES

D. T. Pugh, University of Southampton, Southampton, UK

Introduction

Even the most casual coastal visitor is familiar with marine tides. Slightly more critical observers have noted from early history, relationships between the movements of the moon and sun, and with the phases of the moon. Several plausible and implausible explanations for the links were advanced by ancient civilizations. Apart from basic curiosity, interest in tides was also driven by the seafarer's need for safe and effective navigation, and by the practical interest of all those who worked along the shore.

Our understanding of the physical processes which relate the astronomy with the complicated patterns observed in the regular tidal water movements is now well advanced, and accurate tidal predictions are routine. Numerical models of the ocean responses to gravitational tidal forces allow computations of levels both on- and offshore, and satellite altimetry leads to detailed maps of ocean tides that confirm these. The budgets and flux of tidal energy from the earth–moon dynamics through to final dissipation in a wide range of detailed marine processes has been an active area of research in recent years. For the future, there are difficult challenges in understanding the importance of these processes for many complicated coastal and open ocean phenomena.

Tidal Patterns

Modern tidal theory began when Newton(1642–1727) applied his formulation of the Law of Gravitational Attraction: that two bodies attract each other with a force which is proportional to the product of their masses and inversely proportional to the square of the distance between them. He was able to show why there are two tides for each lunar transit. He also showed why the half-monthly spring-to-neap cycle occurred, why once-daily tides are a maximum when the Moon is furthest from the plane of the equator, and why equinoctial tides are larger than those at the solstices.

The two main tidal features of any sea-level record are the range (measured as the height between successive high and low levels) and the period (the time between one high (or low) level and the next high (or low) level).Spring tides are semidiurnal tides of increased range, which occur approximately twice a month near the time when the moon is either new or full. Neap tides are the semidiurnal tides of small range which occur between spring tides near the time of the first and last lunar quarter. The tidal responses of the ocean to the forcing of the moon and the sun are very complicated and tides vary greatly from one site to another.

Tidal currents, often called tidal streams, have similar variations from place to place. Semidiurnal, mixed, and diurnal currents occur; they usually have the same characteristics as the local tidal changes in sea level, but this is not always so. For example, the currents in the Singapore Strait are often diurnal in character, but the elevations are semidiurnal.

It is important to make a distinction between the popular use of the word 'tide' to signify any change of sea level, and the more specific use of the word to mean only regular, periodic variations. We define tides as periodic movements which are directly related in amplitude and phase to some periodic geophysical force. The dominant geophysical forcing function is the variation of the gravitational field on the surface of the earth, caused by the regular movements of the Moon–Earth and Earth–Sun systems. Movements due to these gravitational forces are termed gravitational tides. This is to distinguish them from the smaller movements due to regular meteorological forces which are called either-meteorological or more usually radiational tides.

Gravitational Potential

The essential elements of a physical understanding of tide dynamics are contained in Newton's Laws of Motion and in the principle of Conservation of Mass. For tidal analysis the basics are Newton's Laws of Motion and the Law of Gravitational Attraction.

The Law of Gravitational Attraction states that for two particles of masses m_1 and m_2, separated by a distance r the mutual attraction is:

$$F = G\,\frac{m_1 m_2}{r^2} \qquad [1]$$

G is the universal gravitational constant.

Use is made of the concept of the gravitational potential of a body; gravitational potential is the work which must be done against the force of attraction to remove a particle of unit mass to an infinite distance from the body. The potential at P on the Earth's surface (**Figure 1**) due to the moon is:

$$\Omega_p = -\frac{Gm}{MP} \qquad [2]$$

This definition of gravitational potential, involving a negative sign, is the one normally adopted in physics, but there is an alternative convention often used in geodesy, which treats the potential in the above equation as positive. The advantage of the geodetic convention is that an increase in potential on the surface of the earth will result in an increase of the level of the free water surface. Potential has units of $L^2 T^{-2}$. The advantage of working with gravitational potential is that it is a scalar property, which allows simpler mathematical manipulation; in particular, the vector, gravitational force on a particle of unit mass is given by $-\text{grad}(\Omega_p)$.

Applying the cosine law to $\triangle OPM$ in **Figure 1**

$$MP^2 = a^2 + r^2 - 2ar\cos\phi \qquad [3]$$

$$\therefore MP = r\left[1 - 2\frac{a}{r}\cos\phi + \frac{a^2}{r^2}\right]^{1/2} \qquad [4]$$

Hence we have:

$$\Omega_p = -\frac{Gm}{r}\left[1 - 2\frac{a}{r}\cos\phi + \frac{a^2}{r^2}\right]^{-1/2} \qquad [5]$$

which may be expanded:

$$\Omega_P = -\frac{Gm}{r}\left[1 + \frac{a}{r}P_1(\cos\phi) + \frac{a^2}{r^2}P_2(\cos\phi) + \frac{a^2}{r^2}P_3(\cos\phi) + ...\right] \qquad [6]$$

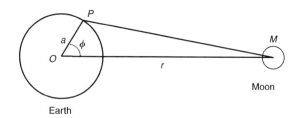

Figure 1 The general position of the point P on the Earth's surface, defined by the angle ϕ.

The terms in $P_n(\cos\phi)$ are the Legendre Polynomials:

$$P_1 = \cos\phi \qquad [7]$$

$$P_2 = \tfrac{1}{2}(3\cos^2\phi - 1) \qquad [8]$$

$$P_3 = \tfrac{1}{2}(5\cos^3\phi - 3\cos\phi) \qquad [9]$$

The tidal forces represented by the terms in this potential are calculated from their spatial gradients $-\text{grad}(P_n)$. The first term in the equation is constant (except for variations in r) and so produces no force. The second term produces a uniform force parallel to OM because differentiating with respect to $(a\cos\phi)$ yields a gradient of potential which provides the force necessary to produce the acceleration in the earth's orbit towards the center of mass of the Moon–Earth system. The third term is the major tide-producing term. For most purposes the fourth term may be neglected, as may all higher terms.

The effective tide-generating potential is therefore written as:

$$\Omega_P = -\tfrac{1}{2}Gm\frac{a^2}{r^3}(3\cos^2\phi - 1) \qquad [10]$$

The force on the unit mass at P may be resolved into two components as functions of ϕ:

vertically upwards : $-\dfrac{\partial\Omega_p}{\partial a} = 2g\Delta_l(\cos^2\phi - \tfrac{1}{3})$ [11]

horizontally in the direction of increasing ϕ.

$$-\frac{\partial\Omega_p}{a\delta\phi} = -g\Delta_l\sin 2\phi \qquad [12]$$

For the Moon:

$$\Delta_l = \frac{3}{2}\frac{m_l}{m_e}\left(\frac{a}{R_l}\right)^3 \qquad [13]$$

m_l is the lunar mass and m_e is the Earth mass. R_1, the lunar distance, replaces r. The resulting forces are shown in **Figure 2**.

To generalize in three dimensions, the lunar angle ϕ must be expressed in suitable astronomical variables. These are chosen to be declination of the Moon north or south of the equator, the north–south latitude of P, ϕ_p, and the hour angle of the moon, which is the difference in longitude between the meridian of P and the meridian of the sublunar point V on the Earth's surface.

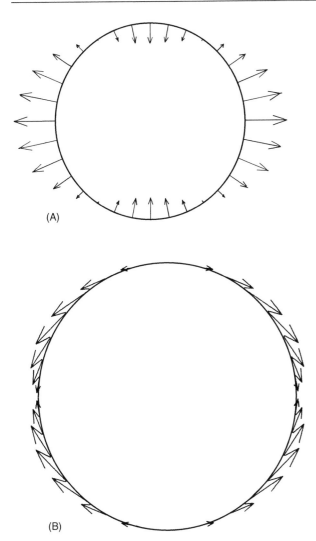

Figure 2 The tide-producing forces at the Earth's surface, due to the Moon. (A) The vertical forces, showing an outward pull at the equator and a smaller downward pull at the poles. (B) The horizontal forces, which are directed away from the poles towards the equator, with a maximum value at 45° latitude.

The Equilibrium Tide

An equilibrium tide can be computed from eqn [10] by replacing $\cos^2\phi$ by the full astronomical expression in terms of d_1, ϕ_p and the hour angle C_1. The equilibrium tide is defined as the elevation of the sea surface that would be in equilibrium with the tidal forces if the Earth were covered with water and the response is instantaneous. It serves as an important reference system for tidal analysis.

It has three coefficients which characterize the three main species of tides: (1) the long period species; (2) the diurnal species at a frequency of one cycle per day ($\cos C$); and (3) the semidiurnal species at two cycles per day ($\cos 2C$).

The equilibrium tide due to the sun is expressed in a form analogous to the lunar tide, but with solar mass, distance, and declination substituted for lunar parameters.

The ratio of the two tidal amplitudes is:

$$\frac{m_s}{m_1}\left(\frac{R_l}{R_s}\right)^3$$

For the semidiurnal lunar tide at the equator when the lunar declination is zero, the equilibrium tidal amplitude is 0.27 m. For the sun it is 0.13 m. The solar amplitudes are smaller by a factor of 0.46 than those of the lunar tide, but the essential details are the same.

The maximum diurnal tidal ranges occur when the lunar declination is greatest. The ranges become very small when the declination is zero. This is because the effect of declination is to produce an asymmetry between the two high- and the two low-water levels observed as a point P rotates on the earth within the two tidal bulges.

The fortnightly spring/neap modulation of semidiurnal tidal amplitudes is due to the various combinations of the separate lunar and solar semidiurnal tides. At times of spring tides the lunar and solar forces combine together, but at neap tides the lunar and solar forces are out of phase and tend to cancel. In practice, the observed spring tides lag the maximum of the tidal forces, usually by one or two days due to the inertia of the oceans and energy losses. This delay is traditionally called the age of the tide.

The observed ocean tides are normally much larger than the equilibrium tide because of the dynamic response of the ocean to the tidal forces. But the observed tides do have their energy at the same frequencies (or periods) as the equilibrium tide. This forms the basis of tidal analysis.

Tidal Analysis

Tidal analysis of data collected by observations of sea levels and currents has two purposes. First, a good analysis provides the basis for predicting tides at future times, a valuable aid for shipping and other operations. Secondly, the results of analyses can be mapped and interpreted scientifically in terms of the hydrodynamics of the seas and their responses to tidal forcing. In tidal analysis the aim is to produce significant time-stable tidal parameters which describe the tidal regime at the place of observation. These parameters are often termed tidal constants on the assumption that the responses of the oceans and seas to tidal forces do not change with time. A good tidal analysis seeks to represent the data by a few significant stable numbers which mean something physically. In general, the longer the period of data included in the analysis, the greater the number of constants which can be independently determined. If possible, an analysis should give some idea of the confidence which should be attributed to each tidal constant determined.

The close relationship between the movement of the moon and sun, and the observed tides, make the lunar and solar coordinates a natural starting point for any analysis scheme. Three basic methods of tidal analysis have been developed. The first, which is now generally of only historical interest, the non-harmonic method, relates high and low water times and heights directly to the phases of the moon and other astronomical parameters. The second method which is generally used for predictions and for scientific work, harmonic analysis, treats the observed tides as the sum of a finite number of harmonic constituents with angular speeds determined from the astronomical arguments. The third method develops the concept, widely used in electronic engineering, of a frequency-dependent response of a system to a driving mechanism. For tides, the driving mechanism is the equilibrium potential. The latter two methods are special applications of the general formalisms of time series analysis.

Analyses of changing sea levels (scalar quantities) are obviously easier than those of currents (vectors), which can be analysed by resolving into two components.

Harmonic Analysis

The basis of harmonic analysis is the assumption that the tidal variations can be represented by a finite number N of harmonic terms of the form:

$$H_n \cos(\sigma_n t - g_n)$$

where H_n is the amplitude, g_n is the phase lag on the equilibrium tide at Greenwich and σ_n is the angular speed. The angular speeds σ_n are determined by an expansion of the equilibrium tide into harmonic terms. The speeds of these terms are found to have the general form:

$$\omega_n = i_a\omega_1 + i_b\omega_2 + i_c\omega_3 + (\omega_4, \omega_5, \omega_6 \text{ terms}) \quad [14]$$

where the values of ω_1 to ω_6 are the angular speeds related to astronomical parameters and the coefficients, i_a to i_c are small integers (normally 0, 1 or 2) (**Table 1**).

The phase lags g_n are defined relative to the phase of the corresponding term in the harmonic expansion of the equilibrium tide.

Full harmonic analysis of the equilibrium tide shows the grouping of tidal terms into species ($_1$;

Table 1 The basic astronomical periods which modulate the tidal forces

	Period	Symbol
Mean solar day (msd)	1.0000 msd	ω_0
Mean lunar day	1.0351 msd	ω_1
Sidereal month	27.3217 msd	ω_2
Tropical year	365.24222 msd	ω_3
Moon's perigee	8.85 years	ω_4
Regression of Moon's nodes	18.61 years	ω_5
Perihelion	20 942 years	ω_6

diurnal, semidiurnal …), groups (ω_2; monthly) and constituents (ω_3; annual).

Response Analysis

The basic ideas involved in response analysis are common to many activities. A system, sometimes called a 'black box', is subjected to an external stimulus or input. The output from a system depends on the input and the system response to that input. The response of the system may be evaluated by comparing the input and output functions at various forcing frequencies. These ideas are common in many different contexts, including mechanical engineering, financial modeling and electronics.

In tidal analysis the input is the equilibrium tidal potential. The tidal variations measured at a particular site may be considered as the output from the system. The system is the ocean, and we seek to describe its response to gravitational forces. This 'response' treatment has the conceptual advantage of clearly separating the astronomy (the input) from the oceanography (the black box). The basic response analysis assumes a linear system, but weak nonlinear interactions can be allowed for with extra terms.

Tidal Dynamics

The equilibrium tide consists of two symmetrical tidal bulges directly opposite the moon or sun. Semidiurnal tidal ranges would reach their maximum value of about 0.5 m at equatorial latitudes. The individual high water bulges would track around the earth, moving from east to west in steady progression. These theoretical characteristics are clearly *not* those of the observed tides.

The observed tides in the main oceans have much larger mean ranges, of about 1 m, but there are considerable variations. Times of tidal high water vary in a geographical pattern which bears no relationship to the simple ideas of a double bulge. The tides spread from the oceans onto the surrounding continental shelves, where even larger ranges are observed. In some shelf seas the spring tidal range may exceed 10 m: the Bay of Fundy, the Bristol Channel and the Argentine Shelf are well-known examples. Laplace (1749–1827) advanced the basic mathematical solutions for tidal waves on a rotating earth. More generally, the reasons for these complicated ocean responses to tidal forcing may be summarized as follows.

1. Movements of water on the surface of the earth must obey the physical laws represented by the hydrodynamic equations of continuity and momentum balance; this means that they must propagate as long waves. Any propagation of a wave, east to west around the earth, is impeded by the north–south continental boundaries.
2. Long waves travel at a speed that is related to the water depth; oceans are too shallow for this to match the tracking of the moon.
3. The various ocean basins have their individual natural modes of oscillation which influence their response to the tide-generating forces. There are many resonant frequencies. However, the whole global ocean system seems to be near to resonance at semidiurnal tidal frequencies, as the observed semidiurnal tides are generally much bigger than the diurnal tides.
4. Water movements are affected by the rotation of the earth. The tendency for water movement to maintain a uniform direction in absolute space means that it performs a curved path in the rotating frame of reference within which our observations are made.
5. The solid earth responds elastically to the imposed gravitational tidal forces, and to the ocean tidal loading. The redistribution of water mass during the tidal cycle affects the gravitational field.

Long-Wave Characteristic, No Rotation

Provided that wave amplitudes are small compared with the depth, and that the depth is small compared with the wavelength, then the speed for the wave propagation is:

$$c = (gD)^{1/2} \qquad [15]$$

where g is gravitational acceleration, and D is the water depth. The currents u are related to the instantaneous level ζ by:

$$u = \zeta(g/D)^{1/2} \qquad [16]$$

Long waves have the special property that the speed c is independent of the frequency, and depends only on the value of g and the water depth; any disturbance which consists of a number of separate harmonic constituents will not change its shape as it propagates – this is nondispersive propagation. Waves at tidal periods are long waves, even in the deep ocean, and so their propagation is nondispersive. In the real ocean, tides cannot propagate endlessly as progressive waves. They undergo reflection at sudden changes of depth and at the coastal boundaries.

Standing Waves and Resonance

Two progressive waves traveling in opposite directions result in a wave motion, called a standing wave. This can happen where a wave is perfectly reflected at a barrier.

Systems which are forced by oscillations close to their natural period have large amplitude responses. The responses of oceans and many seas are close to semidiurnal resonance. In nature, the forced resonant oscillations cannot grow indefinitely because friction limits the response.

Because of energy losses, tidal waves are not perfectly reflected at the head of a basin, which means that the reflected wave is smaller than the ingoing wave. It is easy to show that this is equivalent to a progressive wave superimposed on a standing wave with the progressive wave carrying energy to the head of the basin. Standing waves cannot transmit energy because they consist of two progressive waves of equal amplitude traveling in opposite directions.

Long Waves on a Rotating Earth

A long progressive wave traveling in a channel on a rotating Earth behaves differently from a wave traveling along a nonrotating channel. The geostrophic forces that affect the motion in a rotating system, cause a deflection of the currents towards the right of the direction of motion in the Northern Hemisphere. The build-up of water on the right of the channel gives rise to a pressure gradient across the channel, which in turn develops until at equilibrium it balances the geostrophic force. The resulting Kelvin wave is described mathematically:

$$\zeta(y) = H_o \exp\left(\frac{-fy}{c}\right); u(y) = \left(\frac{g}{D}\right)^{1/2} \zeta(y) \quad [17]$$

where $\zeta(y)$ is the amplitude at a distance y from the right-hand boundary (in the Northern Hemisphere) and H_o is the amplitude of the wave at the boundary. The effect of the rotation appears only in the factor $\exp(-fy/c)$, which gives a decay of wave amplitude away from the boundary with a length scale of

$c/f = [(gD)^{1/2}/f]$, which depends on the latitude and the water depth. This scale is called the Rossby radius of deformation. At a distance $y = c/f$ from the boundary the amplitude has fallen to $0.37H_o$. At 45°N in water of 4000 m depth the Rossby radius is 1900 km, but in water 50 m deep this is reduced to 215 km.

Kelvin waves are not the only solution to the hydrodynamic equations on a rotating Earth: a more general form, called Poincaré waves, gives amplitudes which vary sinusoidally rather than exponentially in the direction transverse to the direction of wave propagation.

The case of a standing-wave oscillation on a rotating Earth is of special interest in tidal studies. Away from the reflecting boundary, tidal waves can be represented by two Kelvin waves traveling in opposite directions. The wave rotates about a nodal point, which is called an amphidrome (**Figure 3**). The cotidal lines all radiate outwards from the amphidrome and the co-amplitude lines form a set of nearly concentric circles around the center at the amphidrome, at which the amplitude is zero. The amplitude is greatest around the boundaries of the basin.

Ocean Tides

Dynamically there are two essentially different types of tidal regime; in the wide and relatively deep ocean basins the observed tides are generated directly by the external gravitational forces; in the shelf seas the tides are driven by co-oscillation with the oceanic tides. The ocean response to the gravitational forcing may be described in terms of a forced linear oscillator, with weak energy dissipation.

A global chart of the principal lunar semidiurnal tidal constituent M_2 shows a complicated pattern of amphidromic systems. As a general rule these conform to the expected behavior for Kelvin wave propagation, with anticlockwise rotation in the Northern Hemisphere, and clockwise rotation in the Southern Hemisphere.

For example, in the Atlantic Ocean the mostfully developed semidiurnal amphidrome is located near 50°N, 39°W. The tidal waves appear to travel around the position in a form which approximates to a Kelvin wave, from Portugal along the edge of the north-west European continental shelf towards Iceland, and thence west and south past Greenland to Newfoundland. There is a considerable leakage of energy to the surrounding continental shelves and to the Arctic Ocean, so the wave reflected in a southerly direction, is weaker than the wave traveling northwards along the European coast.

The patterns of tidal waves on the continental shelf are scaled down as the wave speeds are reduced. In the very shallow water depths (typically less than 20 m) there are strong tidal currents and substantial

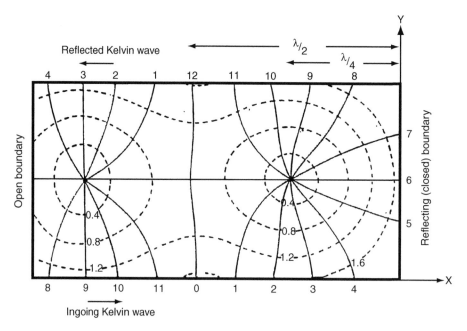

Figure 3 Cotidal and co-amplitude lines for a Kelvin wave reflected without energy loss in a rectangular channel. The incoming wave travels fromleft to right. Continuous lines are cotidal lines at intervals of 1/12 of a full cycle. Broken lines are lines of equal amplitude. Progression of the wave crests in the Northern Hemisphere is anticlockwise for both the amphidromic systems shown. In practice the reflected wave is weaker and so the amphidromes are moved towards the upper wall in the diagram.

energy losses due to bottom friction. Tidal waves are strongly influenced by linear Kelvin wave dynamics and by basin resonances. Energy is propagated to the shallow regions where it is dissipated.

Energy Fluxes and Budgets

The energy lost through tidal friction gradually slows down the rate of rotation of the earth, increasing the length of the day by one second in 41 000 years. Angular momentum of the earth–moon system is conserved by the moon moving away from the earth at 3.7 mm per year. The total rate of tidal energy dissipation due to the M_2 tide can be calculated rather exactly from the astronomic observations at 2.50 ± 0.05 TW, of which 0.1 TW is dissipated in the solid Earth. The total lunar dissipation is 3.0 TW, and the total due to both sun and moon is 4.0 TW. For comparison the geothermal heat loss is 30 TW, and the 1995 total installed global electric capacity was 2.9 TW. Solar radiation input isfive orders of magnitude greater.

Most of the 2.4 TW of M_2 energy lost in the ocean is due to the work against bottom friction which opposes tidal currents. Because the friction increases approximately as the square of current speed, and the energy loses as the cube, tidal energy loses are concentrated in a few shelf areas of strong tidal currents. Notable among these are the north-west European Shelf, the Patagonian Shelf, the Yellow Sea, the Timor and Arafura Seas, Hudson Bay, Baffin Bay, and the Amazon Shelf. It now appears that up to 25% (1 TW) of the tidal energy may be dissipated by internal tidal waves in the deep ocean, where the dissipation processes contribute to vertical mixing and the breakdown of stratification. Again, energy losses may be concentrated in a few areas, for example where the rough topography of midocean ridges and islandarcs create favorable conditions.

One of the main areas of tidal research is increasingly concentrated on gaining a better understanding of the many nonlinear ocean processes that are driven by this cascading tidal energy. Some examples, outlined below, are considered in more detail elsewhere in this Encyclopedia.

- Generation of tidal fronts in shelf seas, where the buoyancy forces due to tidal mixing compete with the buoyancy fluxes due to surface heating; the ratio of the water depth divided by the cube of the tidal currentis a good indicator of the balance between the two factors, and fronts form along lines where this ratio reaches a critical value.
- River discharges to shallow seas near the mouth of rivers. The local input of freshwater buoyancy may be comparable to the buoyancy input fromsummer surface warming. These regions are called ROFIs (regions of freshwater influence). Spring–neap variations in the energy of tidal mixing strongly influence the circulation in these regions.
- The mixing and dispersion of pollutants often driven by the turbulence generated by tides. Maximum turbulent energy occurs some hours after the time of maximum tidal currents.
- Sediment processes of erosion and deposition are often controlled by varying tidal currents, particularly over a spring–neap cycle. The phase delay in suspended sediment concentration after maximum currents may be related to the phase lag in the turbulent energy, which has importantconsequences for sediment deposition and distribution.
- Residual circulation is partly driven by nonlinear responses to tidal currents in shallow water. Tidal flows also induce residual circulation around sandbanks because of the depth variations. In the Northern Hemisphere this circulation is observed to be in a clockwise sense. Near headlands and islands which impede tidal currents, residual eddies can cause marked asymmetry between the time and strength of the tidal ebb and flow currents.
- Tidal currents influence biological breeding patterns, migration, and recruitment. Some types of fish have adapted to changing tidal currents to assist in their migration: they lie dormant on the seabed when the currents are not favorable. Tidal mixing in shallow seas promotes productivity by returning nutrients to surface waters where light is available. Tidal fronts are known to be areas of high productivity. The most obvious example of tidal influence on biological processes is the zonation of species found at different levels along rocky shorelines.
- Evolution of sedimentary shores due to the dynamic equilibrium between waves, tides and other processes along sedimentary coasts which resultsin a wide range of features such as lagoons, sandbars, channels, and islands. These are very complicated processes which are still difficult to understandand model.
- Tidal amphidrome movements. The tidal amphidromes as shown in **Figure 4** only fall along the center line of the channel if the incoming tidal Kelvin wave is perfectly reflected. In reality, the reflected wave is weaker, and the amphidromes are displaced towards the side of the sea along which the outgoing tidal wave travels. Proportionately more energy is removed at spring tides than at neap tides, so the amphidromes can

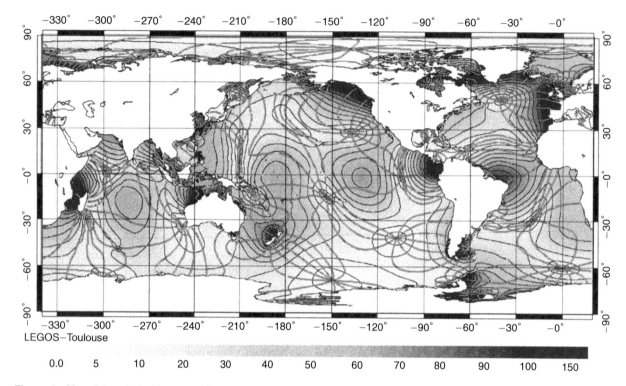

Figure 4 Map of the principal lunar semidiurnal tide produced by computer. Ocean tides observed by satellite and *in situ* now agree very closely with computer modeled tides. The dark areas show regions of high tida lamplitude. Note the convergence of cophase lines at amphidromes. In the Northern Hemisphere the tides normally progress in an anticlockwise sense around the amphidrome; in the Southern Hemisphere the progression is usually clockwise.

move by several tens of kilometers during the-spring–neap cycle.

Nonlinear processes on the basic M_2 tide generate a series of higher harmonics, M_4, M_6..., with corresponding terms such as MS_4 for spring–neap interactions. A better understanding of the significance of the amplitudes and phases of these terms in the analyses of shallow-water tides and tidal phenomena will be an important tool in advancing our overall knowledge of the influence of tides on a wide range of ocean processes.

See also

Beaches, Physical Processes Affecting. Coastal Trapped Waves. Estuarine Circulation. Geomorphology. Intertidal Fishes. Lagoons. River Inputs. Salt Marshes and Mud Flats. Sea Level Change. Shelf Sea and Shelf Slope Fronts. Tidal Energy. Waves on Beaches.

Further Reading

Cartwright DE (1999) *Tides – a Scientific History.* Cambridge: Cambridge University Press.

Garrett C and Maas LRM (1993) Tides and their effects. *Oceanus* 36(1): 27–37.

Parker BB (ed.) (1991) *Tidal Hydrodynamics.* New York: John Wiley.

Prandle D (1997) Tidal characteristics of suspended sediment concentrations. *Journal of Hydraulic Engineering* 123: 341–350.

Pugh DT (1987) *Tides, Surges and Mean Sea Level.* Chichester: John Wiley.

Ray RD and Woodworth PL (eds.) (1997) Special issue on tidal science in honour of David E Cartwright. *Progress in Oceanography* 40.

Simpson JH (1998) Tidal processes in shelf seas. In: Brink KH and Robinson AR (eds.) *The Sea*, Vol. 10. New York: John Wiley.

Wilhelm H, Zurn W, and Wenzel HG (eds.) (1997) *TidalPhenomena: Lecture Notes in Earth Sciences 66.* Berlin: Springer-Verlag.

TIDAL ENERGY

A. M. Gorlov, Northeastern University, Boston, Massachusetts, USA

Introduction

Gravitational forces between the moon, the sun and the earth cause the rhythmic rising and lowering of ocean waters around the world that results in Tide Waves. The moon exerts more than twice as great a force on the tides as the sun due to its much closer position to the earth. As a result, the tide closely follows the moon during its rotation around the earth, creating diurnal tide and ebb cycles at any particular ocean surface. The amplitude or height of the tide wave is very small in the open ocean where it measures several centimeters in the center of the wave distributed over hundreds of kilometers. However, the tide can increase dramatically when it reaches continental shelves, bringing huge masses of water into narrow bays and river estuaries along a coastline. For instance, the tides in the Bay of Fundy in Canada are the greatest in the world, with amplitude between 16 and 17 meters near shore. High tides close to these figures can be observed at many other sites worldwide, such as the Bristol Channel in England, the Kimberly coast of Australia, and the Okhotsk Sea of Russia. Table 1 contains ranges of amplitude for some locations with large tides.

On most coasts tidal fluctuation consists of two floods and two ebbs, with a semidiurnal period of about 12 hours and 25 minutes. However, there are some coasts where tides are twice as long (diurnal tides) or are mixed, with a diurnal inequality, but are still diurnal or semidiurnal in period. The magnitude of tides changes during each lunar month. The highest tides, called spring tides, occur when the moon, earth and sun are positioned close to a straight line (moon syzygy). The lowest tides, called neap tides, occur when the earth, moon and sun are at right angles to each other (moon quadrature). Isaac Newton formulated the phenomenon first as follows: 'The ocean must flow twice and ebb twice, each day, and the highest water occurs at the third hour after the approach of the luminaries to the meridian of the place'. The first tide tables with accurate prediction of tidal amplitudes were published by the British Admiralty in 1833. However, information about tide fluctuations was available long before that time from a fourteenth century British atlas, for example.

Rising and receding tides along a shoreline area can be explained in the following way. A low height tide wave of hundreds of kilometers in diameter runs on the ocean surface under the moon, following its rotation around the earth, until the wave hits a continental shore. The water mass moved by the moon's gravitational pull fills narrow bays and river estuaries where it has no way to escape and spread over the ocean. This leads to interference of waves and accumulation of water inside these bays and estuaries, resulting in dramatic rises of the water level (tide cycle). The tide starts receding as the moon continues its travel further over the land, away from the ocean, reducing its gravitational influence on the ocean waters (ebb cycle).

The above explanation is rather schematic since only the moon's gravitation has been taken into account as the major factor influencing tide fluctuations. Other factors, which affect the tide range are the sun's pull, the centrifugal force resulting from the earth's rotation and, in some cases, local resonance of the gulfs, bays or estuaries.

Energy of Tides

The energy of the tide wave contains two components, namely, potential and kinetic. The potential energy is the work done in lifting the mass of water above the ocean surface. This energy can be calculated as:

$$E = \mathbf{g}\rho A \int z \, \mathrm{d}z = 0.5 \mathbf{g}\rho A b^2,$$

where E is the energy, $\mathbf{g}$ is acceleration of gravity, ρ is the seawater density, which equals its mass per unit

Table 1 Highest tides (tide ranges) of the global ocean

Country	Site	Tide range (m)
Canada	Bay of Fundy	16.2
England	Severn Estuary	14.5
France	Port of Ganville	14.7
France	La Rance	13.5
Argentina	Puerto Rio Gallegos	13.3
Russia	Bay of Mezen (White Sea)	10.0
Russia	Penzhinskaya Guba (Sea of Okhotsk)	13.4

volume, A is the sea area under consideration, z is a vertical coordinate of the ocean surface and h is the tide amplitude. Taking an average $(g\rho) = 10.15 \, \text{kN m}^{-3}$ for seawater, one can obtain for a tide cycle per square meter of ocean surface:

$$E = 1.4h^2, \text{watt-hour}$$

or

$$E = 5.04h^2, \text{kilojoule}$$

The kinetic energy T of the water mass m is its capacity to do work by virtue of its velocity V. It is defined by $T = 0.5 \, \text{m V}^2$. The total tide energy equals the sum of its potential and kinetic energy components.

Knowledge of the potential energy of the tide is important for designing conventional tidal power plants using water dams for creating artificial upstream water heads. Such power plants exploit the potential energy of vertical rise and fall of the water. In contrast, the kinetic energy of the tide has to be known in order to design floating or other types of tidal power plants which harness energy from tidal currents or horizontal water flows induced by tides. They do not involve installation of water dams.

Extracting Tidal Energy: Traditional Approach

People used the phenomenon of tides and tidal currents long before the Christian era. The earliest navigators, for example, needed to know periodical tide fluctuations as well as where and when they could use or would be confronted with a strong tidal current. There are remnants of small tidal hydromechanical installations built in the Middle Ages around the world for water pumping, watermills and other applications. Some of these devices were exploited until recent times. For example, large tidal waterwheels were used for pumping sewage in Hamburg, Germany up to the nineteenth century. The city of London used huge tidal wheels, installed under London Bridge in 1580, for 250 years to supply fresh water to the city. However, the serious study and design of industrial-size tidal power plants for exploiting tidal energy only began in the twentieth century with the rapid growth of the electric industry. Electrification of all aspects of modern civilization has led to the development of various converters for transferring natural potential energy sources into electric power. Along with fossil fuel power systems and nuclear reactors, which create huge new environmental pollution problems, clean renewable energy sources have attracted scientists

and engineers to exploit these resources for the production of electric power. Tidal energy, in particular, is one of the best available renewable energy sources. In contrast to other clean sources, such as wind, solar, geothermal etc., tidal energy can be predicted for centuries ahead from the point of view of time and magnitude. However, this energy source, like wind and solar energy is distributed over large areas, which presents a difficult problem for collecting it. Besides that, complex conventional tidal power installations, which include massive dams in the open ocean, can hardly compete economically with fossil fuel (thermal) power plants, which use cheap oil or coal, presently available in abundance. These thermal power plants are currently the principal component of world electric energy production. Nevertheless, the reserves of oil and coal are limited and rapidly dwindling. Besides, oil and coal cause enormous atmospheric pollution both from emission of green house gases and from their impurities such as sulfur in the fuel. Nuclear power plants produce accumulating nuclear wastes that degrade very slowly, creating hazardous problems for future generations. Tidal energy is clean and not depleting. These features make it an important energy source for global power production in the near future. To achieve this goal, the tidal energy industry has to develop a new generation of efficient, low cost and environmentally friendly apparatus for power extraction from free or ultra-low head water flow.

Four large-scale tidal power plants currently exist. All of them were constructed after World War II. They are the La Rance Plant (France, 1967), the Kislaya Guba Plant (Russia, 1968), the Annapolis Plant (Canada, 1984), and the Jiangxia Plant (China, 1985). The main characteristics of these tidal power plants are given in **Table 2**. The La Rance plant is shown in **Figure 1**.

All existing tidal power plants use the same design that is accepted for construction of conventional river hydropower stations. The three principal structural and mechanical elements of this designare: a water dam across the flow, which creates an artificial water basin and builds up a water head for operation of hydraulic turbines; a number of turbines coupled with electric generators installed at the lowest point of the dam; and hydraulic gates in the dam to control the water flow in and out of the water basin behind the dam. Sluice locks are also used for navigation when necessary. The turbines convert the potential energy of the water mass accumulated on either side of the dam into electric energy during the tide. The tidal power plant can be designed for operation either by double or single action. Double action means that the turbines work in both water

Table 2 Extant large tidal power plants

Country	Site	Installed power (MW)	Basin area (km²)	Mean tide (m)
France	La Rance	240	22	8.55
Russia	Kislaya Guba	0.4	1.1	2.3
Canada	Annapolis	18	15	6.4
China	Jiangxia	3.9	1.4	5.08

flows, i.e. during the tide when the water flows through the turbines, filling the basin, and then, during the ebb, when the water flows back into the ocean draining the basin. In single-action systems, the turbines work only during the ebb cycle. In this case, the water gates are kept open during the tide, allowing the water to fill the basin. Then the gates close, developing the water head, and turbines start operating in the water flow from the basin back into the ocean during the ebb.

Advantages of the double-action method are that it closely models the natural phenomenon of the tide, has least effect on the environment and, in some cases, has higher power efficiency. However, this method requires more complicated and expensive reversible turbines and electrical equipment. The

Figure 1 Aerial view of the La Rance Tidal Power Plant (Source: Electricitéde France).

single action method is simpler, and requires less expensive turbines. The negative aspects of the single action method are its greater potential for harm to the environment by developing a higher water head and causing accumulation of sediments in the basin. Nevertheless, both methods have been used in practice. For example, the La Rance and the Kislaya Guba tidal power plants operate under the double-action scheme, whereas the Annapolis plant uses a single-action method.

One of the principal parameters of a conventional hydropower plant is its power output P (energy per unit time) as a function of the water flow rate Q (volume per time) through the turbines and the water head h (difference between upstream and downstream water levels). Instantaneous power P can be defined by the expression: $P = 9.81\,Qh$, kW, where Q is in $m^3 s^{-1}$, h is in meters and 9.81 is the product (ρg) for fresh water, which has mass density $\rho = 1000$ kg m^{-3} and $g = 9.81$ m s^{-2}. The (ρg) component has to be corrected for applications in salt water due to its different density (see above).

The average annual power production of a conventional tidal power plant with dams can be calculated by taking into account some other geophysical and hydraulic factors, such as the effective basin area, tidal fluctuations, etc. **Tables 2** and **3** contain some characteristics of existing tidal power plants as well as prospects for further development of traditional power systems in various countries using dams and artificial water basins described above.

Extracting Tidal Energy: Non-traditional Approach

As mentioned earlier, all existing tidal power plants have been built using the conventional design developed for river power stations with water dams as their principal component. This traditional river scheme has a poor ecological reputation because the dams block fish migration, destroying their population, and damage the environment by flooding and swamping adjacent lands. Flooding is not an issue for tidal power stations because the water level in the basin cannot be higher than the natural tide. However, blocking migration of fish and other ocean inhabitants by dams may represent a serious environmental problem. In addition, even the highest average global tides, such as in the Bay of Fundy, are small compared with the water heads used in conventional river power plants where they are measured in tens or even hundreds of meters. The relatively low water head in tidal power plants creates a difficult technical problem for designers. The fact is that the very efficient, mostly propeller-type hydraulic turbines developed for high river dams are inefficient, complicated and very expensive for low-head tidal power application.

These environmental and economic factors have forced scientists and engineers to look for a new approach to exploitation of tidal energy that does not require massive ocean dams and the creation of high water heads. The key component of such an approach is using new unconventional turbines, which can efficiently extract the kinetic energy from a free unconstrained tidal current without any dams. One such turbine, the Helical Turbine, is shown in **Figure 2**. This cross-flow turbine was developed in 1994. The turbine consists of one or more long helical blades that run along a cylindrical surface like a screwthread, having a so-called airfoil or 'airplane wing' profile. The blades provide a reaction thrust that can rotate the turbine faster than the water flow itself. The turbine shaft (axis of rotation) must be perpendicular to the water current, and the turbine can be positioned either horizontally or

Table 3 Some potential sites for tidal power installations (traditional approach)

Country	Site	Potential power (MW)	Basin area (km²)	Mean tide (m)
USA	Passamaquoddy	400	300	5.5
USA	Cook Inlet	Up to 18 000	3100	4.35
Russia	Mezen	15 000	2640	5.66
Russia	Tugur	6790	1080	5.38
UK	Severn	6000	490	8.3
UK	Mersey	700	60	8.4
Argentina	San Jose	7000	780	6.0
Korea	Carolim Bay	480	90	4.7
Australia	Secure	570	130	8.4
Australia	Walcott	1750	260	8.4

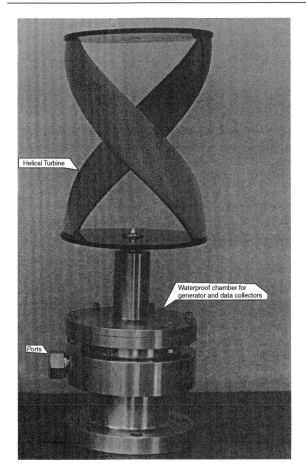

Figure 2 Double-helix turbine with electric generator for underwater installation.

vertically. Due to its axial symmetry, the turbine always develops unidirectional rotation, even in reversible tidal currents. This is a very important advantage, which simplifies design and allows exploitation of the double-action tidal power plants. A pictorial view of a floating tidal power plant with a number of vertically aligned triple-helix turbines is shown in **Figure 3**. This project has been proposed for the Uldolmok Strait in Korea, where a very strong reversible tidal current with flows up to 12 knots (about $6\,\mathrm{m\,s}^{-1}$) changes direction four times a day.

The following expression can be used for calculating the combined turbine power of a floating tidal plant (power extracted by all turbines from a free, unconstrained tidal current): $P_t = 0.5\eta\rho AV^3$, where P_t is the turbine power in kilowatts, η is the turbine efficiency ($\eta = 0.35$ in most tests of the triple-helix turbine in free flow), ρ is the mass water density, A is the total effective frontal area of the turbines in m^2 (cross-section of the flow where the turbines are installed) and V is the tidal current velocity in $\mathrm{m\,s}^{-1}$. Note, that the power of a free water current through

a cross-flow area A is $P_w = 0.5\rho AV^3$. The turbine efficiency η, also called power coefficient, is the ratio of the turbine power output P_t to the power of either the water head for traditional design or unconstrained water current P_w, i.e. $\eta = P_t/P_w$.

The maximum power of the Uldolmok tidal project shown in **Figure 3** is about 90 MW calculated using the above approach for $V = 12$ knots, $A = 2100\,\mathrm{m}^2$ and $\eta = 0.35$.

Along with the floating power farm projects with helical turbines described, there are proposals to use large-diameter propellers installed on the ocean floor to harness kinetic energy of tides as well as other ocean currents. These propellers are, in general, similar to the well known turbines used for wind farms.

Utilizing Electric Energy from Tidal Power Plants

A serious issue that must be addressed is how and where to use the electric power generated by extracting energy from the tides. Tides are cyclical by their nature, and the corresponding power output of a tidal power plant does not always coincide with the peak of human activity. In countries with a well-developed power industry, tidal power plants can be a part of the general power distribution system. However, power from a tidal plant would then have to be transmitted a long distance because locations of high tides are usually far away from industrial and urban centers. An attractive future option is to utilize the tidal power *in situ* for year-round production of hydrogen fuel by electrolysis of the water. The hydrogen, liquefied or stored by another method, can be transported anywhere to be used either as a fuel instead of oil or gasoline or in various fuel cell energy systems. Fuel cells convert hydrogen energy directly into electricity without combustion or moving parts, which is then used, for instance, in electric cars. Many scientists and engineers consider such a development as a future new industrial revolution. However, in order to realize this idea worldwide, clean hydrogen fuel would need to be also available everywhere. At present most hydrogen is produced from natural gases and fossil fuels, which emit greenhouse gases into the atmosphere and harm the global ecosystem. From this point of view, production of hydrogen by water electrolysis using tidal energy is one of the best ways to develop clean hydrogen fuel by a clean method. Thus, tidal energy can be used in the future to help develop a new era of clean industries, for example, to clean up the automotive industry, as well as other energy-consuming areas of human activity.

Electric generators sit above the water

Figure 3 Artist rendition of the floating tidal power plant with vertical triple-helix turbines for Uldolmok Strait (Korean Peninsula).

Conclusion

Tides play a very important role in the formation of global climate as well as the ecosystems for ocean habitants. At the same time, tides are a substantial potential source of clean renewable energy for future human generations. Depleting oil reserves, the emission of greenhouse gases by burning coal, oil and other fossil fuels, as well as the accumulation of nuclear waste from nuclear reactors will inevitably force people to replace most of our traditional energy sources with renewable energy in the future. Tidal energy is one of the best candidates for this approaching revolution. Development of new, efficient, low-cost and environmentally friendly hydraulic energy converters suited to free-flow waters, such as triple-helix turbines, can make tidal energy available worldwide. This type of machine, moreover, can be used not only for multi-megawatt tidalpower farms but also for mini-power stations with turbines generating a few kilowatts. Such power stations can provide clean energy to small communities or even individual households located near continental shorelines, straits or on remote islands with strong tidal currents.

See also

Tides.

Further Reading

Bernshtein LB (ed.) (1996) *Tidal Power Plants*. Seoul: Korea Ocean Research and Development Institute (KORDI).

Gorlov AM (1998) Turbines with a twist. In: Kitzinger U and Frankel EG (eds.) *Macro-Engineering and the Earth: World Projects for the Year 2000 and Beyond*, pp. 1–36. Chichester: Horwood Publishing.

Charlier RH (1982) *Tidal Energy*. New York: Van Nostrand Reinhold.

TOPOGRAPHY AND SEA LEVEL

SEA LEVEL CHANGE

J. A. Church, Antarctic CRC and CSIRO Marine Research, TAS, Australia
J. M. Gregory, Hadley Centre, Berkshire, UK

Introduction

Sea-level changes on a wide range of time and space scales. Here we consider changes in mean sea level, that is, sea level averaged over a sufficient period of time to remove fluctuations associated with surface waves, tides, and individual storm surge events. We focus principally on changes in sea level over the last hundred years or so and on how it might change over the next one hundred years. However, to understand these changes we need to consider what has happened since the last glacial maximum 20 000 years ago. We also consider the longer-term implications of changes in the earth's climate arising from changes in atmospheric greenhouse gas concentrations.

Changes in mean sea level can be measured with respect to the nearby land (relative sea level) or a fixed reference frame. Relative sea level, which changes as either the height of the ocean surface or the height of the land changes, can be measured by a coastal tide gauge.

The world ocean, which has an average depth of about 3800 m, contains over 97% of the earth's water. The Antarctic ice sheet, the Greenland ice sheet, and the hundred thousand nonpolar glaciers/ ice caps, presently contain water sufficient to raise sea level by 61 m, 7 m, and 0.5 m respectively if they were entirely melted. Ground water stored shallower than 4000 m depth is equivalent to about 25 m (12 m stored shallower than 750 m) of sea-level change. Lakes and rivers hold the equivalent of less than 1 m, while the atmosphere accounts for only about 0.04 m.

On the time-scales of millions of years, continental drift and sedimentation change the volume of the ocean basins, and hence affect sea level. A major influence is the volume of mid-ocean ridges, which is related to the arrangement of the continental plates and the rate of sea floor spreading.

Sea level also changes when mass is exchanged between any of the terrestrial, ice, or atmospheric reservoirs and the ocean. During glacial times (ice ages), water is removed from the ocean and stored in large ice sheets in high-latitude regions. Variations in the surface loading of the earth's crust by water and the surface loading of the earth's crust by water and ice change the shape of the earth as a result of the elastic response of the lithosphere and viscous flow of material in the earth's mantle and thus change the level of the land and relative sea level. These changes in the distribution of mass alter the gravitational field of the earth, thus changing sea level. Relative sea level can also be affected by local tectonic activities as well as by the land sinking when ground water is extracted or sedimentation increases. Sea water density is a function of temperature. As a result, sea level will change if the ocean's temperature varies (as a result of thermal expansion) without any change in mass.

Sea-Level Changes Since the Last Glacial Maximum

On timescales of thousands to hundreds of thousands of years, the most important processes affecting sea-level are those associated with the growth and decay of the ice sheets through glacial–interglacial cycles. These are also relevant to current and future sea level rise because they are the cause of ongoing land movements (as a result of changing surface loads and the resultant small changes in shape of the earth – postglacial rebound) and ongoing changes in the ice sheets.

Sea-level variations during a glacial cycle exceed 100 m in amplitude, with rates of up to tens of millimetres per year during periods of rapid decay of the ice sheets (**Figure 1**). At the last glacial maximum (about 21 000 years ago), sea level was more than 120 m below current levels. The largest contribution to this sea-level lowering was the additional ice that formed the North American (Laurentide) and European (Fennoscandian) ice sheets. In addition, the Antarctic ice sheet was larger than at present and there were smaller ice sheets in presently ice-free areas.

Observed Recent Sea-Level Change

Long-term relative sea-level changes have been inferred from the geological records, such as radio-carbon dates of shorelines displaced from present day sea level, and information from corals and sediment cores. Today, the most common method of measuring sea level relative to a local datum is by tide gauges at coastal and island sites. A global data set is maintained by the Permanent Service for Mean Sea Level (PSMSL). During the 1990s, sea level has been measured globally with satellites.

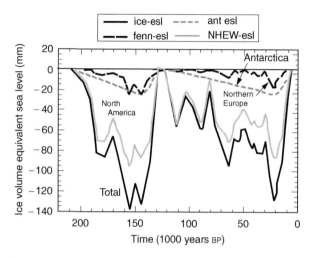

Figure 1 Change in ice sheet volume over the last 200 000 years. Fenn, Fennoscandian; ant, Antarctic; NHEW; North America; esl, equivalent sea level. (Reproduced from Lambeck, 1998.)

Tide-gauge Observations

Unfortunately, determination of global-averaged sea-level rise is severely limited by the small number of gauges (mostly in Europe and North America) with long records (up to several hundred years, **Figure 2**). To correct for vertical land motions, some sea-level change estimates have used geological data, whereas others have used rates of present-day vertical land movement calculated from models of postglacial rebound.

A widely accepted estimate of the current rate of global-average sea-level rise is about $1.8 \, \mathrm{mm \, y^{-1}}$. This estimate is based on a set of 24 long tide-gauge records, corrected for land movements resulting from deglaciation. However, other analyses produce different results. For example, recent analyses suggest that sea-level change in the British Isles, the North Sea region and Fennoscandia has been about $1 \, \mathrm{mm \, y^{-1}}$ during the past century. The various assessments of the global-average rate of sea-level change over the past century are not all consistent within stated uncertainties, indicating further sources of error. The treatment of vertical land movements remains a source of potential inconsistency, perhaps amounting to $0.5 \, \mathrm{mm \, y^{-1}}$. Other sources of error include variability over periods of years and longer and any spatial distribution in regional sea level rise (perhaps several tenths of a millimeter per year).

Comparison of the rates of sea-level rise over the last 100 years ($1.0–2.0 \, \mathrm{mm \, y^{-1}}$) and over the last two millennia ($0.1–0.0 \, \mathrm{mm \, y^{-1}}$) suggests the rate has accelerated fairly recently. From the few very long tide-gauge records (**Figure 2**), it appears that an acceleration of about $0.3–0.9 \, \mathrm{mm \, y^{-1}}$ per century occurred over the nineteenth and twentieth century. However, there is little indication that sea-level rise accelerated during the twentieth century.

Altimeter Observations

Following the advent of high-quality satellite radar altimeter missions in the 1990s, near-global and homogeneous measurement of sea level is possible, thereby overcoming the inhomogeneous spatial sampling from coastal and island tide gauges. However, clarifying rates of global sea-level change requires continuous satellite operations over many

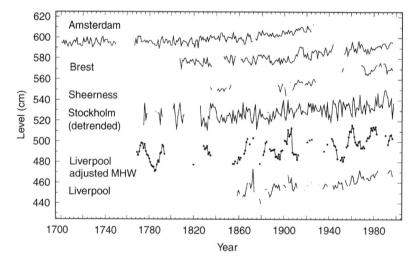

Figure 2 Time series of relative sea level over the last 300 years from several European coastal locations. For the Stockholm record, the trend over the period 1774 to 1873 has been removed from the entire data set. For Liverpool two series are given. These are mean sea level and, for a longer period, the mean high water (MHW) level. (Reproduced with permission from Woodworth, 1999.)

years and careful control of biases within and between missions.

To date, the TOPEX/POSEIDON satellite-altimeter mission, with its (near) global coverage from 66°N to 66°S (almost all of the ice-free oceans) from late 1992 to the present, has proved to be of most value in producing direct estimates of sea-level change. The present data allow global-average sea level to be estimated to a precision of several millimeters every 10 days, with the absolute accuracy limited by systematic errors. The most recent estimates of global-average sea level rise based on the short (since 1992) TOPEX/POSEIDON time series range from $2.1 \, \text{mm} \, \text{y}^{-1}$ to $3.1 \, \text{mm} \, \text{y}^{-1}$.

The alimeter record for the 1990s indicates a rate of sea-level rise above the average for the twentieth century. It is not yet clear if this is a result of an increase in the rate of sea-level rise, systematic differences between the tide-gauge and altimeter data sets or the shortness of the record.

Processes Determining Present Rates of Sea-Level Change

The major factors determining sea-level change during the twentieth and twenty-first century are ocean thermal expansion, the melting of nonpolar glaciers and ice caps, variation in the mass of the Antarctic and Greenland ice sheets, and changes in terrestrial storage.

Projections of climate change caused by human activity rely principally on detailed computer models referred to as atmosphere–ocean general circulation models (AOGCMs). These simulate the global three-dimensional behavior of the ocean and atmosphere by numerical solution of equations representing the underlying physics. For simulations of the next hundred years, future atmospheric concentrations of gases that may affect the climate (especially carbon dioxide from combustion of fossil fuels) are estimated on the basis of assumptions about future population growth, economic growth, and technological change. AOGCM experiments indicate that the global-average temperature may rise by 1.4–5.8°C between 1990 and 2100, but there is a great deal of regional and seasonal variation in the predicted changes in temperature, sea level, precipitation, winds, and other parameters.

Ocean Thermal Expansion

The broad pattern of sea level is maintained by surface winds, air–sea fluxes of heat and fresh water (precipitation, evaporation, and fresh water runoff from the land), and internal ocean dynamics. Mean sea level varies on seasonal and longer timescales. A particularly striking example of local sea-level variations occurs in the Pacific Ocean during El Niño events. When the trade winds abate, warm water moves eastward along the equator, rapidly raising sea level in the east and lowering it in the west by about 20 cm.

As the ocean warms, its density decreases. Thus, even at constant mass, the volume of the ocean increases. This thermal expansion is larger at higher temperatures and is one of the main contributors to recent and future sea-level change. Salinity changes within the ocean also have a significant impact on the local density, and thus on local sea level, but have little effect on the global-average sea level.

The rate of global temperature rise depends strongly on the rate at which heat is moved from the ocean surface layers into the deep ocean; if the ocean absorbs heat more readily, climate change is retarded but sea level rises more rapidly. Therefore, time-dependent climate change simulation requires a model that represents the sequestration of heat in the ocean and the evolution of temperature as a function of depth. The large heat capacity of the ocean means that there will be considerable delay before the full effects of surface warming are felt throughout the depth of the ocean. As a result, the ocean will not be in equilibrium and global-average sea level will continue to rise for centuries after atmospheric greenhouse gas concentrations have stabilized. The geographical distribution of sea-level change may take many decades to arrive at its final state.

While the evidence is still somewhat fragmentary, and in some cases contradictory, observations indicate ocean warming and thus thermal expansion, particularly in the subtropical gyres, at rates resulting in sea-level rise of order $1 \, \text{mm} \, \text{y}^{-1}$. The observations are mostly over the last few decades, but some observations date back to early in the twentieth century. The evidence is most convincing for the subtropical gyre of the North Atlantic, for which the longest temperature records (up to 73 years) and most complete oceanographic data sets exist. However, the pattern also extends into the South Atlantic and the Pacific and Indian oceans. The only areas of substantial ocean cooling are the subpolar gyres of the North Atlantic and perhaps the North Pacific. To date, the only estimate of a global average rate of sea-level rise from thermal expansion is $0.55 \, \text{mm} \, \text{y}^{-1}$.

The warming in the Pacific and Indian Oceans is confined to the main thermocline (mostly the upper 1 km) of the subtropical gyres. This contrasts with the North Atlantic, where the warming is also seen at greater depths.

AOGCM simulations of sea level suggest that during the twentieth century the average rate

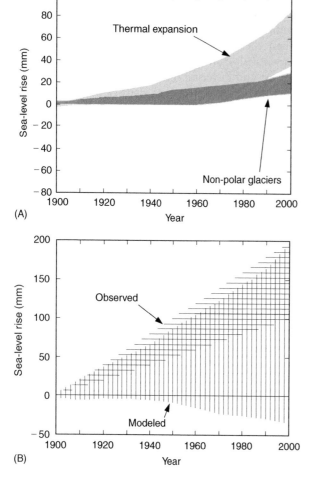

Figure 3 Computed sea-level rise from 1900 to 2000 AD. (A) The estimated thermal expansion is shown by the light stippling; the estimated nonpolar glacial contribution is shown by the medium-density stippling. (B) The computed total sea level change during the twentieth century is shown by the vertical hatching and the observed sea level change is shown by the horizontal hatching.

of change due to thermal expansion was of the order of $0.3–0.8\,\text{mm}\,\text{y}^{-1}$ (**Figure 3**). The rate rises to $0.6–1.1\,\text{mm}\,\text{y}^{-1}$ in recent decades, similar to the observational estimates of ocean thermal expansion.

Nonpolar Glaciers and Ice Caps

Nonpolar glaciers and ice caps are rather sensitive to climate change, and rapid changes in their mass contribute significantly to sea-level change. Glaciers gain mass by accumulating snow, and lose mass (ablation) by melting at the surface or base. Net accumulation occurs at higher altitude, net ablation at lower altitude. Ice may also be removed by discharge into a floating ice shelf and/or by direct calving of icebergs into the sea.

In the past decade, estimates of the regional totals of the area and volume of glaciers have been improved. However, there are continuous mass balance records longer than 20 years for only about 40 glaciers worldwide. Owing to the paucity of measurements, the changes in mass balance are estimated as a function of climate.

On the global average, increased precipitation during the twenty-first century is estimated to offset only 5% of the increased ablation resulting from warmer temperatures, although it might be significant in particular localities. (For instance, while glaciers in most parts of the world have had negative mass balance in the past 20 years, southern Scandinavian glaciers have been advancing, largely because of increases in precipitation.) A detailed computation of transient response also requires allowance for the contracting area of glaciers.

Recent estimates of glacier mass balance, based on both observations and model studies, indicate a contribution to global-average sea level of 0.2 to $0.4\,\text{mm}\,\text{y}^{-1}$ during the twentieth century. The model results shown in **Figure 3** indicate an average rate of 0.1 to $0.3\,\text{mm}\,\text{y}^{-1}$.

Greenland and Antarctic Ice Sheets

A small fractional change in the volume of the Greenland and Antarctic ice sheets would have a significant effect on sea level. The average annual solid precipitation falling onto the ice sheets is equivalent to 6.5 mm of sea level, but this input is approximately balanced by loss from melting and iceberg calving. In the Antarctic, temperatures are so low that surface melting is negligible, and the ice sheet loses mass mainly by ice discharge into floating ice shelves, which melt at their underside and eventually break up to form icebergs. In Greenland, summer temperatures are high enough to cause widespread surface melting, which accounts for about half of the ice loss, the remainder being discharged as icebergs or into small ice shelves.

The surface mass balance plays the dominant role in sea-level changes on a century timescale, because changes in ice discharge generally involve response times of the order of 10^2 to 10^4 years. In view of these long timescales, it is unlikely that the ice sheets have completely adjusted to the transition from the previous glacial conditions. Their present contribution to sea-level change may therefore include a term related to this ongoing adjustment, in addition to the effects of climate change over the last hundred years. The current rate of change of volume of the polar ice sheets can be assessed by estimating the individual mass balance terms or by monitoring surface elevation changes directly (such as by airborne and satellite altimetry during the 1990s). However, these techniques give results with large uncertainties. Indirect methods (including numerical modeling of ice-sheets, observed sea-level changes over the last few millennia, and changes in the earth's rotation parameters) give narrower bounds, suggesting that the present contribution of the ice sheets to sea level is a few tenths of a millimeter per year at most.

Calculations suggest that, over the next hundred years, surface melting is likely to remain negligible in Antarctica. However, projected increases in precipitation would result in a net negative sea-level contribution from the Antarctic ice sheet. On the other hand, in Greenland, surface melting is projected to increase at a rate more than enough to offset changes in precipitation, resulting in a positive contribution to sea-level rise.

Changes in Terrestrial Storage

Changes in terrestrial storage include reductions in the volumes of some of the world's lakes (e.g., the Caspian and Aral seas), ground water extraction in excess of natural recharge, more water being impounded in reservoirs (with some seeping into aquifers), and possibly changes in surface runoff. Order-of-magnitude evaluations of these terms are uncertain but suggest that each of the contributions could be several tenths of millimeter per year, with a small net effect (**Figure 3**). If dam building continues at the same rate as in the last 50 years of the twentieth century, there may be a tendency to reduce sea-level rise. Changes in volumes of lakes and rivers will make only a negligible contribution.

Permafrost currently occupies about 25% of land area in the northern hemisphere. Climate warming leads to some thawing of permafrost, with partial runoff into the ocean. The contribution to sea level in the twentieth century is probably less than 5 mm.

Projected Sea-Level Changes for the Twenty-first Century

Detailed projections of changes in sea level derived from AOCGM results are given in material listed as Further Reading. The major components are thermal expansion of the ocean (a few tens of centimeters), melting of nonpolar glaciers (about 10–20 cm), melting of Greenland ice sheet (several centimeters), and increased storage in the Antarctic (several centimeters).

After allowance for the continuing changes in the ice sheets since the last glacial maximum and the

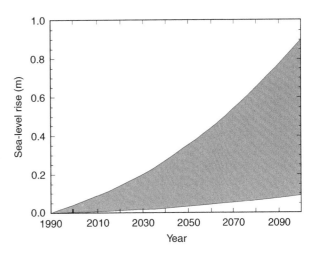

Figure 4 The estimated range of future global-average sea-level rise from 1990 to 2100 AD for a range of plausible projections in atmospheric greenhouse gas concentrations.

melting of permafrost (but not including changes in terrestrial storage), total projected sea-level rise during the twenty-first century is currently estimated to be between about 9 and 88 cm (**Figure 4**).

Regional Sea-Level Change

Estimates of the regional distribution of sea-level rise are available from several AOGCMs. Our confidence in these distributions is low because there is little similarity between model results. However, models agree on the qualitative conclusion that the range of regional variation is substantial compared with the global-average sea-level rise. One common feature is that nearly all models predict less than average sea-level rise in the Southern Ocean.

The most serious impacts of sea-level change on coastal communities and ecosystems will occur during the exceptionally high water levels known as storm surges produced by low air pressure or driving winds. As well as changing mean sea level, climate change could also affect the frequency and severity of these meteorological conditions, making storm surges more or less severe at particular locations.

Longer-term Changes

Even if greenhouse gas concentrations were to be stabilized, sea level would continue to rise for several hundred years. After 500 years, sea-level rise from thermal expansion could be about 0.3–2 m but may be only half of its eventual level. Glaciers presently contain the equivalent of 0.5 m of sea level. If the CO_2 levels projected for 2100 AD were sustained, there would be further reductions in glacier mass.

Ice sheets will continue to react to climatic change during the present millennium. The Greenland ice sheet is particularly vulnerable. Some models suggest that with a warming of several degrees no ice sheet could be sustained on Greenland. Complete melting of the ice sheet would take at least a thousand years and probably longer.

Most of the ice in Antarctica forms the East Antarctic Ice Sheet, which would disintegrate only if extreme warming took place, beyond what is currently thought possible. The West Antarctic Ice Sheet (WAIS) has attracted special attention because it contains enough ice to raise sea level by 6 m and because of suggestions that instabilities associated with its being grounded below sea level may result in rapid ice discharge when the surrounding ice shelves are weakened. However, there is now general agreement that major loss of grounded ice, and accelerated sea-level rise, is very unlikely during the twenty-first century. The contribution of this ice sheet to sea level change will probably not exceed 3 m over the next millennium.

Summary

On timescales of decades to tens of thousands of years, sea-level change results from exchanges of mass between the polar ice sheets, the nonpolar glaciers, terrestrial water storage, and the ocean. Sea level also changes if the density of the ocean changes (as a result of changing temperature) even though there is no change in mass. During the last century, sea level is estimated to have risen by 10–20 cm, as a result of combination of thermal expansion of the ocean as its temperature rose and increased mass of the ocean from melting glaciers and ice sheets. Over the twenty-first century, sea level is expected to rise as a result of anthropogenic climate change. The main contributors to this rise are expected to be thermal expansion of the ocean and the partial melting of nonpolar glaciers and the Greenland ice sheet. Increased precipitation in Antarctica is expected to offset some of the rise from other contributions. Changes in terrestrial storage are uncertain but may also partially offset rises from other contributions. After allowance for the continuing changes in the ice sheets since the last glacial maximum, the total projected sea-level rise over the twenty-first century is currently estimated to be between about 9 and 88 cm.

See also

Glacial Crustal Rebound, Sea Levels, and Shorelines. Ice Shelf Stability.

Further Reading

Church JA, Gregory JM, Huybrechts P, *et al.* (2001) Changes in sea level. In: Houghton JT (ed.) *Climate Change 2001; The Scientific Basis.* Cambridge: Cambridge University Press.

Douglas BC, Keaney M, and Leatherman SP (eds.) (2000) *Sea Level Rise: History and Consequences, 232 pp.* San Diego: Academic Press.

Fleming K, Johnston P, Zwartz D, *et al.* Refining the eustatic sea-level curve since the Last Glacial Maximum using far- and intermediate-field sites. *Earth and Planetary Science Letters* 163: 327–342.

Lambeck K (1998) Northern European Stage 3 ice sheet and shoreline reconstructions: Preliminary results. News 5, Stage 3 Project, Godwin Institute for Quaternary Research, 9 pp.

Peltier WR (1998) Postglacial variations in the level of the sea: implications for climate dynamics and solid-earth geophysics. *Review of Geophysics* 36: 603–689.

Summerfield MA (1991) *Global Geomorphology.* Harlowe: Longman.

Warrick RA, Barrow EM, and Wigley TML (1993) *Climate and Sea Level Change: Observations, Projections and Implications.* Cambridge: Cambridge University Press.

WWW pages of the Permanent Service for Mean Sea Level, http://www.pol.ac.uk/psmsl/

ECONOMICS OF SEA LEVEL RISE

R. S. J. Tol, Economic and Social Research Institute, Dublin, Republic of Ireland

Introduction

The economics of sea level rise is part of the larger area of the economics of climate change. Climate economics is concerned with five broad areas:

- What are the economic implications of climate change, and how would this be affected by policies to limit climate change?
- How would and should people, companies, and government adapt and at what cost?
- What are the economic implications of greenhouse gas emission reduction?
- How much should emissions be reduced?
- How can and should emissions be reduced?

The economics of sea level rise is limited to the first two questions, which can be rephrased as follows:

- What are the costs of sea level rise?
- What are the costs of adaptation to sea level rise?
- How would and should societies adapt to sea level rise?

These three questions are discussed below.

There are a number of processes that cause the level of the sea to rise. Thermal expansion is the dominant cause for the twenty-first century. Although a few degrees of global warming would expand seawater by only a fraction, the ocean is deep. A 0.01% expansion of a column of 5 km of water would give a sea level rise of 50 cm, which is somewhere in the middle of the projections for 2100. Melting of sea ice does not lead to sea level rise – as the ice currently displaces water. Melting of mountain glaciers does contribute to sea level, but these glaciers are too small to have much of an effect. The same is true for more rapid runoff of surface and groundwater. The large ice sheets on Greenland and Antarctica could contribute to sea level rise, but the science is not yet settled. A complete melting of Greenland and West Antarctica would lead to a sea level rise of some 12 m, and East Antarctica holds some 100 m of sea level. Climate change is unlikely to warm East Antarctica above freezing point, and in fact additional snowfall is likely to store more ice on East Antarctica. Greenland and West Antarctica are much warmer, and climate change may

well imply that these ice caps melt in the next 1000 years or so. It may also be, however, that these ice caps are destabilized – which would mean that sea level would rise by 10–12 m in a matter of centuries rather than millennia.

Concern about sea level rise is only one of many reasons to reduce greenhouse gas emissions. In fact, it is only a minor reason, as sea level rise responds only with a great delay to changes in emissions. Although the costs of sea level rise may be substantial, the benefits of reduced sea level rise are limited – for the simple fact that sea level rise can be slowed only marginally. However, avoiding a collapse of the Greenland and West Antarctic ice sheets would justify emission reduction – but for the fact that it is not known by how much or even whether emission abatement would reduce the probability of such a collapse. Therefore, the focus here is on the costs of sea level rise and the only policy considered is adaptation to sea level rise.

What Are the Costs of Sea Level Rise?

The impacts of sea level rise are manifold, the most prominent being erosion, episodic flooding, permanent flooding, and saltwater intrusion. These impacts occur onshore as well as near shore/on coastal wetlands, and affect natural and human systems. Most of these impacts can be mitigated with adaptation, but some would get worse with adaptation elsewhere. For instance, dikes protect onshore cities, but prevent wetlands from inland migration, leading to greater losses.

The direct costs of erosion and permanent flooding equal the amount of land lost times the value of that land. Ocean front property is often highly valuable, but one should not forget that sea level rise will not result in the loss of the ocean front – the ocean front simply moves inland. Therefore, the average land value may be a better approximation of the true cost than the beach property value.

The amount of land lost depends primarily on the type of coast. Steep and rocky coasts would see little impact, while soft cliffs may retreat up to a few hundred meters. Deltas and alluvial plains are at a much greater risk. For the world as a whole, land loss for 1 m of sea level rise is a fraction of a percent, even without additional coastal protection. However, the distribution of land loss is very skewed. Some islands, particularly atolls, would disappear altogether – and this may lead to the disappearance of entire nations and cultures. The number of people involved is small, though. This is different for deltas, which tend to be densely populated and heavily used because of superb soils and excellent transport. Without additional protection, the deltas of the Ganges–Brahmaputra, Mekong, and Nile could lose a quarter of their area even for relative modest sea level rise. This would force tens of millions of people to migrate, and ruin the economies of the respective countries.

Besides permanent inundation, sea level rise would also cause more frequent and more intense episodic flooding. The costs of this are more difficult to estimate. Conceptually, one would use the difference in the expected annual flood damage. In practice, however, floods are infrequent and the stock-at-risk changes rapidly. This makes estimates particularly uncertain. Furthermore, floods are caused by storms, and climate models cannot yet predict the effect of the enhanced greenhouse effect on storms. That said, tropical cyclones kill only about 10 000 people per year worldwide, and economic damages similarly are only a minor fraction of gross national product (GDP). Even a 10-fold increase because of sea level rise, which is unlikely even without additional protection, would not be a major impact at the global scale. Here, as above, the global average is likely to hide substantial regional differences, but there has been too little research to put numbers on this.

Sea level rise would cause salt water to intrude in surface and groundwater near the coast. Saltwater intrusion would require desalination of drinking water, or moving the water inlet upstream. The latter is not an option on small islands, which may lose all freshwater resources long before they are submerged by sea level rise. The economic costs of desalination are relatively small, particularly near the coast, but desalination is energy-intensive and produces brine, which may cause local ecological problems. Because of saltwater intrusion, coastal agriculture would suffer a loss of productivity, and may become impossible. Halophytes (salt-tolerant plants) are a lucrative nice market for vegetables, but not one that is constrained by a lack of brackish water. Halophytes' defense mechanisms against salt work at the expense of overall plant growth. Saltwater intrusion could induce a shift from agriculture to aquaculture, which may be more profitable. Few studies are available for saltwater intrusion.

Sea level rise would erode coastal wetlands, particularly if hard structures protect human occupations. Current estimates suggest that a third of all coastal wetlands could be lost for less than 40-cm sea level rise, and up to half for 70 cm. Coastal wetlands provide many services. They are habitats for fish (also as nurseries), shellfish, and birds (including migratory birds). Wetlands purify water, and protect coast against storms. Wetlands also provide food and recreation. If wetlands get lost, so do these functions – unless scarce resources are reallocated to provide these services. Estimates of the value of wetlands vary between $100 and $10 000 per hectare, depending on the type of wetland, the services it provides, population density, and per capita income. At present, there are about 70 million ha of coastal wetlands, so the economic loss of sea level rise would be measured in billions of dollars per year.

Besides the direct economic costs, there are also higher-order effects. A loss of agricultural land would restrict production and drive up the price of food. Some estimates have that a 25-cm sea level rise would increase the price of food by 0.5%. These effects would not be limited to the affected regions, but would be spread from international trade. Australia, for instance, has few direct effects from sea level rise – and may therefore benefit from increased exports to make up for the reduced production elsewhere. Other markets would be affected too, as farmers would buy more fertilizer to produce more on the remaining land, and as workers elsewhere would demand higher salaries to compensate for the higher cost of living. Such spillovers are small in developed economies, because agriculture is only a small part of the economy. They are much larger in developing countries.

What Are the Costs of Adaptation to Sea Level Rise?

There are three basic ways to adapt to sea level rise, although in reality mixes and variations will dominate. The two extremes are protection and retreat. In between lies accommodation. Protection entails such things as building dikes and nourishing beaches – essentially, measures are taken to prevent the impact of sea level rise. Retreat implies giving up land and moving people and infrastructure further inland – essentially, the sea is given free range, but people and their things are moved out of harm's way. Accommodation means coping with the consequences of sea level rise. Examples include placing houses on stilts and purchasing flood insurance.

Besides adaptation, there is also failure to (properly) adapt. In most years, this will imply that adaptation costs (see below) are less. In some years, a storm will come to kill people and damage property. The next section has a more extensive discussion on adaptation.

Estimating the costs of protection is straightforward. Dike building and beach nourishment are routine operations practised by many engineering companies around the world. The same holds for accommodation. Estimates have that the total annual cost of coastal protection is less than 0.01% of global GDP. Again, the average hides the extremes. In small island nations, the protection bill may be over 1% of GDP per year and even reach 10%. Note that a sacrifice of 10% is still better than 100% loss without protection.

There may be an issue of scale. Gradual sea level rise would imply a modest extension of the 'wet engineering' sector, if that exists, or a limited expense for hiring foreign consultants. However, rapid sea level rise would imply (local) shortages of qualified engineers – and this would drive up the costs and drive down the effectiveness of protection and accommodation. If materials are not available locally, they will need to be transported to the coast. This may be a problem in densely populated deltas. Few studies have looked into this, but those that did suggest that a sea level rise of less than 2 m per century would not cause logistical problems. Scale is therefore only an issue in case of ice sheet collapse.

Costing retreat is more difficult. The obvious impact is land loss, but this may be partly offset by wetland gains (see above). Retreat also implies relocation. There are no solid estimates of the costs of forced migration. Again, the issue is scale. People and companies move all the time. A well-planned move by a handful of people would barely register. A hasty retreat by a large number would be noticed. If coasts are well protected, the number of forced migrants would be limited to less than 10 000 a year. If coastal protection fails, or is not attempted, over 100 000 people could be displaced by sea level rise each year.

Like impacts, adaptation would have economy-wide implications. Dike building would stimulate the construction sector and crowd out other investments and consumption. In most countries, these effects would be hard to notice, but in small island economies this may lead to stagnation of economic growth.

How Would and Should Societies Adapt to Sea Level Rise?

Decision analysis of coastal protection goes back to Von Dantzig, one of the founding fathers of operations research. This has also been applied to additional protection for sea level rise. Some studies simply compare the best guess of the costs of dike building against the best guess of the value of land lost if no dikes were built, and decide to protect or not on the basis of a simple cost–benefit ratio. Other studies use more advanced methods that, however, conceptually boil down to cost–benefit analysis as well. These studies invariably conclude that it is economically optimal to protect most of the populated coast against sea level rise. Estimates go up to 85% protection, and 15% abandonment. The reason for these high protection levels are that coastal protection is relatively cheap, and that low-lying deltas are often very densely populated so that the value per hectare is high even if people are poor.

However, coastal protection has rarely been based on cost–benefit analysis, and there is no reason to assume that adaptation to sea level rise will be different. One reason is that coastal hazards manifest themselves irregularly and with unpredictable force. Society tends to downplay most risks, and overemphasize a few. The selection of risks shifts over time, in response to events that may or may not be related to the hazards. As a result, coastal protection tends to be neglected for long periods, interspersed with short periods of frantic activity. Decisions are not always rational. There are always 'solutions' waiting for a problem, and under pressure from a public that demands rapid and visible action, politicians may select a 'solution' that is in fact more appropriate for a different problem. The result is a fairly haphazard coastal protection policy.

Some policies make matters worse. Successful past protection creates a sense of safety and hence neglect, and attracts more people and business to the areas deemed safe. Subsidized flood insurance has the same effect. Local authorities may inflate their safety record, relax their building standards and land zoning, and relocate their budget to attract more people.

A fundamental problem is that coastal protection is partly a public good. It is much cheaper to build a dike around a 'community' than around every single property within that community. However, that does require that there is an authority at the appropriate level. Sometimes, there is no such authority and homeowners are left to fend for themselves. In other cases, the authority for coastal protection sits at provincial or national level, with civil servants who are occupied with other matters.

A number of detailed studies have been published on decision making on adaptation to coastal and other hazards. The unfortunate lesson from this work is that every case is unique – extrapolation is not possible except at the bland, conceptual level.

Conclusion

The economics of climate change are still at a formative stage, and so are the economics of sea level rise. First estimates have been developed of the order of magnitude of the problem. Sea level rise is not a substantial economic problem at the global or even the continental scale. It is, however, a substantial problem for a number of countries, and a dominant issue for some local economies. The main issue, therefore, is the distribution of the impacts, rather than the impacts themselves.

Future estimates are unlikely to narrow down the current uncertainties. The uncertainties are not so much due to a paucity of data and studies, but are intrinsic to the problem. The impacts of sea level rise are local, complex, and in a distant future. The priority for economic research should be in developing dynamic models of economies and their interaction with the coast, to replace the current, static assessments.

See also

Beaches, Physical Processes Affecting. Coastal Zone Management. Salt Marshes and Mud Flats. Sea Level Change.

Further Reading

Burton I, Kates RW, and White GF (1993) *The Environment as Hazard*, 2nd edn. New York: The Guilford Press.

Nicholls RJ, Wong PP, Burkett VR, *et al.* (2007) Coastal systems and low-lying areas. In: Parry ML, Canziani O, Palutikof J, van der Linden P, and Hanson C (eds.) *Climate Change 2007: Impacts, Adaptation and Vulnerability – Contribution of Working Group II to the Fourth Assessment Report of the Intergovernmental Panel on Climate Change*, pp. 315–356. Cambridge, UK: Cambridge University Press.

COASTAL TOPOGRAPHY, HUMAN IMPACT ON

D. M. Bush, State University of West Georgia, Carrollton, GA, USA

O. H. Pilkey, Duke University, Durham, NC, USA

W. J. Neal, Grand Valley State University, Allendale, MI, USA

Introduction

The trademark of humans throughout time is the modification of the natural landscape. Topography has been modified from the earliest farming to the modern modifications of nature for transportation and commerce (e.g., roads, utilities, mining), and often for recreation, pleasure, and esthetics. While human modifications of the environment have affected vast areas of the continents, and small portions of the ocean floor, nowhere have human intentions met headlong with nature's forces as in the coastal zone.

A most significant change in human behavior since the 1950s has been the dramatic, rapid increase in population and nonessential development in the coastal zone (**Figure 1**). The associated density of development is in an area that is far more vulnerable and likely to be impacted by natural processes (e.g., wind, waves, storm-surge flooding, and coastal erosion) than most inland areas. Not only are more

Figure 1 The coastal population explosion has resulted in too many people and buildings crowded too close to the shoreline. As sea level rises, the shoreline naturally moves back and encounters the immovable structures of human development. In this example from San Juan, Puerto Rico, erosion in front of buildings has necessitated engineering of the shoreline.

people and development in harm's way, but the human modifications of the coastal zone (e.g., dune removal) have increased the frequency and severity of the hazards. Finally, coastal engineering as a means to combat coastal erosion and management of waterways, ports, and harbors has had profound and often deleterious effects on coastal environments. The endproduct is a total interruption of sediment interchange between land and sea, and a heavily modified topography. Natural hazard mitigation is now moving with a more positive, albeit small, approach by restoring natural features, such as beaches and dunes, and their associated interchangeable sediment supply.

The Scope of Human Impact on the Coast

The natural coastal zone is highly dynamic, with geomorphic changes occurring over several time scales. Equally significant changes are made by humans. On Ocean Isle, NC, USA, an interior dune ridge, the only one on the island, was removed to make way for development. The lowered elevation put the entire development in a higher hazard zone, with a corresponding greater risk for property damage from flooding and other storm processes.

Another example of change, impacting on property damage risk, can be seen in Kitty Hawk, North Carolina. A large shorefront dune once extended in front of the entire community. The dune was constructed in the 1930s by the Civilian Conservation Corps to halt shoreline erosion, and provide a 'protected' area along which to build a road. The modification was done before barrier island migration was understood. Erosion was assumed to be permanent land loss. The artificial dune actually increased erosion here by acting as a seawall in a long-term sense, blocking overwash sand which would have raised island elevation and brought sand to the backside of the island, although the dune did afford some protection for development. As a consequence, buildings by the hundreds were built in the lee of the dune. Fifty years on, however, the price is being paid. During the 1980s, the dune began to deteriorate due to storm penetration, and the 1991 Halloween northeaster finished the job by creating large gaps in the dune, resulting in flooding of portions of the community. The dune cannot be rebuilt in place because the old dune location is now occupied by the beach, backed up against the frontal road. Between the time of dune construction and 1991, the community had only experienced major flooding once, in the great 1962 Ash Wednesday storm. Between 1991 and 2000, the community was flooded four times.

The effect of shoreline engineering on a whole-island system is starkly portrayed by the contrast between Ocean City, MD and the next island to the south, Assateague Island, MD. It has taken several decades to be fully realized, but the impact of the jetties is now apparent. Assateague Island has moved back one entire island width due to sand trapping by an updrift jetty. Similar stories abound along the coast. The Charleston lighthouse, once on the

backside of Morris Island, SC, now stands some 650 m at sea; a sentinel that watched Morris Island rapidly migrate away after the Charleston Harbor jetties, built in 1898, halted the supply of sand to the island (**Figure 2**).

Human alterations of the natural environment have direct and indirect effects. Some types of human modifications to the coastal environment include: (1) construction site modification, (2) building and infrastructure construction, (3) hard shoreline stabilization, (4) soft shoreline stabilization, and (5) major coastal engineering construction projects for waterway, port, and harbor management and inlet channel alteration. Each of the modification types impacts the coastal environment in a variety of ways and also

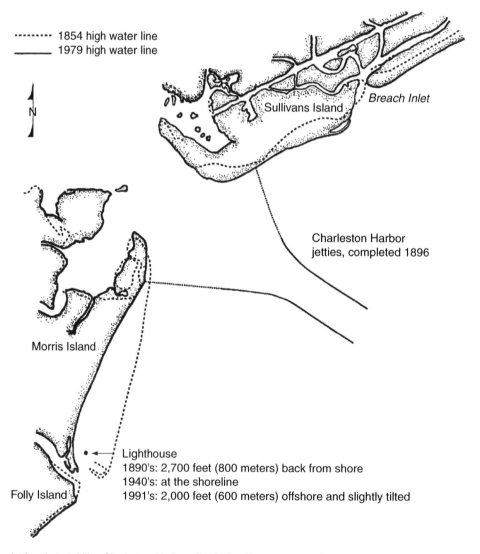

Figure 2 The jetties that stabilize Charleston Harbor, South Carolina, were completed in 1896. The jetties block the southward transport sediment along the beach. As a result, the islands to the north of the jetties have grown seaward slightly, but the islands to the south of the inlet have eroded back more than 1400 m. The Morris Island Lighthouse, once on the back side of the island, is now 650 m offshore.

has several direct and indirect effects. Some of the effects are obvious and intuitive, but many are surprising in that there can be a domino effect as one simple modification creates potential for damage and destruction by increasing the frequency and intensity of natural hazards at individual sites.

Construction Site Modification

Building sites are often flattened and vegetation is removed for ease of construction. Activities such as grading of the natural coastal topography include dune and forest removal. Furthermore, paving of large areas is common, as roads, parking lots, and driveways are constructed. Direct effects of building site modification, in addition to changes in the natural landform configuration, include demobilization of sediment in some places by paving and building footpaths, but also sediment mobilization by removal of vegetation. In either case, rates of onshore–offshore sediment transport and storm-recovery capabilities are changed, which can increase or decrease erosion rates as sediment supply changes.

Other common site modifications include excavating through dunes (dune notching) to improve beach access or sea views. This is particularly common at the ends of streets running toward the beach. After Hurricane Hugo in South Carolina in 1989, shore-perpendicular streets where dunes were notched at their ocean termini were seen to have acted as storm-surge ebb conduits, funneling water back to the sea and increasing scour and property damage. The same effect was noted after Hurricane Gilbert along the northern coast of the Yucatán Peninsula of Mexico in 1988.

Building and Infrastructure Construction

A variety of buildings are constructed in the coastal zone, ranging from single-family homes to high-rise hotels and commercial structures. Some of the common direct effects of building construction are alteration of wind patterns as the buildings themselves interact with natural wind flow, obstruction of sediment movement, marking the landward limit of the beach or dune, channelizing storm surge and storm-surge ebb flow, and reflection of wave energy. Indirect effects result from the simple fact that once there is construction in an area, people tend to want to add more construction, and to increase and improve infrastructure and services. As buildings become threatened by shoreline erosion, coastal engineering endeavors begin.

Roads, streets, water lines, and other utilities are often laid out in the standard grid pattern used inland, cutting through interior and frontal dunes instead of over and around coastal topography (**Figure 3**). Buildings block natural sediment flow (e.g., overwash) while the ends of streets and gaps between rigid buildings funnel and concentrate flow, accentuating the erosive power of flood waters. As noted above, during Hurricane Hugo, water, sand, and debris were carried inland along shore-perpendicular roads in several South Carolina communities. Storm-surge ebb along the shore-perpendicular roads caused scour channels, which undermined roadways and damaged adjacent houses and property. Even something as seemingly harmless as buried utilities may cause a problem as the excavation disrupts the substrate, resulting in a less stable topography after post-construction restorations.

Plugging dune gaps can be a part of nourishment and sand conservation projects. Because dunes are critical coastal geomorphic features with respect to property damage mitigation, they are now often protected, right down to vegetation types that are critical to dune growth. Prior to strict coastal-zone management regulations, however, frontal dunes were often excavated for ocean views or building sites, or notched at road termini for beach access. These artificially created dune gaps are exploited by waves and storm-surge, and by storm surge ebb flows. Wherever dune removal for development has occurred, the probability is increased for the likelihood of complete overwash and possible inlet formation.

Hard Shoreline Stabilization

Hard shoreline stabilization includes various fixed, immovable structures designed to hold an eroding shoreline in place. Hard stabilization is one of the most common modifiers of topography in the coastal zone and is discussed in more detail below. Seawalls, jetties, groins, and offshore breakwaters interrupt sediment exchange and reduce shoreline flexibility to respond to wave and tidal actions. Armoring the shoreline changes the location and intensity of erosion and deposition. Indirectly, hard shoreline stabilization gives a false sense of security and encourages increased development landward of the walls, placing more and more people and property at risk from coastal hazards including waves, storm surge, and wind. Eventual loss of the recreational beach as shoreline erosion continues and catches up with the static line of stabilization is almost a certainty. In addition, structures beget more structures as small walls or groins are replaced by larger and larger walls and groins.

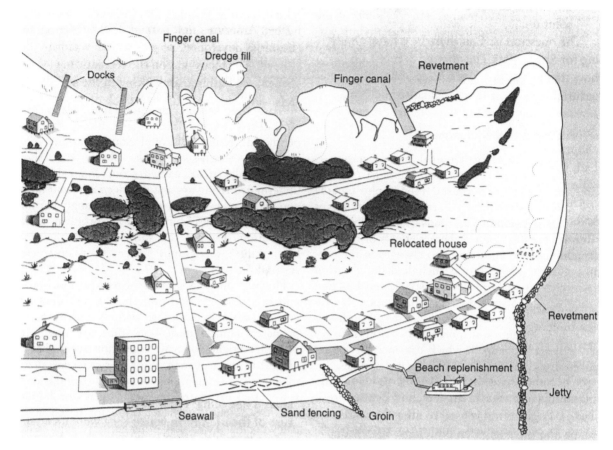

Figure 3 A compilation of many of the impacts humans have on the coastal topography. In this fictional barrier island, roads have been cut through excavated dunes, maritime forest removed for building sites, finger canals dredged, structures built too close to the water, and several types of coastal engineering projects undertaken.

Soft Shoreline Stabilization

The most common forms of soft shoreline stabilization are beach nourishment, dune building, sand fencing, beach bulldozing (beach scraping), and planting of vegetation to grow or stabilize dunes. Direct effects of such manipulations are changes in sedimentation rates and severity of erosion, and interruption of the onshore–offshore sediment transfer, similar in that respect to hard shoreline stabilization. Indirectly, soft shoreline stabilization may make it more difficult to recognize the severity of an erosion problem, i.e., 'masking' the erosion problem. Moreover, as with hard shoreline stabilization, development is actually encouraged in the high-hazard zone behind the beach.

Coastal Engineering Construction Projects

The construction of harbors, port facilities, waterways (e.g., shipping channels, canals) and inlet channel alterations significantly change the coastal outline as well as eliminating land topographic features or erecting artificial shorelines and dredge spoil banks. The Intracoastal Waterway of the Atlantic and Gulf Coasts is one of the longest artificial coastal modifications in the world. Large harbors in many places around the world represent significant alteration of the landscape. Many examples of coastal fill or artificial shorelines exist, but one of the best examples of such a managed shoreline is Chicago's 18 miles of continuous public waterfront. Major canals such as the Suez, Panama, Cape Cod, or Great Dismal Swamp Canal also represent major modifications in the coastal zone. The Houston Ship Channel made the city of Houston, Texas, a major port some 40 miles from the Gulf of Mexico.

Tidal inlets, either on the mainland or between barrier islands, can be altered by dredging, relocation, or artificial closure. Direct effects of dredging tidal inlets are changes in current patterns, which may change the location and degree of erosion and deposition events, and prevention of sand transfer across inlets. In either case, additional shoreline hardening is a common response.

The Scope of Coastal Engineering Impacts

Between 80 and 90% of the American open-ocean shoreline is retreating in a landward direction because of sea-level rise and coastal erosion. Because more static buildings are being sited next to this moving and constantly changing coastline, our society faces major problems. Various coastal engineering approaches to dealing with the coastal erosion problem have been developed (**Figure 3**). More than a century of experience with seawalls and other engineering structures in New Jersey and other coastal developments shows that the process of holding the shoreline in place leads to the loss of the beach, dunes, and other coastal landforms. The real societal issue is how to save both buildings and beaches. The action taken often leads to modifications to the coast that limit the natural flexibility of the coastal zone to respond to storms, that inhibit the natural onshore–offshore exchange of sand, and that interrupt the natural alongshore flow of sand.

Seawalls

Seawalls include a family of coastal engineering structures built either on land at the back of the beach or on the beach, parallel to the shoreline. Strictly defined, seawalls are free-standing structures near the surf-zone edge. The best examples are the giant walls of the northern New Jersey coast, the end result of more than a century of armoring the shoreline (**Figure 4**). If such walls are filled in behind with soil or sand, they are referred to as bulkheads. Revetments, commonly made of piled loose rock, are walls built up against the lower dune-face or land at the back of the beach. For the purpose of considering their alteration of topography both at their construction site and laterally, the distinction between the types of walls is gradational and unimportant, and the general term seawall is used here for all structures on the beach that parallel the shoreline.

Seawalls are usually built to protect the property, not to protect the beach. Sometimes low seawalls are intended only to prevent shoreline retreat, rather than to block wave attack on buildings. Seawalls are successful in preventing property damage if built strongly, high enough to avoid being overtopped, and kept in good repair. The problem is that a very high societal price is paid for such protection. That price is the eventual loss of the recreational beach and steepening of the shoreface or outer beach. This is why several states in the USA (e.g. Maine, Rhode Island, North Carolina, South Carolina, Texas, and Oregon) prohibit or place strict limits on shoreline armoring.

Three mechanisms account for beach degradation by seawalls. Passive loss is the most important. Whatever is causing the shoreline to retreat is unaffected by the wall, and the beach eventually retreats up against the wall. Placement loss refers to the emplacement of walls on the beach seaward of the high-tide line, thus removing part or all of the beach when the wall is constructed (**Figure 5**). Seawall placement was responsible for much of the beach loss in Miami Beach, Florida, necessitating a major beach nourishment project, completed in 1981. Active loss is the least understood of the beach

Figure 4 Cape May, New Jersey has been a popular seaside resort since 1800. Several generations of larger and larger seawalls have been built as coastal erosion caught up with the older structures. Today in many places there is no beach left in front of the seawall.

Figure 5 An example of placement loss in Virginia Beach, Virginia. The seawall was built out on the recreational beach, instantaneously narrowing the beach in front of the wall.

degradation mechanisms. Seawalls are assumed to interact with the surf during storms, which enhances the rate of beach loss. This interaction can occur in a number of ways including seaward reflection of waves, refraction of waves toward the end of the wall, and intensification of surf-zone currents.

By the year 2000, 50% of the developed shoreline on Florida's western (Gulf of Mexico) coast was armored, the same as the New Jersey coast. Similarly 45% of developed shoreline on Florida's eastern (Atlantic Ocean) coast was armored, in contrast to 27% for South Carolina, and only 6% of the developed North Carolina open-ocean shoreline. These figures represent the armored percentage of developed shorelines and do not include protected areas such as parks and National Seashores.

Shoreline stabilization is a difficult political issue because seawalls take as long as five or six decades to destroy beaches, although the usual time range for the beach to be entirely eroded at mid-to-high tide may be only one to three decades. Thus it takes a politician of some foresight to vote for prohibition of armoring. Another issue of political difficulty is that there is no room for compromise. Once a seawall is in place, it is rarely removed. The economic reasoning is that the wall must be maintained and even itself protected, so most walls grow higher and longer.

Figure 6 A groin field along Pawleys Island, South Carolina. Trapping of sand on the updrift side of a groin, and erosion of the beach on the downdrift side usually results in a sawtooth pattern to the beach. Note that in this example the beach is the same width on both sides of each groin, indicating little or no longshore transport of sand.

Groins and Jetties

Groins and jetties are walls or barriers built perpendicular to the shoreline. A jetty, often very long (thousands of feet), is intended to keep sand from flowing into a ship channel within an inlet and to reduce the cost of channel maintenance by dredging. Groins are much shorter structures built on straight stretches of beach away from inlets. Groins are intended to trap sand moving in longshore currents. They can be made of wood, stone, concrete, steel, or fabric bags filled with sand. Some designs are referred to as T-groins because the end of the structure terminates in a short shore-parallel segment.

Both groins and jetties are very successful sand traps. If a groin is working correctly, more sand should be piled up on one side of the groin than on the other. The problem with groins is that they trap sand that is flowing to a neighboring beach. Thus, if a groin is growing the topographic beach updrift, it must be causing downdrift beach loss. Per Bruun, past director of the Coastal Engineering program at the University of Florida, has observed that, on a worldwide basis, groins may be a losing proposition, i.e. more beach may be lost than gained by the use of groins. After one groin is built, the increased rate of erosion effect on adjacent beaches has to be addressed. So other groins are constructed, in self-defense. The result is a series of groins sometimes extending for miles (**Figure 6**). The resulting groin field is a saw-toothed beach in plan view.

Groins fail when continued erosion at their landward end causes the groin to become detached, allowing water and sand to pass behind the groin. When detachment occurs, beach retreat is renewed and additional alteration of the topography occurs.

Jetties, because of their length, can cause major topographic changes. After jetty emplacement, massive tidal deltas at most barrier island inlents will be dispersed by wave activity. In addition, major build-out of the updrift and retreat of the downdrift shorelines may occur. In the case of the Charleston, SC, jetties noted earlier, beach accretion occurred on the updrift Sullivans Island and Isle of Palms.

Offshore Breakwaters

Offshore breakwaters are walls built parallel to the shoreline but at some distance offshore, typically a few tens of meters seaward of the normal surf zone. These structures dampen the wave energy on the 'protected' shoreline behind the breakwater, interrupting the longshore current and causing sand to be deposited and a beach to form. Sometimes these deposits will accumulate out to the breakwater, creating a feature like a natural tombolo. As in the case of groins, the sand trapped behind breakwaters causes a shortage of sediment downdrift in the directions of dominant longshore transport, leading to additional shoreline retreat (e.g. beach and dune loss, scarping of the fastland, accelerated mass wasting).

Beach Nourishment

Beach nourishment consists of pumping or trucking sand onto the beach. The goal of most communities is to improve their recreational beach, to halt shoreline erosion, and to afford storm protection for beachfront buildings. Many famous beaches in developed areas, in fact, are now artificial!

The beach or zone of active sand movement actually extends out to a water depth of 9–12 m below the low-tide line. This surface is referred to as the shoreface. With nourishment, only the upper beach is covered with new sand so that a steeper beach is created, i.e. the topographic profile is modified on land and offshore. This new steepened profile often increases the rate of erosion; in general, replenished beaches almost always disappear at a faster rate than their natural predecessors.

Beach scraping (bulldozing) should not be confused with beach nourishment. Beach sand is moved from the low-tide beach to the upper back beach (independent of building artificial dunes) as an erosion-mitigation technique. In effect this is beach erosion! A relatively thin layer of sand (≤ 30 cm) is removed from over the entire lower beach using a variety of heavy machinery (drag, grader, bulldozer, front-end loader) and spread over the upper beach. The objectives are to build a wider, higher, high-tide dry beach; to fill in any trough-like lows that drain across the beach; and to encourage additional sand to accrete to the lower beach.

The newly accreted sand in turn, can be scraped, leading to a net gain of sand on the manicured beach. An enhanced recreational beach may be achieved for the short term, but no new sand has been added to the system. Ideally, scraping is intended to encourage onshore transport of sand, but most of the sand 'trapped' on the lower beach is brought in by the longshore transport. Removal of this lower beach sand deprives downdrift beaches of their natural nourishment, steepens the beach topographic profile, and destroys beach organisms.

Dune building is often an important part of beach nourishment design, or it may be carried out independently of beach nourishment. Coastal dunes are a common landform at the back of the beach and part of the dynamic equilibrium of barrier beach systems. Although extensive literature exists about dunes, their protective role often is unknown or misunderstood. Frontal dunes are the last line of defense against ocean storm wave attack and flooding from overwash, but interior dunes may provide high ground and protection against penetration of overwash, and against the damaging effects of storm-surge ebb scour.

Human Impact on Sand Supply

In most of the preceding discussion the impact of humans on beaches and shoreline shape and position was emphasized. The beach plays a major role in supplying sand to barrier islands and, in fact, is important in supplying sand and gravel to any kind of upland, mainland, or island. In this sense, any topographic modification, however small, that affects the sand supply of the beach will affect the topography. In beach communities, sand is routinely removed from the streets and driveways after storms or when sand deposited by wind has accumulated to an uncomfortable level for the community. This sand would have been part of the island or coastal evolution process. Often, dunes are replaced by flat, well-manicured lawns. Sand-trapping dune vegetation is often removed altogether.

The previously mentioned Civilian Conservation Corps construction of the large dune line along almost the entire length of the Outer Banks of North Carolina is an example of a major topographic modification that had unexpected ramifications, namely the increased rate of erosion on the beach as well as on the backside of the islands. Prior to dune construction, the surf zone, especially during storms, expended its energy across a wide band of island surface which was overwashed several times a year. After construction of the frontal dune, wave energy was expended in a much narrower zone, leading to increased rates of shoreline retreat, and overwash no longer nourished the backside of the island. Now that the frontal dune is deteriorating, North Carolina Highway-12 is buried by overwash sand in a least a dozen places 1–4 times each year. Overwash sand is an important part of the island migration process, because these deposits raise the elevation of islands, and when sediment extends entirely across an island, widening occurs. If not for human activities, much of the Outer Banks would be migrating at this point in time, but because preservation of the highway is deemed essential to connect the eight villages of the southern Outer Banks, the NC Department of Transportation removes sand and places it back on the beach. As a result, the island fails to gain elevation.

Inlet formation also is an important part of barrier island evolution. Each barrier island system is different, but inlets form, evolve, and close in a manner to allow the most efficient means of moving water in and out of estuaries and lagoons. Humans interfere by preventing inlets from forming, by closing them after they open naturally (usually during storms), or by preventing their natural migration by construction of jetties. The net result is clogging of navigation channels by construction of huge tidal deltas and reduced water circulation and exchange between the sea and estuaries.

Globally shoreline change is being affected by human activity that causes subsidence and loss of sand supply. The Mississippi River delta is a classic example. The sediment discharge from the Mississippi River has been substantially reduced by upstream dam construction on the river and its tributaries. Large flood-control levees constructed along the lower Mississippi River prevent sediment from reaching the marshes and barrier islands along the rim of the delta in the Gulf of Mexico. Natural land subsidence caused by compaction of muds has added to the problem by creating a rapid (1–2 m per century) relative sea-level rise. Finally, maintaining the river channel south of New Orleans has extended the river mouth to the edge of the continental shelf, causing most remaining sediment to be deposited in the deep sea rather than on the delta. The end result is an extraordinary loss rate of salt marshes and very rapid island migration. The face of the Mississippi River delta is changing with remarkable rapidity.

Other deltas around the world have similar problems that accelerate changes in the shape of associated marshes and barrier islands. The Niger and Nile deltas have lost a significant part of their sediment supply because of trapping sand behind dams. Land loss on the Nile delta is permanent and not just migration of the outermost barrier islands. On the Niger delta the lost sediment supply is compounded by the subsidence caused by oil, gas, and water extraction. The barrier islands there are rapidly thinning.

Sand mining is a worldwide phenomenon whose quantitative importance is difficult to guage. Mining dunes, beaches, and river mouths for sand has reduced the sand supply to the shoreface, beaches, and barrier islands. In developing countries beach-sand mining is ubiquitous, while in developed countries beach and dune mining often is illegal and certainly less extensive, although still a problem. For example, sand mining has adversely affected the beaches of many West Indies nations going through the growing pains of development. Dune mining has been going on for so long that many current residents cannot remember sand dunes ever being present on the beaches, although they must have been there at one time, given the sand supply and the strong winds. For example, on the dual-island nation of Antigua and Barbuda, beach ridges – evidence of accumulating sand – can be observed on Barbuda, but are missing on Antigua. The beach ridges of Barbuda have survived to date only because it is much less heavily developed and populated. Sadly, Barbuda's beach ridges are being actively mined.

Puerto Rico is a heavily developed Caribbean island, much larger than Antigua or Barbuda, and with a more diversified economy. Many of Puerto Rico's dunes have been trucked away (**Figure 7**). East of the capital city of San Juan, large sand dunes were mined to construct the International Airport at Isla Verde by filling in coastal wetlands. As a result of removing the dunes, the highway was regularly overwashed and flooded during even moderate winter storms. First an attempt was made to rebuild the dune, then a major seawall was built to protect the lone coastal road.

Dredging or pumping sand from offshore seems like a quick and simple solution to replace lost beach sand; however, such operations must be considered with great care. The offshore dredge hole may allow larger waves to attack the adjacent beach. Offshore sand may be finer in grain size, or it may be

Figure 7 The dunes here near Camuy, Puerto Rico, used to be over 20 m high. After mining for construction purposes, all that remains is a thin veneer of sand over a rock outcrop.

composed of calcium carbonate, which breaks up quickly under wave abrasion. In all of these cases, the new beach will erode faster than the original beach. Dredging also may create turbidity that can kill bottom organisms. Offshore, protective reefs may be damaged by increased turbidity. Loss of reefs will mean faster beach erosion, as well as the obvious loss to the fishery habitat.

Sand can also be brought in from land sources by dump truck, but this may prove to be more expensive. Sand is a scarce resource, and beaches/dunes have been regarded as a source for mining rather than areas that need artificial replenishment. Past beach and dune mining may well be a principal cause of present beach erosion. In some cases, gravel may be better for nourishment than sand, but the recreational value of beaches declines when gravel is substituted.

Sand mining of beaches and dunes accounts for many of Puerto Rico's problem erosion areas. Such sand removal is now illegal, but permits are given to remove sand for highway construction and emergency repair purposes. However, the extraction limits of such permits are often exceeded – and illegal removal of sand for construction aggregate continues. In all cases, the sand removal eliminates natural shore protection in the area of mining, and robs from the sand budget of downdrift beaches, accelerating erosion. Even a small removal operation can set off a sequence of major shoreline changes.

The Caribe Playa Seabeach Resort along the southeastern coast illustrates just such a chain reaction. Located west of Cabo Mala Pascua, the resort has lost nearly 15 m of beach in recent years according to the owner. The problem dates to the days before permits and regulation, when an updrift property owner sold beach sand for 50 cents per dump truck load; a bargain by anyone's standards but a swindle to downdrift property owners. Where the sand was removed the beach eroded, resulting in shoreline retreat and tree kills. In an effort to restabilize the shore, and ultimately protect the highway, a rip-rap groin and seawall were constructed. Today, only a narrow gravel beach remains. Undoubtedly much of the aggregate in the concrete making up the buildings lining the shore, and now endangered by beach erosion, was beach sand. What extreme irony: taking sand from the beach to build structures that were subsequently endangered by the loss of beach sand.

Conclusion

The majority of the world's population lives in the coastal zone, and the percentage is growing. As this trend continues, the coastal zone will see increased impact of humans as more loss of habitats, more inlet dredging and jetties, continued sand removal, topography modification for building, sand starvation from groins and jetties, and the increased tourism and industrial use of coasts and estuaries. Our society's history illustrates the impact of humans as geomorphic agents, and nowhere is that fact borne out as it is in the coastal zone. The ultimate irony is that many of the human modifications on coastal topography actually decrease the esthetics of the area or increase the potential hazards.

See also

Beaches, Physical Processes Affecting. Coastal Zone Management. Sandy Beaches, Biology of. Viral and Bacterial Contamination of Beaches.

Further Reading

Bush DM and Pilkey OH (1994) Mitigation of hurricane property damage on barrier islands: a geological view. *Journal of Coastal Research* Special issue no. 12: 311–326.

Bush DM, Pilkey OH, and Neal WJ (1996) *Living by the Rules of the Sea*. Durham, NC: Duke University Press.

Bush DM, Neal WJ, Young RS, and Pilkey OH (1999) Utilization of geoindicators for rapid assessment of coastal-hazard risk and mitigation. *Ocean and Coastal Management* 42: 647–670.

Carter RWG and Woodroffe CD (eds.) (1994) *Coastal Evolution: Late Quaternary Shoreline Morphodynamics*. Cambridge: Cambridge University Press.

Carter RWG (1988) *Coastal Environments: An Introduction to the Physical, Ecological, and Cultural Systems of Coastlines*. London: Academic Press.

Davis RA Jr (1997) *The Evolving Coast*. New York: Scientific American Library.

French PW (1997) *Coastal and Estuarine Management, Routledge Environmental Management Series*. London: Routledge Press.

Kaufmann W and Pilkey OH Jr (1983) *The Beaches are Moving: The Drowning of America's Shoreline*. Durham, NC: Duke University Press.

Klee GA (1999) *The Coastal Environment: Toward Integrated Coastal and Marine Sanctuary Management*. Upper Saddle River, NJ: Prentice Hall.

Nordstrom KF (1987) Shoreline changes on developed coastal barriers. In: Platt RH, Pelczarski SG, and Burbank BKR (eds.) *Cities on the Beach: Management Issues of Developed Coastal Barriers*, pp. 65–79. University of Chicago, Department of Geography, Research Paper no. 224.

Nordstrom KF (1994) Developed coasts. In: Carter RWG and Woodroffe CD (eds.) *Coastal Evolution: Late Quaternary Shoreline Morphodynamics*, pp. 477–509. Cambridge: Cambridge University Press.

Nordstrom KF (2000) *Beaches and Dunes of Developed Coasts*. Cambridge: Cambridge University Press.

Pilkey OH and Dixon KL (1996) *The Corps and the Shore*. Washington, DC: Island Press.

Platt RH, Pelczarski SG and Burbank BKR (eds.) (1987) *Cities on the Beach: Management Issues of Developed Coastal Barriers*. University of Chicago, Department of Geography, Research Paper no. 224.

Viles H and Spencer T (1995) *Coastal Problems: Geomorphology, Ecology, and Society at the Coast*. New York: Oxford University Press.

GLACIAL CRUSTAL REBOUND, SEA LEVELS, AND SHORELINES

K. Lambeck, Australian National University, Canberra, ACT, Australia

Introduction

Geological, geomorphological, and instrumental records point to a complex and changing relation between land and sea surfaces. Elevated coral reefs or wave-cut rock platforms indicate that in some localities sea levels have been higher in the past, while observations elsewhere of submerged forests or flooded sites of human occupation attest to levels having been lower. Such observations are indicators of the relative change in the land and sea levels: raised shorelines are indicative of land having been uplifted or of the ocean volume having decreased, while submerged shorelines are a consequence of land subsidence or of an increase in ocean volume. A major scientific goal of sea-level studies is to separate out these two effects.

A number of factors contribute to the instability of the land surfaces, including the tectonic upheavals of the crust emanating from the Earth's interior and the planet's inability to support large surface loads of ice or sediments without undergoing deformation. Factors contributing to the ocean volume changes include the removal or addition of water to the oceans as ice sheets wax and wane, as well as addition of water into the oceans from the Earth's interior through volcanic activity. These various processes operate over a range of timescales and sea level fluctuations can be expected to fluctuate over geological time and are recorded as doing so.

The study of such fluctuations is more than a scientific curiosity because its outcome impacts on a number of areas of research. Modern sea level change, for example, must be seen against this background of geologically–climatologically driven change before contributions arising from the actions of man can be securely evaluated. In geophysics, one outcome of the sea level analyses is an estimate of the viscosity of the Earth, a physical property that is essential in any quantification of internal convection and thermal processes. Glaciological modeling of the behavior of large ice sheets during the last cold period is critically dependent on independent constraints on ice volume, and this can be extracted from the sea level information. Finally, as sea level rises and falls, so the shorelines advance and retreat. As major sea level changes have occurred during critical periods of human development, reconstructions of coastal regions are an important part in assessing the impact of changing sea levels on human movements and settlement.

Tectonics and Sea Level Change

Major causes of land movements are the tectonic processes that have shaped the planet's surface over geological time. Convection within the high-temperature viscous interior of the Earth results in stresses being generated in the upper, cold, and relatively rigid zone known as the lithosphere, a layer some 50–100 km thick that includes the crust. This convection drives plate tectonics — the movement of large parts of the lithosphere over the Earth's surface — mountain building, volcanism, and earthquakes, all with concomitant vertical displacements of the crust and hence relative sea level changes. The geological record indicates that these processes have been occurring throughout much of the planet's history. In the Andes, for example, Charles Darwin identified fossil seashells and petrified pine trees trapped in marine sediments at 4000 m elevation. In Papua New Guinea, 120 000-year-old coral reefs occur at elevations of up to 400 m above present sea level (**Figure 1**).

One of the consequences of the global tectonic events is that the ocean basins are being continually reshaped as mid-ocean ridges form or as ocean floor collides with continents. The associated sea level changes are global but their timescale is long, of the order 10^7 to 10^8 years, and the rates are small, less than 0.01 mm per year. **Figure 2** illustrates the global sea level curve inferred for the past 600 million years from sediment records on continental margins. The long-term trends of rising and falling sea levels on timescales of 50–100 million years are attributed to these major changes in the ocean basin configurations. Superimposed on this are smaller-amplitude and shorter-period oscillations that reflect more regional processes such as large-scale volcanism in an ocean environment or the collision of continents. More locally, land is pushed up or down episodically in response to the deeper processes. The associated

Figure 1 Raised coral reefs from the Huon Peninsula, Papua New Guinea. In this section the highest reef indicated (point 1) is about 340 m above sea level and is dated at about 125 000 years old. Elsewhere this reef attains more than 400 m elevation. The top of the present sea cliffs (point 2) is about 7000 years old and lies at about 20 m above sea level. The intermediate reef tops formed at times when the rate of tectonic uplift was about equal to the rate of sea level rise, so that prolonged periods of reef growth were possible. Photograph by Y. Ota.

vertical crustal displacements are rapid, resulting in sea level rises or falls that may attain a few meters in amplitude, but which are followed by much longer periods of inactivity or even a relaxation of the original displacements. The raised reefs illustrated in **Figure 1**, for example, are the result of a large number of episodic uplift events each of typically a meter amplitude. Such displacements are mostly local phenomena, the Papua New Guinea example extending only for some 100–150 km of coastline.

The episodic but local tectonic causes of the changing position between land and sea can usually be identified as such because of the associated seismic activity and other tell-tale geological signatures. The development of geophysical models to describe these local vertical movements is still in a state of infancy and, in any discussion of global changes in sea level,

information from such localities is best set aside in favor of observations from more stable environments.

Glacial Cycles and Sea Level Change

More important for understanding sea level change on human timescales than the tectonic contributions – important in terms of rates of change and in terms of their globality – is the change in ocean volume driven by cyclic global changes in climate from glacial to interglacial conditions. In Quaternary time, about the last two million years, glacial and interglacial conditions have followed each other on timescales of the order of 10^4–10^5 years. During interglacials, climate conditions were similar to those of today and sea levels were within a few meters of their present day position. During the major glacials, such as 20 000 years ago, large ice sheets formed in the northern hemisphere and the Antarctic ice sheet expanded, taking enough water out of the oceans to lower sea levels by between 100 and 150 m. **Figure 3** illustrates the changes in global sea level over the last 130 000 years, from the last interglacial, the last time that conditions were similar to those of today, through the Last Glacial Maximum and to the present. At the end of the last interglacial, at 120 000 years ago, climate began to fluctuate; increasingly colder conditions were reached, ice sheets over North America and Europe became more or less permanent features of the landscape, sea levels reached progressively lower values, and large parts of today's coastal shelves were exposed. Soon after the culmination of maximum glaciation, the ice sheets again disappeared, within a period of about

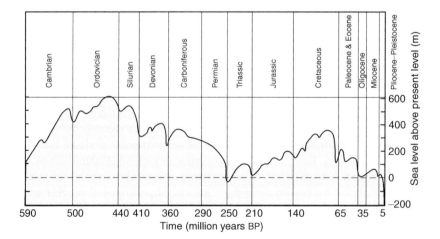

Figure 2 Global sea level variations through the last 600 million years estimated from seismic stratigraphic studies of sediments deposited at continental margins. Redrawn with permission from Hallam A (1984) Pre-Quaternary sea-level changes. *Annual Review of Earth and Planetary Science* 12: 205–243.

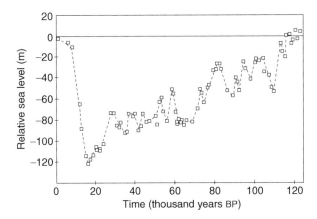

Figure 3 Global sea level variations (relative to present) since the time of the last interglacial 120 000–130 000 years ago when climate and environmental conditions were last similar to those of the last few thousand years. Redrawn with permission from Chapell J *et al.* (1996) Reconciliation of Late Quaternary sea level changes derived from coral terraces at Huon Peninsula with deep sea oxygen isotope records. *Earth and Planetary Science Letters* 141b: 227–236.

10 000 years, and climate returned to interglacial conditions.

The global changes illustrated in **Figure 3** are only one part of the sea level signal because the actual ice–water mass exchange does not give rise to a spatially uniform response. Under a growing ice sheet, the Earth is stressed; the load stresses are transmitted through the lithosphere to the viscous underlying mantle, which begins to flow away from the stressed area and the crust subsides beneath the ice load. When the ice melts, the crust rebounds. Also, the meltwater added to the oceans loads the seafloor and the additional stresses are transmitted to the mantle, where they tend to dissipate by driving flow to unstressed regions below the continents. Hence the seafloor subsides while the interiors of the continents rise, causing a tilting of the continental margins. The combined adjustments to the changing ice and water loads are called the glacio- and hydro-isostatic effects and together they result in a complex pattern of spatial sea level change each time ice sheets wax and wane.

Observations of Sea Level Change Since the Last Glacial Maximum

Evidence for the positions of past shorelines occurs in many forms. Submerged freshwater peats and tree stumps, tidal-dwelling mollusks, and archaeological sites would all point to a rise in relative sea level since the time of growth, deposition, or construction. Raised coral reefs, such as in **Figure 1**, whale bones cemented in beach deposits, wave-cut rock platforms

and notches, or peats formed from saline-loving plants would all be indicative of a falling sea level since the time of formation. To obtain useful sea level measurements from these data requires several steps: an understanding of the relationship between the feature's elevation and mean sea level at the time of growth or deposition, a measurement of the height or depth with respect to present sea level, and a measurement of the age. All aspects of the observation present their own peculiar problems but over recent years a substantial body of observational evidence has been built up for sea level change since the last glaciation that ended at about 20 000 years ago. Some of this evidence is illustrated in **Figure 4**, which also indicates the very significant spatial variability that may occur even when, as is the case here, the evidence is from sites that are believed to be tectonically stable.

In areas of former major glaciation, raised shorelines occur with ages that are progressively greater with increasing elevation and with the oldest shorelines corresponding to the time when the region first became ice-free. Two examples, from northern Sweden and Hudson Bay in Canada, respectively, are illustrated in **Figure 4A**. In both cases sea level has been falling from the time the area became ice-free and open to the sea. This occurred at about 9000 years ago in the former case and about 7000 years later in the second case. Sea level curves from localities just within the margins of the former ice sheet are illustrated in **Figure 4B**, from western Norway and Scotland, respectively. Here, immediately after the area became ice-free, the sea level fell but a time was reached when it again rose. A local maximum was reached at about 6000 years ago, after which the fall continued up until the present. Farther away from the former ice margins the observed sea level pattern changes dramatically, as is illustrated in **Figure 4C**. Here the sea level initially rose rapidly but the rate decreased for the last 7000 years up to the present. The two examples illustrated, from southern England and the Atlantic coast of the United States, are representative of tectonically stable localities that lie within a few thousand kilometers from the former centers of glaciation. Much farther away again from the former ice sheets the sea level signal undergoes a further small but significant change in that the present level was first reached at about 6000 years ago and then exceeded by a small amount before returning to its present value. The two examples illustrated in **Figure 4D** are from nearby localities in northern Australia. At both sites present sea level was reached about 6000 years ago after a prolonged period of rapid rise with resulting highstands that are small in amplitude but geographically variable.

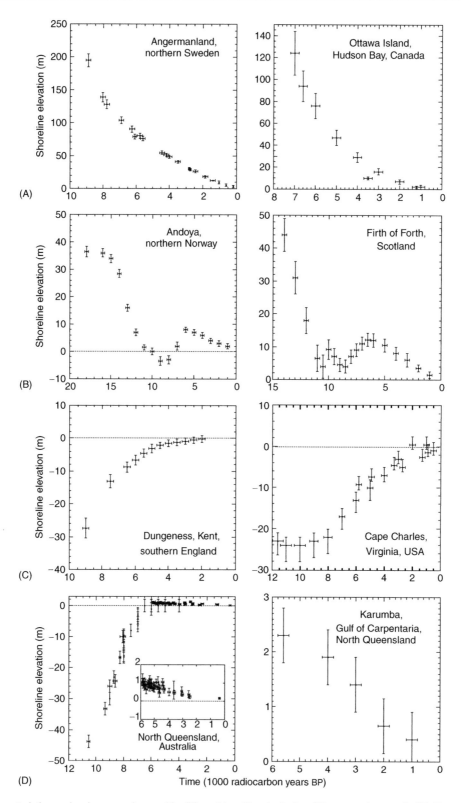

Figure 4 Characteristic sea level curves observed in different localities (note the different scales used). (A) From Ångermanland, Gulf of Bothnia, Sweden, and from Hudson Bay, Canada. Both sites lie close to centers of former ice sheet, over northern Europe and North America respectively. (B) From the Atlantic coast of Norway and the west coast of Scotland. These sites are close to former ice-sheet margins at the time of the last glaciation. (C) From southern England and Virginia, USA. These sites lie outside the areas of former glaciation and at a distance from the margins where the crust is subsiding in response to the melting of the nearby icesheet. (D) Two sites in northern Australia, one for the Coral Sea coast of Northern Queensland and the second from the Gulf of Carpentaria some 200 km away but on the other side of the Cape York Peninsula. At both localities sea level rose until about 6000 years ago before peaking above present level.

The examples illustrated in **Figure 4** indicate the rich spectrum of variation in sea level that occurred when the last large ice sheets melted. Earlier glacial cycles resulted in a similar spatial variability, but much of the record before the Last Glacial Maximum has been overwritten by the effects of the last deglaciation.

Glacio-Hydro-Isostatic Models

If, during the decay of the ice sheets, the meltwater volume was distributed uniformly over the oceans, then the sea level change at time t would be

$$\Delta\zeta_e(t) = (\text{change in ice volume/ocean surface area})$$
$$\times \rho_i/\rho_w \qquad [1a]$$

where ρ_i and ρ_w are the densities of ice and water, respectively. (See end of article – symbols used.) This term is the ice-volume equivalent sea level and it provides a measure of the change in ice volume through time. Because the ocean area changes with time a more precise definition is

$$\Delta\zeta_e(t) = \frac{dV_i}{dt} dt \frac{\rho_i}{\rho_0} \int_t \frac{\rho_i}{A_0(t)} \qquad [1b]$$

where A_0 is the area of the ocean surface, excluding areas covered by grounded ice. The sea level curve illustrated in **Figure 3** is essentially this function. However, it represents only a zero-order approximation of the actual sea level change because of the changing gravitational field and deformation of the Earth.

In the absence of winds or ocean currents, the ocean surface is of constant gravitational potential. A planet of a defined mass distribution has a family of such surfaces outside it, one of which – the geoid – corresponds to mean sea level. If the mass distribution on the surface (the ice and water) or in the interior (the load-forced mass redistribution in the mantle) changes, so will the gravity field, the geoid, and the sea level change. The ice sheet, for example, represents a large mass that exerts a gravitational pull on the ocean and, in the absence of other factors, sea level will rise in the vicinity of the ice sheet and fall farther away. At the same time, the Earth deforms under the changing load, with two consequences: the land surface with respect to which the level of the sea is measured is time-dependent, as is the gravitational attraction of the solid Earth and hence the geoid.

The calculation of the change of sea level resulting from the growth or decay of ice sheets therefore involves three steps: the calculation of the amount of water entering into the ocean and the distribution of this meltwater over the globe; the calculation of the deformation of the Earth's surface; and the calculation of the change in the shape of the gravitational equipotential surfaces. In the absence of vertical tectonic motions of the Earth's surface, the relative sea level change $\Delta\zeta(\varphi, t)$ at a site φ and time t can be written schematically as

$$\Delta\zeta(\varphi, t) = \Delta\zeta_e(t) + \Delta\zeta_I(\varphi, t) \qquad [2]$$

where $\Delta\zeta_I(\varphi, t)$ is the combined perturbation from the uniform sea level rise term [1]. This is referred to as the isostatic contribution to relative sea level change.

In a first approximation, the Earth's response to a global force is that of an elastic outer spherical layer (the lithosphere) overlying a viscous or viscoelastic mantle that itself contains a fluid core. When subjected to an external force (e.g., gravity) or a surface load (e.g., an ice cap), the planet experiences an instantaneous elastic deformation followed by a time-dependent or viscous response with a characteristic timescale(s) that is a function of the viscosity. Such behavior of the Earth is well documented by other geophysical observations: the gravitational attraction of the Sun and Moon raises tides in the solid Earth; ocean tides load the seafloor with a time-dependent water load to which the Earth's surface responds by further deformation; atmospheric pressure fluctuations over the continents induce deformations in the solid Earth. The displacements, measured with precision scientific instruments, have both an elastic and a viscous component, with the latter becoming increasingly important as the duration of the load or force increases. These loads are much smaller than the ice and water loads associated with the major deglaciation, the half-daily ocean tide amplitudes being only 1% of the glacial sea level change, and they indicate that the Earth will respond to even small changes in the ice sheets and to small additions of meltwater into the oceans.

The theory underpinning the formulation of planetary deformation by external forces or surface loads is well developed and has been tested against a range of different geophysical and geological observations. Essentially, the theory is one of formulating the response of the planet to a point load and then integrating this point-load solution over the load through time, calculating at each epoch the surface deformation and the shape of the equipotential surfaces, making sure that the meltwater is appropriately distributed into the oceans and that the total ice–water mass is preserved. Physical inputs into the formulation are the time–space history of the ice sheets, a description of the ocean basins from which the water is extracted or into which it is added, and a description of the rheology, or response parameters, of the Earth. For the last requirement, the elastic properties of the Earth, as well as the density distribution with depth, have been determined from seismological and geodetic studies. Less well determined are the viscous properties of the mantle, and the usual procedure is to adopt a simple parametrization of the viscosity structure and to estimate the relevant parameters from analyses of the sea level change itself.

The formulation is conveniently separated into two parts for schematic reasons: the glacio-isostatic effect representing the crustal and geoid displacements due to the ice load, and the hydro-isostatic effect due to the water load. Thus

$$\Delta\zeta(\varphi,t) = \Delta\zeta_e(t) + \Delta\zeta_i(\varphi,t) + \Delta\zeta_w(\varphi,t) \quad [3]$$

where $\Delta\zeta_i(\varphi,t)$ is the glacio-isostatic and $\Delta\zeta_w(\varphi,t)$ the hydro-isostatic contribution to sea level change. (In reality the two are coupled and this is included in the formulation). If $\Delta\zeta_i(\varphi,t)$ is evaluated everywhere, the past water depths and land elevations $H(\varphi,t)$ measured with respect to coeval sea level are given by

$$H(\varphi,t)H_0(\varphi) - \Delta\zeta(\varphi,t) \quad [4]$$

where $H(\phi)$ is the present water depth or land elevation at location φ.

A frequently encountered concept is eustatic sea level, which is the globally averaged sea level at any time t. Because of the deformation of the seafloor during and after the deglaciation, the isostatic term $\Delta\zeta(\varphi,t)$ is not zero when averaged over the ocean at any time t, so that the eustatic sea level change is

$$\Delta\zeta_{eus}(t) = \Delta\zeta_e(t) + \langle\Delta\zeta_I\{\varphi,t\}\rangle_0$$

where the second term on the right-hand side denotes the spatially averaged isostatic term. Note that $\Delta\zeta_e(t)$ relates directly to the ice volume, and not $\Delta\zeta_{eus}(t)$.

The Anatomy of the Sea Level Function

The relative importance of the two isostatic terms in eqn. [3] determines the spatial variability in the sea level signal. Consider an ice sheet of radius that is much larger than the lithospheric thickness: the limiting crustal deflection beneath the center of the load is $I\rho_i/I\rho_m$ where I is the maximum ice thickness and $I\rho_m$ is the upper mantle density. This is the local isostatic approximation and it provides a reasonable approximation of the crustal deflection if the loading time is long compared with the relaxation time of the mantle. Thus for a 3 km thick ice sheet the maximum

deflection of the crust can reach 1 km, compared with a typical ice volume for a large ice sheet that raises sea level globally by 50–100 m. Near the centers of the formerly glaciated regions it is the crustal rebound that dominates and sea level falls with respect to the land.

This is indeed observed, as illustrated in **Figure 4A**, and the sea level curve here consists of essentially the sum of two terms, the major glacio-isostatic term and the minor ice-volume equivalent sea level term $\Delta\zeta_e(t)$ (**Figure 5A**). Of note is that the rebound continues long after all ice has disappeared, and this is evidence that the mantle response includes a memory of the earlier load. It is the decay time of this part of the curve that determines the mantle viscosity. As the ice margin is approached, the local ice thickness becomes less and the crustal rebound is reduced and at some stage is equal, but of opposite sign, to $\Delta\zeta_e(t)$. Hence sea level is constant for a period (**Figure 5B**) before rising again when the rebound becomes the minor term. After global melting has ceased, the dominant signal is the late stage of the crustal rebound and levels fall up to the present. Thus the oscillating sea level curves observed in areas such as Norway and Scotland (**Figure 4B**) are essentially the sum of two effects of similar magnitude but opposite sign. The early part of the observation contains information on earth rheology as well as on the local ice thickness and the globally averaged rate of addition of meltwater into the oceans. Furthermore, the secondary maximum is indicative of the timing of the end of global glaciation and the latter part of the record is indicative mainly of the mantle response.

At the sites beyond the ice margins it is the meltwater term $\Delta\zeta_e(t)$ that is important, but it is not the sole factor. When the ice sheet builds up, mantle material flows away from the stressed mantle region and, because flow is confined, the crust around the periphery of the ice sheet is uplifted. When the ice sheet melts, subsidence of the crust occurs in a broad zone peripheral to the original ice sheet and at these locations the isostatic effect is one of an apparent subsidence of the crust. This is illustrated in **Figure 5C**. Thus, when the ocean volumes have stabilized, sea level continues to rise, further indicating that the planet responds viscously to the changing surface loads. The early part of the observational record (e.g., **Figure 4C**) is mostly indicative of the rate at which meltwater is added into the ocean, whereas the latter part is more indicative of mantle viscosity.

In all of the examples considered so far it is the glacial-rebound that dominates the total isostatic adjustment, the hydro-isostatic term being present but comparatively small. Consider an addition of water that raises sea level by an amount D. The local

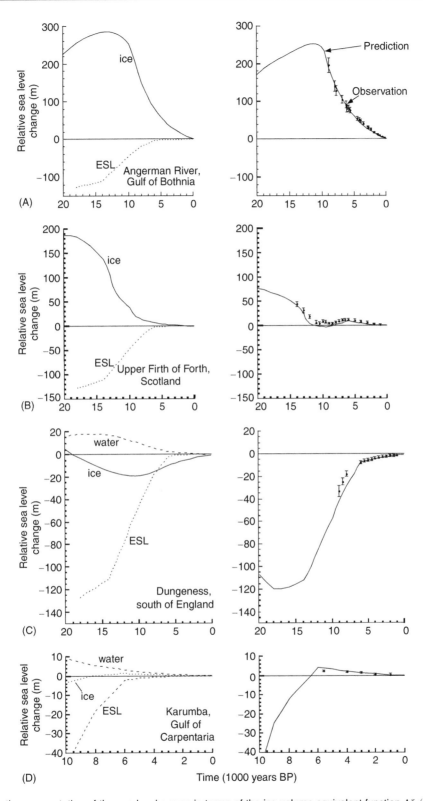

Figure 5 Schematic representation of the sea level curves in terms of the ice-volume equivalent function $\Delta\zeta_e(t)$ (denoted by ESL) and the glacio- and hydro-isostatic contributions $\Delta\zeta_i(\phi, t)$, $\Delta\zeta_w(\phi, t)$ (denoted by ice and water, respectively). The panels on the left indicate the predicted individual components and the panels on the right indicate the total predicted change compared with the observed values (data points). (A) For the sites at the ice center where $\Delta\zeta_i(\phi, t) \gg \Delta\zeta_e(t) \gg \Delta\zeta_w(\phi, t)$. (B) For sites near but within the ice margin where $|\Delta\zeta_i(\phi, t)| \sim |\Delta\zeta_e(t)| \gg |\Delta\zeta_w(\phi, t)|$, but the first two terms are of opposite sign. (C) For sites beyond the ice margin where $|\Delta\zeta_e(t)| > |\Delta\zeta_i(\phi, t)| \gg |\Delta\zeta_w(\phi, t)|$. (D) For sites at continental margins far from the former ice sheets where $|\Delta\zeta_w(\phi, t)| > |\Delta\zeta_i(t)|$. Adapted with permission from Lambeck K and Johnston P (1988). The viscosity of the mantle: evidence from analyses of glacial rebound phenomena. In: Jackson ISN (ed.) *The Earth's Mantle – Composition, Structure and Evolution*, pp. 461–502. Cambridge: Cambridge University Press.

isostatic response to this load is $D\rho_w/\rho_m$, where ρ_w is the density of ocean water. This gives an upper limit to the amount of subsidence of the sea floor of about 30 m for a 100 m sea level rise. This is for the middle of large ocean basins and at the margins the response is about half as great. Thus the hydro-isostatic effect is significant and is the dominant perturbing term at margins far from the former ice sheets. This occurs at the Australian margin, for example, where the sea level signal is essentially determined by $\Delta\zeta_e(t)$ and the water-load response (**Figure 4D**). Up to the end of melting, sea level is dominated by $\Delta\zeta_e(t)$ but there-after it is determined largely by the water-load term $\Delta\zeta_w(\varphi,t)$, such that small highstands develop at 6000 years. The amplitudes of these highstands turn out to be strongly dependent on the geometry of the water-load distribution around the site: for narrow gulfs, for example, their amplitude increases with distance from the coast and from the water load at rates that are particularly sensitive to the mantle viscosity.

While the examples in **Figure 5** explain the general characteristics of the global spatial variability of the sea level signal, they also indicate how observations of such variability are used to estimate the physical quantities that enter into the schematic model (3). Thus, observations near the center of the ice sheet partially constrain the mantle viscosity and central ice thickness. Observations from the ice sheet margin partially constrain both the viscosity and local ice thickness and establish the time of termination of global melting. Observations far from the ice margins determine the total volumes of ice that melted into the oceans as well as providing further constraints on the mantle response. By selecting data from different localities and time intervals, it is possible to estimate the various parameters that underpin the sea level eqn. [3] and to use these models to predict sea level and shoreline change for unobserved areas. Analyses of sea-level change from different regions of the world lead to estimates for the lithospheric thickness of between 60–100 km, average upper mantle viscosity (from the base of the lithosphere to a depth of 670 km) of about $(1-5)10^{20}$ Pa s and an average lower mantle viscosity of about $(1-5)10^{22}$ Pa s. Some evidence exists that these parameters vary spatially; lithosphere thickness and upper mantle viscosity being lower beneath oceans than beneath continents.

Sea Level Change and Shoreline Migration: Some Examples

Figure 6 illustrates the ice-volume equivalent sea level function $\Delta\zeta_e(t)$ since the time of the last maximum glaciation. This curve is based on sea level indicators from a number of localities, all far from the former ice sheets, with corrections applied for the isostatic effects. The right-hand axis indicates the corresponding change in ice volume (from the relation [1b]). Much of this ice came from the ice sheets over North America and northern Europe but a not insubstantial part also originated from Antarctica. Much of the melting of these ice sheets occurred within 10 000 years, at times the rise in sea level exceeding 30 mm per year, and by 6000 years ago most of the deglaciation was completed. With this sea level function, individual ice sheet models, the formulation for the isostatic factors, and a knowledge of the topography and bathymetry of the world, it becomes possible to reconstruct the paleo shorelines using the relation [4].

Scandinavia is a well-studied area for glacial rebound and sea level change since the time the ice retreat began about 18 000 years ago. The observational evidence is quite plentiful and a good record of ice margin retreat exists. **Figure 7** illustrates examples for two epochs. The first (**Figure 7A**), at 16 000 years ago, corresponds to a time after onset of deglaciation. A large ice sheet existed over

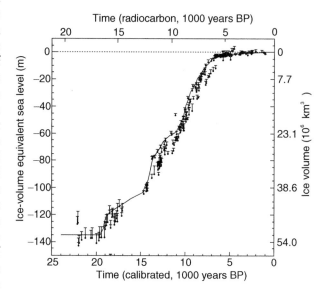

Figure 6 The ice-volume equivalent function $\Delta\zeta_e(t)$ and ice volumes since the time of the last glacial maximum inferred from corals from Barbados, from sediment facies from north-western Australia, and from other sources for the last 7000 years. The actual sea level function lies at the upper limit defined by these observations (continuous line). The upper time scale corresponds to the radiocarbon timescale and the lower one is calibrated to calendar years. Adapted with permission from Fleming K *et al.* (1998) Refining the eustatic sea-level curve since the LGM using the far- and intermediate-field sites. *Earth and Planetary Science Letters* 163: 327–342; and Yokoyama Y *et al.* (2000) Timing of the Last Glacial Maximum from observed sea-level minimum. *Nature* 406: 713–716.

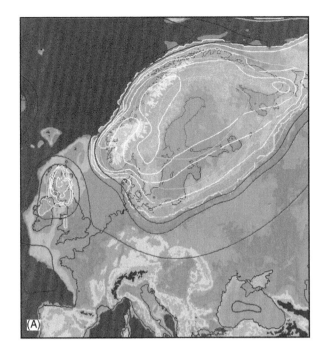

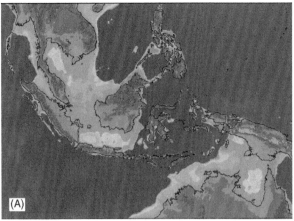

Figure 8 Shoreline reconstructions for South East Asia and northern Australia at the time of the last glacial maximum, (A) at about 18 000 years ago and (B) at 12 000 years ago. The water depth contours are the same as in **Figure 7**. The water depths in the inland lakes at 18 000 years ago are contoured at 25 m intervals relative to their maximum levels that can be attained without overflowing.

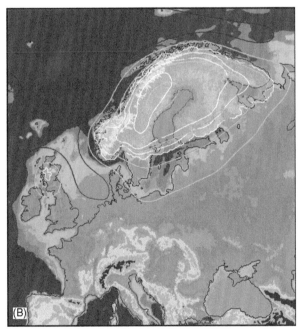

Figure 7 Shoreline reconstructions for Europe (A) at 16 000 and (B) at 10 500 years ago. The contours are at 400 m intervals for the ice thickness (white) and 25 m for the water depths less than 100 m. The orange, yellow, and red contours are the predicted lines of equal sea level change from the specified epoch to the present and indicate where shorelines of these epochs could be expected if conditions permitted their formation and preservation. The zero contour is in yellow; orange contours, at intervals of 100 m, are above present, and red contours, at 50 m intervals, are below present. At 10 500 years ago the Baltic is isolated from the Atlantic and its level lies about 25 m above that of the latter.

Scandinavia with a smaller one over the British Isles. Globally sea level was about 110 m lower than now and large parts of the present shallow seas were exposed, for example, the North Sea and the English Channel but also the coastal shelf farther south such as the northern Adriatic Sea. The red and orange contours indicate the sea level change between this period and the present. Beneath the ice these rebound contours are positive, indicating that if shorelines could form here they would be above sea level today. Immediately beyond the ice margin, a broad but shallow bulge develops in the topography, which will subside as the ice sheet retreats. At the second epoch selected (**Figure 7B**) 10 500 years ago, the ice has retreated and reached a temporary halt as the climate briefly returned to colder conditions; the Younger Dryas time of Europe. Much of the Baltic was then ice-free and a freshwater lake developed at some

25–30 m above coeval sea level. The flooding of the North Sea had begun in earnest. By 9000 years ago, most of the ice was gone and shorelines began to approach their present configuration.

The sea level change around the Australian margin has also been examined in some detail and it has been possible to make detailed reconstructions of the shoreline evolution there. **Figure 8** illustrates the reconstructions for northern Australia, the Indonesian islands, and the Malay–IndoChina peninsula. At the time of the Last Glacial Maximum much of the shallow shelves were exposed and deeper depressions within them, such as in the Gulf of Carpentaria or the Gulf of Thailand, would have been isolated from the open sea. Sediments in these depressions will sometimes retain signatures of these pre-marine conditions and such data provide important constraints on the models of sea level change. Part of the information illustrated in **Figure 6**, for example, comes from the shallow depression on the Northwest Shelf of Australia. By 12 000 years ago the sea has begun its encroachment of the shelves and the inland lakes were replaced by shallow seas.

Symbols used

$\Delta \zeta_e(t)$ Ice-volume equivalent sea level. Uniform change in sea level produced by an ice volume that is distributed uniformly over the ocean surface.

$\Delta \zeta_I(t)$ Perturbation in sea level due to glacioisostatic $\Delta \zeta_i(t)$ and hydro-isostatic $\Delta \zeta_w(t)$ effects.

$\Delta \zeta_{eus}(t)$ Eustatic sea level change. The globally averaged sea level at time t.

See also

Beaches, Physical Processes Affecting. Coral Reefs. Fiordic Ecosystems. Geomorphology. Lagoons. Rocky Shores. Salt Marshes and Mud Flats. Salt Marsh Vegetation. Sandy Beaches, Biology of. Sea Level Change.

Further Reading

Lambeck K (1988) *Geophysical Geodesy: The Slow Deformations of the Earth*. Oxford: Oxford University Press.

Lambeck K and Johnston P (1999) The viscosity of the mantle: evidence from analysis of glacial-rebound phenomena. In: Jackson ISN (ed.) *The Earth's Mantle – Composition, Structure and Evolution*, pp. 461–502. Cambridge: Cambridge University Press.

Lambeck K, Smither C, and Johnston P (1998) Sea-level change, glacial rebound and mantle viscosity for northern Europe. *Geophysical Journal International* 134: 102–144.

Peltier WR (1998) Postglacial variations in the level of the sea: implications for climate dynamics and solid-earth geophysics. *Reviews in Geophysics* 36: 603–689.

Pirazzoli PA (1991) *World Atlas of Holocene Sea-Level Changes*. Amsterdam: Elsevier.

van de Plassche O (ed.) (1986) *Sea-Level Research: A Manual for the Collection and Evaluation of Data*. Norwich: Geo Books.

Sabadini R, Lambeck K, and Boschi E (1991) *Glacial Isostasy, Sea Level and Mantle Rheology*. Dordrecht: Kluwer.

GEOMORPHOLOGY

C. Woodroffe, University of Wollongong, Wollongong, NSW, Australia

Introduction

Geomorphology is the study of the form of the earth. Coastal geomorphologists study the way that the coastal zone, one of the most dynamic and changeable parts of the earth, evolves, including its profile, plan-form, and the architecture of foreshore, backshore, and nearshore rock and sediment bodies. To understand these it is necessary to examine wave processes and current action, but it may also involve drainage basins that feed to the coast, and the shallow continental shelves which modify oceanographic processes before they impinge upon the shore. Morphodynamics, study of the mutual co-adjustment of form and process, leads to development of conceptual, physical, mathematical, and simulation models, which may help explain the changes that are experienced on the coast.

In order to understand coastal variability from place to place there are a number of boundary conditions which need to be considered, including geophysical and geological factors, oceanographic factors, and climatic constraints.

At the broadest level plate-tectonic setting is important. Coasts on a plate margin where oceanic plate is subducted under continental crust, such as along the western coast of the Americas, are known as collision coasts. These are typically rocky coasts, parallel to the structural grain, characterized by seismic and volcanic activity and are likely to be uplifting. They contrast with trailing-edge coasts where the continental margin sits mid-plate, which are the locus of large sedimentary basins. Smaller basins are typical of marginal sea coasts, behind a tectonically active island arc.

The nature of the material forming the coast is partly a reflection of these broad plate-tectonic factors. Whether the shoreline is rock or unconsolidated sediment is clearly important. Resistant igneous or metamorphic rocks are more likely to give rise to rocky coasts than are those areas composed of broad sedimentary sequences of clays or mudstones. Within sedimentary coasts sandstones may be more resistant than mudstones. Rocky coasts tend to be relatively resistant to change, whereas coasts composed of sandy sediments are relatively easily reshaped, and muddy coasts, in low-energy environments, accrete slowly.

The relative position of the sea with respect to the land has changed, and represents an important boundary condition. The sea may have flooded areas which were previously land (submergence), or formerly submarine areas may now be dry (emergence). The form of the coast may be inherited from previously subaerial landforms. Particularly distinctive are landscapes that have been shaped by glacial processes, thus fiords are glacially eroded valleys, and fields of drumlins deposited by glacial processes form a prominent feature on paraglacial coasts.

The oceanographic factors that shape coasts include waves, tides, and currents. Waves occur as a result of wind transferring energy to the ocean surface. Waves vary in size depending on the strength of the wind, the duration for which it has blown and the fetch over which it acts. Wave trains move out of the area of formation and are then known as swell. The swell and wave energy received at a coast may be a complex assemblage generated by specific storms from several areas of origin. Tides represent a large-wavelength wave formed as a result of gravitational attractions of the sun and moon. Tides occur as a diurnal or semidiurnal fluctuation of the sea surface that may translate into significant tidal currents particularly in narrow straits and estuaries. In addition storm-generated surges and tsunamis may cause elevated water levels with significant geomorphological consequences.

The coast is shaped by these oceanographic processes working on the rocks and unconsolidated sediments of the shoreline. Climate is significant in terms of several factors. First, wind conditions lead to generation of waves and swell, and may blow sand into dunes along the backshore. Climate also influences the rate at which weathering and catchment processes operate. In addition, regional-scale climate factors such as monsoonal wind systems and the El Niño Southern Oscillation phenomenon demonstrate oscillatory behavior and may reshape the coast seasonally or interannually. Gradual climate change may mean that shoreline fluctuations do not revolve around stationary boundary conditions but may exhibit gradual change themselves.

In this respect the impact of perceived global climate change during recent decades as a result of human-modified environmental factors is likely to be felt in the coastal zone. Of particular concern is

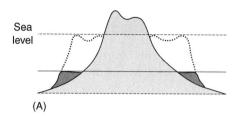

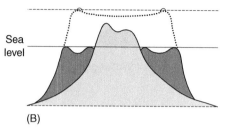

Figure 1 Schematic representation of the original woodcut illustrations by Charles Darwin showing (A) the manner in which he deduced that fringing reefs around a volcanic island would develop into barrier reefs (dotted), and (B) barrier reefs would develop into an atoll, as a result of subsidence and vertical reef growth.

anticipated sea-level rise which may have a range of geomorphological effects, as well as other significant socioeconomic impacts.

History

Geomorphology has its origins in the nineteenth century with the results of exploration, and the realization that the surface of the earth had been shaped over a long time through the operation of processes that are largely in operation today (uniformitarianism).

The observations by Charles Darwin during the voyage of the *Beagle* extended this view, particularly his remarkable deduction that fringing reefs might become barrier reefs which in turn might form atolls as a result of gradual subsidence of volcanic islands combined with vertical reef growth (**Figure 1**). In the first part of the twentieth century geomorphology was dominated by the 'geographical cycle' of erosion of William Morris Davis who anticipated landscape denudation through a series of stages culminating in

peneplanation, and subsequent rejuvenation by uplift. This highly conceptual model, across landscapes in geological time, was also applied to the coast by Davis who envisaged progressive erosion of the coast reducing shoreline irregularities with time (**Figure 2**). Such landscape-scale studies were extended by Douglas Johnson who emphasized the role of submergence or emergence as a result of sea-level change (**Figure 3**).

The Davisian view was reassessed in the second half of the twentieth century with a greater emphasis on process geomorphology whereby studies focused on attempting to measure rates of process operation and morphological responses to those processes, reflecting ideas of earlier researchers such as G. K. Gilbert. The concept of the landscape as a system was examined, in which coastal landforms adjusted to equilibrium, perhaps a dynamic equilibrium, in relation to processes at work on them. Studies of

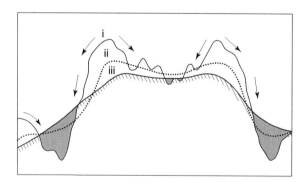

Figure 2 Schematic representation of the planform stages that W. M. Davis conceptualized through which a shoreline would progress from an initial rugged form (1) through maturity (2) to a regularized shoreline (3) that is cliffed (hatched) and infilled with sand (stippled). He envisaged this in parallel to the geographical cycle of erosion by which mountains (like the initial shoreline form, 1) would be reduced to a peneplain (like the solid ultimate shoreline, 3).

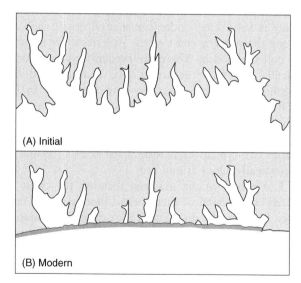

Figure 3 Schematic representation of (A) the initial shoreline form envisaged by Douglas Johnson for an area in Marthas Vineyard, New England, and (B) the modern regularized, form of the shore.

sediment movement and the adjustment of beach shape under different wave conditions were typical.

Scales of Study

Coastal geomorphology now studies landforms and the processes that operate on them at a range of spatial and temporal scales (**Figure 4**). At the smallest scale geomorphology is concerned with an 'instantaneous' timescale where the principles of fluid dynamics apply. It should be possible to determine details of sediment entrainment, complexities of turbulent flow and processes leading to deposition of individual bedform laminae. The laws of physics apply at these scales, though they may operate stochastically. In theory, behavior of an entire embayment could be understood; in practice studies simply cannot be undertaken at that level of detail and broad extrapolations based on important empirical relationships are made.

The next level of study is the 'event' timescale, which may cover a single event such as an individual storm or an aggregation of several lesser events over a year or more. The mechanistic relationships from instantaneous time are scaled up in a deterministic or empirical way to understand the operation of coasts at larger spatial and temporal scales. Thus, stripping of a beach during a single storm, and the more gradual reconstruction to its original state under ensuing calmer periods can be observed. Time taken for reaction to the event, and relaxation back to a more ambient state may be known from surveys of beaches, enabling definition of a 'sweep zone' within which the beach is regularly active.

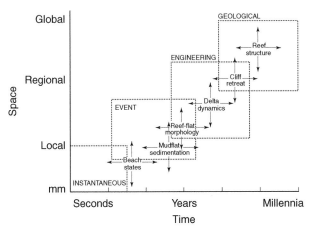

Figure 4 Representation of space and timescales appropriate for the study of coastal geomorphology and schematic representation of some of the examples discussed in the context of instantaneous, event, engineering, and geological scales of enquiry.

At a larger scale of operation, the coastal geomorphologist is interested in the way in which coasts change over timescales that are relevant to societies and at which coastal-zone managers need to plan. This is the 'engineering' timescale, involving several decades. It is perhaps the most difficult timescale on which to understand coastal geomorphology and to anticipate behavior of the shoreline. The largest timescales are geological timescales. Studies over geological timescales are primarily discursive, conceptual models which recognize that boundary conditions, including climate and the rate of operation of oceanographic processes, change. It is clear that sea level itself has changed dramatically over millennia as a result of expansion and contraction of ice sheets during the Quaternary ice ages, and so the position of the coastline has changed substantially between glaciations and interglaciations.

Models of Coastal Evolution

There has been considerable improvement in understanding the long-term development of coasts as a result of significant advances in paleoenvironmental reconstruction and geochronological techniques (especially radiometric dating). Incomplete records of past coastal conditions may be preserved, either as erosional morphology (notches, marine terraces, etc.) or within sedimentary sequences. Although this record is selective, reconstructions of Quaternary paleoenvironments, together with interpretation of geological sequences in older rocks, have enabled the formulation of geomorphological models based on sedimentary evidence. In the case of deltas, where there may be important hydrocarbon reserves, the complementary development of geological models and study of modern deltas has led to better understanding of process and response at longer timescales than those for which observations exist.

Unconsolidated sediments are the key to coastal morphodynamics because coasts change through erosion, transport, and deposition of sediment. Study of coastal systems has led to many insights, particularly in terms of the various pathways through which sediment may move within a coastal compartment or circulation cell. However, it is clear that a completely reductionist approach to coastal geomorphology will not lead to understanding of all components and the way in which they interact. Empirical relationships and the presumed deterministic nature of sediment response to forcing factors remain incomplete and are all too often formulated on the basis of presumed uniform sediment sizes or absence of biotic influence and are ultimately

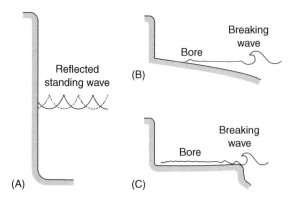

Figure 5 The morphology of cliffs and shore platforms. Plunging cliffs (A) reflect wave energy and are generally not eroded. Shore platforms dissipate wave energy and may be either (B) ramped, or (C) subhorizontal. Change in these systems tends to occur at geological (hard rock) or engineering (soft rock) timescales.

unrealistic. More recently coastal systems have been investigated as nonlinear dynamical systems, because relationships are not linear and behavior is potentially chaotic.

Nonlinear dynamical systems behave in a way that is in part dependent upon antecedent conditions. This is well illustrated by studies of beach state; beach and nearshore sediments are particularly easily reshaped by wave processes, thus a series of characteristics typical of distinct beach states can be recognized (see **Figure 6**). However, the beach is only partly a response to incident wave conditions, its shape being also dependent on the previous shape of the beach which was in the process of adjusting to wave conditions incident at that time.

Coastal systems are inherently unpredictable. However, they may operate within a broad range of conditions with certain states being recurrent. Chaotic systems may tend towards self-organization, for instance patterns of beach cusps characteristic along low-energy beaches which reflect much of the wave energy may adopt a self-organized cuspate morphology in which swash processes, sediment sorting, and form are balanced.

Models may be developed based on patterns of change inferred over geological timescales but consistent with mechanistic processes known to operate over lesser event timescales. Simulation modelling is not intended to reconstruct coastal evolution exactly, but becomes a tool for experimentation and extrapolation within which broad scenarios of change can be modeled and sensitivity to parameterization of variables examined.

Coasts may be divided into rocky coasts, sandy coasts, and muddy coasts, and coral reefs and deltas and estuaries can be differentiated. The geomorphology of each of these behaves differently. Processes operate at different rates, transitions between different states occur over different timescales and the significance of antecedent conditions varies.

Rocky Coasts

Rocky coasts are characterized by sea cliffs, especially on tectonically active, plate-margin coasts where there are resistant rocks. Cliffed coastlines in resistant rock appear to adopt one of two forms (states): plunging cliffs where a vertical cliff extends below sea level, and shore platforms where a broad bench occurs at sea level in front of a cliff (**Figure 5**). Erosion of cliffs occurs where the erosional force of waves exceeds the resistance of the rocks, and where sufficient time has elapsed.

Plunging cliffs occur where the rock is too resistant to be eroded. The vertical face results in a standing wave which reflects wave energy, so that there is little force to erode the cliff at water level. Waves exert a greater force if they break, or if they are already broken. This can only occur if the water depth is shallow offshore from the cliff face in which case the increased energy from the breaking waves is also able to entrain sediment from the floor (or rock fragments quarried from the foot of the cliff). This process of erosion accelerates through a positive feedback cutting a shore platform.

Shore platforms thus develop in those situations where the erosive force of waves exceeds resistance of the rock. A platform widens as a result of erosion at the foot of the cliff behind the platform. The cliff oversteepens, leading to toppling and fall of detritus onto the rear of the platform, which slows further erosion of the cliff face until that talus has been removed. A series of such negative feedbacks slow the rate at which platforms widen over time, and there is often considerable uniformity of platforms up to a maximum width in any particular lithological setting.

Shore platforms adopt either a gradually sloping ramped form, or a subhorizontal form often with a seaward rampart (**Figure 5**). There has been much discussion as to the relative roles of wave and subaerial processes in the formation of these platforms. Platforms in relatively sheltered locations appear to owe their origin to processes of water-layer leveling (physiochemical processes in pools which persist on platforms at low tide) or wetting and drying and its weakening of the rock. In other cases wave quarrying and abrasion are involved. In many cases both processes may be important. Other platforms may be polygenetic with inheritance from former stands of sea level (reflecting antecedent conditions).

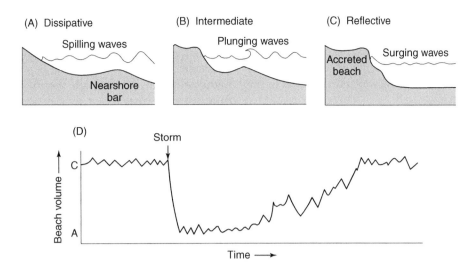

Figure 6 Beach state may be (A) dissipative, where high wave-energy impinges on low-gradient beaches; (B) intermediate (in which several states are possible); or (C) reflective, where low wave-energy surges onto steep or coarse beaches. Change between beach states occurs at event timescales. (D) A beach may react rapidly to a perturbation such as a storm changing from reflective state (C) to more dissipative state (A) but then change much more slowly through intermediate states, readjusting towards state (C) again unless further disrupted.

Coral Reefs

A group of coastlines that are of particular interest to the geomorphologist are those formed by coral reefs. Corals are colonial animals that secrete a limestone exoskeleton that may form the matrix of a reef. Coral reefs flourish in tropical seas in high-energy settings where a significant swell reaches the shoreline. In these circumstances the reef attenuates much of the wave energy such that a relatively quiet-water environment occurs sheltered behind the reef.

Not only is this reef ecosystem of enormous geomorphological significance in terms of the solid reef structure that it forms, but in addition, the carbonate reef material breaks down into calcareous gravels, sands and muds which form the sediments that further modify these coastlines.

On the one hand reefs can be divided into fringing reefs, barrier reefs, and atolls, distinct morphological states that form part of an evolutionary sequence as a result of gradual subsidence of the volcanic basement upon which the reef established (see **Figure 1**). This powerful deduction by Darwin relating to reef structure and operating over geological timescales, has been generally supported by drilling and geochronological studies on mid-plate islands in reef-forming seas. On the other hand, the surface morphology of reefs is extremely dynamic with rapid production of skeletal sediments and their redistribution over event timescales, with the landforms of reefs responding to minor sea-level oscillations, storms, El Niño, coral bleaching, and other perturbations.

Sandy Coasts

Beaches form where sandy (or gravel) material is available forming a sediment wedge at the shoreline. The beach is shaped by incident wave energy, and can undergo modification particularly by formation and migration of nearshore bars which in turn modify the wave-energy spectrum. Various beach states can be recognized across a continuum from beaches which predominantly reflect wave energy, and those which dissipate wave energy across the nearshore zone. Reflective beaches are steep, waves surge up the beach and much of the energy is reflected, and the beach face may develop cusps. Dissipative beaches are much flatter, and waves spill before reaching the shoreline (**Figure 6**).

During a storm, sand is generally eroded from the beach face and deposited in the nearshore, often forming shore-parallel or transverse bars. Waves consequently break on the bars and energy is lost, the form modifying the process in a mutual way. It may be possible to recognize a series of beach states intermediate between reflective and dissipative and any one beach may adopt one or several beach states over time (**Figure 6**). Beach state is clearly modified by incident wave conditions, but the rate of adjustment between states takes time, and a beach is also partly a function of antecedent beach states.

Although erosion and redeposition of sand is the way in which the nearshore adjusts, there may also be long-term storage of sediment. In particular broad, flat beaches may develop dunes behind them.

In other cases a sequence of beach ridges may develop. Geomorphological changes in state over geological timescales represented by beach-ridge plains may relate to variations in supply of sediment to the system (perhaps by rivers) as well as variations in the processes operating.

Deltas and Estuaries

Where rivers bring sediment to the coast, deltas and estuaries can develop. In this case the sediment budget of the shoreline compartment or the receiving basin is augmented and there is generally a positive sediment budget. Deltas are characterized by broad wedges of sediment deposition. Delta morphology tends to reflect the processes that are dominant (**Figure 7**). Where wave energy is low and tides are minor, it is primarily river flows which account for sedimentation patterns. River flow may be likened to a jet, influenced by inertial forces, friction with the basin floor or buoyancy where there are density differences between outflow and the water of the receiving basin.

Wave action tends to smooth the shoreline and a wave-dominated deltaic shoreline will be characterized by shore-parallel bars or ridges. Where river-supplied sediment is relatively low in volume, wave action may form a sandy barrier along the coast and rivers may supply sediment only to lagoons formed behind these barriers. Such is the case for much of the barrier-island shoreline of the eastern shores of the Americas; where barrier islands may be continually reworked landwards during relative sea-level rise. On the other hand if sea level is stable, a stable sand barrier is likely to form closing each

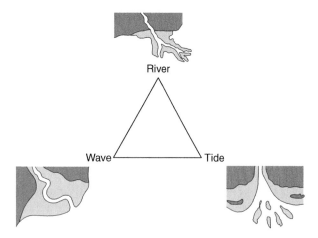

Figure 7 Deltas adopt a variety of forms, but morphology appears to reflect the relative balance of river, wave, and tide action. The broad morphology of the delta tends to be digitate where river-dominated, shore-parallel where wave-dominated, and tapering where tide-dominated.

embayment or creating a barrier estuary as in the case of coastal lagoons in Asia and Australasia.

Estuaries are embayments which are likely to infill incrementally either through the deposition of river-borne sediments or through the influx of sediment from seaward by wave or tidal processes (**Figure 8**). The sediment from seaward may either be derived from the shelf, or from shoreline erosion.

Tidal processes differ from river processes; they are bi-directional, flowing in during flood tide and out during ebb, and the flow is forced by the rising level of the sea. Small embayments are flooded and drained by a tidal prism (the volume of water between low and high tide). Longer estuaries, on the other hand, may have a series of tidal waves which progress up them. Where the tidal range is large this may flow as a tidal bore. Tide-dominated estuarine channels adopt a distinctive tapered form, with width (and depth) decreasing from the mouth upstream (**Figure 8**).

Wave and tidal processes tend to shape deltas and estuaries to varying degrees depending on their relative operation. Many of the deltaic-estuarine processes operate over cycles of change. The hydraulic efficiency of distributaries decreases as they lengthen, until an alternative, shorter course with steeper hydraulic gradient is adopted. The abandoned distributaries of deltas often become tidally dominated with sinuous, tapering tidal creeks dominating what may be a gradually subsiding abandoned delta plain. Wave processes, in wave-dominated settings, smooth and rework the abandoned delta shore, often forming barrier islands.

Muddy Coasts

Muddy coasts are associated with the lowest energy environments. Mud banks may occur in high-energy, wave-exposed settings where large volumes of mud are supplied to the mouth of large rivers. Thus longshore drift north west of the Amazon, and around Bohai Bay downdrift from the Yellow River, enables mud-shoal deposition in open-water settings. Elsewhere mud flats are typical of sheltered settings, within delta interdistributary bays, around coastal lagoons, etc. These muddy environments are likely to be colonized by halophytic vegetation. Mangrove forests occur in tropical settings, whereas salt marsh occurs in higher latitudes, often extending into tropical areas also.

These coastal wetlands further promote retention of fine-grained sediment. The muddy environments are areas of complex hydrodynamics and sedimentation. Sedimentation is likely to occur with a negative feedback such that as the tidal wetlands

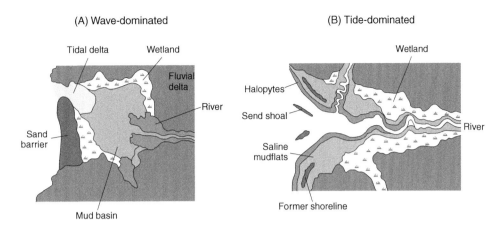

Figure 8 Estuaries are broad embayments which may adopt a wide range of morphologies. They tend to show sand-barrier accumulation where wave-dominated (A), but be prominently tapering where tide-dominated (B). Estuaries are generally sediment sinks with the rates and patterns of infill reflecting the relative dominance of river, wave, and tide processes and sediment sources.

accrete sediment and as the substrate is elevated, they are flooded less frequently and therefore sedimentation decelerates. Boundary conditions, particularly sea level, are likely to vary at rates similar to the rate of sedimentation and prograded coastal plains contain complex sedimentary records of changes in ecological and geomorphological state.

Conclusion

Coastal geomorphology is the study of the evolving form of the shoreline in response to mutually adjusting processes acting upon it. It spans instantaneous and event timescales, over which beaches respond to wave energy, through engineering and geological timescales, over which deltas build seaward or switch distributaries, sea level fluctuates, and cliff morphology evolves. The morphology (state) of the coast changes in response to perturbations, particularly extreme events such as storms, but also thresholds within the system (as when a cliff oversteepens and falls, or a distributary lengthens and then switches). Human action may also represent a perturbation to the system. In each coastal setting, the influence of human modifications is being felt. Thus there are fewer coasts which are not in some way influenced by society. As anthropogenic modification of climate and sea level occurs at a global scale, the human factor increasingly needs to be given prominence in coastal geomorphology.

See also

Beaches, Physical Processes Affecting. Coral Reefs. Rocky Shores. Salt Marshes and Mud Flats.

Further Reading

Boyd R, Dalrymple R, and Zaitlin BA (1992) Classification of clastic coastal depositional environments. *Sedimentary Geology* 80: 139–150.

Carter RWG and Woodroffe CD (eds.) (1994) *Coastal Evolution: Late Quaternary Shoreline Morphodynamics*. Cambridge: Cambridge University Press.

Cowell PJ and Thom BG (1994) Morphodynamics of coastal evolution. In: Carter RWG and Woodroffe CD (eds.) *Coastal Evolution: Late Quaternary Shoreline Morphodynamics*, pp. 33–86. Cambridge: Cambridge University Press.

Darwin C (1842) *The Structure and Distribution of Coral Reefs*. London: Smith Elder.

Davis RA (1985) *Coastal Sedimentary Environments*. New York: Springer-Verlag.

Johnson DW (1919) *Shore Processes and Shoreline Development*. New York: Prentice Hall.

Trenhaile AS (1997) *Coastal Dynamics and Landforms*. Oxford: Clarendon Press.

Wright LD and Thom BG (1977) Coastal depositional landforms: a morphodynamic approach. *Progress in Physical Geography* 1: 412–459.

PLANKTON AND BENTHOS

MARINE PLANKTON COMMUNITIES

G.-A. Paffenhöfer, Skidaway Institute of
Oceanography, Savannah, GA, USA

Introduction

By definition, a community is an interacting population of various kinds of individuals (species) in a
common location (*Webster's Collegiate Dictionary*,
1977).

The objective of this article is to provide general
information on the composition and functioning of
various marine plankton communities, which is accompanied by some characteristic details on their
dynamicism.

General Features of a Plankton Community

The expression 'plankton community' implies that
such a community is located in a water column. It
has a range of components (groups of organisms)
that can be organized according to their size. They
range in size from tiny single-celled organisms such
as bacteria (0.4–1-μm diameter) to large predators
like scyphomedusae of more than 1 m in diameter.
A common method which has been in use for
decades is to group according to size, which here is
attributed to the organism's largest dimension; thus
the organisms range from picoplankton to macroplankton (**Figure 1**). It is, however, the smallest
dimension of an organism which usually determines whether it is retained by a mesh, since in a
flow, elongated particles align themselves with the
flow.

A plankton community is operating/functioning
continuously, that is, physical, chemical, and biological variables are always at work. Interactions
among its components occur all the time. As one
well-known fluid dynamicist stated, "The surface of
the ocean can be flat calm but below that surface
there is always motion of the water at various
scales." Many of the particles/organisms are moving
or being moved most of the time: Those without
flagella or appendages can do so due to processes
within or due to external forcing, for example, from
water motion due to internal waves; and those with
flagella/cilia or appendages or muscles move or create motion of the water in order to exist. Oriented
motion is usually in the vertical which often results

in distinct layers of certain organisms. However,
physical variables also, such as light or density differences of water masses, can result in layering of
planktonic organisms. Such layers which are often
horizontally extended are usually referred to as
patches.

As stated in the definition, the components of a
plankton community interact. It is usually the case
that a larger organism will ingest a smaller one or a
part of it (**Figure 1**). However, there are exceptions.
The driving force for a planktonic community originates from sun energy, that is, primary productivity's
(1) direct and (2) indirect products: (1) autotrophs
(phytoplankton cells) which can range from near 2 to
more than 300-μm width/diameter, or chemotrophs;
and (2) dissolved organic matter, most of which is
released by phytoplankton cells and protozoa as
metabolic end products, and being taken up by bacteria and mixo- and heterotroph protozoa (**Figure 1**).
These two components mainly set the microbial
loop (ML) in motion; that is, unicellular organisms
of different sizes and behaviors (auto-, mixo-, and
heterotrophs) depend on each other – usually,
but not always, the smaller being ingested by the
larger. Most of nutrients and energy are recirculated
within this subcommunity of unicellular organisms in
all marine regions of our planet (for more details,
especially the ML). These processes of the ML dominate the transfer of energy in all plankton communities largely because the processes (rates of
ingestion, growth, reproduction) of unicellular heterotrophs almost always outpace those of phytoplankton, and also of metazooplankton taxa at most
times.

The main question actually could be: "What is the
composition of plankton communities, and how do
they function?" **Figure 1** reveals sizes and relationships within a plankton community including the
ML. It shows the so-called 'bottom-up' and 'top-
down' effects as well as indirect effects like the
above-mentioned labile dissolved organic matter
(labile DOM), released by auto- and also by heterotrophs, which not only drives bacterial growth but
can also be taken up or used by other protozoa.
There can also be reversals, called two-way processes. At times a predator eating an adult metazoan
will be affected by the same metazoan which is able
to eat the predator's early juveniles (e.g., well-grown
ctenophores capturing adult omnivorous copepods
which have the ability to capture and ingest very
young ctenophores).

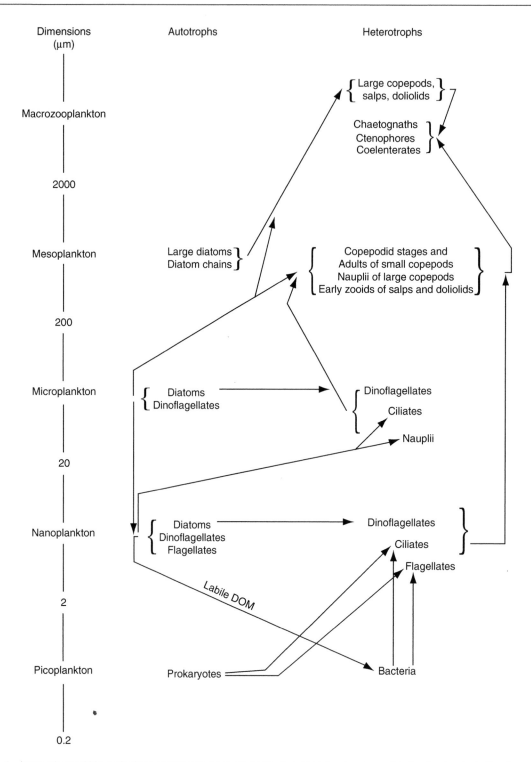

Figure 1 Interactions within a plankton community separated into size classes of auto- and heterotrophs, including the microbial loop; the arrows point to the respective grazer, or receiver of DOM; the figure is partly related to figure 9 from Landry MR and Kirchman DL (2002) Microbial community structure and variability in the tropical Pacific. *Deep-Sea Research II* 49: 2669–2693.

To comprehend the functioning of a plankton community requires a quantitative assessment of the abundances and activities of its components. First, almost all of our knowledge to date stems from *in situ* sampling, that is, making spot measurements of the abundance and distribution of organisms in the water column. The accurate determination of abundance and distribution requires using meshes or devices which quantitatively collect the respective organisms. Because of methodological difficulties and insufficient comprehension of organisms' sizes and activities, quantitative sampling/quantification of a community's main components has been often inadequate. The following serves as an example of this. Despite our knowledge that copepods consist of 11 juvenile stages aside of adults, the majority of studies of marine zooplankton hardly considered the juveniles' significance and this manifested itself in sampling with meshes which often collected merely the adults quantitatively. Second, much knowledge on rate processes comes from quantifying the respective organisms' activities under controlled conditions in the laboratory. Some *in situ* measurements (e.g., of temperature, salinity, chlorophyll concentrations, and acoustic recordings of zooplankton sizes) have been achieved 'continuously' over time, resulting in time series of increases and decreases of certain major community components. To date there are few, if any, direct *in situ* observations on the activity scales of the respective organisms, from bacteria to proto- and to metazooplankton, mainly because of methodological difficulties. In essence, our present understanding of processes within plankton communities is incomplete.

Specific Plankton Communities

We will provide several examples of plankton communities of our oceans. They will include information about the main variables affecting them, their main components, partly their functioning over time, including particular specifics characterizing each of those communities.

In this section, plankton communities are presented for three different types of marine environments: estuaries/inshore, continental shelves, and open ocean regions.

Estuaries

Estuaries and near-shore regions, being shallow, will rapidly take up and lose heat, that is, will be strongly affected by atmospheric changes in temperature, both short- and long-term, the latter showing in the seasonal extremes ranging from 2 to 32 °C in estuaries of North Carolina. Runoff of fresh water, providing continuous nutrient input for primary production, and tides contribute to rapid changes in salinity. This implies that resident planktonic taxa ought to be eurytherm as well as – therm. Only very few metazooplanktonic species are able to exist in such an environment (**Table 1**). In North Carolinian

estuaries, representative of other estuaries, they are the copepod species *Acartia tonsa*, *Oithona oculata*, and *Parvocalanus crassirostris*. In estuaries of Rhode Island, two species of the genus *Acartia* occur. During colder temperatures *Acartia hudsonica* produces dormant eggs as temperatures increase and then is replaced by *A. tonsa*, which produces dormant eggs once temperatures again decrease later in the year. Such estuaries are known for high primary productivity, which is accompanied by high abundances of heterotroph protozoa preying on phytoplankton. Such high abundances of unicellular organisms imply that food is hardly limiting the growth of the above-mentioned copepods which can graze on auto- as well as heterotrophs. However, such estuaries are often nursery grounds for juvenile fish like menhaden which prey heavily on late juveniles and adults of such copepods, especially *Acartia*, which is not only the largest of those three dominant copepod species but also moves the most, and thus can be seen most easily by those visual predators. This has resulted in diurnal migrations mostly of their adults, remaining at the seafloor during the day where they hardly eat, thus avoiding predation by such visual predators, and only entering the water column during dark hours. That then is their period of pronounced feeding. The other two species which are not heavily preyed upon by juvenile fish, however, can be affected by the co-occurring *Acartia*, because from early copepodid stages on this genus can be strongly carnivorous, readily preying on the nauplii of its own and of those other species.

Nevertheless, the usually continuous abundance of food organisms for all stages of the three copepod species results in high concentrations of nauplii which in North Carolinian estuaries can reach $100 \, l^{-1}$, as can their combined copepodid stages. The former is an underestimate, because sampling was done with a 75-μm mesh, which is passed through by most of those nauplii. By comparison, in an estuary on the west coast of Japan (Yellow Sea), dominated also by the genera *Acartia*, *Oithona*, and *Paracalanus* and sampling with 25-μm mesh, nauplius concentrations during summer surpassed $700 \, l^{-1}$, mostly from the genus *Oithona*. And copepodid stages plus adults repeatedly exceeded $100 \, l^{-1}$. Here sampling with such narrow mesh ensured that even the smallest copepods were collected quantitatively.

In essence, estuaries are known to attain among the highest concentrations of proto- and metazooplankton. The known copepod species occur during most of the year, and are observed year after year which implies persistence of those species beyond decades.

Table 1 Some characteristics of marine plankton communities

| | Estuaries | Shelves | Open ocean gyres | | Epipelagic subtropical |
			Subarctic Pacific	Boreal Atlantic	Atlantic/Pacific
Physical variables	Wide range of temperature and salinity	Intermittent and seasonal atmospheric forcing	Steady salinity, seasonal temp. variability	Major seasonal variability of temperature	Steady temperature and salinity, continuous atmospheric forcing
Nutrient supply	Continuous	Episodic	Seasonal	Seasonal	Occasional
Phytoplankton abundance	High from spring to autumn	Intermittently high	Always low	Major spring bloom	Always low
Phytoplankton composition	Flagellates, diatoms	Flagellates, diatoms, dinoflagellates	Nanoflagellates	Spring: diatoms Other: mostly nanoplankton	Mostly prokaryotes, small nano- and dinoflagellates
Primary Productivity	High at most times	Intermittently high	Maximum in spring	Max. in spring and autumn	Always low
No. of metazoan species	≤5	~10–30	>10	>20	>100
Seasonal variability of metazoan abundance	High spring and summer, low winter	Highly variable	High	High	Low
Copepod Ranges	$N^a \sim 10$–$500\,l^{-1}$	<5–$50\,l^{-1}$			3–$10\,l^{-1}$
Abundance	$Cop^b \sim 5$–$100\,l^{-1}$	<3–$30\,l^{-1}$	Up to $1000\,m^{-3}$ Neocalanus	Up to $1000\,m^{-3}$ C. finmarchicus	300–$1000\,m^{-3}$
Dominant metazooplankton taxa	Acartia Oithona Parvocalanus	Oithona Paracalanus Temora Doliolida	Neocalanus Oithona Metridia	Calanus Oithona Oncaea	Oithona Clausocalanus Oncaea

[a] Nauplii.
[b] Copepodids and adult copepods.

Continental Shelves

By definition they extend to the 200-m isobath, and range from narrow (few kilometers) to wide (more than 100-km width). The latter are of interest because the former are affected almost continuously and entirely by the nearby open ocean. Shelves are affected by freshwater runoff and seasonally changing physical variables. Water masses on continental shelves are evaluated concerning their residence time, because atmospheric events sustained for more than 1 week can replace most of the water residing on a wide shelf with water offshore but less so from near shore. This implies that plankton communities on wide continental shelves, which are often near boundary currents, usually persist for limited periods of time, from weeks to months (**Table 1**). They include shelves like the Agulhas Bank, the Campeche Banks/Yucatan Shelf, the East China Sea Shelf, the East Australian Shelf, and the US southeastern continental shelf. There can be a continuous influx year-round of new water from adjacent boundary currents as seen for the Yucatan Peninsula and Cape Canaveral (Florida). The momentum of the boundary current (here the Yucatan Current and Florida Current) passing a protruding cape will partly displace water along downstream-positioned diverging isobaths while the majority will follow the current's general direction. This implies that upstream-produced plankton organisms can serve as seed populations toward developing a plankton community on such wide continental shelves.

Whereas estuarine plankton communities receive almost continuously nutrients for primary production from runoff and pronounced benthic-pelagic coupling, those on wide continental shelves infrequently receive new nutrients. Thus they are at most times a heterotroph community unless they obtain nutrients from the benthos due to storms, or receive episodically input of cool, nutrient-rich water from greater depths of the nearby boundary current as can be seen for the US SE shelf. Passing along the outer shelf at about weekly intervals are nutrient-rich cold-core Gulf Stream eddies which contain plankton organisms from the highly productive Gulf of Mexico. Surface winds, displacing shelf surface water offshore, lead to an advance of the deep cool water onto the shelf which can be flooded entirely by it. Pronounced irradiance and high-nutrient loads in such upwellings result in phytoplankton blooms which then serve as a food source for protozoo- and metazooplankton. Bacteria concentrations in such cool water masses increase within several days by 1 order of magnitude. Within 2–3 weeks most of the smaller phytoplankton (c. <20-μm width) has been greatly reduced, usually due to grazing by protozoa and relatively slow-growing assemblages of planktonic copepods of various genera such as *Temora*, *Oithona*, *Paracalanus*, *Eucalanus*, and *Oncaea*. However, quite frequently, the Florida Current which becomes the Gulf Stream carries small numbers of Thaliacea (Tunicata), which are known for intermittent and very fast asexual reproduction. Such salps and doliolids, due to their high reproductive and growth rate, can colonize large water masses, the latter increasing from ~5 to >500 zooids per cubic meter within 2 weeks, and thus form huge patches, covering several thousands of square kilometers, as the cool bottom water is displaced over much of the shelf. The increased abundance of salps (usually in the warmer and particle-poor surface waters) and doliolids (mainly in the deeper, cooler, particle-rich waters, also observed on the outer East China shelf) can control phytoplankton growth once they achieve bloom concentrations. The development of such large and dense patches is partly due to the lack of predators.

Although the mixing processes between the initially quite cool intruding bottom (13–20 °C) and the warm, upper mixed layer water (27–28 °C) are limited, interactions across the thermocline occur, thus creating a plankton community throughout the water column of previously resident and newly arriving components. The warm upper mixed layer often has an extraordinary abundance of early copepodid stages of the poecilostomatoid copepod *Oncaea*, thanks to their ontogenetical migration after having been released by the adult females which occur exclusively in the cold intruding water. Also, early stages of the copepod *Temora turbinata* are abundant in the warm upper mixed layer; while *T. turbinata*'s late juvenile stages prefer the cool layer because of the abundance of large, readily available phytoplankton cells. As in estuaries, the copepod genus *Oithona* flourishes on warm, temperate, and polar continental shelves throughout most of the euphotic zone.

Such wide subtropical shelves will usually be well mixed during the cooler seasons, and then harbor, due to lower temperatures, fewer metazooplankton species which are often those tolerant of wider or lower temperature ranges. Such wide shelves are usually found in subtropical regions, which explains the rapidity of the development of their plankton communities. They, however, are also found in cooler climates, like the wide and productive Argentinian/Brazilian continental shelf about which our knowledge is limited. Other large shelves, like the southern North Sea, have a limited exchange of water with the open ocean but at the same time considerable influx of runoff, plus nutrient supply from the benthos due to storm events, and thus can maintain identical plankton communities over months and seasons.

In essence, continental shelf plankton communities are usually relatively short-lived, which is largely due to their water's limited residence time.

Open Ocean

The open ocean, even when not including ocean margins (up to 1000-m water column), includes by far the largest regions of the marine environment. Its deep-water columns range from the polar seas to the Tropics. All these regions are under different atmospheric and seasonal regimes, which affect plankton communities. Most of these communities are seasonally driven and have evolved along the physical conditions characterizing each region. The focus here is on gyres as they represent specific ocean communities whose physical environment can be readily presented.

Gyres represent huge water masses extending horizontally over hundreds to even thousands of kilometers in which the water moves cyclonically or anticyclonically. They are encountered in subpolar, temperate, and subtropical regions. The best-studied ones are:

- subpolar: Alaskan Gyre;
- temperate: Norwegian Sea Gyre, Labrador–Irminger Sea Gyre;
- subtropical: North Pacific Central Gyre (NPCG), North Atlantic Subtropical Gyre (NASG).

The Alaskan Gyre is part of the subarctic Pacific (**Table 1**) and is characterized physically by a shallow halocline (~110-m depth) which prevents convective mixing during storms. Biologically it is characterized by a persistent low-standing stock of phytoplankton despite high nutrient abundance, and several species of large copepods which have evolved to persist via a life cycle as shown for *Neocalanus plumchrus*. By midsummer, fifth copepodids (C5) in the upper 100 m which have accumulated large amounts of lipids begin to descend to greater depths of 250 m and beyond undergoing diapause, and eventually molt to females which soon begin to spawn. Spawning females are found in abundance from August to January. Nauplii living off their lipid reserves and moving upward begin to reach surface waters by mid- to late winter as copepodid stage 1 (C1), and start feeding on the abundant small phytoplankton cells (probably passively by using their second maxillae, but mostly by feeding actively on heterotrophic protozoa which are the main consumers of the tiny phytoplankton cells). The developing copepodid stages accumulate lipids

which in C5 can amount to as much as 50% of their body mass, which then serve as the energy source for metabolism of the females at depth, ovary development, and the nauplii's metabolism plus growth. While the genus *Neocalanus* over much of the year provides the highest amount of zooplankton biomass, the cyclopoid *Oithona* is the most abundant metazooplankter; other abundant metazooplankton taxa include *Euphausia pacifica*, and in the latter part of the year *Metridia pacifica* and *Calanus pacificus*.

In the temperate Atlantic (**Table 1**), the Norwegian Sea Gyre maintains a planktonic community which is characterized, like much of the temperate oceanic North Atlantic, by the following physical features. Pronounced winds during winter mix the water column to beyond 400-m depth, being followed by lesser winds and surface warming resulting in stratification and a spring bloom of mostly diatoms, and a weak autumn phytoplankton bloom. A major consumer of this phytoplankton bloom and characteristic of this environment is the copepod *Calanus finmarchicus*, occurring all over the cool North Atlantic. This species takes advantage of the pronounced spring bloom after emerging from diapause at >400-m depth, by moulting to adult, and grazing of females at high clearance rates on the diatoms, right away starting to reproduce and releasing up to more than 2000 fertilized eggs during their lifetime. Its nauplii start to feed as nauplius stage 3 (N3), being able to ingest diatoms of similar size as the adult females, and can reach copepodid stage 5 (C5) within about 7 weeks in the Norwegian Sea, accumulating during that period large amounts of lipids (wax ester) which serve as the main energy source for the overwintering diapause period. Part of the success of C. finmarchicus is found in its ability of being omnivorous. C5s either descend to greater depths and begin an extended diapause period, or could moult to adult females, thus producing another generation which then initiates diapause at mostly C5. Its early to late copepodid stages constitute the main food for juvenile herring which accumulate the copepods' lipids for subsequent overwintering and reproduction. Of the other copepods, the genus *Oithona* together with the poecilostomatoid *Oncaea* and the calanoid *Pseudocalanus* were the most abundant.

Subtropical and tropical parts of the oceans cover more than 50% of our oceans. Of these, the NPCG, positioned between *c.* 10° and 45° N and moving anticyclonically, has been frequently studied. It includes a southern and northern component, the latter being affected by the Kuroshio and westerly winds, the former by the North Equatorial Current and the trade winds. Despite this, the NPCG has been considered as an ecosystem as well as a huge plankton community. The NASG, found between *c.* 15° and 40° N and moving anticyclonically, is of similar horizontal dimensions. There are close relations between subtropical and tropical communities; for example, the Atlantic south of Bermuda is considered close to tropical conditions. Vertical mixing in both gyres is limited. Here we focus on the epipelagic community which ranges from the surface to about 150-m depth, that is, the euphotic zone. The epipelagial is physically characterized by an upper mixed layer of *c.* 15–40 m of higher temperature, below which a thermocline with steadily decreasing temperatures extends to below 150-m depth. In these two gyres, the concentrations of phytoplankton hardly change throughout the year in the epipelagic (**Table 1**) and together with the heterotrophic protozoa provide a low and quite steady food concentration (**Table 1**) for higher trophic levels. Such very low particle abundances imply that almost all metazooplankton taxa depending on them are living on the edge, that is, are severely food-limited. Despite this fact, there are more than 100 copepod species registered in the epipelagial of each of the two gyres. How can that be? Almost all these copepod species are small and rather diverse in their behavior: the four most abundant genera have different strategies to obtain food particles: the intermittently moving *Oithona* is found in the entire epipelagial and depends on moving food particles (hydrodynamic signals); *Clausocalanus* is mainly found in the upper 50 m of the epipelagial and always moves at high speed, thus encountering numerous food particles, mainly via chemosensory; *Oncaea* copepodids and females occur in the lower part of the epipelagial and feed on aggregates; and the feeding-current producing *Calocalanus* perceives particles via chemosensory. This implies that any copepod species can persist in these gyres as long as it obtains sufficient food for growth and reproduction. This is possible because protozooplankton always controls the abundance of available food particles; thus, there is no competition for food among the metazooplankton. In addition, since total copepod abundance (quantitatively collected with a 63-μm mesh by three different teams) is steady and usually $< 1000 \, \text{m}^{-3}$ including copepodid stages (pronounced patchiness of metazooplankton has not yet been observed in these oligotrophic waters), the probability of encounter (only a minority of the zooplankton is carnivorous on metazooplankton) is very low, and therefore the probability of predation low within the metazooplankton. In summary, these steady conditions make it possible that in the epipelagial more than 100 copepod species can coexist, and are in a steady state throughout much of the year.

Conclusions

All epipelagic marine plankton communities are at most times directly or indirectly controlled or affected by the activity of the ML, that is, unicellular organisms. Most of the main metazooplankton species are adapted to the physical and biological conditions of the respective community, be it polar, subpolar, temperate, subtropical, or tropical. The only metazooplankton genus found in all communities mentioned above, and also all other studied marine plankton communities, is the copepod genus *Oithona*. This copepod has the ability to persist under adverse conditions, for example, as shown for the subarctic Pacific. This genus can withstand the physical as well as biological (predation) pressures of an estuary, the persistent very low food levels in the warm open ocean, and the varying conditions of the Antarctic Ocean. Large copepods like the genus *Neocalanus* in the subarctic Pacific, and *C. finmarchicus* in the temperate to subarctic North Atlantic are adapted with respective distinct annual cycles in their respective communities. Among the abundant components of most marine plankton communities from near shore to the open ocean are appendicularia (Tunicata) and the predatory chaetognaths.

Our present knowledge of the composition and functioning of marine planktonic communities derives from (1) oceanographic sampling and time series, optimally accompanied by the quantification of physical and chemical variables; and (2) laboratory/onboard experimental observations, including some time series which provide results on small-scale interactions (microns to meters; milliseconds to hours) among components of the community. Optimally, direct *in situ* observations on small scales in conjunction with respective modeling would provide insights in the true functioning of a plankton community which operates continuously on scales of milliseconds and larger.

Our future efforts are aimed at developing instrumentation to quantify *in situ* interactions of the various components of marine plankton communities. Together with traditional oceanographic methods we would go 'from small scales to the big picture', implying the necessity of understanding the functioning on the individual scale for a comprehensive understanding as to how communities operate.

See also

Chemical Processes in Estuarine Sediments. Estuarine Circulation. Gas Exchange in Estuaries.

Further Reading

Atkinson LP, Lee TN, Blanton JO, and Paffenhöfer G-A (1987) Summer upwelling on the southeastern continental shelf of the USA during 1981: Hydrographic observations. *Progress in Oceanography* 19: 231–266.

Fulton RS, III (1984) Distribution and community structure of estuarine copepods. *Estuaries* 7: 38–50.

Hayward TL and McGowan JA (1979) Pattern and structure in an oceanic zooplankton community. *American Zoologist* 19: 1045–1055.

Landry MR and Kirchman DL (2002) Microbial community structure and variability in the tropical Pacific. *Deep-Sea Research II* 49: 2669–2693.

Longhurst AR (1998) *Ecological Geography of the Sea*, 398pp. San Diego, CA: Academic Press.

Mackas DL and Tsuda A (1999) Mesozooplankton in the eastern and western subarctic Pacific: Community structure, seasonal life histories, and interannual variability. *Progress in Oceanography* 43: 335–363.

Marine Zooplankton Colloquium 1 (1998) Future marine zooplankton research – a perspective. *Marine Ecology Progress Series* 55: 197–206.

Menzel DW (1993) *Ocean Processes: US Southeast Continental Shelf*, 112pp. Washington, DC: US Department of Energy.

Miller CB (1993) Pelagic production processes in the subarctic Pacific. *Progress in Oceanography* 32: 1–15.

Miller CB (2004) *Biological Oceanography*, 402pp. Boston: Blackwell.

Paffenhöfer G-A and Mazzocchi MG (2003) Vertical distribution of subtropical epiplanktonic copepods. *Journal of Plankton Research* 25: 1139–1156.

Paffenhöfer G-A, Sherman BK, and Lee TN (1987) Summer upwelling on the southeastern continental shelf of the USA during 1981: Abundance, distribution and patch formation of zooplankton. *Progress in Oceanography* 19: 403–436.

Paffenhöfer G-A, Tzeng M, Hristov R, Smith CL, and Mazzocchi MG (2003) Abundance and distribution of nanoplankton in the epipelagic subtropical/tropical open Atlantic Ocean. *Journal of Plankton Research* 25: 1535–1549.

Smetacek V, DeBaar HJW, Bathmann UV, Lochte K, and Van Der Loeff MMR (1997) Ecology and biogeochemistry of the Antarctic Circumpolar Current during austral spring: A summary of Southern Ocean JGOFS cruise ANT X/6 of RV *Polarstern*. *Deep-Sea Research II* 44: 1–21 (and all articles in this issue).

Speirs DC, Gurney WSC, Heath MR, Horbelt W, Wood SN, and de Cuevas BA (2006) Ocean-scale modeling of the distribution, abundance, and seasonal dynamics of the copepod *Calanus finmarchicus*. *Marine Ecology Progress Series* 313: 173–192.

Tande KS and Miller CB (2000) Population dynamics of *Calanus* in the North Atlantic: Results from the Trans-Atlantic Study of *Calanus finmarchicus*. *ICES Journal of Marine Science* 57: 1527 (entire issue).

Webber MK and Roff JC (1995) Annual structure of the copepod community and its associated pelagic environment off Discovery Bay, Jamaica. *Marine Biology* 123: 467–479.

BENTHIC ORGANISMS OVERVIEW

P. F. Kingston, Heriot-Watt University, Edinburgh, UK

Introduction

The term benthos is derived from the Greek word βαος (vathos, meaning depth) and refers to those organisms that live on the seabed and the bottom of rivers and lakes. In the oceans the benthos extends from the deepest oceanic trench to the intertidal spray zone. It includes those organisms that live in and on sediments, those that inhabit rocky substrata and those that make up the biodiversity of coral reefs.

The benthic environment, sometimes referred to as the benthal, may be divided up into various well defined zones that seem to be distinguished by depth (**Figure 1**).

Physical Conditions Affecting the Benthos

In most parts of the world the water level of the upper region of the benthal fluctuates, so that the animals and plants are subjected to the influence of the water only at certain times. At the highest level, only spray is involved; in the remainder of the region, the covering of water fluctuates as a result of tides and wind and other atmospheric factors. Below the level of extreme low water, the seafloor is permanently covered in water.

Water pressure increases rapidly with depth. Pressure is directly related to depth and increases by one atmospheric (101 kPa) per 10 m. Thus at 100 m, water pressure would be 11 atmospheres and at 10 000 m, 1001 atmospheres. Water is essentially incompressible, so that the size and shape of organisms are not affected by depth, providing the species does not possess a gas space and move between depth zones.

Light is essential to plants and it is its rapid attenuation with increasing water depth that limits the distribution of benthic flora to the coastal margins. Daylight is reduced to 1% of its surface values between 10 m and 30 m in coastal waters; in addition, the spectral quality of the light changes with depth, with the longer wavelengths being absorbed first. This influences the depth zonation of plant species, which is partially based on photosynthetic pigment type.

Surface water temperatures are highest in the tropics, becoming gradually cooler toward the higher latitudes. Diurnal changes in temperature are confined to the uppermost few meters and are usually quite small (3°C); however, water temperature falls with increasing depth to between 0.5°C and 2°C in the abyssal zone.

The average salinity of the oceans is 34.7 ppt. Salinity values in the deep ocean remain very near to this value. However, surface water salinity may vary considerably. In some enclosed areas, such as the Baltic Sea, salinity may be as low as 14 ppt, while in

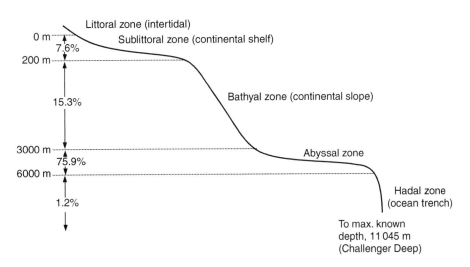

Figure 1 Classification of benthic environment with percentage representation of each depth zone.

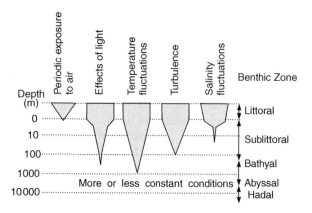

Figure 2 Diagram summarizing the influence of physical factors on the benthos. Relative influence of each factor indicated by width of shaded area.

areas of high evaporation, such as the Arabian Gulf, salinities in excess of 50 ppt may be reached. Salinity profiles may become quite complex in estuarine conditions or in coastal regions where there is a high fresh-water run off from the land (e.g., fiordic conditions) (**Figure 2**).

The composition of the benthos is profoundly affected by the nature of the substratum. Hard substrata tend to be dominated by surface dwelling forms, providing a base for the attachment of sessile animals and plants and a large variety of micro-habitats for organisms of cryptic habits. In contrast, sedimentary substrata are dominated by burrowing organisms and, apart from the intertidal zone and the shallowest waters, they are devoid of plants.

Hard substrata are most common in coastal waters where there are strong tidal currents and surface turbulence. Farther offshore, and in the deep sea, the seabed is dominated by sediments, hard substrata occurring only on the slopes of seamounts, oceanic trenches, and other irregular features where the gradient of the seabed is too steep to permit significant accumulation of sedimentary material. Thus, viewed in its entirety, the seabed can be considered predominantly a level-bottom sedimentary environment (**Figure 3**).

Classification of the Benthos

It was the Danish marine scientist, Petersen working in the early part of the twentieth century, who first defined the two principal groups of benthic animals:

● the epifauna, comprising all animals living on or associated with the surface of the substratum, including sediments, rocks, shells, reefs and vegetation;
● the infauna comprising all animals living within the substratum, either moving freely through it or living in burrows and tubes or between the grains of sediments.

According to the great benthic ecologist Gunnar Thorson, the epifauna occupies less than 10% of the total area of the seabed, reaching its maximum abundance in the shallow waters and intertidal zones of tropical regions. However, the infauna, which Thorson believed occupies more than half the surface area of the planet, is most fully developed sublittorally. Nevertheless, the number of epifaunal species is far greater than the number of infaunal species. This is because the level-bottom habitat

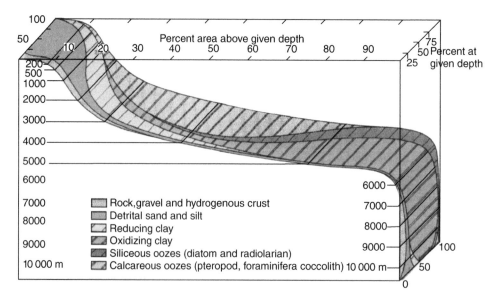

Figure 3 Distribution of sediment types in the ocean. (After Brunn, in Hedgepeth (1957).)

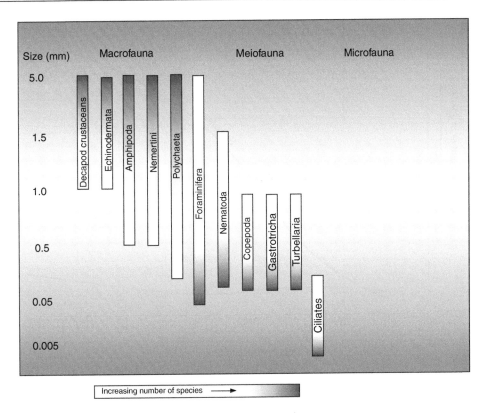

Figure 4 Principal taxa associated with each of the major categories of infauna.

provides a more uniform environment than hard substrata, with fewer potential habitat types to support a wide diversity of species. The epifauna of polar regions are not generally as well developed as those in the lower latitudes because of the effects of low temperatures in shallow waters and the effects of ice and meltwater in the intertidal region. The infauna largely escape these effects, exhibiting less latitudinal variation in number of species (**Figure 4**).

The infauna is further classified on the basis of size, the size categories broadly agreeing with major taxa that characterize the groups (**Figure 5**):

- macrofauna – animals that are retained on a 0.5 mm aperture sieve;
- meiofauna – animals that pass a 0.5 mm sieve but are retained on 0.06 mm sieve;
- microfauna – animals that pass a 0.06 mm sieve.

Petersen was also the first marine biologist to quantitatively sample soft-bottom habitats. After examining hundreds of samples from Danish coastal waters, he was struck by the fact that extensive areas of seabed were occupied by recurrent groups of species. These assemblages differed from area to area, in each case only a few species making up the bulk of the individuals and biomass. This contrasted with nonquantitative epifaunal dredge samples taken over the same range of areas, for which faunal lists

might be almost identical. Petersen proposed the infaunal assemblages that he had distinguished as communities and named them on the basis of the most visually dominant animals (**Table 1**).

Following the publication of Petersen's work in 1911–12, other marine biologists began to investigate benthic infauna quantitatively, and it began to emerge that there existed parallel bottom communities in which similar habitats around the world supported similar communities to those found by Petersen. These communities, although composed of different species, were closely similar, both ecologically and taxonomically, the characteristic species belonging to at least the same genus or a nearby taxon (**Table 2**).

Although the concept of parallel bottom communities appeared to hold good for temperate waters, in tropical regions such communities are not clearly definable, because of the presence of very large numbers of species and the small likelihood of any particular species or group of species dominating.

Not all benthic ecologists believe in Petersen's communities as functional biological units, since it is clear that abiotic factors such as sediment type must play a central role in determining species distribution. Alternative approaches have been proposed in which benthic associations are linked to

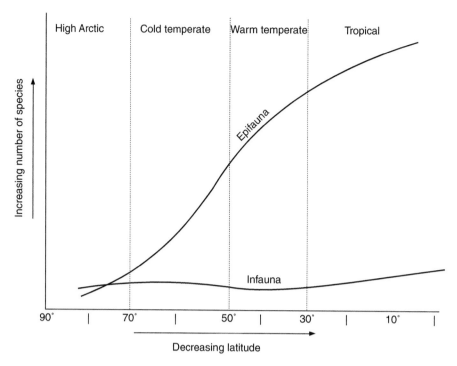

Figure 5 Relationship between numbers of epifaunal/infaunal species and latitude.

substratum (and latitude), suggesting that it is a common environmental requirement that forms the basis of the perceived community rather than an affinity of the members of the assemblage with one another.

Feeding Habits of Benthic Animals

Benthic animals, like all other animals, ultimately rely on the plant kingdom as their primary source of food. Most benthic animals are dependent on the rain of dead or partially decayed material (organic detritus) from above, and it is only in shallow coastal waters that macrophytes and phytoplankton are directly available for grazing or filtering by bottom-living forms. Much of this material consists of cellulose (dead plant cells) and chitin (crustacean exoskeletons). Few animals are able to digest these substances themselves and most rely on the action of bacteria to render them available, either as bacterial biomass or breakdown products.

Food particles are intercepted in the water before they reach the seabed, or are collected from the sediment surface after they have settled out, or are extracted from the sediment after becoming incorporated into it. These three scenarios reflect the three main types of detrivores: suspension feeders, selective deposit feeders, and direct deposit feeders.

Table 1 Petersen's benthic communities

Petersen's community	Typical species	Substratum
Macoma or Baltic community	Macoma balthica, Mya arenaria, Cerastoderma edule	Intertidal mud
Abra community	Abra alba, Macoma calcarea, Corbula gibba, Nephtys spp.	Inshore mud
Venus community	Venus striatula, Tellina fabula, Echinocardium cordatum	Offshore sand
Echinocardium–Filiformis community	E. cordatum, Amphiura filiformis, Cultellus pellucidus, Turritella communis	Offshore muddy sand
Brissopsis–Chiagei community	Brissopsis lyrifera, Amphiura chiagei, Calocaris macandreae	Offshore mud
Brissopsis–Sarsi community	B. lyrifera, Ophiura Sarsi, Abra nitida, Nucula tenuis	Deep mud
Amphilepis–Pecten community	Amphilepis norvegica, Chlamys (= Pecten) vitrea, Thyasira flexuosa	Deep mud
Haploops community	Haploops tubicola, Chlamys septemradiata, Lima loscombi	Offshore clay
Deep Venus community	Venus gallina, Spatangus purpureus, Abra prismatica	Deep sand

Table 2 Examples of parallel bottom communities identified by the Danish ecologist Gunnar Thorson

Species	Genera			
	NE Atlantic	NE Pacific	Arctic	NW Pacific
Macoma	baltica	nasuta	calcarea	incongrua
Cardium	edule	corbis	ciliatum	hungerfordi
Mya	arenaria	arenaria	truncata	
Arenicola	marina	claparedii	marina	

Table 3 Percentage representation of different feeding types from sandy and muddy sediments

Feeding types	Sandy sediment (< 15 silt/clay) (%)	Muddy sediment (> 90% silt/clay) (%)
Predators/omnivores	25.3	22.4
Suspension feeders	27.2	14.9
Selective deposit feeders	41.1	50.8
Direct deposit feeders	6.3	11.9

- Suspension feeders may be passive or active. Passive suspension feeders trap passing particles on extended appendages that are covered in sticky mucus and rely on natural water movements to bring the food to them (e.g., crinoid echinoderms). Active suspension feeders create a strong water current of their own, filtering out particles using specially modified organs (e.g., most bivalve mollusks).
- Selective deposit feeders either consume surface deposits in their immediate vicinity using unspecialized mouth parts, or, where food is less abundant, use extendable tentacles or siphons to pick up particles over a large area (e.g., terebellid polychaetes).
- Direct deposit feeders indiscriminately ingest sediment using organic matter and microbial organisms contained in the sediment as food. Polychaeta, such as the lugworm, *Arenicola*, construct L-shaped burrows and mine sediment from a horizontal gallery, reversing up the vertical shaft to defecate at the surface. Such animals play an important role in the physical turnover of the sediments.

Grazing or browsing animals are most common intertidally or in shallow waters. They are mobile consumers, cropping exposed tissues of sessile prey usually without killing the whole organism. They include animals that feed on macroalgae and those that feed on colonial cnidaria, bryozoans, and tunicates such as gastropod mollusks and echinoids. On the level bottom, the tips of tentacles and siphons of infaunal animals are grazed by demersal fish, providing a route for energy transfer from the seabed back into the pelagic system.

Predatory hunters are common among benthic epifauna and include crustaceans, asteroid echinoderms, and gastropod mollusks. These are mobile animals that seek out and consume individual prey items one at a time. Although less common, such predators are also found in the infauna, moving through the sediment, attacking their prey *in situ* (e.g., the polychaete *Glycera*).

Although, overall, deposit feeders form the largest single group of benthic infauna, the proportion of each trophic group is greatly influenced by the nature of the sediment. Thus in coarser sandy sediments, where water movement is relatively strong, the proportion of suspension feeders increases, while fine silts and muds are usually dominated by deposit feeders (see **Table 3**).

Spatial Distribution of Benthos

Competition between benthic infauna is usually for space. This is because most benthic animals are either suspension or deposit feeders and are competing for access to the same food source. In this respect, benthic communities are similar to those of terrestrial plants since, in both, competition between individuals is for an energy source that originates from above. Indeed, early approaches to the statistical analysis of benthic community structure were often rooted in principles originally developed by terrestrial botanists.

Suspension feeding may take place at more than one level. Some species, such as sabellid worms, extend their feeding organs several centimeters into the water column; some keep a lower profile, with short siphons projecting just a few millimeters above the sediment surface, while others have open burrows, drawing water into galleries below the surface. In addition, selective deposit feeders scour the sediment surface for food particles using palps or tentacles that in some species can extend up to a meter. The result is a contagious horizontal distribution of animals (i.e., neither random nor regular) that is maintained primarily by inter- and intra-specific competition for space.

Benthic infauna also show a marked vertical distribution. This is more a function of the subsurface physical conditions of the sediments and the need for the majority of species to be in communication with the surface than of competition between the animals.

Petersen chose physically large representatives to describe his benthic communities, primarily because they were easy to identify under field conditions. However, most benthic infaunal species are too small to be recognized with the naked eye, with burrows that penetrate no more than a few centimeters into the substratum. The result is that, for most level-bottom communities, some 95–99% of the animals are located within 5 cm of the sediment surface.

Reproduction in Benthic Animals

The act of reproduction offers benthic animals, the majority of which are either sessile or very restricted in their migratory powers, an opportunity to disperse and to colonize new ground. It is therefore not surprising that the majority of benthic species experience at least some sort of pelagic phase during their early development. Most invertebrates have larvae that swim for varying amounts of time before settlement and metamorphosis. The larvae, which develop freely in the surface waters of the ocean, either feed on planktonic organisms (planktotrophic larvae) or develop independently from a self-contained food supply or yolk (lecithotrophic larvae). Pelagic development in temperate waters can take several weeks, during which time developing larvae may be transported over great distances. Where it is within the interests of a particular species to ensure that its offspring are not dispersed (e.g., some intertidal habitats), a free-living larval phase may be dispensed with. In this case eggs may develop directly into miniature adults (oviparity) or may be retained within the body of the adult with the young being born fully developed (viviparity). Reproductive strategies such as these are also common in the deep-sea and polar regions where the supply of phytoplankton for feeding is unreliable or nonexistent. A good example of a latitudinal trend in this respect was demonstrated by Thorson. Analysing the developmental types of prosobranchs, he was able to show that the proportion of species with nonpelagic larvae decreases from the arctic to the tropics, while the proportion with pelagic larvae increases (**Figure 6**).

For many years deep-sea biologists believed that the energetic investment required to produce large numbers of planktotrophic larvae, and the huge distances required to be covered by such larvae in order to reach surface waters, would preclude such a reproductive strategy for deep-sea animals. However, it is now known that several species of ophiuroids living at depths of 2000–3000 m not only exhibit seasonal reproductive behavior but also produce larvae that feed in ocean surface waters.

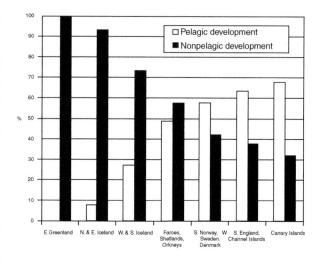

Figure 6 Percentage distribution of prosobranchs with pelagic and nonpelagic development in relation to latitude. (Adapted from Thorson (1950).)

Many benthic invertebrates are able to reproduce asexually. For example, polychaetes from the family the Syllidae are able to reproduce by budding; others, such as the cirratulid *Dodecaceria* or the ctenodrilid *Raphadrilus*, simply fragment, each fragment growing into a new individual. The ability to switch between sexual and vegetative means of propagation provides the potential for such species to rapidly colonize areas that have been disturbed. Where disturbance is accompanied by organic enrichment, for example, from sewage or paper pulp discharge, huge localized populations may result. These are the so-called opportunistic species that are sometimes used as indicators of pollution.

Although planktonic larvae are able to swim, they are very small and, for the most part, are obliged to go where ocean currents take them. The critical time arrives just before the larvae are about to settle. At one time it was thought that the process of settlement was random, with individuals that settled in unfavorable substrata perishing. Although this undoubtedly happens, most species seem to have some sort of behavioral pattern to increase their chances of finding a suitable substratum. The larvae usually pass through one or more stages of photopositive and photonegative behavior. These enable the larvae to remain near the sea surface to feed and then to drop to the bottom to seek a suitable substratum on which to settle. Depending on the species, larvae may cue on the mechanical attributes of the substratum or on its chemical nature. Chemical attraction is also important in gregarious species in which the young are attracted to settle at sites where adults of the same species are already present (e.g., oysters). Most larvae go through a period when, although able to settle permanently, they retain the ability to swim. This allows them to 'test' the substratum, rising back into the water and any prevailing currents should the nature of the ground be unsuitable. After settling, larvae may move a short distance, usually no more than a few centimeters. These early stages in the recruitment of benthic organisms are crucial in the maintenance of benthic community structure and it is now believed that it is at this time that the nature of the community is established.

It is clear that the vast majority of planktonic larvae never make it to adulthood. Mortality from predation and transport away from a suitable habitat are on a massive scale. To compensate, species with planktotrophic larvae produce huge numbers of eggs (e.g., the sea hare *Aplysia californiensis* spawns as many as 450 000 000 eggs at one time). This is possible because there is no need for a large, and energetically expensive, yolk; the larvae hatch at an early embryonic stage and rely almost entirely on plankton-derived food for their development. One consequence of this is that the recruitment varies depending on the success of the plankton production in a particular year and the vagaries of local currents. Thus, populations of benthic species that reproduce by means of planktotrophic larvae tend to fluctuate numerically from year to year, with the potential for heavy recruitment when the combination of environmental factors is favorable, or recruitment failure when they are not. Species reproducing by means of nonpelagic larvae or by direct development tend to produce fewer eggs, since there is a large yolk required to nourish the developing embryo. Although annual recruitment is relatively modest for these species, it is less variable between years, producing populations with a greater temporal stability (**Figure 7**). Because of this, these populations are

likely to be slow to recover from major natural environmental disturbances (e.g., unusual temperature extremes or physical disturbance) or major pollution events.

Deep-sea Benthos

Because of regional differences, it is difficult to define the exact upper limit of the deep-sea benthic environment. However, it is generally regarded as beginning beyond the 200 m depth contour. At this point, where the continental shelf gives way to the continental slope, there is often a marked change in the benthic fauna. In the past there have been many attempts to produce a scheme of zonation for the deep-sea environment. There are three major regions beyond 200 m depth – these are the bathyal region, the abyssal region, and the hadal region (see **Figure 1**). The bathyal region represents the transition region between the edge of the continental shelf and the true deep sea (the continental slope). Its boundary with the abyssal region has been variously defined by workers. It is believed that the 4 °C isotherm limits the depth at which the endemic abyssal organisms can survive. Since this varies in depth according to geographical and hydrographical conditions in the abyssal region, it follows that the upper depth limit of the abyssal fauna will also vary. This depth is usually between 1000 and 3000 m. The abyssal region is by far the most extensive, reaching down to 6000 m depth and accounting for over half the surface area of the planet. The hadal zone, sometimes called the ultra-abyssal zone, is largely restricted to the deep oceanic trenches. The composition of the benthos in these trenches differs from that of nearby abyssal areas. The trenches are geographically isolated from one another and the fauna exhibits a high degree of endemism.

The deep sea is aphotic and so has no primary production except in certain areas where chemosynthetic bacteria are found. Thus, the fauna of the deep sea is almost wholly reliant on organic material that has been generated in the surface layers of the oceans and has sunk to the seabed. Since the likelihood of a particle of food being consumed on its way down through the water column is related to time, the deeper the water the less food is available for the animals that live on the seafloor. This results in a relatively impoverished fauna, in which the density of organisms is low and the size of most is quite small. Paradoxically, the deep sea supports a few rare species that grow unusually large. This phenomenon is known as abyssal gigantism and is found primarily in crustacean species. The reasons for abyssal

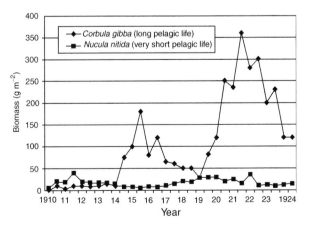

Figure 7 Example of two populations of bivalves showing the influence of type of larvae on population stability. (Adapted from Thorson (1950).)

gigantism are not clear, but it is believed to be the result either of a peculiar metabolism under conditions of high pressure or of the slow growth and time taken to reach sexual maturity.

Although the number of individuals in the deep sea is quite low, there are many species. This combination of many species represented by few individuals results in high calculated diversity values and has led to the suggestion by some that the biodiversity of the deep sea is comparable to that of tropical rain forests. Accepting that diversity is high in the deep sea, there are several theories that attempt an explanation. The earliest is the stability–time hypothesis put forward by Howard Sanders of the Woods Hole Oceanographic Institution in the late 1960s. This suggests that the highly stable environmental conditions of the deep sea that have persisted over geological time might have allowed many species to evolve that are highly specialized for a particular microhabitat or food source. Another theory, the cropper or disturbance theory, suggests that, as a result of the scarcity of food in the deep sea, none of the animals is a food specialist, the animals being forced to feed indiscriminately on anything living or dead that is smaller than itself. High diversity results from intense predation, which allows a large number of species to persist, eating the same food, but never becoming abundant enough to compete with one another. More recently, Grassle and Morse-Porteus have suggested that a combination of factors might be responsible, including the patchy distribution of organically enriched areas in a background of low productivity; the occurrence of discrete, small-scale disturbances (primarily biological) in an area of otherwise great constancy; and the lack of barriers to dispersal among species distributed over a very large area.

Although most animal groups are represented in the deep sea, the fauna is often dominated by Holothuroidea (sea cucumbers) or Ophiuroidea (brittle stars). Crustacea and polychaete worms are also important members of deep-sea communities. For many years it was believed that the deep seafloor was too remote from the surface, and the physical conditions were too constant, for the organisms living there to be influenced by the seasons. Although this may be the case for many deep-sea species, long-term time-lapse photography has shown that cyclical events, such as the accumulation of organic detritus on the deep seabed, do take place. Furthermore, in temperate waters, these appear to correspond with seasonally driven processes such as the spring plankton bloom. Where they occur, these pulses of organic input inevitably influence the deep-sea communities below with the consequence that seasonal life cycles are not uncommon in abyssal animals. Localized areas of organic enrichment can occur in the deep sea as the result of the sinking of large objects such as the carcass of a large sea mammal or fish or waterlogged tree trunk. The surprisingly quick response of deep-sea scavengers to large food-falls such as these, and the frequency with which they have been recorded, has led researchers to believe that they are important contributers to energy flow on the deep-sea floor. Hydrothermal vents are also thought to provide a significant input of energy into the benthic environment. These are a relatively recent discovery and, at first, were thought to be rare and isolated phenomena occurring only on the Galapagos Rift off Ecuador. It is now known that active vents are associated with nearly all areas of tectonic activity that have been investigated in the deep Pacific and Atlantic. These vents provide a nonphotosynthetic source of organic carbon through the medium of chemoautotrophic bacteria. Theses organisms use sulfur-containing inorganic compounds as an oxidizing substrate to synthesize organic carbon from carbon dioxide and methane without the need for sunlight. The chemicals come from hot water that originates deep within the Earth's crust. Some of the bacteria form dense white mats on the surface of the sediments similar to those of *Beggiatoa*, an anaerobic bacterium found in anoxic sediments in shallow water; others enter into symbiotic relationships with the bacteria, either hosting them on their gills (e.g., the mussel *Bathymodiolus*) or within a special internal sac, the trophosome (e.g., *Riftia*). The relationship is very complex as there has to be a compromise between the anaerobic needs of the bacteria and the aerobic needs of the animals. Nevertheless, the arrangement is very successful and the animals, which sometimes occur in huge numbers, often grow to gargantuan size. There is still much to be learned about these vent communities, which can support concentrations of biomass several orders of magnitude greater than that of the nearby seafloor. It remains a mystery how these vents become populated, since they are known to be transient and variable, probably lasting only decades or less. It has been suggested that pelagic larvae of many vent species may be able to delay settlement for months at a time so as to increase their chances of locating a suitable site.

Glossary

Abyssal region That region of the seabed from between 1000 and 3000 m depth reaching down to 6000 m.

Atmosphere Measure of pressure (1 atm = 101 kPa).

Bathyal region Region of the seabed that represents the transition region between the edge of the continental shelf and the true deep sea (the continental slope).

Benthal The benthic environment.

Benthos Those organisms that live on the seabed and the bottoms of rivers and lakes.

Continental shelf Region of the seabed from low-water mark to a depth of 200 m.

Detritivores Animals feeding on organic detritus.

Direct deposit feeders Animals indiscriminately ingesting sediment using organic matter and microbial organisms contained in the sediment as food.

Epifauna All animals living on or associated with the surface of the substratum.

Hadal zone Region of the seabed below 6000 m depth, largely restricted to the deep oceanic trenches.

Infauna All animals living within the substratum or moving freely through it.

Kilopascal (kPa) Measure of pressure (100 kPa = 1 bar = 0.987 atm).

Lecithotrophic larvae Pelagic larvae of marine animals that develop freely in the surface waters of the ocean, developing independently from a self-contained food supply or yolk.

Macrofauna Animals retained on a 0.5 mm aperture sieve.

Meiofauna Animals passing a 0.5 mm sieve but retained on a 0.06 mm sieve.

Microfauna Animals passing a 0.06 mm sieve.

Organic detritus Dead or partially decayed plant and animal material.

Oviparity Eggs laid by the adult develop directly into miniature adults.

Pelagic larvae Larvae of marine animals that swim freely in the water column.

Planktotrophic larvae Pelagic larvae of marine animals that develop freely in the surface waters of the ocean, feeding on planktonic organisms.

Selective deposit feeders Animals feeding on surface particles of organic matter or sediment particles supporting a rich bacterial flora.

Suspension feeders Animals feeding on organisms or organic detritus suspended in the water column.

Viviparity Young are born fully developed either from eggs retained within the body of the mother (oviparity) or after internal embryonic development.

See also

Demersal Species Fisheries. Grabs for Shelf Benthic Sampling. Macrobenthos. Phytobenthos. Tides.

Further Reading

Cushing DH and Walsh JJ (eds.) (1976) *The Ecology of the Seas.* Oxford: Blackwell Scientific Publications.

Friedrich H (1969) *Marine Biology: An Introduction to Its Problems and Results.* London: Sidgwick and Jackson.

Gage JD and Tyler PA (1991) *Deep-sea Biology: A Natural History of Organisms at the Deep-Sea Floor.* Cambridge: Cambridge University Press.

Hedgepeth JW (ed.) (1957) *Treatise on Marine Ecology and Paleoecology,* Vol. 1, *Ecology,* The Geological Society of America, Memoir 67. Washington, DC: Geological Society of America.

Jøgensen CB (1990) *Bivalve Filter-Feeding: Hydrodynamics, Bioenergetics, Physiology and Ecology.* Fredensborg, Denmark: Olsen and Olsen.

Levington JS (1995) *Marine Biology: Function, Biodiversity, Ecology.* Oxford: Oxford University Press.

Nybakken JW (1993) *Marine Biology.* New York: Harper-Collins College Publishers.

Thorson G (1950) Reproduction and larval ecology of marine bottom invertebrates. *Biological Reviews* 25: 1–25.

Webber HH and Thurman HV (1991) *Marine Biology.* New York: Harper Collins.

MACROBENTHOS

J. D. Gage, Scottish Association for Marine Science, Oban, UK

Introduction

The macrobenthos is a size-based category that is the most taxonomically diverse section of the benthos. Only in shallow water does the macrobenthos include both plants and animals. Here, attached macrophytes (including various algae and green vascular plants) may make up a large part of the benthic biomass in coastal areas. Meadows of sea grass (which root into and stabilize coarser sediments) and forests of macroalgae (usually attached to hard bottoms) provide habitat for smaller plant and animal species. In warm, shallow water, stony corals, which flourish as a consequence of symbiotic algae living in their tissues, overgrow large areas and these reefs provide a biogenic habitat for a wealth of other species, and their broken-down skeletons provide much of the sediment in adjacent areas. But the importance of such areas declines rapidly with declining potential for photosynthesis as light quickly vanishes with increasing depth, and the macrobenthos becomes the exclusive domain of heterotrophic life (fueled by breakdown of complex organic material) in soft sediments. Only at deep-sea hydrothermal vents is there an exception to this. These support lush concentrations of benthic biomass that relies not on photosynthetic production ultimately derived from the surface but on the activity of chemoautotrophic bacteria exploiting the emissions of reduced sulfur-containing inorganic compounds.

The animal macrobenthos may be attached or may be able to move over hard surfaces provided by exposed bedrock, or may use as habitat the much larger and quantitatively important areas covered by soft sediment. Areas of rock, exposed as a consequence of water currents and turbulence (or, in the case of the newly formed sea floor at the spreading centers along the mid-ocean ridges, rock that has not had time to become covered in sediment settling from above) provide habitat for epibenthos, or epifauna. Even if epibenthos, both plant and animal, looks conspicuous between the tides (and is certainly important with respect to fouling of colonizers of submerged hard surfaces made by man, such as ships and jetties), it is only a vanishingly small proportion of the huge area of the benthic habitat covering more than half the globe that is not covered by soft sediment.

The term infauna has been used to categorize the organisms inhabiting soft sediment. But many epifauna, such as sea stars, are motile and can forage over the surface of sediments. The activities of the animals of the so-called infauna are usually focused on the sediment–water interface where their detrital food is concentrated, but this should not be taken to imply that sediment fauna is always burrowed out of sight. None the less, some species, particularly among larger crustacea, are capable of burrowing even a meter or more deep into the sediment. There are also many species closely related to epifaunal groups of hard substrata, such as sponges and Cnidaria (including sea pens, soft and stony corals and sea anemones), that anchor into the sediment for a sedentary life style, catching small particles from the bed flow.

Global Pattern in Macrobenthic Biomass

Extensive Russian sampling after World War II established the precipitous decline in benthic biomass with increasing depth into the abyss. This is caused largely by mid-water consumption of particles escaping from the euphotic zone. Globally, the amount leaving the euphotic zone should be equivalent to the so-called 'new production' of roughly $3.4–4.7 \times 10^9$ tonnes $C\,y^{-1}$, about 10% of total surface primary production. But this is distributed very unevenly. Perhaps 25–60% is exported in shallow seas, while in the deep ocean only 1–10% reaches the bottom; this is also influenced by latitude-related differences in depth of the mixed layer and intensity of seasonality in the upper ocean. **Figure 1** shows how influences such as upwelling and inshore surface productivity will also affect local values of benthic biomass. Overall, these range from highs of more than $500\,g\,m^{-2}$ in shallow, productive waters just tens of meters deep, to less than $0.05\,g\,m^{-2}$ (equivalent to $2\,mg\,C\,m^2$) on the abyssal plains. Trenches are deeper still but, by acting as sumps for material washed in from nearby island arcs and land mass, can support higher than expected biomass.

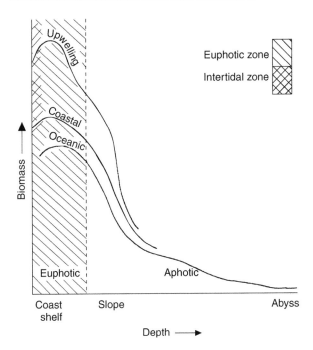

Figure 1 Conceptual model of food availability and benthic biomass in relation to depth. Upwelling areas provide nutrients for enhanced coastal productivity. The 'coastal' curve refers to shelf areas supporting high productivity (usually wider shelves with land inputs, e.g., rivers) compared to the 'oceanic' curve where oceanic effects prevail (usually narrow shelves with little land input). The mismatch in the intertidal between biomass and high food availability is explained by the co-occurrence with the latter of high hydrodynamic disturbance by waves and currents. (Modified from Pearson and Rosenberg (1987).)

History and Size Limits of Macrobenthos

The term macrobenthos dates from the early 1940s when Molly Mare published a study of an area of coastal soft sediment off Plymouth, England. In recognizing that the benthic ecosystem is fueled by a detrital rain of particles derived from photosynthetic production by macrophytes or phytoplankton, she identified the potential importance of the smaller size classes of metazoan and single-celled organism, right down to bacteria, in the decomposition cycle and food chains in the sediment. She differentiated the benthos into several subcategories based on size, or biomass. Before this time a distinct category for macrobenthos was unnecessary because the only part of the benthos generally thought to be worth studying was the animal life large enough to be eaten by fish. We now differentiate these from the very small metazoans and single-celled algae that Mare named meiobenthos. Another category is the hyperbenthos – small metazoan life that can swim off the bottom and form a distinct community in the benthic boundary layer.

The lower size limit of macrobenthos was determined by Mare as that part retained in a 1 mm sieve, but later became reduced downwards to just 0.5 mm as it was realized not only that small, juvenile sizes were being lost in numbers but also that smaller species of groups already being sampled and taken as part of the macrofauna were not always retained adequately. Mare recognized that the limit might depend on habitat and deep-sea benthic biologists found they had to use even finer-meshed screens to collect the same sorts of animals that characteristically make up the macrobenthos in shallow waters. In the 1970s, Hessler found that he had to use a 297 μm mesh sieve to catch sufficient animal macrobenthos for study in box cores from the abyssal central North Pacific (a very oligotrophic area – nutrient-poor and therefore thin in plankton). Sieves with meshes of just 250 μm are now standard in recent large European studies on the deep-sea macrobenthos.

Sources of Food and Feeding Types

Patterns in feeding of macrobenthos have often been used to distinguish ecological zones. Although exact definition of feeding category for individual organisms has been controversial, the simplest classification is into suspension and deposit feeders, carnivores, and herbivores. More detailed categorization has proved difficult because of overlap and behavioral flexibility. Although most macrobenthos feed on detrital particles settling from the water column, such as feces, molts, and dead bodies of plankton, this passive sinking is augmented by currents that may resuspend particles periodically from the bottom. Macrobenthos may gather these particles either by catching them from bottom flow or by ingesting the sediment itself as deposit feeders, either in bulk or more selectively for the most nutritious particles. Where currents vary periodically, some animals can feed on both suspended and settled particles by simply changing the way they use their feeding appendages. Just as the particles caught by suspension feeders may range from inert floating detritus up to small swimming organisms, deposit feeding shades into predation where the particles encountered include smaller living benthos. Whether macrofauna can utilize dissolved organic matter in the sediment porewaters to any great extent is still unclear.

Wildish has provided the most satisfactory classification of macrofaunal feeding types related to environment. This keeps all three categories but separates deposit feeders into surface and burrowing deposit feeders. Each of the five groups is subdivided in terms of motility and also in terms of food-gathering technique, such as use of jaws and particle-entangling structures. These may be arranged along an environmental gradient, such as that illustrated in **Figure 2**, to allow insight into the causal basis of previously described composition of macrofaunal communities.

The relation of feeding to small body size in deep-sea macrobenthos may be important. Thiel thought that small body size is a result of a balance between limited food and metabolic rate that makes larger size more efficient than smaller, and of the effects of small population size on reproductive success. Being small allows organisms to maintain higher population densities that increase the chance of encountering the opposite sex and so of reproducing and maintaining the population. The extent of faunal miniaturization is still debated and, surprisingly, not readily summarized by simply taking the total bulk of

the sample and dividing by the number of animals present. The exceptions seem be those organisms that have overcome the reproductive problem by being highly motile scavengers and that need also to be large enough to allow them to forage for the large food falls that occur very sporadically on the deep ocean bed. This scavenger community is quite well developed and includes close relatives of typical macrofaunal organisms in shallow water that in the deep sea grow to a relatively enormous size (**Figure 3**).

Size Spectra

If the sizes of all individuals from an area of sediment are measured and plotted as frequencies along a logarithmic size axis, a pattern of peaks shows up corresponding to the micro-, meio- and macrobenthic size classes (**Figure 4**). This supports practical intuition but does not explain why such peaks occur (no

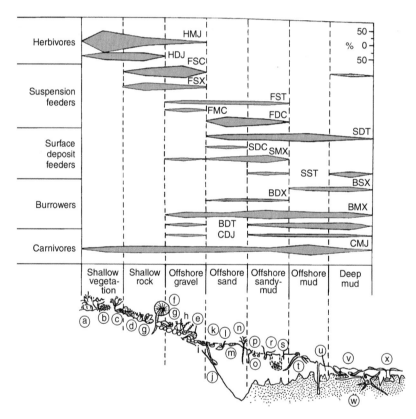

Figure 2 Distribution of functional groups in boreal coastal macrobenthos (compiled from genera listed by N. S. Jones for the Irish Sea) along an environmental gradient of decreasing food availability and water turbulence and increasing depth and sedimentation. Functional groups: H, herbivore; F, suspension feeder; S, surface deposit feeder; B, burrowing deposit feeder; C, carnivore. Motility: M, motile; D, semi-motile; S, sessile. Feeding habit: J, jawed; C, ciliary mechanisms; T, tentaculate; X, other types. In the upper panel, width of line representing each functional group along the gradient indicates proportional composition at that depth. The lower panel gives a diagrammatic representation of typical feeding position of taxa representative of various groups relative to the sediment–water interface along each gradient. Key to taxa: (a) macroalgae; (b) sea urchins, e.g., *Echinus* (HMJ); (c) limpets, e.g., *Patella* (HDL); (d) Barnacles, e.g., *Balanus* (FSX); (e), (f) serpulids, sabellids (FST); (g) epifaunal bivalves, e.g., *Mytilus* (FSC); (h) brittle stars, e.g., *Ophiothrix* (FMC); (i) *Venus* (FSX); (j) *Mya* (FSC); (k) *Cardium* (FDC); (l) *Tellinba* (FDT); (m) *Turritella* (SMX); (n) *Lanice* (SST); (o) *Abra* (SDC); (p) *Spio* (SST); (r) *Amphiura* (FDT); (s) *Echinocardium* (BMX); (t) *Ampharete* (SST); (u) *Maldane* (BSX); (v) *Glycera* (CDJ); (w) Thyasira (BDX); (x) *Amphiura* (SDT). From Pearson and Rosenberg (1987).

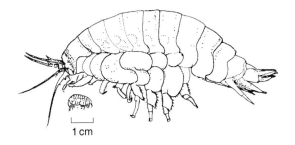

Figure 3 Gigantism in a scavenging amphipod (family Lysianassidae), the cosmopolitan deep-sea species *Eurythenes gryllus*, compared to the size of a typical shallow-water lysiannasid, *Orchomene nana* (bottom left), a northern European shallow-water species. (Redrawn from Gage and Tyler (1991) and Hayward PJ and Ryland JS (1995) *Handbook of the Marine Fauna of North-West Europe*. Oxford: Oxford University Press.)

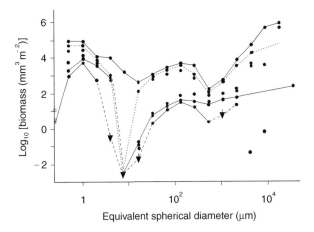

Figure 4 Macrobenthic size spectra measured from an intertidal inlet in Nova Scotia, subtidal Bay of Fundy, and abyssal sediment from the Nares Abyssal Plain south of Bermuda. The median lines (dotted for intertidal and inshore, and double-dashed for abyssal plain) and range (continuous or solid lines) show a coherent pattern with biomass peaks at 1256 and 8192 µm equivalent spherical diameter (ESD). Downward-pointing arrows indicate minimum detectable biomass. Abundance in the bacterial (leftmost) and meiofaunal (middle) peaks averages 5×10^3 mm^3 m^{-2}. The macrofaunal biomass peak is an order of magnitude higher, but shows greater variability. Biomass in the troughs is about 2–3 orders of magnitude less than adjacent peaks. Sediment was a fluid silt-clay. (From Schwinghamer P (1985) Observations on size-structure and pelagic coupling of some shelf and abyssal benthic communities. In: Gibbs PE (ed) Proceedings of the Nineteenth European Marine Biology Symposium, Plymouth, Devon, U.K. 16–21 September 1984, pp. 347–359. Cambridge: Cambridge University Press.)

such peaks occur, for example, in pelagic communities). Schwinghamer thought that these peaks reflect the way the organism perceives its sediment environment: macrofauna as a continuous medium on, or in, which to move and burrow; meiofauna as a series of interstices between sediment particles; while to microbenthos and bacteria each particle is a little world on which to attach and grow. Warwick provided a complementary explanation that the peaks also reflect size-related adaptation in the way the life history of the organism is optimized to its environment. For example, larvae of macrofauna exploit the trough between macro- and meiofauna to escape from meiofaunal predators, and thereafter quickly grow into the size range of the 'macrofaunal' peak. This is generally lower and less defined than the meiofaunal peak, where organism longevity is just a few weeks at most and there is therefore a narrow range in size, while individual macrofauna might grow over several years so that population size distributions are wider. However, subsequent studies have not found that clear peaks in size spectra occur everywhere. In the deep sea, body size miniaturization does not destroy this pattern, even if the trough at 512–1024 µm between meio- and macrofauna may be less than in coastal sediment (**Figure 4**). It seems more likely that low food supply has become more important than anything else, so that macrofauna, although settling at roughly the same size, simply do not grow anything like as large as similar coastal species, rather than their somehow perceiving the sediment environment differently from typical macrofaunal organisms in shallow water.

We cannot therefore reject the idea that size-based differentiation of the benthos occurs; but is it sensible to stick rigidly to the strict size-based divisions that define the macrofauna as only those organisms within a given range of size, or is it better to compare like with like on the basis of higher taxa determining limits rather than size? With the former definition, the lower limit of the macrobenthos will be determined by size at 1.0 or 0.5 mm, even if this excludes smaller specimens belonging to the same higher taxon, or even much lower-level taxa. This assumes that a size-based functional distinction operates that for the purposes of the study (perhaps environmental impact assessment) will be more important in determining variability than taxonomic affinity. The former function-based definition has been referred to as macrofauna *sensu stricto*, while the latter, taxonomic one as macrofauna *sensu lato*.

Composition and Succession

Macrobenthos characteristically includes a huge range of phyla (the major divisions of the animal kingdom). In fact, most higher-level taxa are marine and benthic, with most of these part of the macrobenthos. Of the 35 or so known phyla (the major divisions of the animal kingdom), 22 are exclusively marine, with 11 restricted to the benthic environment. Virtually every known phylum is represented in the macrobenthos except for one, the Chaetognatha, or arrow worms (arguably found only in the plankton, although one bottom-living genus is known). This contrasts with the land (including freshwater environments) where only 12 phyla are found (there is only one small, obscure phylum of worm-like animals, the Onychophora, known only on land). This reflects the marine origins of life and the much shorter time of occupation for life on land (barely 400 million years), compared with 800 million years since metazoan organisms first appeared in the ancient ocean. Only five metazoan phyla are normally regarded as part of the next size group down, the meiofauna.

The proportional representation of major taxa is conservative, and seems to vary little worldwide with depth, latitude, or productivity regime. It is only in stressed soft-sediment environments, such as those with high organic loading and depleted oxygen (often occurring together), where major departures to this pattern are found (**Figure 5**). Inshore studies on effects of pollution have contributed to a concept whereby such stress leads to a modified macrobenthos with fewer species, and these characterized by opportunist forms, mainly polychaetes. In tracing recovery after pollution events, it is not clear to what extent a predictable succession occurs. The modern consensus is that there is a random component imposed on a facultative succession in which 'opportunist' species pioneer colonization and bring about amelioration in sediment conditions. This allows a more diverse set of species that are more highly tuned to particular habitats to become established through progressively deeper and more extensive bioturbation (**Figure 6**).

How Many Macrobenthic Species are There?

Up to a few years ago it was thought that of the 1.4 to 1.8 million or so species recorded on earth there are perhaps only 160 000 or so known marine species, about 10% of the total. A large-scale sampling programme in deep water off the eastern United States has thrown this into doubt. Along a 180 km section of the continental slope at about 2000 m depth, Grassle and Maciolek found 58% of the species – especially among polychaete (bristle) worms and peracarids (small, sandhopper sized crustaceans) – new to science. The curve of the accumulation of species plotted against increasing area sampled showed no sign of tailing off; the steady increment of new (but rare) species encouraging an extrapolation that this will apply over the wider area of the deep ocean.

Depths below the shelf edge cover about 90% of the domain of macrobenthic infauna. But an area of only about $0.5 \, km^2$ out of the almost $335 \times 10^6 \, km^2$ area below 200 m depth has yet to be adequately sampled for macrofauna using grabs or corers. Because of this huge unexplored area, the actual number of marine macrobenthic species present today is unknown. It must be vastly greater than earlier estimates based on shallow seas, and according to Grassle and Maciolek is conservatively greater than one million, and more likely to rise to 10 million as more of the deep sea is sampled.

The overwhelming taxonomic challenge of describing these new species, the painstaking work of sorting samples from the sediment, and the difficulty of seabed experimentation have perhaps slowed progress in understanding the deep-water sediment community. In contrast, much more has been achieved in biological knowledge of hydrothermal vents (and to a lesser extent cold seep communities) since their discovery in 1977.

Large-scale Patterns in Macrobenthic Diversity

Large-scale patterns, other than that for biomass, remain controversial. On the basis largely of sampling of the continental shelf, Thorson pointed out that species richness of the epifauna, occupying less than 10% of the total area, and maximally developed intertidally, rises steeply from low levels in the ice-scoured shallows in the Arctic to high values in the tropics. In contrast, the sediment macrofauna he referred to as 'infauna,' usually found deeper and unaffected by ice and meltwater, show much less change. This lack of a latitudinal gradient is supported in some other studies. However, latitudinal comparisons by Sanders in the 1960s found depressed diversity in shallow boreal macrobenthos stressed by wide seasonal temperature change compared to the tropics. Thorson through there were about four times more epibenthic than infaunal species, the microhabitat complexity and consequently high species diversification of the epibenthic habitat being much less obvious in sediments. The sameness of this habitat regardless of latitude led Thorson to his concept of parallel level-bottom communities related to sediment type. But several recent studies indicate shallow tropical and deep-sea sediments do not follow this pattern, with much more species-rich communities developing, albeit including lots of 'rare' species, in these habitats.

Macrofaunal abundances

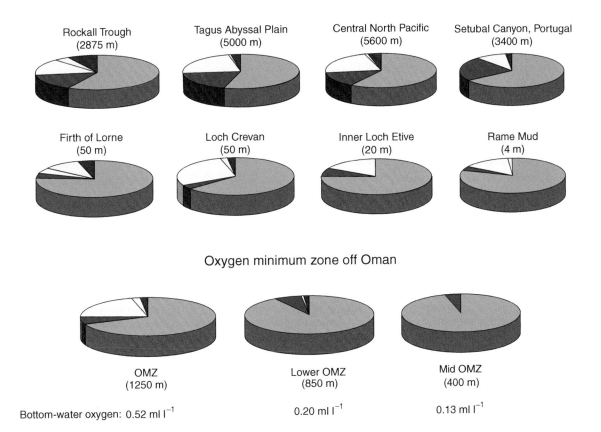

Oxygen minimum zone off Oman

Macrofaunal species

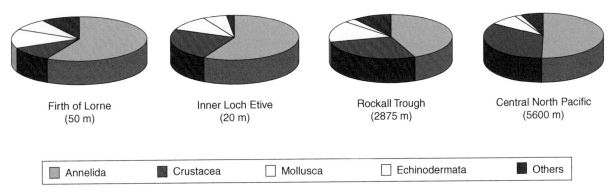

Figure 5 Proportional representation of the major taxonomic groups of macrofauna in differing soft sediment habitats and depths worldwide. The upper diagrams show this in terms of the relative abundance in these groups; the lower ones in terms of the number of species represented. A broadly similar pattern is shown between both sets although Crustacea are often more abundant in the deep sea rather than shallow-water macrobenthos. Representation of Annelida (mostly polychaete worms) shows most obvious variation in relation to organic carbon loading and oxygen, the three samples from the oxygen minimum zone (OMZ) in the Arabian Sea off Oman showing a pattern of increasing dominance by Annelida, and eventually complete loss of all other groups except Crustacea, with increasing oxygen depletion. (Data from Mare (1942); Gage (1972) Community structure of the benthos in Scottish sea-lochs. I. Introduction and species diversity. *Marine Biology* 14: 281–297, Gage J (1977) Structure of the abyssal macrobenthic community in the Rockall Trough. In: Keegan BF, O'Ceidigh P and Boaden PJS (eds) *Biology of Benthic Organisms* (11th European Marine Biology Symposium), pp. 247–260. Oxford: Pergamon Press; Levin LA, Gage JD, Martin C and Lamont PA (2000) Macrobenthic community structure within and beneath the oxygen minimum zone, NW Arabian Sea. *Deep-Sea Research II* 47: 189–226.)

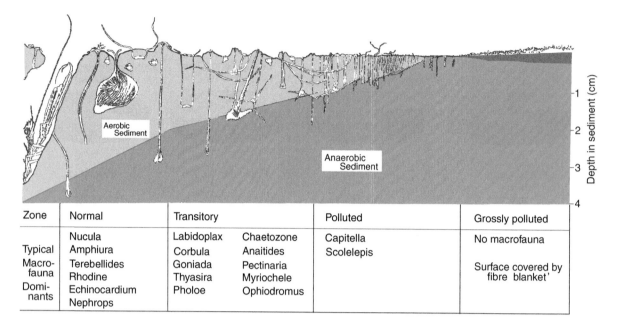

Zone	Normal		Transitory		Polluted	Grossly polluted
Typical Macro-fauna Domi-nants	Nucula Amphiura Terebellides Rhodine Echinocardium Nephrops		Labidoplax Corbula Goniada Thyasira Pholoe	Chaetozone Anaitides Pectinaria Myriochele Ophiodromus	Capitella Scolelepis	No macrofauna Surface covered by fibre blanket'

Figure 6 Changes in macrobenthic fauna along an enrichment–disturbance gradient, such as that associated with pollution. The gradient can be replaced by time in tracing recovery along the *x*-axis to a 'normal' community (left) after a severe pollution event (right). Note the shift in body size (reflecting change from quick-growing and fast-turnover 'pioneer' species to slower-growing, longer-lived population species) from right to left, as well as the increased depth and extent of bioturbation in the sediment. There is also a concomitant increase in macrobenthic diversity. (From Pearson TH and Rosenbeg R (1978).)

Depth-related patterns in macrobenthic composition have been studied, particularly for larger benthic invertebrates (technically megabenthos, see discussion earlier) and demersal (bottom-living) fish. Rates of macrofaunal turnover, and clinal variation in individual species, correspond to the rate of change in depth. It is highest in upper bathyal and slowest and most subtle in the abyssal. Many changes in species composition can be related to trophic strategies along a gradient in food and hydrodynamic energy (see **Figure 2**), but changes in the sort and intensity of biological interactions, such as predation, varying with depth, may also be important, as can life-history characteristics (such as the incidence of planktotropic larval development). Recent studies have also established a degree of pressure adaptation during early development that will further limit vertical range. Such ecological processes must be considered in concert with processes at the evolutionary timescale for understanding of zonation patterns. These processes have been summarized using multivariate statistics. While helping in formulating ideas on causal factors, these may obscure the underlying complexity, which is best understood as the sum of the range and adaptation and evolutionary history of individual species.

Rex postulated a mod-slope peak in macrobenthic diversity from studies of sled samples taken throughout the Atlantic by Sanders. However, there is high

variability among individual sample values compared and there are conflicting results from other sites worked in the north-eastern Atlantic. That deep-sea macrobenthos has high species diversity seems well founded, but the extent to which this contrasts with shallow water is unclear. In his original study off the north-eastern United States, Sanders showed an impoverished species richness in samples of macrofauna compared to the adjacent slope and rise. Gray has pointed out that on the outer continental shelf off Norway macrofaunal diversity may be comparable to that found in Sanders' deep-sea samples, and it is considerably higher still off south-eastern Australia. This suggests not only that the inshore shelf off New England is rather poor in species richness but that the North Atlantic as a whole may be atypical, with perhaps historical factors operating there to restrict macrobenthic diversity compared to the southern hemisphere.

Such factors may determine the differing response shown in a comparison of deep-sea macrobenthic diversity among sites throughout the Atlantic where the depression at high latitudes is absent south of the equator. The reduced levels at high latitudes may simply reflect Quaternary glaciation so that the deep Norwegian Sea, isolated by shallow sills from deep water to the south, has a much more recently diverged and quite distinct macrofauna from that in the Atlantic.

Small-scale Pattern

In sampling macrobenthos from a ship it is easy to assume that the animals in this apparently homogeneous habitat are randomly distributed, but sample replicates may not always provide good estimates of the population mean and its sampling error. When samples are mapped over the sediment, or variability is analysed in large numbers of replicates, clumped distributions of some kind are commonplace (**Figure 7**). Nonrandomly even (regular) dispersions have been detected, but only at the centimeter scale, suggesting that they are actively defined by the ambit (such as the area swept by feeding tentacles) of individual animals. To describe rather than just detect such nonrandom spatial pattern has been a challenging task, not least because most macrofauna are not readily visible in seabed photographs and the very analysis of samples by sieving and mud will destroy fine-scale pattern. Spatial pattern is usually envisaged in the horizontal plane because macrobenthic organisms concentrate their activity on the sediment–water interface in feeding, movement, and reproduction. Pattern is a dynamic expression of this and consequently may change through time but marine sediments provide a three-dimensional habitat so that vertical as well as horizontal spatial patterns may occur. The latter may be best developed at the small scale where smaller macrofauna (and meiofauna) are concentrated around irrigatory or feeding burrows of larger species (**Figure 8**). The problem is that to analyze this pattern it is difficult not also to disrupt the habitat. Yet in order to understand the basis of such pattern it is vital to analyze and map pattern over a range of scales in conjunction with variability in the sediment habitat. Some of the most revealing studies have examined dispersions of individual species over plots measuring tens of meters square. These may reveal the two aspects of pattern, intensity and form. Intensity can relatively easily be measured by the ratio of variance to mean. This will distinguish distributions that are clumped, regular, or not statistically distinguishable from random. The form of pattern is an aspect that classical statistical tests of nonrandomness do not address. Yet a clumped pattern may be very different in form from that shown by another species that shows similar intensity of aggregation.

Although a nuisance for the easy interpretation of sample statistics, an understanding of the biological basis of patterns will provide important insight into the processes maintaining macrobenthic communities.

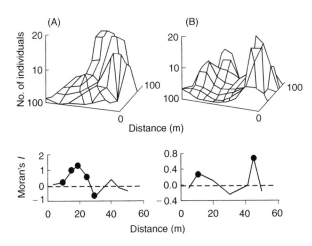

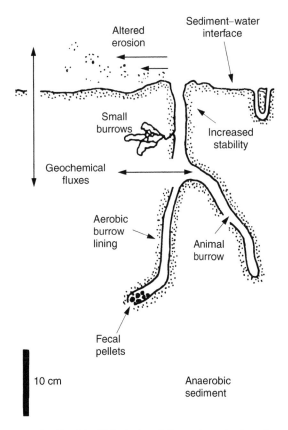

Figure 7 Upper, three-dimensional plots of abundance showing spatial dispersion patterns of two infaunal bivalve molluscs *Nucula hartivigiana* (A) and *Soletellna siliqua* (B) in a 9000 m² area of mid-tide sandflat with no obvious gradients in physicochemical conditions. Both species show quite different spatial patterns. The lower plots show spatial correlograms with significant autocorrelation coefficients (measured as Moran's *I*) denoted by filled circles (From Hall *et al.* (1984). Thrush *et al.*, (1989).)

Figure 8 Benthic biological activity and seabed sediment structure. The diagram shows some of the ways macrobenthos, in conjunction with other size classes, influence sediment fabric, physicochemical properties and solute fluxes (see also **Table 1**). (After Meadows PS (1986) Biologica activity and seabed sediment structure. *Nature* 323: 207.)

Functional Importance of Macrobenthos

Using a grab as a quantitative sampler in the early years of the twentieth century, C. J. J. Petersen hoped to be able to work out from his thousands of samples taken in the North Sea how much food was available to fish such as flounder. Such links were well supported from finds of large numbers of benthic animals in fish guts, even if later work showed that fish are by no means as important predators of macrobenthos as are invertebrates such as sea stars. Petersen also noticed that characteristic and uniform assemblages of macrobenthos were found that could be related to sediment type and that provided him with statistical units that Thorson later used as his descriptive units in his concept of 'parallel level bottom communities.' Ecologists debated whether these existed as anything more than assemblages responding to similar conditions (as originally inferred by Petersen) or reflected functional units or 'biocoenoses' where biological interactions play an important, if unknown, role. However, the importance of biological interactions in the subtidal community is difficult to address experimentally, and most data are available from intertidal mudflats and sandflats where ecological gradients related to tidal exposure pose additional complexity. Manipulative experiments on the effects of predation by caging small areas show that predators like shore crabs can have big effects on prey densities, while other studies show competitive exclusion between worms with different burrowing styles that can be reflected in clumping patterns. This contrasts with the importance of grazers and predators in preventing dominance by fast-growing competitive superior species on rocky shores. Such biological interaction cascades down through the community – so-called 'top-down' control. But, in sediments, effects such as predation are not so marked overall. Perhaps the three-dimensional structure, the uneven distribution of food and irrigatory flows, and the often intense stratification of chemical processes reduce competition. Indirect effects, such an bioturbation, may take the place of competition. By bulk processing large quantities of sediment, large macrofaunal deposit feeders rework the sediment down to the greatest ocean depths and thus exert a major influence on benthic community structure. It has been suggested that this constant process of biogenic disturbance and alteration of the benthic environment by macrofauna (**Figure 8, Table 1**) encourages high species richness among smaller macrobenthos by reducing them to levels where competition is relaxed, so that more species can coexist. It is also argued that the constantly changing micro-landscape created by other, larger species provides a rich niche variety for macrofauna. It is difficult to see this process operating on the vast abyssal plains where faunal densities, and therefore such biogenic effects, are so low but species richness is high. Grassle's spatiotemporal mosaic theory sees the deep-sea bed having patchy and ephemeral food resources that create a relatively small, discrete, and widely separated patch structure promoting coexistence.

Effects of larger-scale disturbances are more difficult to detect let alone manipulate in experiments with the sediment community. Yet the evidence is that physical disturbance such as that caused by storm-driven sediment scour and resuspension may have an important effect on assemblage structure and species richness on the exposed continental shelf and margin. The expectation that, just as on an exposed sandy shore, only a relatively small suite of species will be able to adapt to such conditions is confirmed in the deep sea on the continental rise off Nova Scotia, where benthic storms occur with relatively high frequency. Benthic storms may occasionally occur on the abyssal plains, so it should not be assumed that biogenic structure is simply longer-lasting there because it takes so long to be covered by the very low rate of natural sedimentation.

Perhaps the most important determinant of the macrobenthic assemblage, or community, is the larval stage usually dispersed in the water column. Larvae can test the substratum and swim off until they find conditions suitable for settlement and metamorphosis. On rock this may involve a series of precise cues that can include presence of their own or other species. Less is known about settlement of infaunal species, but it is thought that positive cues such as microtopography may be much less important in sediment dwellers, while negative cues such as the presence of other species or unattractive sediment are more important. Nevertheless, a community will still be very largely constrained by supply of propagules. In a coastal area the access of larvae supply from adjacent breeding populations may be constrained by coastal topography and currents, not to mention barriers formed by features such as estuaries. The patch structure in the deep sea is maintained by water-borne dispersal stages with the resulting metapopulations spatially unautocorrelated (presence of an organism not dependent on other occurrences). In deep water the openness of the system may mean that the sediment is exposed to a much larger pool of species, even if they are at very low densities as larvae. In the tropics a similar effect results from the greater incidence of planktotrophy (larvae feeding in the plankton), when the longer

Table 1 Direct and indirect effects of macrofauna on soft sediments and their ecological consequences

Direct effect	Indirect effects and ecological feedback
Bioresuspension: pelletization of superficial sediment as egesta changes sediment granulometry and increases openness of sediment fabric	Bioresuspension and formation of bottom turbidity layers through decrease in threshold where bed-shear stress will mobilize sediment Alteration in benthic community by reducing suspension feeders (trophic group 'amenalism') Reduced numbers of epifauna, and total community diversity Disruption and rapid obliteration of biogenic traces
Biodeposition: increased stickiness of sediment surface, and sediment structure altering hydrodynamic conditions to trap particles in pits, etc.	Biodeposition increases sediment stability and trapping of fine particles utilizable as food Encourages suspension feeders, and hence increases trophic diversity Results in tighter sediment fabric; increased sediment microstructure, and habitat complexity causing greater niche variety, and thereby increasing community diversity Better preservation of biogenic traces
Sediment irrigation	Spatial and temporal variability in exchange of dissolved gases, dissolved or absorbed ions and complex chemical species nutrients across sediment water interface Alteration in vertical gradients in Eh, pH etc. Transfer of reduced compounds from reducing conditions below the interface to aerobic conditions above Enhances recycling of dissolved nutrients Creates chemical heterogeneity from hotspots of elements concentrated by larger organisms in the sediment
Vertical sediment transport by larger deposit feeders	Particle selection may cause vertical shifts, granulometry Alteration of chemical microenvironment and diagenesis of sediment and smearing of stratigraphic signal
Formation of surface traces, by movement, burrowing or defecation	Increases bottom roughness, affecting near-bed flow Increased complexity in microenvironment creates niche mosaic at sediment surface

larval life will ensure wider dispersal than that of the nonfeeding larvae prevalent in cooler waters. This may help to explain why so many, mostly rare, species can coexist in both environments.

Importance of Macrobenthos in Environmental Assessment

Because the benthic community (unlike fish or plankton) is stationary or at best slow-moving over a small area of bottom, it is useful in monitoring environmental change caused by eutrophication and chemical contamination. Macrobenthos studies have defined the generic effects of such sources of stress by changing representation of major taxa, reduction in diversity, and increasing numerical dominance by small-sized opportunist species causing a downward shift in size structure. This seems to be accompanied by greater patchiness, reflected by increased variability in species abundances in sample replicates. It is also seen as greater variability in local species diversity caused by greater heterogeneity in species identities. This reflects subtle changes in abundance and, particularly in the more species-rich communities, changes in presence/absence of rare species that might be detected earlier at less severe levels of disturbance. It is claimed that in bioassessments comparing species richness using samples of macrobenthos rare species should receive greater attention by taking larger samples because they contribute relatively more to diversity than the abundant community dominants. Other workers argue that very many species, especially rare ones, are interchangeable in the way they characterize samples. This question requires investigation of the way stressors impact the community, and whether it is the dominant or the rare species that are most sensitive, and therefore most rewarding for study in detecting impacts.

Interpretation of impacts also has to proceed against a background of natural changes in benthic communities caused by little-understood, year-to-year differences in annual recruitment. In establishing a baseline there is a need also to take into account the little-understood effects of bottom trawling on coastal benthos. Such disturbance in parts of the North Sea may date back at least 100 years, and now means that virtually every square meter of bottom is trawled over at least once a year. Such monitoring has in the past entailed costly benthic survey and tedious analysis of samples to species level. Consequently, there has been effort to see whether the effects of stress can be detected at higher taxonomic levels, such as families. Higher taxonomic levels may more closely reflect gradients in contamination than they do abundance of individual species because of the statistical noise generated from natural recruitment variability and from seasonal cycles such as reproduction. This hierarchical structure of macrobenthic response means that, as stress increases, the adaptability of first individual animals, then the species, and then genus, family, and so on, is exceeded so that the stress is manifest at progressively higher taxonomic level.

Such new approaches, along with the nascent awareness of conservation of the rich benthic diversity, and with a need for improved environmental impact assessment on the deep continental margin, should ensure a continued active scientific interest in macrobenthos in the years to come.

See also

Benthic Organisms Overview. Coral Reefs. Demersal Species Fisheries. Fiordic Ecosystems. Grabs for Shelf Benthic Sampling. Phytobenthos. Rocky Shores. Sandy Beaches, Biology of.

Further Reading

Gage JD and Tyler PA (1991) *Deep-sea Biology: A Natural History of Organisms at The Deep-sea Floor.* Cambridge: Cambridge University Press.

Graf G and Rosenberg R (1997) Bioresuspension and biodeposition: a review. *Journal of Marine Systems* 11: 269–278.

Gray JS (1981) *The Ecology of Marine Sediments.* Cambridge: Cambridge University Press.

Hall SJ, Raffaelli D, and Thrush SF (1986) Patchiness and disturbance in shallow water benthic assemblages. In: Gee JHR and Giller PS (eds.) *Organization of Communities: Past and Present*, pp. 333–375. Oxford: Blackwell.

Hall SJ, Raffaelli D, and Thrush SF (1994) Patchiness and disturbances in shallow water benthic assemblages. In: Giller PS, Hildrew HG, and Raffaelli DG (eds.) *Aqautic Ecology: Scale, Patterns and Processes*, pp. 333–375. Oxford: Blackwell Scientific Publications.

Mare MF (1942) A study of a marine benthic community with special reference to the micro-organisms. *Journal of the Marine Biological Association of the United Kingdom* 25: 517–554.

McLusky DS and McIntyre AD (1988) Characteristics of the benthic fauna. In: Postma H and Zijlstra JJ (eds.) *Ecosystems of the World 27, Continental Shelves*, pp. 131–154. Amsterdam: Elsevier.

Pearson TH and Rosenberg R (1978) Macrobenthic succession in relation to organic enrichment and pollution of the marine environment. *Oceanography and Marine Biology: an Annual Review* 16: 229–311.

Pearson TH and Rosenberg R (1987) Feast and famine: structuring factors in marine benthic communities. In: Gee JHR and Giller PS (eds.) *Organization of Communities: Past and Present*, pp. 373–395. Oxford: Blackwell Scientific Publications.

Rex MA (1997) Large-scale patterns of species diversity in the deep-sea benthos. In: Ormond RFG, Gage JD, and Angel MV (eds.) *Marine Biodiversity: Patterns and Processes*, pp. 94–121. Cambridge: Cambridge University Press.

Rhoads DC (1974) Organism–sediment relations on the muddy sea floor. *Oceanography and Marine Biology Annual Reviews* 12: 263–300.

Thorson G (1957) Bottom communities (sublittoral or shallow shelf). In: Hedgepeth JW (ed.) *Treatise on Marine Ecology and Paleoecology*, pp. 461–534. New York: Geological Society of America.

Thrush S (1991) Spatial pattern in soft-bottom communities. *Trends in Ecology and Evolution* 6: 75–79.

PHYTOBENTHOS

M. Wilkinson, Heriot-Watt University, Edinburgh, UK

What is Phytobenthos?

'Phytobenthos' means plants of the seabed, both intertidal and subtidal, and both sedimentary and hard. Such plants belong almost entirely to the algae although seagrasses, which form meadows on some subtidal and intertidal areas, are flowering plants or angiosperms. Algae of sedimentary shores are usually microscopic, unicellular or filamentous, and are known as the microphytobenthos or benthic microalgae. Marine algae on hard surfaces can range from microscopic single-celled forms to large cartilaginous plants. Some use the term 'seaweed' for macroscopic forms whereas others also include the smaller algae of rocky seashores. This article is concerned with the nature, diversity, ecology, and exploitation of the marine benthic algae. Other plants of the shore are dealt with elsewhere in this encyclopedia as salt-marshes, mangroves, and seagrasses. (*see* Salt Marsh Vegetation).

What are Algae?

Algae were regarded as the least highly evolved members of the plant kingdom. Nowadays most classifications either regard the microscopic algae as protists, while leaving the macroscopic ones in the plant kingdom, or regard all algae as protists. This distinction is not important for an understanding of the ecological role of these organisms so they will all be called plants in this article. The fundamental feature that algae share in common with the rest of the plant kingdom is photoautotrophic nutrition. In photosynthesis they convert inorganic carbon (as carbon dioxide, carbonate, or bicarbonate) into organic carbon using light energy. Thus they are primary producers, which act as the route of entry of carbon and energy into food chains. They are not the only autotrophs in the sea. Besides the other nonalgal plant communities mentioned earlier, there are chemoautotrophs in hydrothermal vent communities, which use inorganic reactions rather than light as the energy source, and some bacteria are photosynthetic. However, algae are responsible for at least 95% of marine primary production. Algal photosynthesis uses chlorophyll *a* as the principal pigment that traps and converts light energy into chemical energy (although many accessory photosynthetic pigments may also be present) and water is the source of the hydrogen that is used to reduce inorganic carbon to carbohydrate. Oxygen is a by-product so the process is called oxygenic photosynthesis. Broadly the same process occurs throughout the plant kingdom. Those true bacteria that are photosynthetic use alternative pathways, pigments, and hydrogen donors, e.g., hydrogen sulfide.

One group of organisms falls between the algae and the bacteria – the cyanobacteria, until recently regarded as blue–green algae. These perform oxygenic photosynthesis and have similar photosynthetic pigments to algae, but their cell structure is fundamentally different. In common with the bacteria they have the more primitive, prokaryotic cell structure, lacking membrane-bound organelles and organized nuclei. Algae have the more advanced and efficient eukaryotic cell structure, with membrane-bound organelles and defined nuclei, in common with all other plants and animals. Blue–greens also have some physiological affinities with bacteria, particularly nitrogen fixation. This means that they can use elemental nitrogen as a source of nitrogen for biosynthesis of various organic nitrogen compounds, starting from the simple organic compounds formed in photosynthesis. Algae and other plants have lost this ability and require to absorb fixed nitrogen, combined inorganic nitrogen as nitrate and ammonium ions, from solution in sea water. Despite internal cellular differences, blue–greens have similar overall morphology to smaller algae and live indistinguishably in algal communities as primary producers. They will therefore be included with algae in this review.

Algae are therefore similar to the 'true' plants in having oxygenic photosynthesis. They are distinguished from the rest of the plant kingdom only on a rather technical botanical point. Algae have simpler reproductive structures. In all the higher plant groups the reproductive organs are surrounded by walls of sterile cells (i.e., cells that are not gametes or spores) that are formed as a specific part of the reproductive organ. This does not occur in algae. Even in the highly complex large brown seaweeds with apparently complex reproductive structures, the gametangia, which produce eggs and sperm, enclosed within complex reproductive structures, have only membranous walls rather than cellular walls.

Diversity of Algae

Algae as defined in the previous section include a large diversity of organisms from microscopic single-celled ones, as little as about 2 μm in diameter, to the complex giant kelp nearly 70 m long, the largest plant on Earth. We can make sense of this diversity in three ways:

- Structural diversity
- Habitat diversity
- Taxonomic classification

Structural Diversity of Algae

The simplest algae are single cells, which can vary in size from about 2 μm to 1 mm. They can be non-motile, lacking flagella, or motile by means of flagella. An interesting intermediate situation is in the diatoms which lack flagella but are nonetheless motile by gliding over surfaces. Unicells can differ in the presence of external sculpturing and the number and orientation of flagella on each cell. Colonies are aggregations of single cells which can also be flagellate or nonflagellate. The simplest truly multi-cellular algae are filamentous, i.e., hair-like, chains of cells. These can be branched or unbranched and may be only one cell in thickness (uniseriate) or more than one cell in thickness (multiseriate). Heterotrichy is an advanced form of filamentous construction in which two separate branched systems of filaments may be present on one plant – a prostrate system which creeps along the substratum and an erect system which arises into the seawater medium from the prostrate system.

The larger more advanced types of seaweeds can be traced in origin to modifications of the hetero-trichous system. Reduction of one of the two fila-ment systems and elaboration of the other can give rise either to encrusting forms (erect reduced, prostrate elaborated) or to erect plants in which the prostrate system only forms the attachment organ or holdfast. Three further modifications give rise to a wide diversity of large cartilaginous (leathery), foliose (leaf-like) and complex filamentous seaweeds. These three modifications are:

- Presence of meristems, localized areas where cell division is concentrated, which may be apical, at the growing tips of branches, or may be inter-calary, located along the length of the plant (in simpler algae growth is diffuse with cell division occurring anywhere in the plant, not localized to meristems).
- Pseudoparenchyma formation – the aggregation of many separate filaments together to make a massive plant body, as opposed to true par-enchyma formation, where massive tissues result only from multiplication of adjacent cells. This is a different use of the term parenchyma from that in higher plants where it means an unspe-cialized type of cell which acts as packing tissue.
- The occurrence of two or more phases of growth, which may or may not differ in pattern (pseudo-parenchymatous or truly parenchymatous) and may involve formation of a secondary lateral meristem. Various phases of growth can give rise to the different tissues seen in cross-sections of seaweeds which may help in giving the ability to bend in response to water motion and wave action, without breaking.

Some seaweeds are able to secrete calcium car-bonate so that they appear solid. Red calcareous species appear pink and are common throughout the world whereas green calcareous forms are commoner in the tropics and subtropics. Calcareous encrusting red algae form a pink calcareous coating on the rock surface which can be mistaken by the nonspecialist for a geological feature.

The structure of a kelp plant illustrates the life of seaweeds. The plant is attached to the rock surface by a holdfast, which is branched and fits intimately to the microtopography of the rocks. From the holdfast arises the stipe, a stem-like structure which supports the frond in the water column. The frond is a wide flat area which gives a high surface area to volume ratio for light, carbon dioxide, and nutrient absorption. Superficially there is a resemblance to the roots, stem, and leaves of higher plants but there is no real equivalence because of the different life style. The holdfast is not an absorptive root system and does not penetrate the substratum, unlike roots penetrating the soil, since the seaweed can obtain all its requirements by direct absorption over its surface. The stipe is not a stem containing transport systems, as in higher plants, since these are not needed, again because of direct absorption. Similarly the frond does not have the complex structure of leaves with gas exchange and water retention organs such as stomata. Seaweeds do not have resistant phases such as seeds. Development is direct from spores or zygotes.

A seaweed, such as a kelp, can be viewed as a chemical factory taking in light, nutrients, and inorganic carbon from the water and converting them into organic matter. This production can be going on even when the plant does not seem to be increasing in size. The formation of new organic matter is then balanced by the loss of decaying tissue from the tip of the plant and by organic secretions from the frond. Both of these will be contributing to heterotrophic production in the kelp's ecosystem.

Habitat Diversity of Benthic Algae

Microphytobenthos in sedimentary shores can be distinguished according to whether they are epipsammic (attached to sand particles) or epipelic (between mud particles). They are mainly unicellular forms: diatoms, euglenoids, and blue–greens in estuarine muds; diatoms, dinoflagellates, and blue–greens in sand. Many show vertical migration within the top few millimeters of the sediment, photosynthesizing when the tide is out and burrowing before the return of the tide so that some escape being washed away. This is not 100% effective so that resuspended microphytobenthos can be a significant proportion of apparent phytoplankton in some estuaries. Diatoms migrate by gliding motility whereas euglenoids do so by alternate contraction and relaxation of the cell shape (metaboly). In estuaries, which may be turbid environments where photosynthesis by submerged plants may be reduced, and large expanses of intertidal mud flats may be available, microphytobenthos could be important primary producers which have been underestimated because they are not visually obvious. They may also help to stabilize sediments by the mucus secretions which keep them from desiccation when on the mud surface.

Seaweeds do not just grow attached to hard surfaces such as bedrock, boulders, and artificial structures. In Britain, a habitat-diverse, open coast shore is likely to have 70–100 species of seaweed present out of a British total of about 630 species. Such a high total on a shore is only realized because of many habitat variations that harbor the more microscopic species.

Most seaweeds have smaller species that grow attached to them as epiphytes. In turn they have even smaller species attached and this may continue for several orders down to very small microscopic plants. Endophytes are microscopic algae that grow between the cells within the tissues of larger ones. Epizoic algae grow attached to animals and endozoic forms grow inside animals, usually in skeletal parts. These include algae that penetrate calcareous substrata, i.e., shell-boring algae – red, green, and blue–green forms that bore through mollusk and barnacle shells and coral skeletons. They also include forms that inhabit proteinaceous animal skeletons – red and green filamentous species in skeletons of hydroids and bryozoans, and filamentous green seaweeds, mainly *Tellamia*, in the periostracum of periwinkles. Some of the shell-boring algae are also endolithic, boring through chalk rocks and so possibly aiding coastal erosion. Finally, there are endozoic algae that live in soft parts of benthic animals. Zooxanthellae are nonmotile dinoflagellate unicells in coral polyps, where they contribute to the high productivity of coral reefs, and unicellular blue–greens live in the tissues of some sea-slugs.

Taxonomic Classification of Algae

Algae are classified into a number of divisions of the plant kingdom (equivalent to phylum), varying in number from about 8 to 16 depending on author. Distinction is based on fundamental cellular and biochemical features and so is independent of the form of the plant. Each division can contain a range of forms, from unicellular to complex multicellular, although in many divisions the unicellular and colonial forms predominate. General features used to distinguish the divisions are:

- The range of accessory photosynthetic pigments present in addition to chlorophyll *a* (other chlorophylls, carotenoids, and biloproteins);
- The secondary more soluble components of the cell wall present in addition to the main fibrillar component;
- The chemical nature of the insoluble storage products resulting from excess photosynthesis;
- The presence, fine structure, number, and position of flagella on vegetative cells of flagellate organisms or on flagellate reproductive bodies of larger species;
- Specialized aspects of cell structure, peculiar to particular divisions, such as the silica frustules, which encase diatom cells.

The three biochemical features above are relatively uniform in the higher plants, compared with the algae, and similar to those of one algal division, *Chlorophyta* (green algae). This suggested origin of land plants occurred from only this one division, although this is now contested. At a fundamental level the algae are therefore much more diverse than the higher plants. Characteristics for each division can be found in the Further Reading list.

The seaweeds are the macroscopic marine algae in the divisions *Chlorophyta* (green), *Phaeophyta* (brown) and *Rhodophyta* (red algae). Greens are mainly foliose and filamentous seaweeds. Browns have no unicellular forms and include the very large complex and leathery forms such as kelps and rockweeds. Reds include a wide range of heterotrichous forms forming a wide diversity of complex foliose and filamentous forms.

Seaweed Life Cycles

Most seaweeds have more than one phase in their life cycle. They have generally the same pattern as higher plants where a sexually reproducing gametophyte generation gives rise to an asexually reproducing sporophyte generation and vice versa. There is not always an obligate alternation as in higher plants and there is much more diversity in the nature of the different phases in algae, with some red algae even having a third generation. Some only have one generation. In some cases this may reflect lack of experimental culture which is necessary for life cycle determination.

The simplest life cycle with two phases is termed isomorphic where the gametophyte and sporophyte are morphologically identical. Many common green seaweeds are like this, e.g., *Ulva* and *Enteromorpha*. Life cycles with morphologically different phases are heteromorphic. An example is kelp plants which have a massive leathery sporophyte which alternates with a microscopic filamentous gametophyte.

In many cases the two generations in a heteromorphic life cycle may have been known since the nineteenth century by separate names from before the life cycle was determined in culture. For example, the various species of the red seaweed, *Porphyra*, alternate with a filamentous shell-boring sporophyte formerly known as *Conchocelis rosea*.

Life cycles can be under environmental control. In some *Porphyra* species the change between generations is controlled by daylength, bringing about an annual seasonal life cycle. In some simpler life cycles, individual generations may be able to propogate themselves so that the full life cycle is not seen. This can be environmentally controlled so that, for example, in Europe there is a change with latitude of the relative proportions of the sexual and asexual generations of the brown filamentous seaweed, *Ectocarpus*, connected with latitudinal variation of sea temperature.

The Validity of Laboratory Cultures of Phytobenthos

Experimental culture is needed to determine life cycles but it is difficult to simulate all environmental conditions. Wave action and water flow are not usually simulated in the numerous batch culture dishes needed for replicated ecological experiments. Artificial light sources are unlike daylight in spectral composition and intensity, yet light quality may control photomorphogenesis in plants. It is therefore possible that laboratory culture could give false results. Two examples are given below.

The red seaweeds *Asparagopsis armata* and *Falkenbergia rufulanosa*, originally described as separate species, are phases in the same life cycle. During the twentieth century they have been spreading their geographical limit northwards in Europe from the Mediterranean to northern Scotland. Populations in Britain are rarely seen with reproductive organs but have a mode of attachment to the rock surface that suggests they were produced by vegetative reproduction. The two phases seem to have spread independently in Britain although in culture they could be made to participate in the same life cycle. This reflects a wider phenomenon in red seaweeds where by going north and south from the center of geographical distribution, reproductive potential declines.

In ecological experiments aseptic conditions are often not used, to the surprise of microbiologists. This lack of sterility may be desirable. For example, the green foliose seaweed, *Monostroma*, develops abnormally as a filamentous form in aseptic culture because of the need for growth factors from contaminating bacteria. These examples are not meant to decry culture experiments but to counsel their critical interpretation.

Seaweed Ecology

Seaweeds are present on all rocky shores but are more obvious where wave action is less. On temperate shores intertidal zones are generally dominated by brown fucoid seaweeds (wracks or rockweeds), while the shallow subtidal area is occupied by a kelp forest formed of large laminarian seaweeds. A variety of red, green, and brown seaweeds forms an understorey.

Some general rules can be exemplified by consideration of shores in north-west Europe. Firstly, on intertidal rocky shores the number of species is greatest at the lower tidal levels and declines with increasing intertidal height. Shores sheltered from wave action show the greatest cover and biomass of seaweeds. Shores exposed to very strong wave action tend to be dominated by sessile animals rather than seaweeds. Most shores are intermediate in wave action and tend to have a mosaic distribution of organisms, at least on lower and mid-shore. This may include patches of grazing animals interspersed with patches of different seaweed communities. The mosaic may make it hard to see a zonation of organisms with height on the shore. On very sheltered shores there may be a very obvious zonation of large brown seaweeds, in order of descending height on the shore: *Pelvetia canalicaulata*, *Fucus spiralis*,

Fucus vesiculosus and *Ascophyllum nodosum, Fucus serratus, Laminaria digitata.* (Similar zonations, but with different species, may occur on temperate shores outside north-west Europe.) Clear-cut zonations with visually dominant species can give a false impression that discrete communities exist at different tidal heights. Understorey species also have zones but their boundaries do not coincide with the larger species so as not to give sharply delimited communities.

With increase in wave action there is a change of species, e.g., in fucoids *Fucus serratus* is replaced by *Himanthalia elongata* and in laminarians *Laminaria digitata* is replaced by *Alaria esculenta.* Some species change form with wave action, for example, the bladder wrack, *F. vesiculosus,* loses its bladders (such morphological plasticity is a common feature confusing seaweed identification). With increased wave action, zones increase in breadth and height on shore. All these features can be incorporated in biologically defined exposure scales for the comparative description of rocky shores, which place shores on a numerical exposure scale. This facilitates the comparison of similar shores in pollution-monitoring studies to ensure that differences along a pollution gradient are due to human disturbance rather than to wave action.

Rock pools interrupt the gradient of conditions with height on shore. They provide a constantly submerged environment, like the subtidal one, but which is of limited volume and so undergoes physicochemical fluctuations while the tide is out, unlike the open sea. There is a corresponding zonation of dominant seaweed types in rock pools. On the lower shore they are characterized by sublittoral species and on the mid-shore they include species restricted to pools such as *Halidrys siliquosa.* Upper shore pools, where salinity fluctuates, have few species and are characterized by euryhaline opportunists such as *Enteromorpha* spp.

Intertidal zonation is only partly due to the desiccation and salinity tolerance of the seaweeds. Such factor tolerance is particularly important on the upper shore where conditions are most harsh for a marine organism and so few species are present, with few biotic interactions. On mid- and lower-shores biotic factors are important in determining species boundaries. Grazing by limpets and periwinkles on smaller algae, and on the microscopic germlings of larger ones, is important as is biological competition, which narrows down species occurrence to less than their tolerance range.

Subtidally a zonation may also be seen with dense kelp forest, with a large variety of understorey species in the shallowest water, below which is a kelp park with only scattered plants. Below the kelp depth limit may be a red algal zone to the photic depth limit where light becomes insufficient for positive net photosynthesis. Important factors are again both physical and biotic. Light tolerance plays a role but in the shallowest waters, where plant density is greatest, competition is important and zone limits may be set by grazers, this time by sea urchins rather than limpets.

There is an old view that accessory pigment differences between green, brown, and red seaweeds equip them to dominate at different depths according to which spectral quality of light penetrates. This is an oversimplification. Deeper-growing plants are shade plants with lower overall light requirements and can be of any color group.

Distribution into estuaries along a generally decreasing salinity gradient is another modifying factor like wave action. Colonization of hard surfaces by phytobenthos in estuaries is largely by marine species with species number declining going upstream. This occurs by selective attenuation firstly of red, then of brown species. Estuaries have broadly two algal zones: an outer one with fucoid dominated shores, which are a species-poor version of a sheltered open coast shore; and an inner zone dominated by filamentous mat-forming algae, principally greens and blue–greens.

There can be a successional sequence in which a bare area of shore is successively colonized, starting with unicells, with increasingly larger and more complex algae. Patches in a mosaic distribution may be at different stages in such a sequence. The succession involves contrasting types of seaweed as shown in **Table 1**.

Although opportunists are good at colonizing bare rock, in a stable environment they are eventually replaced by more precisely adapted late successional species.

Various conditions may favor the unusual abundance of opportunists. Mistaken conclusions about effects of effluent discharges can be reached by environmentalists who do not realize that there are both natural and artificial causes. Opportunist domination of rocks is favored naturally by sand scour and they may have natural summer outbursts in temperate climates. Pollution, particularly by sewage-derived nutrients, may favor opportunist domination. On tidal flats, 'green tides' may occur where the mudflat is completely covered by thick opportunist mats. This can induce anoxia and ammonia release beneath the mat so inhibiting benthic invertebrate populations and interfering with bird-feeding on tidal flats.

Table 1 Types of seaweed involved in successional colonization

	Opportunist species early in succession, e.g., foliose and filamantous green algae	Late successional species, e.g., large cartilaginous plants such as fucoids
Morphology	Simple	Complex
Size	Smaller	Larger
Growth rate	Faster	Slower
Life span	Shorter	Longer
Reproduction	All year round	Likely to have seasons
Environmental tolerance	Very wide but not precise adaptation to a specific niche	Narrower but good adaptation to a specific niche

Uses of Seaweed

Benthic macroalgae are an important biological resource. Their principal uses are human food, animal feeds, fertilizer, and industrial chemicals.

In the West, human seaweed consumption is small compared with the east. One of the most valuable fisheries listed by the Food and Agriculture Organization (FAO) is a seaweed, *Porphyra*, used extensively for human consumption in Japan. In the West, human consumption is more a health food market with beneficial effects ascribed to high trace element content, because of the high bioaccumulation activity of many seaweeds, but this remains to be rigorously tested.

Aqueous extracts of seaweeds such as fucoids and kelps are used as commercial and domestic fertilizers. Beneficial effects have been claimed in horticulture such as increased growth, faster ripening and increased fruit yield. Effects are usually ascribed to trace element or plant hormone (usually cytokinin) content.

Various high value fine chemicals, such as pigments, can be obtained from seaweeds but the main seaweed chemical industry is extraction of phycocolloids. These are secondary cell wall components of red and brown seaweeds, which have gel-forming and emulsifying properties in aqueous solutions. This gives them hundreds of industrial applications ranging from textile printing to ice cream manufacture. The chemicals concerned are all macromolecular carbohydrates, principally alginates from brown seaweeds and agars and carrageenans from red seaweeds. Alginate is not a single substance but a biopolymer with considerable possibility for structural variation. It is a linear structure of repeating sugar units of two kinds, mannuronic acid (M units) and guluronic (G units). Different alginates vary in the ratio of M and G units and in chain length (total number of units). Structural differences confer different gel-forming and emulsifying properties making different alginates suitable for different industrial applications. In turn, different alginate structures are found in different species so that different seaweeds are required for different applications.

Ensuring Supplies of Commercial Seaweeds

Seaweeds can be harvested from natural populations or farmed in the sea. The approach may depend on the value of the product. In the West, harvesting is preferred for phycocolloids. Despite the wide industrial application they are low-value products and so will not support the high labor costs needed for cultivation. By contrast alginate is successfully produced from farmed kelps in China, where labor costs are lower, and *Porphyra* can be farmed for human consumption in Japan because of its high value.

Sustainable harvesting should consider the ability of the seaweed resource to recover. Mechanized *Macrocystis* (giant kelp) harvesting in California is sustainable yet productive because of the growth pattern of the plant. It grows to almost 70 m long and cropping of the distal few meters by barges floating over the forest canopy allows numerous meristems to remain intact so growth continues. By contrast *Laminaria hyperborea* harvesting in Norway is more destructive because the desired alginate is in the stipe (the supporting 'stem') of the plant. Harvesting by kelp dredges, i.e., large bags with cutters at the front end towed over the seabed on skis just above the rock surface, cuts plants off just above the holdfast, so that the meristematic area is harvested. Sustainability of the *Laminaria* forest is based on the vegetation structure. There are different layers of plants. The dredge takes only the large canopy-forming plants with tough, nonyielding stipes. The smaller plants bend rather than breaking and so survive harvesting. With the absence of the canopy they receive more light and so grow quickly to replace the harvested plants. The forest biomass is regenerated in 3 years so the Norwegian

Government licenses areas of the seabed for harvesting not less than every 4 years. The regenerated kelp is better for alginate extraction as it is not contaminated by epiphytes. However, the sustainability of the kelp forest biodiversity is separate from commercial yield. It is a diverse ecosystem with many invertebrates in the kelp holdfasts, as well as diverse seabed fauna and flora, and pelagic species sheltered by the forest. The diversity of this had not recovered after much more than 4 years so the licensing system does not conserve the full ecosystem.

There are biotic considerations in protecting the phytobenthic resource. In the early 1970s the decline of the sea otter population off California allowed its prey, a sea urchin which was a kelp grazer, to increase. This hindered natural regeneration of the forest, which was already declining due to sewage pollution from Los Angeles. Recovery required manual transplantation. Various species of sea urchin are among the most voracious subtidal grazers. Another one, *Strongylocentrotus*, was able to create barren areas of seabed off the Canadian coast where *Laminaria longicruris* had been harvested, thus preventing regeneration.

Farming of seaweed can require knowledge from laboratory culture of environmental tolerances and the factors controlling life cycles. This allows manipulation of stocks in seawater tanks on land under controlled conditions to produce reproductive bodies. These can be used to seed ropes, nets, canes or other substrata which are then planted out into sheltered sea areas for growing on. This increases the habitat area for attachment in the sea, and so increases yield. In the case of *Porphyra* in Japan, such manipulation of environmental conditions allows up to five generations per year from an area of coast where natural seasonal changes would only give one. It is not practical to grow the plants entirely on land in environmentally controlled seawater tanks because of the high cost of the facilities, but it is feasible to retain a small stock of, for example, shells infected with the *Conchocelis*-phase of *Porphyra* from which spores can be obtained on demand by manipulation of daylength and temperature. The technology exists to cultivate seaweeds entirely in land-based tanks should the economics be favorable in the future. For example, there are different genetic strains of *Chondrus crispus*, whose carrageenan yield and type can be controlled by nutrient and salinity conditions. Such tank culture may be useful in the future to produce high value fine chemicals for medical applications.

Seaweeds are the most obvious type of plant in the sea and are the main component of the phytobenthos but other algae and other marine plants are described elsewhere in this Encyclopedia.

See also

Coral Reefs. Exotic Species, Introduction of. Rocky Shores.

Further Reading

Guiry MD and Blunden G (eds.) (1991) *Seaweed Resources in Europe: Uses and Potential*. Chichester: Wiley.

Hoek C, van den, Mann DG, and Jahns HM (1995) *Algae: An Introduction to Phycology*. Cambridge: Cambridge University Press.

Lembi CA and Waaland JR (eds.) (1988) *Algae and Human Affairs*. Cambridge: Cambridge University Press.

Lobban CS and Harrison PJ (1997) *Seaweed Ecology and Physiology*. Cambridge: Cambridge University Press.

Luning K (1990) *Seaweeds: Their Environment, Biogeography and Ecophysiology*. New York: Wiley.

South GR and Whittick A (1987) *Introduction to Phycology*. Oxford: Blackwell Scientific Publications.

MANAGEMENT, MARICULTURE AND FISHERIES

COASTAL ZONE MANAGEMENT

D. R. Godschalk, University of North Carolina, Chapel Hill, NC, USA

Introduction

Home to over half the world's population, coastal area environments and economies make critical contributions to the wealth and well-being of maritime countries. Market values stem from fisheries, tourism and recreation, marine transportation and ports, energy and minerals, and real estate development. Nonmarket values include life support and climate control through evaporation and carbon absorption, provision of productive marine and estuarine habitats, and enjoyment of nature.

Cumulative effects of human activities and coastal engineering threaten the sustainability of coastal resources. We are depleting coastal fisheries, degrading coastal water quality, draining coastal wetlands, blocking natural beach and barrier island movements, and spoiling recreational areas. Population growth is putting pressure on fragile ecosystems and increasing the number of people exposed to natural hazards, such as hurricanes, typhoons, and tsunamis. Maintaining sustainability demands active intervention in the form of coastal management.

Managing the conservation and development of coastal areas is, however, a challenging enterprise. Coastal land and water resources follow natural system boundaries rather than governmental jurisdiction boundaries, fragmenting management authority and responsibility. Coastal scientists and coastal managers operate in different spheres, fragmenting knowledge creation and dissemination. Coastal resource programs tend to focus on single sectors, such as water quality or land use, fragmenting efforts to manage holistic ecosystems. The field of coastal management has evolved to deal with the challenges of managing fragmented transboundary coastal resources.

Evolution of Coastal Zone Management

Coastal zone management developed to protect coastal resources from threats to their sustainability and to overcome the ineffectiveness of single function management approaches. Unlike ocean management issues, such as freedom of navigation and conservation of migratory species, coastal management traditionally focused on issues related to the land–sea interface, such as shoreline erosion, wetland protection, siting of coastal development, and public access. Important conceptual landmarks in the field's development are the intergovernmental framework of the 1972 US Coastal Zone Management Act, the 1987 Brundtland report proposing sustainable development, and the 1992 Earth Summit recommendation for initiation of integrated coastal management (ICM).

Coastal Zone Management

In 1972, the United States enacted the Coastal Zone Management Act to create a formal framework for collaborative planning by federal, state, and local governments. To receive federal funding incentives, states were required to set coastal zone boundaries, define permissible land and water uses, and designate areas of particular concern, such as hazard areas. An additional incentive was the promise that federal government actions would be consistent with the state's approved coastal zone management program. In practice, state programs tended to focus on land-use planning and regulation. Many other developed countries followed suit, establishing their own coastal zone management programs.

North Carolina offers an example of a state program developed under the Coastal Zone Management Act. Its 1974 Coastal Area Management Act established a coastal resource management program for the 20 coastal counties influenced by tidal waters. Policy is made by the Coastal Resources Commission, whose members are appointed by the governor. The act is administered by the state Division of Coastal Management, which issues coastal development permits, manages coastal reserves, and provides financial and technical assistance for local government planning. Required local land-use plans must meet approved standards, include a post-storm reconstruction policy, and identify watershed boundaries, but localities have considerable implementation flexibility. Four Areas of Environmental Concern are designated: estuarine and ocean, ocean hazard, public water supplies, and natural and cultural resources areas. All major development projects within an Area of Environmental Concern must receive a state permit. Small structures must be set back beyond the 30-year line of estimated annual beach erosion and large structures must be set back beyond the 60-year line. Shore-hardening structures are banned. While the North Carolina program has enjoyed considerable success, in practice many coastal localities continue to put economic development ahead of environmental protection.

The New Paradigm of Sustainable Development

In 1987, the UN World Commission on Environment and Development published its report *Our Common Future*, referred to as the Brundtland report in recognition of its chairman, Norwegian Prime Minister Gro Harland Brundtland. The report called for global sustainability, a concept which is now the dominant paradigm in coastal management. Sustainability requires economic development which meets the needs of the present generation to be achieved without compromising the ability of future generations to meet their own needs. Sustainability demands balance among the elements of a triple bottom line, sometimes referred to as 'the three e's': ecological sustainability, social equity, and economic enterprise. The triple bottom line has become a touchstone for accountability reporting of business as well as government actions. The 2002 UN World Summit on Sustainable Development revalidated the concept as a collective responsibility at local, national, regional, and global levels.

The main ideas of sustainable development can be stated in terms of questions to be asked of every environment and development decision:

● How does it improve the quality of human life?
● How does it affect the environment and natural resources?
● Are its benefits distributed equitably?

The Rise of Integrated Coastal Management

The 1992 United Nations Conference on Environment and Development in Rio de Janeiro – the Earth Summit – recommended principles to guide actions on environment, development, and social issues. It also approved Agenda 21, an action plan for sustainable development. While both the principles and actions are nonbinding, they represent a major shift toward understanding that sustainability must address the interdependence of environment and development in both developed countries (North) and developing countries (South). As shown in **Figure 1**, global environmental problems, such as greenhouse gases, generated in the North threaten the ability of the South to develop, while poverty and overpopulation in the South lead to local environmental stresses, such as air and water pollution.

Interdependence generates the need for integration between environment and development in sectors

Figure 1   Interdependence of environment and development. Source: Cicin-Sain B and Knecht RW (1998) *Integrated Coastal and Ocean Management: Concepts and Practices*, p. 83. Washington, DC: Island Press.

and nations. The ocean and coastal issues chapter of Agenda 21 calls for ocean and coastal management to be integrated in content and precautionary and anticipatory in ambit. The precautionary principle holds that lack of full scientific certainty shall not be used as a reason for postponing cost-effective measures to prevent environmental degradation. In other words, a conservative regulatory and management approach should be taken, in the absence of convincing evidence to the contrary.

The Agenda 21 section on integrated management and sustainable development of coastal areas calls on each coastal state to establish coordination mechanisms at both the local and national levels. It suggests undertaking coastal- and marine-use plans, environmental impact assessment and monitoring, contingency planning for both human-induced and natural disasters, improvement of coastal human settlements, conservation and restoration of critical habitats, and integration of sectoral programs such as fishing and tourism into a coordination framework. It also calls for national guidelines for ICM and actions to maintain biodiversity and productivity of marine species and habitats. The section highlights the need for information on coastal and marine systems and natural science and social science variables, along with education and training and capacity building.

The European Commission defines integrated coastal zone management (ICZM) as "a dynamic, multidisciplinary and iterative process to promote sustainable management of coastal zones. It covers the full cycle of information collection, planning (in its broadest sense), decision making, management and monitoring of implementation. ICZM uses the informed participation and cooperation of all stakeholders to assess the societal goals in a given coastal area, and to take actions towards meeting these objectives. ICZM seeks, over the long-term, to balance environmental, economic, social, cultural and recreational objectives, all within the limits set by natural dynamics. Integrated in ICZM refers to the integration of objectives and also to the integration of the many instruments needed to meet these objectives. It means integration of all relevant policy areas, sectors, and levels of administration. It means integration of the terrestrial and marine components of the target territory, in both time and space."

Coastal managers worldwide face many of the same problems: environmental degradation, marine pollution, fishery depletion, and loss of marine habitat. However, each nation's coastal management programs differ due to unique geography, development issues, and political system. Still, similar packages of tools and techniques are seen in many countries. The packages include combinations of regulations, national policies, planning, diagnostic studies of natural and socioeconomic systems and government capacity, incentives provision, and consensus building and participation to respond to conflicts. The ICM influence often is less evident in those developed countries with coastal programs already in place, than in less-developed countries where external funding assistance is needed to generate new coastal management institutions and to sustain traditional coastal livelihoods.

How well does ICM work? Evaluation efforts typically take the form of best practice stories, assessments of process outputs, or descriptions of levels of implementation. Program success indicators based on management evaluations indicate that successful programs tend to be comprehensive (based on ecosystem boundaries, such as watersheds or river basins), participatory (involving stakeholders), cooperative (networking among organizations), contingent (allowing for uncertainty and change), precautionary (acting conservatively to preserve the environment in the absence of conclusive evidence), long-term (with time horizons beyond project deadlines), focused (aimed at perceived issues or problems), incremental (taking small steps toward objectives, as in managed retreat from the shore), and adaptive (using learning to redirect program efforts). When more experience has been acquired, it may be possible to evaluate program outcomes or long-term, large-scale successes.

What types of ICM actions have been taken? Practices may take the form of innovative programs or institutions, as well as innovative applications to particular problems or content areas. Examples of both types of successful ICM best practices are illustrated in **Table 1**.

Regional Integrated Coastal Management Initiatives

With the maturing of ICM, its scope has expanded to include regional or transnational projects and programs. Inherent linkages among terrestrial and ocean processes have led to a number of regional ocean initiatives; 16 regional action plans have been formally adopted as of 2001 under the UN's Regional Seas Program. Each plan focuses on the unique problems of its region. For example, the East Asian Seas region has tackled problems of coastal aquaculture, fisheries exploitation, coral reef restoration, and resource extraction, using an ecoregion approach. Nations in the Western Indian Ocean region, where 70% of the African continent earn their living from natural resources, focus on simultaneously alleviating poverty and conserving natural resources. These developing regions promote sustainable

Table 1 Successful ICM practices

Long-range planning and marine zoning for Australia's Great Barrier Reef Marine Park, where use and protection zones were mapped and participatory visioning was used to prepare a 25-year long-range plan

Bringing ocean and coastal management together in the Republic of Korea's new Ministry of Maritime Affairs and Fisheries, which integrated coastal zone management at the national level with navigation, ports, and fisheries programs

Involving publics in operation of the special area management zones in Ecuador, through education, training, and outreach

Determining the value of coastal ocean utilization in the pilot ICM program in Xiamen, China, through valuing the use of ocean space and charging users for mariculture and anchorage space

Incorporating tradition management practices in American Samoa's villages, which are responsible for monitoring and enforcing ICM measures on their land and waters

Controlling coastal erosion in Sri Lanka through establishment of a coastal zone development permit system, prohibition of coral mining except for research, and recruiting coastal communities into the development of management plans

Control of nonpoint marine pollution in the Chesapeake Bay, where a partnership among three US states (Virginia, Maryland, and Pennsylvania) and the federal government applies science to restore America's largest and most productive estuary

Protecting coastal resources in Turkey though designating specially protected areas to address rampant tourism development and its impacts on coastal waters and fisheries

Community-based coral reef protection in Phuket Island, Thailand, with a bottom-up program of education and outreach to protect the main tourist attraction, 60% of which has been seriously damaged or degraded

Excerpted from Cicin-Sain B and Knecht RW (1998) *Integrated Coastal and Ocean Management: Concepts and Practices*, pp. 297–299. Washington, DC: Island Press.

livelihoods in order to meet both conservation and development goals.

The example of the European Union illustrates the evolution of a regional program. They conducted a 1996 demonstration with 35 projects in member countries and six thematic analyses of key ICM factors, including legislation, information, EU policy, territorial and sectoral cooperation, technical solutions, and participation. Their ICM strategy, approved in 2000, recommended that member states commit to a common vision for the future of their coastal zones, based on durable economic opportunities, functioning social and cultural systems in local communities, adequate open space, and maintenance of ecosystem integrity. It set out principles for a long-term holistic perspective, adaptive management, local specificity, working with natural processes, participatory planning, involvement of relevant administrative bodies, and use of a combination of instruments.

The EU strategy was able to draw on the experience gained from the 1975 Mediterranean Action Plan (MAP), one of the regional seas programs of the United Nations Environment Program (UNEP). Originally focused on marine pollution control, the MAP widened to deal with the land-based sources of 80% of pollution sources. It created coastal area management programs and, following Agenda 21, shifted them to an integrated basis. The Mediterranean region faces rapid coastal urbanization, growing tourist activities, high water consumption, concentrated pollution hot spots, biodiversity losses, and soil erosion. Lessons learned from the MAP experience emphasize the need for building an evaluation and monitoring mechanism at the beginning. Programs start out oriented to specific issues and problems, but must become more comprehensive at later stages in order to deal with complex linkages and provide integrated solutions. Strong political commitment at all levels to the preparation and initiation of initiatives, along with participation of stakeholders and end-users from program design through implementation, are of utmost importance. While ICM has provided a sturdy conceptual framework, full integration remains an elusive goal, as discovered in evaluation studies.

Ten MAP projects implemented during 1987–2001 were evaluated. The projects were located in three entire national coastal areas (Albania, Syria, and Israel), two areas of large parts of the national coast (Malta and Lebanon), two semi-enclosed bays polluted by urban and industrial expansion (Kastela Bay in Croatia and Izmir Bay in Turkey), two islands (Rhodes and Malta), a polluted industrialized urban coast (Sfax, Tunisia), and a relatively virgin coastal area under threat of uncontrolled development (Fuka-Matrouh, Egypt). The evaluation found that many changes occurred in the region during the multiyear implementation period, requiring the ICM projects to adapt to the new conditions, institutional structures, and coastal area priorities. Initial pilot projects focused on data collection, studies of pollution and ecosystem protection, and capacity building. First- and second-cycle projects focused on sectoral studies, resource management, implementation and its basic tools (e.g., environmental impact assessment and geographic information system or GIS), and introducing carrying capacity assessment. Third-cycle projects focused on sustainable development, applying ICM, and newer tools (strategic environmental assessment, systemic sustainability analysis, sustainability indicators, resource valuation, and socio-economic evaluation).

The evaluation reported that the projects had an overall significant impact on national systems in the

Mediterranean region, with a more significant impact on the developing countries. The impacts directly influenced the practices, capacities, and awareness of the national and local authorities, institutions, teams, and stakeholders, as well as the resident coastal populations. The result was a strengthened awareness of the importance of coastal areas and the need for sustainable development, as well as an improvement in the ICM concepts and practices. Three of the 10 projects were also evaluated by a World Bank/Mediterranean Environmental Technical Assistance team in 1997. They found that, despite weaknesses in preparation and financial support, the overall impact of the projects was good, especially in terms of increased institutional capacity, impact on decision makers, and catalytic role of the projects. Performance problems were weak sectoral integration, weak participation and absence of domestic financial support, and uncertain project follow-up.

Assessments of ICM bring out both its strengths and weaknesses. Its strengths are based on the rationality of an integrated attack on interdependent coastal problems aimed at balancing ecological, economic, and equity values. No single government agency or program is equipped to deal with the complex interactions of oceans and coasts, yet it is precisely these interactions that pose such serious challenges to global sustainability. Its weaknesses stem from its need to reform and reinvent long-standing institutions, political turfs, and development practices. The times necessary to achieve such ambitious reforms far exceed the times of the usual 4–6-year ICM project. In absence of a crisis event to catalyze, and a huge infusion of funds to pay for, the necessary institutional change, progress proceeds slowly and long-term coordination takes a back seat to short-term demands. These same weaknesses plague efforts to manage all transboundary resources: air, water, and land.

Future Directions

All indications are that the need for ICM will continue to grow along with continued integration efforts. Serious global environmental problems are on the rise and more and more people are flocking to coastal areas where they will be exposed to potential catastrophes. The natural response by scientists and planners, governments and nongovernmental organizations, and other concerned stakeholders will be to call for further integration of coastal and ocean programs, institutions, and strategies. Increasing support for sustainable development in the form of green buildings and smart growth by professional organizations, private businesses, and governmental bodies should provide a positive foundation for strengthening ICM around the world. Still the questions remain: Can we expect the emergence of sufficient political will to reform decades of territorial and institutional balkanization? Can we overcome the perception that ICM actions are an obstacle to economic development of the coast?

Indicators of a Looming Coastal Crisis

One possible catalyst for a worldwide surge in ICM is the impacts of the dangerous combination of climate change, natural hazards, resource depletion, poverty, and coastal population growth. Scientific studies have made it increasingly difficult for politicians to deny the reality of global warming. The transformation of the Earth's atmosphere through carbon dioxide emissions is driving up global temperature, which is expected to cause a string of disasters, including hurricanes, droughts, glacial melting, sea level rise, and ocean acidification. Since the turn of the century, the world experienced two of history's most catastrophic coastal disasters – the 2004 Indian Ocean tsunami with a death toll of 230 000 people triggered by an earthquake off the coast of Indonesia and the 2005 hurricane Katrina which devastated New Orleans and coastal Mississippi causing 1836 deaths and $81.2 billion in property damage. Yet people, rich and poor alike, continue to move to the ocean's edge, fueling the growth of vulnerable settlements and megacities. An estimated 40% of world cities of 1–10 million population and 75% of cities over 10 million population were located on the coast in 1995 and the number is growing. Clearly, continuation of these trends is a recipe for future disasters unless coastal settlement patterns resilient to natural hazards can be ensured.

Increasing Integration Efforts

One example of a major new ICM initiative is the 2004 report of the US Commission on Ocean Policy, *An Ocean Blueprint for the 21st Century*. The report acknowledges the magnitude of the crisis but insists that all is not lost. It envisions a future in which the coasts are attractive places to live, work, and play with clean water, public access, strong economies, safe harbors, adequate roads and services, and special protection for sensitive habitats and threatened species. Future management boundaries coincide with ecosystem regions, managers balance competing considerations and proceed with caution, and ocean governance is effective, participatory, and well coordinated. The report recommends creation of a National Ocean Council within the Executive Office of the President to coordinate federal policy, encourages groups of states to form networked regional ocean councils, and calls for federal laws and programs to be consolidated and modified to improve sustainability, including: amending the Coastal Zone Management Act to add measurable goals and performance measures and to implement watershed-based coastal zone boundaries; changing federal funding and infrastructure programs to discourage growth in fragile or hazard-prone coastal areas; and revising the National Flood Insurance Program to reduce incentives to develop in high-hazard areas. Response to the report has been generally positive, but it remains to be seen as to whether the recommendations will be implemented.

Possible Futures

The future of ICZM is closely tied to the future directions of global development. The Millennium Ecosystem Assessment has attempted to imagine possible future scenarios of global development pathways between 2000 and 2050. The scenarios represent different combinations of governance and economic development (global vs. regional) and of ecosystem management (reactive vs. proactive). They focus on ecosystem services and the effects of ecosystems on human well-being. The four scenarios are:

- Global Orchestration (socially conscious globalization, with emphasis on equity, economic growth, and public goods, and reactive toward ecosystems);
- Order from Strength (regionalized, with emphasis on security and economic growth, and reactive toward ecosystems);
- Adapting Mosaic (regionalized, with proactive management of ecosystems, local adaptation, and flexible governance); and
- TechnoGarden (globalized, with emphasis on using technology to achieve environmental outcomes and proactive management of ecosystems).

None of the scenarios represents a best or worst path, although ecosystems do better in the Techno-Garden scenario. However, they illuminate the potential consequences of choices in institutional design and management which could inform future decisions about coastal zone management.

Whether the catalysts of environmental threats, integrated management reforms, and scenarios of potential ecosystem outcomes can turn the tide toward fully integrated and effective coastal

management is an open question. What is clear is that a strong worldwide movement is actively pursuing this goal.

See also

Coastal Topography, Human Impact on. Economics of Sea Level Rise. Glacial Crustal Rebound, Sea Levels, and Shorelines. Tsunami.

Further Reading

Barbiere J and Li H (2001) *Third Millennium Special Issue on Megacities. Ocean and Coastal Management* 44: 283–449.

Beatley T, Brower DJ, and Schwab AK (2002) *An Introduction to Coastal Zone Management*, 2nd edn. Washington, DC: Island Press.

Belfiore S (ed.) (2002) *From the 1992 Earth Summit to the 2002 World Summit on Sustainable Development: Continuing Challenges and New Opportunities for Capacity Building in Ocean and Coastal Management. Special Issue on Capacity Building Ocean and Coastal Management* 45: 541–718.

Burbridge P and Humphrey S (eds.) (2003) *Special Issue: The European Demonstration Programme on Integrated Coastal Zone Management. Coastal Management* 31: 121–212.

Cicin-Sain B and Knecht RW (1998) *Integrated Coastal and Ocean Management: Concepts and Practices.* Washington, DC: Island Press.

Cicin-Sain B, Pavlin I, and Belfiore S (eds.) (2002) *Sustainable Coastal Management: A Transatlantic and Euro-Mediterranean Perspective.* Dordrecht: Kluwer.

Francis J, Tobey J, and Torell E (eds.) (2006) *Special Issue: Balancing Development and Conservation Needs in the Western Indian Ocean Region. Ocean and Coastal Management* 49: 789–888.

Glavovic B (2006) Coastal sustainability – an elusive pursuit: Reflections on South Africa's coastal policy experience. *Coastal Management* 34: 111–132.

Heilerman S (ed.) (2006) *IOC Manuals and Guides 46, ICAM Dossier 2: A Handbook for Measuring the Progress and Outcomes of Integrated Coastal and Ocean Management.* Paris: UNESCO.

Kay R and Alder J (2005) *Coastal Planning and Management*, 2nd edn. London: Taylor and Francis.

Mageau C (ed.) (2003) *Special Issue: The Role of Indicators in Integrated Coastal Management. Ocean and Coastal Management* 46: 221–390.

Sain B and Knecht RW (1998) *Integrated Coastal and Ocean Management: Concepts and Practices.* Washington, DC: Island Press.

Stojanovic T, Ballinger RC, and Lalwani CS (2004) Successful integrated coastal management: Measuring it with research and contributing to wise practice. *Ocean and Coastal Management* 47: 273–298.

Tobey J and Lowry K (eds.) (2002) *Learning from the practice of integrated coastal management. Coastal Management* 30: 283–345.

US Commission on Ocean Policy (2004) *An Ocean Blueprint for the 21st Century*. Final Report. Washington, DC: US Commission on Ocean Policy.

Vallega A (2002) The regional approach to the ocean, the ocean regions, and ocean regionalization – a post-modern dilemma. *Ocean and Coastal Management* 45: 721–760.

Williams E, Mcglashan D, and Finn J (2006) Assessing socioeconomic costs and benefits of ICZM in the European Union. *Coastal Management* 34: 65–86.

Relevant Websites

http://web.worldbank.org
  – Coastal and Marine Management, World Bank.
http://www.unesco.org
  – Coastal Regions and Small Islands (CSI) Portal, UNESCO.
http://www.coastalguide.org
  – EUCC: Coastal Guide.
http://ec.europa.eu
  – EUROPA Coastal Zone Policy.
http://www.keysheets.org
  – Integrated Coastal Management Keysheet.
http://www.coastalmanagement.com
  – Integrated Coastal Management Websites.
http://www.icriforum.org
  – International Coral Reef Initiative.

http://ioc3.unesco.org
  – IOC Marine Sciences and Observations for Integrated Coastal Management.
http://www.millenniumassessment.org
  – Millennium Ecosystem Assessment.
http://www.netcoast.nl
  – NetCoast: A Guide to Integrated Coastal Management.
http://www.oneocean.org
  – OneOcean: Coastal Resources and Fisheries Management of the Philippines.
http://www.pemsea.org
  – PEMSEA Integrated Coastal Management.
http://www.unep.org
  – Regional Seas Programme, United Nations Environment Programme.
http://www.deh.gov.au
  – The Contribution of Science to Integrated Coastal Management.
http://www.globaloceans.org
  – The Global Forum on Oceans, Coasts, and Islands.
http://www.csiwisepractices.org
  – Wise Coastal Practices for Sustainable Human Development Forum.
http://www.ngdc.noaa.gov
  – World Data Center for Marine Geology and Geophysics: National Geophysical Data Center.
http://earthtrends.wri.org
  – World Resources Institute: Earth Trends.

MARICULTURE OVERVIEW

M. Phillips, Network of Aquaculture Centres in Asia-Pacific (NACA), Bangkok, Thailand

Introduction: Mariculture – A Growing Ocean Industry of Global Importance

Global production from aquaculture has grown substantially in recent years, contributing in evermore significant quantities to the world's supply of fish for human consumption. According to FAO statistics, in 2004, aquaculture production from mariculture was 30.2 million tonnes, representing 50.9% of the global total of farmed aquatic products. Freshwater aquaculture contributed 25.8 million tonnes, or 43.4%. The remaining 3.4 million tonnes, or 5.7%, came from production in brackish environments (**Figure 1**). Mollusks and aquatic plants (seaweeds), on the other hand, almost evenly make up most of mariculture at 42.9% and 45.9%, respectively. These statistics, while accurately reflecting overall trends, should be viewed with some caution as the definition of mariculture is not adopted consistently across the world. For example, when reporting to FAO, it is known that some countries report penaeid shrimp in brackish water, and some in mariculture categories. In this article, we focus on the culture of aquatic animals in the marine environment.

Nevertheless, the amount of aquatic products from farming of marine animals and plants is substantial, and expected to continue to grow. The overall production of mariculture is dominated by Asia, with seven of the top 10 producing countries within Asia. Other regions of Latin America and Europe however also produce significant and growing quantities of farmed marine product. The largest producer of mariculture products by far is China, with nearly 22 million tonnes of farmed marine species. A breakdown of production among the top 15 countries, and major commodities produced, is given in **Table 1**.

Commodity and System Descriptions

A wide array of species and farming systems are used around the world for farming aquatic animals and plants in the marine environment. The range of marine organisms produced through mariculture currently include seaweeds, mollusks, crustaceans, marine fish, and a wide range of other minor species such as sea cucumbers and sea horses. Various containment or holding facilities are common to marine ecosystems, including sea-based pens, cages, stakes, vertical or horizontal lines, afloat or bottom set, and racks, as well as the seabed for the direct broadcast of clams, cockles, and similar species.

Mollusks

Many species of bivalve and gastropod mollusks are farmed around the world. Bivalve mollusks are the major component of mariculture production, with most, such as the commonly farmed mussel, being high-volume low-value commodities. At the other end of the value spectrum, there is substantial production of pearls from farming, an extremely low-volume but high-value product.

Despite the fact that hatchery production technologies have been developed for many bivalves and gastropods, much bivalve culture still relies on collection of seedstock from the wild. Artificial settlement substrates, such as bamboo poles, wooden stakes, coconut husks, or lengths of frayed rope, are used to collect young bivalves, or spat, at settlements. The spat are then transferred to other grow-out substrates ('relayed'), or cultured on the settlement substrate.

Some high-value species (such as the abalone) are farmed in land-based tanks and raceways, but most mollusk farming takes place in the sea, where three major systems are commonly used:

- Within-particulate substrates – this system is used to culture substrate-inhabiting cockles, clams, and other species. Mesh covers or fences may be used to exclude predators.

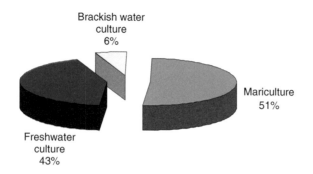

Figure 1 Aquaculture production by environment in 2004. From FAO statistics for 2004.

Table 1 Top 15 mariculture producers

Country	Production (tonnes)	Major species/commodities farmed in mariculture
China	21 980 595	Seaweeds (kelp, wakame), mollusks (oysters, mussels, scallops, cockles, etc.) dominate, but very high diversity of species cultured, and larger volumes of marine fish and crustaceans
Philippines	1 273 598	Seaweeds (*Kappaphycus*), with small quantities of fish (milkfish and groupers) and mollusks
Japan	1 214 958	Seaweeds, mollusks, marine fish
Korea, Republic of	927 557	Seaweeds, mollusks, marine fish
Chile	688 798	Atlantic salmon, other salmonid species, smaller quantities of mollusks, and seaweed
Norway	637 993	Atlantic salmon, other salmonid species
Korea, Dem. People's Rep.	504 295	Seaweeds, mollusks
Indonesia	420 919	Seaweeds, smaller quantities of marine fish, and pearls
Thailand	400 400	Mollusks (cockle, oyster, green mussel), small quantities of marine fish (groupers)
Spain	332 062	Blue mussel dominates, but high diversity of fish and mollusks
United States of America	242 937	Mollusks (oysters) with small quantities of other mollusks and fish (salmon)
France	198 625	Mollusks (mussel and oyster) with small quantities of other mollusks and fish (salmon)
United Kingdom	192 819	Atlantic salmon and mussels
Vietnam	185 235	Seaweeds (*Gracilaria*) and mollusks
Canada	134 699	Atlantic salmon and other salmonid species, and mollusks
Total (all countries)	30 219 472	

From FAO statistics for 2004.

- On or just above the bottom – this culture system is commonly used for culture of bivalves that tolerate intertidal exposure, such as oysters and mussels. Rows of wooden or bamboo stakes are arranged horizontally or vertically. Bivalves may also be cultured on racks above the bottom in mesh boxes, mesh baskets, trays, and horizontal wooden and asbestos-cement battens.
- Surface or suspended culture – bivalves are often cultured on ropes or in containers, suspended from floating rafts or buoyant long-lines.

Management of the mollusk cultures involves thinning the bivalves where culture density is too high to support optimal growth and development, checking for and controlling predators, and controlling biofouling. Mollusk production can be very high, reaching 1800 tonnes per hectare annually. With a cooked meat yield of around 20%, this is equivalent to 360 tonnes of cooked meat per hectare per year, an enormous yield from a limited water area. Farmed bivalves are commonly sold as whole fresh product, although some product is simply processed, for example, shucked and sold as fresh or frozen meat. There has been some development of longer-life products, including canned and pickled mussels.

Because of their filter-feeding nature, and the environments in which they are grown, edible bivalves are subject to a range of human health concerns, including accumulation of heavy metals, retention of human-health bacterial and viral pathogens, and accumulation of toxins responsible for a range of shellfish poisoning syndromes. One option to improve the product quality of bivalves is depuration, which is commonly practiced with temperate mussels, but less so in tropical areas.

Seaweeds

Aquatic plants are a major production component of mariculture, particularly in the Asia-Pacific region. About 13.6 million tonnes of aquatic plants were produced in 2004. China is the largest producer, producing just less than 10 million tonnes. The dominant cultured species is Japanese kelp *Laminaria japonica*. There are around 200 species of seaweed used worldwide, or which about 10 species are intensively cultivated – including the brown algae *L. japonica* and *Undaria pinnatifida*; the red algae *Porphyra*, *Eucheuma*, *Kappaphycus*, and *Gracilaria*; and the green algae *Monostrema* and *Enteromorpha*.

Seaweeds are grown for a variety of uses, including direct consumption, either as food or for medicinal purposes, extraction of the commercially valuable polysaccharides alginate and carrageenan, use as fertilizers, and feed for other aquaculture commodities, such as abalone and sea urchins.

Because cultured seaweeds reproduce vegetatively, seedstock is obtained from cuttings. Grow-out is undertaken using natural substrates, such as long-lines, rafts, nets, ponds, or tanks.

Production technology for seaweeds is inexpensive and requires only simple equipment. For this reason, seaweed culture is often undertaken in relatively undeveloped areas where infrastructure may limit the development of other aquaculture commodities, for example, in the Pacific Island atolls.

Seaweeds can be grown using simple techniques, but are also subject to a range of physiological and pathological problems, such as 'green rot' and 'white rot' caused by environmental conditions, 'ice-ice' disease, and epiphyte growth. In addition, cultured seaweeds are often consumed by herbivores, particularly sea urchins and herbivorous fish species, such as rabbitfish.

Selective breeding for specific traits has been undertaken in China to improve productivity, increase iodine content, and increase thermal tolerance to better meet market demands. More recently, modern genetic manipulation techniques are being used to improve temperature tolerance, increase agar or carrageenan content, and increase growth rates. Improved growth and environmental tolerance of cultured strains is generally regarded as a priority for improving production and value of cultured seaweeds in the future.

Seaweed aquaculture is well suited for small-scale village operations. Seaweed fisheries are traditionally the domain of women in many Pacific island countries, so it is a natural progression for women to be involved in seaweed farming. In the Philippines and Indonesia, seaweed provides much-needed employment and income for many thousands of farmers in remote coastal areas.

Marine Finfish

Marine finfish aquaculture is well established globally, and is growing rapidly. A wide range of species is cultivated, and the diversity of culture is also steadily increasing.

In the Americas and northern Europe, the main species is the Atlantic salmon (*Salmo salar*), with smaller quantities of salmonids and species. Chile, in particular, has seen the most explosive growth of salmon farming in recent years, and is poised to become the number-one producer of Atlantic salmon. In the Mediterranean, a range of warmer water species are cultured, such as seabass and seabream.

Asia is again the major producer of farmed marine fish. The Japanese amberjack *Seriola quinqueradiata* is at the top of the production tables, with around 160 000 tonnes produced in 2004, but the region is characterized by the extreme diversity of species farmed, in line with the diverse fish-eating habits of

the people living in the region. Seabreams are also common, with barramundi or Asian seabass (*Lates calcarifer*) cultured in both brackish water and mariculture environments. Grouper culture is expanding rapidly in Asia, driven by high prices in the live-fish markets of Hong Kong SAR and China, and the decreasing availability of wild-caught product due to overfishing. Southern bluefin tuna (*Thunnus mccoyii*) is cultured in Australia using wild-caught juveniles. Although production of this species is relatively small (3500–4000 tonnes per annum in 2001–03), it brings very high prices in the Japanese market and thus supports a highly lucrative local industry in South Australia. The 2003 production of 3500 tonnes was valued at US$65 million.

Hatchery technologies are well developed for most temperate species (such as salmon and seabream) but less well developed for tropical species such as groupers where the industry is still reliant on collection of wild fingerlings, a concern for future sustainability of the sector.

The bulk of marine fish are presently farmed in net cages located in coastal waters. Most cultured species are carnivores, leading to environmental concerns over the source of feed for marine fish farms, with most still heavily reliant on wild-caught so-called 'trash' fish. Excessive stocking of cages in coastal waters also leads to concern over water and sediment pollution, as well as impacts from escapes and disease transfer on wild fish populations.

Crustaceans

Although there is substantial production of marine shrimps globally, this production is undertaken in coastal brackish water ponds and thus does not meet the definition of mariculture. There has been some experimental culture of shrimp in cages in the Pacific, but this has not yet been commercially implemented. Tropical spiny rock lobsters and particularly the ornate lobster *Panulirus ornatus* are cultured in Southeast Asia, with the bulk of production in Vietnam and the Philippines. Lobster aquaculture in Vietnam produces about 1500 tonnes valued at around US$40 million per annum. Tropical spiny rock lobsters are cultured in cages and fed exclusively on fresh fish and shellfish.

In the medium to long term, it is necessary to develop hatchery production technology for seedstock for tropical spiny rock lobsters. There is currently considerable research effort on developing larval-rearing technologies for tropical spiny rock lobsters in Southeast Asia and in Australia. As in the case of tropical marine fish farming, there is also a need to develop less-wasteful and less-polluting diets to replace the use of wild-caught fish and shellfish as diets.

Other Miscellaneous Invertebrates

There are a range of other invertebrates being farmed in the sea, such as sea cucumbers, sponges, corals, sea horses, and others. Farming of some species has been ongoing for some time, such as the well-developed sea cucumber farming in northern China, but others are more recent innovations or still at the research stage. Sponge farming, for example, is generating considerable interest in the research community, but commercial production of farmed sponges is low, mainly in the Pacific islands. This farming is similar to seaweed culture as sponges can be propagated vegetatively, with little infrastructure necessary to establish farms. The harvested product, bath sponges, can be dried and stored and, like seaweed culture, may be ideal for remote communities, such as those found among the Pacific islands.

Environmental Challenges

Environmental Impacts

Mariculture is an important economic activity in many coastal areas but is facing a number of environmental challenges because of the various environmental 'goods' and 'services' required for its development. The many interactions between mariculture and the environment include impacts of: (1) the environment on mariculture; (2) mariculture on the environment; and (3) mariculture on mariculture.

The environment impacts on mariculture through its effects on water, land, and other resources necessary for successful mariculture. These impacts may be negative or positive, for example, water pollution may provide nutrients which are beneficial to mariculture production in some extensive culture systems, but, on the other hand, toxic pollutants and pathogens can be extremely damaging. An example is the farming of oysters and other filter-feeding mollusks which generally grow faster in areas where nutrient levels are elevated by discharge of wastewater from nearby centers of human population. However, excessive levels of human and industrial waste cause serious problems for mollusk culture, such as contamination with pathogens and toxins from dinoflagellates. Aquaculture is highly sensitive to adverse environmental changes (e.g., water quality and seed quality) and it is therefore in the long-term interests of mariculture farmers and governments to work toward protection and enhancement of environmental quality. The effects of global climate change, although poorly understood in the fishery

sector, are likely to have further significant influences on future mariculture development.

The impacts of mariculture on the environment include the positive and negative effects farming operations may have on water, land, and other resources required by other aquaculturists or other user groups. Impacts may include loss or degradation of natural habitats, water quality, and sediment changes; overharvesting of wild seed; and introduction of disease and exotic species and competition with other sectors for resources. In increasingly crowded coastal areas, mariculture is running into more conflicts with tourism, navigation, and other coastal developments.

Mariculture can have significant positive environmental impacts. The nutrient-absorbing properties of seaweeds and mollusks can help improve coastal water quality. There are also environmental benefits from restocking of overfished populations or degraded habitats, such as coral reefs. For example, farming of high-value coral reef species is being seen as one means of reducing threats associated with overexploitation of threatened coral reef fishes traditionally collected for food and the ornamental trade.

Finally, mariculture development may also have an impact on itself. The rapid expansion in some areas with limited resources (e.g., water and seed) has led to overexploitation of these resources beyond the capacity of the environment to sustain growth, followed by an eventual collapse. In mariculture systems, such problems have been particularly acute in intensive cage culture, where self-pollution has led to disease and water-quality problems which have undermined the sustainability of farming, from economic and environmental viewpoints. Such problems emphasize the importance of environmental sustainability in mariculture management, and the need to minimize overharvesting of resources and hold discharge rates within the assimilative capacity of the surrounding environment.

The nature and the scale of the environmental interactions of mariculture, and people's perception of their significance, are also influenced by a complex interaction of different factors, such as follows:

- The technology, farming and management systems, and the capacity of farmers to manage technology. Most mariculture technology, particularly in extensive and semi-intensive farming systems, such as mollusk and seaweed farming, and well-managed intensive systems, is environmentally neutral or low in impact compared to other food production sectors.

- The environment where mariculture farms are located (i.e., climatic, water, sediment, and biological features), the suitability of the environment for the cultured animals and the environmental conditions under which animals and plants are cultured.
- The financial and economic feasibility and investment, such as the amount invested in proper farm infrastructure, short- versus long-term economic viability of farming operations, and investment and market incentives or disincentives, and the marketability of products.
- The sociocultural aspects, such as the intensity of resource use, population pressures, social and cultural values, and aptitudes in relation to aquaculture. Social conflicts and increasing consumer perceptions all play an important role.
- The institutional and political environment, such as government policy and the legal framework, political interventions, plus the scale and quality of technical extension support and other institutional and noninstitutional factors.

These many interacting factors make both understanding environmental interactions and their management (as in most sectors – not just mariculture) both complex and challenging.

Environmental Management of Mariculture

The sustainable development of mariculture requires adoption of management strategies which enhance positive impacts (social, economic, and environmental impacts) and mitigate against environmental impacts associated with farm siting and operation. Such management requires consideration of: (1) the farming activity, for example, in terms of the location, design, farming system, investment, and operational management; (2) the 'integration' of mariculture into the surrounding coastal environment; and (3) supporting policies and legislation that are favorable toward sustainable development.

Technology and Farming Systems Management

The following factors are of crucial importance in environmental management at the farm level:

- *Farm siting.* The sites selected for aquaculture and the habitat at the farm location play one of the most important roles in the environmental and social interactions of aquaculture. Farm siting is also crucial to the sustainability of an investment; incorrect siting (e.g., cages located in areas with unsuitable water quality) often lead to increased investment costs associated with operation and amelioration of environmental problems. Farms are better sited away from sensitive habitats (e.g., coral reefs) and in areas with sufficient water exchange to maintain environmental conditions. Problems of overstocking of mollusk culture beds are recognized in the Republic of Korea, for example, where regulations have been developed to restrict the areas covered by mollusk culture. For marine cage culture, one particularly interesting aspect of siting is the use of offshore cages, and new technologies developed in European countries are now attracting increasing interest in Asia.

- *Farm construction and design features.* Farm construction and system design has a significant influence on the impact of mariculture operations on the environment. Suitable design and construction techniques should be used when establishing new farms, and as far as possible seek to cause minimum disturbance to the surrounding ecosystems. The design and operation of aquaculture farms should also seek to make efficient use of natural resources used, such as energy and fuel. This approach is not just environmentally sound, but also economic because of increasing energy costs.

- *Water and sediment management.* Development of aquaculture should minimize impacts on water resources, avoiding impacts on water quality caused by discharge of farm nutrients and organic material. For sea-based aquaculture, where waste materials are discharged directly into the surrounding environment, careful control of feed levels and feed quality is the main method of reducing waste discharge, along with good farm siting. In temperate aquaculture, recent research has been responsible for a range of technological and management innovations – low-pollution feeds and novel self-feeding systems, lower stocking densities, vaccines, waste-treatment facilities – that have helped reduce environmental impacts. Complex models have also been developed to predict environmental impacts, and keep stocking levels within the assimilative capacity of the surrounding marine environment. In mariculture, there are also examples of integrated, polyculture, and alternate cropping farming systems that help to reduce impacts. For example in China and Korea, polyculture on sea-based mollusk and seaweed farms is practiced and for more intensive aquaculture operations, effluent rich in nutrients and microorganisms, is potentially suitable for culturing fin fish, mollusks, and seaweed.

- *Suitable species and seed.* A supply of healthy and quality fish, crustacean, and mollusk seed is essential for the development of mariculture. Emphasis should be given to healthy and quality hatchery-reared stock, rather than collection from the wild. Imports of alien species require import risk assessment and management, to reduce risks to local aquaculture industries and native biodiversity.

- *Feeds and feed management.* Access to feeds, and efficient use of feeds is of critical importance for a cost-effective and environmentally sound mariculture industry. This is due to many factors, including the fact that feeds account for 50% or more of intensive farming costs. Waste and uneaten feed can also lead to undesirable water pollution. Increasing concern is also being expressed about the use of marine resources (fish meal as ingredients) for aquaculture feeds. One of the biggest constraints to farming of carnivorous marine fish such as groupers is feed. The development of sustainable supplies of feed needs serious consideration for future development of mariculture at a global level.
- *Aquatic animal health management.* Aquatic animal and plant diseases are a major cause of unsustainability, particularly in more intensive forms of mariculture. Health management practices are necessary to reduce disease risks, to control the entry of pathogens to farming systems, maintain healthy conditions for cultured animals and plants, and avoid use of harmful disease control chemicals.
- *Food safety.* Improving the quality and safety of aquaculture products and reducing risks to ecosystems and human health from chemical use and microbiological contamination is essential for modern aquaculture development, and marketing of products on domestic and international markets. Normally, seafood is considered healthy food but there are some risks associated with production and processing that should be minimized. The two food-safety issues, that can also be considered environmental issues, are chemical and biological. The chemical risk is associated with chemicals applied in aquaculture production and the biological is associated with bacteria or parasites that can be transferred to humans from the seafood products. Increasing calls for total traceability of food products are also affecting the food production industry such that consumers can be assured that the product has been produced without addition of undesirable or harmful chemicals or additives, and that the environments and ecosystems affected by the production facilities have not been compromised in any way.
- *Economic and social/community aspects.* The employment generated by mariculture can be highly significant, and globally aquaculture has become an important employer in remote and poor coastal communities. Poorly planned mariculture can also lead to social conflicts, and the future development and operation of mariculture farms must also be done in a socially responsible manner, as far as possible without compromising local traditions and communities, and ideally contributing positively. The special traditions of many coastal people and their relation with the sea in many places deserve particularly careful attention in planning and implementation of mariculture.

Planning, Policy, and Legal Aspects

Integrated Coastal Area Management

Effective planning processes are essential for sustainable development of mariculture in coastal areas. Integrated coastal area management (ICAM) is a concept that is being given increasing attention as a result of pressures on common resources in coastal areas arising from increasing populations combined with urbanization, pollution, tourism, and other changes. The integration of mariculture into the coastal area has been the subject of considerable recent interest, although practical experience in implementation is still limited in large measure because of the absence of adequate policies and legislation and institutional problems, such as the lack of unitary authorities with sufficiently broad powers and responsibilities.

Zoning of aquaculture areas within the coastal area is showing some success. In China, Korea, Japan, Hong Kong, and Singapore, there are now well-developed zoning regulations for water-based coastal aquaculture operations (marine cages, mollusks, and seaweeds). For example, Hong Kong has 26 designated 'marine fish culture zones' within which all marine fish-culture activities are carried out. In the State of Hawaii, 'best areas' for aquaculture have been identified, and in Europe zoning laws are being strictly applied to many coastal areas where aquaculture is being developed. Such an approach allows for mariculture to be developed in designated areas, reducing risks of conflicts with other coastal zone users and uses.

Policy and Legal Issues

While much can be done at farm levels and by integrated coastal management, government involvement through appropriate policy and legal instruments is important in any strategy for mariculture sustainability. Some of the important issues include legislation, economic incentives/disincentives, private sector/community participation in policy formulation, planning processes, research and knowledge transfer, balance between food and export earnings, and others.

While policy development and most matters of mariculture practice have been regarded as purely national concerns, they are coming to acquire an increasingly international significance. The implication of this is that, while previously states would look merely to national priorities in setting mariculture policy, particularly legislation/standards, for the future it will be necessary for such activities to take account of international requirements, including various bilateral and multilateral trade policies. International standards of public health for aquaculture products and the harmonization of trade controls are examples of this trend.

Government regulations are an important management component in maintaining environmental quality, reducing negative environmental impacts, and allocating natural resources between competing users and integration of aquaculture into coastal area management. Mariculture is a relative newcomer among many traditional uses of natural resources and has commonly been conducted within an amalgam of fisheries, water resources, and agricultural and industrial regulations. It is becoming increasingly clear that specific regulations governing aquaculture are necessary, not least to protect aquaculture development itself. Key issues to be considered in mariculture legislation are farm siting, use of water area and bottom in coastal and offshore waters; waste discharge, protection of wild species, introduction of exotic or nonindigenous species, aquatic animal health; and use of drugs and chemicals.

Environmental impact assessment (EIA) can also be an important legal tool which is being more widely applied to mariculture. The timely application of EIA (covering social, economic, and ecological issues) to larger-scale coastal mariculture projects can be one way to properly identify environmental problems at an early phase of projects, thus enabling proper environmental management measures to be incorporated in project design and management. Such measures will ultimately make the project more sustainable. A major difficulty with EIAs is that they are difficult (and generally impractical) to apply to smaller-scale mariculture developments, common throughout many parts of Asia, and do not easily take account of the potential cumulative effects of many small-scale farms. Strategic environmental assessment (SEA) can provide a broader means of assessing impacts.

Conclusion

Mariculture is and will increasingly become an important producer of aquatic food in coastal areas, as well as a source of employment and income for many coastal communities. Well-planned and -managed mariculture can also contribute positively to coastal environmental integrity. However, mariculture's future development will occur, in many areas, with increasing pressure on coastal resources caused by rising populations, and increasing competition for resources. Thus, considerable attention will be necessary to improve the environmental management of aquaculture through environmentally sound technology and better management, supported by effective policy and planning strategies and legislation.

See also

Mariculture, Economic and Social Impacts.

Further Reading

Clay J (2004) *World Aquaculture and the Environment. A Commodity by Commodity Guide to Impacts and Practices*. Washington, DC: Island Press.

FAO/NACA/UNEP/WB/WWF (2006) *International Principles for Responsible Shrimp Farming*, 20pp. Bangkok, Thailand: Network of Aquaculture Centres in Asia-Pacific (NACA). http://www.enaca.org/uploads/international-shrimp-principles-06.pdf (accessed Apr. 2008).

Hansen PK, Ervik A, Schaanning M, *et al.* (2001) Regulating the local environmental impact of intensive, marine fish farming-II. The monitoring programme of the MOM system (Modelling-Ongrowing fish farms-Monitoring). *Aquaculture* 194: 75–92.

Hites RA, Foran JA, Carpenter DO, Hamilton MC, Knuth BA, and Schwager SJ (2004) Global assessment of organic contaminants in farmed salmon. *Science* 303: 226–229.

Joint FAO/NACA/WHO Study Group (1999) Food safety issues associated with products from aquaculture. *WHO Technical Report Series 883*. http://www.who.int/foodsafety/publications/fs_management/en/aquaculture.pdf (accessed Apr. 2008).

Karakassis I, Pitta P, and Krom MD (2005) Contribution of fish farming to the nutrient loading of the Mediterranean. *Scientia Marina* 69: 313–321.

NACA/FAO (2001) Aquaculture in the third millennium. In: Subasinghe RP, Bueno PB, Phillips MJ, Hough C, McGladdery SE, and Arthur JR (eds.) *Technical*

Proceedings of the Conference on Aquaculture in the Third Millennium. Bangkok, Thailand, 20–25 February 2000, 471pp. Bangkok, NACA and Rome: FAO.

Naylor R, Hindar K, Flaming IA, *et al.* (2005) Fugitive salmon: Assessing the risks of escaped fish from net-pen aquaculture. *BioScience 55*: 427–473.

Neori A, Chopin T, Troell M, *et al.* (2004) Integrated aquaculture: Rationale, evolution and state of the art emphasizing sea-weed biofiltration in modern mariculture. *Aquaculture* 231: 361–391.

Network of Aquaculture Centres in Asia-Pacific (2006) Regional review on aquaculture development. 3. Asia and the Pacific – 2005. *FAO Fisheries Circular No. 1017/3*, 97pp. Rome: FAO.

Phillips MJ (1998) Tropical mariculture and coastal environmental integrity. In: De Silva S (ed.) *Tropical Mariculture*, pp. 17–69. London: Academic Press.

Pillay TVR (1992) *Aquaculture and the Environment*, 158pp. London: Blackwell.

Secretariat of the Convention on Biological Diversity (2004) Solutions for sustainable mariculture – avoiding the adverse effects of mariculture on biological diversity, *CBD Technical Series No. 12*. http://www.biodiv.org/doc/publications/cbd-ts-12.pdf (accessed Apr. 2008).

Tacon AJC, Hasan MR, and Subasinghe RP (2006) Use of fishery resources as feed inputs for aquaculture development: Trends and policy implications. *FAO Fisheries Circular No. 1018*. Rome: FAO.

World Bank (2006) *Aquaculture: Changing the Face of the Waters. Meeting the Promise and Challenge of Sustainable Aquaculture*. Report no. 36622. Agriculture and Rural Development Department, the World Bank. http://siteresources.worldbank.org/INTARD/Resources/Aquaculture_ESW_vGDP.pdf (accessed Apr. 2008).

Relevant Websites

http://www.pbs.org
  – Farming the Seas, Marine Fish and Aquaculture Series, PBS.

http://www.fao.org/fi
  – Food and Agriculture Organisation of the United Nations.

http://www.cbd.int
  Jakarta Mandate, Marine and Coastal Biodiversity: Mariculture, Convention on Biological Diversity.

http://www.enaca.org
  – Network of Aquaculture Centres in Asia-Pacific.

http://www.seaplant.net
  – The Southeast Asia Seaplant Network.

http://www.oceansatlas.org
  – UN Atlas of the Oceans.

MARICULTURE DISEASES AND HEALTH

A. E. Ellis, Marine Laboratory, Aberdeen, Scotland, UK

Introduction

As with all forms of intensive culture where a single species is reared at high population densities, infectious disease agents are able to transmit easily between host individuals and large economic losses can result from disease outbreaks. Husbandry methods are designed to minimize these losses by employing a variety of strategies, but central to all of these is providing the cultured animal with an optimal environment that does not jeopardize the animal's health and well-being. All animals have innate and acquired defenses against infectious agents and when environmental conditions are good for the host, these defense mechanisms will provide protection against most infections. However, animals under stress have less energy available to combat infections and are therefore more prone to disease. Although some facilities on a farm may be able to exclude the entry of pathogens, for example hatcheries with disinfected water supplies, it is impossible to exclude pathogens in an open marine situation. Under these conditions, stress management is paramount in maintaining the health of cultured animals. Even then, because of the close proximity of individuals in a farm, if certain pathogens do gain entry they are able to spread and multiply extremely rapidly and such massive infectious burdens can overcome the defenses of even healthy animals. In such cases some form of treatment, or even better, prophylaxis, is required to prevent crippling losses. This article describes some of the management strategies available to fish and shellfish farmers in avoiding or reducing the losses from infectious diseases and some of the prophylactic measures and treatments. The most important diseases encountered in mariculture are summarized in **Table 1**.

Health Management

Facility Design

Farms and husbandry practices can be designed in such a way as to avoid the introduction of pathogens and to restrict their spread within a farm in a variety of ways.

Isolate the hatchery Infectious agents can be excluded from hatcheries by disinfecting the incoming water using filters, ultraviolet lamps or ozone treatments. It is also important not to introduce infections from other parts of the farm that may be contaminated. The hatchery then should stand apart and strict hygiene standards applied to equipment and personnel entering the hatchery. Some diseases cause major mortalities in young fry while older fish are more resistant. For example, infectious pancreatic necrosis virus (IPNV) causes mass mortality in halibut fry, but juveniles are much more resistant. It is vitally important therefore, to exclude the entry of IPNV into the halibut hatchery and as this virus has a widespread distribution in the marine environment, disinfection of the water supply may be necessary.

Hygiene practice Limiting the spread of disease agents on a farm include having hand nets for each tank and disinfecting the net after each use, disinfectant foot-baths at the farm entrance and between buildings, and restricted movement of staff, their protective clothing and equipment. Prompt removal of dead and moribund stock is essential as large numbers of pathogens are shed into the water from such animals. In small tanks this can easily be done using a hand net. In large sea cages, lifting the net can be stressful to fish and divers are expensive. Special equipment such as air-lift pumps, or specially designed cages with socks fitted in the bottom in which dead fish collect and which can be hoisted out on a pulley are more practical. Proper disposal of dead animals is essential. Methods such as incineration, rendering, ensiling and, on a small scale, in lime pits are recommended.

Husbandry and Minimizing Stress

Animals under stress are more prone to infectious diseases. However, it is not possible to eliminate all the procedures that are known to induce stress in aquaculture animals, as many are integral parts of aquaculture, e.g., netting, grading, and transport. Nevertheless, it is possible for farming practices to minimize the effects of these stressors and others, e.g., overcrowding and poor water quality, can be avoided by farmers adhering to the recommended

Table 1 Principal diseases of fish and shellfish in mariculture

Disease agent	Host	Prevention/treatment
Shellfish pathogens		
Protozoa		
Bonamia ostreae	European flat oyster	Exclusion
Marteilia refringens	Oyster, mussel	Exclusion
Bacteria:		
Aerococcus viridans (Gaffkaemia).	Lobsters	Improve husbandry
Fin-fish pathogens		
Viruses		
Infectious pancreatic necrosis	Salmon, turbot, halibut, cod, sea bass, yellowtail	Sanitary precautions, vaccinate
Infectious salmon anemia	Atlantic salmon	Sanitary precautions, eradicate
Pancreas disease	Atlantic salmon	Avoid stressors, vaccinate
Viral hemorrhagic septicemia	Turbot	Sanitary precautions
Viral nervous necrosis	Stripped jack, Japanese flounder, barramundi, sea bass, sea bream, turbot, halibut	Sanitary precautions
Bacteria		
Vibriosis (*Vibrio anguillarum*)	All marine species	Vaccinate, antibiotics
Cold water vibriosis (*Vibrio salmonicida*)	Salmon, trout, cod	Vaccinate, antibiotics
Winter ulcers (*Vibrio viscosus*)	Salmon	Avoid stress, vaccinate
Typical furunculosis (*Aeromonas salmonicida*)	Salmon	Vaccinate, antibiotics
Atypical furunculosis (*Aeromonas salmonicida achromogenes*)	Turbot, halibut, flounder, salmon	Vaccinate, antibiotics, avoid stress
Flexibacter maritimus	Sole, flounder, turbot, salmon, sea bass	Avoid stress, antibiotics
Enteric redmouth (*Yersinia ruckeri*)	Atlantic salmon	Vaccinate
Bacterial kidney disease (*Renibacterium salmoninarum*)	Pacific salmon	Exclude
Mycobacteriosis (*Mycobacterium marinum*)	Salmonids, sea bass, sea bream, stripped bass, cod, red drum, tilapia	Sanitary
Pseudotuberculosis (*Photobacterium damselae piscicida*)	Yellowtail, sea bass, sea bream	Vaccinate, sanitary
Piscirickettsiosis (*Piscirickettsia salmonis*)	Salmonids	Sanitary
Parasites		
Protozoan: many species; *Amyloodinium, Cryptobia, Ichthyobodo, Cryptocaryon, Tricodina*	Most species	Avoid stress, sanitary, chemical baths, e.g., formalin
Paramebic gill disease	Salmonids	Low salinity bath, H_2O_2
Crustacea: Sea lice	Salmonids, sea bass, sea bream	In-feed insecticides, H_2O_2, cleaner fish

limits for stocking densities, water flow rates and feeding regimes. In cases where stressors are unavoidable, farmers can adopt certain strategies to minimize the stress.

Withdrawal of food prior to handling Following feeding the oxygen requirement of fish is increased. Withdrawal of food two or three days prior to handling the fish will therefore minimize respiratory stress. It also avoids fouling of the water during transportation with fecal material and regurgitated food.

Use of anesthesia Although anesthetics can disturb the physiology of fish, light anesthesia can have a calming effect on fish during handling and transport and so reduce the stress resulting from these procedures.

Avoidance of stressors at high temperatures High temperatures increase the oxygen demand of animals and stress-induced mortality can result from respiratory failure at high temperatures. It is therefore safer to carry out netting, grading and transport at low water temperatures.

Avoidance of multiple stressors and permitting recovery The effects of multiple stressors can be additive or even synergistic so, for instance, sudden changes of temperature should be avoided during or after transport. Where possible, recovery from stress should be facilitated. Generally, the duration of the recovery period is proportional to the duration of the stressor. Thus, reducing the time of netting, grading or transport will result in recovery in a shorter time. The duration required for recovery to occur may be from a few days to two weeks.

Selective breeding In salmonids, it is now established that the magnitude of the stress response is a heritable characteristic and programs now exist for selecting broodstock which have a low stress response to handling stressors. This accelerates the process of domestication to produce stocks, which are more tolerant of aquaculture procedures with resultant benefits in increased health, survival and productivity. Such breeding programs have been conducted in Norway for some years and have achieved improvements in resistance to furunculosis and infectious salmon anemia virus (ISAV) in Atlantic salmon.

Management of the Pathogen

Breaking the pathogen's life cycle If a disease is introduced on to a farm, it is important to restrict the horizontal spread especially to different year classes of fish/shellfish. Hence, before a new year class of animals is introduced to a part of the farm, all tanks, equipment etc. should be thoroughly cleaned and disinfected. In sea-cage sites it is a useful technique to physically separate year classes to break the infection cycle.

Many pathogens do not survive for long periods of time away from their host and allowing a site to be fallow for a period of time may eliminate or drastically reduce the pathogen load. This practice has been very effective in controlling losses from furunculosis in marine salmon farms and also significantly reduces the salmon lice populations particularly in the first year after fallowing.

Eliminate vertical transmission Several pathogens may persist as an asymptomatic carrier state and be present in the gonadal fluids of infected broodstock and can infect the next generation of fry. In salmon farming IPNV may persist in or on the ova and disinfection of the eggs with iodine-based disinfectants (Iodophores) is recommended immediately after fertilization.

The testing of gonadal fluids for the presence of IPNV can also be carried out and batches of eggs

from infected parents destroyed. IPNV has been associated with mass mortalities in salmon and halibut fry and has been isolated from a wide range of marine fish and shellfish, including sea bass, turbot, striped bass, cod, and yellowtail.

Avoid infected stock Many countries employ regulations to prevent the movement of eggs or fish that are infected with certain 'notifiable' diseases from an infected to a noninfected site. These policies are designed to limit the spread of the disease but require specialized sampling procedures and laboratory facilities to perform the diagnostic techniques. By testing the stock frequently they can be certified to be free from these diseases. Such certification is required for international trade in live fish and eggs but within a country it is widely practiced voluntarily because stock certified to be 'disease-free' command premium prices.

Eradication of infected stock Commercially this is a drastic step to take especially when state compensation is not usually available even when state regulations might require eradication of stock. This policy is usually only practiced rarely and when potential calamitous circumstances may result, for instance, the introduction of an important exotic pathogen into an area previously free of that disease. This has occurred in Scotland where the European Commission directives have required compulsory slaughter of turbot infected with viral hemorrhagic septicemia virus (VHSV) and Atlantic salmon infected with ISAV.

Treatments

Viral Diseases

There are no treatments available for viral diseases in aquaculture. These diseases must be controlled by husbandry and management strategies as described above, or by vaccination (see below).

Bacterial Diseases

Antibiotics can be used to treat many bacterial infections in aquaculture. They are usually mixed into the feed. Before the advent of vaccines against many bacterial diseases of fin fish, antibiotic treatments were commonly used. However, after a few years, the bacterial pathogens developed resistance to the antibiotics. Furthermore, there was a growing concern that the large amounts of antibiotics being used in aquaculture would have damaging effects on the environment and that antibiotic residues in fish flesh may have dangerous consequences for consumers by promoting the development of antibiotic resistant strains of human bacterial pathogens. These concerns have led to many restrictions on the use of antibiotics in aquaculture, especially in defining long withdrawal periods to ensure that carcasses for consumption are free of residues. These regulations have made the use of antibiotic treatments impractical for fish that are soon to be harvested for consumption but their use in the hatchery is still an important method of controlling losses from bacterial pathogens.

For most bacterial diseases of fish, vaccines have become the most important means of control (see below) and this has led to drastic reductions in the use of antibiotics in mariculture.

Parasite Diseases

Sea lice The most economically important parasitic disease in mariculture of fin fish is caused by sea lice infestation of salmon. These crustacean parasites normally infest wild fish and when they enter the salmon cages they rapidly multiply. The lice larvae and adults feed on the mucus and skin of the salmon and heavy infestations result in large haemorrhagic ulcers especially on the head and around the dorsal fin. These compromise the fish's osmoregulation and allow opportunistic bacterial pathogens to enter the tissues. Without treatment the fish will die.

A range of treatments are available and recently very effective and environmental friendly in-feed treatments such as 'Slice' and 'Calicide' have replaced the highly toxic organophosphate bath treatments. Hydrogen peroxide is also used. As a biological control method, cleaner fish are used but they are not a complete method of control.

Vaccination

Use of Vaccines

Vertebrates can be distinguished from invertebrates in their ability to respond immunologically in a specific manner to a pathogen or vaccine. Invertebrates, such as shellfish, only possess nonspecific defense mechanisms. In the strict definition, vaccines are used only as a prophylactic measure in vertebrates because a particular vaccine against a particular disease induces protection that is specific for that particular disease and does not protect against other diseases. Vaccines induce long-term protection and in aquaculture a single administration is usually sufficient to induce protection until the fish are harvested. However, vaccines also have nonspecific immunostimulatory properties that can also activate

many nonspecific defense mechanisms. These can increase disease resistance levels but only for a short period of time. Thus, in their capacity to induce such responses they are also used in shellfish culture, especially of shrimps.

Current Status of Vaccination

In Atlantic salmon mariculture, vaccination has been very successful in controlling many bacterial diseases and has almost replaced the need for antibiotic treatments. In recent years vaccination of sea bass and sea bream has become common practice. Most of the commercial vaccines are against bacterial diseases because these are relatively cheap to produce. Obviously the cost per dose of vaccine for use in aquaculture must be very low and it is inexpensive to culture most bacteria in large fermenters and to inactivate the bacteria and their toxins chemically (usually with formalin). It is much more expensive to culture viruses in tissue culture and this has been a major obstacle in commercializing vaccines against virus diseases of fish. However, modern molecular biology techniques have made it possible to transfer viral genes to bacteria and yeasts, which are inexpensive to culture and produce large amounts of viral vaccine cheaply. A number of vaccines against viral diseases of Atlantic salmon are now becoming available. Currently available commercial vaccines for use in mariculture are summarized in **Table 2**.

Methods of Vaccination

There are two methods of administering vaccines to fish: immersion in a dilute suspension of the vaccine or injection into the body cavity. For practical

Table 2 Vaccines available for fish species in mariculture

Vaccines against	Maricultured species
Bacterial diseases	
Vibriosis	Salmon, sea bass, sea bream, turbot, halibut
Winter ulcers	Salmon
Coldwater Vibriosis	Salmon
Enteric redmouth	Salmon
Furunculosis	Salmon
Pseudotuberculosis	Sea bass
Viral diseases	
Infectious pancreatic necrosis (IPNV)	Salmon
Pancreas disease	Salmon
Infectious salmon anemia (ISAV)	Salmon

reasons the latter method requires the fish to be over about 15 g in weight.

Immersion vaccination is effective for some, but not all vaccines. The vaccine against the bacterial disease vibriosis is effective when administered by immersion. It is used widely in salmon and sea bass farming and probably could be administered by this route to most marine fish species. The vaccine against pseudotuberculosis can also be administered by immersion to sea bass. With the exception of the vaccine against enteric redmouth, which is delivered by immersion to fish in freshwater hatcheries, all the other vaccines must be delivered by injection in order to achieve effective protection.

Injection vaccination induces long-term protection and the cost per dose is very small. However, it is obviously very labor intensive. Atlantic salmon are usually vaccinated several months before transfer to sea water so that the protective immunity has time to develop before the stress of transportation to sea and exposure to the pathogens encountered in the marine environment.

Conclusions

It is axiomatic that intensive farming of animals goes hand in hand with culture of their pathogens. The mariculture of fish and shellfish has had severe problems from time to time as a consequence of infectious diseases. During the 1970s, *Bonamia* and *Marteilia* virtually eliminated the culture of the European flat oyster in France and growers turned to production of the more resistant Pacific oyster. In Atlantic salmon farming, Norway was initially plagued with vibriosis diseases and Scotland suffered badly from furunculosis in the late 1980s. These bacterial diseases have been very successfully brought under control by vaccines. However, there are still many diseases for which vaccines are not available and the susceptibility of Pacific salmon to bacterial kidney disease has markedly restricted the development of the culture of these fish species on the Pacific coast of North America. As new industries grow, new diseases come to the foreground, for instance piscirikettsia in Chilean salmon culture, paramoebic gill disease in Tasmanian salmon culture, pseudotuberculosis in Mediterranean sea bass and sea bream and Japanese yellowtail culture. Old

diseases find new hosts, for example IPNV long known to affect salmon hatcheries, has in recent years caused high mortality in salmon postsmolts and has devastated several halibut hatcheries.

To combat these diseases and to ensure the sustainability of aquaculture great attention must be paid to sanitation and good husbandry (including nutrition). In some cases these are insufficient in themselves and the presence of certain enzootic diseases, or following their introduction, have made it impossible for certain species to be cultured, for example, the European flat oyster in France. The treatment of disease by chemotherapy, which was performed widely in the 1970s and 1980s, resulted in the induction of antibiotic-resistant strains of bacteria and chemoresistant lice. Furthermore, the growing concern for the environment and the consumer about the increasing usage of chemicals and antibiotics in aquaculture, led to increasing control and restrictions on their usage. This stimulated much research in the 1980s and 1990s into development of more environmentally and consumer friendly methods of control such as vaccines and immunostimulants. These have achieved remarkable success and the pace of current research in this area using biotechnology to produce vaccines more cheaply, suggests that this approach will allow continued growth and sustainability of fin-fish mariculture into the future.

See also

Mariculture   Overview.   Molluskan   Fisheries. Salmonids.

Further Reading

Bruno DW and Poppe TT (1996) *A Colour Atlas of Salmonid Diseases.* London: Academic Press.

Ellis AE (ed.) (1988) *Fish Vaccination.* London: Academic Press.

Lightner DV (1996) *A Handbook of Pathology and Diagnostic Procedures for Diseases of Cultured Penaeid Shrimps.* The World Acquaculture Society.

Roberts RJ (ed.) (2000) *Fish Pathology*, 3rd edn. London: W.B. Saunders.

MARICULTURE, ECONOMIC AND SOCIAL IMPACTS

C. R. Engle, University of Arkansas at Pine Bluff, Pine Bluff, AR, USA

Introduction

Mariculture is a broad term that encompasses the cultivation of a wide variety of species of aquatic organisms, including both plants and animals. These different products are produced across the world with a wide array of technologies. Each technology in each situation will entail various price and cost structures within different social contexts. Thus, each aquaculture enterprise will have distinct types and levels of economic and social impacts. Moreover, there are as many management philosophies, strategies, and business plans as there are aquaculture entrepreneurs. Each of these will result in different economic and social impacts. As an example, consider two shrimp farms located in a developing country that utilize the same production technology. One farm hires and trains local people as both workers and managers, invests in local schools and health centers, while paying local and national taxes. This farm will likely have a large positive social and economic impact. On the other hand, another farm that imports managers, pays the lowest wages possible, displaces local families through land acquisitions, pays few taxes, and exports earnings to developed countries may have negative social and perhaps even economic impacts.

Economic and social structure, interactions, and impacts are complex and interconnected, even in rural areas with seemingly simple economies. This article discusses a variety of types of impacts that can occur from mariculture enterprises.

Economic Impacts

Economic impacts begin with the direct effects from the sale of product produced by the mariculture operation. However, the impacts extend well beyond the effect of sales revenue to the farm. As the direct output of marine fish, shellfish, and seaweed production increases, the demand for supply inputs such as feed, fingerlings, equipment, repairs, transportation services, and processing services also increases.

These activities represent indirect effects. Subsequent increased household spending will follow. As the industry grows, employment in all segments of the industry also grows and these new jobs create more income that generates additional economic activity. Thus, growth of the mariculture industry results in greater spending that multiplies throughout the economy.

Mariculture is an important economic activity in many parts of the world. **Table 1** lists the top 15 mariculture-producing countries in the world, both in terms of the volume of metric tons produced and the value in 2005. Its economic importance can be measured in total sales volume, total employment, or total export volume for large aquaculture industry sectors. Macroeconomic effects include growth that promotes trade and domestic resource utilization. Fish production in the Philippines, for example, accounted for 3.9% of gross domestic product (GDP) in 2001. In India, it contributed 1.4% of national GDP and 5.4% of agricultural GDP.

On the microlevel, incomes and livelihoods of the poor are enhanced through mariculture production. Small-scale and subsistence mariculture provides high-quality protein for household consumption,

Table 1 Top 15 mariculture-producing countries, with volumes (metric tons) produced in 2005 and value (in US)

Country	Volume of production (metric ton)	Value of production ($1000 US)
China	22 677 724	14 981 801
The Philippines	1 419 727	165 335
Japan	1 211 959	3 848 906
Korea, Rep.	1 042 142	1 317 250
Indonesia	950 819	338 093
Chile	889 867	3 069 169
Korea, Dem.	703 292	283 362
Norway	504 295	2 072 562
Thailand	656 636	58 587
France	347 750	571 543
Canada	216 103	503 974
Spain	136 724	262 394
United Kingdom	190 426	584 152
United States	161 339	235 912
Vietnam	134 937	158 800
Others	173 800	3 268 283
Total	31 417 540	31 720 123

Source: FishStat Plus (http://www.fao.org).

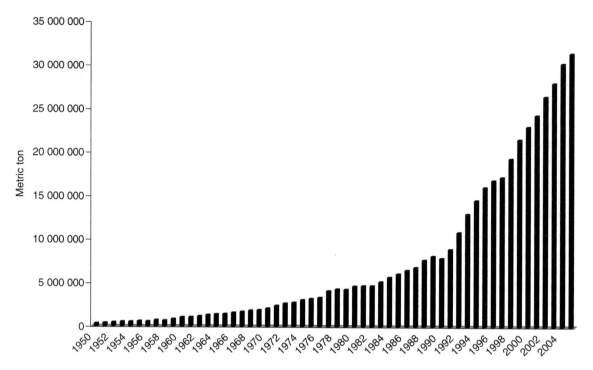

Figure 1   Growth of mariculture production worldwide, 1950–2005. Source: Food and Agriculture Organization of the United Nations.

generates supplemental income from sales to local markets, and can serve as a savings account to meet needs for cash during difficult financial times.

Income from Sales Revenue

According to the Food and Agriculture Organization of the United Nations, mariculture production grew from just over 300 000 metric ton in 1950 to 31.4 million metric ton valued at $31.7 billion in 2005 (**Figure 1**). **Figure 1** also shows that the total volume of mariculture production has tripled over the past decade alone. Sales revenue received by operators of mariculture businesses has increased rapidly over this same time period. The top mariculture product category in 2005, based on quantity produced, was Japanese kelp, followed in descending order of quantities produced by oysters, Japanese carpetshell, wakame, Yesso scallop, and salmon (**Table 2**). In terms of value, salmon was the top mariculture product produced in 2005, followed by oysters, kelp, and Japanese carpetshell. The top five marine finfish raised in 2005 (in terms of value) were Atlantic salmon, Japanese amberjack, rainbow trout, seabreams, and halibut.

The increase in production and sales of mariculture products has resulted not only from the expansion of existing products and technologies (i.e., salmon, Japanese amberjack) but also from the rapid development of technologies to culture new species

Table 2   Top 15 mariculture species cultured, volume (metric tons), and value ($1000 US), 2005

Species	Volume (metric ton)	Value ($1000 US)
Atlantic salmon	1 216 791	4 659 841
Blood cockle	436 924	436 772
Blue mussel	391 210	385 131
Constricted tagelus	713 846	589 836
Green mussel	280 267	26 679
Japanese carpetshell	2 880 687	2 358 586
Japanese kelp	4 911 256	2 941 148
Laver (nori)	1 387 990	1 419 130
Pacific cupped oyster	4 496 196	3 003 831
Rainbow trout	183 575	762 707
Red seaweeds	866 383	121 294
Sea snails	238 331	131 188
Wakame	2 739 753	1 101 507
Warty gracilaria	985 667	394 257
Yesso scallop	1 239 811	1 677 870
Others	8 448 853	11 710 346
Total	31 417 540	31 720 123

Source: FishStat Plus (http://www.fao.org).

(i.e., cobia, grouper, and tuna). New mariculture farms have created new markets and opportunities for local populations, such as the development of backyard hatcheries in countries such as Indonesia. In Bali, for example, the development of small-scale hatcheries has resulted in substantial increases in income as compared to the more traditional coconut crops.

The immediate benefit to a mariculture entrepreneur is the cash income received from sale of the aquatic products produced. This income then is spent to pay for cash expenditures related to feed, fingerlings, repairs to facilities and equipment, fuel, labor, and other operating costs. Payments on loans will be made to financial lenders involved in providing capital for the facilities, equipment, and operation of the business. Profit remaining after expenses can be spent by the owner on other goods and services, invested back into the business, or invested for the future benefit of the owner.

Mariculture of certain species has begun to exceed the value of the same species in capture fisheries. Salmon is a prime example, in which the value of farmed salmon has exceeded that of wild-caught salmon for a number of years. A similar trend can be observed for the rapidly developing capture-based mariculture technologies. In the Murcia region of Spain, for example, the economic value of tuna farming now represents 8 times the value of regional fisheries.

Employment

Mariculture businesses generate employment for the owner of the business, for those who serve as managers and foremen, and those who constitute the principal workforce for the business. Studies in China, Vietnam, Philippines, Indonesia, and Thailand suggest that shrimp farming uses more labor per hectare than does rice farming.

Employment is generated throughout the market supply chain. Development of mariculture businesses increases demand for fry and fingerlings, feeds, construction, equipment used on the farm (trucks, tractors, aerators, nets, boats, etc.), and repairs to facilities and equipment. In Bangladesh, it is estimated that about 300 000 people derive a significant part of their annual income from shrimp seed collection.

Mariculture businesses also create jobs for fish collectors, brokers, and vendors. These intermediaries provide important marketing functions that are difficult for smaller-scale producers to handle. With declines in catch of some previously important fish products, fish vendors need new fish products to sell. Increasing volumes from mariculture can keep these marketers in business.

The employment stimulated by development of mariculture businesses is especially significant in economically depressed areas with high rates of unemployment and underemployment. Many mariculture activities develop in rural coastal communities

where jobs are limited and mariculture can constitute a critical source of employment. Many jobs in fishing communities have been reduced as fishing opportunities have become more scarce. Given its relationship to the fishing industry and some degree of similarity in the skills required, mariculture businesses are often a welcome alternative for the existing skilled workforce.

Capture-based mariculture operations in particular provide job opportunities that require skill sets that are easily met by those who have worked in the fishing industry. The rapid development of capture-based tuna farms in the Mediterranean Sea provides a good example. In Spain, fishermen have become active partners in tuna farms. As a result, the number of specialized bluefin tuna boats has increased to supply the capture-based tuna farms in the area. These boats capture the small pelagic fish that are used as feed. In Croatia, trawlers transport live fish or feed to tuna farms off shore. This has generated important sources of employment in heavily depopulated Croatian islands. Tuna farming is labor-intensive, but offers opportunities for younger workers to develop new skills. Many of the employees on the tuna farms are young, 25–35 years old, who work as divers. The divers are used to crowd the tuna for hand-harvesting without stressing the fish, inspect for mortalities, and check the integrity of moorings. Working conditions on tuna farms are preferable to those on tuna boats because the hours and salaries are more regular. Workers can spend weekends on shore as compared to spending long periods of time at sea on tuna boats. The improved working conditions have improved social stability.

Formal studies of economic impacts have measured the effects on employment for fish farm owners and from the secondary businesses such as supply companies, feed mills, processing plants, transportation services, etc. In the Philippines, fish production accounted for 4.4% of total national employment in 2001. In the United States, catfish production accounted for nearly half of all employment in one particular county.

Tax Revenue

Tax policies vary considerably from country to country. Nevertheless, there is some form of tax structure in most countries that is used to finance local, regional, and national governments. Tax structures typically include some form of business tax in addition to combinations of property, income, value-added, or sales taxes. Tax revenue is the major source of revenue used for investments in roads and bridges, schools, health, and public sanitation facilities.

Mariculture businesses contribute to the tax base in that particular region or market. Moreover, the wages paid to managers and workers in the business are subject to income, sales, and property taxes. As the business grows, it pays more taxes. Increasing payrolls from increased employment by the company results in greater tax revenue from increased incomes, increased spending (sales taxes), and increased property taxes paid (as people purchase larger homes and/or more land).

Export Revenue and Foreign Exchange

The largest volume of mariculture production is in developing countries whereas the largest markets for mariculture products are in the more developed nations of the world. Thus, much of the flow of international trade in mariculture products is from the developing to the developed world. This trade results in export revenue for the companies involved. Exporters in the live grouper trade in Asia have been shown to earn returns on total costs as high as 94%.

Export volumes and revenue further generate foreign exchange for the exporting country. This is particularly important for countries that are dependent on other countries for particular goods and services. In order to import products not produced domestically, the country needs sufficient foreign exchange. Moreover, if the country exports products to countries whose currency is strong (i.e., has a high value relative to other forms of currency), the country can then import a relatively greater amount of product from countries whose currency does not have as high a value. Many low-income food-deficient countries use fish exports to pay for imports of low-value food commodities. For example, in 2000, Indonesia, China, the Philippines, India, and Bangladesh paid off 63%, 62%, 22%, 54%, and 32%, respectively, of their food import bills by exporting fish and fish products.

Economic Growth

Economic growth is generated from increased savings, investment, and money supply. The process starts with profitable businesses. Profits can either be saved by the owner or invested back into the business. Savings deposited in a bank or invested in stocks or bonds increase the amount of capital available to be loaned out to other potential investors. With increasing amounts of capital available, the cost of capital (the interest rate) is reduced, making it easier for both individuals and businesses to borrow capital. As a result, spending on housing, vehicles, property, and new and expanded businesses increases. This in turn increases demand for housing, vehicles, and other goods which increases demand for employment. As the economy grows (as measured by the GDP), the standard of living rises as income levels rise and citizens can afford to purchase greater varieties of goods and services. The availability of goods and services also increases with economic growth. Economic growth increases demand which also tends to increase price levels. As prices increase, businesses become more profitable and wages rise.

Economic growth is necessary for standards of living to increase. However, continuous economic growth means that prices also rise continuously. Continuous increases in prices constitute inflation that, if excessive, can create economic difficulties. It is beyond the scope of this article to discuss the mechanisms used by different countries to manage the economy on a national scale and how to manage both economic growth and inflation.

Economic Development

Economic development occurs as economies diversify, provide a greater variety of goods and services, and as the purchasing power of consumers increases. This combination results in a higher standard of living for more people in the country. Economic development results in higher levels of education, greater employment opportunities, and higher income levels. Communities are strengthened with economic development because increasing numbers of jobs result in higher income levels. Higher standards of living provide greater incentives for young people to stay in the area rather than outmigrate in search of better employment and income opportunities.

Studies of the economic impacts of aquaculture have highlighted the importance of backwardly linked businesses. These are the businesses that produce fingerlings, feed, and sell other supplies to the aquaculture farms. Studies of the linkages in the US catfish industry show that economic output of and the value added by the secondary businesses are greater than those of the farms themselves.

Economic Multiplier Effects

Each time a dollar exchanges hands in an economy, its impact 'multiplies'. This occurs because that same dollar can then be used to purchase some other item. Each purchase represents demand for the product being purchased. Each time that same dollar is used to purchase another good or service, it generates additional demand in the economy. Those businesses that have higher expenditures, particularly those with high initial capital expenditures, tend to have higher multiplier effects in the economy.

Aquaculture businesses tend to have high capital expenditures and, thus, tend to have relatively higher multipliers than other types of businesses, particularly other types of agricultural businesses. These high capital expenditures come from the expense of building ponds, constructing net pens or submerged cages, and the equipment costs associated with aerators, boats, trucks, tractors, and other types of equipment. As an example of how economic multipliers occur, consider a new shrimp farm enterprise. Construction of the ponds will require either purchasing bulldozers and bulldozer operators or contracting with a company that has the equipment and expertise to build ponds. The shrimp farm owner pays the pond construction company for the ponds that were built. The dollars paid to the construction company are used to pay equipment operators, fuel companies, for water supply and control pipes and valves, and repairs to equipment. The pond construction workers use the dollars received as wages to purchase more food, clothing for the family, school costs, healthcare costs, and other items. The dollars spent for food in the local market become income for the vendors who pay the farmers who raised the food. Those farmers can now pay for their seed and fertilizer, buy more clothing, and pay school and healthcare expenses, etc. In this expenditure chain, the dollar 'turned over' five times, creating additional economic demand each time.

Several formal studies of economic impacts from aquaculture have been conducted. These studies have shown large amounts of total economic output, employment, and value-added effects from aquaculture businesses. On the local area, aquaculture can generate the majority of economic output in a given district. Multipliers as high as 6.1 have been calculated for aquaculture activities on the local level.

Potential Negative Economic Impacts

Mariculture has been criticized by some for what economists call 'externalities'. Externalities refer to an effect on an individual or community other than the individual or community that created the problem. Pollution is often referred to as an externality. For example, if effluents discharged from a mariculture business create water-quality problems for another farm or community downstream, that farm or community will have to spend more to clean up that pollution. Yet, those costs are 'external' to the business that created the problem.

To the extent that a mariculture business creates an externality that increases costs for another business or community, there will be negative economic impacts. The primary potential sources of negative economic impacts from mariculture include: (1) discharge of effluents that may contain problematic levels of nutrients, antibiotics, or nonnative species; (2) spread of diseases; (3) use of high levels of fish meal in the diets fed to the species cultured; (4) clearing mangroves from coastal areas; and (5) consequences of predator control measures. The history of the development of mariculture includes cases in which unregulated discharge of untreated effluents from shrimp farms resulted in poor water-quality conditions in bays and estuaries. Since this same water also served as influent for other shrimp farms, its poorer quality stressed shrimp and facilitated spread of viral diseases. Economic losses resulted.

Concerns have been noted over the increasing demand for fish meal as mariculture of carnivorous species grows. The fear is that this increasing use will result in a decline in the stocks of the marine species that serve as forage for wild stocks and that are used in the production of fish meal. If this were to occur, economic losses could occur from declines in capture fisheries.

Loss of mangrove forests in coastal areas results in loss of important nursery grounds for a number of species and less protection during storms. The loss of mangrove areas is due to many uses, including for construction, firewood, for salt production, and others. If the construction of ponds for mariculture reduces mangrove areas, additional negative impacts will occur.

Control of predators on net pens and other types of mariculture operations often involves lethal methods. The use of lethal methods raises concerns related to biodiversity and viability of wild populations. While there is no direct link to other economic activities in many of these cases, the loss of ecosystem services from reductions in biodiversity may at some point result in other economic problems.

Social Impacts

Food Security

Aquaculture is an important food security strategy for many individuals and countries. Domestic production of food provides a buffer against interruptions in the food supply that can result from international disputes, trade embargoes, war, transportation accidents, or natural disasters. There are many examples of fish ponds being used to feed households during times of strife. Examples include fish pond owners in Vietnam during the war with the United States and during the Contra War in Nicaragua, among others.

Aquaculture has been shown to function as a food reserve for many subsistence farming households. Many food crops are seasonal in nature while fish in ponds or cages can be available year-round and used particularly during periods of prolonged drought or in between different crops. Pond levees also can provide additional land area to raise vegetable crops for household consumption and sale.

Nutritional Benefits

Fish are widely recognized to be a high-quality source of animal protein. Fish have long been viewed as 'poor people's food'. Subsistence farmers who raise fish, shellfish, or seaweed have a ready source of nutritious food for home consumption throughout the year. Particularly in Asia, fish play a vital role in supplying inexpensive animal protein to poorer households. This is important because the proportion of the food budget allocated to fish is higher among low-income groups. Moreover, it has been shown that rural people consume more fish than do urban dwellers and that fish producers consume more fish than do nonproducers. Those farmers who also supply the local market with aquatic products provide a supply of high-quality protein to other households in the area. Increasing supplies of farmed products result in lower prices in seafood markets. Tuna prices in Japan, for example, have been falling as a result of the increased farmed supply. This has resulted in making tuna and its nutritional benefits more readily available to middle-income buyers in Japan.

Health Benefits

In addition to the protein content, a number of aquatic products provide additional health benefits. Farmed fish such as salmon have high levels of omega-3 fatty acids that reduce the risk of heart disease. Products such as seaweeds are rich in vitamins. Mariculture can enable the poorest of the poor and even the landless to benefit from public resources. For example, cage culture of fish in public waters, culture of mollusks and seaweeds along the coast, and culture-based fisheries in public water bodies provide a source of healthy food for subsistence families.

Poverty Alleviation

As aquaculture has grown, its development has frequently occurred in economically depressed areas.

Poverty alleviation is accompanied by a number of positive social impacts. These include improved access to food (that results in higher nutritional and health levels), improved access to education (due to higher income levels and ability to pay for fees and supplies), and improved employment opportunities. In Sumatra, Indonesia, profits from grouper farming enable members of Moslem communities to make pilgrimages to Mecca, enriching both the individuals and the community. In Vietnam, income from grouper hatcheries contributes 10–50% of the annual income of fishermen.

Enhancement of Capture Fisheries

Marine aquaculture can improve the condition of fisheries through supplemental stockings from aquaculture. This will help to meet the global demand for fish products. Increased mariculture supply relieves pressure on traditional protein sources for species such as live cod and haddock. Given that fish supplies from capture fisheries are believed to have reached or be close to maximum sustainable yield, mariculture can reduce the expected shortage of fish.

Potential Negative Social Impacts

Mariculture has been criticized for creating negative social impacts. These criticisms have tended to focus on displacement of people if large businesses buy land and move people off that land. Economic development is accompanied by social change and individuals and households frequently are affected in various ways by construction of new infrastructure and production and processing facilities. In developed countries, federal laws typically provide for some degree of compensation for land taken through eminent domain for construction of highways or power lines. However, developing countries rarely have similar mechanisms for compensating those who lose land. In situations in which the resources involved are in the public domain (such as coastal areas), granting a concession for a mariculture operation along the coast may result in reduced access for poor and landless individuals to collect shellfish and other foods for their households.

Resource ownership or use rights in coastal zones frequently are ambiguous. Many of these areas traditionally have been common access areas, but are now under pressure for development for mariculture activities. Displaced and poor migrant people are frequently marginalized in coastal lands and often depend to some degree on common resources. However, the extent of conflict over the use of resources is variable. Surveys in Asia reported less than 10% of farms experiencing conflicts in most countries, but higher incidents of conflict were reported in India (29%) and China (94%).

In the Mediterranean, capture-based mariculture has resulted in conflicts between local tuna fishermen who fish with long lines, and cage towing operations. Other conflicts have occurred in Croatia due to the smell and pollution during the summer from bluefin tuna farms. Uncollected fat skims on the sea surface that results from feeding oily trash fish spread outside the licensed zones onto beaches frequented by tourists. On the positive side, tourism in Spain was enhanced by offering guided tours to offshore cages.

Additional negative impacts could potentially include: (1) marine mammal entanglements; (2) threaten genetic makeup of wild fish stocks; (3) privatize what some think of as free open-access resource; (4) esthetically undesirable; and (5) negative impacts on commercial fishermen. For example, excessive collection of postlarvae and egg-laden female shrimp can result in loss of income for fishermen and reduction of natural shrimp and fish stocks.

Conclusions

The economic and social impacts of mariculture are variable throughout the world and are based on the range of technologies, cultures, habitats, land types, and social, economic, and political differences. Positive economic impacts include increased revenue, employment, tax revenue, foreign exchange, economic growth and development, and multiplicative effects through the economy. Negative economic effects could occur through excessive discharge of effluents, spread of disease, and through excessive use of fish meal. Mariculture enhances food security of households and countries and provides nutritional and health benefits while alleviating poverty. However, if earnings are exported and access to public domain resources restricted to poorer classes, negative social impacts can occur.

See also

Mariculture Diseases and Health. Mariculture Overview.

Further Reading

Aguero M and Gonzalez E (1997) Aquaculture economics in Latin America and the Caribbean: A regional assessment. In: Charles AT, Agbayani RF, Agbayani EC, *et al.* (eds.) Aquaculture Economics in Developing Countries: Regional Assessments and an Annotated Bibliography. *FAO Fisheries Circular No. 932.* Rome: FAO.

Dey MM and Ahmed M (2005) *Special Issue: Sustainable Aquaculture Development in Asia. Aquaculture Economics and Management* 9(1/2): 286pp.

Edwards P (2000) Aquaculture, poverty impacts and livelihoods. *Natural Resource Perspectives* 56 (Jun. 2000).

Engle CR and Kaliba AR (2004) *Special Issue: The Economic Impacts of Aquaculture in the United States. Journal of Applied Aquaculture* 15(1/2): 172pp.

Ottolenghi F, Silvestri C, Giordano P, Lovatelli A, and New MB (2004) *Capture-Based Aquaculture. The Fattening of Eels, Groupers, Tunas and Yellowtails*, 308pp. Rome: FAO.

Tacon GJ (2001) Increasing the contribution of aquaculture for food security and poverty alleviation. In: Subasinghe RP, Bueno P, Phillips MJ, Hough C, McGladdery SE, and Arthur JR (eds.) Aquaculture in the Third Millennium. *Technical Proceedings of the Conference on Aquaculture in the Third Millennium,* pp. 63–72, Bangkok, Thailand, 20–25 February 2000. Bangkok and Rome: NACA and FAO.

Relevant Websites

http://www.fao.org
 – Food and Agriculture Organization of the United Nations.

http://www.worldfishcenter.org
 – WorldFish Center.

SEAWEEDS AND THEIR MARICULTURE

T. Chopin and M. Sawhney, University of New Brunswick, Saint John, NB, Canada

Introduction: What are Seaweeds and their Significance in Coastal Systems

Before explaining how they are cultivated, it is essential to try to define this group of organisms commonly referred to as 'seaweeds'. Unfortunately, it is impossible to give a short definition because this heterogeneous group is only a fraction of an even less natural assemblage, the 'algae'. In fact, algae are not a closely related phylogenetic group but a diverse group of photosynthetic organisms (with a few exceptions) that is difficult to define, by either a lay person or a professional botanist, because they share only a few characteristics: their photosynthetic system is based on chlorophyll *a*, they do not form embryos, they do not have roots, stems, leaves, nor vascular tissues, and their reproductive structures consist of cells that are all potentially fertile and lack sterile cells covering or protecting them. Throughout history, algae have been lumped together in an unnatural group, which now, especially with the progress in molecular techniques, is emerging as having no real cohesion with representatives in four of the five kingdoms of organisms. During their evolution, algae have become a very diverse group of photosynthetic organisms, whose varied origins are reflected in the profound diversity of their size, cellular structure, levels of organization and morphology, type of life history, pigments for photosynthesis, reserve and structural polysaccharides, ecology, and habitats they colonize. Blue-green algae (also known as Cyanobacteria) are prokaryotes closely related to bacteria, and are also considered to be the ancestors of the chloroplasts of some eukaryotic algae and plants (endosymbiotic theory of evolution). The heterokont algae are related to oomycete fungi. At the other end of the spectrum (one cannot presently refer to a typical family tree), green algae (Chlorophyta) are closely related to vascular plants (Tracheophyta). Needless to say, the systematics of algae is still the source of constant changes and controversies, especially recently with new information provided by molecular techniques. Moreover, the fact that the roughly 36 000 known species of algae represent only about 17% of the existing algal species is a measure of our still rudimentary knowledge of this group of organisms despite their key role on this planet: indeed, approximately 50% of the global photosynthesis is algal derived. Thus, every second molecule of oxygen we inhale was produced by an alga, and every second molecule of carbon dioxide we exhale will be re-used by an alga.

Despite this fundamental role played by algae, these organisms are routinely paid less attention than the other inhabitants of the oceans. There are, however, multiple reasons why algae should be fully considered in the understanding of oceanic ecosystems: (1) The fossil record, while limited except in a few phyla with calcified or silicified cell walls, indicates that the most ancient organisms containing chlorophyll *a* were probably blue-green algae 3.5 billion years ago, followed later (900 Ma) by several groups of eukaryotic algae, and hence the primacy of algae in the former plant kingdom. (2) The organization of algae is relatively simple, thus helping to understand the more complex groups of plants. (3) The incredible diversity of types of sexual reproduction, life histories, and photosynthetic pigment apparatuses developed by algae, which seem to have experimented with 'everything' during their evolution. (4) The ever-increasing use of algae as 'systems' or 'models' in biological or biotechnological research. (5) The unique position occupied by algae among the primary producers, as they are an important link in the food web and are essential to the economy of marine and freshwater environments as food organisms. (6) The driving role of algae in the Earth's planetary system, as they initiated an irreversible global change leading to the current oxygen-rich atmosphere; by transfer of atmospheric carbon dioxide into organic biomass and sedimentary deposits, algae contribute to slowing down the accumulation of greenhouse gases leading to global warming; through their role in the production of atmospheric dimethyl sulfide (DMS), algae are believed to be connected with acidic precipitation and cloud formation which leads to global cooling; and their production of halocarbons could be related to global ozone depletion. (7) The incidence of algal blooms, some of which are toxic, seems to be on the increase in both freshwater and marine habitats. (8) The ever-increasing use of algae in pollution control, waste treatment, and biomitigation by developing balanced management practices such as integrated multi-trophic aquaculture (IMTA).

This chapter restricts itself to seaweeds (approximately 10 500 species), which can be defined as marine benthic macroscopic algae. Most of them are members of the phyla Chlorophyta (the green seaweeds of the class Ulvophyceae (893 species)), Ochrophyta (the brown seaweeds of the class Phaeophyceae (1749 species)), and Rhodophyta (the red seaweeds of the classes Bangiophyceae (129 species) and Florideophyceae (5732 species)). To a lot of people, seaweeds are rather unpleasant organisms: these plants are very slimy and slippery and can make swimming or walking along the shore an unpleasant experience to remember! To put it humorously, seaweeds do not have the popular appeal of 'emotional species': only a few have common names, they do not produce flowers, they do not sing like birds, and they are not as cute as furry mammals! One of the key reasons for regularly ignoring seaweeds, even in coastal projects is, in fact, this very problem of identification, as very few people, even among botanists, can identify them correctly. Reasons for this include: a very high morphological plasticity; taxonomic criteria that are not always observable with the naked eye but instead are based on reproductive structures, cross sections, and, increasingly, ultra-structural and molecular arguments; an existing classification of seaweeds that is in a permanent state of revisions; and algal communities with very large numbers of species from different algal taxa that are not always well defined. The production of benthic seaweeds has probably been underestimated, since it may approach 10% of that of all the plankton while only occupying 0.1% of the area used by plankton; this area is, however, crucial, as it is the coastal zone.

The academic, biological, environmental, and economic significance of seaweeds is not always widely appreciated. The following series of arguments emphasizes the importance of seaweeds, and why they should be an unavoidable component of any study that wants to understand coastal biodiversity and processes: (1) Current investigations about the origin of the eukaryotic cell must include features of present-day algae/seaweeds to understand the diversity and the phylogeny of the plant world, and even the animal world. (2) Seaweeds are important primary producers of oxygen and organic matter in coastal environments through their photosynthetic activities. (3) Seaweeds dominate the rocky intertidal zone in most oceans; in temperate and polar regions they also dominate rocky surfaces in the shallow subtidal zone; the deepest seaweeds can be found to depths of 250 m or more, particularly in clear waters. (4) Seaweeds are food for herbivores and, indirectly, carnivores, and hence are part of the foundation of the food web. (5) Seaweeds participate naturally in nutrient recycling and waste treatment (these properties are also used 'artificially' by humans, for example, in IMTA systems). (6) Seaweeds react to changes in water quality and can therefore be used as biomonitors of eutrophication. Seaweeds do not react as rapidly to environmental changes as phytoplankton but can be good indicators over a longer time span (days vs. weeks/months/years) because of the perennial and benthic nature of a lot of them. If seaweeds are 'finally' attracting some media coverage, it is, unfortunately, because of the increasing report of outbreaks of 'green tides' (as well as 'brown and red tides') and fouling species, which are considered a nuisance by tourists and responsible for financial losses by resort operators. (7) Seaweeds can be excellent indicators of natural and/or artificial changes in biodiversity (both in terms of abundance and composition) due to changes in abiotic, biotic, and anthropogenic factors, and hence are excellent monitors of environmental changes. (8) Around 500 species of marine algae (mostly seaweeds) have been used for centuries for human food and medicinal purposes, directly (mostly in Asia) or indirectly, mainly by the phycocolloid industry (agars, carrageenans, and alginates). Seaweeds are the basis of a multibillion-dollar enterprise that is very diversified, including food, brewing, textile, pharmaceutical, nutraceutical, cosmetic, botanical, agrichemical, animal feed, bioactive and antiviral compounds, and biotechnological sectors. Nevertheless, this industry is not very well known to Western consumers, despite the fact that they use seaweed products almost daily (from their orange juice in the morning to their toothpaste in the evening!). This is due partly to the complexity at the biological and chemical level of the raw material, the technical level of the extraction processes, and the commercial level with markets and distribution systems that are difficult to understand and penetrate. (9) The vast majority of seaweed species has yet to be screened for various applications, and their extensive diversity ensures that many new algal products and processes, beneficial to mankind, will be discovered.

Seaweed Mariculture

Seaweed mariculture is believed to have started during the Tokugawa (or Edo) Era (AD 1600–1868) in Japan. Mariculture of any species develops when society's demands exceed what natural resources can supply. As demand increases, natural populations frequently become overexploited and the need for the cultivation of the appropriate species emerges. At present, 92% of the world seaweed supply comes from cultivated species. Depending on the selected species, their biology, life history, level of tissue specialization, and the socioeconomic situation of the region where it is developed, cultivation technology (**Figures 1–6**) can be low-tech (and still extremely successful with highly efficient and simple culture techniques, coupled with intensive labour at low costs) or can become highly advanced and mechanized, requiring on-land cultivation systems for seeding some phases of the life history before growth-out at open-sea aquaculture sites. Cultivation and seed-stock improvement techniques have been refined over centuries, mostly in Asia, and can now be highly sophisticated. High-tech on-land cultivation systems (**Figure 7**) have been developed in a few rare cases, mostly in the Western World; commercial viability has only been reached when high value-added products have been obtained, their markets secured (not necessarily in response to a local demand, but often for export to Asia), and labor costs reduced to balance the significant technological investments and operational costs.

Because the mariculture of aquatic plants (11.3 million tonnes of seaweeds and 2.6 million tonnes of unspecified 'aquatic plants' reported by the Food and Agriculture Organization of the United Nations) has developed essentially in Asia, it remains mostly unknown in the Western World, and is often neglected or ignored in world statistics ... a situation we can only explain as being due to a deeply rooted zoological bias in marine academics, resource managers, bureaucrats, and policy advisors! However, the seaweed aquaculture sector represents 45.9% of the biomass and 24.2% of the value of the world mariculture production, estimated in 2004 at 30.2 million tonnes, and worth US$28.1 billion (**Table 1**). Mollusk aquaculture comes second at 43.0%, and the finfish aquaculture, the subject of many debates, actually only represents 8.9% of the world mariculture production.

The seaweed aquaculture production, which almost doubled between 1996 and 2004, is estimated at 11.3 million tonnes, with 99.7% of the biomass being

Figure 1 Long-line aquaculture of the brown seaweed, *Laminaria japonica* (kombu), in China. Reproduced by permission of Max Troell.

Figure 2 Long-line aquaculture of the brown seaweed, *Undaria pinnatifida* (wakame), in Japan. Photo by Thierry Chopin.

Figure 3 Net aquaculture of the red seaweed, *Porphyra yezoensis* (nori), in Japan. Photo by Thierry Chopin.

cultivated in Asia (**Table 2**). Brown seaweeds represent 63.8% of the production, while red seaweeds represent 36.0%, and the green seaweeds 0.2%. The seaweed aquaculture production is valued at US$5.7 billion (again with 99.7% of the value being provided by Asian countries; **Table 3**). Brown seaweeds dominate with 66.8% of the value, while red seaweeds contribute 33.0%, and the green seaweeds 0.2%. Approximately 220 species of seaweeds are cultivated worldwide; however, six genera (*Laminaria* (kombu; 40.1%), *Undaria* (wakame; 22.3%), *Porphyra* (nori; 12.4%), *Eucheuma/Kappaphycus* (11.6%), and *Gracilaria* (8.4%)) provide 94.8% of the seaweed aquaculture production (**Table 4**), and four genera (*Laminaria* (47.9%), *Porphyra* (23.3%), *Undaria* (17.7%), and *Gracilaria* (6.7%)) provide 95.6% of its value (**Table 5**). Published world statistics, which regularly mention 'data exclude aquatic plants' in their tables, indicate that in 2004 the top ten individual species produced by the global aquaculture (50.9% mariculture, 43.4% freshwater aquaculture, and 5.7% brackishwater aquaculture) were Pacific cupped oyster (*Crassostrea gigas* – 4.4 million tonnes), followed by three species of carp – the silver

Figure 4 Off bottom-line aquaculture of the red seaweed, *Eucheuma denticulatum*, in Zanzibar. Photo by Thierry Chopin.

Figure 5 Bottom-stocking aquaculture of the red alga, *Gracilaria chilensis*, in Chile. Photo by Thierry Chopin.

carp (*Hypophthalmichthys molitrix* – 4.0 million tonnes), the grass carp (*Ctenopharyngodon idellus* – 3.9 million tonnes), and the common carp (*Cyprinus carpio* – 3.4 million tonnes). However, in fact, the kelp, *Laminaria japonica*, was the first top species, with a production of 4.5 million tonnes.

Surprisingly, the best-known component of the seaweed-derived industry is that of the phyco-colloids, the gelling, thickening, emulsifying, binding, stabilizing, clarifying, and protecting agents known as carrageenans, alginates, and agars. However, this component represents only a minor volume (1.26 million tonnes or 11.2%) and value (US$650 million or 10.8%) of the entire seaweed-derived

industry (**Table 6**). The use of seaweeds as sea vegetable for direct human consumption is much more significant in tonnage (8.59 million tonnes or 76.1%) and value (US$5.29 billion or 88.3%). Three genera – *Laminaria* (or kombu), *Porphyra* (or nori), and *Undaria* (or wakame) – dominate the edible seaweed market. The phycosupplement industry is an emerging component. Most of the tonnage is used for the manufacturing of soil additives; however, the agrichemical and animal feed markets are comparatively much more lucrative if one considers the much smaller volume of seaweeds they require. The use of seaweeds in the development of pharmaceuticals and nutraceuticals, and as a source of pigments and

Figure 6 Net aquaculture of the green alga, *Monostroma nitidum*, in Japan. Photo by Thierry Chopin.

Figure 7 Land-based tank aquaculture of the red alga, *Chondrus crispus* (Irish moss), in Canada for high value-added sea-vegetable (Hana-nori) production. Photo by Acadian Seaplants Limited.

bioactive compounds is in full expansion. Presently, that component is difficult to evaluate accurately; the use of 3000 tonnes of raw material to obtain 600 tonnes of products valued at US$3 million could be an underestimation.

The Role of Seaweeds in the Biomitigation of Other Types of Aquaculture

One may be inclined to think that, on the world scale, the two types of aquaculture, fed and extractive,

environmentally balance each other out, as 45.9% of the mariculture production is provided by aquatic plants, 43.0% by mollusks, 8.9% by finfish, 1.8% by crustaceans, and 0.4% by other aquatic animals. However, because of predominantly monoculture practices, economics, and social habits, these different types of aquaculture production are often geographically separate, and, consequently, rarely balance each other out environmentally, on either the local or regional scale (**Figure 8**). For example, salmon aquaculture in Canada represents 68.2% of the tonnage of the aquaculture industry and 87.2% of its farmgate value. In Norway, Scotland, and Chile,

Table 1 World mariculture production and value from 1996 to 2004 according to the main groups of cultivated organisms

	Production (%)		% of value in	
	1996	2000	2004	2004
Mollusks	48	46.2	43.0	34.0
Aquatic plants	44	44.0	45.9	24.2
Finfish	7	8.7	8.9	34.0
Crustaceans	1	1.0	1.8	6.8
Other aquatic animals		0.1	0.4	1.0

Source: FAO (1998) *The State of World Fisheries and Aquaculture 1998*. Italy: Food and Agriculture Organization of United Nations. http://www.fao.org/docrep/w9900e/w9900e00.htm; FAO (2002) *The State of World Fisheries and Aquaculture 2002*. Italy: Food and Agriculture Organization of the United Nations. http://www.fao.org/docrep/005/y7300e/y7300e00.htm; and FAO (2006) *The State of World Fisheries and Aquaculture 2006*. Italy: Food and Agriculture Organization of the United Nations. http://www.fao.org/docrep/009/a0699e/A0699E00.htm (accessed Mar. 2008).

Table 2 Seaweed aquaculture production (tonnage) from 1996 to 2004 according to the main groups of seaweeds (the brown, red, and green seaweeds) and contribution from Asian countries

	Production (tonnes)		
	1996	2000	2004
Brown seaweeds			
World	4 909 269	4 906 280	7 194 316
Asia (%)	4 908 805	4 903 252	7 194 075
	(99.9)	(99.9)	(99.9)
Red seaweeds			
World	1 801 494	1 980 747	4 067 028
Asia (%)	1 678 485	1 924 258	4 035 783
	(93.2)	(97.2)	(99.2)
Green seaweeds			
World	13 418	33 584	19 046
Asia (%)	13 418	33 584	19 046
	(100)	(100)	(100)
Total			
World	6 724 181	6 920 611	11 280 390
Asia (%)	6 600 708	6 861 094	11 248 904
	(98.2)	(99.1)	(99.7)

Source: FAO (1998) *The State of World Fisheries and Aquaculture 1998*. Italy: Food and Agriculture Organization of United Nations. http://www.fao.org/docrep/w9900e/w9900e00.htm; FAO (2002) *The State of World Fisheries and Aquaculture 2002*. Italy: Food and Agriculture Organization of the United Nations. http://www.fao.org/docrep/005/y7300e/y7300e00.htm; and FAO (2006) *The State of World Fisheries and Aquaculture 2006*. Italy: Food and Agriculture Organization of the United Nations. http://www.fao.org/docrep/009/a0699e/A0699E00.htm (accessed Mar. 2008).

salmon aquaculture represents 88.8%, 93.3%, and 81.9% of the tonnage of the aquaculture industry, and 87.3%, 90.9%, and 95.5% of its farmgate value, respectively. Conversely, while Spain (Galicia) produces only 8% of salmon in tonnage (16% in farmgate

Table 3 Seaweed aquaculture production (value) from 1996 to 2004 according to the main groups of seaweeds (the brown, red, and green seaweeds) and contribution from Asian countries

	Value (×US$1000)		
	1996	2000	2004
Brown seaweeds			
World	3 073 255	2 971 990	3 831 445
Asia (%)	3 072 227	2 965 372	3 831 170
	(99.9)	(99.8)	(99.9)
Red seaweeds			
World	1 420 941	1 303 751	1 891 420
Asia (%)	1 367 625	1 275 090	1 875 759
	(96.3)	(97.8)	(99.2)
Green seaweeds			
World	7263	5216	12 751
Asia (%)	7263	5216	12 751
	(100)	(100)	(100)
Total			
World	4 501 459	4 280 957	5 735 615
Asia (%)	4 447 115	4 245 678	5 719 680
	(98.8)	(99.2)	(99.7)

Source: FAO (1998) *The State of World Fisheries and Aquaculture 1998*. Italy: Food and Agriculture Organization of United Nations. http://www.fao.org/docrep/w9900e/w9900e00.htm; FAO (2002) *The State of World Fisheries and Aquaculture 2002*. Italy: Food and Agriculture Organization of the United Nations. http://www.fao.org/docrep/005/y7300e/y7300e00.htm; and FAO (2006) *The State of World Fisheries and Aquaculture 2006*. Italy: Food and Agriculture Organization of the United Nations. http://www.fao.org/docrep/009/a0699e/A0699E00.htm (accessed Mar. 2008).

value), it produces 81% of its tonnage in mussels (28% in farmgate value). Why should one think that the common old saying "Do not put all your eggs in one basket", which applies to agriculture and many other businesses, would not also apply to aquaculture? Having too much production in a single species leaves a business vulnerable to issues of sustainability because of low prices due to oversupply, and the possibility of catastrophic destruction of one's only crop (diseases, damaging weather conditions). Consequently, diversification of the aquaculture industry is imperative to reducing the economic risk and maintaining its sustainability and competitiveness.

Phycomitigation (the treatment of wastes by seaweeds), through the development of IMTA systems, has existed for centuries, especially in Asian countries, through trial and error and experimentation. Other terms have been used to describe similar systems (integrated agriculture-aquaculture systems (IAAS), integrated peri-urban aquaculture systems (IPUAS), integrated fisheries-aquaculture systems (IFAS), fractionated aquaculture, aquaponics); they can, however, be considered to be variations on the IMTA concept. 'Multi-trophic' refers to the incorporation of species from different trophic or nutritional levels in

Table 4 Production, from 1996 to 2004, of the eight genera of seaweeds that provide 96.1% of the biomass for the seaweed aquaculture in 2004

Genus	Production (tonnes)			% of production in 2004
	1996	2000	2004	
Laminaria (kombu)[a]	4 451 570	4 580 056	4 519 701	40.1
Undaria (wakame)[a]	434 235	311 125	2 519 905	22.3
Porphyra (nori)[b]	856 588	1 010 778	1 397 660	12.4
Kappaphycus/ Eucheuma[b]	665 485	698 706	1 309 344	11.6
Gracilaria[b]	130 413	65 024	948 292	8.4
Sargassum[a]	0	0	131 680	1.2
Monostroma[c]	8277	5288	11 514	0.1

[a]Brown seaweeds.
[b]Red seaweeds.
[c]Green seaweeds.
Source: FAO (1998) The State of World Fisheries and Aquaculture 1998. Italy: Food and Agriculture Organization of United Nations. http://www.fao.org/docrep/w9900e/w9900e00.htm; FAO (2002) The State of World Fisheries and Aquaculture 2002. Italy: Food and Agriculture Organization of the United Nations. http://www.fao.org/docrep/005/y7300e/y7300e00.htm; and FAO (2006) The State of World Fisheries and Aquaculture 2006. Italy: Food and Agriculture Organization of the United Nations. http://www.fao.org/docrep/009/a0699e/A0699E00.htm (accessed Mar. 2008).

Table 5 Value, from 1996 to 2004, of the eight genera of seaweeds that provide 99.0% of the value of the seaweed aquaculture in 2004

Genus	Value ($\times$ US$1000)			% of value in 2004
	1996	2000	2004	
Laminaria (kombu)[a]	2 875 497	2 811 440	2 749 837	47.9
Porphyra (nori)[b]	1 276 823	1 183 148	1 338 995	23.3
Undaria (wakame)[a]	178 290	148 860	1 015 040	17.7
Gracilaria[b]	60 983	45 801	385 794	6.7
Kappaphycus/ Eucheuma[b]	67 883	51 725	133 324	2.3
Sargassum[a]	0	0	52 672	0.9
Monostroma[c]	6622	1849	9937	0.2

[a]Brown seaweeds.
[b]Red seaweeds.
[c]Green seaweeds.
Source: FAO (1998) The State of World Fisheries and Aquaculture 1998. Italy: Food and Agriculture Organization of United Nations. http://www.fao.org/docrep/w9900e/w9900e00.htm; FAO (2002) The State of World Fisheries and Aquaculture 2002. Italy: Food and Agriculture Organization of the United Nations. http://www.fao.org/docrep/005/y7300e/y7300e00.htm; and FAO (2006) The State of World Fisheries and Aquaculture 2006. Italy: Food and Agriculture Organization of the United Nations. http://www.fao.org/docrep/009/a0699e/A0699E00.htm (accessed Mar. 2008).

the same system. This is one potential distinction from the age-old practice of aquatic polyculture, which could simply be the co-culture of different fish species from the same trophic level. In this case, these organisms may all share the same biological and chemical processes, with few synergistic benefits, which could potentially lead to significant shifts in the ecosystem. Some traditional polyculture systems may, in fact, incorporate a greater diversity of species, occupying several niches, as extensive cultures (low intensity, low management) within the same pond. The 'integrated' in IMTA refers to the more intensive cultivation of the different species in proximity to each other (but not necessarily right at the same location), connected by nutrient and energy transfer through water.

The IMTA concept is very flexible. IMTA can be land-based or open-water systems, marine or freshwater systems, and may comprise several species combinations. Ideally, the biological and chemical processes in an IMTA system should balance. This is achieved through the appropriate selection and proportioning of different species providing different ecosystem functions. The co-cultured species should be more than just biofilters; they should also be organisms which, while converting solid and soluble nutrients from the fed organisms and their feed into

biomass, become harvestable crops of commercial value and acceptable to consumers. A working IMTA system should result in greater production for the overall system, based on mutual benefits to the co-cultured species and improved ecosystem health, even if the individual production of some of the species is lower compared to what could be reached in monoculture practices over a short-term period.

While IMTA likely occurs due to traditional or incidental, adjacent culture of dissimilar species in some coastal areas (mostly in Asia), deliberately designed IMTA sites are, at present, less common. There has been a renewed interest in IMTA in Western countries over the last 30 years, based on the age-old, common sense, recycling and farming practice in which the by-products from one species become inputs for another: fed aquaculture (fish or shrimp) is combined with inorganic extractive (seaweed) and organic extractive (shellfish) aquaculture to create balanced systems for environmental sustainability (biomitigation), economic stability (product diversification and risk reduction), and social acceptability (better management practices). They are presently simplified systems, like fish/seaweed/shellfish. Efforts to develop such IMTA systems are currently taking place in Canada, Chile, China, Israel, South Africa, the USA, and several European countries. In the future, more advanced

Table 6 Biomass, products, and value of the main components of the world's seaweed-derived industry in 2006

Industry component	Raw material (wet tonnes)	Product (tonnes)	Value (US$)
Sea vegetables	8.59 million	1.42 million	5.29 billion
Kombu (*Laminaria*)[a]	4.52 million	1.08 million	2.75 billion
Nori (*Porphyra*)[b]	1.40 million	141 556	1.34 billion
Wakame (*Undaria*)[a]	2.52 million	166 320	1.02 billion
Phycocolloids	1.26 million	70 630	650 million
Carrageenans[b]	528 000	33 000	300 million
Alginates[a]	600 000	30 000	213 million
Agars[b]	127 167	7630	137 million
Phycosupplements	1.22 million	242 600	53 million
Soil additives	1.10 million	220 000	30 million
Agrichemicals (fertilizers, biostimulants)	20 000	2000	10 million
Animal feeds (supplements, ingredients)	100 000	20 000	10 million
Pharmaceuticals nutraceuticals, botanicals, cosmeceuticals, pigments, bioactive compounds, antiviral agents, brewing, etc.	3000	600	3 million

[a] Brown seaweeds.
[b] Red seaweeds.
Source: McHugh (2003); Chopin and Bastarache (2004); and FAO (2004) *The State of World Fisheries and Aquaculture 2004*. Italy: Food and Agriculture Organization of the United Nations. http://www.fao.org/docrep/007/y5600e/y5600e00.htm (accessed Mar. 2008); FAO (2006) *The State of World Fisheries and Aquaculture 2006*. Italy: Food and Agriculture Organization of the United Nations. http://www.fao.org/docrep/009/a0699e/A0699E00.htm (accessed Mar. 2008).

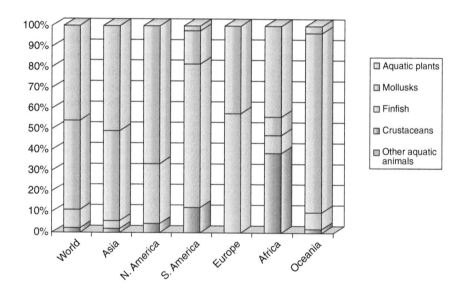

Figure 8 Distribution (%) of mariculture production among the main farmed groups of organisms, worldwide and on the different continents. The world distribution is governed by the distribution in Asia; however, large imbalances between fed and extractive aquacultures exist on the other continents. Source: FAO (2006) *FAO Fisheries Technical Paper 500: State of World Aquaculture 2006*. Rome: Food and Agriculture Organization of the United Nations. http://www.fao.org/docrep/009/a0874e/a0874e00.htm (accessed Mar. 2008).

systems with several other components for different functions, or similar functions but different size brackets of particles, will have to be designed.

Recently, there has been a significant opportunity for repositioning the value and roles seaweeds have in coastal ecosystems through the development of IMTA systems. Scientists working on seaweeds, and industrial companies producing and processing them, have an important role to play in educating the animal-dominated aquaculture world, especially in the Western World, about how to understand and take advantage of the benefits of such extractive organisms, which will help bring a balanced ecosystem approach to aquaculture development.

It is difficult to place a value on the phycomitigation industry, inasmuch as no country has yet implemented guidelines and regulations regarding nutrient discharge into coastal waters. Because the 'user pays' concept is expected to gain momentum as a tool in integrated coastal management, one should soon be able to put a value on the phycomitigation services of IMTA systems for improving water quality and coastal health. Moreover, the conversion of fed aquaculture by-products into the production of salable biomass and biochemicals used in the sea-vegetable, phycocolloid, and phycosupplement sectors should increase the revenues generated by the phycomitigation component.

See also

Mariculture Overview.

Further Reading

Barrington K, Chopin T, and Robinson S (in press) Integrated multi-trophic aquaculture in marine temperate waters. FAO case study review on integrated multi-trophic aquaculture. Rome: Food and Agriculture Organization of the United Nations.

Chopin T, Buschmann AH, Halling C, *et al.* (2001) Integrating seaweeds into marine aquaculture systems: A key towards sustainability. *Journal of Phycology* 37: 975–986.

Chopin T and Robinson SMC (2004) Proceedings of the integrated multi-trophic aquaculture workshop held in Saint John, NB, 25–26 March 2004. *Bulletin of the Aquaculture Association of Canada* 104(3): 1–84.

Chopin T, Robinson SMC, Troell M, *et al.* (in press) Multi-trophic integration for sustainable marine aquaculture. In: Jorgensen SE (ed.) *Encyclopedia of Ecology.* Oxford, UK: Elsevier.

Costa-Pierce BA (2002) *Ecological Aquaculture: The Evolution of the Blue Revolution*, 382pp. Oxford, UK: Blackwell Science.

Critchley AT, Ohno M, and Largo DB (2006) *World Seaweed Resources An Authoritative Reference System.* Amsterdam: ETI BioInformatics Publishers (DVD ROM).

FAO (1998) *The State of World Fisheries and Aquaculture 1998.* Italy: Food and Agriculture Organization of the United Nations. http://www.fao.org/docrep/w9900e/w9900e00.htm (accessed Jul. 2008).

FAO (2002) *The State of World Fisheries and Aquaculture 2002.* Italy: Food and Agriculture Organization of the United Nations. http://www.fao.org/docrep/005/y7300e/y7300e00.htm (accessed Jul. 2008)

FAO (2004) *The State of World Fisheries and Aquaculture 2004.* Italy: Food and Agriculture Organization of the United Nations. http://www.fao.org/docrep/007/y5600e/y5600e00.htm (accessed Mar. 2008).

FAO (2006) *FAO Fisheries Technical Paper 500: State of World Aquaculture 2006.* Rome: Food and Agriculture Organization of the United Nations. http://www.fao.org/docrep/009/a0874e/a0874e00.htm (accessed Mar. 2008).

FAO (2006) *The State of World Fisheries and Aquaculture 2006.* Italy: Food and Agriculture Organization of the United Nations. http://www.fao.org/docrep/009/a0699e/A0699E00.htm (accessed Mar. 2008).

Graham LE and Wilcox LW (2000) *Algae*, 699pp. Upper Saddle River, NJ: Prentice-Hall.

Lobban CS and Harrison PJ (1994) *Seaweed Ecology and Physiology*, 366pp. Cambridge, UK: Cambridge University Press.

McHugh DJ (2001) *Food and Agriculture Organization Fisheries Circular No. 968 FIIU/C968 (En): Prospects for Seaweed Production in Developing Countries.* Rome: FAO. http://www.fao.org/DOCREP/004/Y3550E/Y3550E00.HTM (accessed Mar. 2008).

McVey JP, Stickney RR, Yarish C, and Chopin T (2002) Aquatic polyculture and balanced ecosystem management: New paradigms for seafood production. In: Stickney RR and McVey JP (eds.) *Responsible Marine Aquaculture*, pp. 91–104. Oxon, UK: CABI Publishing.

Neori A, Chopin T, Troell M, *et al.* (2004) Integrated aquaculture: Rationale, evolution and state of the art emphasizing seaweed biofiltration in modern mariculture. *Aquaculture* 231: 361–391.

Neori A, Troell M, Chopin T, Yarish C, Critchley A, and Buschmann AH (2007) The need for a balanced ecosystem approach to blue revolution aquaculture. *Environment* 49(3): 36–43.

Ridler N, Barrington K, Robinson B, *et al.* (2007) Integrated multi-trophic aquaculture. Canadian project combines salmon, mussels, kelps. *Global Aquaculture Advocate* 10(2): 52–55.

Ryther JH, Goldman CE, Gifford JE, *et al.* (1975) Physical models of integrated waste recycling-marine polyculture systems. *Aquaculture* 5: 163–177.

Troell M, Halling C, Neori A, *et al.* (2003) Integrated mariculture: Asking the right questions. *Aquaculture* 226: 69–90.

Troell M, Rönnbäck P, Kautsky N, Halling C, and Buschmann A (1999) Ecological engineering in aquaculture: Use of seaweeds for removing nutrients from intensive mariculture. *Journal of Applied Phycology* 11: 89–97.

Relevant Website

http://en.wikipedia.org
  – Integrated Multi-Trophic Aquaculture, Wikipedia.

MOLLUSKAN FISHERIES

V. S. Kennedy, University of Maryland, Cambridge, MD, USA

Introduction

The Phylum Molluska, the second largest phylum in the animal kingdom with about 100 000 named species, includes commercially important gastropods, bivalves, and cephalopods. All are soft-bodied invertebrates and most have a calcareous shell secreted by a phylum-specific sheet of tissue called the mantle. The shell is usually external (most gastropods, all bivalves), but it can be internal (most cephalopods). Gastropods live in salt water, fresh water, and on land; bivalves live in salt water and fresh water; cephalopods are marine. Gastropods include land and sea slugs (these have a reduced or no internal shell and are rarely exploited by humans) and snails (including edible periwinkles, limpets, whelks, conchs, and abalone). Bivalves include those that cement to hard substrates (oysters); that attach to hard substrates by strong, beard-like byssus threads (marine mussels, some pearl oysters); that burrow into hard or soft substrates (clams, cockles); or that live on the sea bottom but can move into the water column if disturbed (scallops). Cephalopods are mobile and include squid, cuttlefish, octopus, and the chambered nautilus (uniquely among cephalopods, the nautilus has a commercially valuable external shell). Most gastropods are either carnivores or grazers on algae, most commercial bivalves 'filter' suspended food particles from the surrounding water, and cephalopods are carnivores.

Humans have exploited aquatic mollusks for thousands of years, as shown worldwide by shell mounds or middens produced by hunter-gatherers on sea coasts, lake margins, and riverbanks. Some mounds are enormous. In the USA, Turtle Mound in Florida's Canaveral National Seashore occupies about a hectare of land along about 180 m of shoreline, contains nearly 27 000 m³ of oyster shell, and was once an estimated 23 m high (humans have mined such mounds worldwide for the shells, which serve as a source of agricultural and building lime, as a base for roads, and for crumbling into chicken grit). A group of oystershell middens near Damariscotta, Maine, USA stretches along 2 km of shoreline. One mound was about 150 m long, 70 m wide, and 9 m high before it was mined; a smaller mound was estimated to contain about 340 million individual shells. Studies by anthropologists have shown that many middens were started thousands of years ago (e.g. 4000 years ago in Japan, 5000 years ago in Maine, 12 000 years ago in Chile, and perhaps up to 30 000 years ago in eastern Australia; sea level rise may have inundated even older sites).

Mollusks have been exploited for uses other than food. More than 3500 years ago, Phoenicians extracted 'royal Tyrian purple' dye from marine whelks in the genus *Murex* to color fabrics reserved for royalty, and other whelks in the Family Muricidae have long been used in Central and South America to produce textile dyes. From Roman times until the early 1900s, the golden byssus threads of the Mediterranean pen shell *Pinna nobilis* were woven into fine, exceedingly lightweight veils, gloves, stockings, and shawls. Wealth has been displayed worldwide on clothing or objects festooned with cowry (snail) shells. Cowries, especially the money cowry *Cypraea moneta*, have been widely used as currency, perhaps as early as 700 BC in northern China and the first century AD in India. Their use spread to Africa and North America. Also in North America, East Coast Indians made disk-shaped beads, or wampum, from the shells of hard clams *Mercenaria mercenaria* for use as currency.

Mollusk shells have been used as fishhooks and octopus lures (the latter often made of cowry shells), household utensils (bowls, cups, spoons, scrapers, knives, boring devices, adzes, chisels, oil lamps), weapons (knives, axes), signaling devices (trumpets made from large conch shells), and decorative objects (beads, lampshades made of windowpane oyster shell, items made from iridescent 'mother-of-pearl' shell of abalone). Many species of mollusks produce 'pearls' when they cover debris that is irritating their mantle with layers of nacre (mother-of-pearl; an aragonite form of calcium carbonate). The pearl oyster (not a true oyster) and some freshwater mussels (not true mussels) use an especially iridescent nacre that results in commercially valuable pearls. Some religious practices involve shells, including scallop shells used as symbols by pilgrims trekking to Santiago de Compostela in Spain to honor St James (the French name for the scallop is Coquille St-Jacques).

As human populations and the demand for animal protein have grown, harvests of wild mollusks have expanded. However, overharvesting and pollution of

Figure 1 Post-harvest activities on shallow-draft tonging boats in Chesapeake Bay, Maryland, USA. Note the array of tongs, and the sorting or culling platform in the boat in the left foreground. (Photograph by Skip Brown, courtesy of Maryland Sea Grant.)

mollusk habitat have depleted many wild populations. To meet the demand for protein and to combat these losses, aquaculture has increased to supplement wild harvests. Unfortunately, some harvesting and aquaculture practices can have detrimental effects on the environment (see below).

Harvesting Natural Populations

Historical exploitation of mollusks occurred worldwide in shallow coastal and freshwater systems, and artisanal fishing still takes place there. The simplest fisheries involve harvesting by hand or with simple tools. Thus marine mussels and various snails exposed on rocky shores at low tides are harvested by hand, with oysters, limpets, and abalone pried from the rocks. Low tide on soft-substrate shores allows digging by hand to capture many shallow-dwelling species of burrowing clams. Some burrowing clams live more deeply or can burrow quickly when disturbed. Harvesting these requires a shovel, or a modified rake with long tines that can penetrate sediment quickly and be rotated so the tines retain the clam as the rake is pulled to the surface.

Mollusks living below low tide are usually captured by some sort of tool, the simplest being rakes and tongs. For example, oyster tongs in Chesapeake Bay (**Figure 1**) have two wooden shafts, each with a half-cylinder, toothed, metal-rod basket bolted at one end and with a metal pin or rivet holding the shafts together like scissors. A harvester standing on the side of a shallow-draft boat lets the shafts slip through his hands (almost all harvesters are male) until the baskets reach the bottom and then moves the upper ends of the shafts back and forth to scrape oysters and shells into a pile. He closes the baskets on the pile and hoists the contents to the surface manually or by a winch, opening the baskets to dump the scraped material onto a sorting platform on the boat (see **Figure 1**). Harvesters use hand tongs at depths up to about 10 m. Also in Chesapeake Bay, harvesters exploiting oysters living deeper than 10 m deploy much larger, heavier, and more efficient tongs from their boat's boom, using a hydraulic system to raise and lower the tongs and to close them on the bottom. In addition to capturing bivalves, rakes and hand tongs are used worldwide to harvest gastropods like whelks, conchs, and abalone (carnivorous gastropods like whelks and conchs can also be captured in pots baited with dead fish and other animals).

Mollusks can be captured by dredges towed over the bottom by boats, some powered by sail (**Figure 2**) and others by engines (**Figure 3**). Dredges harvest attached mollusks like oysters and marine mussels, as well as buried clams, scallops lying on or swimming

Figure 2 Oyster dredge coming on board a sailboat ('skipjack') in Chesapeake Bay, Maryland, USA. Note the small 'push-boat' or yawl hoisted on the stern on this sailing day. (Photograph by Michael Fincham, courtesy of Maryland Sea Grant.)

Figure 3 Harvesting vessel *Mytilus* with blue mussel dredges in Conwy Bay, Wales. (Photograph courtesy of Dr Eric Edwards.)

just above the sea bottom, and some gastropods (whelks, conchs). Dredges are built of metal and usually have teeth on the leading edge that moves over the bottom. Captured material is retained in a sturdy mesh bag made of wear-resistant heavy metal rings linked together and attached to the dredge frame. Mesh size is usually regulated so that small mollusks can fall out of the bag and back onto the sea bottom.

Some dredges use powerful water jets to blow buried mollusks out of soft sediment and into a metal-mesh bag. Where the water is shallow enough, subtidal clams (and oysters in some regions) are harvested by such water jets, but instead of being captured in a mesh bag the clams are blown onto a wire-mesh conveyor belt that carries them to the surface alongside the boat. Harvesters pick legal-sized clams off the mesh as they move past on the belt. Mesh size is such that under-sized clams and small debris fall through the belt and return to the bottom while everything else continues up the belt and falls back into the water if not removed by the harvester.

Commercial harvesting of freshwater mussels in the USA since the late 1800s has taken advantage of the propensity of bivalves to close their shells tightly when disturbed. Harvest vessels tow 'brails' over the bottom. Brails are long metal rods or galvanized pipe with eyebolts at regular intervals. Wire lines are attached to the eyebolts by snap-swivels. Each line holds a number of 'crowfoot' hooks of various sizes and numbers of prongs, depending on the species being harvested. Small balls are formed on the end of each prong so that, when the prong tip enters

between the partially opened valves of the mussel, the valves close on the prong and the ball keeps the prong from pulling free of the shell. When the brails are brought on deck the clinging mussels are removed. This method works best in river systems with few snags (tree stumps, rocks, trash) that would catch and hold the hooks.

Cephalopods are captured by trawls, drift nets, seines, scoop and cast nets, pots and traps, and hook and line. A traditional gear is the squid jig, which takes various shapes but which has an array of barbless hooks attached. Jigs are moved up and down in the water to attract squid, which grab the jig and ensnare their tentacles, allowing them to be hauled into the boat. In oceanic waters, large vessels using automated systems to oscillate the jig in the water may deploy over 100 jig lines, each bearing 25 jigs. Such vessels fish at night, with lights used to attract squid to the fishing boat. A typical vessel may carry 150 metal-halide, incandescent lamps that together produce 300 kW of light. Lights from concentrations of vessels in the global light-fishing fleet off China and south-east Asia, New Zealand, the Peruvian coast, and southern Argentina can be detected by satellites. With a crew of 20, a vessel as described above may catch 25–30 metric tons of squid per night.

Some harvesters dive for mollusks, especially solitary organisms of high market value such as pearl oysters and abalone. Breath-hold diving has been used for centuries, but most divers now use SCUBA or air-delivery (hooka) systems. Diving is efficient because divers can see their prey, whereas most other capture methods fish 'blind'. Unfortunately, although diving allows for harvesting with minimal damage to the habitat, it has led to the depletion or extinction of some mollusk populations such as those of abalone.

A variety of measures are in place around the world to regulate mollusk fisheries. A common regulation involves setting a minimum size for captured animals that is larger than the size at which individuals of the species become capable of reproducing. This regulation ensures that most individuals can spawn at least once before being captured. Size selection is often accomplished by use of a regulated mesh size in dredge bags or conveyor-belts as described earlier. If the animals are harvested by a method that is not size selective, such as tonging for oysters or brailing for freshwater mussels, then the harvester is usually required to cull undersized individuals from the accumulated catch (see **Figure 1** for the culling platform used by oyster tongers) and return them to the water, usually onto the bed from which they were taken. Oysters are measured with a metal ruler; freshwater mussels are culled by attempting to pass them through metal rings of legal diameter and keeping those that cannot pass through.

Other regulatory mechanisms include limitations as to the number of harvesters allowed to participate in the fishery, the season when harvesting can occur, the type of harvest gear that can be used, or the total catch that the fishery is allowed to harvest. There may be areas of a species' range that are closed to harvest, perhaps when the region has many undersized juveniles or when beds of large adults are thought to be in need of protection so that they can serve as a source of spawn for the surrounding region. Restrictions may spread the capture effort over a harvest season to prevent most of the harvest from occurring at the start of the season, with a corresponding market glut that depresses prices. Finally, managers may protect a fishery by regulations mandating inefficiencies in harvest methods. Thus, in Maryland's Chesapeake Bay, diving for oysters has been strictly regulated because of its efficiency. Similarly, the use of a small boat (**Figure 2**) to push sailboat dredgers ('power dredging') is allowed only on 2 days per week (the days chosen – usually days without wind – are at the captain's discretion). Dredging must be done under sail on the remaining days of the week (harvesting is not allowed on Sunday). Around the world, inspectors ('marine police') are empowered to ensure that regulations are followed, either by boarding vessels at sea or when they dock with their catch.

A great hindrance to informed management of molluskan (and other) fisheries is the lack of data on the quantity of organisms taken by noncommercial (recreational) harvesters. For example, in Maryland's Chesapeake Bay one can gather a bushel (around 45 l) of oysters per day in season without needing a license if the oysters are for personal use. Clearly it is impossible to determine how many of these bushels are harvested during a season in a sizeable body of water. If such harvests are large, the total fishing mortality for the species can be greatly underestimated, complicating efforts to use fishery models to manage the fishery.

Processing mollusks involves mainly shore-based facilities, except for deep-sea cephalopod (mostly squid) fisheries where processing, including freezing, is done on board, and for some scallop species (the large muscle that holds scallop shells shut is usually cut from the shell at sea, with the shell and remaining soft body parts generally discarded overboard). Thus the catch may be landed on the same day it is taken (oysters, freshwater and marine mussels, many clams and gastropods) or within a few days (some scallop and clam fisheries). If the catch is not frozen, ice is used to prevent spoilage at sea.

Suspension-feeding mollusks (mostly bivalves) can concentrate toxins from pollutants or poisonous algae and thereby become a threat to human health. If such mollusks are harvested, they have to be held in clean water for a period of time to purge themselves of the toxin (if that is possible). Thus they may be relaid on clean bottom or held in shore-based systems ('depuration facilities') that use ozone or UV light to sterilize the water that circulates over the mollusks. Relaying and reharvesting the mollusks or maintaining the land-based systems adds to labor, energy, and capital costs.

Aquaculture

Aquaculture involves using either natural 'seed' (small specimens or juveniles that will be moved to suitable habitat for further growth) that is harvested from the wild and reared in specialized facilities, or producing such seed in hatcheries. Most commercial bivalves and abalone can be spawned artificially in hatcheries. The techniques involve either taking adults ripe with eggs or sperm (gametes) from nature or assisting adults to ripen by providing algal food in abundance at temperatures warm enough to support gamete production. Most ripe adults will spawn when provided with a stimulus such as an increase in temperature or food or both, or by the addition to the ambient water of gametes dissected from sacrificed adults. The spawned material is washed through a series of screens to separate the gametes from debris. Eggs are washed into a container and their numbers are estimated by counting samples, then the appropriate density of sperm is added to fertilize the eggs. Depending on the species of mollusk, an adult female may produce millions of eggs in one spawning event, so hundreds of millions of larvae produced by a relatively small number of females may be available for subsequent rearing.

Larvae are reared in specialized containers that allow culture water to be changed every few days and the growing larvae to be captured on screens, counted, and replaced into clean water until they become ready to settle. As they grow, larvae are fed cultured algae at appropriate concentrations (this requires extremely large quantities of algae to support the heavily feeding larvae). Depending on the species, after 2 or more weeks many larvae are ready to settle ('set') onto a solid surface (e.g., oysters, marine mussels, scallops, pearl oysters, abalone) or onto sediment into which they will burrow (e.g. clams). The solid material on which setting occurs is called 'cultch'. Settled larvae are called 'spat'.

Spat can be reared in the hatchery if sufficient algal food is available (usually an expensive proposition given the large quantity of food required). Thus most production facilities move spat (or seed) into nature soon after they have settled. However, this exposes the seed to diverse natural predators – including flatworms, boring sponges, snails, crabs, fish, and birds. Consequently, the seed may need to be protected (e.g. by removing predators by hand, poisoning them, and providing barriers to keep them from the seed), which is labor-intensive and expensive. Mortalities of larvae, spat, and seed are high at all stages of aquaculture operations.

The energy and monetary costs of maintaining hatchery water at suitable temperatures and of rearing enormous quantities of algae means that molluskan hatcheries are expensive to operate profitably. Thus only 'high-value' mollusks are cultured in hatcheries. One option for those who cannot afford to maintain a hatchery is to purchase larvae that are ready to settle. In the supply hatchery, larvae are screened from the culture water onto fine mesh fabric in enormous densities. The densely packed larvae withdraw their soft body parts into their shells, closing them. The larvae can then be shipped by an express delivery service in an insulated container that keeps them cool and moist until they reach the purchaser, who gently rinses the larvae off the fabric into clean water of the appropriate temperature and salinity in a setting tank. Also in the tank is the cultch that the larvae will attach to or the sediment into which they will burrow. This procedure of setting purchased larvae is called remote setting.

Among bivalves, oyster larvae settle on hard surfaces, preferably the shell of adults of the species, but also on cement materials, wood, and discarded trash. The larva cements itself in place and is immobile for the rest of its life. Thus oysters have traditionally been cultured by allowing larvae to cement to cultch like wooden, bamboo, or concrete stakes; stones and cement blocks; or shells (such as those of oysters and scallops) strung on longlines hanging from moored rafts (**Figure 4C**). As noted above, the settling larvae may be either those produced in a hatchery or those living in nature. The spat may be allowed to remain where they settle, or the cultch may be moved to regions where algal production is high and spat growth can accelerate. Scallop, marine mussel, and pearl oyster larvae do not cement to a substrate, but secrete byssus threads for attachment. For these species, fibrous material like hemp rope is provided in hatcheries or is deployed in nature where larvae are abundant. As these settled spat grow, they may be transferred to containers such as single- or multi-compartment nets (**Figure 4A, B**) or to flexible mesh tubes (used for marine mussels). Most clams are burrowers and will settle onto sediment. This may be provided in ponds, sometimes including those used to grow shrimp and other crustaceans (simultaneous culture of various species is called polyculture). Abalone settle on hard surfaces to which their algal food is attached, so they are usually cultured initially on corrugated plastic plates coated with diatoms. As they grow, they may be moved to well-flushed raceways or held suspended in nature (**Figure 4C**) in net cages.

There are risks associated with aquaculture, just as with harvesting natural populations. A major problem is the one that affects many agricultural monocultures and that is the increased susceptibility to disease outbreaks among densely farmed organisms. Such diseases now affect cultured abalone and scallops in China. Another problem involves the reliance on a relatively few animals for spawning purposes, which can lead to genetic deficiencies and inbreeding, further endangering a culture program. The coastal location of aquaculture facilities makes them susceptible to damage by storms, and if such storms increase or intensify with global warming, aquaculturists risk losing their animals, facilities, and investment. Ice can cause damage in cold-winter regions, although this is more predictable than are intense storms and preventative steps can be taken (not always successfully).

Extensive use of rafts and other systems for suspension culture (**Figure 4C**) may interfere with shipping and recreational uses of the water, and in some regions, laws against navigational hazards prevent water-based culture systems. If land

Figure 4 (A) Single-compartment pearl oyster cages attached to intertidal longlines. (B) Multi-compartment lantern net used to culture scallops. (C) Suspended longlines for growing Pacific oyster, scallops, and abalone off Rongchen, China. (Photographs courtesy of Dr Ximing Guo; reproduced with permission from Guo *et al.* (1999) *Journal of Shellfish Research* 18: 19–31.)

contiguous to the water is too expensive because of development or other land-use practices, land-based culture systems may not be economically feasible. Finally, coastal residents may consider rafts to be an eyesore that lowers the value of their property and they may seek to have them banned from the region.

Comparisons of Wild and Cultured Production

Of the 86–93 million metric tons of aquatic species harvested from wild stocks in 1996–1998 (capture landings), fish comprised 86%, mollusks 8%, and crustaceans (shrimp, lobsters, crabs) 6% (**Table 1**). Among the mollusks, the average relative proportions over the 3 years were cephalopods, 44%; bivalves, 32%; gastropods, 2%; and miscellaneous, 21%. Within the bivalves, the relative proportions over the 3 years were clams, 38% of all bivalves; freshwater mussels, 25%; scallops, 22%; marine mussels, 9%; and oysters, 6%.

Aquaculture had a greater effect on total landings (wild harvest plus aquaculture production) of mollusks than of fish and crustaceans. Of FAO's estimated 27–31 million metric tons of aquatic animals cultured worldwide from 1996 to 1998, fish comprised 64% and crustaceans comprised 5% by weight, declines from their proportions of the wild harvest (**Table 1**). In contrast, mollusks comprised 31% of culture production, about four times their proportion of the wild harvest. In addition, 20–40% more mollusks by weight were produced by aquaculture than were harvested from nature; by contrast, the quantities of cultured fish and crustaceans were a small fraction of quantities harvested in nature (**Table 1**). Of the cultured mollusks produced from 1996 to 1998, bivalves represented 87% by weight, followed by miscellaneous mollusks at 13%;

Table 1 Worldwide capture landings (wild harvest) and aquaculture production and value of fish, mollusks, and crustaceans from 1996 to 1998

Category	1996 (%)	1997 (%)	1998 (%)	Average percentage
Capture landings (million metric tons)				
Fish	80.9 (87)	79.7 (86)	72.7 (85)	86
Mollusks	6.6 (7)	7.3 (8)	6.6 (8)	8
Crustaceans	5.5 (6)	5.9 (6)	6.4 (7)	6
Total	93.0	92.9	85.7	
Aquaculture production (million metric tons)				
Fish	17.0 (63)	18.8 (65)	20.0 (65)	64
Mollusks	8.6 (32)	8.7 (30)	9.2 (30)	31
Crustaceans	1.2 (4)	1.4 (5)	1.6 (5)	5
Total	26.8	28.9	30.8	
Aquaculture value (billion $US)				
Fish	26.5 (62)	28.3 (62)	27.8 (61)	62
Mollusks	8.6 (20)	8.7 (19)	8.5 (19)	19
Crustaceans	7.8 (18)	8.5 (19)	9.2 (20)	19

From UN Food and Agriculture Organization statistics as at 21 March 2000.

cultured gastropods and cephalopods represented fractions of a percent of produced weight. When the FAO's harvest values of the wild fisheries and aquaculture production from 1984 to 1998 are combined for bivalves, cephalopods, and gastropods (**Figure 5**), the production of bivalves is seen to have risen greatly in the 1990s, with cephalopod production having increased modestly and gastropod production not at all. The differences can be attributed to the relatively greater yield of cultured bivalves compared with the other two molluskan groups.

The economic value of cultured animals ranged from 43 to 46 billion US dollars over the period 1996–1998, with fish comprising an average of 62%, mollusks 19%, and crustaceans 19% of this amount (**Table 1**). Thus, although production (weight) of cultured mollusks was six to seven times that of cultured crustaceans, their value just equaled that of crustaceans (this was a result of the premium value of cultured shrimps and prawns). Of the cultured mollusks, bivalves represented 93% by economic value followed by miscellaneous mollusks at 6%; cultured gastropods and cephalopods represented fractions of a percent of overall economic value.

Among cultured bivalves, the relative proportions by weight and by economic value were: Oysters, 42% and 42% respectively; clams, 26% and 32%; scallops, 15% and 20%; marine mussels, 17% and 6%; and freshwater mussels, <1% in both categories. Thus, although oysters were the least important bivalve in terms of wild harvests (6% of bivalve landings, see above), they were the most important cultured bivalve. On the other hand, freshwater mussels represented 25% of wild harvests but were relatively insignificant as a cultured item.

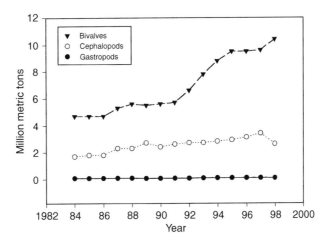

Figure 5 Worldwide production of bivalve, cephalopod, and gastropod mollusks from wild fisheries plus aquaculture activities in the period from 1984 to 1998. (From UN Food and Agriculture Organization statistics as at 21 March 2000.)

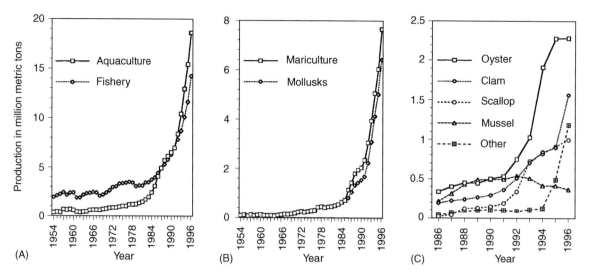

Figure 6 Production (wet, whole-body weight) in the natural fishery and in aquaculture in China. (A) Natural fishery and aquaculture. (B) Total mariculture (marine aquaculture) production and the molluskan component. (C) Comparative production among molluskan groups. (Adapted with permission from Guo *et al.* (1999) *Journal of Shellfish Research* 18: 19–31.)

China provides an excellent example of increased efforts in aquaculture, with its industry producing about 50% of the world's aquacultural output in the early 1990s, rising to about 67% by 1997 and 1998. Oyster and clam culture has been practiced in China for about two millennia, but growth of aquaculture accelerated in the early 1950s, with production outstripping wild fishery harvests by 1988 (**Figure 6A**). Mollusk culture was a substantial factor in this growth (**Figure 6B**). The intensified culture involved new species beyond the traditional oysters and clams, beginning with marine mussels, then scallops, then abalone. Over 30 species of marine mollusks are farmed along China's coasts, including 3 species of oysters, 14 of clams, 4 of marine mussels, 4 of scallops, 2 of abalone, 2 of snails, and 1 of pearl oyster. Marine mussel production has declined since 1992 (**Figure 6C**), apparently because of increased production of 'high-value' species (oysters, scallops, abalone).

Detrimental Effects of Molluskan Fisheries

Harvesting Natural Populations

Harvesting can have direct and indirect effects on local habitats. Direct effects include damage caused by harvest activities, such as when humans trample rocky shore organisms while harvesting edible mollusks. Similarly, heavy dredges with strong teeth can damage and kill undersized bivalves as well as associated inedible species. Dredges that use high-pressure water jets churn up bottom sediments and

displace undersized clams and nontarget organisms that may not be able to rebury before predators find them exposed on the surface. When commercial shellfish are concentrated within beds of aquatic grasses, dredging can uproot the plants.

On the other hand, some harvesting techniques are relatively benign. For example, lures like squid jigs attract the target species with little or no 'by-catch' of nontarget species. Size selectivity can be attained by using different sizes of jigs. However, as with most fishing methods, overharvesting of the target species can still occur.

From the perspective of population dynamics, intense harvesting pressure commonly results in smaller average sizes of individual organisms and eventually most survivors may be juveniles. Care must be taken to ensure that a suitable proportion of individuals remains to reproduce before they are harvested, thus the widespread regulations on minimal sizes of individuals that can be harvested.

In terms of indirect effects, if some molluskan species are overharvested, inedible associated organisms that depend on the harvested species in some way can be affected. For example, oysters are 'ecosystem engineers' (like corals) and produce large structures (beds, reefs) that are exploited by numerous other organisms. Oyster shells provide attachment surfaces for large numbers of barnacles, mussels, sponges, sea squirts, and other invertebrates, as well as for the eggs of some species of fish and snails. Waterborne sediments as well as feces from all the reef organisms settle in interstices among the shells and become habitat for worms and other burrowing invertebrates. The presence of these organisms attracts predators, including crustaceans and fish. Consequently, the diversity of species on and around oyster reefs is higher than it is for adjacent soft-bottom systems. The same is true for marine mussel beds. Harvesting these bivalves can result in lowered biomass and perhaps lowered diversity of associated organisms.

Depleted populations of commercial bivalves may have other ecological consequences. These bivalves are 'suspension feeders' and pump water inside their shell and over their gills to extract oxygen and remove suspended food particles. When Europeans sailed into Chesapeake Bay in the 1600s, oyster reefs broke the Bay's surface and were a navigational hazard. Today, many oyster beds have been scraped almost level with the Bay bottom and annual harvests in Maryland have dropped from an estimated 14 million bushels of oysters in the 1880s to a few hundred thousand bushels today. Knowing the pumping capacity of individual oysters and the estimated abundances of pre-exploitation and present-day populations, scientists can calculate the filtering ability of those populations. Estimates are that a quantity of water equivalent to the total amount contained in the Bay could have been filtered in about a week in the early eighteenth century; today's depleted populations are thought to require nearly a year to filter the same amount of water.

This diminished filtering capacity has implications for food web dynamics. A feeding oyster ingests and digests food particles and expels feces as packets of material larger than the original food particles. Nonfood particles are trapped in mucous strings and expelled into the surrounding water as pseudofeces. Thus, oysters take slowly sinking microscopic particles from the water and expel larger packets of feces and pseudofeces that sink more rapidly to the Bay bottom. When oyster numbers are greatly diminished, this filtering and packaging activity is also diminished and more particles remain suspended in the water column. Thus ecological changes in Chesapeake Bay over the last century may have been enormous. Many scientists believe the Bay has shifted from a 'bottom-dominated' system in which oyster reefs were ubiquitous and the water was clearer than now to one dominated by water-column organisms (plankton) in which light levels are diminished. Smaller plankton serve as food for larger plankton, including jellyfish, and some scientists believe that the large populations of jellyfish present in the Bay in warmer months may be a result of overharvesting of oysters over the past century.

The reverse of this phenomenon is seen in the North American Great Lakes, where huge populations of zebra mussels *Dreissena polymorpha* and quagga mussels *Dreissena bugensis* have developed from invaders inadvertently introduced from Europe in ships' ballast water. The mussels filter the lake water so efficiently that the affected lakes are clearer than they have been for decades – they have become 'bottom-dominated' and plankton populations have decreased in abundance. Unintended consequences resulting from overharvesting other mollusk populations remain to be elucidated.

Aquaculture Practices

A number of environmental problems affect molluskan aquaculture. For one, heavy suspension feeding by densely farmed mollusks in coastal bays may outstrip the ability of the environment to supply algae, so the mollusks may starve or cease to grow. Another problem is that large concentrations of cultured organisms produce fecal and other wastes in quantities that can overwhelm the environment's ability to recycle these wastes. When this happens, eutrophication occurs, inedible species of algae may appear, and abundances of algal species that support molluskan growth may be reduced. To counter these problems, countries like Australia are using Geographic Information System technology to pinpoint suitable and unsuitable locations for aquaculture.

Introductions of exotic species of shellfish for aquaculture purposes have sometimes been counterproductive. A variety of diseases that have depleted native mollusk stocks have been associated with some introductions (e.g. MSX disease in the eastern oyster *Crassostrea virginica* in Chesapeake Bay is thought to be linked to attempts to import the Pacific oyster *Crassostrea gigas* to the bay). 'Hitch-hiking' associates that are carried to the new environment by imported mollusks have sometimes caused problems. For example, the gastropod *Crepidula fornicata* that accompanied oysters of the genus *Crassostrea* that were brought to Europe from Asia has become so abundant that it competes with the European oyster *Ostrea edulis* and blue mussel *Mytilus edulis* for food and may foul oyster and mussel beds with its wastes. Recently, the veined rapa whelk *Rapana venosa* has appeared in lower Chesapeake Bay, apparently arriving from the Black Sea or from Japan in some unknown fashion. It is a carnivore and may pose a threat to the indigenous oyster and hard clam fisheries (although it might prove to be a commercial species itself if a market can be found). As a result of these and other problems, many countries have developed stringent rules to govern movement of mollusks locally and worldwide.

See also

Coral Reef and Other Tropical Fisheries. Corals and Human Disturbance. Exotic Species, Introduction of. Rocky Shores.

Further Reading

Andrews JD (1980) A review of introductions of exotic oysters and biological planning for new importations. *Marine Fisheries Review* 42(12): 1–11.

Attenbrow V (1999) Archaeological research in coastal southeastern Australia: A review. In: Hall J and McNiven IJ (eds.) *Australian Coastal Archaeology*, pp. 195–210. ANH Publications, RSPAS. Canberra, Australia: Australian National University.

Caddy JF (ed.) (1989) *Marine Invertebrate Fisheries: Their Assessment and Management.* New York: John Wiley Sons.

Caddy JF and Rodhouse PG (1998) Cephalopod and groundfish landings: Evidence for ecological change in global fisheries. *Reviews in Fish Biology and Fisheries* 8: 431–444.

Food and Agriculture Organization of the United Nations' website for fishery statistics (www.fao.org).

Guo X, Ford SE, and Zhang F (1999) Molluscan aquaculture in China. *Journal of Shellfish Research* 18: 19–31.

Kaiser MJ, Laing I, Utting SD, and Burnell GM (1998) Environmental impacts of bivalve mariculture. *Journal of Shellfish Research* 17: 59–66.

Kennedy VS (1996) The ecological role of the eastern oyster, *Crassostrea virginica*, with remarks on disease. *Journal of Shellfish Research* 15: 177–183.

MacKenzie CL Jr, Burrell VG Jr, Rosenfield A, and Hobart WL (eds.) (1997) *The History, Present Condition, and Future of the Molluscan Fisheries of North and Central America and Europe. Volume 1, Atlantic and Gulf Coasts. (NOAA Technical Report NMFS 127); Volume 2, Pacific Coast and Supplemental Topics (NOAA Technical Report NMFS 128); Volume 3, Europe (NOAA Technical Report NMFS 129).* Seattle, Washington: US Department of Commerce.

Menzel W (1991) *Estuarine and Marine Bivalve Mollusk Culture.* Boca Raton, FL: CRC Press.

Newell RIE (1988) Ecological changes in Chesapeake Bay: Are they the result of overharvesting of the American oyster, *Crassostrea virginica*. In: Lynch MP and Krome EC (eds.) *Understanding the Estuary: Advances in Chesapeake Bay Research*, pp. 536–546. Chesapeake Research Consortium Publication 129. Solomons, MD: Chesapeake Research Consortium.

Rodhouse PG, Elvidge CD, and Trathan PN (2001) Remote sensing of the global light-fishing fleet: An analysis of interactions with oceanography, other fisheries and predators. *Advances in Marine Biology* 39: 261–303.

Safer JF Gill FM (1982) *Spirals from the Sea. An Anthropological Look at Shells.* New York: Clarkson N Potter.

Sanger D and Sanger M (1986) Boom and bust in the river. The story of the Damariscotta oyster shell heaps. *Archaeology of Eastern North America* 14: 65–78.

OYSTERS – SHELLFISH FARMING

I. Laing, Centre for Environment Fisheries and Aquaculture Science, Weymouth, UK

As I ate the oysters with their strong taste of the sea and their faint metallic taste that the cold white wine washed away, leaving only the sea taste and the succulent texture, and as I drank their cold liquid from each shell and washed it down with the crisp taste of the wine, I lost the empty feeling and began to be happy and to make plans. (Ernest Hemingway in *A Moveable Feast*).

A Long History

Oysters have been prized as a food for millennia. Carbon dating of shell deposits in middens in Australia show that the aborigines took the Sydney rock oyster for consumption in around 6000 BC. Oyster farming, nowadays carried out all over the world, has a very long history, although there are various thoughts as to when it first started. It is generally believed that the ancient Romans and the Greeks were the first to cultivate oysters, but some maintain that artificial oyster beds existed in China long before this.

It is said that the ancient Chinese raised oysters in specially constructed ponds. The Romans harvested immature oysters and transferred them to an environment more favorable to their growth. Greek fishermen would toss broken pottery dishes onto natural oyster beds to encourage the spat to settle.

All of these different cultivation methods are still practiced in a similar form today.

A Healthy Food

Oysters (**Figure 1**) have been an important food source since Neolithic times and are one of the most nutritionally well balanced of foods. They are ideal for inclusion in low-cholesterol diets. They are high in omega-3 fatty acids and are an excellent source of vitamins. Four or five medium-size oysters supply the recommended daily allowance of a whole range of minerals.

The Oyster

Oysters are one among a number of bivalve mollusks that are cultivated. Others include clams, cockles, mussels, and scallops.

Current Status

Oysters form the second most important group, after cyprinid fishes, in world aquaculture production.

Figure 1 A dish of Pacific oysters.

In 2004, 4.6 million tonnes were produced (Food and Agriculture Organization (FAO) data). Asia and the Pacific region produce 93.4% of this total production.

The FAO of the United Nations lists 15 categories of cultivated oyster species. The Pacific oyster (*Crassostrea gigas*) is by far the most important. Annual production of this species now exceeds 4.4 million metric tonnes globally, worth US$2.7 billion, and accounts for 96% of the total oyster production (see **Table 1**). China is the major producer.

The yield from wild oyster fisheries is small compared with that from farming and has declined steadily in recent years. It has fallen from over 300 000 tonnes in 1980 to 152 000 tonnes in 2004. In contrast to this, Pacific oyster production from aquaculture has increased by 51% in the last 10 years (see **Figure 2**).

General Biology

As might be expected, given a long history of cultivation, there is a considerable amount known about the biology of oysters.

The shell of bivalves is in two halves, or valves. Two muscles, called adductors, run between the inner surfaces of the two valves and can contract rapidly to close the shell tightly. When exposed to the air, during the tidal cycle, oysters close tightly to prevent desiccation of the internal tissues. They can respire anaerobically (i.e., without oxygen) when out of water but have to expel toxic metabolites when reimmersed as the tide comes in. They are known to be able to survive for long periods out of water at low temperatures such as those used for storage after collection.

Within the shell is a fleshy layer of tissue called the mantle; there is a cavity (the mantle cavity) between the mantle and the body wall proper. The mantle secretes the layers of the shell, including the inner nacreous, or pearly, layer. Oysters respire by using both gills and mantle. The gills, suspended within the mantle cavity, are large and function in food gathering (filter feeding) as well as in respiration. As water passes over the gills, organic particulate material, especially phytoplankton, is strained out and is carried to the mouth.

Table 1 The six most important oyster species produced worldwide

Oyster species	Number of producing countries	Major producing countries (% of total)	Production (metric tonnes)
Pacific oyster	30	China (85), Korea (5), Japan (5), France (2.5)	4 429 337
American oyster	4	USA (95), Canada (4.5)	110 770
Slipper oyster	1	Philippines (100)	15 915
Sydney oyster	1	Australia (100)	5600
European oyster	17	Spain (50), France (29), Ireland (8)	5071
Mangrove oyster	3	Cuba (99)	1184

Tonnages are 2004 figures (FAO data). All are cupped oyster species, apart from the European oyster, which is a flat oyster species.

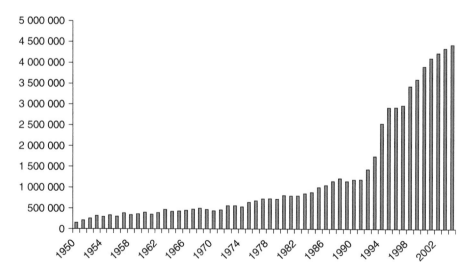

Figure 2 The increase in world production, in metric tonnes, of the Pacific oysters (1950–2004).

Oysters have separate sexes, but they may change sex one or more times during their life span, being true hermaphrodites. In most species, the eggs and sperm are shed directly into the water where fertilization occurs. Larvae are thus formed and these swim and drift in the water, feeding on natural phytoplankton. After 2–3 weeks, depending on local environmental conditions, the larvae are mature and they develop a foot. At this stage they sink to the seabed and explore the sediment surface until they find a suitable surface on which to settle and attach permanently, by cementation. Next, they go through a series of morphological and physiological changes, a process known as metamorphosis, to become immature adults. These are called juveniles, spat, or seed.

Cultivated species of flat oysters brood the young larvae within the mantle cavity, releasing them when they are almost ready to settle. Fecundity is usually related to age, with older and larger females producing many more larvae.

Growth of juveniles is usually quite rapid initially, before slowing down in later years. The length of time that oysters take to reach a marketable size varies considerably, depending on local environmental conditions, particularly temperature and food availability. Pacific oysters may reach a market size of 70–100 g live weight (shell-on) in 18–30 months.

Methods of Cultivation – Seed Supply

Oyster farming is dependent on a regular supply of small juvenile animals for growing on to market size. These can be obtained primarily in one of two ways, either from naturally occurring larvae in the plankton or artificially, in hatcheries.

Wild Larvae Collection

Most oyster farmers obtain their seed by collecting wild set larvae. Special collection materials, generically known as 'cultch', are placed out when large numbers of larvae appear in the plankton. Monitoring of larval activity is helpful to determine where and when to put out the cultch. It can be difficult to discriminate between different types of bivalve larvae in the plankton to ensure that it is the required species that is present but recently, modern highly sensitive molecular methods have been developed for this.

Various materials can be used for collection. Spat collected in this way are often then thinned prior to growing on (**Figure 3**). In China, coir rope is widely used, as well as straw rope, flax rope, and ropes woven by thin bamboo strips. Shells, broken tiles, bamboo, hardwood sticks, plastics, and even old tires can also be used. Coatings are sometimes applied.

Hatcheries

Restricted by natural conditions, the amount of wild spat collected may vary from year to year. Also, where nonnative oyster species are cultivated, there may be no wild larvae available. In these circumstances, hatchery cultivation is necessary. This also allows for genetic manipulation of stocks, to rear and maintain lines specifically adapted for certain

Figure 3 A demonstration of the range of oyster spat collectors used in France.

traits. Important traits for genetic improvement include growth rate, environmental tolerances, disease resistance, and shell shape.

Methods of cryopreservation of larvae are being developed and these will contribute to maintaining genetic lines. Furthermore, triploid oysters, with an extra set of chromosomes, can only be produced in hatcheries. These oysters often have the advantage of better growth and condition, and therefore marketability, during the summer months, as gonad development is inhibited.

Techniques for hatchery rearing were first developed in the 1950s and today follow well-established procedures.

Adult broodstock oysters are obtained from the wild or from held stocks. Depending on the time of year, these may need to be bought into fertile condition by providing a combination of elevated temperature and food (cultivated phytoplankton diets) over several weeks (see **Figure 4**). Selection of the appropriate broodstock conditioning diet is very important. An advantage of hatchery production is that it allows for early season production in colder climates and this ensures that seeds have a maximum growing period prior to their first overwintering.

For cultivated oyster species, mature gametes are usually physically removed from the gonads. This involves sacrificing some ripe adults. Either the gonad can be cut repeatedly with a scalpel and the gametes washed out with filtered seawater into a part-filled container, or a clean Pasteur pipette can be inserted into the gonad and the gametes removed by exerting gentle suction.

Broodstock can also be induced to spawn. Various stimuli can be applied. The most successful methods are those that are natural and minimize stress. These include temporary exposure to air and thermal cycling (alternative elevated and lowered water temperatures). Serotonin and other chemical triggers can also be used to initiate spawning but these methods are not generally recommended as eggs liberated using such methods are often less viable.

Flat oysters, of the genera *Ostrea* and *Tiostrea*, do not need to be stimulated to spawn. They will spawn of their own accord during the conditioning process as they brood larvae within their mantle cavities for varying periods of time depending on species and temperature.

The fertilized eggs are then allowed to develop to the fully shelled D-larva veliger stage, so called because of the characteristic 'D' shape of the shell valves (**Figure 5**).

These larvae are then maintained in bins with gentle aeration. Static water is generally used and this is exchanged daily or once every 2 days. Through-flow systems, with meshes to prevent loss of larvae, are also employed. Cultured microalgae are added into the tanks several times per day, at an appropriate

Figure 4   The SeaCAPS continuous culture system for algae. An essential element for a successful oyster hatchery is the means to cultivate large quantities of marine micro algae (phytoplankton) food species.

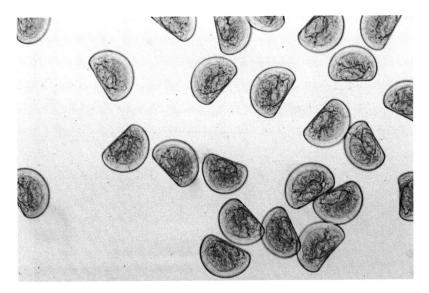

Figure 5 Pacific oyster D larvae. At this stage shell length is about 70 microns.

daily ration according to the number and size of the larvae.

The larvae eventually become competent to settle and for this, surfaces must be provided. The area of settlement surface is important. Types of materials in common usage to provide large surface areas for settlement include sheets of slightly roughened polyvinyl chloride (PVC), layers of shell chips and particles prepared by grinding aged, clean oyster shell spread over the base of settlement trays or tanks, bundles, bags, or strings of aged clean oyster shells dispersed throughout the water column, usually in settlement tanks or various plastic or ceramic materials coated with cement (lime/mortar mix).

For some of these methods, the oysters are subsequently removed from the settlement surface to produce 'cultchless' spat. These can then be grown as separate individuals through to marketable size and sold as whole or half shell, usually live.

Provision of competence to settle Pacific oyster larvae for remote setting at oyster farms is common practice on the Pacific coast of North America. Hatcheries provide the mature larvae and the farmers themselves set them and grow the spat for seeding oyster beds or in suspended culture.

In other parts of the world, hatcheries set the larvae, as described above, and grow the spat to a size that growers are able to handle and grow.

Oyster juveniles from hatchery-reared larvae perform well in standard pumped upwelling systems and survival is usually good, although some early losses may occur immediately following metamorphosis. Diet, ration, stocking density, and water flow rate are all important in these systems (**Figure 6**). They are only suitable for initial rearing of small seed. As the spat grow, food is increasingly likely to become limiting in these systems and they must be transferred to the sea for on-growing.

The size at which spat is supplied is largely dictated by the requirements and maturity of the grow-out industry. Seed native oysters are made available from commercial hatcheries at a range of sizes up to 25–30 mm. The larger the seed, the more expensive they are but this is offset by the higher survival rate of larger seed. Larger seed should also be more tolerant to handling.

Ponds

Pond culture offers a third method to provide seed. This was the method that was originally developed in Europe in early attempts to stimulate production following the decline of native oyster stocks in the late nineteenth century. Ponds of 1–10 ha in area and 1–3 m deep were built near to high water spring tides, filled with seawater, and then isolated for the period of time during which the oysters are breeding naturally, usually May to July. Collectors put into the ponds encourage and collect the settlement of juvenile oysters. There is an inherent limited amount of control over the process and success is very variable. In France, where spat collectors were also deployed in the natural environment, it was relatively successful and became an established method for a time. In the UK, spat production from ponds built in the early 1900s was insufficiently regular to provide a reliable supply of seed to the industry and the method was largely abandoned in the middle of the twentieth century in favor of the more controlled conditions available in hatcheries.

(a)

(b)

Figure 6   Indoor (a) and outdoor (b) oyster nursery system, in which seawater and algae food are pumped up through cylinders fitted with mesh bases and containing the seed.

Methods of Cultivation – On-growing

Various methods are available for on-growing oyster seed once this has been obtained, from either wild set larvae, hatcheries, or ponds.

Oysters smaller than 10 g need to be held in trays or bags attached to a superstructure on the foreshore until they are large enough to be put directly on to the substrate and be safe from predators, strong tidal and wave action, or siltation (**Figure 7**).

Tray cultivation of oysters can be successful in water of minimal flow, where water exchange is driven only by the rise and fall of the tide and gentle wave action. In such circumstances it may even be possible in open baskets.

Figure 7 The traditional bag and trestle method for on-growing Pacific oysters.

Figure 8 Open baskets for on-growing, as developed in Tasmania for Pacific oysters.

In more exposed sites, systems developed in Australia and employing plastic mesh baskets attached to or suspended from wires (see **Figure 8**) are becoming increasingly popular. It is claimed that an oyster with a better shape, free from worm (*Polydora*) infestation, will result from using these systems. *Polydora* can cause unsightly brown blemishes on the inner surface of the shell and decrease marketability of the stock. Less labor is required with these systems but they are more expensive to purchase initially.

In all of the above systems, the mesh size can be increased as the oysters grow, to improve the flow of water and food through the animals.

Holding seed oysters in trays or attached to the cultch material onto which they were settled, and suspended from rafts, pontoons, or long lines is an alternative method in locations where current speed will allow. The oysters should grow more quickly because they are permanently submerged but the shell may be thinner and therefore more susceptible to damage. Early stages of predator species such as crabs and starfish can settle inside containers, where they can cause significant damage unless containers are opened and checked on a regular basis. In the longer term the oysters will perform better on the seabed, although they can be reared to market size in the containers.

Protective fences can be put up around ground plots to give some degree of protection to smaller oysters from shore crabs. Potting crabs in the area of the lays is another method of control. The steps in the oyster cultivation process are shown as a diagram in **Figure 9**.

The correct stocking density is important so as not to exceed the carrying capacity of a body of water. When carrying capacity is exceeded, the algal population in the water is insufficient and growth declines. Mortalities also sometimes occur.

Yields in extensively cultivated areas are usually about 25 tonnes per hectare per year but this can increase to 70 tonnes where individual plots are well separated.

In France, premium quality oysters are sometimes finished by holding them in fattening ponds known as claires (see **Figure 10**), in which a certain type of algae is encouraged to bloom, giving a distinctive green color to the flesh.

In the UK, part-grown native flat oysters from a wild fishery are relayed in spring into areas in which conditions are favorable to give an increase in meat yield over a period of a few months, prior to marketing in the winter.

There are proposals to establish standards for organic certification of mollusk cultivation but many oyster growers consider that the process is intrinsically organic.

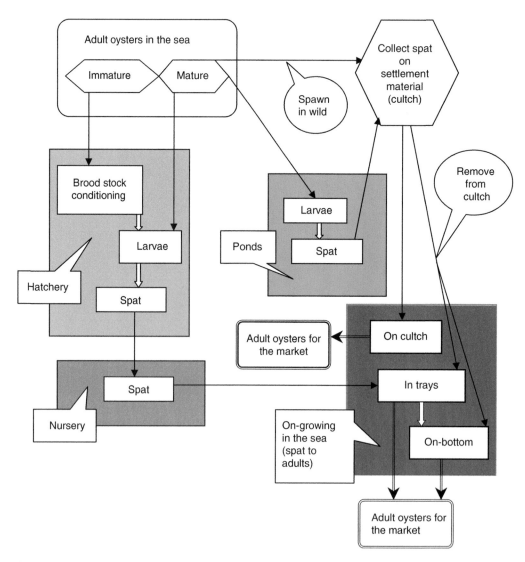

Figure 9 The steps and processes of oyster farming.

Figure 10 Oyster fattening ponds, or Claires, in France.

Pearl Oysters

There is an important component of the oyster farming industry, located mainly in Japan, Australia, and the South Sea Islands, devoted to the production of pearls. The process involves inserting into the oyster a nucleus, made with shell taken from a North American mussel, and a tiny piece of mantle cut from another oyster. The oysters are cultivated in carefully tended suspended systems while pearls develop around the nucleus. Once the pearls have been taken out of the oysters, they are seeded anew. A healthy oyster can be reseeded as many as 4 times with a new nucleus. As the oyster grows, it can accommodate progressively bigger pearl nuclei. Therefore, the biggest pearls are most likely to come from the oldest oysters. There has been a significant increase in demand for pearl jewelry in the last 10 years. The US is the biggest market, with estimated sales of US$1.5 billion of pearl jewelry per year (**Table 2**).

Site Selection for On-growing

A range of physical, biological, and chemical factors will influence survival and growth of oysters. Many of these factors are subject to seasonal and annual variation and prospective oyster farmers are usually advised to monitor the conditions and to see how well oysters grow and survive at their chosen site for at least a year before any commercial culture begins.

Seawater temperature has a major effect on seasonal growth and may be largely responsible for any differences in growth between sites. Changes in

Table 2 Estimated values of pearl production (2004)

	Market share in value (%)	Production value (2004)
White South Sea cultured pearls (Australia, Indonesia, the Philippines, Myanmar)	35	US$220 million
Freshwater cultured pearls (China)	24	US$150 million
Akoya cultured pearls (Japan, China)	22	US$135 million
Tahitian cultured pearls (French Polynesia)	19	US$120 million
Total		US$625 million

Source: Pearl World, The International Pearling Journal (January/February/March 2006).

salinity do not affect the growth of bivalves by as much as variation in temperature. However, most bivalves will usually only feed at higher salinities. Pacific oysters prefer salinity levels nearer to 25 psu, conditions typical of many estuaries and inshore waters.

Growth rates are strongly influenced by the length of time during which the animals are covered by the tide. Growth of oysters in trays stops when they are exposed to air for more than about 35% of the time. However, this fact can be used to advantage by the cultivator who may wish, for commercial reasons, to slow down or temporarily stop the growth of the stock. This can be achieved by moving the stock higher up the beach. It is a practice that is routinely adopted in Korea and Japan for 'hardening off' wild-caught spat prior to sale.

Other considerations are related to access and harvesting. It is important to consider the type of equipment likely to be used for planting, maintenance, and harvesting, particularly at intertidal sites. Some beaches will support wheeled or tracked vehicles, while others are too soft and will require the use of a boat to transport equipment.

Risks

Oyster farming can be at risk from pollutants, predators, competitors, diseases, fouling organisms, and toxic algae species.

Pollutants

Some pollutants can be harmful to oysters at very low concentrations. During the 1980s, it was found that tributyl tin (TBT), a component of marine antifouling paints, was highly toxic to bivalve mollusks at extremely low concentrations in the seawater. Pacific oysters cultivated in areas in which large numbers of small vessels were moored showed stunted growth and thickening of the shell, and natural populations of flat oysters failed to breed (see **Figure 11**). The use of this compound on small vessels was widely banned in the late 1980s, and since then the oyster industry has recovered. The International Maritime Organization has since announced a ban for larger vessels as well. When marketed for consumption, oysters must meet a number of 'end product' standards. These include a requirement that the shellfish should not contain toxic or objectionable compounds such as trace metals, organochlorine compounds, hydrocarbons, and polycyclic aromatic hydrocarbons (PAHs) in such quantities that the calculated dietary intake exceeds the permissible daily amount.

Predators

In some areas oyster drills or tingles, which are marine snails that eat bivalves by rasping a hole through the shell to gain access to the flesh, are a major problem.

Competitors

Competitors include organisms such as slipper limpets (*Crepidula fornicata*), which compete for food and space. In silt laden waters, they also produce a muddy substrate, which is unsuitable for cultivation. They can be a significant problem. An example is around the coast of Brittany, where slipper limpets have proliferated in the native flat oyster areas during the last 30 years. In order to try to control this problem, approximately 40 000–50 000 tonnes of slipper limpets are harvested per year. These are taken to a factory where they are converted to a calcareous fertilizer.

Diseases

The World Organisation for Animal Health (OIE) lists seven notifiable diseases of mollusks and six of

Figure 11   Compared with a normal shell (upper shell section), TBT from marine antifouling paints causes considerable thickening of Pacific oyster shells (lower section). These compounds are now banned.

these can infect at least one of the cultivated oyster species. Some of these diseases, often spread through national and international trade, have had a devastating effect on oyster stocks worldwide. For example, Bonamia, a disease caused by the protist parasitic organism *Bonamia ostreae*, has had a significant negative impact on *Ostrea edulis* production throughout its distribution range in Europe. Mortality rates in excess of 80% have been noted. The effect this can have on yields can be seen in the drastic (93%) drop in recorded production in France, from 20 000 tonnes per year in the early 1970s to 1400 tonnes in 1982. A major factor in the introduction and success of the Pacific oyster as a cultivated species has been that it is resistant to major diseases, although it is not completely immune to all problems.

There are reports of summer mortality episodes, especially in Europe and the USA, and infections by a herpes-like virus have been implicated in some of these. *Vibrio* bacteria are also associated with larval mortality in hatcheries.

The Fisheries and Oceans Canada website lists 53 diseases and pathogens of oysters. Juvenile oyster disease is a significant problem in cultivated *Crassostrea virginica* in the Northeastern United States. It is thought to be caused by a bacterium (*Roseovarius crassostreae*).

Fouling

Typical fouling organisms include various seaweeds, sea squirts, tubeworms, and barnacles. The type and degree of fouling varies with locality. The main effect is to reduce the flow of water and therefore the supply of food to bivalves cultivated in trays or on ropes and to increase the weight and drag on floating installations. Fouling organisms grow in response to the same environmental factors as are desirable for good growth and survival of the cultivated stock, so this is a problem that must be controlled rather than avoided.

Toxic Algae

Mortalities of some marine invertebrates, including bivalves, have been associated with blooms of some alga species, including *Gyrodinium aureolum* and *Karenia mikimoto*. These so-called red tides cause seawater discoloration and mortalities of marine organisms. They have no impact on human health.

Stock Enhancement

If we accept the FAO definition of aquaculture as "The farming of aquatic organisms in inland and coastal areas, involving intervention in the rearing process to enhance production and the individual or corporate ownership of the stock being cultivated," then in many cases stock enhancement can be included as a type of oyster farming.

Natural beds can be managed to encourage the settlement of juvenile oysters and sustain a fishery. Beds can be raked and tilled on a regular basis to remove silt and ensure that suitable substrates are available for the attachment of the juvenile stages. Adding settlement material (cultch) is also beneficial.

In some areas of the world, there has been a dramatic reduction in stocks of the native oyster species. This is attributed mainly to overexploitation, although disease is also implicated. Native oyster beds form a biotope, with many associated epifaunal and infaunal species. Loss of this habitat has resulted in a major decline in species richness in the coastal environment.

Considerable effort has been put into restoring these beds, using techniques that might fall under the definition of oyster farming. A good example is the Chesapeake Bay Program for the native American oyster *Crassostrea virginica*. Overfishing followed by the introduction of two protozoan diseases, believed to be inadvertently introduced to the Chesapeake through the importation of a nonnative oyster, *Crassostrea gigas*, in the 1930s, has combined to reduce oyster populations throughout Chesapeake Bay to about 1% of historical levels. The Chesapeake Bay Program is committed to the restoration and creation of aquatic reefs. A key component of the strategy to restore oysters is to designate sanctuaries, that is, areas where shellfish cannot be harvested. It is often necessary within a sanctuary to rehabilitate the bottom to make it a suitable oyster habitat. Within sanctuaries aquatic reefs are created primarily with oyster shell, the preferred substrate of spat. Alternative materials including concrete and porcelain are also used. These permanent sanctuaries will allow for the development and protection of large oysters and therefore more fecund and potentially disease-resistant oysters. Furthermore, attempts have been made at seeding with disease-free hatchery oysters.

Nonnative Species

The International Council for the Exploration of the Sea (ICES) Code of Practice sets forth recommended procedures and practices to diminish the risks of detrimental effects from the intentional introduction and transfer of marine organisms. The Pacific oyster, originally from Asia, has been introduced around the world and has become invasive in parts of Australia and New Zealand, displacing the native rock oyster in some areas. It is also increasingly becoming a problem in northern European coastal regions. In parts of France naturally recruited stock is competing with cultivated stock and has to be controlled.

The possibility of farming the nonnative Suminoe oyster (*Crassostrea ariakensis*) to restore oyster stocks in Chesapeake Bay is being examined and there are differing opinions about the environmental risks involved, with concerns of potential harmful effects on the local ecology.

It should also be noted that aquaculture has been responsible for the introduction of a whole range of passenger species, including pest species such as the slipper limpet and oyster drills, throughout the world.

Food Safety

Bivalve mollusks filter phytoplankton from the seawater during feeding; they also take in other small particles, such as organic detritus, bacteria, and viruses. Some of these bacteria and viruses, especially those originating from sewage outfalls, can cause serious illnesses in human consumers if they remain in the bivalve when it is eaten. The stock must be purified of any fecal bacterial content in cleansing

(depuration) tanks (**Figure 12**) before sale for consumption. There are regulations governing this, which are based on the level of contamination of the mollusks.

Viruses are not all removed by normal depuration processes and they can cause illness if the bivalves are eaten raw or only lightly cooked, as oysters often are. These viruses can only be detected by using sophisticated equipment and techniques, although research is being carried out to develop simpler methods.

Finally, certain types of naturally occurring algae produce toxins, which can accumulate in the flesh of oysters. People eating shellfish containing these toxins can become ill and in exceptional cases death can result. Cooking does not denature the toxins responsible nor does cleansing the shellfish in depuration tanks eliminate them. The risks to consumers are usually minimized by a requirement for samples to be tested regularly. If the amount of toxin exceeds a certain threshold, the marketing of shellfish for consumption is prohibited until the amount falls to a safe level.

See also

Molluskan Fisheries.

Further Reading

Andersen RA (ed.) (2005) *Algal Culturing Techniques.* New York: Academic Press.

Bueno P, Lovatelli A, and Shetty HPC (1991) Pearl oyster farming and pearl culture, Regional Seafarming Development and Demonstration Project. *FAO Project Report No. 8.* Rome: FAO.

Dore I (1991) *Shellfish – A Guide to Oysters, Mussels, Scallops, Clams and Similar Products for the Commercial User.* London: Springer.

Gosling E (2003) *Bivalve Mollusks; Biology, Ecology and Culture.* London: Blackwell.

Helm MM, Bourne N, and Lovatelli A (2004) Hatchery culture of bivalves; a practical manual. *FAO Fisheries Technical Paper 471.* Rome: FAO.

Huguenin JE and Colt J (eds.) (2002) *Design and Operating Guide for Aquaculture Seawater Systems,* 2nd edn. Amsterdam: Elsevier.

Lavens P and Sorgeloos P (1996) Manual on the production and use of live food for aquaculture. *FAO Fisheries Technical Paper 361.* Rome: FAO.

Mann R (1979) Exotic species in mariculture. Proceedings of Symposium on Exotic Species in Mariculture: Case Histories of the Japanese Oyster, *Crassostrea gigas* (Thunberg), with implications for other fisheries. Woods Hole Oceanographic Institution, Cambridge Press.

Matthiessen G (2001) *Fishing News Books Series: Oyster Culture.* New York: Blackwell.

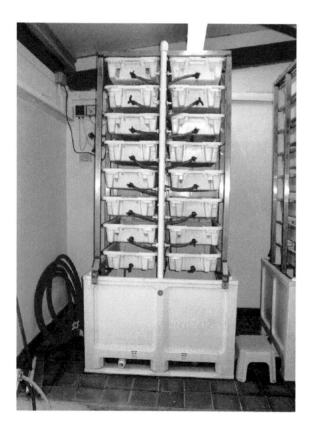

Figure 12 A stacked depuration system, suitable for cleansing oysters of microbiological contaminants. The oysters are held in trays in a cascade of UV-sterlized water.

Spencer BE (2002) *Molluscan Shellfish Farming*. New York: Blackwell.

Relevant Websites

http://www.crlcefas.org
 – Cefas is the EU community reference laboratory for bacteriological and viral contamination of oysters.
http://www.oie.int
 – Definitive information on oyster diseases can be found on the website of The World Organisation for Animal Health (OIE).
http://www.pac.dfo-mpo.gc.ca
 – Further information on diseases can be found on the Fisheries and Oceans Canada website.
http://www.cefas.co.uk
 – The online magazine 'Shellfish News', although produced primarily for the UK industry, has articles of general interest on oyster farming. It can be found on the News/Newsletters area of the Cefas web site. A booklet on site selection for bivalve cultivation is also available at this site.

http://www.vims.edu
 – The Virginia Institute of Marine Science has information on oyster genetics and breeding programs and the proposals to introduce the non-native *Crassostrea ariakensis* to Chesapeake Bay.
http://www.was.org
 – The World Aquaculture Society website has a comprehensive set of links to other relevant web sites, including many national shellfish associations.
http://www.fao.org
 – There is a great deal of information on oyster farming throughout the world on the website of The Food and Agriculture Organization of the United Nations (FAO). This website has statistics on oyster production, datasheets for various cultivated species, including Pacific oysters, and online manuals on the production and use of live food for aquaculture (*FAO Fisheries Technical Paper 361*), Hatchery culture of bivalves (*FAO Fisheries Technical Paper. No. 471*) and Pearl oyster farming and pearl culture (*FAO Project Report No. 8*).

SALMONIDS

D. Mills, Atlantic Salmon Trust, UK

Introduction

The Atlantic and Pacific salmon and related members of the Salmonidae are anadromous fish, breeding in fresh water and migrating to sea as juveniles at various ages where they feed voraciously and grow fast. Survival at sea is dependent on exploitation, sea surface temperature, ocean climate and predation. Their return migration to breed reveals a remarkable homing instinct based on various guidance mechanisms. Some members of the family, however, are either not anadromous or have both anadromous and nonanadromous forms.

Taxonomy

The Atlantic salmon (*Salmo salar*) and the seven species of Pacific salmon (*Oncorhynchus*) are members of one of the most primitive superorders of the teleosts, namely the Protacanthopterygii. The family Salmonidae includes the Atlantic and Pacific salmon, the trout (*Salmo* spp.), the charr (*Salvelinus* spp.) and huchen (*Hucho* spp.). The anatomical features that separate the genera *Salmo* and *Oncorhynchus* from the genus *Salvelinus* are the positioning of the teeth. In the former two genera the teeth form a double or zigzag series over the whole of the vomer bone, which is flat and not boat-shaped, whereas in the latter the teeth are restricted to the front of a boat-shaped vomer. In the genus *Salmo* there is only a small gap between the vomerine and palatine teeth but this gap is wide in adult *Oncorhynchus* and not in *Salmo*. A specialization occurring in *Oncorhynchus* and not in *Salmo* is the simultaneous ripening of all the germ cells so that these fish can only spawn once (semelparity). There are a number of anatomical features that help in the identification of the various species of *Salmo* and *Oncorhynchus*; these include scale and fin ray counts, the number and shape of the gill rakers on the first arch and the length of the maxilla in relation to the eye.

Origin

There has been much debate as to whether the Salmonidae had their origins in the sea or fresh water. Some scientists considered that the Salmonidae had a marine origin with an ancestor similar to the Argentinidae (argentines) which are entirely marine and, like the salmonids and smelts (Osmeridae), bear an adipose fin. Other scientists considered the salmonids to have had a freshwater origin, supporting their case by suggesting that since the group has both freshwater-resident and migratory forms within certain species there has been recent divergence. Furthermore, there are no entirely marine forms among modern salmonids so they can not have had a marine origin. The Salmonidae have been revised as relatively primitive teleosts of probable marine pelagic origin whose specializations are associated with reproduction and early development in fresh water. The hypothesis of the evolution of salmonid life histories through penetration of fresh water by a pelagic marine fish, and progressive restrictions of life history to the freshwater habitat, involves adaptations permitting survival, growth and reproduction there. The salmonid genera show several ranges of evolutionary progression in this direction, with generally greater flexibility among *Salmo* and *Salvelinus* than among *Oncorhynchus* species (**Table 1**). Evidence for this evolutionary progression is perhaps even greater if one starts with the Argentinidae and Osmeridae which are basically marine coastal fishes. Some enter the rivers to breed, some live in fresh water permanently, and others such as the capelin (*Mallotus villosus*) spawn in the gravel of the seashore.

Life Histories

Members of the Salmonidae have a similar life history pattern but with varying degrees of complexity. A typical life history involves the female excavating a hollow in the river gravel into which the large yolky eggs are deposited and fertilized by the male. Because of egg size and the protection afforded them in the gravel the fecundity of the Salmonidae is low when compared with species such as the herring and the cod which are very fecund but whose eggs have no protection. On hatching the salmonid young (alevins) live on their yolk sac within the gravel for some weeks depending on water temperature. On emerging the fry may remain in the freshwater environment for a varying length of time (**Table 2**) changing as they grow into the later stages of parr and then smolt, at which stage they go to sea (**Figure 1**). Not all the Salmonidae have a prolonged freshwater life before entering the sea, and the juveniles of some

Table 1 Examples of flexibility of life history patterns in salmonid genera: anadromy implies emigration from fresh water to the marine environment as juveniles and return to fresh water as adults; nonanadromy implies a completion of the life cycle without leaving fresh water, although this may involve migration between a river and a lake habitat

Genus	Species		
	Anadromous form only	Both anadromous and nonanadromous forms	Nonanadromous form only
Oncorhynchus	gorbuscha (pink salmon) keta (chum salmon) tschawytscha (chinook salmon)	nerka (sockeye salmon) kisutch (coho salmon) masou (cherry salmon) rhodurus (amago salmon) mykiss (steelhead/rainbow trout) clarki (cut throat trout)	aguabonito (golden trout)
Salmo	none	salar (Atlantic salmon) trutta (sea/brown trout)	
Salvelinus	none	alpinus (arctic charr) fontinalis (brook trout) malma (Dolly Varden) leucomanis	namaycush (lake trout)
Hucho	none	perryi	hucho (Danube salmon)

Adapted from **Thorpe (1988)**.

species such as the pink salmon (*Oncorhynchus gorbuscha*) and chum salmon (*O. keta*) migrate to sea on emerging from the gravel. Others such as the sockeye salmon (*O. nerka*) have specialized freshwater requirements, namely the need for a lacustrine environment to which the young migrate on emergence (**Table 2**).

Distribution

Atlantic Salmon

The Atlantic salmon occurs throughout the northern Atlantic Ocean and is found in most countries whose rivers discharge into the North Atlantic Ocean and Baltic Sea from rivers as far south as Spain and Portugal to northern Norway and Russia and one river in Greenland. It has been introduced to some countries in the Southern Hemisphere, including New Zealand where they only survive as a land-locked form.

Sea Trout

The marine distribution of the anadromous form of *S. trutta* is confined to coastal and near-offshore waters and is not found in the open ocean. It has a more limited distribution than the salmon being confined mainly to the eastern seaboard of the North Atlantic, although it has been introduced to some North American rivers. Its natural range includes Iceland, and the Faroe Islands, Scandinavia, the Cheshkaya Gulf in the north, throughout the Baltic and down the coast of Europe to northern Portugal. It occurs as a subspecies in the Black Sea (*S. trutta labrax*) and Caspian Sea (*S. trutta caspius*). It has been introduced to countries in the Southern Hemisphere including Chile and the Falkland Islands.

Sockeye Salmon

The natural range of the sockeye, as other species of Pacific salmon, is the temperate and subarctic waters

Table 2 Life histories

Species	Length of freshwater life (years)	Particular features
Oncorhynchus nerka	1–3	Lake environment required for juveniles
O. gorbuscha	Migrate to estuarine waters on emergence	Tend to spawn closer to sea than other oncorhynchids, and may frequent smaller river systems
O. keta	Migrate to estuarine waters on emergence	Spawning takes place in lower reaches of rivers
O. tschawytscha	Some migrate to estuary on emergence, others remain in fresh water for one or more years	Two races: *anadromous*, long freshwater residence
		semelparous, short freshwater residence
O. kisutch	1–2	Tend to utilize small coastal streams
O. masou	1–2	Large parr become smolts and go to sea
		Small to medium-sized parr remain in fresh water
Salmo salar	1–7	A small percentage spawn more than once. 'Land-locked' forms live in lakes and spawn in afferent or efferent rivers
S. trutta	1–4 (anadromous form)	May spawn frequently, the anadromous form after repeat spawning migrations
S. alpinus	2–6	Anadromous form only occurs in rivers lying north of 60°N

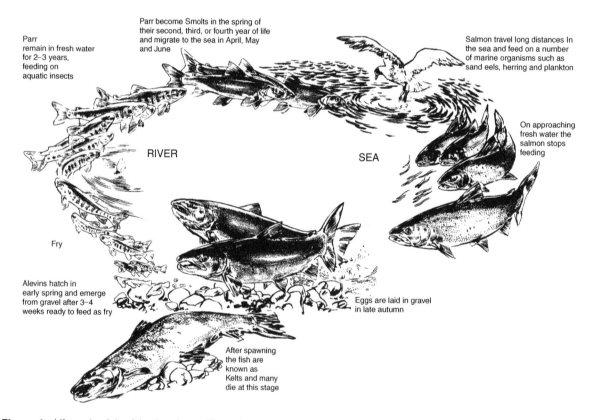

Parr become Smolts in the spring of their second, third, or fourth year of life and migrate to the sea in April, May and June

Parr remain in fresh water for 2–3 years, feeding on aquatic insects

Salmon travel long distances In the sea and feed on a number of marine organisms such as sand eels, herring and plankton

RIVER

SEA

On approaching fresh water the salmon stops feeding

Fry

Alevins hatch in early spring and emerge from gravel after 3–4 weeks ready to feed as fry

Eggs are laid in gravel in late autumn

After spawning the fish are known as Kelts and many die at this stage

Figure 1 Life cycle of the Atlantic salmon. (Reproduced from Mills, 1989.)

of the North Pacific Ocean and the northern adjoining Bering Sea and Sea of Okhotsk. However, because sockeye usually spawn in areas associated with lakes, where their juveniles spend their freshwater existence before going to sea, their spawning distribution is related to north temperate rivers with lakes in their systems. The Bristol Bay watershed in southwestern Alaska and the Fraser River system are therefore the major spawning areas for North American sockeye.

Pink Salmon

The natural freshwater range of pink salmon embraces the Pacific coast of Asia and North America north of 40°N and during the ocean feeding and maturation phase they are found throughout the North Pacific north of 40°N. The pink is the most abundant of the Pacific salmon species followed by the chum and sockeye in that order. Pink salmon have been transplanted outside their natural range to the State of Maine, Newfoundland, Hudson Bay, the North Kola Peninsula and southern Chile.

Chum Salmon

The chum occurs throughout the North Pacific Ocean in both Asian and North American waters north of 40°N to the Arctic Ocean and along the western and eastern arctic seaboard to the Lena River in Russia and the Mackenzie River in Canada.

Chinook Salmon

The chinook has a more southerly distribution than the other Pacific salmon species, extending as far south as the Sacramento–San Joaquin River system in California as well as into the northerly waters of the Arctic Ocean and Beaufort Sea. It has been transplanted to the east coast of North America, the Great Lakes, south Chile and New Zealand.

Coho Salmon

The coho is the least abundant of the Pacific salmon species but has a similar distribution as the other species. It has been introduced to the Great Lakes, some eastern states of North America and Korea and Chile.

Masu Salmon

This species only occurs in Far East Asia, in the Sea of Japan and Sea of Okhotsk. Some have been transported to Chile. The closely related amago salmon mainly remain in fresh water but some do migrate to coastal waters but few to the open sea.

Arctic Charr

The anadromous form of this species has a wide distribution throughout the subarctic and arctic regions of the North Atlantic north of 40°N. It enters rivers in the late summer and may remain in fresh water for some months.

Migrations

Once Atlantic salmon smolts enter the sea and become postsmolts they move relatively quickly into the ocean close to the water surface. Patterns of movement are strongly influenced by surface water currents, wind direction and tidal cycle. In some years postsmolts have been caught in the near-shore zone of the northern Gulf of St Lawrence throughout their first summer at sea, and in Iceland some postsmolts, mainly maturing males, forage along the shore following release from salmon-ranching stations. These results indicate that the migratory behavior of postsmolts can vary among populations. Pelagic trawl surveys conducted in the Iceland/Scotland/Shetland area during May and June and in the Norwegian Sea from 62°N to 73°N in July and August, have shown that postsmolts are widely distributed throughout the sampled area, although they do not reach the Norwegian Sea until July and August. Over much of the study area, catches of postsmolts are closely linked to the main surface currents, although north of about 64°N, where the current systems are less pronounced, postsmolts appear to be more diffusely distributed.

Young salmon from British Columbia rivers descend in discontinuous waves, and it has been suggested that this temporal pattern has evolved in response to short-term fluctuations in the availability of zooplankton prey. Although zooplankton production along the British Columbia coast is adequate to meet this seasonal demand, there is debate about the adequacy of the Alaska coastal current to support the vast populations of growing salmon in the summer. Density-dependent growth has been shown for the sockeye salmon populations at this time, suggesting that food can be limiting.

The movements of salmon in offshore waters are complex and affected by physical factors such as season, temperature and salinity and biological factors such as maturity, age, size and food availability and distribution of food organisms and stock-of-origin (i.e. genetic disposition to specific migratory patterns). Through sampling of stocks of the various Pacific salmon at various times of the year over many years scientists have been able to construct oceanic migration patterns of some of the major stocks of North American sockeye, chum and pink salmon (**Figure 2**) as well as for stocks of chinook and masu salmon. Similarly, as a result of ocean surveys and tagging experiments it has been possible to determine the migration routes of North American Atlantic salmon very fully (**Figures 3** and **4**). A picture of the approximate migration routes of Atlantic salmon in the North Atlantic area as a whole has been achieved from tag recaptures (**Figure 5**).

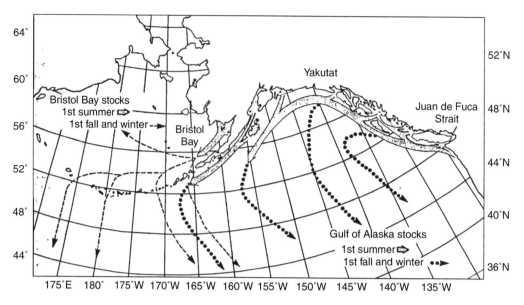

Figure 2 Diagram of oceanic migration patterns of some major stocks of North American sockeye, chum and pink salmon during their first summer at sea, plus probable migrations during their first fall and winter. (Reproduced from Burgner, 1991.)

Movements

Vertical and horizontal movements of Atlantic salmon have been investigated using depth-sensitive tags and data storage tags. Depth records show that salmon migrate mostly in the uppermost few meters and often show a diel rhythm in vertical movements. The salmon are closest to the surface at mid-day and go deeper at night. They have been recorded diving to a depth of 110 m.

Available information from research vessels and operation of commercial Pacific salmon fisheries suggest that Pacific salmon generally occur in near-surface waters.

Details of rates of travel are given in **Table 3**.

Food

The marine diet of both Pacific and Atlantic salmon comprises fish and zooplankton. The proportion of fish and zooplankton in their diet varies with season, availability and area. Among the fish species sockeye eat capelin (*Mallotus villosus*), sand eels (*Ammodytes hexapterus*), herring (*Clupea harengus pallasi*) and pollock (*Theragra chalcogramma*). Zooplankton organisms include euphausiids, squid, copepods and pteropods.

The diet of pink salmon includes fish eggs and larvae, squid, amphipods, euphausiids and copepods, whereas chum salmon were found to take pteropods, salps, euphausiids and amphipods. Chinook salmon eat herring, sand eels, pilchards and anchovies and in some areas zooplankton never exceeds 6% of the diet. However, the diet varies considerably from area

to area and up to 21 different taxonomic groups have been recorded in this species' diet. Similarly, the diet of coho salmon is a varied one, with capelin, sardines, lantern fish (myctophids), other coho salmon, being eaten along with euphausiids, squid, goose barnacles and jellyfish.

Masu salmon eat mainly small-sized fish and squid and large zooplankton such as amphipods and euphausiids. Fish species taken include capelin, herring, Dolly Varden charr, Japanese pearlside (*Maurolicus japonicus*), saury (*Cololabis saira*), sand eels, anchovies, greenlings (*Hexagrammos otakii*) and sculpins (*Hemilepidotus* spp.).

There is no evidence of selective feeding among sockeye, pink and chum salmon.

The food of Atlantic salmon postsmolts is mainly invertebrate consisting of chironomids and gammarids in the early summer in inshore waters and in the late summer and autumn the diet changes to one of small fish such as sand eels and herring larvae. A major dietary study of 4000 maturing Atlantic salmon was undertaken off the Faroes and it confirmed the view that salmon forage opportunistically, but that they demonstrate a preference for fish rather than crustaceans when both are available. They are also selective when feeding on crustaceans, preferring hyperiid amphipods to euphausiids. Feeding intensity and feeding rate of Atlantic salmon north of the Faroes have been shown to be lower in the autumn than in the spring, which might suggest that limited food is available at this time of year. Similar results have been found for Atlantic salmon in the Labrador Sea and in the Baltic. Salmon in the north Atlantic also rely on amphipods in the diet in the

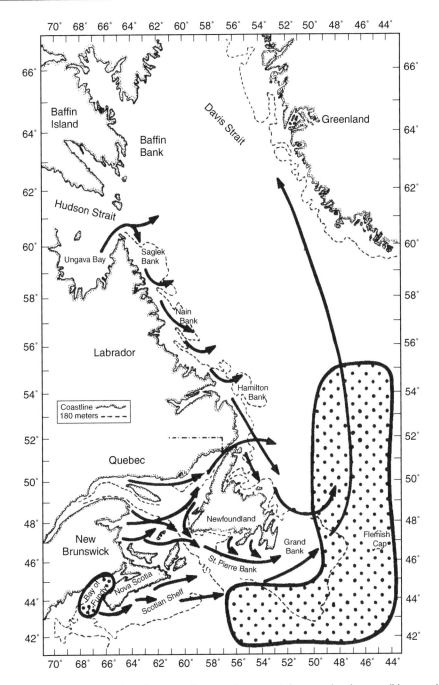

Figure 3 The migration routes for Atlantic salmon smolts away from coastal areas showing possible overwintering areas and movement of multi-sea winter salmon into the West Greenland area. Arrows indicate the path of movement of the salmon, and dotted area indicates the overwintering area. (Reproduced from Reddin, 1988.)

autumn, whereas at other times fish are the major item. Fish taken include capelin, herring, sprat (*Clupea sprattus*), lantern fishes, barracudinas and pearlside (*Maurolicus muelleri*).

Predation

Marine mammals recorded predating on Pacific salmon include harbour seal (*Phoca vitulina*), fur seal (*Callorhinus uresinus*), Californian sea lion (*Zalophus grypus*), humpbacked whale (*Megaptera novaeangliae*) and Pacific white-sided dolphin (*Lagenorhynchus obliquideus*). Pinniped scar wounds on sockeye salmon, caused by the Californian sea lion and harbor seal, increased from 2.8% in 1991 to 25.9% in 1996 and on spring-run chinook salmon they increased from 10.5% in 1991 to 31.8% in 1994.

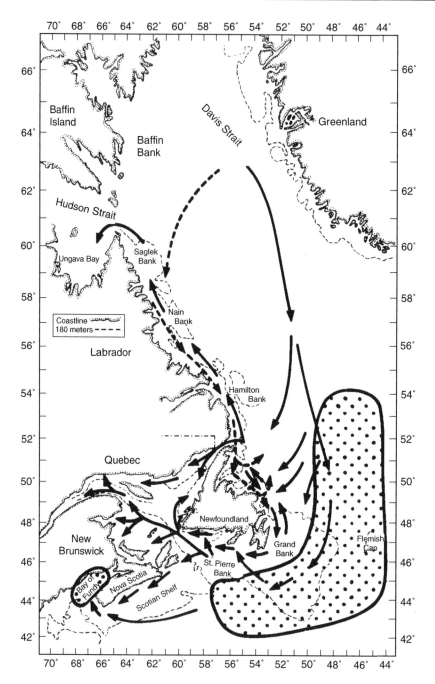

Figure 4 The migration routes of salmon from West Greenland and overwintering areas on return routes to rivers in North America. Solid arrows indicate migration in mid-summer and earlier; broken arrows indicate movement in late summer and fall; dotted areas are the wintering areas. (Reproduced from Reddin, 1988.)

Predators of Atlantic salmon postsmolts include gadoids, bass (*Dicentrarchus labrax*), gannets (*Sula bassana*), cormorants (*Phalacrocorax carbo*) and Caspian terns (*Hydroprogne tschegrava*). Adult fish are taken by a number of predators including the grey seal (*Halichoerus grypus*), the common seal (*Phoca vitulina*), the bottle-nosed dolphin (*Tursiops truncatus*), porbeagle shark (*Lamna cornubica*), Greenland shark (*Somniosus microcephalus*) and ling (*Molva molva*).

Environmental Factors

Surface Salinity

Salinity may have an effect on fish stocks and it has been shown that the great salinity anomaly of the 1970s in the North Atlantic adversely affected the spawning success of eleven of fifteen stocks of fish whose breeding grounds were traversed by the anomaly. In the North Pacific there was found to be

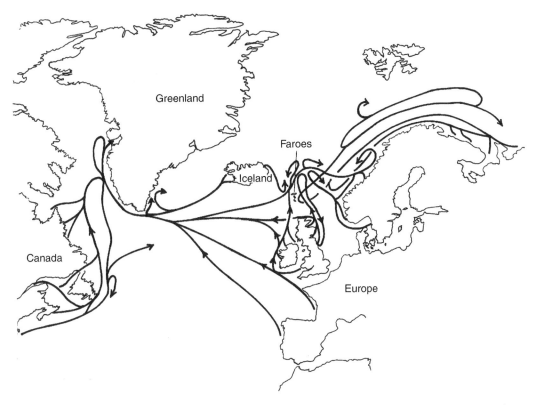

Figure 5 Approximate migration routes of Atlantic salmon in the North Atlantic area. (Reproduced from Mills, 1989.)

little relation between the high seas distribution of sockeye salmon and surface salinity. Sockeye are distributed across a wide variety of salinities, with low salinities characterizing 'salmon waters' of the Subarctic Pacific Region.

Sea Surface Temperature

Sockeye salmon are found over a wide variety of conditions. Sea surface temperature (SST) has not been found to be a strong and consistent determinant of sockeye distribution, but it definitely influences distribution and timing of migrations. Sockeye tend to prefer cooler water than the other Pacific salmon species.

Temperature ranges of waters yielding catches of various salmon species in the northwest Pacific in winter are: sockeye, 1.5–6.0°C; pink, 3.5–8.5°C; chum, 1.5–10.0°C; coho, 5.5–9.0°C.

A significant relationship was found for SST and Atlantic salmon catch rates in the Labrador Sea, Irminger sea and Grand Banks (**Figure 6**), with the greatest abundance of salmon being found in SSTs between 4 and 10°C. This significant relationship suggests that salmon may modify their movements at sea depending on SST. It has also been suggested that the number of returning salmon is linked to environmental change and that the abundance of

salmon off Newfoundland and Labrador was linked to the amount of water of <0°C on the shelf in summer. In years when the amount of cold water was large and the marine climate tended to be cold, there were fewer salmon returning to coastal waters. A statistically significant relationship has been found between the area of ice off Labrador and northern Newfoundland and the number of returning salmon in one of the major rivers on the Atlantic coast of Nova Scotia. The timing and geographical distribution of Atlantic salmon along the Newfoundland and Labrador coasts have been shown to be dependent on the arrival of the 4°C water, salmon arriving earlier during warmer years.

In two Atlantic salmon stocks that inhabit rivers confluent with the North Sea a positive correlation was found between the area of 8–10°C water in May and the survival of salmon. An analysis of SST distribution for periods of good versus poor salmon survival showed that when cool surface waters dominated the Norwegian coast and the North Sea during May salmon survival has been poor. Conversely, when the 8°C isotherm has extended northward along the Norwegian coast during May, survival has been good.

Temperature may also be linked to sea age at maturity. An increase in temperature in the northeast Atlantic subarctic was found to be associated

Table 3 Rates of travel in the sea of Atlantic and Pacific salmon

Species	Rate of travel (km day^{-1})	Conditions
Atlantic salmon	19.5–24	Icelandic postsmolts from ranching stations
	15.2–20.7	Postsmolts based on smolt tag recaptures
	22–52	Grilse and large salmon
	32	Average for maturing salmon of all sea ages
	26	For previous spawners migrating to Newfoundland and Greenland
	28.8–43.2	Icelandic coastal waters
	10–50	Baltic Sea
Sockeye salmon	46–56	During their final 30–60 days at sea
Pink salmon	17.2–19.8	Juveniles during first 2–3 months at sea
	45–54.3	Fish recovered at sea and in coastal waters
	43.3–60.2	For eastern Kamchatka stocks tagged in Aleutian Island passes or the Bering Sea
Coho salmon	30	
	55	Could be maintained over long distances

with large numbers of older (multi-sea winter) salmon and fewer grilse (one-sea winter salmon) returning to the Aberdeenshire Dee in Scotland.

Ocean Climate

Ocean climate appears to have a major influence on mortality and maturation mechanisms in salmon. Maturation, as evidenced by returns and survival of salmon of varying age, has been correlated with a number of environmental factors. The climate over the North Pacific is dominated by the Aleutian Low Pressure System. The long-term pattern of the Aleutian Low Pressure System corresponds with trends in salmon catch, with copepod production and with other climatic indices, indicating that climate and the marine environment may play an important role in salmon production. Survival of Pacific salmon species varies with fluctuations in large-scale circulation patterns such as El Niño Southern Oscillation (ENSO) events and more localized upwelling circulation that would be expected to affect local productivity and juvenile salmon growth. Runs of Pacific salmon in rivers along the western margin of North America stretching from Alaska to California vary on a decadal scale. When catches of a species are high in one region (e.g. Oregon) they may be low in another (e.g. Alaska). These changes are in part caused by marked interdecadal changes in the size and distribution of salmon stocks in the northeast Pacific, which are in turn associated with important ecosystem shifts forced by hydroclimatic changes linked to El Niño and possibly climate change.

Similarly, in the North Atlantic there have been annual and decadal changes in the North Atlantic Oscillation (NAO) index. Associated with these changes is a wide range of physical and biological responses, including effects on wind speed, ocean circulation, sea surface temperature, prevalence and intensity of Atlantic storms and changes in zooplankton production. For example, during years of positive NAO index, the eastern and western North Atlantic display increases in temperature while temperatures in central North Atlantic and in the Labrador Basin decline. The extremely low temperatures in the Labrador Sea during the 1980s coincided with a decline in salmon abundance. Similarly, in Europe, years of low NAO index were associated with high catches, whereas stocks have declined dramatically during high index years.

In the northeast Atlantic significant correlations were obtained between the variations in climate and hydrography with declines in primary production, standing crop of zooplankton and with reduced abundance and altered distribution of pelagic forage fishes and salmon catches.

By analogy with the North Pacific it is likely that changes in salmon abundance in the northeast Atlantic are linked to alterations in plankton productivity and/or structural changes in trophic transfer, each forced by hydrometeorological variability and possibly climate change as a result of the North Atlantic Oscillation and the Gulf Stream indexes.

Homing

It is suggested that homeward movement involves directed navigation. Fish can obtain directional information from the sun, polarized light, the earth's magnetic field and olfactory clues. Chinook and sockeye salmon have been shown capable of

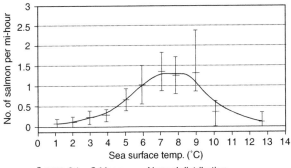

Figure 6 The relationship between sea surface temperatures and salmon catch rates from Labrador Sea, Irminger Sea and Grand Banks, 1965–91. (Reproduced from Reddin and Friedland, 1993.)

detecting changes in magnetic field. It has been suggested that migration from the feeding areas at sea to the natal stream must be accomplished without strictly retracing the outward migration using a variety of geopositioning mechanisms including magnetic and celestial navigation coordinated by an endogenous clock. Olfactory and salinity clues take over from geopositioning (bi-coordinate navigation) once in proximity of the natal stream because even small changes in declination during the time at sea correspond to a large search area for the home stream.

Further Reading

Burgner RL (1991) Life history of sockeye salmon (Oncorhynchus nerka). In: Groot C and Margolis L (eds.) *Pacific Salmon Life Histories*, pp. 2–117. Vancouver: UBC Press.

Foerster RE (1968) *The sockeye salmon, Oncorhynchus nerka. Bulletin of the Fisheries Research Board of Canada* 162:422.

Groot C and Margolis L (eds.) (1991) *Pacific Salmon Life Histories*. Vancouver: UBC Press.

Healey MC (1991) Life history of chinook salmon (Oncorhynchus tschawytscha). In: Groot C and Margolis L (eds.) *Pacific Salmon Life Histories*, pp. 312–393. Vancouver: UBC Press.

Heard WR (1991) Life history of pink salmon (Oncorhynchus gorbuscha). In: Groot C and Margolis L (eds.) *Pacific Salmon Life Histories*, pp. 120–230. Vancouver: UBC Press.

Kals F (1991) Life histories of masu and amago salmon (Oncorhynchus masou and Oncorhynchus rhodurus). In: Groot C and Margolis L (eds.) *Pacific Salmon Life Histories*, pp. 448–520. Vancouver: UBC Press.

McDowell RM (1988) *Diadromy in Fishes*. London: Croom Helm.

Mills DH (ed.) (1989) *Ecology and Management of Atlantic Salmon*. London: Chapman and Hall.

Mills DH (ed.) (1993) *Salmon in the Sea and New Enhancement Strategies*. Oxford: Fishing News Books.

Mills DH (ed.) (1999) *The Ocean Life of Atlantic Salmon*. Oxford: Fishing News Books.

Reddin D (1988) Ocean Life of Atlantic Salmon (Salmo salar L.) in the northwest Atlantic. In: Mills D and Piggins D (eds.) *Atlantic Salmon: Planning for the Future*, pp. 483–511. London and Sydney: Croom Helm.

Reddin D and Friedland K (1993) Marine environmental factors influencing the movement and survival of Atlantic salmon. In: Mills D (ed.) *Salmon in the Sea and New Enhancement Strategies*, pp. 79–103. Oxford: Fishing News Books.

Salo EO (1991) Life history of chum salmon (Oncorhynchus keta). In: Groot C and Margolis L (eds.) *Pacific Salmon Life Histories*, pp. 232–309. Vancouver: UBC Press.

Sandererock FK (1991) Life history of coho salmon (Oncorhynchus kisutch). In: Groot C and Margolis L (eds.) *Pacific Salmon Life Histories*, pp. 396–445. Vancouver: UBC Press.

Thorpe JE (1988) Salmon migration. *Science Progress (Oxford)* 72: 345–370.

SALMONID FARMING

L. M. Laird[†], Aberdeen University, Aberdeen, UK

Introduction

All salmonids spawn in fresh water. Some of them complete their lives in streams, rivers, or lakes but the majority of species are anadromous, migrating to sea as juveniles and returning to spawn as large adults after one or more years feeding. The farmed process follows the life cycle of the wild fish; juveniles are produced in freshwater hatcheries and smolt units and transferred to sea for ongrowing in floating sea cages. An alternative form of salmonid mariculture, ocean ranching, takes advantage of their accuracy of homing. Juveniles are released into rivers or estuaries, complete their growth in sea water and return to the release point where they are harvested.

The salmonids cultured in seawater cages belong to the genera *Salmo*, *Oncorhynchus*, and *Salvelinus*. The last of these, the charrs are currently farmed on a very small scale in Scandinavia; this article concentrates on the former two genera. The Atlantic salmon, *Salmo salar* is the subject of almost all production of fish of the genus *Salmo* (1997 worldwide production 640 000 tonnes) although a small but increasing quantity of sea trout (*Salmo trutta*) is produced (1997 production 7000 tonnes). Three species of *Oncorhynchus*, the Pacific salmon are farmed in significant quantities in cages, the chinook salmon (also known as the king, spring or quinnat salmon), *O. tshawytscha* (1997, 10 000 tonnes), the coho (silver) salmon, *O. kisutch* (1997, 90 000 tonnes) and the rainbow trout, *O. mykiss*. The rainbow trout (steelhead) was formerly given the scientific name *Salmo gairdneri* but following studies on its genetics and native distribution was reclassified as a Pacific salmon species. Much of the world rainbow trout production (1997, 430 000 tonnes) takes place entirely in fresh water although in some countries such as Chile part-grown fish are transferred to sea water in the same way as the salmon species.

Here, the history of salmonid culture leading to the commercial mariculture operations of today is reviewed. This is followed by an overview of the requirements for successful operation of marine salmon farms, constraints limiting developments and prospects for the future.

[†] Deceased

History

Salmonids were first spawned under captive conditions as long ago as the fourteenth century when Dom Pinchon, a French monk from the Abbey of Reome stripped ova from females, fertilized them with milt from males, and placed the fertilized eggs in wooden boxes buried in gravel in a stream. At that time, all other forms of fish culture were based on the fattening of juveniles captured from the wild. However, the large (4–7 mm diameter) salmonid eggs were much easier to handle than the tiny, fragile eggs of most freshwater or marine fish. By the nineteenth century the captive breeding of salmonids was well established; the main aim was to provide fish to enhance river stocks or to transport around the world to provide sport in countries where there were no native salmonids. In this way, brown trout populations have become established in every continent except Antarctica, sustaining game fishing in places such as New Zealand and Patagonia.

A logical development of the production of eggs and juveniles for release was to retain the young fish in captivity, growing them until a suitable size for harvest. The large eggs hatch to produce large juveniles that readily accept appropriate food offered by the farmer. In the early days, the fish were first fed on finely chopped liver and progressed to a diet based on marine fish waste. The freshwater rainbow trout farming industry flourished in Denmark at the start of the twentieth century only to be curtailed by the onset of World War I when German markets disappeared. The success of the Danish trout industry encouraged a similar venture in Norway. However, when winter temperatures in fresh water proved too low, fish were transferred to pens in coastal sea water. Although these pens broke up in bad weather, the practice of seawater salmonid culture had been successfully demonstrated. The next major steps in salmonid mariculture came in the 1950s and 1960s when the Norwegians developed the commercial rearing of rainbow trout and then Atlantic salmon in seawater enclosures and cages. Together with the development of dry, manufactured fishmeal-based diets this led to the industry in its present form.

Salmonid Culture Worldwide

Fish reared in seawater pens are subject to natural conditions of water quality and temperature. Optimum water temperatures for growth of most

Table 1 Production of four species of salmonids reared in seawater cages

	1988 production (t)	1997 production (t)
Atlantic salmon	112 377	638 951
Rainbow trout[a]	248 010	428 963
Coho salmon	25 780	88 431
Chinook salmon	4 698	9774

[a]Includes freshwater production

salmonid species are in the range 8–16°C. Such temperatures, together with unpolluted waters are found not only around North Atlantic and North Pacific coasts within their native range but also in the southern hemisphere along the coastlines of Chile, Tasmania, and New Zealand. Salmon and trout are thus farmed in seawater cages where conditions are suitable both within and outwith their native ranges.

Seawater cages are used for almost the entire sea water production of salmonids. A very small number of farms rear fish in large shore-based silo-type structures into which sea water is pumped. Such structures have the advantage of better protection against storms and predators and the possibility of control of environmental conditions and parasites such as sea lice. However, the high costs of pumping outweigh these advantages and such systems are generally now used only for broodfish that are high in value and benefit from controlled conditions.

The production figures for the four species of salmonid reared in seawater cages are shown in **Table 1**

Norway, the pioneering country of seawater salmonid mariculture, remains the biggest producer of Atlantic salmon. The output figures for 1997 show the major producing countries to be Norway (331 367 t), Scotland, UK (99 422 t), Chile (96 675 t), Canada (51 103 t), USA (18 005 t). Almost the entire farmed production of chinook salmon comes from Canada and New Zealand with Chile producing over 70 000 tonnes of coho salmon (1997 figures). Most of the Atlantic salmon produced in Europe is sold domestically or exported to other European countries such as France and Spain. Production in Chile is exported to North America and to Japan.

Seawater Salmonid Rearing

Smolts

Anadromous salmonids undergo physiological, anatomical and behavioral changes that preadapt them for the transition from fresh water to sea water. At this stage one of the most visible changes in the young fish is a change in appearance from mottled brownish to silver and herring-like. The culture of farmed salmonids in sea water was made possible by the availability of healthy smolts, produced as a result of the technological progress in freshwater units. Hatcheries and tank farms were originally operated to produce juveniles for release into the wild for enhancement of wild stocks, often where there had been losses of spawning grounds or blockage of migration routes by the construction of dams and reservoirs. One of the most significant aspects of the development of freshwater salmon rearing was the progress in the understanding of dietary requirements and the production of manufactured pelleted feed. The replacement of a diet based on wet trash fish with a dry diet also benefitted the freshwater rainbow trout farming industry, by improving growth and survival and reducing disease and the pollution of the watercourses receiving the outflow water from earth ponds.

It was found possible to transfer smolts directly to cages moored in full strength sea water. If the smolts are healthy and the transfer stress-free, survival after transfer is high and feeding begins within 1–3 days (Atlantic salmon).

The smolting process in salmonids is controlled by day length; natural seasonal changes regulate the physiological processes, resulting in the completion of smolting and seaward migration of wild fish in spring. For the first two decades of seawater Atlantic salmon farming, producers were constrained by the annual seasonal availability of smolts. These fish are referred to as 'S1s', being approximately one year post-hatching. This had consequences for the use of equipment (nonoptimal use of cages) and for the timing of harvest. Most salmon reached their optimum harvest size or began to show signs of sexual maturation at the same time of year; thus large quantities of fish arrived on the market together for biological rather than economic reasons. These fish competed with wild salmonids and missed optimum market periods such as Christmas and Easter.

Research on conditions controlling the smolting process enabled smolt producers to alter the timing of smolting by manipulating photoperiod. Compressing the natural year by shortening day length and giving the parr an early 'winter' results in S1/2s or S3/4s, smolting as early as six months after hatch. Similarly, by delaying winter, smolting can be postponed. Thus it is now possible to have Atlantic salmon smolts ready for transfer to sea water throughout the year. Although this benefits marketing it makes site fallowing (see below) more difficult than when smolt input is annual.

The choice of smolts for seawater rearing is becoming increasingly important with the establishment of controlled breeding programs. Few species of fish can be said to be truly domesticated. The only examples approaching domestication are carp species and, to a lesser degree, rainbow trout. Other salmon species have been captive bred for no more (and usually far less than) ten generations; the time between successive generations of Atlantic salmon is usually a minimum of three years which prevents rapid progress in selection for preferred characters although this is countered by the fact that many thousand eggs are produced by each female. Trials carried out mainly in Norway have demonstrated that several commercially important traits can be improved by selective breeding. These include growth rate, age at sexual maturity, food conversion efficiency, fecundity, egg size, disease resistance and survival, adaptation to conditions in captivity and harvest quality, including texture, fat, and color. All of these factors can also be strongly influenced by environmental factors and husbandry.

Sexual maturation before salmonids have reached the desired size for harvest has been a problem for salmonid farmers. Pacific salmon species (except rainbow trout) die after spawning; Atlantic salmon and rainbow trout show increased susceptibility to disease, reduced growth rate, deterioration in flesh quality and changes in appearance including coloration. Male salmonids generally mature at a smaller size and younger age than the females. One solution to this problem, routinely used in rainbow trout culture, is to rear all-female stocks, produced as a result of treating eggs and fry of potential broodstock with methyl testosterone to give functional males which are in fact genetically female. When crossed with normal females, all-female offspring are produced as the Y, male, sex chromosome has been eliminated. Sexual maturation can be eliminated totally by subjecting all-female eggs to pressure or heat shock to produce triploid fish. This is common practice for rainbow trout but used little for other salmonids, partly because improvements in stock selection and husbandry are overcoming the problem but also because of adverse press comment on supposedly genetically modified fish. This same reaction has limited the commercial exploitation of fast-growing genetically modified salmon, produced by the incorporation into eggs of a gene from ocean pout.

Site Selection

The criteria for the ideal site for salmonid cage mariculture have changed with the development of stronger cage systems, use of automatic feeders with a few days storage capacity and generally bigger and stronger boats, cranes and other equipment on the farm. The small, wooden-framed cages (typically $6 \text{ m} \times 6 \text{ m}$ frame, 300 m^3 capacity) with polystyrene flotation required sheltered sites with protection from wind and waves greater than $1-2 \text{ m}$ high. Recommended water depth was around three times the depth of the cage to ensure dispersal of wastes. This led to the siting of cages in inshore sites such as inner sea lochs and fiords. These sheltered sites had several disadvantages, notably variable water quality caused by runoff of fresh water, silt, and wastes from the land and susceptibility to the accumulation of feces and waste feed on the seabed because of poor water exchange. In addition, cage groups were often sited near public roads in places valued for their scenic beauty, attracting adverse public reaction to salmon farming.

Cages in use today are far larger (several thousand m^3 volume) and stronger. Frames are made from either galvanized steel or flexible plastic or rubber and can be designed to withstand waves of 5 m or more. Flotation collars are stronger and mooring systems designed to match cages to sites. Such sites are likely to provide more constant water quality than inshore sites; an ideal salmonid rearing site has temperatures of $6-16°C$ and salinities of $32-35‰$ (parts per thousand). Rearing is thus moving into deeper water away from sheltered lochs and bays. However, there are still advantages to proximity to the coast; these include ease of access from shore bases, proximity to staff accommodation, reduction in costs of transport of feed and stock and ability to keep sites under regular surveillance. Other factors to be taken into account in siting cage groups are the avoidance of navigation routes and the presence of other fish or shellfish farms. Maintaining a minimum separation distance from other fish farms is preferred to minimize the risk of disease transfer; if this is not possible, farms should enter into agreements to manage stock in the same way to reduce risk. Models have been developed in Norway and Scotland to determine the carrying capacity of cage farm sites.

Current speed is an important factor in site selection. Water exchange through the cage net ensures the supply of oxygen to the stock and removal of dissolved wastes such as ammonia as well as feces and waste feed. Salmon have been shown to grow and feed most efficiently in currents with speeds equivalent to $1-2$ body lengths per second. In an ideal site this current regime should be maintained for as much of the tidal cycle as possible. At faster current speeds the salmon will use more energy in swimming and cage nets will tend to twist, sometimes forming pockets and trapping fish, causing scale removal.

Some ideal sites may be situated near offshore islands; access from the mainland may require crossing open water with strong tides and currents making access difficult on stormy days. However, modern workboats and feeding barges with the capacity to store several days supply of feed make the operation of such sites possible.

The presence of predators in the vicinity is often taken as a criterion for site selection. Unprotected salmon cage farms are likely to be subject to predation from seals or, in Chile, sea lions, and birds, such as herons and cormorants. Such predators not only remove fish but also damage others, tear holes in nets leading to escapes and stress stock making it more susceptible to disease. Protection systems to guard against predators include large mesh nets surrounding cages or cage groups, overhead nets and acoustic scaring devices. When used correctly these can all be effective in preventing attacks. Attacks from predators are frequently reported to involve nonlocal animals, attracted to a food source. Because of this and the possibility of excluding and deterring predators, it seems that proximity to colonies is not necessarily one of the most important factors in determining site selection.

A further factor, which must be taken into account in the siting of cage salmonid farms, is the occurrence of phytoplankton blooms. Phytoplankton can enter surface-moored cages and can physically damage gills, cause oxygen depletion or produce lethal toxins that kill fish. Historic records may indicate prevalence of such blooms and therefore sites to be avoided although some cages are now designed to be lowered beneath the surface and operated as semisubmersibles, keeping the fish below the level of the bloom until it passes.

Farm Operation

Operation of the marine salmon farm begins with transfer of stock from freshwater farms. Where possible, transfer in disinfected bins suspended under helicopters is the method of choice as it is quick and relatively stress-free. For longer journeys, tanks on lorries or wellboats are used. The latter require particular vigilance as they may visit more than one farm and have the potential to transfer disease. Conditions in tanks and wellboats should be closely monitored to ensure that the supply of oxygen is adequate (minimum 6 mg l^{-1}).

The numbers of smolts stocked into each cage is a matter for the farmer; some will introduce a relatively small number, allowing for growth to achieve a final stocking density of 10–15 kg^{-3} whereas others stock a greater number and split populations between cages during growth. This latter method makes better use of cage space but increases handling and therefore stress. Differential growth may make grading into two or three size groups necessary.

Stocking density is the subject of debate. It is essential that oxygen concentrations are maintained and that all fish have access to feed when it is being distributed. Fish may not distribute themselves evenly within the water column; because of crowding together the effective stocking density may therefore be a great deal higher than the theoretical one.

As with all farmed animals the importance of vigilance of behavior and health and the maintenance of accurate, useful records cannot be overemphasized. When most salmon farms were small, producing one or two hundred tonnes of salmon a year rather than thousands, hand feeding was normal; observation of stock during feeding provided a good indication of health. Today, fish are often fed automatically using blowers attached to feed storage systems. The best of these systems incorporate detectors to monitor consumption of feed and underwater cameras to observe the stock.

All of the nutrients ingested by cage-reared salmonids are supplied in the feed distributed. Typically, manufactured diets for salmonids will contain 40% protein (mainly obtained from fishmeal) and up to 30% oil, providing the source of energy, sparing protein for growth. Although very poorly digested by salmonids, carbohydrate is necessary to bind other components of the diet. Vitamins and minerals are also added, as are carotenoid pigments such as astaxanthin, necessary to produce the characteristic pink coloration of the flesh of anadromous salmonids. The feed used on marine salmon farms is nowadays almost exclusively a pelleted or extruded fishmeal-based diet manufactured by specialist companies. Feed costs make up the biggest component of farm operating costs, sometimes reaching 50%. It is therefore important to make optimum use of this valuable input by minimizing wastes. This is accomplished by ensuring that feed is delivered to the farm in good condition and handled with care to prevent dust formation, increasing the size of pellets as the fish grow and distributing feed to satisfy the appetites of the fish. Improvements in feed manufacture and in feeding practices have reduced feed conversion efficiency (feed input : increase in weight of fish) from 2 : 1 to close to 1 : 1. Such figures may seem improbable but it must be remembered that they represent the conversion of a nearly dry feed to wet fish flesh and other tissues.

The importance of maintaining a flow of water through the net mesh of the cages has been emphasized. Mesh size is generally selected to be the maximum capable of retaining all fish and preventing escapes. Any structure immersed in the upper few meters of coastal or marine waters will quickly be subjected to colonization by fouling organisms including bacteria, seaweeds, mollusks and sea squirts. Left unchecked, such fouling occludes the mesh, reducing water exchange and may place a burden on the cage reducing its resistance to storm damage. One of the most effective methods of preventing fouling of nets and moorings is to treat them with antifouling paints and chemicals prior to installation. However, one particularly effective treatment used in the early 1980s, tributyl tin, has been shown to have harmful effects on marine invertebrates and to accumulate in the flesh of the farmed fish; its use in aquaculture is now banned. Other antifoulants are copper or oil based; alternative, preferred methods of removing fouling organisms include lifting up sections of netting to dry in air on a regular basis or washing with high pressure hoses or suction devices to remove light fouling.

The aim of the salmonid farmer is to produce maximum output of salable product for minimum financial input. To do this, fish must grow efficiently and a high survival rate from smolt input to harvest must be achieved. Minimizing stress to the fish by reducing handling, maintaining stable environmental conditions and optimizing feeding practices will reduce mortalities. Causes of mortality in salmonid and other farms are reviewed elsewhere (see Mariculture Diseases and Health). It is vital to keep accurate records of mortalities; any increase may indicate the onset of an outbreak of disease. It is also important that dead fish are removed; collection devices installed in the base of cages are often used to facilitate this. Treatment of diseases or parasitic infestations such as sea lice (Lepeophtheirus salmonis, Caligus elongatus) is difficult in fish reared in sea cages because of their large volumes and the high numbers of fish involved. Some treatments for sea lice involve reducing the cage volume and surrounding with a tarpaulin so that the fish can be bathed in chemical. After the specified time the tarpaulin is removed and the chemical disperses into the water surrounding the cage. Newer treatments incorporate the chemicals in feed and are therefore simpler to apply. In the future, vaccines are increasingly likely to replace chemicals.

The health of cage-reared salmonids can be maintained by a site management system incorporating a period of fallowing when groups of cages are left empty for a period of at least three months and preferably longer. This breaks the life cycle of parasites such as sea lice and allows the seabed to recover from the nutrient load falling from the cages. Ideally a farmer will have access to at least three sites; at any given time one will be empty and the other two will contain different year classes, separated to prevent cross-infection.

Harvesting

Most of the farmed salmonids reared in sea water reach the preferred harvest size (3–5 kg) 10 months or more after transfer to sea water. Poor harvesting and handling methods can have a devastating effect on flesh quality, causing gaping in muscle blocks and blood spotting. After a period of starvation to ensure that guts are emptied of feed residues the fish are generally killed by one of two methods. One of these involves immersion in a tank of sea water saturated with carbon dioxide, the other an accurate sharp blow to the cranium. Both methods are followed by excision of the gill arches; the loss of blood is thought to improve flesh quality. It is important that water contaminated with blood is treated to kill any pathogens which might infect live fish.

Ocean Ranching

The anadromous behavior of salmonids and their ability to home to the point of release has been exploited in ocean ranching programs which have been operated successfully with Pacific salmon. Some of these programs are aimed at enhancing wild stocks and others are operated commercially. The low cost of rearing Pacific salmon juveniles, which are released into estuaries within weeks of hatching, makes possible the release of large numbers. In Japan over two billion juveniles are released annually; overall return rates have increased to 2%, 90% of which are chum (Oncorhynchus keta) and 8% pink (Oncorhynchus gorbuscha) salmon. The success of the operation depends on cooperation between those operating and financing the hatcheries and those harvesting the adult fish. The relatively high cost of producing Atlantic salmon smolts and the lack of control over harvest has restricted ranching operations.

See also

Mariculture Diseases and Health.

Further Reading

Anon (ed.) (1999) *Aquaculture Production Statistics 1988–1997*. Rome: Food and Agriculture Organization.

Black KD and Pickering AD (eds.) (1998) *Biology of Farmed Fish*. Sheffield Academic Press.

Heen K, Monahan RL, and Utter F (eds.) (1993) *Salmon Aquaculture*. Oxford: Fishing News Books.

Pennell W and Barton BA (eds.) (1996) *Principles of Salmonid Culture*. Amsterdam: Elsevier.

Stead S and Laird LM (In press) *Handbook of Salmon Farming Praxis*. Chichester: Springer-Praxis.

Willoughby S (1999) *Manual of Salmonid Farming*. Oxford: Blackwell Science.

DEMERSAL SPECIES FISHERIES

K. Brander, International Council for the Exploration of the Sea (ICES), Copenhagen, Denmark

Introduction

Demersal fisheries use a wide variety of fishing methods to catch fish and shellfish on or close to the sea bed. Demersal fisheries are defined by the type of fishing activity, the gear used and the varieties of fish and shellfish which are caught. Catches from demersal fisheries make up a large proportion of the marine harvest used for human consumption and are the most valuable component of fisheries on continental shelves throughout the world.

Demersal fisheries have been a major source of human nutrition and commerce for thousands of years. Models of papyrus pair trawlers were found in Egyptian graves dating back 3000 years. The intensity of fishing activity throughout the world, including demersal fisheries, has increased rapidly over the past century, with more fishing vessels, greater engine power, better fishing gear and improved navigational and fish finding aids. Many demersal fisheries are now overexploited and all are in need of careful assessment and management if they are to provide a sustainable harvest.

Demersal fisheries are often contrasted with pelagic fisheries, which use different methods to catch fish in midwater and close to the water surface. Demersal species are also contrasted with pelagic species (see relevant sections), but the distinction between them is not always clear. Demersal species frequently occur in mid-water and pelagic species occur close to the seabed, so that 'demersal' species are frequently caught in 'pelagic' fisheries and 'pelagic' species in demersal fisheries. For example Atlantic cod (*Gadus morhua*), a typical 'demersal' species, occurs close to the seabed, but also throughout the water column and in some areas is caught equally in 'demersal' and 'pelagic' fishing gear. Atlantic herring (*Clupea harengus*), a typical pelagic species, is frequently caught on the seabed, when it forms large spawning concentrations as it lays its eggs on gravel banks.

Total marine production rose steadily from less than 20 million tonnes (Mt) in 1950 to around 100 Mt during the late 1990s (**Figure 1**). The demersal fish catch rose from just over 5 Mt in 1950 to around 20 Mt by the early 1970s and has since fluctuated around that level (**Figure 2**). The proportion of demersal fish in this total has therefore declined over the period 1970–1998.

The products of demersal fisheries are mainly used for human consumption. The species caught tend to be relatively large and of high value compared with typical pelagic species, but there are exceptions to such generalizations. For example the industrial (fishmeal) fisheries of the North Sea, which take over half of the total fish catch, are principally based on

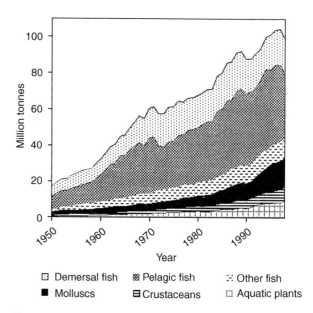

Figure 1 Total marine landings.

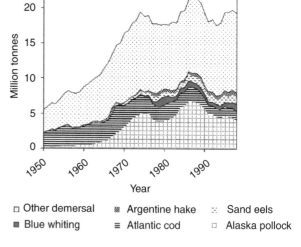

Figure 2 Total demersal fish landings.

low-value demersal species which live in or close to the seabed (sand eels, small gadoids).

Demersal fisheries are also often known as groundfish fisheries, but the terms are not exact equivalents because 'groundfish' excludes shellfish, which can properly be considered a part of the demersal catch. Shellfish such as shrimps and lobsters constitute the most valuable component of the demersal trawl catch in some areas.

Principal Species Caught in Demersal Fisheries

For statistical, population dynamics and fisheries management purposes the catch of each species (or group of species) is recorded separately by FAO (UN Food and Agriculture Organization). The FAO definitions of species, categories and areas are used here. FAO groups the major commercial species into a number of categories, of which, the flatfish (flounders, halibuts, soles) and the shrimps and prawns are entirely demersal. The gadiforms (cods, hakes, haddocks) include some species which are entirely demersal (haddock) and others which are not (blue whiting). The lobsters are demersal, but most species are caught in special fisheries using traps. An exception is the Norway lobster, which is caught in directed trawl fisheries or as a by-catch, as well as being caught in traps.

The five demersal marine fish with the highest average catches over the decade 1989–98 are Alaska (walleye) pollock, Atlantic cod, sand eels, blue whiting and Argentine hake. These species all spend a considerable proportion of their time in mid-water. Sand eels spend most of their lives on or in the seabed, but might not be regarded as a typical demersal species, being small, relatively short-lived and of low value. Sand eels and eight of the other 20 top 'species' in fact consist of more than one biological species, which are not identified separately in the FAO classification (Table 1).

The vast majority of demersal fisheries take place on the continental shelves, at depths of less than 200 m. Fisheries for deep-sea species, down to several thousand meters, have only been undertaken for the past few decades, as the technology to do so developed and it became profitable to exploit other species.

The species caught in demersal fisheries are often contrasted with pelagic species in textbooks and described as large, long-lived, high-value fish species, with relatively slow growth rates, low variability in recruitment and low mortality. There are so many exceptions to such generalizations that they are likely to be misleading. For example tuna and salmon are large, high-value pelagic species. Three of the main pelagic species in the North Atlantic (herring, mackerel and horse mackerel) are longer lived, have lower mortality rates and lower variability of recruitment than most demersal stocks in that area (Table 2).

Table 1 Total world catch of 20 top demersal fish species (averaged from 1989–1998)

Common name	Scientific name	Tonnes
Alaska pollock	*Theragra chalcogramma*	3 182 645
Atlantic cod	*Gadus morhua*	2 317 261
Sand eels	*Ammodytes* spp.	1 003 343
Blue whiting	*Micromesistius poutassou*	628 918
Argentine hake	*Merluccius hubbsi*	526 573
Croakers, drums nei	Sciaenidae	492 528
Pacific cod	*Gadus macrocephalus*	425 467
Saithe (Pollock)	*Pollachius virens*	385 227
Sharks, rays, skates, etc. nei	Elasmobranchii	337 819
Atlantic redfishes nei	*Sebastes* spp.	318 383
Norway pout	*Trisopterus esmarkii*	299 145
Flatfishes nei	Pleuronectiformes	289 551
Haddock	*Melanogrammus aeglefinus*	273 459
Cape hakes	*Merluccius capensis, M. paradox*	266 854
Blue grenadier	*Macruronus novaezelandiae*	255 421
Sea catfishes nei	Ariidae	242 815
Atka mackerel	*Pleurogrammus azonus*	237 843
Filefishes	*Cantherhines* (= *Navodon*) spp.	234 446
Patagonian grenadier	*Macruronus magellanicus*	230 221
South Pacific hake	*Merluccius gayi*	197 911
Threadfin breams nei	*Nemipterus* spp.	186 201

	Probability of being caught during next year	Coefficient of variation of recruitment
Demersal species		
Cod	33–64%	38–65%
Haddock	19–52%	70–151%
Hake	27%	33%
Plaice	32–48%	35–56%
Saithe	29–42%	45–56%
Sole	28–37%	15–94%
Whiting	47–56%	41–61%
Pelagic species		
Herring	12–40%	56–63%
Horse mackerel	16%	40%
Mackerel	21%	41%

Fishing Gears and Fishing Operations

A very wide range of fishing gear is used in demersal fisheries, the main ones being bottom trawls of different kinds, which are dragged along the seabed behind a trawler. Other methods include seine nets, trammel nets, gill nets, set nets, baited lines and longlines, temporary or permanent traps and barriers.

Some fisheries and fishermen concentrate exclusively on demersal fishing operations, but many alternate seasonally, or even within a single day's fishing activity, between different methods. Fishing vessels may be designed specifically for demersal or pelagic fishing or may be multipurpose.

Effects of Demersal Fisheries on the Species They Exploit

Most types of demersal fishing operation are non-selective in the sense that they catch a variety of different sizes and species, many of which are of no commercial value and are discarded. Stones, sponges, corals and other epibenthic organisms are frequently caught by bottom trawls and the action of the fishing gear also disturbs the seabed and the benthic community on and within it. Thus in addition to the intended catch, there is unintended disruption or destruction of marine life.

The fact that demersal fishing methods are nonselective has important consequences when trying to limit their impact on marine life. There are direct impacts, when organisms are killed or disturbed by fishing, and indirect impacts, when the prey or predators of an organism are removed or its habitat is changed.

The resilience or vulnerability of marine organisms to demersal fishing depends on their life history. In areas where intensive demersal fisheries have been operating for decades to centuries the more vulnerable species will have declined a long time ago, often before there were adequate records of their occurrence. For example, demersal fisheries caused a decline in the population of common skate (*Raia batis*) in the north-east Atlantic and barndoor skate (*Raia laevis*) in the north-west Atlantic, to the point where they are locally extinct in areas where they were previously common. These large species of elasmobranch have life histories which, in some respects, resemble marine mammals more than they do teleost fish. They do not mature until 11 years old, and lay only a small number of eggs each year. They are vulnerable to most kinds of demersal fishery, including trawls, seines, lines, and shrimp fisheries in shallow water.

The selective (evolutionary) pressure exerted by fisheries favors the survival of species which are resilient and abundant. It is difficult to protect species with vulnerable life histories from demersal fisheries and they may be an inevitable casualty of fishing. Some gear modifications, such as separator panels may help and it may be possible to create refuges for vulnerable species through the use of large-scale marine protected areas. Until recently fisheries management ignored such vulnerable species and concentrated on the assessment and management of a few major commercial species.

In areas with intensive demersal fisheries the probability that commercial-sized fish will be caught within one year is often greater than 50% and the fisheries therefore have a very great effect on the level and variability in abundance (**Table 2**). The effect of fishing explains much of the change in abundance of commercial species which has been observed during the few decades for which information is available and the effects of the environment, which are more difficult to estimate, are regarded as introducing 'noise', particularly in the survival of young fish. As the length of the observational time series increases and information about the effects of the environment on fish accumulates, it is becoming possible to turn more of the 'noise' into signal. It is no longer credible or sensible to ignore environmental effects when evaluating fluctuations in demersal fisheries, but a considerable scientific effort is still needed in order to include such information effectively.

Effects of the Environment on Demersal Fisheries

The term 'environment' is used to include all the physical, chemical and biological factors external to the fish, which influence it. Temperature is one of the main environmental factors affecting marine species. Because fish and shellfish are ectotherms, the temperature of the water surrounding them (ambient temperature) governs the rates of their molecular, physiological and behavioral processes. The relationship between temperature and many of these rates processes (growth, reproductive output, mortality) is domed, with an optimum temperature, which is species and size specific (**Figure 3**). The effects of variability in temperature are therefore most easily detected at the extremes and apply to

populations and fisheries as well as to processes within a single organism. Temperature change may cause particular species to become more or less abundant in the demersal fisheries of an area, without necessarily affecting the aggregate total yield.

The cod (*Gadus morhua*) at Greenland is at the cold limit of its thermal range and provides a good example of the effects of the environment on a demersal fishery; the changes in the fishery for it during the twentieth century are mainly a consequence of changes in temperature (**Figure 4**). Cod were present only around the southern tip of Greenland until 1917, when a prolonged period of warming resulted in the poleward expansion of the range by about 1000 km during the 1920s and 1930s. Many other boreal marine species also extended their range at the same time and subsequently retreated during the late 1960s, when colder conditions returned.

Changes in wind also affect demersal fish in many different ways. Increased wind speed causes mixing of the water column which alters plankton production. The probability of encounter between fish larvae and their prey is altered as turbulence increases. Changes in wind speed and direction affect the transport of water masses and hence of the planktonic stages of fish (eggs and larvae). For example, in some years a large proportion of the fish larvae on Georges Bank are transported into the Mid-Atlantic Bight instead of remaining on the Bank. In some areas, such as the Baltic, the salinity and oxygen levels are very dependent on inflow of oceanic water, which is largely wind driven. Salinity and oxygen in turn affect the survival of cod eggs and larvae, with major consequences for the biomass of cod in the area. These environmental effects on the early life stages of demersal fish affect their survival

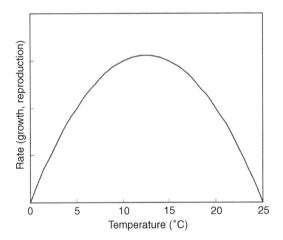

Figure 3 The relationship between temperature and many rate processes (growth, reproductive output, mortality) is domed. The optimum temperature is species and size specific.

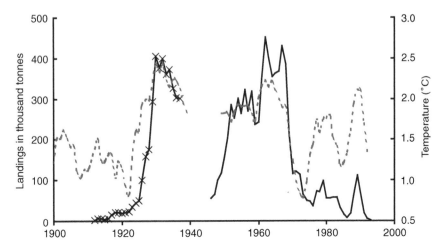

Figure 4 Cod catch and water temperature at West Greenland. Temperature is the running five-year mean of upper layer (0–40 m) values. ×, local catch × 20; ——, international catch; – – –, temperature.

and hence the numbers which recruit to the adult population.

World Catches from Demersal Fisheries and the Limits

Demersal fisheries occur mainly on the continental shelves (i.e., at depths less than 200 m). This is because shelf areas are much more productive than the open oceans, but also because it is easier to fish at shallower depths, nearer to the coast. The average catch of demersal fish per unit area on northern hemisphere temperate shelves is twice as high as on southern hemisphere temperate shelves and more than five times higher than on tropical shelves (**Figure 5**). The difference is probably due to nutrient supply. The effects of differences in productive capacity of the biological system on potential yield from demersal fisheries are dealt with elsewhere.

Demersal fisheries provide the bulk of fish and shellfish for direct human consumption. The steady increase in the world catch of demersal fish species ended in the early 1970s and has fluctuated around 20 Mt since then (**Figure 2**). Many of the fisheries are overexploited and yields from them are declining. In a few cases it would seem that the decline has been arrested and the goal of managing for a sustainable harvest may be closer.

A recent analysis classified the top 200 marine fish species, accounting for 77% of world marine fish production, into four groups – undeveloped, developing, mature and declining (senescent). The proportional change in these groups over the second half of the twentieth century (**Figure 6**) shows how

fishing has intensified, so that by 1994 35% of the fish stocks were in the declining phase, compared with 25% mature and 40% developing. Other analyses reach similar conclusions – that roughly two-thirds of marine fish stocks are fully exploited or overexploited and that effective management is needed to stabilize current catch levels.

Fish farming (aquaculture) is regarded as one of the principal means of increasing world fish production, but one should recall that a considerable proportion of the diet of farmed fish is supplied by demersal fisheries on species such as sand eel and Norway pout. Fishmeal is also used to feed terrestrial farmed animals.

Management of Demersal Fisheries

The purposes of managing demersal fisheries can be categorized as biological, economic and social. Biological goals used to be set in terms of maximum sustainable yield of a few main species, but a broader and more cautious approach is now being introduced, which includes consideration of the ecosystem within which these species are produced and which takes account of the uncertainty in our assessment of the consequences of our activities. The formulation of biological goals is evolving, but even the most basic, such as avoiding extinction of species, are not being achieved in many cases. At a global level it is evident that economic goals are not being achieved, because the capital and operating costs of marine fisheries are about 1.8 times higher than the gross revenue. There are innumerable examples of adverse social impacts of changes in fisheries, often caused by the effects of

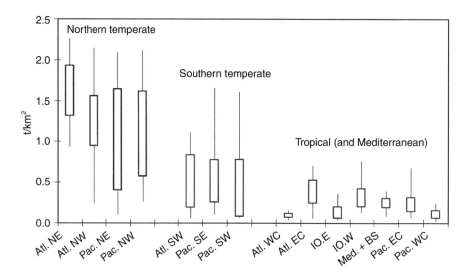

Figure 5 Demersal fish landings per unit area of continental shelf <200 m deep for the main temperate and tropical areas. The boxes show the spread between the upper and lower quartiles of annual landings and the whiskers show the highest and lowest annual landings 1950–1998. Atl, Atlantic; Pac, Pacific; IO, Indian Ocean; Med, Mediterranean; BS, Black Sea.

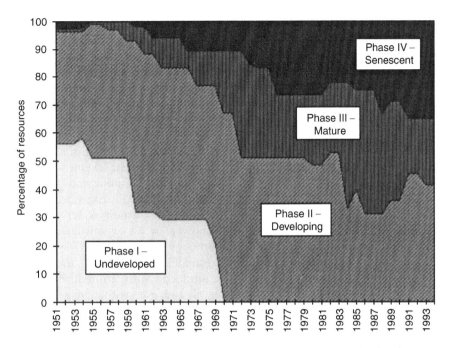

Figure 6 Temporal change in the level of exploitation of the top 200 marine fish species, showing the progression from mainly undeveloped or developing in 1951 to mainly fully exploited or overexploited in 1994.

larger, industrial fishing operations on the quality of life and standard of living of small-scale fishing communities. Clearly there is scope for improvement in fisheries management.

From this rather pessimistic analysis of where fisheries management has got us to date, it follows that a description of existing management regimes is a record of current practice rather than a record of successful practice.

Biological management of demersal fisheries has developed mainly from a single species 'yield-per-recruit' model of fish stocks. The output (yield) is controlled by adjusting the mortality and size of fish that are caught. Instruments for limiting fishing mortality are catch quotas (TAC, total allowable catch) and limits on fishing effort. Since it is much easier to define and measure catch than effort, the former is more widely used. In many shared, international fisheries, such as those governed by the European Union, the annual allocation of catch quotas is the main instrument of fisheries management. This requires costly annual assessment of many fish stocks, which must be added to the operating costs when looking at the economic balance for a fishery. Because annual assessments are costly, only the most important species, which tend to be less vulnerable, are assessed. For example the US National Marine Fisheries Service estimate that the status of 64% of the stocks in their area of responsibility is unknown.

The instruments for limiting the size of fish caught are mesh sizes, minimum landing sizes and various kinds of escape panels in the fishing gear. These instruments can be quite effective, particularly where catches are dominated by a single species. In multi-species fisheries, which catch species with different growth patterns, they are less effective because the optimal mesh size for one species is not optimal for all.

There are two classes of economic instruments for fisheries management: (1) property rights and (2) corrective taxes and subsidies. The former demands less detailed information than the latter and may also lead to greater stakeholder participation in the management process, because it fosters a sense of ownership.

The management of demersal fisheries will always be a complex problem, because the marine ecosystem is complicated and subject to change as the global environment changes. Management will always be based on incomplete information and understanding and imperfect management tools. A critical step towards better management would be to monitor performance in relation to the target objectives and provide feedback in order to improve the system.

One of the changes which has taken place over the past few years is the adoption of the precautionary approach. This seeks to evaluate the quality of the evidence, so that a cautious strategy is adopted when the evidence is weak. Whereas in the past such balance of evidence arguments were sometimes applied in order to avoid taking management action unless the evidence was strong (in order to avoid possible unnecessary disruption to the fishing industry), the presumption now is that in case of doubt it is the fish stocks rather than the short-term interests of the fishing industry which should be protected. This is a very significant change in attitude, which gives some grounds for optimism in the continuing struggle to achieve sustainable fisheries and healthy ecosystems.

Demersal Fisheries by Region

The demersal fisheries of the world vary greatly in their history, fishing methods, principal species and management regimes and it is not possible to review all of these here. Instead one heavily exploited area with a long history (New England) and one less heavily exploited area with a short history (the south-west Atlantic) will be described.

New England Demersal Fisheries

The groundfish resources of New England have been exploited for over 400 years and made an enormous contribution to the economic and cultural development of the USA since the time of the first European settlements. Until the early twentieth century, large fleets of schooners sailed from New England ports to fish, mainly for cod, from Cape Cod to the Grand Banks. The first steam-powered otter trawlers started to operate in 1906 and the introduction of better handling, preservation and distribution changed the market for fish and the species composition of the catch. Haddock became the principal target and their landings increased to over 100 000 t by the late

1920s. The advent of steam trawling raised concern about discarding and about the damage to the seabed and to benthic organisms.

From the early 1960s fishing fleets from European and Asian countries began to take an increasing share of the groundfish resources off New England. The total groundfish landings rose from 200 000 t to 760 000 t between 1960 and 1965. This resulted in a steep decline in groundfish abundance and in 1970 a quota management scheme was introduced under the International Commission for Northwest Atlantic Fisheries (ICNAF). Extended jurisdiction ended the activities of distant water fleets, but was quickly followed by an expansion of the US fleet, so that although the period 1974–78 saw an increase in groundfish abundance, the decline subsequently continued. Most groundfish species remain at low levels and, even with management measures intended to rebuild the stocks, are likely to take more than a decade to recover.

The changes in abundance of 'traditional' groundfish stocks (cod, haddock, redfish, winter flounder, yellowtail flounder) have to some extent been offset by increases in other species, including some elasmobranchs (sharks and rays). However, prolonged high levels of fishing have resulted in severe declines in the less resilient species of both elasmobranchs (e.g. barndoor skate) and teleosts (halibut, redfish). This is covered more fully elsewhere.

The second half of the twentieth century saw major changes in the species composition of the US

Table 3 Average US catch from the north-west Atlantic region

Species	Scientific name	Average catch (t)	
		1950–59	1989–98
Atlantic redfishes nei	Sebastes spp	77 923	518
Haddock	Melanogrammus aeglefinus	64 489	1487
Silver hake	Merluccius bilinearis	47 770	16 469
Atlantic cod	Gadus morhua	18 748	24 112
Scup	Stenotomus chrysops	17 904	3846
Saithe (Pollock)	Pollachius virens	11 127	6059
Yellowtail flounder	Limanda ferruginea	9103	5085
Winter flounder	Pseudopleuronectes americanus	7849	5700
Dogfish sharks nei	Squalidae	502	17 655
Raja rays nei	Raja spp.	73	10 304
American angler	Lophius americanus	41	19 136
American sea scallop	Placopecten magellanicus	77 650	81 320
Northern quahog (Hard clam)	Mercenaria mercenaria	52 038	25 607
Atlantic surf clam	Spisula solidissima	43 923	160 795
Blue crab	Callinectes sapidus	28 285	57 955
American lobster	Homarus americanus	12 325	29 829
Sand gaper	Mya arenaria	12 324	7739
Ocean quahog	Arctica islandica	1139	179 312

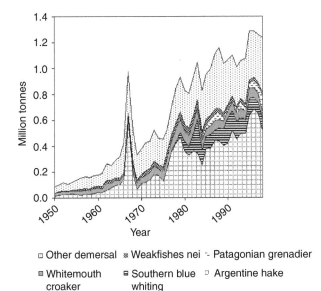

Legend:
- □ Other demersal
- ⊠ Weakfishes nei
- -- Patagonian grenadier
- ⊞ Whitemouth croaker
- ⊟ Southern blue whiting
- ⊡ Argentine hake

Figure 7 Total demersal fish landings from the south-west Atlantic (FAO Statistical area 41).

demersal fisheries. By the last decade of the century catches of the two top fish species during the decade 1950–59, redfish and haddock, had declined to 0.6% and 2.3% of their previous levels, respectively. Shellfish had become the main element of the demersal catch (**Table 3**).

Demersal Fisheries of the South-West Atlantic (FAO Statistical Area 41)

The catch from demersal fisheries in the SW Atlantic increased steadily from under 90 000 tonnes in 1950 to over 1.2 Mt in 1998 (**Figure 7**). The demersal catch consists mainly of fish species, of which Argentine hake has been predominant throughout the 50-year record. The catch of shrimps, prawns, lobsters and crabs is almost 100 000 t per year and they have a relatively high market value.

Thirty countries have taken part in demersal fisheries in this area, the principal ones being Argentina, Brazil, and Uruguay, with a substantial and continuing component of East European effort. A three year 'pulse' of trawling by the USSR fleet resulted in a catch over 500 000 t of Argentine hake and over 100 000 t of demersal percomorphs in 1967. The fisheries in this area are mostly industrialized and long range. As the stocks on the continental shelf have become fully exploited, the fisheries have extended into deeper water, where they take pink cusk eel and Patagonian toothfish. The main coastal demersal species are whitemouth croaker, Argentine croaker and weakfishes.

The Argentine hake fishery extends over most of the Patagonian shelf. It is now fully exploited and possibly even overexploited. Southern blue whiting and Patagonian grenadier are also close to full exploitation. Hake and other demersal species are regulated by annual TAC and minimum mesh size regulations.

Conclusions

Demersal fisheries have been a major source of protein for people all over the world for thousands of years. World catches increased rapidly during the first three-quarters of the twentieth century. Since then the catches from some stocks have declined, due to overfishing, while other previously underexploited stocks have increased their yields. The limits of biological production have probably been reached in many areas and careful management is needed in order to maintain the fisheries and to protect the ecosystems which support them.

Further Reading

Cochrane KL (2000) Reconciling sustainability, economic efficiency and equity in fisheries: the one that got away? *Fish and Fisheries* 1: 3–21.

Cushing DH (1996) *Towards a Science of Recruitment in Fish Populations*. Ecology Institute, D-21385 Oldendorf/Luhe, Germany.

Gulland JA (1988) *Fish Population Dynamics: The Implications for Management*, 2nd edn Chichester: John Wiley.

Kurlansky M (1997) *Cod. A Biography of the Fish That Changed the World*. London: Jonathan Cape.

INTERTIDAL FISHES

R. N. Gibson, Scottish Association for Marine Science, Argyll, Scotland

Introduction and Classification of Intertidal Fishes

The intertidal zone is the most temporally and spatially variable of all marine habitats. It ranges from sand and mud flats to rocky reefs and allows the development of a wide variety of plant and animal communities. The members of these communities are subject to the many and frequent changes imposed by wave action and the ebb and flow of the tide. Consequently, animals living permanently in the intertidal zone have evolved a variety of anatomical, physiological and behavioral adaptations that enable them to survive in this challenging habitat. The greater motility of fishes compared with most other intertidal animals allows them greater flexibility in combating these stresses and they adopt one of two basic strategies. The first is to remain in the zone at low tide. This strategy used by the 'residents', requires the availability of some form of shelter to alleviate the dangers of exposure to air and to predators. 'Visitors' or 'transients', that is species not adapted to cope with large changes in environmental conditions, only enter the intertidal zone when it is submerged and leave as the tide ebbs. The extent to which particular species employ either of these strategies varies widely. Many species found in the intertidal zone spend most of their lives there and are integral parts of the intertidal ecosystem. At the other extreme, others simply use the intertidal zone at high tide as an extension of their normal subtidal living space. In between these extremes are species that spend seasons of the year or parts of their life history in the intertidal zone and use it principally as a nursery or spawning ground. The different behavior patterns used by residents and visitors mean that few fishes are accidentally stranded by the outgoing tide.

Habitats, Abundance, and Systematics

Fishes can be found in almost all intertidal habitats and in all nonpolar regions. Most shelter is found on rocky shores in the form of weed, pools, crevices and spaces beneath boulders and it is on rocky shores that resident intertidal fishes are usually most numerous. If, however, fishes are capable of constructing their own shelter in the form of burrows in the sediment, as in the tropical mudskippers (Gobiidae), they may be abundant in such habitats. Visiting species may also be extremely numerous on occasions and are particularly common in habitats such as sandy beaches, mudflats and saltmarshes. Few fishes occupy gravel beaches although species like the Pacific herring (*Clupea pallasii*), capelin (*Mallotus villosus*, Osmeridae) and some pufferfishes (Tetraodontidae) may spawn on such beaches. Estimating abundance in terms of numbers per unit area can be difficult because of the cryptic nature of the fishes and the patchiness of the habitat. The difficulty is particularly acute on rocky shores where fishes may be highly concentrated in areas such as rock pools but absent elsewhere. Nevertheless, fish densities can be relatively high, particularly at the time of recruitment from the plankton, and on occasions may exceed 10 individuals per m^2.

Over 700 species of fishes from 110 families have so far been recorded in the intertidal zone worldwide. This figure represents less than 3% of known fish species but is certainly an underestimate because it is based only on species recorded on rocky shores. Species on soft sediment shores are not included and many areas of the world have yet to be studied in detail. The final count is therefore likely to be much higher. Intertidal fish faunas are frequently dominated by members of a few families (**Table 1**). Generally speaking, more species are found in the tropics than in temperate zones and each area of the world tends to have its own characteristic fauna. The Atlantic coast of South Africa, for example, is characterized by large numbers of clinid species, New Zealand by triplefins, the northeast Pacific by sculpins and pricklebacks (Stichaeidae) and many other areas by blennies, gobies and clingfishes.

Characteristics of Intertidal Fishes as Adaptations to Intertidal Life

Resident intertidal fishes are probably descended from subtidal ancestors and have few, if any, characters that are truly unique. They are thus representatives of families that have convergently evolved morphological, behavioral and physiological traits that enable them to survive in shallow turbulent habitats. The distribution of many resident intertidal

Table 1 Analysis of 47 worldwide collections of rocky shore intertidal fishes to show the 10 families with the largest number of species. Based on Prochazka *et al.* in *Horn et al.* (1999). Note that abundance of species does not necessarily imply that families are also numerically abundant

Family	Common name	Number of species
Blenniidae	Blennies and rockskippers	55
Gobiidae	Gobies and mudskippers	54
Labridae	Wrasses	44
Clinidae	Clinids, kelpfishes, klipfishes	33
Pomacentridae	Damselfishes	30
Tripterygiidae	Triplefin blennies	30
Cottidae	Sculpins	26
Labrisomidae	Labrisomids	26
Scorpaenidae	Scorpionfishes	25
Gobiesocidae	Clingfishes	24

species also extends below low water mark but they mainly differ from their fully subtidal relatives in the degree to which they can withstand exposure to air and are capable of terrestrial locomotion. Nevertheless, a few species can be considered truly intertidal in their distribution because they never occur below low water mark. Examples are the

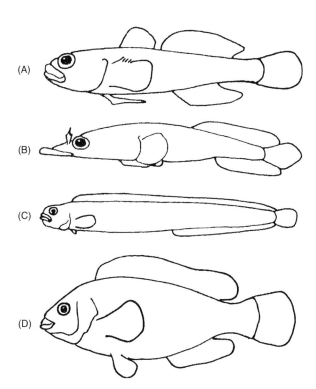

Figure 1 Sketches of four intertidal fish species to demonstrate the basic body shapes. (A) Terete (*Gobius paganellus*, Gobiidae); (B) dorsoventrally flattened (*Lepadogaster lepadogaster*, Gobiesocidae); (C) elongate (*Pholis gunnellus*, Pholidae); (D) laterally compressed (*Symphodus melops*, Labridae).

mudskippers of the genus *Periophthalmus* and blennies of the genera *Alticus* and *Coryphoblennius*.

Morphology

All the common families of intertidal fishes possess the characteristic morphological features of fishes adapted for benthic life in turbulent waters. They are cryptically colored, are rarely more than 15 cm long and are negatively buoyant because they lack a swimbladder or possess one that is reduced in volume. Four basic body shapes can be recognized: elongate, dorsoventrally flattened (depressed), smoothly cylindrical (terete), or laterally compressed (**Figure 1**). In many species the fins are modified to act as attachment devices to prevent dislodgement by turbulence or to assist movement over rough surfaces. In most blennies, the rays of the paired and ventral fins are hooked at their distal ends and may be covered with a thick cuticle to minimize wear. In the gobies and clingfishes, the pelvic fins are fused to form suction cups and allow the fish to attach themselves firmly to the substratum. There seem to be few 'typical' sensory adaptations to intertidal life although many species have reduced olfactory and lateral line systems. These sensory systems would be of limited use for species living in turbulent waters or in those that frequently emerge from the water.

Behavior

Intertidal fishes also show characteristic behavior that enables them to cope with the rigors of intertidal life. Their modified fins and relatively high density allow them to remain on or close to the bottom with the minimum of effort and to resist displacement by surge. Most are also thigmotactic, a behavior that keeps as much of their body touching the substratum as possible and ensures that they come to rest in contact with solid objects when inactive. Their mode of locomotion also reflects this bottom-dwelling lifestyle. Few excursions are made into open water and those species with large pectoral fins use them as much as the tail for forward movement and swim in a series of short hops. Clingfishes and gobies can progress slowly over horizontal and vertical surfaces using their sucker. Elongate forms, which usually have reduced paired fins, creep along the bottom using sinuous movements of their body or alternate lateral flexions of the tail. Strong lateral flips of the tail are also used by some blennies and gobies that can jump between rock pools at low tide and by the semiterrestrial mudskippers in their characteristic 'skipping' movements over the surface of the mud. When progressing more slowly mudskippers use the muscular pectoral fins as 'crutches' to move over the substratum.

Resident species are generally active at a particular state of the tide, although these tidally phased movements can be modulated by the day/night cycle so that some species may only be active, for example, on high tides that occur during the night. Visitors present a more complicated picture because, although their movements are also basically of tidal frequency, they are modulated by a wider range of cycles of lower frequency. Individuals may migrate intertidally on each tide, on every other tide depending on whether they are diurnal or nocturnal, or only on day or night spring tides. They may enter the zone as juveniles in spring or summer, stay there for several months, during which time they are tidally active, and then leave when conditions become unsuitable in winter or as they grow and mature.

The distances over which fishes move in the course of their intertidal movements are dependent on several factors. At one end of the scale are relatively gradual but ultimately extensive shifts in position related to the seasons. At the other end are local, short-term, tidally related foraging excursions. Residents tend to be very restricted in their movements and are often territorial, whereas visitors regularly enter and leave the intertidal zone and may cover considerable distances in each tide. Body size also determines the scale of movement. Residents are small and have limited powers of locomotion whereas visitors usually possess good locomotory abilities and can travel greater distances more rapidly.

The small size and poor swimming abilities of resident species partially account for the restricted extent of their movements but there is good evidence that some species also possess good homing abilities. Most evidence comes from experiments in which individuals are experimentally displaced short distances from their 'home' pools and subsequently reappear in these pools a short time later. Experiments with the goby *Bathygobius soporator* suggest that this species acquires a knowledge of its surroundings by swimming over them at high tide. It can remember this knowledge for several weeks and use its knowledge to return to its pool of origin. Homing is also known in some species of blennies and sculpins and is presumably based on the use of visual clues in the environment. Displacement experiments with the sculpin *Oligocottus maculosus*, however, suggest that this species at least may also use olfactory clues to find its way back to its home pool.

The energy expended in these movements at the various temporal and spatial scales described suggest that they play an important part in the ecology of both resident and visiting species alike. Several functions have been proposed for these movement patterns of which the most obvious is feeding. Visitors move into the intertidal zone on the rising tide to take advantage of the food resources that are only accessible at high water and move out again as the tide ebbs. Residents, on the other hand, simply move out of their low-tide refuge, forage while the tide is high, and return to the refuge before low tide. Following the flooding tide into the intertidal zone may have the added benefit of providing protection from larger predators in deeper water. Movements at both short and long time scales may also be in response to changing environmental conditions. Visitors avoid being stranded above low water mark at low tide because of the lack of refuges or because they are not adapted for the low tide conditions that may arise in possible refuges such as rock pools. Longer term seasonal movements into deeper water can be viewed as responses to changes in such physical factors as temperature, salinity and turbulence. Finally, several species whose distribution is basically subtidal move into the intertidal zone to spawn (see Life histories and reproduction below).

In order to synchronize their behavior with the constantly changing environment fishes must be able to detect and respond to the cues produced by these changes. The cues that fish actually use in timing and directing their tidally synchronized movements are mostly unknown. Synchronization could be achieved by a direct response to change. The flooding of a tide pool or the changing pressure associated with the rising tide, for example, could be used to signal the start of activity. In addition, behavior may be synchronized with the external environment by reference to an internal timing mechanism. The possession of such a 'biological clock' that is phased with, but operates independently of, external conditions is a feature common to all intertidal fishes in which it has been investigated. In the laboratory the presence of the 'clock' can be demonstrated by recording the activity of fish in the absence of external cues. Under these conditions fish show a rhythm of swimming activity in which periods of activity alternate with periods of rest (**Figure 2**). In most cases the period of greatest activity appears at the time of predicted high tide on the shore from which the fish originated. These 'circatidal' activity rhythms, so called because in constant conditions the period of the rhythm only approximates the average period of the natural tidal cycle (12.4 h), can persist in the laboratory for several days without reinforcement by external cues. After this time activity becomes random but in the blenny *Lipophrys pholis* the rhythm can be restarted (entrained) by exposing fish to

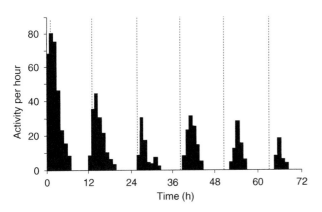

Figure 2 The 'circatidal' activity pattern shown by the blenny *Lipohrys pholis* in constant laboratory conditions. Peaks of activity correspond initially to the predicted times of high tide (dotted lines) but gradually occur later because the 'biological clock' has a period greater than that of the natural tidal cycle (12.4 h). In the sea the clock would be continually synchronized by local tidal conditions.

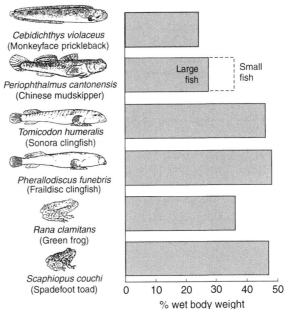

Figure 3 Tolerance of water loss as a percentage of wet body weight in four intertidal fishes compared with two amphibians. (Reproduced with permission from Horn and Gibson, 1988.)

experimental cycles of wave action or hydrostatic pressure or by replacing them in the sea. The probable function of this circatidal 'clock' is that it enables a fish to anticipate future changes in tidal state and regulate its activity and physiology accordingly. If activity is prevented in the wild, by storms for example, then the persistent nature of the clock allows fish to resume activity that is appropriate to the tidal state at the next opportunity.

Physiology

The ebbing tide exposes the intertidal zone to air and so the location of a fish is critical to its survival. Over the low tide period resident fish face not only the danger of exposure to air but also to marked changes in other physical and chemical conditions. On rocky shores, pools act as low-tide refuges but even here temperature, salinity, pH and oxygen content of the water can change markedly for the few hours that the pool is isolated from the sea. Consequently, many resident species are usually more tolerant of changes in these factors than subtidal species. Exposure to air could result in desiccation but, surprisingly, resident intertidal fishes show no major physiological or anatomical adaptations for resisting desiccation. Instead, desiccation is minimized by behavior patterns that ensure fish hide in pools, or in wet areas under stones and clumps of weed at low tide. Nevertheless, many species can survive out of water in moist conditions for many hours and tolerate water losses of more than 20% of their body weight, equivalent to that of some amphibians (**Figure 3**). The ability to tolerate water loss is generally correlated with position on the shore; those fish that occupy higher

levels are the most resistant. Prolonged emersion could also present fish with problems of nitrogen excretion and osmoregulation but those species that have been investigated seem to be able to cope with any changes in their internal medium caused by the absence of water surrounding them.

A further consequence of emersion is the change in the availability of oxygen. Although air contains a greater percentage of oxygen than water its density is much lower causing the gill filaments to collapse and reducing the area of the primary respiratory surface. Unlike some freshwater fishes, intertidal fishes that leave the water or inhabit regions where the water is likely to become hypoxic have no specialized air-breathing organs but maximize aerial gas exchange in other ways. Some species have reduced secondary gill lamellae, thickened gill epithelia and their gills are stiffened with cartilaginous rods. Such features reduce the likelihood of gill collapse when the fish is out of water. The skin is also used as an efficient respiratory surface because it is in contact with air and is close to surface blood vessels. In order to be effective, however, the skin must be kept moist and fish that are active out of water frequently roll on their sides or return to the sea to wet the skin. It is probable that some species use vascularized linings of the mouth, opercular cavities, and, possibly, the esophagus as respiratory surfaces.

The ability to respire in air is currently known for at least 60 species from 12 families and many of these voluntarily leave the water. Fish capable of respiring out of water have been classified into three main types. 'Skippers' are commonly seen out of water and actively feed, display and defend territories on land. They are typified by the tropical and subtropical mudskippers (Gobiidae) and rockskippers (Blenniidae). The second much less terrestrially active group, the 'tidepool emergers', crawl or jump out of tide pools mainly in response to hypoxia. Species from several families show this behavior but it has been best studied in the sculpins. The third group, the 'remainers', comprises many species that sit out the low-tide period emersed beneath rocks and in crevices or weed clumps. Some may be guarding egg masses and they are not active out of water unless disturbed.

Feeding Ecology and Predation Impact

Intertidal fishes are no more specialized or generalized in their diets and feeding ecology than subtidal fishes. Most are carnivorous or omnivorous and a few are herbivores. Herbivores appear to be less common in higher latitudes but a satisfactory explanation of this phenomenon is still awaited. In some cases the diet changes with size so that the youngest stages are carnivorous but larger amounts of algae are included in the diet as the fish grow. The small size of most resident fishes also means that their diet is composed of small items such as copepods and amphipods. In common with many other small fishes, they may also feed on parts of larger animals such as the cirri of barnacles or the siphons of bivalve molluscs, a form of browsing that does not destroy the prey. The extent to which intertidal fishes have an impact on the abundance of their prey and on the structure of intertidal communities is not clear. In some areas no impact has been detected whereas in others fish may have a marked effect, particularly on the size and species composition of intertidal algae. Those species that only enter the intertidal zone to feed contribute to the export of energy from this area into deeper water.

Life Histories and Reproduction

The majority of resident intertidal fishes rarely live longer than two to three years although some temperate gobies and blennies have a maximum life span of up to 10 years. Maturity is achieved in the first or second year of life and the females of longer-lived species may spawn several times a year for each year thereafter. Representatives of 25 teleost families are known to spawn in the intertidal zone. Of these,

residents and visitors make up about equal proportions. Intertidal spawning has both costs and benefits. For resident species, intertidal spawning reduces the likelihood of dispersal of the offspring from the adult habitat. It also obviates the need for movement to distant spawning grounds; a process that would be energetically costly for small-bodied demersal fishes with limited powers of locomotion and would at the same time expose them to greater risks of predation. These advantages do not apply to subtidal species many of which are good swimmers and whose adult habitat is offshore. For these species, the benefits are considered to be reduced egg predation rates and possibly faster development if the eggs become emersed. In both residents and visitors alike eggs spawned intertidally may be subject to the costs of increased mortality caused by desiccation and temperature stress. In addition, visitors may be vulnerable to avian and terrestrial predators during the spawning process.

The rugose topography of rocky shores and the cryptic sites chosen for spawning has led to the development of complex mating behavior in many species, particularly the blennies and gobies. In these species, courtship displays by the male include elements of mate attraction and a demonstration of the location of the chosen spawning site. Observing these displays is difficult in most groups but field observations have been made on several Mediterranean blennies that live in holes in rock walls and on mudskippers that perform their mating behavior out of water on the surface of the mud.

All species spawn relatively few (range approximately 10^2–10^5) large eggs (~ 1 mm diameter) that are laid on or buried in the substratum. Large eggs produce large larvae which may reduce dispersal from shallow water by minimizing the amount of time spent in the planktonic stage. On hard substrata the eggs are laid under stones, in holes and crevices and in or under weed. They may be attached individually to the substratum surface in a single layer (blennies, gobies, clingfishes) or in a clump (sculpins). In the gunnels and pricklebacks the eggs adhere to each other in balls but not to the surface. In soft sediments the eggs may be buried by the female as in the grunions (Atherinidae), or laid in burrows as in the mudskippers and some gobies. Killifishes (*Fundulus*) lay their eggs in salt-marsh vegetation.

In temperate latitudes spawning usually takes place in the spring and early summer, but spawning rhythms of shorter frequency may be superimposed on this annual seasonality. Subtidal species such as the grunions, capelin and pufferfishes that use the intertidal zone as a spawning ground mostly take advantage of spring tides to deposit their eggs in the sediment high on the shore, usually at night. During the reproductive season such fish, therefore, spawn at fortnightly intervals at the times of the new and full moons. The larvae develop over the intervening weeks and hatch when the eggs are next immersed.

Buried eggs are left unattended but eggs laid in layers, clumps or balls are always cared for by the parent. The sex of the individual that undertakes this parental care varies between species. In some it is the male, in others the female and in yet others both sexes participate. In some members of the families Embiotocidae, Clinidae and Zoarcidae fertilization is internal and the young are produced live. Parental care by oviparous species takes a variety of forms but all have the function of increasing egg survival rates and removal of the guardian parent greatly increases mortality. Most species guard the eggs against predators but some also clean and fan the eggs to maintain a good supply of oxygen and reduce the possibility of attack by pathogens.

Development time of the larvae depends on species and temperature but when fully formed the eggs hatch to release free-swimming planktonic larvae. In only one case, the plainfin midshipman (*Porichthys notatus*, Batrachoididae) does the male parent also care for the larvae, which in this species remain attached to the substratum near the nest site. The factors stimulating hatching are mostly unknown although it has been suggested that wave shock and temperature change associated with the rising tide may be involved. Until recently it was assumed that the hatched larvae were dispersed randomly by currents and turbulence. It has been shown, however, that at least some species minimize this dispersion by forming schools close to the bottom and are rarely found offshore (**Figure 4**). On completion of the larval phase the larvae metamorphose into the benthic juvenile phase and settle on the bottom. The clues used by settling larvae to select the appropriate substratum are poorly known but there is some evidence to suggest that individuals can discriminate between substratum types and settle on their preferred type.

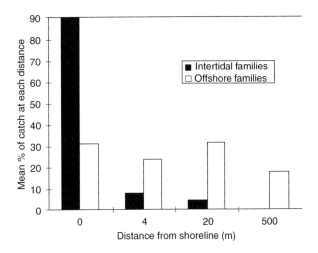

Figure 4 Comparison of the distribution of fish larvae from four intertidal and five offshore families caught in plankton nets at four distances from the shoreline in Vancouver harbour, British Columbia. (Based on data in Marliave JB (1986) *Transactions of the American Fisheries Society* 115: 149–154.)

See also

Rocky Shores. Salt Marshes and Mud Flats. Sandy Beaches, Biology of.

Further Reading

Gibson RN (1996) Intertidal fishes: life in a fluctuating environment. In: Pitcher TJ (ed.) *The Behaviour of Teleost Fishes*, 2nd edn, pp. 513–586. London: Chapman Hall.

Horn MH and Gibson RN (1988) Intertidal fishes. *Scientific American* 256: 64–70.

Horn MH, Martin KLM, and Chotkowski MA (eds.) (1999) *Intertidal Fishes: Life in Two Worlds*. San Diego: Academic Press.

EXOTIC SPECIES, INTRODUCTION OF

D. Minchin, Marine Organism Investigations, Killaloe, Republic of Ireland

Introduction

Exotic species, often referred to as alien, nonnative, nonindigenous, or introduced species, are those that occur in areas outside of their natural geographic range. Vagrant species are those that appear from time to time beyond their normal range and are often confused with exotic species. Since marine science evolved following periods of human exploration and worldwide trade, there are species that may have become introduced at an early time but cannot clearly be ascribed as native or exotic; these are known as cryptogenic species. The full contribution of exotic species among native assemblages remains, and probably will continue to remain, unknown, but these add to the diversity of an area. There are no documented accounts of an introduced species resulting in the extinction of native species in marine habitats as has occurred in freshwater systems. Nevertheless, exotic species can result in habitat modifications that may reduce native species abundance and restructure communities. The greatest numbers of exotic species are inadvertently distributed by shipping either attached to the hull or carried in the large volumes of ballast water used for ship stability. Introductions may also be deliberate. The dependence for food in developing countries and expansion of luxury food products in the developed world have led to increases in food production by cultivation of aquatic plants, invertebrates, and fishes. Many native species do not perform as well as the desired features of some introduced organisms now in widespread cultivation, for example, the Pacific oyster *Crassostrea gigas* and Atlantic salmon *Salmo salar*. Unfortunately, unwanted organisms that have been unintentionally introduced with stock movements can reduce production. Consequently, care is needed when introductions are made. An accidentally introduced exotic species may remain unnoticed until such time as it either becomes abundant or causes harmful effects, whereas larger organisms are normally recognized sooner. Little is known about the movement of the smallest organisms, yet these must be in transit in great numbers every day.

History

Species have been moved for several hundreds of years and some have almost certainly been carried with the earliest human expeditions. Evidence for early movement is scant; this can normally be determined only from hard remains, from a well-understood biogeography of a taxonomic group or perhaps from genetic comparisons. Although the soft-shell clam *Mya arenaria* was present in northern Europe during the late Pliocene, it disappeared during the last glacial period. In the thirteenth century their shells were found in north European middens; this was considered evidence of its introduction by returning Vikings from North America. It may have been used as fresh food during long sea journeys, as it is easily collected and perhaps was maintained in the bilges of their vessels, or may have unintentionally been carried with sediment used as ballast discharged on return to Europe. Certainly it became sufficiently abundant in the Baltic at about this time to be referred to by quaternary geologists as the Mya Sea. This species was also introduced to the Black Sea in the 1960s and rapidly became abundant, resembling its sudden apparent Baltic expansion.

The South American coral *Oculina patagonica*, established in southern Europe, was most probably introduced on sailing ship hulls returning from South America during the sixteenth or seventeenth centuries. Indeed these sailing vessels probably carried a wide range of organisms attached to their hulls. Predictions of the most likely fouling species during these times are based on settlements on panels attached to the hull of a reconstructed sailing ship. Hulls of wooden vessels were fouled with complex communities and excavated by boring organisms whose vacant galleries could provide refugia for a wide range of species including mobile animals. The drag imposed by fouling and the structural damage caused by the boring organisms played its role in the outcome of naval engagements and in the duration of the working life of these vessels, particularly in tropical regions where boring activities and fouling communities evolve rapidly. The periods of stay by sailing ships in ports could provide opportunities for creating new populations arising from drop-off or from reproduction while remaining attached. Some vessels never returned but would have provided instant reefs by sinking, wreckage, or abandonment, thereby distributing a wide assemblage of exotic species. Wooden vessels are still in widespread use and continue to endure problems of drag and structural damage. Many preparations have been used for controlling this, with the most effective being developed over the last 150 years. Wooden vessels became sheathed with thin copper plates and were vulnerable only where these became displaced or damaged. Ironclad sailing vessels also evolved and normal hull life became prolonged because damage was considerably reduced. The building of iron and then steel ship hulls eliminated opportunities for boring and cryptic fauna, yet hull surfaces needed protection from rusting and fouling. The development of protective and also toxic coatings then evolved. The incorporation of copper and other salts in these coatings considerably reduced fouling, and therefore drag, and protected the underlying surfaces from corrosion. Antifouling coatings, in particular, organotins, such as tributyltin compounds, have been very effective in fouling control, and have enabled most vessels to remain in operation before reentering dry dock for servicing, about every 5 years. Unfortunately, tributyltin is a powerful biocide harmful to aquatic organisms in port regions. Its use during the 1980s and 1990s became restricted or banned in many countries on vessels <25 m. And between 2003 and 2008 all merchant shipping are expected to phase out its use. New products are being developed with an emphasis on less toxic and nontoxic applications.

Most exotic species, used in mariculture or for developing fisheries, were spread following the development of steam transport. An early example is the transport from the eastern coast of North America of the striped bass *Morone saxatilis* fingerlings to California in the 1880s by train. Within 10 years, it became regularly sold in the fish markets of San Francisco. Attempts at culturing different species developed slowly due to a poor knowledge of the full life cycle. Planktonic stages were especially misunderstood and pond systems for rearing oyster larvae, for example, were not fully successful until the larval biology became fully known. There were strong economic pressures to develop this knowledge at an early stage. The progressive understanding of behavior and the physical requirements of organisms, and development of algae, brine shrimp, and other cultures and the use of synthetic materials, such as plastics, enabled a rapid expansion of marine culture products ranging from pearls, chemical compounds, to food.

The Vectors of Exotic Species

Organisms are deliberately introduced for culture, to create fisheries, or as ornamental species. Future

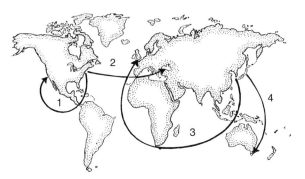

Figure 1 Some examples of primary inoculations: 1, *Carcinus maenas*, the shore crab, a predator of mollusks (vector, probably with shore macroalgae used for transporting fishing baits from the North Atlantic); 2, *Mnemiopsis leidyi* a comb jelly, planktonic predator of larval fishes and crustaceans (vector, ballast water); 3, *Styela clava* an Asian tunicate that causes trophic competition by filtering the water in docks and estuaries (vector, hull fouling); 4, *Asterias amurensis*, an Asian sea star, avid predator of mollusks (vector, ballast water or hull fouling).

introductions may involve species producing products used in pharmacy, as food additives, in the management of diseases, for biological control, or for water quality management.

A primary inoculation is where the first population of an exotic becomes established in a new biogeographic region (**Figure 1**). A secondary inoculation arises by means of further established populations from the first population to other nearby regions by a wide range of activities and so the risk of spread is increased. In many cases, these range extensions are inevitable, either because of human behavior and marketing patterns or because of their mode of life. For example, the Indo-Pacific tubeworm *Ficopomatus enigmaticus* can extensively foul brackish areas and once transported by shipping is easily carried to other estuaries by leisure craft or aquaculture activities, or both.

Species are introduced by a wide range of vectors that can include attachment to seaweeds used for packing material or releases of imported angling baits. The principal vectors are discussed below.

Aquaculture

Useful species have a high tolerance of handling, can endure a wide range of physical conditions, are easily induced to reproduce, and have high survival throughout their production phase. Because of capital costs to retain stock in cages and/or trays, the species must, most usually, be cultivated at densities higher than occurs naturally. The species most likely to adapt to these conditions will be those living under variable conditions at high densities, such as mussels and oysters and some shoaling fishes. Species that do not naturally thrive under these conditions may require more attention and will tend to depend on speciality markets, such as some scallops, tuna, and grouper. Further, aquaculture efforts target those species which grow fast to marketable size.

Mollusks High-value species that are tolerant to handling as well as a wide range of physical conditions are preferred. Some have a long shelf life that allows for live sales. Mollusks, such as clams (e.g., *Ruditapes philippinarum, Mercenaria mercenaria*), scallops (e.g., *Patinopecten yessoensis, Argopecten irradians*), oysters (e.g., *Ostrea edulis, C. gigas*), and abalone (e.g., *Haliotis discus hannai, Haliotis rufescens*), have been introduced for culture (**Figure 2**). Because oysters survive under cool damp conditions for several days, large consignments are easily transported long distances and so have become widely distributed since the advent of steam transportation. With the exception of the east coast of North America, the Pacific oyster *C. gigas*, for example, is now widespread in the Northern Hemisphere and accounts for 80% of world oyster production. Its wide tolerance of salinity, temperature, and turbid conditions and rapid growth make it a desirable candidate for culture. In many cases, it has replaced the native oyster production where this declined either because of overexploitation, diseases of former stocks, or to develop new culture areas. However, pests, parasites, and diseases can be carried within the tissues, mantle cavity, and both in and on the shells of mollusks when they are moved, and can compromise molluskan culture and wild fisheries and may also modify ecosystems. Some examples of pests moved with oysters include the slipper limpet *Crepidula fornicata* and the sea squirt *Styela clava* (**Figure 2**).

Crustaceans The worldwide expansion of penaeid shrimp culture has led to a series of unregulated movements resulting in serious declines of shrimp production caused by pathogenic viruses, bacteria, protozoa, and fungi. Viruses have caused the most serious mortalities of farmed shrimps' broodstock and pose a deterrent to developing culture projects because of casual broodstock movements rather than using those certified as specific-pathogen-free. One parvovirus is the infectious hypodermal and hematopoietic necrosis virus found in wild juvenile and adult prawn stages. This virus was endemic to Southeast Asia, Indonesia, and the Philippines and is now widespread at farms in the tropical regions of the Indo-Pacific and the Americas. About six serious viruses of penaeid shrimp are known. Production may be limited unless disease-resistant

stocks for viruses can be developed together with vaccination against bacterial diseases.

Fishes Of the thousands of fish species only a small number generate a market price high enough to cover production costs. Such fish need to have a high flesh to body weight and high acceptance. While in culture, they should be tolerant of handling and to the wide range of seasonal and diurnal conditions and should not be competitive. The North Atlantic salmon *S. salar* is one of the very few exotics in culture in both hemispheres in the cool to warm temperate climates of North America, Japan, Chile, New Zealand, and Tasmania (**Figure 2**). Fertilized eggs of improved stock are in constant movement between these countries. However, various diseases such as infectious salmon anemia (ISA) may limit the extent of future movements.

It is likely that intensive shore culture facilities for marine species will become more common as the physiological and behavioral requirements of promising species become better understood. For example, in Japan, the bastard sole *Paralichthys olivaceus* is extensively cultivated in shore tanks (and has been introduced to Hawaii) and the sole *Solea senegalensis* in managed lagoons in Portugal. Cultivation under these circumstances is likely to lead to better control of pests, parasites, and diseases, whereas cage culture under more exposed conditions off shore may lead to unexpected mortalities from siphonophores, medusae, algal blooms, and epizootic infestations from parasites, some perhaps introduced.

Shipping

Much of the world trade depends on shipping; the scale and magnitude of these vessels are seldom appreciated. To travel safely, they must either carry cargo or water (as ballast) so that the vessel is correctly immersed to provide more responsive steerage, by allowing better propulsion (without cavitation), rudder bite, and greater stability. The amount of ballast water carried can amount to 30% or more of the overall weight of the ship. Ships also have a large immersed surface area to which organisms attach and result in increased drag that results in higher fuel costs. Nevertheless, despite the best efforts of management, ships carry a great diversity and large numbers of species throughout the world. Unfortunately the risk of introducing further species increases because more ships are in transit and many travel faster than before and operate a wide trading network with new evolving commercial links.

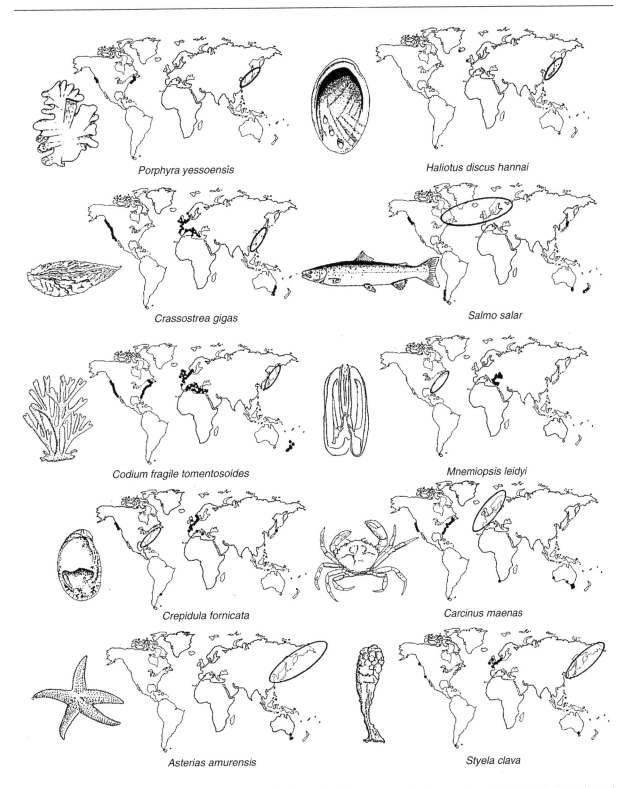

Figure 2 Examples of exotic species. Useful species in culture: *Porphyra yessoensis*, Japanese laver to 30 cm; *H. discus hannai*, Japanese abalone to 18 cm; *C. gigas*, Pacific oyster to 16 cm; *S. salar*, Atlantic salmon to 30 kg. Unwanted invasive species: *Codium fragile tomentosoides*, a green alga to 45 cm; *Mnemiopsis leidyi*, a comb jelly to 11 cm; *Crepidula fornicata*, the slipper limpet to 5 cm; *Carcinus maenas*, the shore or green crab to 9 cm; *Asterias amurensis*, the Asian sea star to 25 cm; *Styela clava*, an Asian tunicate to 18 cm. Open rings, native range; solid dots, established in culture or in the environment.

Ships' ballast water Ballast water is held on board in specially constructed tanks used only for holding water. The design and size of these tanks depend on the type and size of vessel (**Figure 3**). In some vessels that carry bulk products, a cargo hold may also be flooded with ballast water, but these vessels will also have ballast tanks. Ballast tanks are designed to add structural support to the ship, and will have access ladders, frames, and perforated platforms and baffles to reduce water slopping, making them complex engineering structures. All vessels carry ballast for trim; some of this may be permanent ballast (in which case it is not exchanged at any time) or it may be 'all' released (there is always a small portion of unpumpable water that remains) before taking on cargo. There are several tanks on a ship, and because there may be partial discharges of water, the water within nonpermanent ballast tanks can contain a mixture of water from different ports.

Pumping large volumes of water when ballasting is time-consuming and costly. Should a vessel be ballasted incorrectly, the structural forces produced could compromise the hull. Ships have broken up in port because of incorrect deballasting procedures in relation to the loading/unloading of cargo. Other vessels are known to have capsized due to incorrect ballast water operations. For these reasons, there are special guidelines for the correct ballasting of tanks in relation to cargo load and weather conditions.

Ballasting of water normally occurs close to a port, often in an estuary or shallow harbor; this can include turbid water, sewage discharges, and pollutants. The particulates settle inside the tanks to form sediment accumulations that may be 30 cm in depth. Worms, crustaceans, mollusks, protozoa, and the resting stages of dinoflagellates, as well as other species, can live in these muds. Resting stages may remain dormant for months or years before 'hatching'. Ballast sediments and biota are thus an important component of ballast water uptake adding to the diversity of carried organisms released in new regions through larval releases or from sediment resuspension following journeys with strong winds.

Fresh ballast water normally includes a wide range of different animal and plant groups. However, most of these expire over time, so that on long journeys fewer species survive, and those that do survive, do so in reduced numbers. Because ballast water is held in darkness the contained phytoplankton are unable to photosynthesize. Animals, dependent on these microscopic plants, during vulnerable stages in their development may expire because of insufficient food, unless the tanks are frequently flushed. Scavengers and predators may have better opportunities to survive. To date, observations of organisms surviving in ballast water are incomplete because of the inability or poor efficiency in obtaining adequate samples. Some ballast tanks may only be sampled through a narrow sounding pipe to provide a limited sample of the less active species from one region near the tank floor. Nevertheless, all studies to date point to a wide range of organisms surviving ballast transport, ranging from bacteria to adult fishes. The invasions of the comb jelly *Mnemiopsis leidyi* (from the eastern coast of North America to the Black Sea), the Asian sea star *Asterias amurensis* (from Japan to Tasmania), and the shore crab *Carcinus maenas* (to Australia, the Pacific and Atlantic coasts of North America, South Africa, and the Patagonian coast from Europe) probably arose from ballast water transport (**Figure 2**).

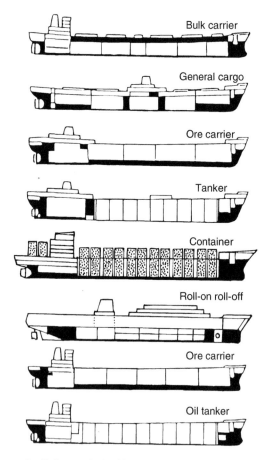

Figure 3 Ballast tanks in ships are dedicated for carrying water for the necessary stabilization of ships. The ballast tanks are outlined in black and show the variation of ballast tank position and size in relation to the type of vessel. Wing tanks (ballast tanks situated along the sides) are not shown in this diagram.

Ships' hulls Ships are dry-docked for inspection, structural repairs, and recoating the hull with antifouling paint about every 3–5 years. The interdocking time varies according to the type of vessel and its age. While in dry dock, the ship is supported on wooden blocks. The areas beneath the blocks, which may in total cover a ship surface of >100 m^2, are not painted and here fouling may freely develop once the ship is returned to service. The hull is cleaned by shot blasting or using powerful water jets, then recoated. Many coatings have a toxic surface that over time becomes leached or worn from the hull. Unfortunately, some toxic compounds, once released, have caused unexpected and unwanted environmental effects. The use of organotin paints such as tributyltin effectively reduced hull fouling since the 1970s, but it has caused sexual distortion in snails and reproductive and respiratory impairment in other taxa, especially in coastal waters along highly frequented shipping routes. Vessels trading in tropical waters have a higher rate of loss of the active coat and on these vessels a greater fouling can occur at an earlier time.

Most organisms carried on ships' hulls are not harmful but may act as the potential carriers of a wide range of pests, parasites, and diseases. Oysters and mussels frequently attach to hulls and may transmit their diseases to aquaculture sites in the vicinity of shipping ports. Small numbers of disease organisms, once released, may be sufficient to become established if a suitable host is found. Should the host be a cultivated species, it may rapidly spread to other areas in the course of normal trade before the disease becomes recognized.

Many marine organisms spawn profusely in response to sea temperature changes. Spawning could arise following entry of a ship to a port, during unloading of cargo or ballast in stratified water, or from diurnal temperature changes. A ship may turn around in port in a few hours to leave behind larvae that may ultimately settle to form a new population.

Vessels moored for long periods normally acquire a dense and complex fouling community that includes barnacles. Once of sufficient size, these can provide toxic-free surfaces for further attachment, in particular for other barnacles (e.g., the Australasian barnacle *Elminius modestus*, present in many Northern European ports), mollusks, hydroids, bryozoa, and tunicates (e.g., the Asian sea squirt *S. clava*, widely distributed in North America and northern Europe (**Figure 2**)). Vacant shells of barnacles and oysters cemented to the hull can provide shelter for mobile species such as crustaceans and nematodes, and some mobile species such as anemones and flatworms can attach directly to the hull.

Small vessels such as yachts and motorboats also undertake long journeys and may become fouled and thereby extend the range of attached biota. Normally small vessels would be involved in secondary inoculations from port regions to small inlets and lagoons (areas where shipping does not normally have access) and spread species such as the barnacle *Balanus improvisus* to other brackish areas.

Aquarium Species

Aquarium releases seldom become established in the sea whereas in fresh water this is common. The attractive green feather-like alga *Caulerpa taxifolia* was probably released into the Mediterranean Sea from an aquarium in Monaco and by 2000 was present on the French Mediterranean coast, Italy, the Balearic Islands, and the Adriatic Sea. It has a 'root' system that grows over rock, gravel, or sand to form extensive meadows in shallows and may occur to depths of 80 m. It excludes seagrasses and most encrusting fauna and so changes community structure. Although it possesses mild toxins that deter some grazing invertebrates, several fishes will browse on it. This invasive plant is still expanding its range and may extend throughout much of the Mediterranean coastline and perhaps to some Atlantic coasts.

Trade Agreements and Guaranteed Product Production

Trade agreements do not normally take account of the biogeographical regions among trading partners, so introductions of unwanted species are almost inevitable. Often, a population of a harmful species is further spread by trading and thereby has the ability to compromise the production of useful species. Trade in the same species from regions where quarantine was not undertaken may compromise cultured exotic species introduced through the expensive quarantine process which are disease-free. The risk of movement of serious diseases from one country to another is usually recognized, whereas pests are not normally regulated by veterinary regulations, and so are more likely to become freely distributed.

Harvesting of cultured species may be prohibited in areas following the occurrence of toxin-producing phytoplankton. Toxins naturally occur in certain phytoplankton species (most usually dinoflagellates) and are filtered by the mollusk, and these become concentrated within molluskan tissues, with some organs storing greater amounts. Some of these toxins may subsequently accumulate within the tissues of molluskan-feeding crustaceans, such as crabs. If contaminated shellfish is consumed by humans, symptoms such as diahorritic shellfish poisoning, paralytic shellfish poisoning, neurological shellfish poisoning, and amnesic shellfish poisoning may occur. These conditions are caused by different toxins, and new toxins continue to be described. As they all have different breakdown rates in the shellfish tissues, there are varying periods when the products are prohibited from sale. There are monitoring programs for evaluating the levels of these contaminants in most countries. On occasions when harvesting of mollusks is suspended, consignments from abroad may be imported to maintain production levels. If these consignments arrive in poor condition, or are unsuitable because of heavy fouling, they may be relaid or dumped in the sea or on the shore. Such actions can extend the range of unwanted exotic species.

The Increasing Emergence of Unexplained Events

Natural eruptions of endemic or introduced pathogens may be responsible for chronic mortalities or unwanted phenomena. The die-off of the sea urchin *Diadema antillarum* in the western Atlantic may be due to a virus. The unsightly fibropapillomae of the green turtle *Chelonia mydas*, previously known only in the Atlantic Ocean, are now found in the Indo-Pacific, where it was previously unknown. The great mortalities of pilchard *Sardinops sagax neopilchardus* off Australia in the 1990s and mortalities of the bay scallop *A. irradians* in China may be a result of introduced microorganisms. In the case of introduced culture species, mortalities may ensue from a lack of resistance to local pathogens. *Bonamia ostreae*, a protozoan parasite in the blood of some oyster species, was unknown until found in the European flat oyster *O. edulis* following its importation to France from the American Pacific coast, where it had been introduced several years earlier. Very often harmful species first become noticed when aquaculture species are cultivated at high densities, for example, the rhizocephalan-like *Pectinophilus inornata* found in the scallop *Pa. yessoensis* in Japanese waters. Often careful studies of native biota reveal new species to science, such as the generally harmful protozoans, *Perkensis* species found in clams, oysters, and scallops that can be transmitted to each generation by adhering to eggs. Pests, parasites, and diseases in living products used in trade are likely sources of transmission. However, viruses or resting stages of some species may also be transmitted in frozen and dried products. Changes in global climate may be contributing to some of these appearances aided by high levels of human mobility.

Vulnerable Regions

Some areas favor exotic species with opportunities for establishment and so enable their subsequent spread to nearby regions. These areas are normally shipping ports within partly enclosed harbors with low tidal amplitudes and/or with good water retention and a large number of arriving vessels. Although it may be possible to predict which ports are the main sites for primary introductions, the factors involved are not clearly understood and information on the exotic species component is presently only available for a small number of ports. Nevertheless, ports with many exotic species are areas where further exotics will be found. Such regions are likely sites for introductions because ships carry a very large number of a wide range of species from different taxonomic groups. Port regions with known concentrations of exotic species include San Francisco Bay, Prince William Sound, Chesapeake Bay, Port Phillip Bay, Derwent Estuary, Brest Harbor, Cork Harbour, and The Solent (**Figure 4**). In the Baltic Sea, there are several ports that receive ballast water from ships operated via canals from the Black and Caspian Seas as well as arising from direct overseas trade. The component of the exotic species biomass in this region is high. In some areas such as the Curonian Lagoon, Lithuania, exotic species comprise the main biomass. The Black Sea has similar conditions to the Baltic Sea where established species have modified the economy of the region. The effects vary from dying clams *Mya arenaria* creating a stench on tourist beaches, poor recruitment of pilchard and anchovy due to a combination of high exploitation of the fisheries, changes in water quality and an increased predation of their larvae by an introduced comb jelly *M. leidyi*, to the development

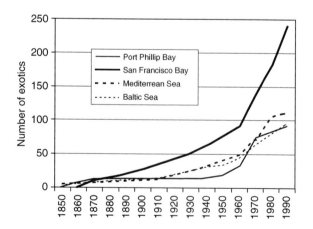

Figure 4 Accumulative numbers of known exotic species in vulnerable areas. The real numbers of exotic species are probably much greater than shown.

of an industry based on harvesting and export of the large introduced predatory snail *Rapana venosa*. The Mediterranean Sea has increasing numbers of unintended introductions arising from trade, mainly by shipping, and from expansion of their range from the Red Sea via the Suez Canal (**Figure 4**).

The majority of exotic species is found in temperate regions of the world. Whether this is a surmise because of a lack of a full understanding, or whether this is due to a real effect, is not known. It may be that tropical species are well dispersed because of natural vectors and that further transmission by shipping is of little consequence to the overall biota present. However, it may be that there is such a daunting diversity of species present in tropical regions, with many still to be described, that it is difficult to grasp the complexity and so be able to understand exotic species' movements in this zone. In the contrasting colder climates, the likely slower development of species may inhibit an introduction from being successful. Temperate ports on either side of the same ocean appear to share several species in common, whereas species from temperate regions from other hemispheres or oceans are less common. This suggests that species that do not undergo undue physiological stresses and those present in shorter voyages have a greater probability of becoming established.

For many species, although opportunities for their dispersal in the past may have taken place, their successful establishment has not succeeded. If an inoculation is to succeed, it must pass through a series of challenges. On most occasions, the populations are unable to be maintained at the point of release in sufficient numbers capable of establishment, even though some may survive. Any sightings of exotic individuals may often pose as a warning that this same species, under different conditions, could become established. The increasing volume and speed of shipping, and of air transport (in the case of foods for human consumption and aquarium species) will provide new opportunities. A successful transfer for a species will depend on: appropriate season, life-history stage in transit, survival time without food, re-immersion, temperature and salinity tolerance, rate of dispersal at the point of release, inoculation size and frequency, presence/absence of predators in the receiving environment, and many unknown factors.

Mode of Life

The dispersal of a species from its point of introduction, and the speed at which it expands its range,

will depend on the behavior of its life-history stages and hydrodynamic conditions in the recipient habitat. Species with very short or no planktonic stages, and with limited mobility, are likely to remain close to the site of introduction, unless carried elsewhere by other means. Tunicates have short larval stages and a sessile adult life and so are normally confined to inlets. Buoyant and planktonic species, which include seaweeds with air bladders and many invertebrates, may be carried by combinations of wind and current and become rapidly dispersed in a directional pattern ruled by the principal vectors. Most species have dispersal potentials that lie between these extremes and some, such as active crustaceans and fishes, may become distributed over a wide range as a result of their own activities. The attempt to establish the pink salmon *Onchorhynchus gorbuscha* on the northern coast of Russia resulted in its capture as far south in the North Atlantic as Ireland. A recent and successful introduction of the king crab *Paralithodes camtschaticus* to the Barents Sea has resulted in its rapid expansion to Norway aided by its planktonic larval stages carried by currents and an ability to travel great distances by walking. With a better knowledge of the behavior of organisms during their life-history stages and of the prominent physical vectors from a point of release, theoretical models of dispersal should be possible. Such models would also be valuable tools for predicting the plumes arising from discharges of ships' ballast water and for evaluating relative risk scenarios of transmitting or receiving exotic species.

The reproductive capability of a species is also important. Those that release broods that are confined to the benthos such as the Chinese hat snail *Calyptraea chinensis*, are likely to remain in one region, and once mature will have a good opportunity for effective reproduction. Species such as *Littorina saxatilis* have the theoretical capability of establishing themselves from the release of a single female by producing miniature crawlers without a planktonic life stage. In contrast, species with long planktonic stages requiring stable conditions are unlikely to succeed. It is doubtful whether spiny lobsters will become inadvertently transferred in ballast water.

Impacts on Society

Exotic species have a wide range of effects: some provide economic opportunities whereas others impose unwanted consequences that can result in serious financial loss and unemployment. In agriculture, the main species utilized in temperate environments for food production were introduced and have taken some thousands of years to develop. In the marine environment, the cultivation of species is relatively new and comparatively few exotic species are utilized. This suggests that in future years an assemblage of exotic species, some presently in cultivation, and others yet to be developed, will form a basis for significant food production. Few exotics in marine cultivation form the basis for subsistence, whereas this is common in freshwater systems. The cultivation of marine species is normally for specialist markets where food quality is an important criterion. Expanding ranges of harmful exotics could erode opportunities by impairing the quality in some way or interfering with production targets.

Unfortunately, many shipping port regions are in areas where conditions for cultivation are suitable, either for the practical reason of lower capital costs for management, or for the ease of operation and/or shelter, or since the conditions favor optimal growth and/or the nearby market. The proximity of shipping to aquaculture activities poses the unquantifiable threat that some imported organisms will impair survival, compromise growth, or render a product unmarketable.

Diseases of organisms and humans are spreading throughout the world. In the marine environment, a large bulk of biota is in transit in ballast water. Ballasting by ships in port may result in loading untreated discharges of human sewage containing bacteria and viruses that may have consequences for human health once discharged elsewhere. In 1991, at the time of the South American cholera epidemic, caused by *Vibrio cholerae*, oysters and fish in Mobile Bay, Alabama (USA), were found with the same strain of this infectious bacterium. Ballast water was considered as a possible source of this event. Subsequently five of 19 ships sampled in Gulf of Mexico ports arriving from Latin America were found with this same strain. The epidemic in South America may have been originally sourced from Asia, and may also have been transmitted by ships.

Of grave concern is the discovery of new algal toxins and the apparent spread of amnesic shellfish poisoning, diahorritic shellfish poisoning, paralytic shellfish poisoning, and neurological shellfish poisoning throughout the world. The associated algae can form dense blooms that, with onshore winds, can form aerosols that may be carried ashore to influence human health. The apparent increase in the frequency of these events may be due to poor historical knowledge of previous occurrences or to a real expansion of the phenomena. There is good evidence that ballast water may be distributing some of these harmful species. Some of the 'bloom-forming' species, such as the naked dinoflagellate *Karenia*

mikimotoi, are almost certainly introduced and cause sufficiently dense blooms to impair respiration in fishes by congestion of the gills, and can also purge the water column of many zooplankton species and cause mortalities of the benthos.

Some introduced invertebrates may act as an intermediate host for human and livestock diseases. The Chinese mitten crab *Eriocheir sinensis*, apart from being a nuisance species, acts as the second intermediate host for the lung fluke *Paragonimus westermanii*. The first intermediate stage appears in snails. The Chinese mitten crab has been introduced to the Mediterranean and Black Seas, North America, and northern Europe. Should the lung fluke be introduced, the ability for it to become established now exists where it may cause health problems for mammals, including humans.

Management of Exotic Species

Aquaculture

There is an expanding interest in aquaculture as an industry to provide employment, revenue, and food. Already several species in production worldwide contribute to these aims. However, any introduction may be responsible for unwanted and harmful introductions of pests, parasites, and diseases. Those involved in future species introductions should consider the International Council for the Exploration of the Sea's (ICES) Code of Practice on Introductions and Transfers of Marine Organisms (**Table 1**). This code takes into account precautionary measures so that unwanted species are unlikely to become unintentionally released. The code also includes provisions for the release of genetically modified organisms (GMOs). These are treated in the same way as if they are exotic species introductions intended for culture. The code is updated from time to time in the light of recent scientific findings. Should this code be ignored the involved parties could be accused of acting inappropriately. By using the code, introductions may take several years before significant production can be achieved. This is because the original broodstock are not released to the wild, only a generation arising from them that is disease-free. This generation must be examined closely for ecological interactions before the species can be freely cultivated. In developing countries it is important that all reasonable precautions are taken to reduce obvious risks. It has been shown historically that direct introductions, even when some precautions have been taken, may lead to problems that can compromise the intended industry or influence other industries, activities, and the environment.

Table 1 Main features of the ICES Code of Practice in relation to an introduction

Conduct a desk evaluation well in advance of the introduction, to include:
  previous known introductions of the species elsewhere;
  review the known diseases, parasites, and pests in the native environment;
  understand its physical tolerances and ecological interactions in its native environment;
  develop a knowledge of its genetics; and
  provide a justification for the introduction
Determine the likely consequences of the introduction and undertake a hazard assessment
Introduce the organisms to a secure quarantine facility and treat all wastewater and waste materials effectively
Cultivate F1 generation in isolation in quarantine and destroy broodstock
Disease-free filial generation may be used in a limited pilot project with a contingency withdrawal plan
Development of the species for culture

At all stages the advice of the ICES Working Group on the Introductions and Transfers of Marine Organisms is sought. Organisms with deliberately modified heritable traits, such as genetically modified native organisms, are considered as exotic species and are required to follow the ICES Code of Practice.

Table 2 Treatment measures of ballast water

Disinfection: Tank wall coatings, biocides, ozone, raised temperature, electrical charges and microwaves, deoxygenation, filtration, ultraviolet light, ultrasonification, mechanical agitation, exchanges with different salinity
By management: Special shore facilities or lighters (transfer vessels) for discharges, provision of clean water (fresh water) by port authorities, no ballasting when organisms are abundant (i.e., during algal blooms, at night) or of turbid water (i.e., during dredging, in shallows), removal of sediments and disposal ashore. Specific port management plans taking account of local port conditions and seasonality of the port as a donor area
Passive effects: Increase time to deballasting, long voyages, reballast at sea.

Experience has shown that good water quality, moderate stocking densities, and meteorological and oceanographic conditions, within the normal limits of species, or the culture system, are of importance for successful cultivation. Maintenance of production in deteriorating conditions, following high sedimentation or pollution, can rarely be achieved by using other introduced species.

Exotic species generally used in aquaculture may not always prove to be beneficial. Although the Pacific oyster *C. gigas* is generally accepted as a useful species, in some parts of the Adriatic Sea it fouls metal ladders, rocks, and stones causing cuts to bathers' skin, this in a region where revenue from tourism exceeds that from aquaculture. This same species is unwelcome in New South Wales because of competition with the Sydney rock oyster *Saccostrea commercialis* and there have been attempts to eradicate it.

Ecomorphology Organisms can respond to changes in their environment by adapting specific characteristics that provide them with advantage. Sometimes these changes can be noted within a single lifetime, but more usually this takes place over many generations and may ultimately lead to species separation. When considering a species for introduction its morphology may provide clues as to whether it will compete with native species or whether its feeding capabilities or range is likely to be distinct and separate, overlapping or coinciding. However, it is not possible to evaluate the overall impacts of a species for introduction in advance of the introduction using morphological features alone.

Ships' Ballast Water

Sterilization techniques of ballast water are difficult, and most ideas are not cost-effective or practical, either because the great volumes of water require large amounts of chemicals or because of the added corrosion to tanks or because the treated water, when discharged, has now become an environmental hazard. Reballasting at sea is the current requirement by the International Maritime Organization (IMO), the United Nations body which deals with shipping. Ballast tanks cannot be completely drained and so three exchanges are required to remove >95% of the original ballast water. However, it is not possible for ships to reballast in mid-ocean in every case. Ships that deballast in bad weather can be structurally compromised or become unstable; this could lead to the loss of the vessel and its crew. A further method under consideration that does not compromise the safety of the vessel is the continuous flushing of water while in passage. Several further techniques have either been considered, researched, or are in development (**Table 2**). Exotic species management in ballast water is likely to become a major research area into the twenty-first century.

Ships' Hulls

The use of the toxic yet effective organotins as antifouling agents is likely to become phased out over the first decade of the twenty-first century. Replacement coatings will need to be as, or more, effective, if ships are going to manage fuel costs at current levels. Nontoxic coatings or paint coatings containing deterrents to settling organisms are likely to evolve, rather than coatings containing biocides. However, the effectiveness of these coatings will need to take account the normal interdocking times for ships. In some cases robots may be required for reactivating coat surfaces and removing undue fouling. Locations and/or special management procedures where these activities take place need to be carefully planned to avoid establishment of species from the 'rain' of detritus from cleaning operations.

Biocontrol

Biological control is the release of an organism that will consume or attack a pest species resulting in a population decrease to a level where it is no longer considered a pest. Although there are many effective examples of biological control in terrestrial systems, this has not been practiced in the marine environment using exotic species. However, biological control has been considered in a number of cases.

1. The green alga *C. taxifolia* has become invasive in the Mediterranean following its likely release from an aquarium; the species presently ranges from the Adriatic Sea to the Balearic Islands and forms meadows over rock, gravels, and sands, displacing many local communities. The introduction of a Caribbean saccoglossan sea slug that does not have a planktonic stage and avidly feeds on this alga has been under consideration for release. These sea slugs were cultured in southern France, but their release to the wild has not been approved.

2. The comb jelly *M. leidyi* became abundant in the Black Sea in the mid-1980s. It readily feeds on larval fishes and stocks of anchovy and pilchard declined in concert with its expansion. The introduction of either cod *Gadus morhua* from the Baltic Sea or chum salmon *Onchorhynchus gorbusha* from North America was considered. Also considered for control was a related predatory comb jelly. However, the predatory comb jelly *Beroe* became introduced to the Black Sea, possibly in ballast water, and the *M. leidyi* abundance has since declined. More recently, *M. leidyi* has entered the Caspian Sea and is repeating the same pattern.

3. The European green crab *C. maenas* has been introduced to South Africa, Western Australia, and Tasmania as well as to the Pacific and Atlantic coasts of North America. In Tasmania, it avidly feeds on shellfish in culture and on wild mollusks.

Here the introduction of a rhizocephalan *Sacculina carcini*, commonly found within the crab's home range and which reduces reproductive output in populations, was considered. However, because there was evidence that this parasite was not host-specific and may infect other crab species, the project did not proceed.

4. The North Pacific sea star *A. amurensis* has become abundant in eastern Tasmania and may have been introduced there as a result of either ships' fouling or ballast water releases taken from Japan. This species feeds on a wide range of benthic organisms, including cultured shellfish. A Japanese ciliate *Orchitophyra* sp. that castrates its host was considered.

Exotic species used in biocontrol need to be species-specific. Generalist predators, parasites, and diseases should be avoided. Complete eradication of a pest species may not be possible or cost-effective except under very special circumstances. In order to evaluate the potential input of the control organism, a good knowledge of the biological system is needed in order to avoid predictable effects that may result in a cascade of changes through the trophic system. Native equivalent species of the biological control organism should be sought for first when introducing a biocontrol species. The ICES Code of Practice should be considered and an external panel of consultants should be involved in all discussions. In some cases, control may be possible by developing a fishery for the pest species as has happened in Turkey following the introduction of the rapa whelk *R. venosa* to the Black Sea and in Norway to control the population of the red king crab *P. camtschaticus*.

Management

Shipping is seen as the most widespread means of disseminating species worldwide and the IMO has held conventions that relate to improved management of ships' ballast water and sediments and hull-fouling management. Aquaculture and the trade in living products also has been responsible for many species transmissions. To this end, ICES has developed a Code of Practice, providing wise advice on deliberate movements of marine species that should be consulted. Species may be released in ignorance to the wild with potential consequences for native communities, such as what may have happened with the introduction of lionfish *Pterois volitans* to the western North Atlantic. Elimination of an impacting species can rarely be achieved unless found soon after arrival. Regular surveys as well as public participation and awareness may reveal an early arrival. However, this involves up-to-date information. The development of rapid assessment surveys for known target species may greatly reduce the duration of surveys and provide early information. Such approaches work. The byssate bivalve *Mytilopsis sallei* was found during a survey in Darwin docks, Australia, and the Mediterranean form of the marine alga *C. taxifolia* was found in a small bay in California, USA; it was first recognized by a member of the public as the result of an awareness campaign. Such management requires up-to-date dissemination of information to convey current affairs to all stakeholders. Rapid dissemination by free-access online journals, such as *Aquatic Invaders*, as well as regular scientific meetings, for example, the International Aquatic Invasive Species Conferences, greatly aid in distributing current knowledge. Management of impacting species may take place in different ways according to their mode of life; it will also depend on the vectors involved in their spread. Where natural vectors disperse a species rapidly, controls may be futile. Programs such as the European Union's specialist projects Assessing Large Scale Risks for Biodiversity with Tested Methods (ALARM) and Delivery Alien Invasive Species Inventories for Europe (DAISIE) to map and manage invading species are likely to lead to new management advances in understanding vector processes and through the development of risk assessments. It is certain that further exotic species will spread and that some will have unpredictable consequences.

Conclusions

Exotic species are an important component of human economic affairs and when used in culture or for sport fisheries, etc., require careful management at the time of introduction. They should pass through a quarantine procedure to reduce transmission of any pests, parasites, and diseases. Aquaculture activities, where practicable, should be sited away from port regions, because there is a risk that shipping may introduce unwanted organisms that may compromise aquaculture production.

Movements of aquarium species also need careful attention and should be sold together with advice not to release these to the wild. Transported fish are normally stressed and many of these have been shown to carry pathogenic bacteria and parasites. There is a risk that serious diseases of fishes could be transmitted outside of the Indo-Pacific region by the aquarium trade.

Trading networks need to consider ways in which organisms transferred alive or organisms and disease agents that may be transferred in or with products, are not released to the wild in areas that lie beyond their normal range.

More effective and less toxic antifouling agents are needed to replace the effective but highly toxic organotin paint applications on ships. This may lead to less toxic port regions which in turn may now become more suitable for invasive species to become established. Despite widespread usage of new antifouling agents on ships' hulls, unpainted or worn or damaged regions of the hull are areas where fouling organisms will continue to colonize.

Ships' ballast water and its sediments pose a serious threat of transmitting harmful organisms. Future designs of ballast tanks that facilitate complete exchanges, with reduced sediment loading and options for sterilization, could greatly reduce the volume of distributed biota.

Exotic species are becoming established at an apparently increasing rate. Some of these will have serious implications for human health, industry, and the environment.

Further Reading

Cohen AN and Carlton JT (1995) *Nonindigenous Aquatic Species in a United States Estuary: A Case Study of the Biological Invasions of the San Francisco Bay and Delta*, 246pp. Washington, DC: United States Fish and Wildlife Service.

Davenport J and Davenport JL (2006) *Environmental Pollution, Vol. 10: The Ecology of Transportation: Managing Mobility for the Environment*, 392pp. Dordrecht: Springer (ISBN 1-4020-4503-4).

Drake LA, Choi KH, Ruiz GM, and Dobbs FC (2001) Global redistribution of bacterioplankton and virioplankton communities. *Biological Invasions* 3: 193–199.

Hewitt CL (2002) The distribution and diversity of tropical Australian marine bio-invasions. *Pacific Science* 56(2): 213–222.

ICES (2005) Vector pathways and the spread of exotic species in the sea. *ICES Co-Operative Report No. 271*, 25pp. Copenhagen: ICES.

Leppäkoski E, Gollasch S, and Olenin S (2002) *Invasive Aquatic Species of Europe: Distribution, Impacts and Management*, 583pp. Dordrecht: Kluwer Academic Publishers (ISBN 1-4020-0837-6).

Minchin D and Gollasch S (2003) Fouling and ships' hulls: How changing circumstances and spawning events may result in the spread of exotic species. *Biofouling* 19(supplement): 111–122.

Padilla DK and Williams SL (2004) Beyond ballast water: Aquarium and ornamental trades as sources of invasive species in aquatic systems. *Frontiers in Ecology and Environment* 2(3): 131–138.

Ruiz GM, Carlton JT, Grozholtz ED, and Hunes AH (1997) Global invasions of marine and estuarine habitats by non-indigenous species: Mechanisms, extent and consequences. *American Zoologist* 37: 621–632.

Williamson AT, Bax NJ, Gonzalez E, and Geeves W (eds.) (2002) Development of a regional risk management framework for APEC economies for use in the control and prevention of introduced pests. *APEC MRC-WG Final Report: Control and Prevention of Introduced Marine Pests*. Singapore: Asia-Pacific Economic Cooperation Secretariat.

Relevant Websites

http://www.alarmproject.net
    – ALARM (Assessing Large Scale Risks for Biodiversity with Tested Methods).
http://www.aquaticinvasions.ru
    – *Aquatic Invasions* (online journal).
http://www.daisie.se
    – Delivering Alien Invasive Species Inventories for Europe (DAISIE).
http://www.ices.dk
    – International Council for the Exploration of the Sea.
http://www.imo.org
    – International Maritime Organization.

GRABS FOR SHELF BENTHIC SAMPLING

P. F. Kingston, Heriot-Watt University, Edinburgh, UK

Introduction

The sedimentary environment is theoretically one of the easiest to sample quantitatively and one of the most convenient ways to secure such samples is by means of grabs. Grab samplers are used for both faunal samples, when the grab contents are retained in their entirety and then sieved to remove the biota from the sediment, and for chemical/physical samples when a subsample is usually taken from the surface of the sediment obtained. In both cases, the sampling program is reliant on the grab sampler taking consistent and relatively undisturbed sediment samples.

Conventional Grab Samplers

The forerunner of the grab samplers used today is the Petersen grab, designed by C.G.J. Petersen to conduct benthic faunal investigations in Danish fiords in the early part of the twentieth century. It consisted of two quadrant buckets that were held in an open position and lowered to the seabed (**Figure 1**). On the bottom, the relaxing of the tension on the lowering warp released the buckets and subsequent hauling caused them to close before they left the bottom. The instrument is still used today but is seriously limited in its range of usefulness, working efficiently only in very soft mud.

Petersen's grab formed the basis for the design of many that came after. One enduring example is the van Veen grab, a sampler that is in common use today (**Figure 2**). The main improvement over Petersen's design is the provision of long arms attached to the buckets to provide additional leverage to the closing action. The arms also provided a means by which the complex closing mechanism of the Petersen grab could be simplified with the hauling warp being attached to chains on the ends of the arms. The mechanical advantage of the long arms can be improved further by using an endless warp rig; this has the added advantage of helping to prevent the grab being jerked off the bottom if the ship rolls as the grab is closing. The van Veen grab was designed in 1933 and is still widely used in benthic infaunal studies owing to its simple design, robustness, and digging efficiency. The van Veen grab

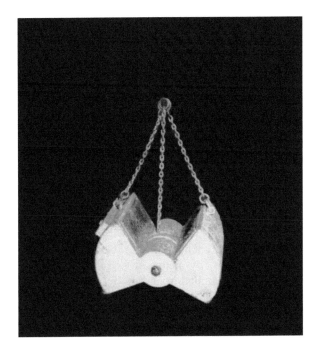

Figure 1 Petersen grab.

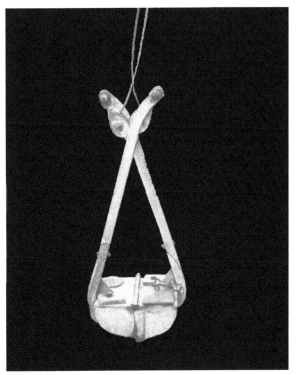

Figure 2 Van Veen grab.

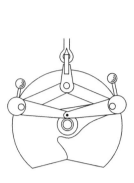
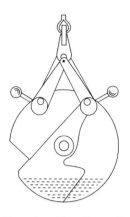

Figure 3 Diagram of a Hunter grab. Reproduced from Hunter B and Simpson AE (1976) A benthic grab designed for easy operation and durability. *Journal of the Marine Biological Association* 56: 951–957.

Figure 4 Smith–McIntyre grab.

typically covers a surface area of $0.1\,m^2$, although instruments of twice this size are sometimes used.

A more recent design of frameless grab is the Hunter grab (**Figure 3**). This is of a more compact design than the van Veen. The jaws are closed by levers attached to the buckets in a parallelogram configuration, giving the mechanism a good overall mechanical advantage. The closing action requires no chains or pulleys and the instrument can be operated by one person. Its disadvantage is that the bucket design does not encourage good initial penetration of the sediment, which is important in hard-packed sediments.

A disadvantage of the grab samplers discussed so far is that there is little latitude for horizontal movement of the ship while the sample is being secured: the smallest amount of drift and the sampler is likely to be pulled over. The Smith–McIntyre grab was designed to reduce this problem by mounting the grab buckets in a stabilizing frame (**Figure 4**). Initial penetration of the leading edge of the buckets is assisted by the use of powerful springs and the buckets are closed by cables pulling on attached short arms in a similar way to that on the van Veen grab. The driving springs are released by two trigger plates, one on either side of the supporting frame to ensure that the sampler is resting flat on the seabed before the sample is taken. In firm sand the Smith–McIntyre grab penetrates to about the same depth of sediment as the van Veen grab. Its main disadvantage is the need to cock the spring mechanism on deck before deploying the sampler, a process that can be quite hazardous in rough weather.

The Day grab is a simplified form of the Smith–McIntyre instrument in which the trigger and closing mechanism remains the same, but without spring assistance for initial penetration of the buckets

Figure 5 Day grab.

(**Figure 5**). The Day grab is widely used, particularly for monitoring work, despite its poor performance in hard-packed sandy sediments.

Most of the grabs thus far discussed have been designed to take samples with a surface area of 0.1 or $0.2\,m^2$. The Baird grab, however, takes samples of $0.5\,m^2$ by means of two inclined digging plates that are pulled together by tension on the warp (**Figure 6**). The grab is useful where a relatively large surface area needs to be covered, but has the disadvantage of taking a shallow bite and having the surface of the sample exposed while it is being hauled in.

Warp Activation

All the grabs described above use the warp acting against the weight of the sampler to close the jaws. However, direct contact with the vessel on the surface during the closure of the grab mechanism poses several problems.

Warp Heave

As tension is taken up by the warp to close the jaws, there is a tendency for the grab to be pulled up off the bottom, resulting in a shallower bite than might be expected from the geometry of the sampler. This tendency is related to the total weight of the sampler and the speed of hauling and is exacerbated by firm sediments. For example, the theoretical maximum depth of bite of a 120-kg Day grab is 13 cm (based

Figure 6   Baird grab.

on direct measurements of the sampler); however, in medium sand, the digging performance is reduced to a maximum depth of only 8 cm (**Figure 7(e)**). The influence of warp action on the digging efficiency of a grab sampler can also depend on the way in which the sampler is rigged. This is particularly true of the van Veen grab. **Figure 7(b)** shows the bite profile of the chain-rigged sampler in which the end of each arm is directly connected to the warp by a chain. The vertical sides of the profile represent the initial penetration of the grab while the central rise shows the upward movement of the grab as the jaws close. **Figure 7(c)** shows the bite profile of a van Veen of similar size and weight (30 kg) rigged with an endless warp in which the arms are closed by a loop of wire passing through a block at the end of each arm (as in **Figure 2**). The vertical profile of the initial penetration is again apparent; however, in this case, the overall depth of the sampler in the sediment is maintained as the jaws close. The endless warp rig increases the mechanical advantage of the pull of the warp while decreasing the speed at which the jaws are closed. The result is that the sampler is 'insulated' from surface conditions to a greater extent than when chain-rigged, giving a better digging efficiency.

Grab 'Bounce'

In calm sea conditions it is relatively easy to control the rate of warp heave and obtain at least some consistency in the volume of sediment secured. However, such conditions are seldom experienced in the open sea where it is more usual to encounter wave action. Few ships used in offshore benthic

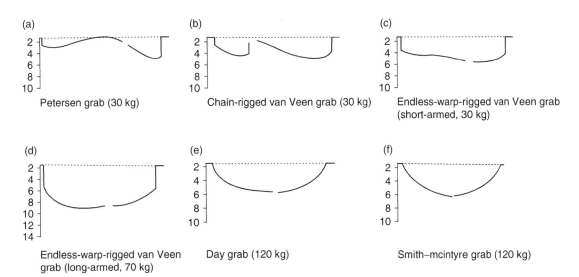

Figure 7   Digging profiles of a range of commonly used benthic grab samplers obtained in a test tank using a fine sand substratum. (a) Peterson grab (30 kg); (b) chain-rigged van Veen grab (30 kg); (c) endless-warp-rigged van Veen grab (short-armed, 30 kg); (d) endless-warp-rigged van Veen grab (long-armed, 70 kg); (e) Day grab (120 kg); (f) Smith–McIntyre grab (120 kg).

studies are fitted with winches with heave compensators so that the effect of ship's roll is to introduce an erratic motion to the warp. This may result in the grab 'bouncing' off the bottom where the ship rises just as bottom contact is made, or in the grab being snatched off the bottom where the ship rises just as hauling commences. In the former instance, it is unlikely that any sediment is secured; in the latter, the amount of material and its integrity as a sample will vary considerably, depending on the exact circumstances of its retrieval.

The intensity of this effect will depend on the severity of the weather conditions. **Figure 8** shows the relationship between wind speed and grab failure rate, which is over 60% of hauls at wind force 8. What is of more concern to the scientist attempting to obtain quantitative samples is the dramatic increase in variability with increase in wind speed with a coefficient of variation between 20 and 30 at force 7. The high cost of ship-time places considerable pressure on operators to work in as severe weather conditions as possible and it is not unusual for sampling to continue in wind force 7 conditions with all its disadvantages.

Drift

For a warp-activated grab sampler to operate efficiently it should be hauled with the warp positioned vertically above. Where there is a strong wind or current, these conditions may be difficult to achieve. The result is that the grab samplers are pulled on to their sides. This is a particular problem with samplers, such as the van Veen grab, that do not have stabilizing frames. Diver observations have shown, however, that at least in shallow water, where the drift effect is at its greatest on the bottom, even the framed heavily weighted Day and Smith–McIntyre grabs can be toppled.

Initial Penetration

It is clear that the weight of the sampler is an important element in determining the volume of the sample secured. Much of the improved digging efficiency of the van Veen grab shown in **Figure 7(d)** can be attributed to the addition of an extra 40 kg of weight which increased the initial penetration of the sampler on contact with the sediment surface.

Initial penetration is one of the most important factors in the sequence of events in grab operation, determining the final volume of sediment secured. **Figure 9** shows the relationship between initial penetration and final sample volume obtained for a van Veen grab. Over 70% of the final volume is determined by the initial penetration. Subsequent digging of the sampler is hampered, as already shown, by the pull of the warp.

For most benthic faunal studies it is important for the sampler to penetrate at least 5 cm into the sediment (for a 0.1 m² surface area sample this gives 5 l of sample). In terms of number of species and individuals, over 90% of benthic macrofauna are found in the top 4–5 cm of sediment. **Figure 10** shows how the number of individuals relates to average sample

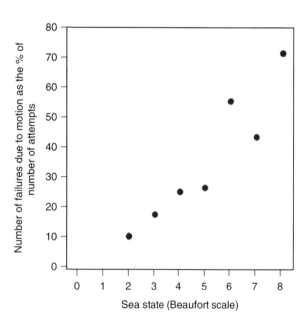

Figure 8 Relationship between wind speed and grab failure rate.

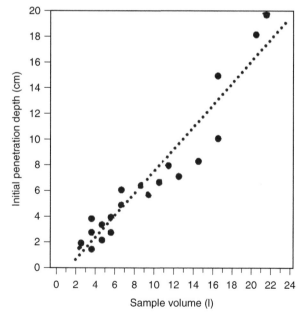

Figure 9 Relationship between initial penetration of a van Veen grab sampler and volume of sediment secured.

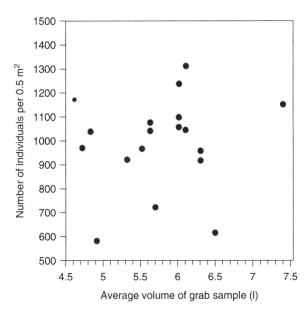

Figure 10 Relationship between number of benthic fauna individuals captured and sample volume for a boreal offshore sand substratum.

Figure 11 Shipek grab.

volume for 18 stations in the southern North Sea (each liter recorded represents 1 cm of penetration). Although there is a considerable variation in the numbers of individuals between stations, there is no significant trend linking increased abundance with increased sample volume penetration. No sample volumes of less than 4.5 l were taken, indicating that at that level of penetration most of the fauna were being captured.

Samplers in which the jaws are held rigidly in a frame have no initial penetration if the edges of the jaw buckets, when held in the open position, are on a level with the base of the frame. The lack of any initial penetration in such instruments has the added disadvantage in benthic fauna work of under-sampling at the edges of the bite profile (see **Figures 7(e)** and **7(f)**), although the addition of weight will usually increase the sample volume obtained.

Pressure Wave Effect

The descent of the grab necessarily creates a bow wave. Under field conditions, it is usually im-practicable to lower the grab at a rate that will eliminate a preceding bow wave, even if the sea were flat calm. There have been several investigations of the effects of 'downwash' both theoretical, using artificially placed surface objects, and *in situ*. The effects of downwash can be reduced by replacing the upper surface of the buckets with an open mesh. Although there is still a considerable effect on the surface flock layer (rendering the samples of dubious

value for chemical contamination studies), the effect on the numbers of benthic fauna is generally very small.

Self-Activated Bottom Samplers

There can be little doubt that one of the most im-portant factors responsible for sampler failure or sample variability in heavy seas is the reliance of most presently used instruments on warp-activated closure. The most immediate and obvious answer to this problem is to make the closing action in-dependent of the warp by incorporating a self-powering mechanism.

Spring-Powered Samplers

One solution to the problem is to use a spring to actuate the sampler buckets. Such instruments are in existence, possibly the most widely used being the Shipek grab, a small sampler ($0.04 \, \text{m}^2$) consisting of a spring-loaded scoop (**Figure 11**). This instrument is widely used where small superficial sediment samples are required for physical or chemical analysis. The use of a pretensioned spring unfortunately sets practical limits on the size of the sampler, since to cock a spring in order to operate a sampler capable of taking a $0.1 \, \text{m}^2$ sample would require a force that would be impracticable to apply routinely on deck. In addition, in rough weather conditions, a loaded sampler of this size would be very hazardous to deploy.

Compressed-Air-Powered Samplers

Another approach has been to use compressed air power. In the 1960s, Flury fitted a compressed air ram to a modified Petersen grab with success. However, the restricted depth range of the instru-ment and the inconvenience of having to recharge the air reservoir for each haul limited its potential for routine offshore work.

Hydraulically Powered Samplers

Hydraulically powered grabs are commonly used for large-scale sediment shifting operations such as sea-bed dredging. The Bedford Institute of Oceanography, Nova Scotia, successfully scaled down this technology to that of a practical benthic sampler. Their instrument is relatively large, standing 2.5 m high and weighing some 1136 kg. It covers a surface area of $0.5 \, m^2$ and samples to a maximum sediment depth of 25 cm. At full penetration, the sediment volume taken is about 100 l. The buckets are driven closed by hydraulic rams powered from the surface. The grab is also fitted with an underwater television camera which allows the operator to visually select the precise sampling area on the seabed, close and open the bucket remotely, and verify that the bucket closed properly prior to recovery. The top of the buckets remain open during descent to minimize the effect of downwash and close on retrieval to reduce washout of the sample on ascent. The current operating depth of the instrument is 500 m. The instrument has been successfully used on several major offshore studies, but does require the use of a substantial vessel for its deployment.

Hydrostatically Powered Samplers

Hydrostatically powered samplers use the potential energy of the difference in hydrostatic pressure at the sea surface and the seabed. The idea of using this power source is not new. In the early part of the twentieth century, a 'hydraulic engine' was in use by marine geologists that harnessed hydrostatic pressure to drive a rock drill. Hydrostatic power has also been used to drive corers largely for geological studies. However, these instruments were principally concerned with deep sediment corers and were not designed to collect macrofauna or material at the sediment–water interface.

A more recent development has been that of a grab built by Heriot-Watt University, Edinburgh. The sampler uses water pressure difference to operate a hydraulic ram that is activated when the grab reaches the seabed. **Figure 12** shows the general layout of the instrument. Water enters the upper chamber of the cylinder when the sampler is on the seabed, forcing down a piston that is connected to a system of levers that close the jaws. The actuating valve is held shut by the weight of the sampler and there is a delay mechanism to prevent premature closure of the jaws resulting from 'bounce'. Back on the ship, the jaws are held shut by an overcenter locking mechanism

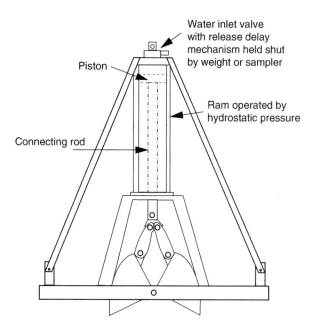

Figure 12 Diagram of a grab sampler using hydrostatic pressure to close the jaws.

and, on release, are drawn open by reversal of the piston motion from air pressure built up on the underside of the piston during its initial power stroke. Since the powering of the grab jaws is independent of the warp, the sampler may be used successfully in a much wider range of surface weather conditions than conventional grabs.

Alternatives to Grab Samplers

Ideally a benthic sediment sample for faunal studies should be straight-sided to the maximum depth of its excavation and should retain the original stratification of the sediment. Grab samplers by the very nature of their action will never achieve this end.

Suction Samplers

One answer to this problem is to employ some sort of corer designed to take samples of sufficient surface area to satisfy the present approaches to benthic studies. The Knudsen sampler is such a device and is theoretically capable of taking the perfect benthic sample. It uses a suction technique to drive a core tube of $0.1 \, m^2$ cross-sectional area 30 cm into the sediment. Water is pumped out of the core tube on the seabed by a pump that is powered by unwinding a cable from a drum. The sample is retrieved by pulling the core out sideways using a wishbone arrangement and returning it to the surface bottom-side up (**Figures 13** and **14**). Under ideal conditions, the device will take a straight-sided sample to a depth

Figure 13   Knudsen sampler in descent position.

of 30 cm. However, conditions have to be flat calm in order to allow time for the pump to operate on the seabed and evacuate the water from the core. This limits the use of the Knudsen sampler and it is generally not suitable for sampling in unsheltered conditions offshore. Mounting the sampler in a stabilizing frame can improve its success rate and it is used regularly for inshore monitoring work where it is necessary to capture deep burrowing species.

Spade Box Samplers

Another approach to the problem is to drive an open-ended box into the sediment, using the weight of the sampler, and arrange for a shutter to close off the bottom end. The most widespread design of such an instrument is that of the spade box sampler, first described by Reineck in the 1950s and later subjected to various modifications. The sampler consists of a removable steel box open at both ends and driven into the sediment by its own weight. The lower end of the box is closed by a shutter supported on an arm pivoted in such a way as to cause it to slide through the sediment and across the mouth of the box (**Figure 15**). As with the grab samplers previously described, the shutter is driven by the act of hauling on the warp with all the attendant disadvantages. Nevertheless, box corers are very successful and are used widely for obtaining relatively undisturbed samples of up to $0.25 \, \text{m}^2$ surface area (**Figure 16**). One big advantage of the box sampler is that the box can usually be removed with the sample and its overlying water left intact, allowing detailed studies of the sediment surface. Furthermore, it is possible to subsample using small-diameter corers for studies of chemical and physical characteristics. Despite their potential of securing the 'ideal' sediment sample, box corers are rarely used for routine benthic monitoring work. This is largely because of their size (a box corer capable of taking a $0.1 \, \text{m}^2$

Figure 14   Knudsen sampler in ascent position.

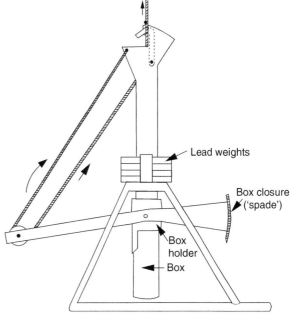

Figure 15   Diagram of a Reineck spade box sampler.

Figure 16   A 0.25 m² spade box sampler.

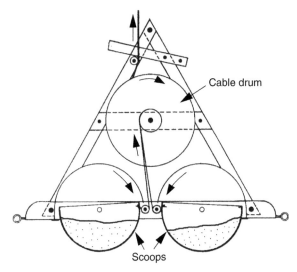

Figure 17   Diagram of a Holme scoop. Reproduced from Holme NA (1953) The biomass of the bottom fauna in the English Channel off Plymouth. *Journal of the Marine Biological Association* 32: 1–49.

macrobenthos sampling. At present, there is no instrument that fulfils the requirements for a quick turnaround precision multiple corer for offshore sampling.

Sampling Difficult Sediments

Most of the samplers so far discussed operate reasonably well in mud or sand substrata. Few operate satisfactorily in gravel or stony mixed ground either because the bottom is too hard for the sampler to penetrate the substratum or because of the increased likelihood of a stone holding the jaws open when they are drawn together. To get around this problem various types of scoops have been devised. The Holme grab has a double scoop action with two buckets rotating in opposite directions to minimize any lateral movement during digging. The scoops are closed by means of a cable and pulley arrangement (**Figure 17**) and simultaneously take two samples of 0.05 m² surface area.

The Hamon grab, which has proved to be very effective in coarse, loose sediments, takes a single rectangular scoop of the substratum covering a surface area of about 0.29 m². The scoop is forced into the sediment by a long lever driven by pulleys that are powered by the pull of the warp (**Figure 18**). Although the samples may not always be as consistent as those from a more conventional grab sampler, the Hamon grab has found widespread use where regular sampling on rough ground is impossible by any other means.

sample weighs over 750 kg and stands 2 m high) and the difficulty in deployment and recovery in heavy seas.

Precision Corers

For chemical monitoring, it is important that the sediment–water interface is maintained intact, for it is the surface flock layer that will contain the most recently deposited material. Unfortunately, such undisturbed samples are rarely obtained using grab samplers or box corers. Precision corers are capable of securing undisturbed surface sediment cores; however, they are unsuitable for routine offshore work because of the time taken to secure a sample on the seabed and dependence on warp-activated closure. Additionally, the cross-sectional area of the core (0.002–0.004 m²) would necessitate the taking of large numbers of replicate samples in order to capture sufficient numbers of benthic macrofauna to be useful. This would be impracticable given the time taken to take a single sample. Large multiple precision corers have been constructed; these are usually too large and difficult to deploy for routine

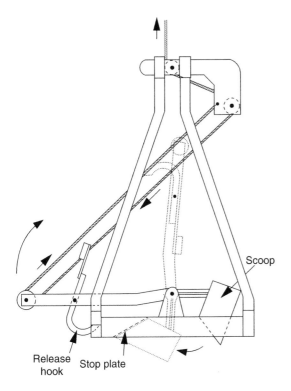

Figure 18 Diagram of a Hamon scoop.

Present State of Technology

It is perhaps surprising that given the high state of technology of survey vessels, position fixing, and analytical equipment, the most commonly used samplers are relatively primitive (being designed some 40 or more years ago). Yet the quality of the sample is of fundamental importance to any research or monitoring work. Currently the most popular instruments are grab samplers, probably because of their wide operational weather window and apparent reliability. Samplers such as the Day, van Veen, and Smith–McIntyre grab samplers are still routinely used for sampling sediment for chemical and biological analysis despite their well-documented shortcomings. As discussed earlier, the most important of these are the substantial downwash that precedes the sampler as it descends and the disturbance of the trapped sediment layers by the closing action of the jaws. Both chemical and meiofaunal studies are particularly vulnerable to these.

Although box corers go some way to reducing disturbance of the sediment strata, the all-important surface flocculent layer is invariably washed away. A big disadvantage of the box corer for routine offshore work is that it is sensitive to weather conditions; in addition, the larger instruments do not perform well on sand substrata.

A generic disadvantage of most samplers presently in use is that they rely on slackening of the warp to trigger the action and the heave of the warp to drive the closing mechanism. In calm conditions, this presents no great problem, but with increasing sea state the vertical movement of the warp decreases reliability dramatically until the variability and finally failure rate of hauls make further sampling effort fruitless.

Specific Problems, Requirements, and Future Developments

Chemical Studies

Studies involving sediment chemistry require precision sampling if undisturbed samples at the sediment/water interface are to be obtained. This can be critical, particularly when the results of recent sedimentation are of interest. The impracticality of using existing hydraulically damped corers such as the Craib corer for offshore work has led to the widespread use of less 'weather sensitive' devices such as spade box corers and grabs for routine monitoring purposes. However, studies carried out have shown that these samplers produce a considerable 'downwash effect', blowing the surface flocculent layer away before the sample is secured. This can have serious consequences if any meaningful estimation of the surface chemistry of the sediment is desired. Repetitive and accurate sampling is also a prerequisite for determining spatial and temporal change in sediment chemistry.

Meiofauna Studies

Meiofauna has increasingly been shown to have potential as an important tool in benthic monitoring. One of the major factors limiting its wider adoption is the lack of a suitable sampler. Although instruments such as the Craib corer and its multicorer derivatives are capable of sampling the critically important superficial sediment layer, these designs provide a poor level of success on harder sediments and in anything but near-perfect weather conditions. They also have a slow turnaround time. Box corers are widely used as an alternative; however, they are known to be unreliable in their sampling of meiofauna. As with the macrobenthos, meiobenthic patchiness results in low levels of precision of abundance estimates unless large numbers of samples are taken.

Macrobenthic Studies

The measurement and prediction of spatial and temporal variation in natural populations are of great importance to population biologists, both for fundamental research into population dynamics and productivity and in the characterization of benthic communities for determining change induced by environmental impact. Though always an important consideration, cost-effectiveness of sampling and sample processing is not so crucial in fundamental research, since time and funding may be tailored to fit objectives. This is rarely the case in routine monitoring work where often resolution and timescales have to fit the resources available.

Benthic fauna are contagiously distributed and to sample such communities with a precision that will enable distinction between temporal variation and incipient change resulting from pollution effects, it is generally accepted that five replicate 0.1 m^2 hauls from each station are necessary (giving a precision at which the standard error is no more than 20% of the mean). This frequency of sampling requires approximately 10–15 man-days of sediment faunal analysis per sample station. While this may be acceptable in community structure studies in which time and manpower (and thus cost) are not a primary consideration, this high cost of analyzing samples is of importance in routine monitoring studies and has led monitoring agencies and offshore operators to reduce sampling frequency on cost grounds to as few as two replicates per station. This reduces the precision with which faunal abundance can be estimated to a level at which only gross change can be demonstrated. However, sampling to an acceptable precision may be achieved from an area equivalent to 0.1 m^2 if a smaller sampling unit is used. For example, 50 5-cm core samples (with a similar total surface area) have been shown to give a similar precision to that of 5 to 12 0.1-m^2 grab samples. Thus a similar degree of precision may be obtained for around one-fifth to one-twelfth the analytical costs using a conventional approach.

The problem is to be able to secure the 50 core samples per site that would be needed for the macrofaunal monitoring in a timescale that would be realistic offshore. At present, there is no instrument capable of supporting such a sampling demand and operating in the range of sediment types that wide-scale monitoring studies demand. Clearly a single core sampler would be impracticable, and one must look to the future development of a multiple corer that is capable of flexibility in its operation, which allows a quick on-deck turnaround between hauls.

See also

Benthic Organisms Overview.

Further Reading

Ankar S (1977) Digging profile and penetration of the van Veen grab in different sediment types. *Contributions from the Askö Laboratory, University of Stockholm, Sweden* 16: 1–12.

Beukema JJ (1974) The efficiency of the van Veen grab compared with the Reineck box sampler. *Journal du Conseil Permanent International pour l'Exploration de la Mer* 35: 319–327.

Eleftheriou A and McIntyre AD (eds.) (2005) *Methods for the Study of the Marine Benthos.* Oxford, UK: Blackwell.

Flury JA (1967) Modified Petersen grab. *Journal of the Fisheries Research Board of Canada* 20: 1549–1550.

Holme NA (1953) The biomass of the bottom fauna in the English Channel off Plymouth. *Journal of the Marine Biological Association* 32: 1–49.

Hunter B and Simpson AE (1976) A benthic grab designed for easy operation and durability. *Journal of the Marine Biological Association* 56: 951–957.

Riddle MJ (1988) Bite profiles of some benthic grab samplers. *Estuarine, Coastal and Shelf Science* 29(3): 285–292.

Thorsen G (1957) Sampling the benthos. In: Hedgepeth JW (ed.) *Treatise on Marine Ecology and Paleoecology, Vol. 1: Ecology.* Washington, DC: The Geological Society of America.

POLLUTION

THERMAL DISCHARGES AND POLLUTION

T. E. L. Langford, University of Southampton,
Southampton, UK

Sources of Thermal Discharges

The largest single source of heat to most water
bodies, including the sea, is the sun. Natural thermal
springs also occur in many parts of the world, almost
all as fresh water, some of which discharge to the sea.
In the deep oceans hydrothermal vents discharge
mineral-rich hot water at temperatures greatly exceeding any natural temperatures either at depth or
at the surface. To add to these natural sources of
heat, industrial processes have discharged heated
effluents into coastal waters in many parts of the
world for at least 150 years. By far the largest volumes of these heated effluents reaching the sea in the
past 60 years have originated from the electricity
generation industry (power industry). Indeed more
than 80% of the volume of heated effluents to the sea
originate from the power industry compared with 3–
5% from the petroleum industries and up to 7% (in
the USA) from chemical and steel industries.

The process known as 'thermal' power generation,
in which a fuel such as oil or coal or the process of
nuclear fission is used to heat water to steam to drive
turbines, requires large volumes of cooling water to
remove the waste heat produced in the process.
Where power stations are sited on or near the coast
all of this waste heat, representing some 60–65% of
that used in the process, is discharged to the sea. The
heat is then dissipated through dilution, conduction,
or convection. In a few, atypical coastal situations,
where the receiving water does not have the capacity
to dissipate the heat, artificial means of cooling the
effluent such as ponds or cooling-towers are used.
Here the effluent is cooled prior to discharge and
much of the heat dissipated to the air.

The waste heat is related to the theoretical thermal
efficiency of the Rankine cycle, which is the modification of the Carnot thermodynamic cycle on which
the process is based. This has a maximum theoretical
efficiency of about 60% but because of environmental temperatures and material properties the
practical efficiency is around 40%. Given this level of
efficiency and the normal operating conditions of a
modern coal- or oil-fired power station, namely
steam at 550°C and a pressure of 10.3×10^6 kg cm^{-2}

with corresponding heat rates of 2200 kg cal kWh^{-1}
of electricity, some 1400 kg cal kWh^{-1} of heat is
discharged to the environment, usually in cooling
water at coastal sites. This assumes a natural water
temperature of 10°C. Nuclear power stations usually
reject about 50% more heat per unit of electricity
generated because they operate at lower temperatures and pressures. Since the 1920s efficiencies have
increased from about 20% to 38–40% today with a
corresponding reduction by up to 50% of the rate of
heat loss. The massive expansion of the industry
since the 1920s has, however, increased the total
amounts of heat discharged to the sea.

Thus for each conventional modern power station
of 2000 MW capacity some 63 m^3 s^{-1} of cooling
water is required to remove the heat. Modern developments such as the combined cycle gas turbine
(CCGT) power stations with increased thermal efficiencies have reduced water requirements and heat
loss further so that a 500 MW power station may
require about 9–10 m^3 s^{-1} to remove the heat, a reduction of over 30% on conventional thermal
stations.

Water Temperatures

Natural sea surface temperatures vary widely both
spatially and temporally throughout the world with
overall ranges recorded from -2°C to 30°C in open
oceans and -2°C to 43°C in coastal waters. Diurnal
fluctuations at the sea surface are rarely more than
1°C though records of up to 1.9°C have been made in
shallow seas. The highest temperatures have been
recorded in sheltered tropical embayments where
there is little exchange with open waters. Most
thermal effluents are discharged into coastal waters
and these are therefore most exposed to both physical and biological effects. In deeper waters vertical
thermal stratification can often exceed 10°C and a
natural maximum difference of over 23°C between
surface and bottom has been recorded in some tropical waters.

The temperatures of thermal discharges from
power stations are typically 8–12°C higher than the
natural ambient water temperature though at some
sites, particularly nuclear power stations, temperature rises can exceed 15°C (**Figure 1**). Maximum
discharge temperatures in some tropical coastal
waters have reached 42°C though 35–38°C is more
typical. There are seasonal and diurnal fluctuations
at many sites related to the natural seasonal

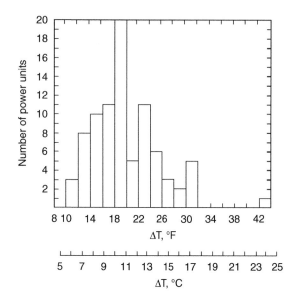

Figure 1 Frequency distribution of designed temperature rises for thermal discharges in US power station cooling water systems. (Reproduced with permission from Langford, 1990.)

temperature cycles and to the operating cycles of the power station.

Thermal Plumes and Mixing Zones

Once discharged into the sea a typical thermal effluent will spread and form a three-dimensional layer with the temperature decreasing with distance from the outfall. The behavior and size of the plume will depend on the design and siting of the outfall, the tidal currents, the degree of exposure and the volume and temperature of the effluent itself. Very few effluents are discharged more than 1 km from the shore. Effects on the shore are, however, maximized by shoreline discharge (**Figure 2**).

The concept of the mixing zone, usually in three components, near field, mid-field and far-field, related to the distance from the outfall, has mainly been used in determining legislative limits on the effluents. Most ecological studies have dealt with near and mid-field effects. The boundary of the mixing zone is, for most ecological limits, set where the water is at 0.5°C above natural ambients though this tends to be an arbitrary limit and not based on ecological data. Mixing zones for coastal discharges can be highly variable in both temperature and area of effect (**Figure 3**).

In addition to heat effects, thermal discharges also contain chemicals, mainly those used for the control of marine fouling in pipework and culverts. Chlorine compounds are the most common and because of its strong biocidal properties, the effects of chlorine are

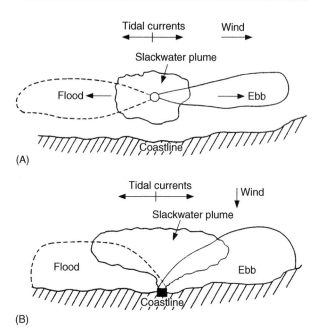

Figure 2 Movements of thermal plumes in tidal waters. (A) Offshore outfall; (B) onshore outfall. (Reproduced with permission from Langford, 1990.)

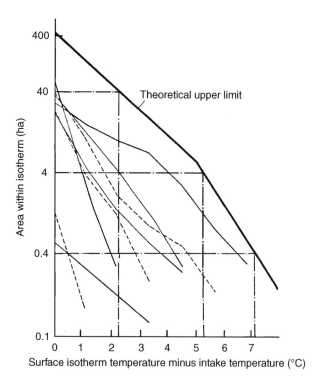

Figure 3 Relationship between surface-temperature elevation and surface area affected for nine different surveys at Moro Bay Power Plant, California. (Reproduced with permission from Langford, 1990.)

of primary concern in many coastal discharges, irrespective of the temperature. Measured chlorine residuals immediately after application vary between

0.5 and 10 mg l^{-1} throughout the world but most are within the 0.5–1.0 mg l^{-1} range. Because of the complex chemical reactions in sea water free chlorine residuals in discharges are usually a factor of 10 lower than the initial dosing rate. Even so, concentrations of chlorine compounds can exceed lethal limits for organisms at some sites.

Biological Effects of Temperature

The biological effects of temperature on marine and coastal organisms have been reviewed by a number of authors. Most animals and plants can survive ranges of temperature which are essentially genetically determined. The ultimate lethal temperature varies within poikilothermic groups but there is a trend of tolerance which is inversely related to the structural complexity of the organism (Table 1). Thus groups of microorganisms tend to contain species which have much higher tolerances than invertebrates, fish, or vascular plants. Because the life processes and survival of many organisms is so dependent on water temperature many species have developed physiological or behavioral strategies for optimizing temperature exposure and for survival in extremes of heat or cold. Examples are to be found among intertidal species and in polar fish.

The effects of temperature on organisms can be classified mainly from experimental data as lethal, controlling, direct, and indirect and all are relevant to the effects of thermal discharges in the marine environment. These effects can be defined briefly as follows.

- *Lethal*: high or low temperatures which will kill an organism within a finite time, usually less than its normal life span. The lethal temperature for any one organism depends on many factors within genetic limits. These include acclimatization, rate of change of temperature, physiological state (health) of the organism and any adaptive mechanisms.
- *Controlling*: temperatures below lethal temperatures which affect life processes, i.e., growth, oxygen consumption, digestive rates, or reproduction. There is a general trend for most organisms to show increases in metabolic activity with increasing temperature up to a threshold after which it declines sharply.
- *Direct*: temperatures causing behavioral responses such as avoidance or selection, movements, or migrations. Such effects have been amply demonstrated in experiments but for some work *in situ* the effects are not always clear.
- *Indirect*: where temperatures do not act directly but through another agent, for example poisons or oxygen levels or through effects on prey or predators. Temperature acts synergistically with toxic substances which can be important to its effects on chlorine toxicity *in situ* in thermal plumes. Where temperature immobilizes or kills prey animals they can become much more vulnerable to predation.

Biological Effects of Thermal Discharges

The translation of data obtained from experimental studies to field sites is often not simple. The complexity of the natural environment can mask or exacerbate effects so that they bear little relation to experimental conditions and this has occurred in many studies of thermal discharges *in situ*. Further, factors other than that being studied may be responsible for the observed effects. Examples relevant to thermal discharges are discussed later in this article.

Entrainment

The biological effects of any thermal discharge on marine organisms begin before the effluent is discharged. Cooling water abstracted from the sea usually contains many planktonic organisms notably bacteria, algae, small crustacea, and fish larvae. Within the power station cooling system these

Table 1 Upper temperature limits for aquatic organisms. Data from studies of geothermal waters[a]

Group	Temperature (°C)
Animals	
Fish and other aquatic vertebrates	38
Insects	45–50
Ostracods (crustaceans)	49–50
Protozoa	50
Plants	
Vascular plants	45
Mosses	50
Eukaryotic algae	56
Fungi	60
Prokaryotic microorganisms	
Blue–green algae	70–73
Photosynthetic bacteria	70–73
Nonphotosynthetic bacteria	>99

[a]Reproduced with permission from Langford (1990).

organisms experience a sudden increase in temperature (10–20°C, depending on the station) as they pass through the cooling condensers. They will also experience changes in pressure (1–2 atm) and be dosed with chlorine (0.5–5 mg l^{-1}) during this entrainment, with the effect that many organisms may be killed before they are discharged to the receiving water. Estimates for power stations in various countries have shown that if the ultimate temperatures are less than 23°C the photosynthesis of entrained planktonic marine algae may increase, but at 27–28°C a decrease of 20% was recorded. At 29–34°C the rates decreased by 61–84% at one US power station. Only at temperatures exceeding 40°C has total mortality been recorded. Concentrations of chlorine (total residual) of 1.0 mg l^{-1} have been found to depress carbon fixation in entrained algae by over 90% irrespective of temperature (**Figure 4**). Diatoms were less affected than other groups and the effects in open coastal waters were less marked than in estuarine waters. Unless the dose was high enough to cause complete mortality many algae showed evidence of recovery.

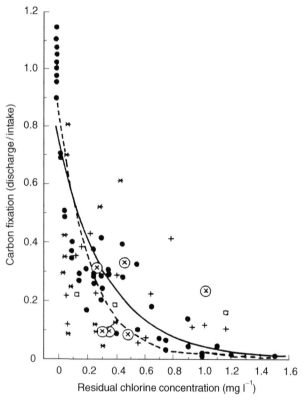

Figure 4 The effect of cooling water chlorination on carbon fixation in marine phytoplankton. ⊗, San Onofre; +, Morgantown; ∗, Hudson River; ●, Fawley; □, Dunkerque. (Reproduced with permission from Langford, 1990.)

The mean mortality rates for marine zooplankton entrained through cooling-water systems were shown to be less than 30% except in unusual cases where extreme temperatures and high chlorine doses caused 100% mortality. High mortalities can also occur where the entrained organisms are exposed to high temperatures in cooling ponds after discharge from the cooling system. In general, zooplankton do not suffer percentage mortalities as high as those of phytoplankton under typical cooling-water conditions. After passing through the cooling system at a US power station the dead or dying zooplankton were observed being eaten by large numbers of fish gathered at the outfall and hence passed into the food chain.

There are few published observations of marine macro-invertebrates entrained in cooling-water systems though at a site in the UK, field experiments showed that specimens of the prawn *Palaemonetes varians* were killed by mechanical damage as they passed through a cooling system. Larval fish are probably most vulnerable to the effects of entrainment, mostly killed by the combination of mechanical, chemical, and temperature effects. Mortalities of ichthyoplankton have varied from 27 to 100% at sites in the US and UK. Many of these were, however, on estuaries or tidal reaches of rivers. At a coastal site in California the mortality rate increased from 10 to 100% as temperatures rose from 31 to 38°C. Some 13% mortalities occurred with no heat, mainly as a result of pressure and abrasion in the system. The significance of 20% larval mortalities to the natural populations of flounders (*Platichthyes americanus*) calculated for a site on Long Island Sound indicated that the annual mortality was estimated at a factor of 0.01 which might cause a reduction of 9% of the adult population over 35 years provided the fish showed no compensatory reproductive or survival mechanism, or no immigration occurred.

Effects of Discharges in Receiving Waters

Algae

At some US power stations the metabolism of phytoplankton was found to be inhibited by chlorine as far as 200 m from the outfall. Also, intermittent chlorination caused reductions of 80–90% of the photosynthetic activity some 50 m from the outfall at a site on the Californian coast. From an assessment of the total entrainment and discharge effects, however, it was reported that where dilution was 300 times per second the effect of even a 100% kill of phytoplankton would not be detectable in the receiving water. In Southampton Water in the UK, mortality rates of 60% were estimated as causing about 1.2–3% reduction in the productivity of the tidal exchange volume where a power station used 6% of this for cooling. The main problem with most assessments is that replication of samples was typically low and estimates suggest that 88 samples would be needed from control and affected areas to detect a difference of 5% in productivity at a site, 22 samples for a 10% change and at least six for a 20% change. Such replication is rarely recorded in site studies.

Temperatures of 35–36°C killed shore algae at a coastal site in Florida, particularly *Halimeda* sp. and *Penicillus* spp. but factors other than temperature, most likely chlorine and scour, removed macro-algae at another tropical site. Blue–green algae were found where temperatures reached 40°C intermittently and *Enteromorpha* sp. occurred where temperatures of 39°C were recorded. In more temperate waters the algae *Ascophyllum* and *Fucus* were eliminated where temperatures reached 27–30°C at an outfall but no data on chlorination were shown. Replacing the shoreline outfall with an offshore diffuser outfall (which increased dilution and cooling rates) allowed algae to recolonize and recover at a coastal power station in Maine (US). On the Californian coast one of the potentially most vulnerable algal systems, the kelps *Macrocystis*, were predicted to be badly affected by power station effluents, but data suggest that at one site only about 0.7 ha was affected near the outfall.

The seagrass systems (*Thalassia* spp.) of the Florida coastal bays appeared to be affected markedly by the effluents from the Turkey Point power stations and a long series of studies indicated that within the +3 to +4°C isotherm in the plume, seagrass cover declined by 50% over an area of 10–12 ha. However, the results from two sets of studies were unequivocal as to the effects of temperature. The data are outlined briefly in the following summary. First, the effluent was chlorinated. Second it contained high levels of copper and iron. Third, the main bare patch denuded of seagrass, according to some observations, may have been caused by the digging of the effluent canal. Although one set of data concluded that the threshold temperature for adverse effects on seagrass was +1.5°C (summer 33–35°C) a second series of observations noted that *Thalassia* persisted apparently unharmed in areas affected by the thermal discharge, though temperatures rarely exceed 35°C. From an objective analysis, it would appear that the effects were caused mainly by a combination of thermal and chemical stresses.

Zooplankton and Microcrustacea

In Southampton Water in the UK, the barnacle *Elminius modestus* formed large colonies in culverts at the Marchwood power station and discharged large numbers of nauplii into the effluent stream increasing total zooplankton densities. Similar increases occurred where fouling mussels (*Mytilus* sp.) released veliger larvae into effluent streams. Data from 10 US coastal power stations were inconclusive about the effects of thermal discharges on zooplankton in receiving waters with some showing increases and others the reverse. Changes in community composition in some areas receiving thermal discharges were a result of the transport of species from littoral zones to offshore outfalls or vice versa. At Tampa Bay in Florida no living specimens of the benthic ostracod *Haplocytheridea setipunctata* were found when the temperature in the thermal plume exceeded 35°C. Similarly the benthic ostracod *Sarsiella zostericola* was absent from the area of a power effluent channel in the UK experiencing the highest temperature range.

Macro-invertebrates

As with other organisms there is no general pattern of change in invertebrate communities associated with thermal discharges to the sea which can be solely related to temperature. Some of the earliest studies in enclosed temperate saline waters in the UK showed that no species was consistently absent from areas affected by thermal plumes and the studies at Bradwell power station on the east coast showed no evidence of a decline in species richness over some 20 years though changes in methodology could have obscured changes in the fauna. No changes in bottom fauna were recorded at other sites affected by thermal plumes in both Europe and the US. The polychaete *Heteromastus filiformis* was found to be common to many of the thermal plume zones in several countries. In these temperate waters temperatures rarely exceeded 33–35°C.

In contrast, in tropical coastal waters data suggest that species of invertebrates are excluded by thermal plumes. For example in Florida, surveys showed that some 60 ha of the area affected by the Turkey Point thermal plume showed reduced diversity and abundance of benthos in summer, but there was marked recovery in winter. The 60 ha represented 0.0023% of the total bay area. A rise of 4–5°C resulted in a dramatic reduction in the benthic community. Similarly, at Tampa Bay, 35 of the 104 indigenous invertebrate species were excluded from the thermal plume area. Removal of the vegetation was considered to be the primary cause of the loss of benthic invertebrates. In an extreme tropical case few species of macro-invertebrates survived in a thermal effluent where temperatures reached 40°C, 10 species occurred at the 37°C isotherm and the number at the control site was 87. Scour may have caused the absence of species from some areas nearer the outfall (**Figure 5**). The effluent was chlorinated but no data on chlorine concentrations were published. Corals suffered severe mortalities at the Kahe power station in Hawaii but the bleaching of the colonies suggested that again chlorine was the primary cause of deaths, despite temperatures of up to 35°C. It has been suggested that temperature increases of as little as 1–2°C could cause damage to tropical ecosystems but detailed scrutiny of the data indicate that it would be difficult to come to that conclusion from field data, especially where chlorination was used for antifouling.

The changes in the invertebrate faunas of rocky shores in thermal effluents have been less well studied. Minimal changes were found on breakwaters in the paths of thermal plumes at two Californian sites. Any measurable effects were within 200–300 m of the outfalls. In contrast in southern France a chlorinated thermal discharge reduced the numbers of species on rocks near the outfall though 11 species of Hydroida were found in the path of the effluent. In most of the studies, chlorine would appear to be the primary cause of reductions in species and abundance though where temperatures exceeded 37°C both factors were significant. There is some evidence that species showed advanced reproduction and growth in the thermal plumes areas of some power stations where neither temperature nor chlorine were sufficient to cause mortality. Also behavioral effects were demonstrated for invertebrates at a Texas coastal power station. Here, blue crabs (*Callinectes sapidus*) and shrimps (*Penaeus aztecus* and *P. setiferus*) avoided the highest temperatures (exceeding 38°C) in the discharge canal but recolonized as temperatures fell below 35°C. At another site in tropical waters, crabs (*Pachygrapsus transversus*) avoided the highest temperatures (and possibly chlorine) by climbing out of the water on to mangrove roots.

Fish

There are few records of marine fish mortalities caused by temperature in thermal discharges except where fish are trapped in effluent canals. For example mortalities of Gulf menhaden (*Brevoortia petronus*), sea catfish (*Arius felis*) occurred in the canal of a Texas power station when temperatures reached 39°C. Also a rapid rise of 15°C killed menhaden (*Alosa* sp.) in the effluent canal of the Northport power station in the US. Avoidance behavior prevents mortalities where fish can move freely.

The apparent attraction of many fish species to thermal discharges, widely reported, was originally associated with behavioral thermoregulation and the selection of preferred temperatures. Perhaps the best recorded example is the European seabass (*Dicentrarchus labrax*) found associated with cooling-water outfalls in Europe. Temperature selection is, however, not now believed to be the cause of the aggregations. Careful analysis and observations indicate clearly that the main cause of aggregation is the large amounts of food organisms discharged either dead or alive in the discharge. Millions of shrimps and small fish can be passed through into effluents and are readily consumed by the aggregated fish. Further where fish have gathered, usually in cooler weather, they remain active in the warmer water and are readily caught by anglers unlike the individuals outside the plume. This also gives the impression that there are more fish in the warmer water. Active tracking of fish has shown mainly short-term association with outfalls though some species have been

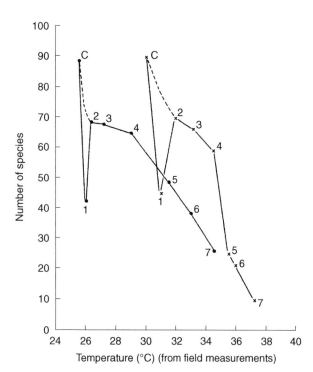

Figure 5 Numbers of invertebrate species recorded at various temperatures, taken from two separate surveys (×, October; ●, winter) at Guyanilla Bay. C, control sampling; 1–7, effluent sampling sites. (Reproduced with permission from Langford, 1990.)

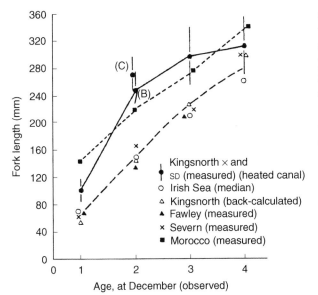

Figure 6 Growth of bass (*Dicentrarchus labrax*) in a thermal discharge canal in comparison to other locations. Note that back calculated lengths are smaller because they probably relate to cold water growth. (Reproduced with permission from Langford, 1990.)

shown as entering water at temperatures above their lethal maximum for very short periods to collect food. There is clear evidence, however, that fish avoid adversely high temperatures for most of the time and will return to a discharge area once the temperatures cool. Avoidance behavior is also apparent at high chlorine concentrations.

Where water temperatures and chemical conditions allow consistent residence, fish in thermal discharge canals show increased growth (**Figure 6**). At the Kingsnorth power station in the UK seabass (*D. labrax*) grew at twice the rate as in cold water, particularly in the first year. Fish showed varying residence times and sequential colonization of the canal at various ages. Winter growth occurred and the scales of older fish with long-term association with the discharge showed no evidence of annual winter growth checks. The fish were able to move into and out of the canal freely.

Occurrence of Exotic Species

Exotic or introduced marine invertebrate species have been recorded from thermal discharges in various parts of the world. Some of the earliest were from the enclosed docks heated by power station effluents near Swansea and Shoreham in the UK. The exotic barnacle *Balanus amphitrite* var. *denticulata*

and the woodborer *Limnoria tripunctata* replaced the indigenous species in the heated areas but declined in abundance when the effluent ceased. The polyzoan *Bugula neritina* originally a favored immigrant species in the heated water disappeared completely as the waters cooled. In New Jersey (USA) subtropical species of *Teredo* bred in a heated effluent and both the ascidian *Styela clavata* and the copepod *Acartia tonsa* have both been regarded as immigrant species favored by heated effluents. However many immigrant species have survived in temperate waters without being associated with heated waters. In UK waters, only the crab *Brachynotus sexdentatus* and the barnacle *B. amphitrite* may be regarded as the only species truly associated with thermal discharges despite records of various other species.

The decline of the American hard shell clam fishery (*Mercenaria mercenaria*) introduced to Southampton Water in the UK was reportedly caused by the closure of the Marchwood power station combined with overfishing. Recruitment of young clams and their early growth were maximized in the heated water and reproduction was advanced but all declined as the thermal discharge ceased.

Aquaculture in Marine Thermal Discharges

The use of marine thermal discharges from power stations for aquaculture has not been highly successful in most parts of the world. Although it is clear that some species will grow faster in warmer water, the presence and unpredictability of chlorination has been a major obstacle. It is not generally economically viable to allow a large power station to become fouled such that efficiency declines merely to allow fish to grow. In Japan some farming ventures are regarded as profitable at power stations but in Europe and the USA such schemes are rarely profitable. At the Hunterston power station in Scotland plaice (*Pleuronectes platessa*) and halibut (*Hippoglossus hippoglossus*) grew almost twice as fast in warm water as in natural ambient but the costs of pump maintenance and capital equipment caused the system to be uneconomic irrespective of chlorination problems. Optimization of temperature is also a problem especially where temperatures fluctuate diurnally or where they exceed optimal growth temperatures. The general conclusion is that commercial uses of heated effluents in marine systems are not yet proven and are unlikely to become large-scale global ventures.

Thermal Discharges and Future Developments

The closure of many older, less-efficient power stations, has led to an increase in the efficiency of the use of water and a decline in the discharge of heat to the sea per unit of electricity generated. However, the increasing industrialization of the developing countries, China, Malaysia, India and the African countries is leading to the construction of new, large power stations in areas not previously developed. The widespread use of CCGT stations can reduce the localized problems of heat loss and water use further as well as reducing emissions of carbon dioxide and sulphur dioxide to the air but the overall increase in power demand and generation will lead to an increase in the total aerial and aquatic emissions in some regions.

In some tropical countries the delicate coastal ecosystems will be vulnerable not only to heat and higher temperature but much more importantly to the biocides used for antifouling. There is as yet no practical alternative that is as economic as chlorine though different methods have been tried with varying success in some parts of the world. There is little doubt that the same problems will be recognized in the areas of new development but as in the past after they have occurred.

Conclusions

It is clear that the problems of the discharge of heated effluents are essentially local and depend on many factors. Although temperatures of over 37°C are lethal to many species which cannot avoid exposure, there are species which can tolerate such temperatures for short periods. Indeed it can be concluded for open coastal waters that discharge temperatures may exceed the lethal limits of mobile species at least for short periods. This, of course, would not apply if vulnerable sessile species were involved, though again some provisos may be acceptable. For example, an effluent which stratified at the surface in deep water would be unlikely to affect the benthos. On the other hand an effluent which impinges on the shore may need strict controls to protect the benthic community. From all the data it is clear that blanket temperature criteria intended to cover all situations would not protect the most vulnerable ecosystems and may be too harsh for those that are more tolerant or less exposed. Constraints should therefore be tailored to each specific site and ecosystem.

Irrespective of temperature it is also very clear that chlorination or other biocidal treatment has been responsible for many of the adverse ecological effects originally associated with temperature. The solution to fouling control and the reduction of chlorination of other antifouling chemicals is therefore probably more important than reducing heat loss and discharge temperatures particularly where vulnerable marine ecosystems are at risk.

See also

Demersal Species Fisheries.

Further Reading

Barnett PRO and Hardy BLS (1984) Thermal deformations. In: Kinne O (ed.) *Marine Ecology*, vol. V. *Ocean Management*, part 4, *Pollution and Protection of the Seas, Pesticides, Domestic Wastes and Thermal Deformations*, pp. 1769–1926. New York: Wiley.

Jenner HA, Whitehouse JW, Taylor CJL and Khalanski M (1998) *Cooling Water Management in European Power Stations: Biology and Control of Fouling.* Hydroecologie Appliquee. Electricité de France.

Kinne O (1970) *Marine Ecology, vol. 1, Environmental Factors, part 1.* New York: Wiley-Interscience.

Langford TE (1983) *Electricity Generation and the Ecology of Natural Waters.* Liverpool: Liverpool University Press.

Langford TE (1990) *Ecological Effects of Thermal Discharges.* London: Elsevier Applied Science.

Newell RC (1970) *The Biology of Intertidal Animals.* London: Logos Press.

OIL POLLUTION

J. M. Baker, Clock Cottage, Shrewsbury, UK

Introduction

This article describes the sources of oil pollution, composition of oil, fate when spilt, and environmental effects. The initial impact of a spill can vary from minimal to the death of nearly everything in a particular biological community, and recovery times can vary from less than one year to more than 30 years. Information is provided on the range of effects together with the factors which help to determine the course of events. These include oil type and volume, local geography, climate and season, species and biological communities, local economic and amenity considerations, and clean-up methods. With respect to clean-up, decisions sometimes have to be made between different, conflicting environmental concerns. Oil spill response is facilitated by pre-spill contingency planning and assessment including the production of resource sensitivity maps.

Oil: A High Profile Pollutant

Consider some of the worst and most distressing effects of an oil spill: dying wildlife covered with oil; smothered shellfish beds on the shore; unusable amenity beaches. It is not surprising that ever since the *Torrey Canyon* in 1967 (the first major tanker accident) oil spills have been media events. Questions about environmental effects and adequacy of response arise time and time again, but to help answer these it is now possible to draw upon decades of experience from three main types of activity: post-spill case studies and long-term monitoring; field experiments to test clean-up methods; and laboratory tests to investigate toxicities of oils and dispersants.

Oil is a complex substance containing hundreds of different compounds, mainly hydrocarbons (compounds consisting of carbon and hydrogen only). The three main classes of hydrocarbons in oil are alkanes (also known as paraffins), cycloalkanes (cycloparaffins or naphthenes) and arenes (aromatics). Compounds within each class have a wide range of molecular weights. Different refined products ranging from petrol to heavy fuel oil obviously differ in physical properties such as boiling range and evaporation rates, and this is related to the molecular weights of the compounds they contain. Crude oils are also variable in their chemical composition and physical properties, depending on the field of origin. Chemical and physical properties are factors which partly determine environmental effects of spilt oil.

Tanker Accidents Compared With Chronic Inputs

Tanker accidents represent a small, but highly visible, proportion of the total oil inputs to the world's oceans each year (**Figure 1**). The incidence of large tanker spills has, however, declined since the 1970s. Irrespective of accidental spills, background hydrocarbons are ubiquitous, though at low concentrations. Water in the open ocean typically contains 1–10 parts per billion but higher concentrations are found in nearshore waters. Sediments or organisms may accumulate hydrocarbons in higher concentrations than does water, and it is common for sediments in industrialized bays to contain several hundred parts per million.

Sources of background hydrocarbons include:

- Operational discharges (e.g. bilge water) from ships and boats;
- Land-based discharges (e.g. industrial effluents, rainwater run-off from roads, sewage discharges);
- Natural seeps of petroleum hydrocarbons, such as occur, for example, along the coasts of Baffin Island and California;

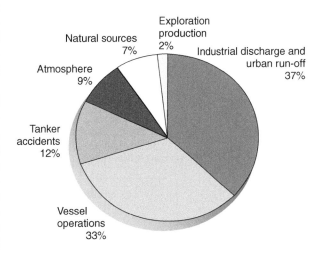

Figure 1 Major inputs of oil to the marine environment.

- Airborne combustion products, either natural (e.g. from forest fires) or artificial (e.g. from the burning of fossil fuels);
- Organisms, e.g. leaf waxes and hydrocarbons synthesized by algae.

Notwithstanding the relatively great total amounts of oil from operational and land-based sources these discharges usually comprise diluted, dissolved and dispersed oil. The sections below focus on accidental spills because it is these which produce visible slicks which may coat wildlife and shores, and which typically require a clean-up response.

Natural Fate of Oil Slicks

When oil is spilt on water, a series of complex interactions of physical, chemical, and biological processes takes place. Collectively these are called 'weathering'; they tend to reduce the toxicity of the oil and in time they lead to natural cleaning. The main physical processes are spreading, evaporation, dispersion as small drops, solution, adsorption onto sediment particles, and sinking of such particles. Degradation occurs through chemical oxidation (especially under the influence of light) and biological action – a large number of different species of bacteria and fungi are hydrocarbon degraders. Case history evidence gives a reasonable indication of natural cleaning timescales in different conditions. For open water sites, half-lives (the time taken for natural removal of 50% of the oil from the water surface) typically range from about half a day for the lightest oils (e.g. kerosene) to seven days or more for heavy oils (e.g. heavy fuel oil). However, for large spills near coastlines, some oil typically is stranded on the shore within a few days; once oil is stranded, the natural cleaning timescale may be prolonged. On the shore observed timescales range from a few days (some very exposed rocky shores) to more than 30 years (some very sheltered shores notably salt marshes). It is estimated that natural shore cleaning may take several decades in extreme cases.

Shore cleaning timescales are affected by factors including:

- exposure of the shore to wave energy (**Figure 2**) from very exposed rocky headlands to sheltered tidal flats, salt marshes and mangroves. This in turn depends on a number of variables that include fetch; speed, direction, duration and frequency of winds; and open angle of the shore;
- localized exposure/shelter – even on an exposed shore, cracks, crevices, and spaces under boulders can provide sheltered conditions where oil may persist;
- steepness/shore profile – extensive, gently sloping shores dissipate wave energy;

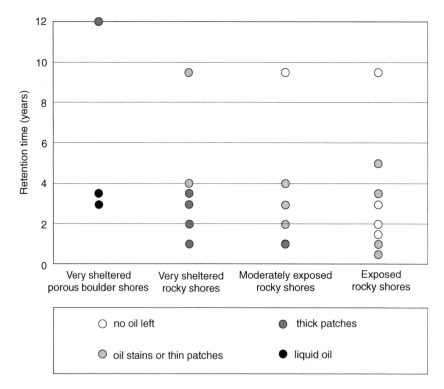

Figure 2 Oil residence times on a variety of rocky shores where no clean up has been attempted.

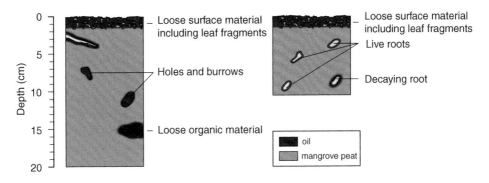

Figure 3 Two core samples taken from a mangrove swamp, showing penetration of oil down biological pathways.

- substratum – oil does not easily penetrate fine sediments (especially if they are waterlogged) unless they have biological pores such as crab and worm burrows or root holes (**Figure 3**). Penetration of shingle, gravel and some sand beaches can take place relatively easily, sometimes to depths of more than one metre;
- clay-oil flocculation – this process (first noticed on some Alaskan shores after the *Exxon Valdez* spill) reduces adherence of oil to shore substrata and facilitates natural cleaning;
- height of the stranded oil on the shore – oil spots taken into the supratidal zone by spray can persist for many years, conversely, oil on the middle and lower shore is more likely to be removed by water action. It is common to have stranded oil concentrated in the high tide area;
- oil type, e.g. viscosity affects movement into and out of sediment shores.

Oil Spill Response

Aims

The aims of oil spill response are to minimize damage and reduce the time for environmental recovery by:

- guiding or re-distributing the oil into less sensitive environmental components (e.g. deflecting oil away from mangroves onto a sandy beach) and/or
- removing an appropriate amount of oil from the area of concern and disposing of it responsibly.

Initiation of a response, or decision to stop cleaning or leave an area for natural clean-up, needs to be focused on agreed definitions of 'how clean is clean', otherwise there is no yardstick for determining whether the response has achieved the desired result.

The main response options are described below.

Booms and Skimmers

Booms and skimmers can be successful if the diversion and containment of the oil starts before it has had too much time to spread over the water surface. Booms tend to work well under calm sea conditions, but they are ineffective in rough seas. When current speeds are greater than 0.7–1.0 knots (1.3–1.85 km h^{-1}), with the boom at right angles to the current, the oil is entrained into the water, passes under the boom and is lost. In some cases it may be possible to angle the boom to prevent this. Booms can also be used for shoreline protection, for example to stop oil from entering sheltered inlets with marshes or mangroves. The time available for protective boom deployment depends on the position and movement of the slick, and can vary from hours to many days. The efficiency of skimmers depends on the oil thickness and viscosity, the sea state and the storage capacity of the skimmer. Skimmers normally work best in sheltered waters. Because of their limitations, recovering 10% of the oil at a large spill in open seas is considered good for these mechanisms.

Dispersants

Dispersants, which contain surfactants (surface active agents), reduce interfacial tension between oil and water. They promote the formation of numerous tiny oil droplets and retard the recoalescence of droplets into slicks. They do not clean oil out of the water, but can improve biodegradation by increasing the oil surface area, thereby increasing exposure to bacteria and oxygen. Information about dispersant effectiveness from accidental spills is limited for various reasons, such as inadequate monitoring and the difficulty of distinguishing between the contributions of different response methods to remove oil from the water surface. However, in at least some situations dispersants appear to remove a greater proportion of oil from the water surface than mechanical methods. Moreover, they can be used relatively quickly and under sea conditions where mechanical collection is impossible. Dispersants, however, do not work well in all circumstances (e.g. on heavy fuel oil) and even for initially dispersable oils, there is only a short 'window of opportunity', typically 1–2 days, after which the oil becomes too weathered for dispersants to be effective. The main environmental concern is that the dispersed oil droplets in the water column may in some cases affect organisms such as corals or fish larvae.

In situ Burning

Burning requires the use of special fireproof containment booms. It is best achieved on relatively fresh oil and is most effective when the sea is fairly calm. It generates a lot of smoke and as with dispersants, the window of opportunity is short, typically a few hours to a day or two, depending on the oil type and the environmental conditions at the time of the spill. When it is safe and logistically feasible, *in situ* burning is highly efficient in removing oil from the water surface.

Nonaggressive Shore Cleaning

Nonaggressive methods of shore cleaning, methods with minimal impact on shore structure and shore organisms, include:

- physical removal of surface oil from sandy beaches using machinery such as front-end loaders (avoiding removal of underlying sediment);
- manual removal of oil, asphalt patches, tar balls etc., by small, trained crews using equipment such as spades and buckets;
- collection of oil using sorbent materials (followed by safe disposal);
- low-pressure flushing with seawater at ambient temperature;
- bioremediation using fertilizers to stimulate indigenous hydrocarbon-degrading bacteria.

In appropriate circumstances these methods can be effective, but they may also be labor-intensive and clean-up crews must be careful to minimize trampling damage.

Nonaggressive methods do not work well in all circumstances. Low-pressure flushing, for example, is ineffective on weathered firmly adhering oil on rocks; and bioremediation is ineffective for subsurface oil in poorly aerated sediments.

Aggressive Shore Cleaning

Aggressive methods of shore cleaning, those that are likely to damage shore structure and/or shore organisms, include:

- removal of shore material such as sand, stones, or oily vegetation together with underlying roots and mud (in some cases the material may be washed and returned to the shore);
- water flushing at high pressure and/or high temperature;
- sand blasting.

In some cases these methods are effective at cleaning oil from the shore, e.g. hot water was more effective than cold water for removing weathered, viscous *Exxon Valdez* oil from rocks. However, heavy machinery, trampling and high-pressure water all can force oil into sediments and make matters worse.

Net Environmental Benefit Analysis for Oil Spill Response

In many cases a possible response to an oil spill is potentially damaging to the environment. The public perception of disaster has sometimes been heightened by headlines such as 'Clean-up makes things worse'. The advantages and disadvantages of different responses need to be weighed up and compared both with each other and with the advantages and disadvantages of entirely natural cleaning. This evaluation process is sometimes known as net environmental benefit analysis. The approach accepts that some response actions cause damage but may be justifiable because they reduce the overall problems resulting from the spill and response.

Example 1 Consider sticky viscous fuel oil adhering to rocks which are an important site for seals. If effective removal of oil could only be achieved by high-pressure hot-water washing or sand blasting, prolonged recovery times of shore organisms such as algae, barnacles and mussels might be accepted because the seals were given a higher priority. A consideration here and in similar cases is that populations of wildlife species are likely to be smaller, more localized, and slower to recover if affected by oil than populations of abundant and widespread shore algae and invertebrates.

Example 2 Consider a slick moving over shallow nearshore water in which there are coral reefs of particular conservation interest. The slick is moving towards sandy beaches important for tourism but of low biological productivity. Dispersant spraying will minimize pollution of the beaches, but some coral reef organisms may be damaged by dispersed oil. From an ecological point of view, it is best not to use dispersants but to allow the oil to strand on the beaches, from where it may be quickly and easily cleaned.

Effects and Recovery

Range of Effects

The range of oil effects after a spill can encompass:

- physical and chemical alteration of habitats, e.g. resulting from oil incorporation into sediments;
- physical smothering effects on flora and fauna;
- lethal or sublethal toxic effects on flora and fauna;
- short- and longer-term changes in biological communities resulting from oil effects on key organisms, e.g. increased abundance of intertidal algae following death of limpets which normally graze the algae;
- tainting of edible species, notably fish and shelfish, such that they are inedible and unmarketable (even though they are alive and capable of self-cleansing in the long term);
- loss of use of amenity areas such as sandy beaches;
- loss of market for fisheries products and tourism because of bad publicity (irrespective of the actual extent of the tainting or beach pollution);
- fouling of boats, fishing gear, slipways and jetties;
- temporary interruption of industrial processes requiring a supply of clean water from sea intakes (e.g. desalination).

The extent of biological damage can vary from minimal (e.g. following some open ocean spills) to the death of nearly everything in a particular biological community. Examples of extreme cases of damage following individual spills include deaths of thousands of sea birds, the death of more than 100 acres of mangrove forest, and damage to fisheries and/or aquaculture with settlements in excess of a million pounds.

Extent of damage, and recovery times, are influenced by the nature of the clean-up operations, and by natural cleaning processes. Other interacting factors include oil type, oil loading (thickness), local geography, climate and season, species and biological communities, and local economic and amenity considerations. With respect to oil type, crude oils and products differ widely in toxicity. Severe toxic effects are associated with hydrocarbons with low boiling points, particularly aromatics, because these hydrocarbons are most likely to penetrate and disrupt cell membranes. The greatest toxic damage has been caused by spills of lighter oil particularly when

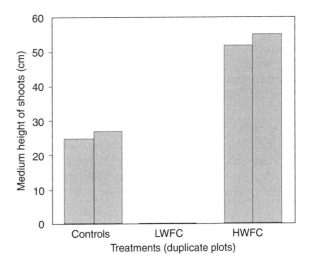

Figure 4 The effects of experimental oil treatments (duplicate plots) on shoot heights of the common salt marsh grass *Spartina anglica*. The measurements shown were taken four months after treatment. Lightly weathered Forties crude (LWFC) killed most of the grass shoots. Heavily weathered Flotta crude (HWFC) stimulated growth.

confined to a small area. Spills of heavy oils, such as some crudes and heavy fuel oil, may blanket areas of shore and kill organisms primarily through smothering (a physical effect) rather than through acute toxic effects. Oil toxicity is reduced as the oil weathers. Thus a crude oil that quickly reaches a shore is more toxic to shore life than oil that weathered at sea for several days before stranding. There have been cases of small quantities of heavy or weathered oils stimulating the growth of salt marsh plants (**Figure 4**).

Geographical factors which have a bearing on the course of events include the characteristics of the water body (e.g. calm shallow sea or deep rough sea), wave energy levels along the shoreline (because these affect natural cleaning) and shoreline sediment characteristics. Temperature and wind speeds influence oil weathering, and according to season, vulnerable groups of birds or mammals may be congregated at breeding colonies, and fish may be spawning in shallow nearshore waters.

Vulnerable Natural Resources

Salt marshes Salt marshes are sheltered 'oil traps' where oil may persist for many years. In cases where perennial plants are coated with relatively thin oil films, recovery can take place through new growth from underground stems and rootstocks. In extreme cases of thick smothering deposits, recovery times may be decades.

Mangroves Mangrove forests are one of the most sensitive habitats to oil pollution. The trees are easily killed by crude oil, and with their death comes loss of habitat for the fish, shellfish and wildlife which depend on them. Mangrove estuaries are sheltered 'oil trap' areas into which oil tends to move with the tide and then remain among the prop roots and breathing roots, and in the sediments (**Figure 5**).

Coral reefs   Coral reef species are sensitive to oil if actually coated with it. There is case-history evidence of long-term damage when oil was stranded on a reef flat at low tide. However, the risk of this type of scenario is quite low – oil slicks will float over coral reefs at most stages of the tide, causing little damage. Deep water corals will escape direct oiling at any stage of the tide.

Fish   Eggs, larvae and young fish are comparatively sensitive but there is no definitive evidence

which suggests that oil pollution has significant effects on fish populations in the open sea. This is partly because fish can take avoiding action and partly because oil-induced mortalities of young life stages are often of little significance compared with huge natural losses each year (e.g. through predation). There is an increased risk to some species and life stages of fish if oil enters shallow near-shore waters which are fish breeding and feeding grounds. If oil slicks enter into fish cage areas there may be some fish mortalities, but even if this is not the case there is likely to be tainting. Fishing nets, fish traps and aquaculture cages are all sensitive because adhering oil is difficult to clean and may taint the fish.

Turtles   Turtles are likely to suffer most from oil pollution during the breeding season, when oil at

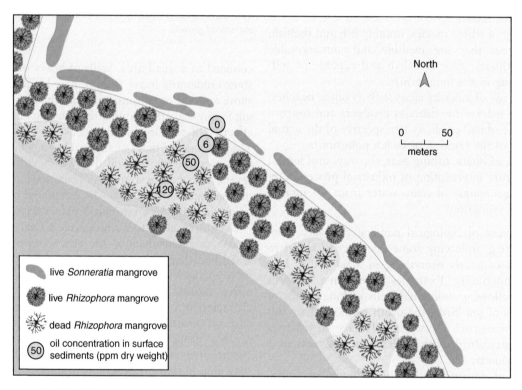

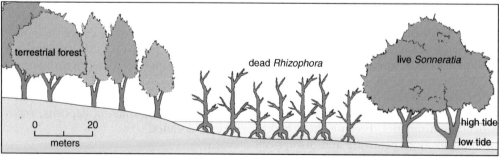

Figure 5   Plan and profile showing mangrove patches killed by small oil slicks.

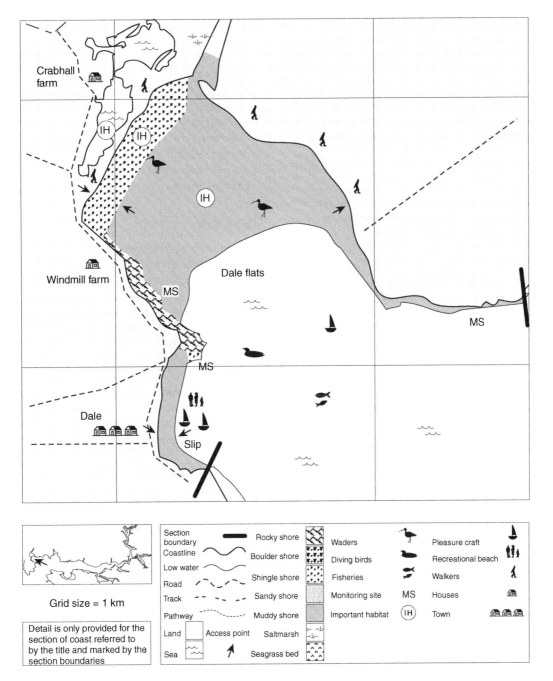

Figure 6 An example of a sensitivity map. This large-scale map is part of an atlas covering the oil port of Milford Haven, Wales. It facilitates response when oil has come near shore or onshore, and the protection or clean-up of specific locations is being planned.

egg-laying sites could have serious effects on eggs or hatchlings. If oiling occurs, the effects from the turtle conservation point of view could be serious, because the various turtle species are endangered.

Birds Seabirds are extremely sensitive to oiling, with high mortality rates of oiled birds. Moreover, there is experimental evidence that small amounts of oil transferred to eggs by sublethally oiled adults can significantly reduce hatching success. Shore birds, notably waders, are also at risk. For them, a worst-case scenario would be oil impacting shore feeding grounds at a time when large numbers of migratory birds were coming into the area.

Mammals Marine mammals with restricted coastal distributions are more likely to encounter oil than wide-ranging species moving quickly through an area. Species at particular risk are those which rely on fur for conservation of body heat (e.g. sea otters). If the fur becomes matted with oil, they rapidly lose body heat and die from hypothermia. At sea, whales, dolphins and seals are at less risk because they have a layer of insulating blubber under the skin. However, there have been oil-related mortalities of young seals at breeding colonies.

Recovery

It is unrealistic to define recovery as a return to prespill conditions. This is partly because quantitative information on prespill conditions is only rarely available and, more importantly, because marine ecosystems are in a constant state of flux due to natural causes. These fluctuations can be as great as those caused by the impact of an oil spill. The following definition takes these problems into account.

Recovery is marked by the re-establishment of a healthy biological community in which the plants and animals characteristic of that community are present and functioning normally. It may not have the same composition or age structure as that which was present before the damage, and will continue to show further change and development. It is impossible to say whether an ecosystem that has recovered from an oil spill is the same as, or different from, that which would have persisted in the absence of the spill.

Assessment

Before a Spill

Before a spill it is important to identify what the particular sensitivities are for the area covered by any particular oil spill contingency plan, and to put the information on a sensitivity map which will be available to response teams. An example is shown in **Figure 6**. Maps should include information on the following.

* Shoreline sensitivity. Shorelines may be ranked using the basic principles that sensitivity to oil increases with increasing shelter of the shore from wave action, penetration of oil into the substratum, natural oil retention times on the shore, and biological productivity of shore organisms. Typically, the least sensitive shorelines are exposed rocky headlands, and the most sensitive are marshes and mangroves forests.
* Other ecological resources such as coral reefs, seagrass and kelp beds, and wildlife such as turtles, birds and mammals.

* Socioeconomic resources, for example fishing areas, shellfish beds, fish and crustacean nursery areas, fish traps and aquaculture facilities. Other features include boat facilities such as harbors and slipways, industrial water intakes, recreational resources such as amenity beaches, and sites of cultural or historical significance. Sensitivities are influenced by many factors including ease of protection and clean-up, recovery times, importance for subsistence, economic value and seasonal changes in use.

After a Spill

The response options need to be reviewed and fine-tuned throughout the response period, in the light of information being received about distribution and degree of oiling and resources affected. In extreme cases this process can be lengthy, for example over three years for the shoreline response to the *Exxon Valdez* spill in Prince William Sound, Alaska. In this case information was provided by shoreline clean-up assessment teams who carried out postspill surveys with the following objectives: assessment of the presence, distribution, and amount of surface and subsurface oil, and collection of information needed to make environmentally sound decisions on clean-up techniques. The standardized methods developed have subsequently been used as a model for other spills.

Acknowledgments

The International Petroleum Industry Environmental Conservation Association is gratefully acknowledged for permission to use material from its Report series.

See also

Coral Reefs.

Further Reading

American Petroleum Institute, Washington, DC. *Oil Spill Conference Proceedings*, published biennially from 1969 onwards. The primary source detailed papers on all aspects of oil pollution.

IPIECA Report Series, Vol. 1 (1991) *Guidelines on Biological Impacts of Oil Pollution*; vol. 2 (1991) *A Guide to Contingency Planning for Oil Spills on Water*; vol. 3 (1992) *Biological Impacts of Oil Pollution: Coral Reefs*; vol. 4 (1993) *Biological Impacts of Oil Pollution: Mangroves*; vol. 5 (1993) *Dispersants and their Role in Oil Spill Response*; vol. 6 (1994) *Biological Impacts of*

Oil Pollution: Saltmarshes; vol. 7 (1995) *Biological Impacts of Oil Pollution: Rocky Shores*; vol. 8 (1997) *Biological Impacts of Oil Pollution: Fisheries*; vol. 9 (1999) *Biological Impacts of Oil Pollution: Sedimentary Shores*; vol. 10 (2000) Choosing Spill Response Options to Minimize Damage. London: International Petroleum Industry Environmental Conservation Association.

ITOPF (1987) *Response to Marine Oil Spills. International Tanker Owners Pollution Federation Ltd, London*. London: Witherby.

POLLUTION, SOLIDS

C. M. G. Vivian and L. A. Murray, The Centre for Environment, Fisheries and Aquaculture Sciences, Lowestoft, UK

Introduction

A very wide variety of solid wastes have been dumped at sea or discharged into the oceans via pipelines or rivers, either deliberately or as a consequence of other activities. The main solid materials involved are dredged material, particulate wastes from sand/gravel extraction and land reclamation, industrial wastes, including mining wastes and munitions, and plastics and litter.

Regulation

The London Convention 1972 has regulated dumping at sea of these materials on a global basis since it came into force in 1974 and currently has 83 contracting parties (mid-2008). In 1996, following a detailed review that began in 1993, the Contracting Parties to the London Convention 1972 agreed a Protocol to the Convention to update and modernize it. This came into force for those states that had ratified the Protocol on 24 March 2006 after the 27th state ratified it and currently (mid-2008) has 34 contracting parties. Regional conventions also regulate the dumping of these materials in some parts of the world, for example, the OSPAR Convention 1992 for the North-East Atlantic, the Helsinki Convention for the Baltic (updated in 2004), the Barcelona Convention 1995 for the Mediterranean, the Cartagena Convention 1983 for the Caribbean, and the Noumea Convention 1986 for the South Pacific. Where materials are discharged into the oceans via pipelines or rivers, this usually falls under national regulation but regional controls or standards may strongly influence this regulation, for example, within the countries of the European Union. When human activities introduce solid materials as a by-product of those activities, this is commonly subject only to national regulation, if any at all.

Impacts

All solid wastes have significant physical impacts at the point of disposal when disposed of in bulk. These impacts can include covering of the seabed, shoaling, and short-term local increases in the levels of water-suspended solids. Physical impacts may also result from the subsequent transport, particularly of the finer fractions, by wave and tidal action and residual current movements. All these impacts can lead to interference with navigation, recreation, fishing, and other uses of the sea. In relatively enclosed waters, the materials with a high chemical or biological oxygen demand (e.g., organic carbon-rich) can adversely affect the oxygen regime of the receiving environment while materials with high levels of nutrients can significantly affect the nutrient flux and thus potentially cause eutrophication.

Biological consequences of these physical impacts can include smothering of benthic organisms, habitat modification, and interference with the migration of fish or shellfish due to turbidity or sediment deposition. The toxicological effects of the constituents of these materials may be significant. In addition, constituents may undergo physical, chemical, and biochemical changes when entering the marine environment and these changes have to be considered in the light of the eventual fate and potential effects of the material.

Dredged Material

Dredging

Dredging is undertaken for a variety of reasons including navigation, environmental remediation, flood control, and the emplacement of structures (e.g., foundations, pipelines, and tunnels). These activities can generate large volumes of waste requiring disposal. Dredging is also undertaken to win materials, for example, for reclamation or beach nourishment. Types of material dredged can include sand, silt, clay, gravel, coral, rock, boulders, and peat. Mining for ores is considered further within the section on industrial wastes.

Most dredging activities, particularly hydraulic dredging, generate an overspill of fine solid material as a consequence of the dredging activity. The particle size characteristics of this material and the hydrodynamics of the site will determine how far it may drift before settling and this will influence the extent and severity of any physical impact outside the immediate dredging site. The chemical characteristics of this material have to be taken into consideration in any risk assessment of the consequences to the environment.

Dredged Material

Waterborne transport is vital to domestic and international commerce. It offers an economical, energy efficient and environmentally friendly transportation for all types of cargo. Dredging is essential to maintain navigable depths in ports and harbors and their approach channels in estuaries and coastal waters (maintenance) as well as for the development of port facilities (capital). Maintenance dredged materials tend generally to consist of sands and silts, whereas capital dredged material may include any of the materials mentioned under dredging. Dredged material can be used for land reclamation or other beneficial uses but most of that generated in marine areas is disposed of in estuaries and coastal waters by dumping from ships or barges. The disposal of dredged material is the largest mass input of wastes deposited directly into the oceans. However, it does not follow that dredged material is the most environmentally significant waste deposited in the ocean, since much dredged material disposal is relocation of sediments already in the marine environment, rather than fresh input.

London Convention data indicate that Contracting Parties dumped around 400 million tonnes dredged material in 2003. However, not all coastal states have signed and ratified the Convention and less than half of the Contracting Parties regularly report on their dumping activities. Data from the International Association of Ports and Harbours indicated that in 1981 around 580 million tonnes of dredged material were being dumped each year and that survey only had data from half the countries contacted. Thus, it would seem likely that actual quantities of dredged material dumped in the world's ocean could be up to 1000 million tonnes per annum.

Regulation

The Specific Guidelines for Assessment of Dredged Material (SGADM) of the global London Convention 1972 is widely accepted as the standard guidance for the management of dredged material disposal at sea. These guidelines were last updated in 2000 and are available on the London Convention website. Regional conventions will often have their own guidance documents that will usually be more specific to take account of local circumstances, for example, The Guidelines for the Management of Dredged Material of the OSPAR Convention last revised in 2004.

Characterization

Dredged material needs to be characterized before disposal as part of the risk assessment procedure if environmental harm is to be minimized. This will usually require consideration of its physical and chemical characteristics and may require assessment of its biological impacts, either inferred from the chemical data, or directly through the conduct of tests for toxicity, bioaccumulation, etc. The physical properties and chemical composition of dredged materials are highly variable depending on the nature of the material concerned and its origin, for example, grain size, mineralogy, bulk properties, organic matter content, and exposure to contamination.

Impacts

In considering the impacts of dredged material disposal, a distinction may be made between natural geological materials not generally exposed to anthropogenic influences, for example, material excavated from beneath the seafloor in deepening a navigation channel and those sediments that have been contaminated by human activities. Although physical, and associated biological, impacts may arise from both types of materials, chemical impacts and their associated biological consequences are usually of particular concern from the latter.

In common with the other materials covered in this article, the physical impact of dredged material disposal is most often the most obvious and significant impact on the aquatic environment. The seabed impacts can be aggravated if the sediment characteristics of the material are very different from that of the sediment at the disposal site or if the material is contaminated with debris such as pieces of cable, scrap metal, and wood.

Most of the material dredged from coastal waters and estuaries is, by its nature, either uncontaminated or only lightly contaminated by human activity (i.e., at, or close to, natural background levels). Sediments in enclosed docks and waterways are often subject to more intense contamination, both because of local practices, and the reduced rate of flushing of such areas. This contamination may be by metals, oil, synthetic organics (e.g., polychlorinated biphenyls, polyaromatic hydrocarbons, and pesticides) or organometallic compounds such as tributyl tin. The latter has been a major concern in recent years due to its previous widespread use in antifouling paints, its consequential occurrence in estuarine sediments in particular, and its wide range of harmful effects to many marine organisms. In recognition of this concern, the international community agreed the International Convention on the Control of Harmful Anti-fouling Systems on Ships, which was adopted on 5 October 2001. This will prohibit the use of harmful organotins in antifouling paints used on ships and will establish a mechanism to prevent the potential future use of other harmful substances in antifouling systems. The convention will enter into force on 17 September 2008.

However, it is generally the case that only a small proportion of dredged material is contaminated to an extent that either major environmental constraints need to be applied when depositing these sediments or the material has to be removed to land for treatment or to specialized confinement facilities.

Beneficial Uses

Dredged material is increasingly being regarded as a resource rather than a waste. Worldwide, around 90% of dredged material is either uncontaminated or only lightly contaminated by human activity so that it can be acceptable for a wide range of uses. The London Convention SGADM recognizes this and requires that possible beneficial uses of dredged material be considered as the first step in examining dredged material management options. Beneficial uses of dredged materials can include beach nourishment, coastal protection, habitat development or enhancement and land reclamation. Operational feasibility is a crucial aspect of many beneficial uses, that is, the availability of suitable material in the required amounts at the right time sufficiently close to the use site. PIANC produced practical guidance on beneficial uses of dredged material in 1992 that is currently (2008) in the process of being revised.

Sand/Gravel Extraction

Introduction

Sand and gravel is dredged from the seabed in various parts of the world for, among other things, land reclamation, concreting aggregate, building sand, beach nourishment, and coastal protection. The largest producer in the world is Japan at around $80-100 \, \text{Mm}^3$ per year, Hong Kong at $25-30 \, \text{Mm}^3$ per year, the Netherlands and the UK regularly producing some $20-30 \, \text{Mm}^3$ per year, and Denmark, the Republic of Korea, and China lesser amounts.

Reclamation with Marine Dredged Sand and Gravel

In recent years, some very large land reclamations have taken place for port and airport developments particularly in Asia. For example, in Hong Kong some $170 \, \text{Mm}^3$ was placed for the new artificial island airport development with another $80 \, \text{Mm}^3$ used for port developments over the period 1990–98. Demand projections indicate a need for a further $300 \, \text{Mm}^3$ by 2010. Also, in Singapore the Jurong

Island reclamation project initiated in 1999 was projected to require $220 \, \text{Mm}^3$ of material over 3 years. However, in 2003 Singapore estimated that it needed 1.8 billion cubic meters of sand over the following 8 years for reclamation works including Tuas View, Jurong Island, and Changi East. The planned Maasvlakte 2 extension of Rotterdam Port in the Netherlands will reclaim some 2000 ha and require $400 \, \text{Mm}^3$ of sand with construction due to take place between 2008 and 2014.

Regulation

National authorities generally carry out the regulation of sand and gravel extraction and they may have guidance on the environmental assessment of the practice. There is currently no accepted international guidance other than for the Baltic Sea area under the Helsinki Convention and the North-East Atlantic under the OSPAR Convention that has adopted the ICES guidance.

Impacts

The environmental impacts of sand and gravel dredging depend on the type and particle size of the material being dredged, the dredging technique used, the hydrodynamic situation of the area and the sensitivity of biota to disturbance, turbidity, or sediment deposition. Screening of cargoes as they are loaded is commonly employed when dredging for sand or gravel to ensure specific sand:gravel ratios are retained in the dredging vessel. Exceptionally, this can involve the rejection of up to 5 times as much sediment over the side of the vessel as is kept as cargo. When large volumes of sand or gravel are used in land reclamation, the runoff from placing the material may contain high levels of suspended sediment. The potential effects of this runoff are very similar to the impacts from dredging itself. The *ICES Cooperative Research Report No. 247* published in 2001 deals with the effects of extraction of marine sediments on the marine ecosystem. It summarizes the impacts of dredging for sand and gravel. An update of this report is due for completion in 2007.

Physical impacts   The most obvious and immediate impacts of sand and gravel extraction are physical ones arising from the following.

• *Substrate removal and alteration of bottom topography.* Trailer suction dredgers leave a furrow in the sediment of up to 2 m wide by about 30 cm deep but stationary (or anchor) suction dredgers may leave deep pits of up to 5 m deep or more. Infill of the pits or furrows created depends on the natural stability of the sediment and the rate of sediment movement due to tidal currents or wave action. This can take many years in some instances. A consequence of significant depressions in the seabed is the potential for a localized drop in current strength resulting in the deposition of finer sediments and possibly a localized depletion in dissolved oxygen.

- *Creation of turbidity plumes in the water column.* This results mainly from the overflow of surplus water/sediment from the spillways of dredgers, the rejection of unwanted sediment fractions by screening, and the mechanical disturbance of the seabed by the draghead. It is generally accepted that the latter is of relatively small significance compared to the other two sources. Recent studies in Hong Kong and the UK indicate that the bulk of the discharged material is likely to settle to the seabed within 500 m of the dredger but the very fine material (<0.063 mm) may remain in suspension over greater distances due to the low settling velocity of the fine particles.
- *Redeposition of fines from the turbidity plumes and subsequent sediment transport.* Sediment that settles out from plumes will cover the seabed within and close to the extraction site. It may also be subject to subsequent transport away from the site of deposition due to wave and tidal current action since it is liable to have less cohesion and may be finer (due to screening) than undredged sediments.

Chemical impacts   The chemical effects of sand and gravel dredging are likely to be minor due to the very low organic and clay mineral content of commercial sand and gravel deposit from geological deposits not generally exposed to anthropogenic influences and to the generally limited spatial and temporal extent of the dredging operations.

Biological impacts   The biological impacts of sand and gravel dredging derive from the physical impacts described above and the most obvious impacts are on the benthic biota. The consequences are as follows.

- *Substrate removal and alteration of bottom topography.* This is the most obvious and immediate impact on the ecosystem. Few organisms are likely to survive intact as a result of passage through a dredger but damaged specimens may provide a food source to scavenging invertebrates and fish. Studies in Europe on dredged areas show very large depletions in species abundance, number of taxa, and biomass of benthos immediately

following dredging. The recolonization process following cessation of dredging depends primarily on the physical stability of the seabed sediments and that is closely related to the hydrodynamic situation. Significant recovery toward the original faunal state of the benthos depends on having seabed sediment of similar characteristics to that occurring before dredging. The biota of mobile sediments tends to be more resilient to disturbance and able to recolonize more quickly than the more long-lived biota of stable environments. Generally, sands and sandy gravels have been found to have recovery times of 2–4 years. However, recovery may take longer where rare slow-growing animals were present prior to dredging. The fauna of coarser deposit are likely to recover more slowly, taking from 5 to 10 years. Recolonizing fauna may move in from adjacent undredged areas or result from larval settlement from the water column.

- *Creation of turbidity plumes in the water column.* As these tend to be limited in space and time, they are not likely to be of great significance for water column fauna unless dredging takes place frequently or continuously.
- *Redeposition of fines from the turbidity plumes and subsequent sediment transport.* Deposition of fines may smother the benthos in and around the area actually dredged but may not cause mortality where the benthos is adapted to dealing with such a situation naturally, for example, in areas with mobile sediments. The deposited fines may be transported away from the site of deposition and affect more distant areas, particularly if the deposited sediment is finer than that naturally on the seabed as a result of screening.

Industrial Solids

Introduction

Industries based around coasts and estuaries have historically used those waters for waste disposal whether by direct discharge or by dumping at sea from vessels. Many of these industries do not generate significant quantities of solid wastes but some, particularly mining and those processing bulk raw materials, may do so.

The sea may be impacted by mining taking place on land, where waste materials are disposed into estuaries and the sea by river inputs, direct tipping, dumping at sea or through discharge pipes. Examples include china clay waste discharged into coastal waters of Cornwall (southwest England) via rivers, colliery waste dumped off the coast of northeast England, and tailings from metalliferous mines in Norway and Canada tipped into fjords and coastal waters. Alternatively, mining may itself be a water-borne activity, such as the mining of diamonds in Namibian waters or tin in Malaysian waters, with waste material discharged directly into the sea.

Fly ash has been dumped at sea off the coast of northeast England since the early 1950s, although this ceased in 1992, and was also carried out off the coast of Israel in the Mediterranean Sea between 1988 and 1998. Fly ash is derived from 'pulverized fuel ash' from the boilers of coal power stations but is mainly 'fly ash' extracted from electrostatic filters and other air pollution control equipment of coal- and oil-fired power stations. This fine material is mainly composed of silicon dioxide with oxides of aluminum and iron but may carry elemental contaminants condensed onto its surfaces following the combustion process.

Surplus munitions, including chemical munitions, were disposed of at sea in various parts of the world in large quantities after both World Wars I and II. In addition, a number of countries have made regular but smaller disposals at sea since 1945. In earlier times much of this material was deposited on continental shelves, but since around 1970 most disposals have taken place in deep ocean waters. OSPAR has collated information from its contracting parties on munitions disposal into a database that is available on its website.

Regulation

The London Convention 1972 has regulated the dumping of industrial wastes at sea since it came into force in 1974. Annex I of the Convention was amended with effect from 1 January 1996 to ban the dumping at sea of industrial wastes by its Contracting Parties. However, not all coastal states have signed and ratified the Convention and less than half of the Contracting Parties regularly report on their dumping activities. The 1996 Protocol to the London Convention restricts the categories of wastes permitted to be considered for disposal at sea to the 7 listed in Annex 2 of the Protocol, which is very similar to the amended Annex I of the Convention referred to above. The protocol was amended in late 2006 to allow the sequestration of carbon dioxide in sub-seabed geological structures as one of a range of climate change mitigation options. In the period 2000–04 Japan dumped some 1.0 million tonnes per annum of bauxite residues into the sea. This activity was controversial as some contracting parties to the Convention regarded the material as industrial waste, and thus banned from disposal, while Japan maintained that it was an Inert Material of Natural Origin and thus allowed to be disposed of at sea. However, Japan announced in 2005 that it would phase out this disposal by 2015.

Characterization

Like dredged material, these materials need to be characterized before disposal can be permitted to ensure that significant environmental harm is not caused as a result of its disposal.

Impacts

The impacts of the disposal of solid industrial wastes into the sea are as described in the introduction, although chemical and biological impacts may be more significant than for some other solid wastes due to the constituents of the materials. However, physical impacts are usually dominant due to the bulk properties of these materials. Fly ash has a special property (pozzolanic activity) that causes the development of strong cohesiveness between the particles on contact with water. This leads to aggregation of the particles and the formation of concretions that may cover all or part of the seabed.

Beneficial Uses

Beneficial uses can often be found for these materials but depend strongly on the characteristics of the material and the opportunities for use in the area of production. Inorganic materials may be used for example for fill, land reclamation, road foundations, or building blocks, and organic materials may be composted, used as animal feed or as a fertilizer.

Plastics and Litter

Introduction

There has been growing concern over the last decade or two about the increasing amounts of persistent plastics and other debris found at sea and on the coastline. Since the 1940s there has been an enormous increase in the use of plastics and other synthetic materials that have replaced many natural and more degradable materials. These materials are generally very durable and cheap but his means that they tend to be both very persistent and readily discarded. They are often buoyant so that they can accumulate in shoreline sinks.

Regulation

A number of international conventions including the London Convention 1972, MARPOL 73/78, and regional seas conventions ban the disposal into the sea of persistent plastics. Annex V of the International Convention for the Prevention of Pollution from Ships 1973, as modified by the 1978 Protocol (known as MARPOL 73/78), states that "the disposal into the sea of all plastics, including but not limited to synthetic ropes, synthetic fishing nets and plastic garbage bags, is prohibited." An exception is made for the accidental loss of synthetic fishing nets provided all reasonable precautions have been taken to prevent such loss. This Annex came into force in 1988. Guidelines for implementing Annex V of MARPOL 73/78 suggest best practice for waste handling on vessels. Many parties to MARPOL 73/78 have built additional waste reception facilities at ports and/or expanded their capacity in order to encourage waste materials to be landed rather than disposed of into the sea. Annex I of the London Convention 1972 prohibits the dumping of "persistent plastics and other persistent synthetic materials, for example netting and ropes, which may float or remain in suspension in the sea in such a manner as to interfere materially with fishing, navigation or other legitimate uses of the sea." Since these materials are not listed in Annex 2 of the 1996 Protocol to the London Convention, this instrument also bans them. All land-based sources of these materials fall under national regulation.

Impacts

Persistent plastics and other synthetic materials present a threat to marine wildlife, as well as being very unsightly and potentially affecting economic interests in recreation areas. They are a visible reminder that the ocean is being used as a dump for plastic and other wastes. These materials also present a threat to navigation through entangling propellers and blocking of cooling water systems of vessels. In 1975 the US Academy of Sciences estimated that 6.4 million tonnes of litter was being discarded each year by the shipping and fishing industries. According to other estimates in a 2005 UNEP report some 8 million items of marine litter are dumped into seas and oceans everyday, approximately 5 million of which (solid waste) are thrown overboard or lost from ships. It has also been estimated that over 13 000 pieces of plastic litter are floating on every square kilometer of ocean today. The UNEP Regional Seas Programme is currently (2006) developing a series of global and regional activities aimed at controlling, reducing, and abating the problem. The debris that is most likely to be disposed of or lost can be divided into three groups.

Fishing gear and equipment Nets that are discarded or lost can continue to trap marine life whether floating on the surface, on the bottom, or at some intermediate level. Marine mammals, fish, seabirds, and turtles are among the animals caught. In 1975 the UN Food and Agriculture Organization estimated that 150 000 tonnes of fishing gear was lost annually worldwide.

Strapping bands and synthetic ropes These are used to secure cargoes, strap boxes, crates, or packing cases and hold materials on pallets. When pulled off rather than cut, the discarded bands may encircle marine mammals or large fish and become progressively tighter as the animal grows. They also entangle limbs, jaws, heads, etc., and affect the animal's ability to move or eat. Plastic straps for four or six packs of cans or bottles can affect smaller animals in a similar way.

Miscellaneous other debris This covers a wide variety of materials including plastic bags or sheeting, packing material, plastic waste materials, and containers for beverages and other liquids. There are numerous studies of turtles, whales, and other marine mammals that were apparently killed by ingesting plastic bags or sheeting. Turtles appear particularly vulnerable to this type of pollution, perhaps by mistaking it for their normal food, for example, jellyfish. A potentially more serious problem may be the increasing quantities of small plastic particles widely found in the ocean, probably from plastics production and insulation, and packing materials. The principal impacts of this material are via ingestion affecting animals' feeding digestion processes. These plastic particles have been found in 25% of the world's sea bird species in one study and up to 90% of the chicks of a single species in another study.

See also

Benthic Organisms Overview.

Further Reading

Arnaudo R (1990) The problem of persistent plastics and marine debris in the oceans. In: *Technical Annexes to the GESAMP Report on the State of the Marine Environment*, UNEP Regional Seas Reports and Studies No. 114/1, Annex I, 20pp. London: UNEP

Duedall IW, Ketchum BH, Park PK, and Kester DR (1983) Global inputs, characteristics and fates of ocean-dumped industrial and sewage wastes: An overview. In: Duedall IW, Ketchum BH, Park PK, and Kester DR (eds.) *Wastes in the Ocean, Industrial and Sewage Wastes in the Ocean*, vol. 1, pp. 3–45. New York, NY: Wiley.

IADC/CEDA (1996–99) *Environmental Aspects of Dredging Guides 1–5*. The Hague: International Association of Dredging Companies.

IADC/IAPH (1997) *Dredging for Development*. The Hague: International Association of Dredging Companies.

ICES (2001) Effects of extraction of marine sediments on the marine ecosystem. *International Council for the Exploration of the Sea, Cooperative Research Report* 247.

Kester DR, Ketchum BH, Duedall IW, and Park PK (1983) The problem of dredged material disposal. In: Kester DR, Ketchum BH, Duedall IW, and Park PK (eds.) *Wastes in the Ocean, Dredged-Material Disposal in the Ocean*, vol. 2, pp. 3–27. New York: Wiley.

Newell RC, Seiderer LJ, and Hitchcock DR (1998) The impact of dredging on biological resources of the seabed. *Oceanography and Marine Biology Annual Review* 36: 127–178.

Thompson RC, Olsen Y, Mitchell RP, *et al.* (2004) Lost at sea: Where is all the plastic? *Science* 304: 838.

United Nations Environment Programme (2005) *Marine Litter: An Analytical overview*. Nairobi: UNEP.

Relevant Websites

http://www.helcom.fi
 – Helsinki Commission: Baltic Marine Environment Protection Commission.

http://www.ices.dk
 – International Council for the Exploration of the Sea.

http://www.londonconvention.org
 – London Convention 1972.

http://www.ospar.org
 – OSPAR Commission.

http://www.pianc-aipcn.org
 – PIANC.

http://www.unep.org
 – United Nations Environment Programme (UNEP).

http://www.woda.org
 – World Organisation of Dredging Associations.

NUCLEAR FUEL REPROCESSING AND RELATED DISCHARGES

H. N. Edmonds, University of Texas at Austin, Port Aransas, TX, USA

Introduction

Oceanographers have for some decades made use of human perturbations to the environment as tracers of ocean circulation and of the behavior of similar substances in the oceans. The study of the releases of anthropogenic radionuclides – by-products of the nuclear industry – by nuclear fuel reprocessing plants is an example of such an application. Other examples of the use of anthropogenic substances as oceanographic tracers addressed in this Encyclopedia include tritium and radiocarbon, largely resulting from atmospheric nuclear weapons testing in the 1950s and 1960s, and chlorofluorocarbons (CFCs), a family of gases which have been used in a variety of applications such as refrigeration and polymer manufacture since the 1920s. As outlined in this article, the source function of nuclear fuel reprocessing discharges, which is very different from that of either weapons test fallout or CFCs, is both a blessing and a curse, defining the unique applicability of these tracers but also requiring much additional and careful quantification.

The two nuclear fuel reprocessing plants of most interest to oceanographers are located at Sellafield, in the UK, which has been discharging wastes into the Irish Sea since 1952, and at Cap de la Hague, in north-western France, which has been discharging wastes into the English Channel since 1966. These releases have been well documented and monitored in most cases. It should be noted, however, that the releases of some isotopes of interest to oceanographers have not been as well monitored throughout the history of the plants because they were not of radiological concern, were difficult to measure, or both. This is particularly true for ^{99}Tc and ^{129}I, which are the isotopes currently of greatest interest to the oceanographic community but for which specific, official release data are only available for the last two decades or less. The radionuclides most widely studied by ocean scientists have been ^{137}Cs, ^{134}Cs, ^{90}Sr, ^{125}Sb, ^{99}Tc, ^{129}I, and Pu isotopes. The Sellafield plant has dominated the releases of most of these, with the exception of ^{125}Sb and, particularly in the 1990s, ^{129}I. In general, the Sellafield releases have been particularly well documented and are the easiest for interested scientists and other parties to obtain. The figures of releases presented in this article are based primarily on the Sellafield data, except where sufficient data are available for Cap de la Hague.

The location of the Sellafield and Cap de la Hague plants is important to the oceanographic application of their releases. As discussed below, the releases enter the surface circulation of the high latitude North Atlantic and are transported northwards into the Norwegian and Greenland Seas and the Arctic Ocean. They have been very useful as tracers of ventilation and deep water formation in these regions, which are of great importance to global thermohaline circulation and climate. In addition, the study of the dispersion of radionuclides from the reprocessing plants at Sellafield and Cap de la Hague serves as an analog for understanding the ultimate distribution and fate of other wastes arising from industrial activities in north-western Europe.

Much of the literature on reprocessing releases in the oceans has focused on, or been driven by, concerns relating to contamination and radiological effects, and they have not found the same broad or general oceanographic application as some other anthropogenic tracers despite their great utility and the excellent quality of published work. This may be attributable to a variety of factors, including perhaps (1) the contaminant-based focus of much of the work and its publication outside the mainstream oceanographic literature, (2) the complications of the source function as compared with some other tracers, and (3) the comparative difficulty of measuring many of these tracers and the large volumes of water that have typically been required. This article addresses the historic and future oceanographic applications of these releases.

A Note on Units

In most of the literature concerning nuclear fuel reprocessing releases, amounts (both releases and ensuing environmental concentrations) are expressed in activity units. The SI unit for activity is the becquerel (Bq), which replaced the previously common curie (Ci), a unit which was based on the activity of one gram of radium. These two units are related as follows:

$$1 Bq = 1 \text{disintegration per second}$$
$$1 Ci = 3.7 \times 10^{10} Bq$$

The conversion from activity to mass or molar units is dependent on the half-life of the isotope in question. Activity (A) is the product of the number of atoms of the isotope present (N) and its decay constant (λ): $A = \lambda N$. The decay constant, λ, is related to the half-life by $t_{1/2} = (ln2)/\lambda$. Using this equation, it is simple to convert from Bq to mol, given the decay constants in units of s^{-1}, and Avogadro's number of 6.023×10^{23} atoms mol^{-1}.

It is also worth noting that the total activity of naturally occurring radionuclides in sea water is approximately $12 kBq \ m^{-3}$. Most of this activity is from the long-lived naturally occuring isotope, $^{40}K(t_{1/2} = 1.25 \times 10^9$ years). Activities contributed from anthropogenic radionuclides greatly exceed this (millions of Bq m^{-3}) in the immediate vicinity of the Sellafield and La Hague outfall pipes. However, most oceanographic studies of these releases involve measurements of much lower activities: up to 10s of Bq m^{-3} in the case of ^{137}Cs, and generally even lower for other radionuclides, for instance mBq m^{-3} or less for plutonium isotopes and ^{129}I.

Nuclear Fuel Reprocessing and Resulting Tracer Releases

Origin and Description of Reprocessing Tracers

Nuclear fuel reprocessing involves the recovery of fissile material (plutonium and enriched uranium) and the separation of waste products from 'spent' (used) fuel rods from nuclear reactors. In the process, fuel rods, which have been stored for a time to allow short-lived radionuclides to decay, are dissolved and the resulting solution is chemically purified and separated into wastes of different composition and activity. Routine releases from the plants to the environment occur under controlled conditions and are limited to discharge totals dictated by the overseeing authorities of each country. The discharges have varied over the years as a function of the amount and type of fuel processed and changes in the reprocessing technology. The discharge limits themselves have changed, in response to monitoring efforts and also spurred by technological advances in waste-treatment capabilities. As a general rule, these changes have resulted in decreasing releases for most nuclides (most notably cesium and the actinide elements), but

there are exceptions. For instance, the Enhanced Actinide Removal Plant (EARP) was constructed at the Sellafield site in the 1990s in order to enable the additional treatment and subsequent discharge of a backlog of previously stored wastes. Although the new technology enabled the removal of actinide elements from these wastes, it is not effective at removing ^{99}Tc. Therefore an allowance was made for increased discharge of ^{99}Tc, up to 200 TBq per year. The resulting pulse of increased ^{99}Tc discharges from Sellafield beginning in 1994 is currently being followed with great interest, as is discussed in more detail below.

The end result of these processes is the availability of a suite of oceanographic tracers with different discharge histories (e.g., in terms of the timing and magnitude of spikes) and a range of half-lives, and thus with a range of utility and applicability to studies of oceanographic processes at a variety of spatial and temporal scales. A brief summary of the reprocessing radionuclides most widely applied in oceanography is shown in **Table 1**, and examples of discharge histories are given in **Figure 1**. It has also been particularly useful in some cases to measure the ratio of a pair of tracers, for instance $^{134}Cs/^{137}Cs$, $^{137}Cs/^{90}Sr$, or potentially, $^{99}Tc/^{129}I$. The use of isotope ratios can (1) provide temporal information and aid the estimation of rates of circulation, (2) mitigate the effects of mixing which will often alter individual concentrations more than ratios, and (3) aid in distinguishing the relative contributions of different sources of the nuclides in question. Finally, differences in the chemical behavior of the different elements released can be exploited to study different processes. Although most of the widely applied tracers are largely conservative in sea water and therefore serve as tracers of water movement, the actinides, particularly Pu, have a high affinity for particulate material and accumulate in the sediments. These tracers are then useful for studying sedimentary processes.

The Reprocessing Tracer Source Function

The primary difference between the north-western European reprocessing releases and other anthropogenic tracers used in oceanography is the nature of their introduction to the oceans. Reprocessing releases enter the oceans essentially at a point source, rather than in a more globally distributed fashion as is the case for weapons test fallout or the chlorofluorocarbons. Thus, reprocessing tracers are excellent, specific tracers for waters originating from north-western Europe. Because these waters are transported to the north into the Nordic Seas and

Table 1 Summary of the major reprocessing tracers and their applications

Isotope	Half-life (years)	Sources	Applications
^{137}Cs	30	Weapons testing, reprocessing (mostly Sellafield), Chernobyl	The 'signature' reprocessing tracer, used in the earliest studies of Sellafield releases. Has been applied in European coastal waters, the Nordic Seas, Arctic Ocean, and deep North Atlantic
^{90}Sr	28	Weapons testing, reprocessing (mostly Sellafield)	In combination with ^{137}Cs, early tracer for Sellafield discharges. The ^{137}Cs/^{90}Sr ratio was used to distinguish reprocessing and weapons sources
^{134}Cs	2.06	Reprocessing and Chernobyl	In combination with ^{137}Cs, this short-lived isotope has been used in the estimation of transit times, e.g., from Sellafield to the northern exit of the Irish Sea
^{125}Sb	2.7	Reprocessing, mostly Cap de la Hague	Circulation in the English Channel and North Sea, into the Skaggerak and Norwegian coastal waters
^{99}Tc	213 000	Reproducing – large pulse from Sellafield beginning in 1994. Some weapons testing	Surface circulation of European coastal waters, the Nordic Seas, Arctic Ocean, and East and West Greenland Currents
^{129}I	15.7×10^6	Reprocessing, mostly Cap de la Hague since 1990, some weapons testing	Deep water circulation in the Nordic Seas, Arctic Ocean, and Deep Western Boundary Current of the Atlantic Ocean. It has also been measured in European coastal waters and the Gulf of Mexico
^{241}Pu	14	Weapons testing, reprocessing (mostly Sellafield)	Not widely used as circulation tracers but often measured in conjunction with other reprocessing radionuclides. Other transuranic elements that have been measured include ^{241}Am and ^{237}Np
^{239}Pu	24 000		
^{240}Pu	6000		
^{238}Pu	88		

Adapted in part from Dahlgaard (1995).

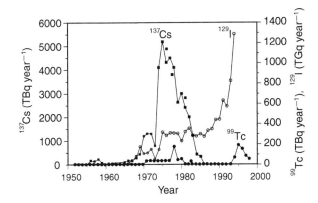

Figure 1 Examples of the source functions of reprocessing tracers to the oceans, illustrating some of the differences in magnitude and timing of the releases. ^{137}Cs data, from Gray et al. (1995) and ^{99}Tc data, courtesy of Peter Kershaw, are for liquid discharges from Sellafield. ^{129}I release information is courtesy of G. Raisbeck and F. Yiou, and combines available information for Sellafield from 1966 and Cap de la Hague from 1975. Note the different scales used for the releases of the three isotopes.

Arctic Ocean, reprocessing releases are particularly sensitive tracers of the climatically important deep water formation processes that occur in these regions.

There are several complications associated with this point source tracer introduction, however. Comparison with CFCs indicates some of these difficulties. Within each hemisphere (northern or southern), CFCs are well mixed throughout the troposphere, and the time history of their concentrations is well known. Their entry into the oceans occurs by equilibration with surface waters, and the details of their solubilities as a function of temperature and salinity have been well characterized. The primary complication is that in some areas equilibrium saturation is not reached, and so assumptions must be made about the degree of equilibration. Where they have been necessary, these assumptions appear to be fairly robust. In the case of reprocessing tracers, the discharge amounts have not always been as well known, although this situation has improved continuously. One major difficulty arises in translating a discharge amount, in kg or Bq per year, or per month, to a concentration in sea water some time later. In order to do this, the surface circulation of the coastal regions must be very well known. This circulation is highly variable, on daily, seasonal, interannual and decadal timescales, further complicating the problem. Also, other sources of the radionuclides under consideration must often be taken into account. Depending on the tracer, these sources may be important and/or numerous, and due to the secrecy involved in many aspects of the operation of nuclear installations the necessary information may not always be available.

Other Sources of Anthropogenic Radionuclides

Other sources of radionuclides to the oceans complicate the interpretation of Sellafield and Cap de la Hague tracers to varying extents, depending on the isotope under consideration, the location, and the time. First there is the need to understand the mixing of signals from these two plants. Other sources may include fallout from nuclear weapons tests, uncontrolled releases due to nuclear accidents, dumping on the seabed, other reprocessing plants, atmospheric releases from Sellafield and Cap de la Hague, and unknown sources. Many of these other sources are small compared to the Sellafield and Cap de la Hague releases. The primary complicating source for many isotopes is nuclear weapons test fallout, which peaked in the 1950s and again, more strongly, in 1962–63. This source has been particularly important for ^{137}Cs and ^{90}Sr. The Chernobyl accident in 1986 also released significant amounts of radioactive materials which have themselves found use as oceanographic tracers. In terms of comparison to releases from Sellafield, the Chernobyl accident has been most important with respect to ^{134}Cs and ^{137}Cs.

In addition, it has recently been noted that in the case of some nuclides, particularly ^{137}Cs and Pu isotopes, the sediments of the Irish Sea have become a significant ongoing source of tracers to the North Atlantic. Although generally considered a conservative tracer, ^{137}Cs exhibits some affinity for particulate material and a significant amount has accumulated in the sediments around Sellafield. With the continuing reduction of ^{137}Cs activities in liquid effluents from Sellafield, release from the sediments, either by resuspension of the sediments or by reequilibration with the reduced seawater concentrations, has become a relatively large (though still small in an absolute sense compared to the liquid discharges of the 1970s) contributor to the current flux out of the Irish Sea, and will continue to be such for years to come.

A further complication of the use of some tracers derived from Sellafield and Cap de la Hague, but one that cannot be lamented, is the continuing reduction of the releases, as well as the radioactive decay of those with the shorter half-lives, such as ^{137}Cs which was released in large quantities over 20 years ago. With respect to many of these complications, and for

several other reasons, the reprocessing radionuclides attracting the most attention in the oceanographic community today are ^{99}Tc and ^{129}I. Both are long lived and have fairly small relative contributions from weapons testing and other sources. Unlike most radionuclides, their releases increased in the 1990s, and the recent releases of each are largely dominated by a single source: ^{99}Tc by Sellafield and ^{129}I by Cap de la Hague. Significant advances have been and are being made in the measurement of these isotopes, allowing their measurement on smaller sample volumes. The activity of ^{129}I is so low that it is measured by accelerator mass spectrometry. This technique allows measurement of ^{129}I on 1 liter seawater samples. Advances in ^{99}Tc measurement, using both radiochemical and mass-spectrometric (ICP-MS) techniques, are continuing. The primary limitation of ^{99}Tc studies continues to be the comparative difficulty of its measurement. This is also true to some extent for ^{129}I, for although the sample sizes have been greatly reduced the measurement requires highly specialized technology. A further complication for ^{129}I is its volatility, and the fact that as much as 10% of the reprocessing discharges have been released directly to the atmosphere. Nevertheless, the promise of both tracers is such that these difficulties are likely to be overcome.

Regional Setting and Circulation of Reprocessing Discharges

A summary of the regional circulation into which the liquid effluents from Sellafield and Cap de la Hague are released is presented in **Figure 2**. It is particularly important to note that studies of the reprocessing discharges have contributed greatly to the development of this detailed picture of the regional circulation. Briefly, from the Sellafield site the waste stream is carried north out of the Irish Sea, around the coast of Scotland, through the North Sea, and into the northward-flowing Norwegian Coastal Current (NCC). Transport across the North Sea occurs at various latitudes: some fraction of the reprocessing releases 'short-circuits' across the northern part of the North Sea, while some flows farther south along the eastern coast of the UK before turning east and north. Recent studies of the EARP ^{99}Tc pulse from Sellafield have suggested that the rate and preferred transport path of Sellafield releases across the North Sea into the NCC may vary in relation to climatic conditions in the North Atlantic such as the North Atlantic Oscillation (NAO).

Radionuclides discharged from the Cap de la Hague reprocessing plant flow north-east through the English Channel and into the North Sea,

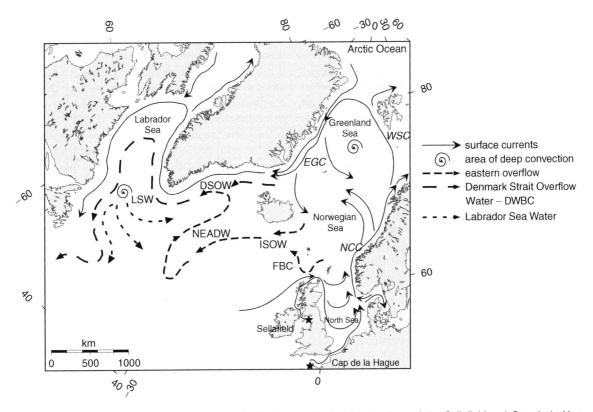

Figure 2 Map of the northern North Atlantic and Nordic Seas, indicating the locations of the Sellafield and Cap de la Hague reprocessing plants (stars), and the major circulation pathways relevant to the discussion of the dispersal of reprocessing wastes. Surface currents are indicated with thin solid lines, and deep waters, (deep overflows from the Nordic Seas, the resulting Deep Western Boundary Current (DWBC) and Labrador Sea Water (LSW)), with heavy dashed lines. Curled arrows indicate the two major areas where deep waters are formed by convective processes. Additional abbreviations: NCC, Norwegian Coastal Current; WSC, West Spitsbergen Current; EGC, East Greenland Current; FBC, Faroe Banks Channel; ISOW, Iceland Scotland Overflow Water; NEADW, Northeast Atlantic Deep Water; DSOW, Denmark Strait Overflow Water.

following the coast and joining the Sellafield releases in the NCC. A small amount of the Sellafield releases and some of the Cap de la Hague releases, which flow closer to the coast, flow east through the Skaggerak and Kattegat to enter the Baltic Sea.

The NCC is formed of a mixture of coastal waters (containing the reprocessing tracers) and warm, saline North Atlantic surface waters. Dilution of the reprocessing signal with Atlantic water continues along the northward flow path of the NCC. There is evidence from reprocessing tracers that turbulent eddies between the NCC and Atlantic water result in episodic transport westward into the surface waters of the Norwegian Sea, in addition to a fairly well-defined westward advective transport towards Jan Mayen Island. The NCC branches north of Norway, with one branch, the North Cape (or Nordkap) Current, flowing eastward through the Barents Sea and thence into the Kara Sea and Arctic Ocean, and the remainder flowing north and west as part of the West Spitsbergen Current (WSC). This latter flow branches in the Fram Strait west of Spitsbergen, with some recirculation to the west and south joining the

southward flowing East Greenland Current (EGC), and the remainder entering the Arctic Ocean. The majority of surface outflow from the Arctic is through the Fram Strait into the EGC, thus the bulk of the reprocessing nuclides entering the Arctic Ocean will eventually exit to the Nordic (Greenland, Iceland, and Norwegian) Seas and the North Atlantic. The presence of high concentrations of radionuclides derived from reprocessing in the surface waters of the Barents and Nordic Seas is also an indication of their utility as tracers of deep-water ventilation and formation in these regions, as discussed below.

The Use of Reprocessing Releases in Oceanography

Historical Background

The first papers reporting measurements of Sellafield-derived ^{137}Cs in coastal waters appeared in the early 1970s. Much of this early work arose from monitoring efforts by the Division of Fisheries

Research of the UK's Ministry of Agriculture, Fisheries, and Food, which holds a joint oversight role on the Sellafield discharges and which continues to this day, now as the Centre for Environment, Fisheries and Aquaculture Science, to be a leader in studies of the distribution and oceanographic application of reprocessing radionuclides. In the late 1970s scientists at the Woods Hole Oceanographic Institution published the first of a number of papers detailing the unique utility of the reprocessing tracers and demonstrating their application. Leading studies of reprocessing tracers in the oceans have been undertaken by researchers in numerous other countries affected by the radiological implications of the releases, including France, Germany, Denmark, Canada, Norway, and the former Soviet Union.

Coastal and Surface Circulation

A great deal of work has been published using the documented releases of radioisotopes from Sellafield and Cap de la Hague to study the local circulations of the Irish Sea, North Sea, and English Channel. Early studies of the Sellafield releases examined a variety of isotopes, including ^{134}Cs, ^{137}Cs, ^{90}Sr, and Pu isotopes. Most attention focused on ^{137}Cs and ^{90}Sr, and their activity ratio, because: (1) the releases of these two isotopes were well documented, (2) they had been studied extensively since the 1950s and 1960s in weapons test fallout; and (3) the ^{137}Cs/^{90}Sr ratio in reprocessing releases (particularly those from Sellafield) was significantly higher than in global fallout and thus could be used to distinguish the sources of these isotopes in a given water sample. The use of the ^{134}Cs/^{137}Cs ratio enabled estimates of transit times, assuming the initial ratio in the releases was constant and making use of the short half-life of ^{134}Cs. The short-lived isotope ^{125}Sb has been used as a specific tracer of the circulation of Cap de la Hague discharges through the English Channel and North Sea, into the Baltic and the Norwegian Coastal Current.

Summaries of transit times and dilution factors for the transport of Sellafield and Cap de la Hague discharges to points throughout the North Sea, Norwegian Coastal Current, Barents and Kara Seas, Greenland Sea, and East and West Greenland Currents have been published in recent reviews. Numbers are not included in this article because they are currently under revision. In terms of transport to the NCC, where the two waste streams are generally considered to merge, the consensus has been that the transit time from Sellafield to about 60°N is three to four years, and that from Cap de la Hague to the same area is one to two years. Compilations of 'transfer factors,' which relate observed concentrations to the discharge amounts, and factor in the transit times, suggest that the two reprocessing waste streams meet in approximately equal proportions in the NCC. In other words, if Sellafield and Cap de la Hague released equal amounts of a radionuclide, with the Cap de la Hague release two years later, they would make equal contributions to the NCC.

Recently, detailed studies have been undertaken of the dispersal of the EARP ^{99}Tc pulse from Sellafield, which began in 1994. These studies have suggested substantially shorter circulation times for Sellafield releases to northern Scottish coastal waters and across the North Sea to the NCC than previously accepted. For instance, the ^{99}Tc pulse reached the NCC within 2.5 years, rather than three to four. It has been suggested that this is a real difference between sampling periods, resulting from climatically induced circulation changes in the North Atlantic. Some of the difference may also be related to the fact that much more detailed data, both in terms of seawater sampling and regarding the releases, are available on this recent event. The continuing passage of the EARP ^{99}Tc signal promises to be very useful in the Nordic Seas and Arctic Ocean as well.

Deep-water Formation in the North Atlantic and Arctic Oceans

In addition to surface water flows, deep-water formation processes within the Arctic Ocean and Nordic Seas have been elucidated through the study of reprocessing releases. In a classic presentation of reprocessing tracer data, Livingston showed that the surface water distribution of ^{137}Cs in the Nordic Seas in the early 1980s was marked by high concentrations at the margins of the seas, highlighting the delivery of the isotope in the northward-flowing NCC and WSC to the west and the return flow in the southward-flowing EGC to the east. The opposite distribution was found in the deep waters, with higher concentrations in the center of the Greenland Basin than at the margins, as a result of the ventilation of the Greenland Sea Deep Water by deep convective processes in the center of the gyre.

In the Arctic Ocean, elevated ^{137}Cs and ^{90}Sr concentrations and ^{137}Cs/^{90}Sr ratios at 1500 m at the LOREX ice station near the North Pole in 1979 indicated that deep layers of the Arctic Ocean were ventilated from the shelves. Similar observations in deep water north of Fram Strait provided early evidence suggesting a contribution of dense brines from the Barents Sea shelf to the bottom waters of the Nansen Basin. Reprocessing tracers, particularly those, like ^{129}I and ^{99}Tc, which have only a small contribution from other sources such as weapons fallout, hold great promise for illuminating eastern Arctic Ocean deep water ventilation processes from the Barents and Kara Sea shelves. The NCC delivers reprocessing tracers to the Barents and Kara relatively rapidly ($\sim$ 5 years) and more importantly in high concentration. With high tracer concentrations in the area of interest as a source water, the reprocessing tracers may be particularly sensitive tracers of a contribution of dense shelf waters to the deep Arctic Ocean.

In addition to the deep waters formed through convection in the polar regions (most notably in the Greenland Sea) which fill the deep basins of the Nordic Seas, intermediate waters are formed which subsequently overflow the sills between Greenland, Iceland, and Scotland and ventilate the deep North Atlantic. The presence of ^{137}Cs and ^{90}Sr from Sellafield was reported in the overflow waters immediately south of the Denmark Straits sampled during the Transient Tracers in the Ocean (TTO) program in 1981. Later, it was demonstrated that reprocessing cesium and strontium could be distinguished in the deep waters as far as TTO Station 214, off the Grand Banks of Newfoundland. No samples were taken for reprocessing radionuclides further south as part of that study, but it was clear in retrospect that the reprocessing signal had traveled even further in the Deep Western Boundary Current (DWBC). Recent work on ^{129}I has highlighted the utility of reprocessing radionuclides as tracers of northern source water masses and the DWBC of the Atlantic. Profiles in stations south of the overflows show much clearer tracer signals for ^{129}I than for the CFCs, and ^{129}I has been detected in the DWBC as far south as Cape Hatteras. As with the cesium studies of the 1980s, it is likely that sampling further south will reveal the tracer there as well.

Conclusions

Releases of radionuclides from the nuclear fuel-reprocessing plants at Sellafield and Cap de la Hague have provided tracers for detailed studies of the circulations of the local environment into which they are released, namely the Irish Sea, English Channel, and North Sea, and for the larger-scale circulation processes of the North Atlantic and Arctic Oceans. These tracers have very different source functions for their introduction into the oceans compared to other widely used anthropogenic tracers, and in some cases compared to each other. The fact that they are released at point sources makes them highly specific

tracers of several interesting processes in ocean circulation, but the nature of the releases has complicated their quantitative interpretation to some extent.

Recent advances have been made in the measurement of ^{129}I and ^{99}Tc, long-lived tracers whose releases have increased in recent years and which experience little complication from other sources. These two tracers hold great promise for elucidating deep water formation and ventilation processes in the North Atlantic, Nordic Seas, and Arctic Ocean in the years to come.

See also

Radioactive Wastes.

Further Reading

Aarkrog A, Dahlgaard H, Hallstadius L, Hansen H, and Hohm E (1983) Radiocaesium from Sellafield effluents in Greenland waters. *Nature* 304: 49–51.

Dahlgaard H (1995) Transfer of European coastal pollution to the Arctic: radioactive tracers. *Marine Pollution Bulletin* 31: 3–7.

Gray J, Jones SR, and Smith AD (1995) Discharges to the environment from the Sellafield Site, 1951–1992. *Journal of Radiological Protection* 15: 99–131.

Jefferies DF, Preston A, and Steele AK (1973) Distribution of caesium-137 in British coastal waters. *Marine Pollution Bulletin* 4: 118–122.

Kershaw P and Baxter A (1995) The transfer of reprocessing wastes from north-west Europe to the Arctic. *Deep-Sea Research II* 42: 1413–1448.

Kershaw PJ, McCubbin D, and Leonard KS (1999) Continuing contamination of north Atlantic and Arctic waters by Sellafield radionuclides. *Science of Total Environment* 237/238: 119–132.

Livingston HD (1988) The use of Cs and Sr isotopes as tracers in the Arctic Mediterranean Seas. *Philosophical Transactions of the Royal Society of London* A 325: 161–176.

Livingston HD, Bowen VT, and Kupferman SL (1982) Radionuclides from Windscale discharges I: non-equilibrium tracer experiments in high-latitude oceanography. *Journal of Marine Research* 40: 253–272.

Livingston HD, Bowen VT, and Kupferman SL (1982) Radionuclides from Windscale discharges II: their dispersion in Scottish and Norwegian coastal circulation. *Journal of Marine Research* 40: 1227–1258.

Livingston HD, Swift JH, and Östlund HG (1985) Artificial radionuclide tracer supply to the Denmark Strait Overflow between 1972 and 1981. *Journal of Geophysical Research* 90: 6971–6982.

RADIOACTIVE WASTES

L. Føyn, Institute of Marine Research, Bergen, Norway

Introduction

The discovery and the history of radioactivity is closely connected to that of modern science. In 1896 Antoine Henri Becquerel observed and described the spontaneous emission of radiation by uranium and its compounds. Two years later, in 1898, the chemical research of Marie and Pierre Curie led to the discovery of polonium and radium.

In 1934 Frédéric Joliot and Irène Curie discovered artificial radioactivity. This discovery was soon followed by the discovery of fission and the enormous amounts of energy released by this process. However, few in the then limited community of scientists working with radioactivity believed that it would be possible within a fairly near future to establish enough resources to develop the fission process for commercial production of energy or even think about the development of mass-destruction weapons.

World War II made a dramatic change to this. The race that began in order to be the first to develop mass-destruction weapons based on nuclear energy is well known. Following this came the development of nuclear reactors for commercial production of electricity. From the rapidly growing nuclear industry, both military and commercial, radioactive waste was produced and became a problem. As with many other waste problems, discharges to the sea or ocean dumping were looked upon as the simplest and thereby the best and final solution.

The Sources

Anthropogenic radioactive contamination of the marine environment has several sources: disposal at sea, discharges to the sea, accidental releases and fallout from nuclear weapon tests and nuclear accidents. In addition, discharge of naturally occurring radioactive materials (NORM) from offshore oil and gas production is a considerable source for contamination.

The marine environment receives in addition various forms of radioactive components from medical, scientific and industrial use. These contributions are mostly short-lived radionuclides and enter the local marine environment through diffuse outlets like muncipal sewage systems and rivers.

Disposal at Sea

The first ocean dumping of radioactive waste was conducted by the USA in 1946 some 80 km off the coast of California. The International Atomic Energy Agency (IAEA) published in August 1999 an 'Inventory of radioactive waste disposal at sea' according to which the disposal areas and the radioactivity can be listed as shown in **Table 1**.

Figure 1 shows the worldwide distribution of disposal-points for radioactive waste.

The majority of the waste disposed consists of solid waste in various forms and origin, only 1.44% of the total activity is contributed by low-level liquid waste. The disposal areas in the north-east Atlantic and the Arctic contain about 95% of the total radioactive waste disposed at sea.

Most disposal of radioactive waste was performed in accordance with national or international regulations. Since 1967 the disposals in the northeast Atlantic were for the most part conducted in accordance with a consultative mechanism of the Organization for Economic Co-operation and Development/Nuclear Energy Agency (OECD/NEA).

The majority of the north-east Atlantic disposals were of low-level solid waste at depths of 1500–5000 m, but the Arctic Sea disposals consist of various types of waste from reactors with spent fuel to containers with low-level solid waste dumped in fairly shallow waters ranging from about 300 m depth in the Kara Sea to less than 20 m depth in some fiords on the east coast of Novaya Zemlya.

Most of the disposals in the Arctic were carried out by the former Soviet Union and were not reported internationally. An inventory of the USSR disposals was presented by the Russian government in 1993. Already before this, the good collaboration between Russian and Norwegian authorities had led

Table 1 Worldwide disposal at sea of radioactive waste[a]

North-west Atlantic Ocean	2.94 PBq
North-west Atlantic Ocean	2.94 PBq
Arctic Ocean	38.37 PBq
North-east Pacific Ocean	0.55 PBq
West Pacific Ocean	0.89 PBq

[a]PBq (petaBq) = 10^{15} Bq (1 Bq1 = disintegration s^{-1}). The old unit for radioactivity was Curie (Ci); 1 Ci3.7 × 10^{10} Bq.

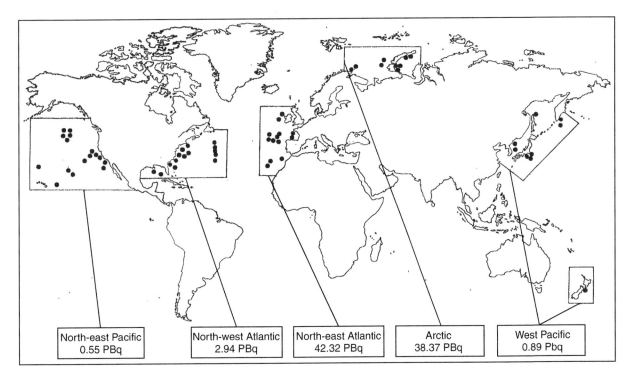

Figure 1 The worldwide location for disposal of radioactive waste at sea.

to a joint Norwegian–Russian expedition to the Kara Sea in 1992 followed by two more joint expeditions, in 1993 and 1994. The main purpose of these expeditions was to locate and inspect the most important dumped objects and to collect samples for assessing the present environmental impact and to assess the possibility for potential future leakage and environmental impacts.

These Arctic disposals differ significantly from the rest of the reported sea disposals in other parts of the world oceans as most of the dumped objects are found in shallow waters and some must be characterized as high-level radioactive waste, i.e. nuclear reactors with fuel. Possible releases from these sources may be expected to enter the surface circulation of the Kara Sea and from there be transported to important fisheries areas in the Barents and Norwegian Seas. **Figures 2** and **3** give examples of some of the radioactive waste dumped in the Arctic. Pictures were taken with a video camera mounted on a ROV (remote operated vehicle). The ROV was also equipped with a NaI-detector for gamma-radiation measurements and a device for sediment sampling close to the actual objects.

The Global Convention on the Prevention of Marine Pollution by Dumping of Wastes and Other Matter was adopted by an Intergovernmental Conference in London in 1972. The convention named the London Convention 1972, formerly the London Dumping Convention (LDC), addressed from the

very beginning the problem of radioactive waste. But it was not until 20 February 1994 that a total prohibition on radioactive waste disposal at sea came into force.

Discharges to the Sea

Of the total world production of electricity about 16% is produced in nuclear power plants. In some countries nuclear energy counts for the majority of the electricity produced, France 75% and Lithuania 77%, and in the USA with the largest production of nuclear energy of more than 96 000 MWh this accounts for about 18% of the total energy production.

Routine operations of nuclear reactors and other installations in the nuclear fuel cycle release small amounts of radioactive material to the air and as liquid effluents. However, the estimated total releases of ^{90}Sr, ^{131}I and ^{137}Cs over the entire periods of operation are negligible compared to the amounts released to the environment due to nuclear weapon tests.

Some of the first reactors that were constructed used a single-pass cooling system. The eight reactors constructed for plutonium production at Hanford, USA, between 1943 and 1956, pumped water from Columbia River through the reactor cores then delayed it in cooling ponds before returning it to the river. The river water and its contents of particles and components were thereby exposed to a great neutron

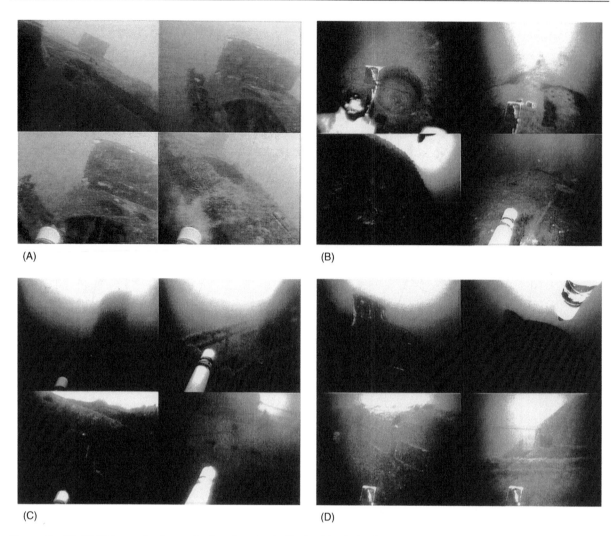

(A)

(B)

(C)

(D)

Figure 2   (A)–(D) Pictures of a disposed submarine at a depth of *c.* 30 m in the Stepovogo Fiord, east coast of Novaya Zemyla. The submarine contains a sodium-cooled reactor with spent fuel. Some of the hatches of the submarine are open which allows for 'free' circulation of water inside the vessel.

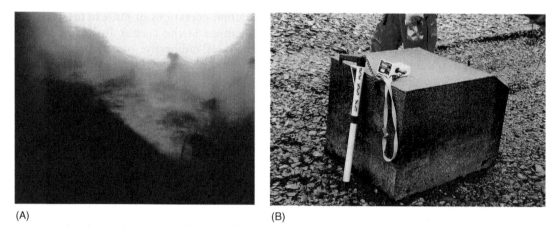

(A)

(B)

Figure 3   (A) Containers of low level solid waste at the bottom of the Abrosimov Fiord at a depth of *c.* 15 m and (B) a similar container found washed ashore.

flux and various radioactive isotopes were created. In addition corrosion of neutron-activated metal within the reactor structure contributed to the radioactive contamination of the cooling water. Only a limited number of these radionuclides reached the river mouth and only ^{32}P, ^{51}Cr, ^{54}Mn and ^{65}Zn were detected regularly in water, sediments and marine organisms in the near-shore coastal waters of the US Pacific Northwest.

Reactors operating today all have closed primary cooling systems that do not allow for this type of contamination. Therefore, under normal conditions production of electricity from nuclear reactors does not create significant amounts of operational discharges of radionuclides. However, the 434 energy-producing nuclear plants of the world in 1998 created radioactive waste in the form of utilized fuel. Utilized fuel is either stored or reprocessed.

Only 4–5% of the utilized nuclear fuel worldwide is reprocessed. Commercial, nonmilitary, reprocessing of nuclear fuel takes place in France, Japan, India and the United Kingdom. Other reprocessing plants defined as defense-related are in operation and producing waste but without discharges. For example in the USA, at the Savannah River Plant and the Hanford complex, about 83 000 m^3 and 190 000 m^3, respectively, of high-level liquid waste was in storage in 1985.

Reprocessing plants and the nuclear industry in the former Soviet Union have discharged to the Ob and Yenisey river systems ending up in the Arctic ocean. In 1950–51 about 77×10^6 m^3 liquid waste of 100 PBq was discharged to the River Techa. The Techa River is connected to the River Ob as is the Tomsk River where the Tomsk-7, a major production site for nuclear weapons plutonium, is situated. Other nuclear plants, such as the Krasnoyarsk industrial complex, have discharged to the Yenisey river. Large amounts of radioactive waste are also stored at the sites.

Radioactive waste stored close to rivers has the potential of contaminating the oceans should an accident happen to the various storage facilities.

The commercial reprocessing plants in France at Cap de la Hague and in the UK at Sellafield have for many years, and still do, contributed to the radioactive contamination of the marine environment. They both discharge low-level liquid radioactive effluents to the sea. Most important, however, these discharges and their behavior in the marine environment have been and are still thoroughly studied and the results are published in the open literature. The importance of these discharges is extensive as radionuclides from Sellafield and la Hague are traced throughout the whole North

Table 2 Total discharges of some radionuclides from Sellafield 1952–92

^{3}H	39 PBq
^{90}Sr	6.3 PBq
^{134}Cs	5.8 PBq
^{137}Cs	41.2 PBq
^{238}Pu	0.12 PBq
^{239}Pu	0.6 PBq
^{241}Pu	21.5 PBq
^{241}Am	0.5 PBq

Atlantic. Most important is Sellafield; **Table 2** summarizes the reported discharge of some important radionuclides.

In addition a range of other radionuclides have been discharged from Sellafield, but prior to 1978 the determination of radionuclides was, for many components, not specific. Technetium (^{99}Tc) for instance was included in the 'total beta' determinations with an estimated annual discharge from 1952 to 1970 below 5 TBq and from 1970 to 1977 below 50 TBq. Specific determination of ^{99}Tc in the effluents became part of the routine in 1978 when about 180 TBq was discharged followed by about 50 TBq in 1979 and 1980 and then an almost negligible amount until 1994.

The reason for mentioning ^{99}Tc is that this radionuclide, in an oceanographic context, represents an almost ideal tracer in the oceans. Technetium is most likely to be present as pertechnetate, TcO_4^-, totally dissolved in seawater; it acts conservatively and moves as a part of the water masses. In addition the main discharges of technetium originate from point sources with good documentation of time for and amount of the release. The discharges from Sellafield are a good example of this. From 1994 the UK authorities have allowed for a yearly ^{99}Tc discharge of up to 200 TBq.

Based on surveys before and after the discharges in 1994, 30 TBq (March–April) and 32 TBq (September–October), the transit time for technetium from the Irish Sea to the North Sea was calculated to be considerably faster than previous estimations of transit times for released radionuclides. This faster transport is demonstrated by measurements indicating that the first discharge plume of ^{99}Tc had reached the south coast of Norway before November 1996 in about 2.5 years compared to the previously estimated transit time of 3–4 years.

Other reprocessing plants may have discharges to the sea, but without a particular impact in the world oceans. The reprocessing plant at Trombay, India, may, for example, be a source for marine contamination.

Accidental Releases

Accidents resulting in direct radioactive releases to the sea are not well known as most of them are connected to wreckage of submarines. Eight nuclear submarines with nuclear weapons have been reported lost at sea, two US and six former USSR. The last known USSR wreck was the submarine *Komsomolets* which sank in the Norwegian Sea southwest of Bear Island, on 7 April 1989. The activity content in the wreck is estimated by Russian authorities to be 1.55–2.8 PBq ^{90}Sr and 2.03–3 PBq ^{137}Cs and the two nuclear warheads on board may contain about 16 TBq ^{239,240}Pu equivalent to 6–7 kg plutonium. Other estimates indicate that each warhead may contain 10 kg of highly enriched uranium or 4–5 kg plutonium.

On August 12th 2000, the Russian nuclear submarine *Kursk* sank at a depth of 108 meters in the Barents Sea north of the Kola peninsula. Vigorous explosions in the submarine's torpedo-chambers caused the wreckage where 118 crew-members were entrapped and lost their lives. *Kursk*, and Oscar II attack submarine, was commissioned in 1995 and was powered by two pressurized water reactors. *Kursk* had no nuclear weapons on board. Measurements close to the wreck in the weeks after the wreckage showed no radioactive contamination indicating that the primary cooling-systems were not damaged in the accident. A rough inventory calculation estimates that the reactors at present contain about 56 000 TBq. Russian authorities are planning for a salvage operation where the submarine or part of the submarine will be lifted from the water and transported to land. Both a possible salvage operation or to leave the wreck where it is will be create a demand for monitoring as the location of the wreck is within important fishing grounds.

The wreckage of *Komsomolets* in 1989 and the attempts to raise money for an internationally financed Russian led salvage operation became very public. The Russian explanation for the intensive attempts of financing the salvage was said to be the potential for radioactive pollution. The wreck of the submarine is, however, located at a depth of 1658 m and possible leaching of radionuclides from the wreck will, due to the hydrography of the area, hardly have any vertical migration and radioactive components will spread along the isopycnic surfaces gradually dispersing the released radioactivity in the deep water masses of the Nordic Seas. An explanation for the extensive work laid down for a salvage operation and for what became the final solution, coverage of the torpedo-part of the hull, may be that this submarine was said to be able to fire its torpedo missiles with nuclear warheads from a depth of 1000 m.

In 1990, the Institute of Marine Research, Bergen, Norway, started regular sampling of sediments and water close to the wreck of *Komsomolets*. Values of ^{137}Cs were in the range 1–10 Bq per kg dry weight sediment and 1–30 Bq per m^3 water. No trends were found in the contamination as the variation between samples taken at the same date were equal to the variation observed from year to year. Detectable amounts of ^{134}Cs in the sediment samples indicate that there is some leaching of radioactivity from the reactor.

Accidents with submarines and their possible impact on the marine environment are seldom noticed in the open literature and there is therefore little common knowledge available. An accident, however, that is well known is the crash of a US B-52 aircraft, carrying four nuclear bombs, on the ice off Thule air base on the northwest coast of Greenland in January 1968. Approximately 0.4 kg plutonium ended up on the sea floor at a depth of 100–300 m. The marine environment became contaminated by about 1 TBq ^{239,240}Pu which led to enhanced levels of plutonium in benthic animals, such as bivalves, sea-stars and shrimps after the accident. This contamination has decreased rapidly to the present level of one order of magnitude below the initial levels.

Fallout from Nuclear Weapon Tests and Nuclear Accidents

Nuclear weapon tests in the atmosphere from 1945 to 1980 have caused the greatest man-made release of radioactive material to the environment. The most intensive nuclear weapon tests took place before 1963 when a test-ban treaty signed by the UK, USA and USSR came into force. France and China did not sign the treaty and continued some atmospheric tests, but after 1980 no atmospheric tests have taken place.

It is estimated that 60% of the total fallout has initially entered the oceans, i.e. 370 PBq ^{90}Sr, 600 PBq ^{137}Cs and 12 PBq ^{239,240}Pu. Runoff from land will slightly increase this number. As the majority of the weapon tests took place in the northern hemisphere the deposition there was about three times as high as in the southern hemisphere.

Results from the GEOSECS expeditions, 1972–74, show a considerable discrepancy between the measured inventories in the ocean of 900 PBq ^{137}Cs, 600 PBq ^{90}Sr and 16 PBq ^{239,240}Pu and the estimated input from fallout. The measured values are far higher than would be expected from the assumed fallout data. Thus the exact input of anthropogenic radionuclides may be partly unknown or the geographical coverage of the measurements in the oceans were for some areas not dense enough for accurate calculations.

Another known accident contributing to marine contamination was the burn-up of a US satellite (SNAP 9A) above the Mozambique channel in 1964 which released 0.63 PBq ^{238}Pu and 0.48 TBq ^{239}Pu; 73% was eventually deposited in the southern hemisphere.

The Chernobyl accident in 1986 in the former USSR is the latest major event creating fallout to the oceans. Two-thirds of the c.100 PBq ^{137}Cs released was deposited outside the Soviet Union. The total input to the world oceans of ^{137}Cs from Chernobyl is estimated to be from 15–20 Pbq, i.e. 4.5 PBq in the Baltic Sea; 3–5 PBq in the Mediterranean Sea, 1.2 PBq in the North Sea and about 5 PBq in the northeast Atlantic.

Natural Occurring Radioactive Material

Oil and gas production mobilize naturally occurring radioactive material (NORM) from the deep underground reservoir rock. The radionuclides are primarily ^{226}Ra, ^{228}Ra and ^{210}Pb and appear in sludge and scales and in the produced water. Scales and sludge containing NORM represent an increasing amount of waste. There are different national regulations for handling this type of waste. In Norway, for example, waste containing radioactivity above 10 Bq g^{-1} is stored on land in a place specially designed for this purpose. However, there are reasons to believe that a major part of radioactive contaminated scales and sludge from the worldwide offshore oil and gas production are discharged to the sea.

Reported NORM values in scales are in the ranges of 0.6–57.2 Bq g^{-1} ^{226}Ra + ^{228}Ra (Norway), 0.4–3700 Bq g^{-1} (USA) and 1–1000 Bq g^{-1} ^{226}Ra (UK).

Scales are an operational hindrance in oil and gas production. Frequent use of scale-inhibitors reduce the scaling process but radioactive components are released to the production water adding to its already elevated radioactivity. More than 90% of the radioactivity in produced water is due to ^{226}Ra and ^{228}Ra having a concentration 100–1000 times higher than normal for seawater.

The discharge of produced water is a continuous process and the amount of water discharged is considerable and increases with the age of the production wells. As an example, the estimated amount of produced water discharged to the North Sea in 1998 was 340 million m^3 and multiplying by an average value of 5 Bq^{-1} of ^{226}Ra in produced water, the total input of ^{226}Ra to the North Sea in 1998 was 1.7 TBq.

Discussion

The total input of anthropogenic radioactivity to the world's oceans is not known exactly, but a very rough estimate gives the following amounts: 85 PBq dumped, 100 PBq discharged from reprocessing and 1500 PBq from fallout. Some of the radionuclides have very long half-lives and will persist in the ocean, for example ^{99}Tc has a half-life of 2.1×10^5 years, ^{239,240}Pu, 2.4×10^4 years and ^{226}Ra, 1600 years. ^{137}Cs, ^{90}Sr and ^{228}Ra with half-lives of 30 years, 29 years and 5.75 years, respectively, will slowly decrease depending on the amount of new releases.

In an oceanographic context it is worth mentioning the differences in denomination between radioactivity and other elements in the ocean. The old denomination for radioactivity was named after Curie (Ci) and 1 g radium was defined to have a radioactivity of 1 Ci; 1 Ci 3.7×10^{10} Bq and 1 PBq 27 000 Ci. Therefore released radioactivity of 1 PBq can be compared to the radioactivity of 27 kg radium.

The common denominations for major and minor elements in seawater are given in weight per volume. For comparison if 1 PBq or 27 kg radium were diluted in 1 km^3 of seawater, this would give a radium concentration of 0.027 µg l^{-1} or 1000 Bq l^{-1}. Calculations like this clearly visualize the sensitivity of the analytical methods used for measuring radioactivity. In the Atlantic Ocean for example radium (^{228}Ra) has a concentration of 0.017–3.40 mBq l^{-1}, whereas ^{99}Tc measured in surface waters off the southwest coast of Norway is in the range of 0.9–6.5 mBq l^{-1}.

Measured in weight the total amount of radionuclides do not represent a huge amount compared to the presence of nonradioactive components in seawater. The radioisotopes of cesium and strontium are both important in a radioecological context since they have chemical behavior resembling potassium and calcium, respectively. Cesium follows potassium in and out of the soft tissue cells whereas strontium follows calcium into bone cells and stays. Since uptake and release in organisms is due to the chemical characteristics and rarely if the element is radioactive or not, radionuclides such as ^{137}Cs and ^{90}Sr have to compete with the nonradioactive isotopes of cesium and strontium.

Oceanic water has a cesium content of about 0.5 l µg^{-1} and a strontium content of about 8000 µg l^{-1}. Uptake in a marine organism is most likely to be in proportion to the abundance of the radioactive and the nonradioactive isotopes of the actual element. This can be illustrated by the following example. The sunken nuclear submarine *Komsomolets* contained an estimated (lowest) amount of 1.55 PBq ^{90}Sr (about 300 g) and 2.03 PBq

^{137}Cs (about 630 g). If all this was released at once and diluted in the immediate surrounding 1 km^3 of water the radioactive concentration would have been 1550 Bq l^{-1} for ^{90}Sr and 2030 Bq l^{-1} for ^{137}Cs, the concentration in weight per volume would have been 0.000 3 μg l^{-1} ^{90}Sr and 0.000 63 μg l^{-1} ^{137}Cs. This means that even if the radioactive material was kept in the extremely small volume of 1 km^3, compared to the volume of the deep water of the Norwegian Sea available for a primary dilution, the proportion of radioactive to nonradioactive isotopes of strontium and cesium, available for uptake in marine organisms, would have been about 2.7×10^6 and 7.9×10^{-5}, respectively.

From the examples above it can be seen that if uptake, and thereby impact, in marine organisms follows regular chemical–physiological rules there is a 'competition' in seawater in favor of the nonradioactive isotopes for elements normally present in seawater. Measurable amount of radionuclides of cesium and strontium are detected in marine organisms but at levels far below the concentrations in freshwater fish. Average concentrations of ^{137}Cs in fish from the Barents Sea during the period with the most intensive nuclear weapon tests in that area, 1962–63, never exceeded 90 Bq kg^{-1} fresh weight, whereas fallout from Chernobyl resulted in concentrations in freshwater fish in some mountain lakes in Norway far exceeding 10 000 Bq kg^{-1}.

For radionuclides like technetium and plutonium, which will persist in the marine environment, uptake will be based only on the actual concentrations in seawater of radionuclide. The levels of ^{99}Tc, for example, increased in seaweed (*Fucus vesiculosus*) from 70 Bq per kg dry weight (December 1997) to 124 Bq kg^{-1} in January 1998 in northern Norway which reflected the increased concentration in the water as the peak of the technetium plume from Sellafield reached this area.

Previously the effects of anthropogenic radioactivity have been based on the possible dose effect to humans. Most of the modeling work has been concentrated on assessing the dose to critical population groups eating fish and other marine organisms. But even if the radiation from anthropogenic radionuclides to marine organisms is small compared to natural radiation from radionuclides like potassium, ^{40}K, the presence of additional radiation may give a chronic exposure with possible effects, at least on individual marine organisms.

The input of radioactivity, NORM, from the offshore oil and gas production may also give reason for concern. The input will increase as it is a continuous part of the production. Even if radium as the main radionuclide is not likely to be taken up by marine organisms the use of chemicals like scale inhibitors may change this making radium more available for marine organisms.

Conclusion

The sea began receiving radioactive waste from anthropogenic sources in 1946, in a rather unregulated way in the first decades. Both national and international regulations controlling disposals have now slowly come into force. Considerable amounts are still discharged regularly from nuclear industries and the practice of using the sea as a suitable wastebasket is likely to continue for ever. In 1994 an international total prohibition on radioactive waste disposal at sea came into force, but the approximately 85 PBq of solid radioactive waste that has already been dumped will sooner or later be gradually released to the water masses.

Compared to other wastes disposed of at sea the amount of radioactive waste by weight is rather diminutive. However, contrary to most of the 'ordinary' wastes in the sea, detectable amounts of anthropogenic radioactivity are found in all parts of the world oceans and will continue to contaminate the sea for many thousands of years to come. This means that anthropogenic radioactive material has become an extra chronic radiation burden for marine organisms. In addition, the release of natural occurring radionuclides from offshore oil and gas production will gradually increase the levels of radium, in particular, with a possible, at present unknown, effect.

However, marine food is not, and probably never will be, contaminated at a level that represents any danger to consumers. The ocean has always received debris from human activities and has a potential for receiving much more and thereby help to solve the waste disposal problems of humans. But as soon as a waste product is released and diluted in the sea it is almost impossible to retrieve. Therefore, in principal, no waste should be disposed of in the sea without clear documentation that it will never create any damage to the marine environment and its living resources. This means that with present knowledge no radioactive wastes should be allowed to be released into the sea.

See also

Nuclear   Fuel   Reprocessing   and   Related Discharges.

Further Reading

Guary JC, Guegueniat P, and Pentreath RJ (eds.) (1988) *Radionuclides: A Tool for Oceanography.* London, New York: Elsevier Applied Science.

Hunt GJ, Kershaw PJ, and Swift DJ (eds.) (1998) Radionuclides in the oceans (RADOC 96–97). Distribution, Models and Impacts. *Radiation Protection Dosimetry* 75: 1–4.

IAEA (1995) *Environmental impact of radioactive releases*; Proceedings of an International Symposium on Environmental Impact of Radioactive Releases. Vienna: International Atomic Energy Agency.

IAEA (1999) *Inventory of Radioactive Waste Disposals at Sea.* IAEA-TECDOC-1105 Vienna: International Atomic Energy Agency. pp. 24 A.1–A.22.

ATMOSPHERIC INPUT OF POLLUTANTS

R. A. Duce, Texas A&M University, College Station, TX, USA

Introduction

For about a century oceanographers have tried to understand the budgets and processes associated with both natural and human-derived substances entering the ocean. Much of the early work focused on the most obvious inputs – those carried by rivers and streams. Later studies investigated sewage outfalls, dumping, and other direct input pathways for pollutants. Over the past decade or two, however, it has become apparent that the atmosphere is also not only a significant, but in some cases dominant, pathway by which both natural materials and contaminants are transported from the continents to both the coastal and open oceans. These substances include mineral dust and plant residues, metals, nitrogen compounds from combustion processes and fertilizers, and pesticides and a wide range of other synthetic organic compounds from industrial and domestic sources. Some of these substances carried into the ocean by the atmosphere, such as lead and some chlorinated hydrocarbons, are potentially harmful to marine biological systems. Other substances, such as nitrogen compounds, phosphorus, and iron, are nutrients and may enhance marine productivity. For some substances, such as aluminum and some rare earth elements, the atmospheric input has an important impact on their natural chemical cycle in the sea.

In subsequent sections there will be discussions of the input of specific chemicals via the atmosphere to estuarine and coastal waters. This will be followed by considerations of the atmospheric input to open ocean regions and its potential importance. The atmospheric estimates will be compared with the input via other pathways when possible. Note that there are still very large uncertainties in all of the fluxes presented, both those from the atmosphere and those from other sources. Unless otherwise indicated, it should be assumed that the atmospheric input rates have uncertainties ranging from a factor of 2 to 4, sometimes even larger.

Estimating Atmospheric Contaminant Deposition

Contaminants present as gases in the atmosphere can exchange directly across the air/sea boundary or they may be scavenged by rain and snow. Pollutants present on particles (aerosols) may deposit on the ocean either by direct (dry) deposition or they may also be scavenged by precipitation. The removal of gases and/or particles by rain and snow is termed wet deposition.

Direct Deposition of Gases

Actual measurement of the fluxes of gases to a water surface is possible for only a very few chemicals at the present time, although extensive research is underway in this area, and analytical capabilities for fast response measurements of some trace gases are becoming available. Modeling the flux of gaseous compounds to the sea surface or to rain droplets requires a knowledge of the Henry's law constants and air/sea exchange coefficients as well as atmospheric and oceanic concentrations of the chemicals of interest. For many chemicals this information is not available. Discussions of the details of these processes of air/sea gas exchange can be found in other articles in this volume.

Particle Dry Deposition

Reliable methods do not currently exist to measure directly the dry deposition of the full size range of aerosol particles to a water surface. Thus, dry deposition of aerosols is often estimated using the dry deposition velocity, v_d. For dry deposition, the flux is then given by:

$$F_d = v_d \cdot C_a \qquad [1]$$

where F_d is the dry deposition flux (e.g., in $g\,m^{-2}\,s^{-1}$), v_d is the dry deposition velocity (e.g., in $m\,s^{-1}$), and C_a is the concentration of the substance on the aerosol particles in the atmosphere (e.g., in $g\,m^{-3}$). In this formulation v_d incorporates all the processes of dry deposition, including diffusion, impaction, and gravitational settling of the particles to a water surface. It is very difficult to parameterize accurately the dry deposition velocity since each of these processes is acting on a particle population, and they are each dependent upon a number of factors, including wind speed, particle size, relative humidity, etc. The following are dry deposition velocities that have been used in some studies of atmospheric deposition of particles to the ocean:

- Submicrometer aerosol particles, $0.001 \, \text{m s}^{-1} \pm$ a factor of three
- Supermicrometer crustal particles not associated with sea salt, $0.01 \, \text{m s}^{-1} \pm$ a factor of three
- Giant sea-salt particles and materials carried by them, $0.03 \, \text{m s}^{-1} \pm$ a factor of two

Proper use of eqn [1] requires that information be available on the size distribution of the aerosol particles and the material present in them.

Particle and Gas Wet Deposition

The direct measurement of contaminants in precipitation samples is certainly the best approach for determining wet deposition, but problems with rain sampling, contamination, and the natural variability of the concentration of trace substances in precipitation often make representative flux estimates difficult using this approach. Studies have shown that the concentration of a substance in rain is related to the concentration of that substance in the atmosphere. This relationship can be expressed in terms of a scavenging ratio, S:

$$S = C_r \cdot \rho \cdot C_{a/g}^{-1} \qquad [2]$$

where C_r is the concentration of the substance in rain (e.g., in g kg^{-1}), ρ is the density of air ($\sim 1.2 \, \text{kg m}^{-3}$), $C_{a/g}$ is the aerosol or gas phase concentration in the atmosphere (e.g., in g m^{-3}), and S is dimensionless. Values of S for substances present in aerosol particles range from a few hundred to a few thousand, which roughly means that $1 \, \text{g}$ (or $1 \, \text{ml}$) of rain scavenges $\subseteq 1 \, \text{m}^3$ of air. For aerosols, S is dependent upon such factors as particle size and chemical composition. For gases, S can vary over many orders of magnitude depending on the specific gas, its Henry's law constant, and its gas/water exchange coefficient. For both aerosols and gases, S is also dependent upon the vertical concentration distribution and vertical extent of the precipitating cloud, so the use of scavenging ratios requires great care, and the results have significant uncertainties. However, if the concentration of an atmospheric substance and its scavenging ratio are known, the scavenging ratio approach can be used to estimate wet deposition fluxes as follows:

$$F_r = P \cdot C_r = P \cdot S \cdot C_{a/g} \cdot \rho^{-1} \qquad [3]$$

where F_r is the wet deposition flux (e.g., in $\text{g m}^{-2} \text{year}^{-1}$) and P is the precipitation rate (e.g., in m year^{-1}), with appropriate conversion factors to translate rainfall depth to mass of water per unit area. Note that $P \cdot S \cdot \rho^{-1}$ is equivalent to a wet deposition velocity.

Atmospheric Deposition to Estuaries and the Coastal Ocean

Metals

The atmospheric deposition of certain metals to coastal and estuarine regions has been studied more than that for any other chemicals. These metals are generally present on particles in the atmosphere. Chesapeake Bay is among the most thoroughly studied regions in North America in this regard. Table 1 provides a comparison of the atmospheric and riverine deposition of a number of metals to Chesapeake Bay. The atmospheric numbers represent a combination of wet plus dry deposition directly onto the Bay surface. Note that the atmospheric input ranges from as low as 1% of the total input for manganese to as high as 82% for aluminum. With the exception of Al and Fe, which are largely derived from natural weathering processes (e.g., mineral matter or soil), most of the input of the other metals is from human-derived sources. For metals with anthropogenic sources the atmosphere is most important for lead (32%).

There have also been a number of investigations of the input of metals to the North Sea, Baltic Sea, and Mediterranean Sea. Some modeling studies of the North Sea considered not only the direct input pathway represented by the figures in Table 1, but also considered Baltic Sea inflow, Atlantic Ocean inflow and outflow, and exchange of metals with the

Table 1 Estimates of the riverine and atmospheric input of some metals to Chesapeake Bay

Metal	Riverine input (10^6 g year^{-1})	Atmospheric input (10^6 g year^{-1})	% Atmospheric input
Aluminum	160	700	81
Iron	600	400	40
Manganese	1300	13	1
Zinc	50	18	26
Copper	59	3.5	6
Nickel	100	4	4
Lead	15	7	32
Chromium	15	1.5	10
Arsenic	5	0.8	14
Cadmium	2.6	0.4	13

Data reproduced with permission from Scudlark JR, Conko KM and Church TM (1994) Atmospheric wet desposition of trace elements to Chesapezke Bay: (CBAD) study year 1 results. *Atmospheric Environment* 28: 1487–1498.

sediments, as well as the atmospheric contribution to all of these inputs. **Figure 1** shows schematically some modeling results for lead, copper, and cadmium. Note that for copper, atmospheric input is relatively unimportant in this larger context, while atmospheric input is somewhat more important for cadmium, and it is quite important for lead, being approximately equal to the inflow from the Atlantic Ocean, although still less than that entering the North Sea from dumping. As regards lead, note that approximately 20% of the inflow from the Atlantic to the North Sea is also derived from the atmosphere. This type of approach gives perhaps the most accurate and in-depth analysis of the importance of

atmospheric input relative to all other sources of a chemical in a water mass.

Nitrogen Species

The input of nitrogen species from the atmosphere is of particular interest because nitrogen is a necessary nutrient for biological production and growth in the ocean. There has been an increasing number of studies of the atmospheric input of nitrogen to estuaries and the coastal ocean. Perhaps the area most intensively studied is once again Chesapeake Bay. **Table 2** shows that approximately 40% of all the nitrogen contributed by human activity to Chesapeake Bay enters via

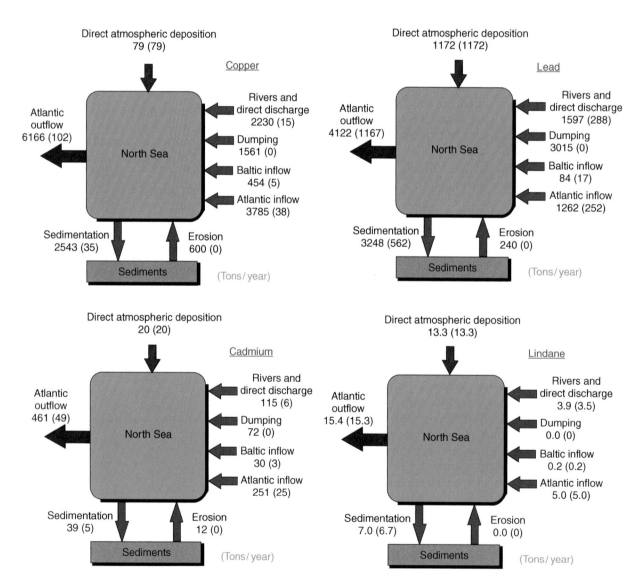

Figure 1 Input of copper, lead, cadmium, and lindane to the North Sea. Values in parentheses denote atmospheric contribution. For example, for copper the atmospheric contribution to rivers and direct discharges is 15 tons per year. (Figure reproduced with permission from Duce, 1998. Data adapted with permission from van den Hout, 1994.)

Table 2 Estimates of the input of nitrogen to Chesapeake Bay

Source	Total input $(10^9 g\ year^{-1})$	Areal input rate $(g\ m^{-2}year^{-1})$	% of the total
Animal waste	5	0.4	3
Fertilizers	48	4.2	34
Point sources	33	2.9	24
Atmospheric precipitation			
nitrate	35	3.1	25
ammonium	19	1.7	14
Total	140	12.3	100

Data reproduced with permission from Fisher D, Ceroso T, Mathew T and Oppenheimer M (1988) *Polluted Coastal Waters: The Role of Acid Rain*. New York: Environmental Defense Fund.

precipitation falling directly on the Bay or its watershed. These studies were different from most earlier studies because the atmospheric contributions were considered not only to be direct deposition on the water surface, but also to include that coming in via the atmosphere but falling on the watershed and then entering the Bay. Note from **Table 2** that atmospheric input of nitrogen exceeded that from animal waste, fertilizers, and point sources. In the case of nitrate, about 23% falls directly on the Bay, with the remaining 77% falling on the watershed. These results suggest that studies that consider only the direct deposition on a water surface (e.g., the results shown in **Table 1**) may significantly underestimate the true contribution of atmospheric input. The total nitrogen fertilizer applied to croplands in the Chesapeake Bay region is $\sim 5.4\ g\ m^{-2}\ year^{-1}$, while the atmospheric nitrate and ammonium nitrogen entering the Bay is $\sim 4.8\ g\ m^{-2}\ year^{-1}$. Chesapeake Bay is almost as heavily fertilized from atmospheric nitrogen, largely anthropogenic, as the croplands are by fertilizer in that watershed!

Results from studies investigating nitrogen input to some other estuarine and coastal regions are summarized in **Table 3**. In this table atmospheric sources for nitrogen are compared with all other sources, where possible. The atmospheric input ranges from 10% to almost 70% of the total. Note that some estimates compare only direct atmospheric deposition with all other sources and some include as part of the atmospheric input the portion of the deposition to the watershed that reaches the estuary or coast.

Synthetic Organic Compounds

Concern is growing about the input of a wide range of synthetic organic compounds to the coastal ocean. To date there have been relatively few estimates of the atmospheric fluxes of synthetic organic compounds to the ocean, and these estimates have significant uncertainties. These compounds are often both persistent and toxic pollutants, and many have relatively high molecular weights. The calculation of the atmospheric input of these compounds to the

Table 3 Estimates of the input of nitrogen to some coastal areas[a]

Region	Total atmospheric input[b] $(10^9 g\ year^{-1})$	Total input all sources $(10^9 g\ year^{-1})$	% Atmospheric input
North Sea	400[c]	1500	27[c]
Western Mediterranean Sea	400[c]	577[d]	69[c]
Baltic Sea	500	~1200	42
Chesapeake Bay	54	140	39
New York Bight	–	–	13[c]
Long Island Sound	11	49	22
Neuse River Estuary, NC	1.7	7.5	23

[a]Data from several sources in the literature.
[b]Total from direct atmospheric deposition and runoff of atmospheric material from the watershed.
[c]Direct atmospheric deposition to the water only.
[d]Total from atmospheric and riverine input only.

Table 4 Estimates of the input of synthetic organic compounds to the North Sea

Organic compound	Atmospheric input (10^6 g year^{-1})	Input from other sources (10^6 g year^{-1})	% Atmospheric input
PCB	40	3	93
Lindane	36	3	92
Polycyclic aromatic hydrocarbons	80	90	47
Benzene	400	500	44
Trichloroethene	300	80	80
Trichloroethane	90	60	94
Tetrachloroethene	100	10	91
Carbon tetrachloride	6	40	13

Data reproduced with permission from Warmerhoven JP, Duiser JA, de Leu LT and Veldt C (1989) *The Contribution of the Input from the Atmosphere to the Contamination of the North Sea and the Dutch Wadden Sea.* Delft, The Netherlands: TNO Institute of Environmental Sciences.

coastal ocean is complicated by the fact that many of them are found primarily in the gas phase in the atmosphere, and most of the deposition is related to the wet and dry removal of that phase. The atmospheric residence times of most of these compounds are long compared with those of metals and nitrogen species. Thus the potential source regions for these compounds entering coastal waters can be distant and widely dispersed.

Figure 1 shows the input of the pesticide lindane to the North Sea. Note that the atmospheric input of lindane dominates that from all other sources. **Table 4** compares the atmospheric input to the North Sea with that of other transport paths for a number of other synthetic organic compounds. In almost every case atmospheric input dominates the other sources combined.

Atmospheric Deposition to the Open Ocean

Studies of the atmospheric input of chemicals to the open ocean have also been increasing lately. For many substances a relatively small fraction of the material delivered to estuaries and the coastal zone by rivers and streams makes its way through the near shore environment to open ocean regions. Most of this material is lost via flocculation and sedimentation to the sediments as it passes from the freshwater environment to open sea water. Since aerosol particles in the size range of a few micrometers or less have atmospheric residence times of one to several days, depending upon their size distribution and local precipitation patterns, and most substances of interest in the gas phase have similar or even longer atmospheric residence times, there is ample opportunity

for these atmospheric materials to be carried hundreds to thousands of kilometers before being deposited on the ocean surface.

Metals

Table 5 presents estimates of the natural and anthropogenic emission of several metals to the global atmosphere. Note that ranges of estimates and the best estimate are given. It appears from **Table 5** that anthropogenic sources dominate for lead, cadmium, and zinc, with essentially equal contributions for copper, nickel, and arsenic. Clearly a significant fraction of the input of these metals from the atmosphere to the ocean could be derived largely from anthropogenic sources.

Table 6 provides an estimate of the global input of several metals from the atmosphere to the ocean and compares these fluxes with those from rivers. Estimates are given for both the dissolved and particulate forms of the metals. These estimates suggest that rivers are generally the primary source of particulate

Table 5 Emissions of some metals to the global atmosphere

Metal	Anthropogenic emissions (10^9 g year^{-1})		Natural emissions (10^9 g year^{-1})	
	Range	Best estimate	Range	Best estimate
Lead	289–376	332	1–23	12
Cadmium	3.1–12	7.6	0.15–2.6	1.3
Zinc	70–194	132	4–86	45
Copper	20–51	35	2.3–54	28
Arsenic	12–26	18	0.9–23	12
Nickel	24–87	56	3–57	30

Data reproduced with permission from Duce *et al.*, 1991.

Table 6   Estimates of the input of some metals to the global ocean

Metal	Atmospheric input		Riverine input	
	Dissolved (10^9 g year^{-1})	Particulate (10^9 g year^{-1})	Dissolved (10^9 g year^{-1})	Particulate (10^9 g year^{-1})
Iron	1600–4800	14 000–42 000	1100	110 000
Copper	14–45	2–7	10	1 500
Nickel	8–11	14–17	11	1 400
Zinc	33–170	11–55	6	3 900
Arsenic	2.3–5	1.3–3	10	80
Cadmium	1.9–3.3	0.4–0.7	0.3	15
Lead	50–100	6–12	2	1 600

Data reproduced with permission from Duce *et al.*, 1991.

metals in the ocean, although again a significant fraction of this material may not get past the coastal zone. For the dissolved phase atmospheric and riverine inputs are roughly equal for metals such as iron, copper, and nickel; while for zinc, cadmium, and particularly lead atmospheric inputs appear to dominate. These estimates were made based on data collected in the mid-1980s. Extensive efforts to control the release of atmospheric lead, which has been primarily from the combustion of leaded gasoline, are now resulting in considerably lower concentrations of lead in many areas of the open ocean. For example, **Figure 2** shows that the concentration of dissolved lead in surface sea water near Bermuda has been decreasing regularly over the past 15–20

years, as has the atmospheric lead concentration in that region. This indicates clearly that at least for very particle-reactive metals such as lead, which has a short lifetime in the ocean (several years), even the open ocean can recover rather rapidly when the anthropogenic input of such metals is reduced or ended. Unfortunately, many of the other metals of most concern have much longer residence times in the ocean (thousands to tens of thousands of years).

Figure 3 presents the calculated fluxes of several metals from the atmosphere to the ocean surface and from the ocean to the seafloor in the 1980s in the tropical central North Pacific. Note that for most metals the two fluxes are quite similar, suggesting the potential importance of atmospheric input to the

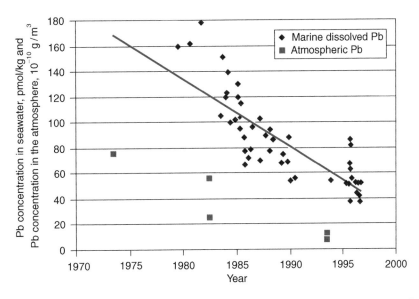

Figure 2   Changes in concentration of atmospheric lead at Bermuda and dissolved surface oceanic lead near Bermuda from the mid-1970s to the mid-1990s. (Data reproduced with permission from Wu J and Boyle EA (1997) Lead in the western North Atlantic Ocean: Completed response to leaded gasoline phaseout. *Geochimica et Cosmochimica Acta* 61: 3279–3283; and from Huang S, Arimoto R and Rahn KA (1996) Changes in atmospheric lead and other pollution elements at Bermuda. *Journal of Geophysical Research* 101: 21 033–21 040.)

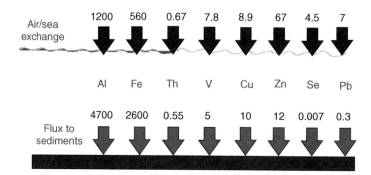

Figure 3 A comparison of the calculated fluxes of aluminum (Al), iron (Fe), thorium (Th), vanadium (V), copper (Cu), zinc (Zn), selenium (Se), and lead (Pb) (in 10^{-9} g cm^{-2} year^{-1}) from the atmosphere to the ocean and from the ocean to the sediments in the central tropical North Pacific. For each metal note the relative similarity in the two fluxes, except for lead and selenium. (Reproduced with permission from Duce, 1998.)

marine sedimentation of these metals in this region. Lead and selenium are exceptions, however, as the atmospheric flux is much greater than the flux to the seafloor. The fluxes to the seafloor represent average fluxes over the past several thousand years, whereas the atmospheric fluxes are roughly for the present time. The atmospheric lead flux is apparently much larger than the flux of lead to the sediments, primarily because of the high flux of anthropogenic lead from the atmosphere to the ocean since the introduction of tetraethyllead in gasoline in the 1920s. (The atmospheric flux is much lower now than in the 1980s, as discussed above.) However, in the case of selenium the apparently higher atmospheric flux is an artifact, because most of the flux of selenium from the atmosphere to the ocean is simply marine-derived selenium that has been emitted from the ocean to the atmosphere as gases, such as dimethyl selenide (DMSe). DMSe is oxidized in the atmosphere and returned to the ocean, i.e., the selenium input is simply a recycled marine flux. Thus, care must be taken when making comparisons of this type.

Nitrogen Species

There is growing concern about the input of anthropogenic nitrogen species to the global ocean. This issue is of particular importance in regions where nitrogen is the limiting nutrient, e.g., the oligotrophic waters of the central oceanic gyres. Estimates to date suggest that in such regions atmospheric nitrogen will in general account for only a few percent of the total 'new' nitrogen delivered to the photic zone, with most of the 'new' nutrient nitrogen derived from the upwelling of nutrient-rich deeper waters and from nitrogen fixation in the sea. It is recognized, however, that the atmospheric input is highly episodic, and at times it may play a much more important role as a source for nitrogen in

surface waters. **Table 7** presents a recent estimate of the current input of fixed nitrogen to the global ocean from rivers, the atmosphere, and nitrogen fixation. From the numbers given it is apparent that all three sources are likely important, and within the uncertainties of the estimates they are roughly equal. In the case of rivers, about half of the nitrogen input is anthropogenic for atmospheric input perhaps the most important information in **Table 7** is that the organic nitrogen flux appears to be equal to or perhaps significantly greater than the inorganic (i.e., ammonium and nitrate) nitrogen flux. The source of the organic nitrogen is not known, but there are indications that a large fraction of it is anthropogenic in origin. This is a form of atmospheric nitrogen input to the ocean that had not been considered until very recently, as there had been few measurements of organic nitrogen input to the ocean before the mid-1990s. The chemical forms of this organic nitrogen are still largely unknown.

Of particular concern are potential changes to the input of atmospheric nitrogen to the open ocean in

Table 7 Estimates of the current input of reactive nitrogen to the global ocean

Source	Nitrogen input (10^{12} g year^{-1})
From the atmosphere	
Dissolved inorganic nitrogen	28–70
Dissolved organic nitrogen	28–84
From rivers (dissolved inorganic + organic nitrogen)	
Natural	14–35
Anthropogenic	7–35
From nitrogen fixation within the ocean	14–42

Data reproduced with permission from Cornell S, Rendell A and Jickells T (1995) Atmospheric inputs of dissolved organic nitrogen to the oceans. *Nature* 376: 243–246.

Table 8   Estimates of anthropogenic reactive nitrogen production, 1990 and 2020

Region	Energy (NO$_x$)					Fertilizer				
	1990 (10^{12}g N year^{-1})	2020	Δ	Factor	% of total increase	1990 (10^{12}g N year^{-1})	2020	Δ	Factor	% of total increase
USA/Canada	7.6	10.1	2.5	1.3	10	13.3	14.2	0.9	1.1	1.6
Europe	4.9	5.2	0.3	1.1	1	15.4	15.4	0	1.0	0
Australia	0.3	0.4	0.1	1.3	0.4	—	—	—	—	—
Japan	0.8	0.8	0	1.0	0	—	—	—	—	—
Asia	3.5	13.2	9.7	3.8	39	36	85	49	2.4	88
Central/South America	1.5	5.9	4.4	3.9	18	1.8	4.5	2.7	2.5	5
Africa	0.7	4.2	3.5	6.0	15	2.1	5.2	3.1	2.5	6
Former Soviet Union	2.2	5.7	3.5	2.5	15	10	10	0	1.0	0
Total	21	45	24	2.1	100	79	134	55	1.7	100

Data adapted with permission from Galloway et al., 1995.

the future as a result of increasing human activities. The amount of nitrogen fixation (formation of reactive nitrogen) produced from energy sources (primarily as NO$_x$, nitrogen oxides), fertilizers, and legumes in 1990 and in 2020 as a result of human activities as well as the current and predicted future geographic distribution of the atmospheric deposition of reactive nitrogen to the continents and ocean have been evaluated recently. **Table 8** presents estimates of the formation of fixed nitrogen from energy use and production and from fertilizers, the two processes which would lead to the most important fluxes of reactive nitrogen to the atmosphere. Note that the most highly developed regions in the world, represented by the first four regions in the table, are predicted to show relatively little increase in the formation of fixed nitrogen, with none of these areas having a predicted increase by 2020 of more than a factor of 1.3 nor a contribution to the overall global increase in reactive nitrogen exceeding 10%. However, the regions in the lower part of **Table 8** will probably contribute very significantly to increased anthropogenic reactive nitrogen formation in 2020. For example, it is predicted that the production of reactive nitrogen in Asia from energy sources will increase ≤ fourfold, and that Asia will account for almost 40% of the global increase, while Africa will have a sixfold increase and will account for 15% of the global increase in energy-derived fixed nitrogen. It is predicted that production of reactive nitrogen from the use of fertilizers in Asia will increase by a factor of 2.4, and Asia will account for ∼88% of the global increase from this source. Since both energy sources (NO$_x$, and ultimately nitrate) and fertilizer (ammonia and nitrate) result in the extensive release of reactive nitrogen to the atmosphere, the predictions above indicate that there

should be very significant increases in the atmospheric deposition to the ocean of nutrient nitrogen species downwind of such regions as Asia, Central and South America, Africa, and the former Soviet Union.

This prediction has been supported by numerical modeling studies. These studies have resulted in the generation of maps of the 1980 and expected 2020 annual deposition of reactive nitrogen to the global ocean. **Figure 4** shows the expected significant increase in reactive nitrogen deposition from fossil fuel combustion to the ocean to the east of all of Asia, from Southeast Asia to the Asian portion of the former Soviet Union; to the east of South Africa, northeast Africa and the Mideast and Central America and southern South America; and to the west of northwest Africa. This increased reactive nitrogen transport and deposition to the ocean will provide new sources of nutrient nitrogen to some regions of the ocean where biological production is currently nitrogen-limited. There is thus the possibility of significant impacts on regional biological primary production, at least episodically, in these regions of the open ocean.

Synthetic Organic Compounds

The atmospheric residence times of many synthetic organic compounds are relatively long compared with those of the metals and nitrogen species, as mentioned previously. Many of these substances are found primarily in the gas phase in the atmosphere, and they are thus very effectively mobilized into the atmosphere during their production and use. Their long atmospheric residence times of weeks to months leads to atmospheric transport that can often be hemispheric or near hemispheric in scale. Thus

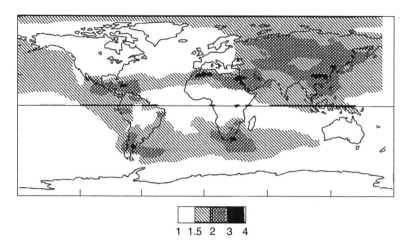

Figure 4 The ratio of the estimated deposition of reactive nitrogen to ocean and land surfaces in 2020 relative to 1980. (Figure reproduced with permission from Watson AJ (1997) Surface Ocean–Lower Atmosphere Study (SOLAS). *Global Change Newsletter, IGBP* no. 31, 9–12; data in figure adapted with permission from Galloway JN, Levy H and Kasibhatla PS (1994) Year 2020: Consequences of population growth and development on deposition of oxidized nitrogen. *Ambio* 23: 120–123.).

Table 9 Estimates of the atmospheric input of organochlorine compounds to the global ocean

Ocean	ΣHCH $(10^6 g\ year^{-1})$	ΣDDT $(10^6 g\ year^{-1})$	ΣPCB $(10^6 g\ year^{-1})$	HCB $(10^6 g\ year^{-1})$	Dieldrin $(10^6 g\ year^{-1})$	Chlordane $(10^6 g\ year^{-1})$
North Atlantic	850	16	100	17	17	8.7
South Atlantic	97	14	14	10	2.0	1.0
North Pacific	2600	66	36	20	8.9	8.3
South Pacific	470	26	29	19	9.5	1.9
Indian	700	43	52	11	6.0	2.4
Global input via the atmosphere	~4700	~170	~230	~80	~40	~22
Global input via rivers	~60	~4	~60	~4	~4	~4
% Atmospheric input	~99%	~98%	~80%	~95%	~91%	~85%

(Data reproduced with permission from Duce *et al.*, 1991.)

atmospheric transport and deposition in general dominates all other sources for these chemicals in sea water in open ocean regions.

Table 9 compares the atmospheric and riverine inputs to the world ocean for a number of synthetic organochlorine compounds. Note that the atmosphere in most cases accounts for 90% or more of the input of these compounds to the ocean. **Table 9** also presents estimates of the input of these same organochlorine compounds to the major ocean basins. Since most of these synthetic organic compounds are produced and used in the northern hemisphere, it is not surprising that the flux into the northern hemisphere ocean is greater than that to southern hemisphere marine regions. There are some differences for specific compounds in different ocean basins. For example, HCH (hexachlorocyclohexane) and DDT have a higher input rate to the North Pacific than the North Atlantic, largely because of the greater use of these compounds in Asia than in North America or

Europe. On the other hand, the input of PCBs (polychlorinated biphenyls) and dieldrin is higher to the North Atlantic than the North Pacific, primarily because of their greater use in the continental regions adjacent to the North Atlantic.

Conclusions

The atmosphere transports materials to the ocean that are both harmful to marine life and that are essential for marine biological productivity. It is now apparent that atmospheric transport and deposition of some metals, nitrogen species, and synthetic organic compounds can be a significant and in some cases dominant pathway for these substances entering both estuarine and coastal waters as well as some open ocean regions. Atmospheric input clearly must be considered in any evaluation of material fluxes to marine ecosystems. However, the uncertainties in the atmospheric fluxes of these materials to the ocean are large. The primary reasons for these large uncertainties are:

- The lack of atmospheric concentration data over vast regions of the coastal and open ocean, particularly over extended periods of time and under varying meteorological conditions;
- The episodic nature of the atmospheric deposition to the ocean;
- The lack of accurate models of air/sea exchange, particularly for gases;
- The inability to measure accurately the dry deposition of particles; and
- The inability to measure accurately the air/sea exchange of gases.

Further Reading

Duce RA (1998) *Atmospheric Input of Pollution to the Oceans*, pp. 9–26. Proceedings of the Commission for Marine Meteorology Technical Conference on Marine Pollution, World Meteorological Organization TD-No. 890, Geneva, Switzerland.

Duce RA, Liss PS, Merrill JT, *et al.* (1991) The atmospheric input of trace species to the world ocean. *Global Biogeochemical Cycles* 5: 193–259.

Galloway JN, Schlesinger WH, Levy H, Michaels A, and Schnoor JL (1995) Nitrogen fixation: anthropogenic enhancement – environmental reponse. *Global Biogeochemical Cycles* 9: 235–252.

Jickells TD (1995) Atmospheric inputs of metals and nutrients to the oceans: their magnitude and effects. *Marine Chemistry* 48: 199–214.

Liss PS and Duce RA (1997) *The Sea Surface and Global Change*. Cambridge: Cambridge University Press.

Paerl HW and Whitall DR (1999) Anthropogenically-derived atmospheric nitrogen deposition, marine eutrophication and harmful algal bloom expansion: Is there a link? *Ambio* 28: 307–311.

Prospero JM, Barrett K, Church T, *et al.* (1996) Nitrogen dynamics of the North Atlantic Ocean – Atmospheric deposition of nutrients to the North Atlantic Ocean. *Biogeochemistry* 35: 27–73.

van den Hout KD (ed.) (1994) *The Impact of Atmospheric Deposition of Non-Acidifying Pollutants on the Quality of European Forest Soils and the North Sea*. Report of the ESQUAD Project, IMW-TNO Report No. R 93/329.

VIRAL AND BACTERIAL CONTAMINATION OF BEACHES

J. Bartram, World Health Organization, Geneva, Switzerland
H. Salas, CEPIS/HEP/Pan American Health Organization, Lima, Peru
A. Dufour, United States Environmental Protection Agency, OH, USA

Introduction

Interest in the contamination of beaches by microbes is driven by concern for human health. The agents of concern are human pathogens, microorganisms capable of causing disease. Most are derived from human feces; therefore disposal of excreta and water-borne sewage are of particular importance in their control. Pathogens derived from animal feces may also be significant in some circumstances. The human population of concern constitutes primarily the recreational users, whether local residents, visitors, or tourists. Recreational use of natural waters (including coastal waters) is common worldwide and the associated tourism may be an important component of local and/or national economy.

Scientific underpinning and insight into public health concern for fecal pollution of beaches developed rapidly from around 1980. Approaches to regulation and control (including monitoring) have yet to respond to the increased body of knowledge, although some insights into potential approaches are available.

This article draws heavily on two recent substantial publications: the World Health Organization *Guidelines for Safe Recreational Water Environment*, released as a 'draft for consultation' in 1998, and *Monitoring Bathing Waters* by Bartram and Rees, published in 2000.

Public Health Basis for Concern

Recreational waters typically contain a mixture of pathogenic (i.e., disease-causing) and nonpathogenic microbes derived from multiple sources. These sources include sewage effluents, non-sewage excreta disposals (such as septic tanks), the recreational user population (through defecation and/or shedding), industrial processes (including food processing, for example), farming activities (especially feed lots and animal husbandry), and wildlife, in addition to the indigenous aquatic microflora. Exposure to pathogens in recreational waters may lead to adverse health effects if a suitable quantity (infectious dose) of a pathogen is ingested and colonizes (infects) a suitable site in the body and leads to disease.

What constitutes an infectious dose varies with the agent (pathogen) concerned, the form in which it is encountered, the conditions (route) of exposure, and host susceptibility and immune status. For some viruses and protozoa, this may be very few viable infectious particles (conceptually one). The infectious dose for bacteria varies widely from few particles (e.g., some *Shigella* spp., the cause of bacillary dysentery) to large numbers (e.g. 10^8 for *Vibrio cholerae*, the cause of cholera). In all cases it is important to recall that microorganisms rarely exist as homogeneous dispersions in water and are often aggregated on particles, where they may be partially protected from environmental stresses and as a result of which the probability of ingestion of an infectious dose is increased.

Transmission of disease through recreational water use is biologically plausible and is supported by a generalized dose–response model and the overall body of evidence. For infectious disease acquired through recreational water use, most attention has been paid to diseases transmitted by the fecal–oral route, in which pathogens are excreted in feces, are ingested by mouth, and establish infection in the alimentary canal.

Other routes of infectious disease transmission may also be significant as a result of exposure though recreational water use. Surface exposure can lead to ear infections and inhalation exposure may result in respiratory infections.

Sewage-polluted waters typically contain a range of pathogens and both individuals and recreational user populations are rarely limited to exposure to a single encounter with a single pathogen. The effects of multiple and simultaneous or consecutive exposure to pathogens remain poorly understood.

Water is not a natural ambient medium for the human body, and use of water (whether contaminated or not) for recreational purposes may compromise the body's natural defenses. The most obvious example of this concerns the eye. Epidemiological studies support the logical inference that recreational water use involving repeated immersion will increase the likelihood of eye infection through compromising natural resistance mechanisms, regardless of the quality of the water.

On the basis of a review of all identified and accessible publications concerning epidemiological studies on health outcomes associated with recreational water exposure, the WHO has recently concluded the following:

- The rate of occurrence of certain symptoms or symptom groups is significantly related to the count of fecal indicator bacteria. An increase in outcome rate with increasing indicator count is reported in most studies.
- Mainly gastrointestinal symptoms (including 'highlycredible' or 'objective' gastrointestinal symptoms) are associated with fecal indicator bacteria such as enterococci, fecal streptococci, thermotolerant coliforms and *Escherichia coli*.
- Overall relative risks for gastroenteric symptoms of exposure to relatively clean water lie between 1.0 and 2.5.
- Overall relative risks of swimming in relatively polluted water versus swimming in clean water vary between 0.4 and 3.

- Many studies suggest continuously increasing risk models with thresholds for various indicator organisms and health outcomes. Most of thesuggested threshold values are low in comparison with the water qualities often encountered in coastal waters used for recreation.
- The indicator organisms that correlate best with health outcome are enterococci/fecal streptococci for marine and freshwater, and *E.coli* for freshwater. Other indicators showing correlation are fecal coliforms and staphylococci. The latter may correlate with density of bathers and were reported to be significantly associated with ear, skin, respiratory, and enteric diseases.

In assessing the adequacy of the overall body of evidence for the association of bathing water quality and gastrointestinal symptoms, WHO referred to Bradford Hill's criteria forcausation in environmental studies (Table 1). Seven of the nine criteria were fulfilled. The criterion on specificity of association was considered inapplicable because the etiological agents were suspected to be numerous and relatively outcome-nonspecific. Results of experiments on the impact of preventive actions on health outcome frequency have not been reported.

Table 1 Criteria for causation in environmental studies (according to Bradford Hill, 1965). Application to bathing water quality and gastrointestinal symptoms

Criterion	Explanation	Fulfillment
1. Strength of association	Difference in illness rate between exposed and nonexposed groups, measured as a ratio	Yes Significant associations have been found; the ratios are relatively low (usually < 3)
2. Consistency	Has it been observed by different people at different places?	Yes In several countries and by various authors
3. Specificity of association	A particular type of exposure is linked with a particular site of infection or a particular disease	No
4. Temporality	Does the exposure precede the disease rather than following it?	Yes Most studies indicate temporal relationship
5. Biological gradient	A dose–response curve can be detected	Yes Most of the selected studies show significant exposure–response relationships
6. Plausibility	Does the present relationship seem likely in terms of present knowledge?	Yes For example, the results are in line with findings on ingestion of infective doses of pathogens
7. Coherence	Cause-and-effect interpretation of the data should not conflict with knowledge of natural history and biology of the disease	Yes
8. Experiment	Did preventive actions change the disease frequency?	Preventive actions have not yet been described in the studies
9. Analogy	Are similar agents known to cause similar diseases in similar circumstances?	Yes Similar to ingestion of recreational water, gastrointestinal symptoms are known to be caused by fecally polluted drinking water

This degree of fulfillment suggests that the association is causal.

Because of the study areas used, especially for the available randomized controlled trials, the results are primarily indicated for adult populations in temperate climates. Greater susceptibility among younger age groups has been shown and the overall roles of endemicity and immunity in relation to exposure and response are inadequately understood.

The overall conclusions of the work of WHO concerning fecal contamination of recreational waters and the different potential adverse health outcomes among user groups were as follows:

- The overall body of evidence suggests a casual relationship between increasing exposure to fecal contamination and frequency of gastroenteritis. Limited information concerning the dose–response relationships narrows the ability to apply cost–benefit approaches to control. Misclassification of exposure is likely to produce artificially lowthreshold values in observational studies. The one randomized trial indicated a higher threshold of 33 fecal streptococci per 100 ml for gastrointestinal symptoms.
- A cause–effect relationship between fecal pollution or bather-derived pollution and acute febrile respiratory illness is biologically plausible since associations have been reported and a significant exposure–response relationship with a threshold of 59 fecal streptococci per 100 ml was reported.
- Associations between ear infections and microbiological indicators of fecal pollution and bather load have been reported. A significant dose–response effect has been reported in one study. A cause–effect relationship between fecal or bather derived pollution and ear infection is biologically plausible.
- Increased rates of eye symptoms have been reported among bathers and evidence suggests that bathing, regardless of water quality, compromises the eye's immune defenses. Despite biological plausibility, no credible evidence is available for increased rates of eye ailments associated with water pollution.
- No credible evidence is available for an association of skin disease with either water exposure or microbiological water quality.
- Most investigations have either not addressed severe health outcomes such as hepatitis, enteric fever, or poliomyelitis or have not been undertaken in areas of low or zero endemicity. By inference, transmission of enteric hepatitis viruses and of poliomyelitis – should exposure of susceptible persons occur – is biologically plausible, and one study reported enteric fever (typhoid) causation.

The WHO work of 1998 led to the derivation of draft guideline values as summarized in **Table 2**.

Sources and Control

The principal sources of fecal pollution are sewage (and industrial) discharges, combined sewer overflows, urban runoff, and agriculture. These may lead to pollution remote from their source or point of discharge because of transport in rivers or through currents in coastal areas or lakes. The public health significance of any of these sources may be modified by a number of factors, some of which provide management opportunities for controlling human health risk.

With regard to public health, most attention has, logically, been paid to sewage as the source of fecal pollution. Pollution abatement measures for sewage may be grouped into three disposal alternatives, although there is some variation within and overlap between these: treatment, dispersion through sea outfalls, and discharge not to surface water bodies(e.g., to agriculture or ground water injection).

Where significant attention has been paid to sewage management, it has often been found that other sources of fecal contamination are also significant. Most important among these are combined sewer overflows (and 'sanitary sewer' overflows) and riverine discharges to coastal areas and lakes. Combined sewer overflows(CSOs) generally operate as a result of rainfall. Their effect is rapid and discharge may be directly to areas used for recreation. Riverine discharge may derive from agriculture, from upstream sewage discharges (treated or otherwise), and from upstream CSOs. The effect may be continuous (e.g., from upstream sewage treatment) or rainfall-related (agricultural runoff, urban runoff, CSOs). Where it is rainfall-related, the effect on downstream recreational water use areas may persist for several days. In river systems the decrease in microbiological concentrations downstream of a source (conventionally termed 'die-off') largely reflects sedimentation. After settlement in riverbed sediments, survival times are significantly increased and re-suspension will occur when river flow increases. Because of this and the increased inputs from sources such as CSOs and urban and agricultural runoff during rainfall events, rivers may demonstrate a close correlation between flow and bacterial indicator concentration.

Table 2 Draft guideline values for microbiological quality of marine recreational waters (fecal streptococci per 100 ml)

95th centile value of fecal streptococci per 100 ml	Basis of derivation	Estimated disease burden
10	This value is below the no-observed-adverse-effect level (NOAEL) in most epidemiological studies that have attempted to define a NOAEL	Using the indicator level/burden of disease relationship it corresponds to the 95th centile value that is associated with less than a single excess incidence of enteric symptoms for a family of four healthy adult bathers having 80 exposures per bathing season (rounded value), over a 5-year period, making a total of 400 exposures
50	This value is above the threshold and lowest-observed-adverse-effect level (LOAEL) for gastroenteritis in most epidemiological studies that have attempted to define a LOAEL	Using the indicator level/burden of disease relationship it corresponds to the 95th centile value that is associated with a single excess incidence of enteric symptoms for a family of four healthy adult bathers having 80 exposures per bathing season (rounded value)
200	This value is above the threshold and lowest-observed-adverse-effect level for all adverse health outcomes in most epidemiological studies	Using the indicator level/burden of disease relationship it corresponds to the 95th centile value that is associated with a single excess incidence of enteric symptoms for a healthy adult bather having 20 exposures per bathing season (rounded value)
1000	Derived from limited evidence regarding transmission of typhoid fever in areas of low-level typhoid endemicity and of paratyphoid. These are used in this context as indicators of severe health outcome	The exceedence of this level should be considered a public health risk leading to immediate investigation by the competent authorities. Such an interpretation should generally be supported by evidence of human fecal contamination (e.g., a sewage outfall)

Notes
1. This table would produce protection of 'healthy adult bathers' exposed to marine waters in temperate north European waters.
2. It does not relate to children, the elderly, or the immunocompromised who would have lower immunity and might require a greater degree of protection. There are no available data with which to quantify this and no correction factors are therefore applied.
3. Epidemiological data on fresh waters or exposure other than bathing (e.g., high exposures activities such as surfing or whitewater canoeing) are currently inadequate to present a parallel analysis for defined reference risks. Thus a single guideline value is proposed, at this time, for all recreational uses of water because insufficient evidence exists at present to do otherwise. However, it is recommended that the severity and frequency of exposure encountered by special-interest groups (such as body-, board-, and wind-surfers, subaqua divers, canoeists, and dinghy sailors) be taken into account.
4. Where disinfection is used to reduce the density of indicator bacteria in effluents and discharges, the presumed relationship between fecal streptococci (as indicators of fecal contamination) and pathogen presence may be altered. This alteration is, at present, poorly understood. In water receiving such effluents and discharges, fecal streptococci counts may not provide an accurate estimate of the risk of suffering from mild gastrointestinal symptoms.
5. The values calculated here assume that the probability on each exposure is additive.
Reproduced with permission from WHO (1998).

The efficiency of removal of major groups of microorganisms of concern in various types of treatment processes is described in **Table 3**.

Advanced sewage treatment (for instance based upon ultrafiltration or nanofiltration) can also be effective in removal of viruses and other pathogens. The role and efficiency of ultraviolet light, ozone, and other disinfectants are being critically re-evaluated. Treatment in oxidation ponds may remove significant numbers of pathogens, especially the larger protozoan cysts and helminth ova. However, short-circuiting due to poor design, thermal gradients, or hydraulic overload may reduce residence time from the typical design range of 30–90 days.

During detention in oxidation ponds, pathogens are removed or inactivated by sedimentation, sunlight, temperature, predation, and time.

Disposal of sewage through properly designed-long-sea outfalls provides a high degree of protection for human health, minimizing the risk that bathers will come into contact with sewage. In addition, long-sea outfalls reduce demand on land area in comparison with treatment systems, but they may be considered to have unacceptable environmental impacts (for instance, nutrient discharge into areas wheredilution or flushing is limited). They tend to have high capital costs, although these are comparable to those of land-based treatment systems

Table 3   Pathogen removal during sewage treatment

Treatment	Enterococci (cfu l^{-1})[a]	Enteric viruses	Salmonella	C. perfringens[b]
Raw sewage (l^{-1})	2 800 000	100 000–1 000 000	5000–80 000	100 000
Primary treatment[c]				
Percentage removal	32	50–98.3	95.5–99.8	30
Number remaining (l^{-1})	1 900 000	1700–500 000	10–3600	70 000
Secondary treatment[d]				
Percentage removal	96	53–99.92	98.65–99.996	98
Number remaining (l^{-1})	110 000	80–470 000	< 1–1080	2000
Tertiary treatment[e]				
Percentage removal	99.6	99.983–99.999 9998	99.99–99.999 995	99.9
Numbers remaining (l^{-1})	11 000	< 1–170	< 1–8	100

[a]Miescier JJ and Cabelli VJ (1982) Enterococci and other microbial indicators in municipal wastewater effluents, *Journal of Water Pollution Control Federation* 54: 1599–1606.
[b]Long and Ashbolt (1994) *Microbiological Quality of Sewage Treatment Plant Effluents*, AWT-Science and Environmental Report No. 94/123, Water Board, Sydney.
[c]Secondary = primary sedimentation, trickling filter/activated sludge and disinfection.
[d]Tertiary = primary sedimentation, trickling filter/activated sludge, disinfection, coagulation–sand filtration, and disinfection; note that tertiary does not involve coagulation–sand filtration and second disinfection steps for *C. perfringens*.
[e]Primary = physical sedimentation.
Adapted from Yates and Gerba (1998) *Microbial Considerations in Wastewater Reclamation and Reuse*. Vol. 10, Water Quality Management Library, Technomic Publishing Co., Inc. Lancaster, PA, pp. 437–488.

depending on the degree of treatment, whereas recurrent costs are relatively much lower. Ludwig (1988) has presented a comparison of costs and ecological impacts of long-sea outfalls versus treatment levels. Diffuser length, depth, and orientation, as well as the area and spacing of ports are key design considerations. Pathogens are diluted and dispersed and suffer die-off in the marine environment. These are major considerationsin length of outfall and outfall locations. Pretreatment by screening removes large particulates and 'floatables'. Grease and oil removal are also often undertaken.

Re-use of wastewater and groundwater recharge are two methods of sewage disposal that have minimal impact upon recreational waters. Especially in arid areas, sewage can be a safe and important resource (of water and nutrients) used for agricultural purposes such as crop irrigation. Direct injection or infiltration of sewagefor ground water recharge generally presents very low risk for human healththrough recreational water use.

Control of human health hazards associated with recreational use of the water environment may be achieved through control of the hazard itself (that is, pollution control) or through control of exposure. Fecal pollution of recreational waters may be subject to substantial variability whether temporally (e.g., time-limited changes in response to rainfall) or spatially (e.g., because, as aresult of the effects of discharge and currents, one part of a beach may behighly contaminated while another part is of good quality). This temporaland spatial variability

provides opportunities to reduce human exposure while pollution control is planned or implemented or in areas where pollution control cannot or will not be implemented for reasons such as cost. The measures used may include public education, control/limitation of access, or posting of advisory notices; they are often relatively affordable and can be implemented relatively rapidly.

Monitoring, Assessment and Regulation

Present regulatory schemes for the microbiological quality of recreational water are primarily or exclusively based upon percentage compliance with fecal indicator counts(Table 4).

These regulations and standards have had some success in driving cleanup, increasing public awareness, and contributing to improved personal choice. Not withstanding these successes, a number of constraints are evident in established approaches to regulation and standardsetting:

- Management actions are retrospective and can only be deployed after human exposure to the hazard.
- The risk to human health is primarily from human feces, the traditional indicators of which may also derive from other sources.
- There is poor interlaboratory and international comparability of microbiological analytical data.

Table 4 Microbiological quality of water guidelines/standards per 100 ml[i]

Country	Primary contact recreation			References
	TC[a]	FC[b]	Other	
Brazil	80% < 5000[c]	80% < 1000[c]		Brazil, Ministerio del Interior (1976)
Colombia	1000	200		Colombia, Ministerio de Salud (1979)
Cuba	1000[d]	200[d] 90% < 400		Cuba, Ministerio de Salud (1986)
EEC[e], Europe	80% < 500[f] 95% < 10 000[g]	80% < 100[f] 95% < 2000[g]	Fecal streptococci 100[f] Salmonella 0 l^{-1g} Enteroviruses 0 pfu l^{-1} Enterococci 90% < 100	EEC (1976) CEPPOL (1991)
Ecuador	1000	200		Ecuador, Ministerio de Salud Publica (1987)
France	< 2000	< 500	Fecal streptococci < 100	WHO (1977)
Israel	80% < 1000[h]			Argentina, INCYTH (1984)
Japan	1000			Japan, Environmental Agency (1981)
Mexico	80% < 1000[j] 100% < 10 000[k]			Mexico, SEDUE (1983)
Peru	80% < 5000[j]	80% < 1000[j]		Peru, Ministerio de Salud (1983)
Poland			E. coli < 1000	WHO (1975)
Puerto Rico		200[l] 80% < 400		Puerto Rico, JCA (1983)
California	80% < 1000[m,n] 100% < 10 000[k]	200[d,n] 90% < 400[o]		California State Water Resources Board (no date)
United States, USEPA			Enterococci 35[d] (marine), 33[d] (fresh) E. coli 126[d] (fresh)	USEPA (1986)
Former USSR			E. coli < 100	WHO (1977)
UNEP/WHO		50% < 100[p] 90% < 1000[p]		WHO/UNEP (1978)
Uruguay		< 500[f] < 1000[q]		Uruguay, DINAMA (1998)
Venezuela	90% < 1000 100% < 5000	90% < 200 100% < 400		Venezuela (1978)
Yugoslavia	2000			Argentina, INCYTH (1984)

[a]Total coliforms.
[b]Fecal or thermotolerant coliforms.
[c]'Satisfactory' waters, samples obtained in each of the preceding 5 weeks.
[d]Logarithmic average for a period of 30 days of at least five samples.
[e]Minimum sampling frequency – fortnightly.
[f]Guide.
[g]Mandatory.
[h]Minimum 10 samples per month.
[i]Monthly average.
[j]At least 5 samples per month.
[k]Not a sample taken during the verification period of 48 hours should exceed 10 000/100 ml.
[l]At least 5 samples taken sequentially from the waters in a given instance.
[m]Period of 30 days.
[n]Within a zone bounded by the shoreline and a distance of 1000 feet from the shoreline or the 30-foot depth contour, whichever is further from the shoreline.
[o]Period of 60 days.
[p]Geometric mean of at least 5 samples.
[q]Not to be exceeded in at least 5 samples.
Reproduced with permission from Bartram and Rees (2000).
Salas H (1998) History and application of microbiological water quality standards in the marine environment. Pan-American Center for Sanitary Engineering and Environmental Sciences (CEPIS)/Pan-American Health Organization, Lima, Peru.

Table 5 Risk potential to human health through exposure tosewage

Treatment	Discharge type		
	Directly on beach	Short outfall[a]	Effective outfall[b]
None[c]	Very high	High	NA
Preliminary	Very high	High	Low
Primary (including septic tanks)	Very high	High	Low
Secondary	High	High	Low
Secondary plus disinfection	Medium	Medium	Very Low
Tertiary	Medium	Medium	Very Low
Tertiary plus disinfection	Very Low	Very Low	Very Low
Lagoons	High	High	Low

[a]The relative risk is modified by population size. Relative risk is increased for discharges from large populations and decreased for discharges from small populations.
[b]This assumes that the design capacity has not been exceeded and that climatic and oceanic extreme conditions are considered in the design objective (i.e., no sewage on the beach zone).
[c]Includes combined sewer overflows.
NA, not applicable.
Reproduced with permission from Bartram and Rees (2000).

Table 6 Risk potential to human health through exposure to sewage through riverine flow and discharge

Dilution effect[a, b]	Treatment level				
	None	Primary	Secondary	Secondary plus disinfection	Lagoon
High population with low river flow	Very high	Very high	High	Low	Medium
Low population with low river flow	Very high	High	Medium	Very low	Medium
Medium population with medium river flow	High	Medium	Low	Very low	Low
High population with high river flow	High	Medium	Low	Very low	Low
Low population with high river flow	High	Medium	Very low	Very low	Very low

[a]The population factor includes all the population upstream from the beach to be classified and assumes no instream reduction in hazard factor used to classify the beach.
[b]Stream flow is the 10% flow during the period of active beach use. Stream flow assumes no dispersion plug flow conditions to the beach.
Reproduced with permission from Bartram and Rees (2000).

- While beaches are classified as 'safe' or 'unsafe', there is a gradient of increasing frequency and variety of adverse health effects with increasing fecal pollution and it is desirable to promote incremental improvements by prioritizing 'worst failures'.

The present form of regulation also tends to focus attention upon sewage, treatment, and outfall management as the principal or only effective solutions. Owing to high costs of these measures, local authorities may be effectively disenfranchised and few options for effective local intervention in securing bather safety appear to be available.

A modified approach to regulation of recreational water quality could provide for improved protection of public health, possibly with reduced monitoring effort and greater scope for interventions, especially within the scope for local authority intervention. This was discussed in detail at an international meeting of experts in 1998 leading to the development of the 'Annapolis Protocol'.

Table 7 Risk potential to human health through exposure to sewage from bathers

Bather shedding	Category
High bather density, high dilution[a]	Low
Low bather density, high dilution	Very low
High bather density, low dilution[a, b]	Medium
Low bather density, low dilution[b]	Low

[a]Move to next higher category if no sanitary facilities are available at beach site.
[b]If no water movement. Reproduced with permission from Bartram and Rees (2000).

Table 8 Possible sewage contamination indicators and their functions

Indicator/use	Function	
	Pros	Cons
Fecal streptococci/ enterococci	Marine and potentially freshwater human health indicator More persistent in water and sediments than coliforms Fecal streptococci may be cheaper than enterococci to assay	May not be valid for tropical waters, due to potential growth in soils
Thermotolerant coliforms	Indicator of recent fecal contamination	Possibly not suitable for tropical waters owing to growth in soils and waters Confounded by non-sewage sources (e.g., *Klebsiella* spp. in pulp and paper wastewaters)
E. coli	Potentially a freshwater human health indicator. Indicator of recent fecal contamination. Potential for typing *E. coli* to aid identifying sources of fecal contamination. Rapid identification possible if defined as β-glucuronidase-producing bacteria	Possibly not suitable for tropical waters owing to growth in soils and waters
Sanitary plastics	Immediate assessment can be made for each bathing day Can be categorized Little training of staff required	May reflect old sewage contamination and be of little health significance Subjective and prone to variable description
Rainfall in preceding 12, 24, 48 or 72 h	Simple regressions may account for 30–60% of the variation in microbial indicators for a particular beach	Each beach catchment may need to have its rainfall response assessed Response may depend on the period before the event
Sulfite-reducing clostridia/*Clostridium perfringens*	Always in sewage impacted waters Possibly correlated with enteric viruses and parasitic protozoa Inexpensive assay with H_2S production	May also come from dog feces May be too conservative an indicator Enumeration requires anaerobic culture
Somatic coliphages	Standard method well established Similar physical behavior to human enteric viruses	Not specific to sewage May not be as persistent as human enteric viruses Host may grow in the environment
F-specific RNA phages	Standard ISO method available More persistent than some coliphages Host does not grow in environmental waters below 30°C	Not specific to sewage WG49 host may lose plasmid (although F-amp is more stable) Not as persistent in marine waters
Bacteroides fragilis phages	Appear to be specific to sewage ISO method recently published More resistant than other phages in the environment and similar to hardy human enteric viruses	Requires anaerobic culture Numbers in sewage are lower than other phages, and many humans do not excrete this phage (hence no value for small populations)
Fecals sterols	Coprostanol largely specific to sewage Coprostanol degradation in water similar to die-off of thermotolerant coliforms Ratio of $5\beta/5\alpha$ stanols > 0.5 is indicative of fecal contamination; i.e., coprostanol/5α-cholestanol > 0.5 indicates human fecal contamination; while C_{29} 5β (24-ethylcoprostanol)/5α stanol ratio > 0.5 indicates herbivore feces Ratio of coprostanol/24-ethylcoprostanol can be used to indicate the proportion of human fecal contamination, which can be further supported by ratios with fecal indicator bacteria	Requires gas-chromatographic analysis and is expensive (about \$100/sample) Requires up to 10 litres of sample to be filtered through a glass fiber filter (Whatman) to concentrate particulate stanols

(Continued)

Table 8 *Continued*

Indicator/use	Function	
	Pros	*Cons*
Caffeine	May be specific to sewage, but unproven to date Could be developed into a dipstick assay	Yet to be proven as a reliable method
Detergents (calcufluors)	Relatively routine methods available	May not be related to sewage (e.g., industrial pollution)
Turbidity	Simple, direct, and inexpensive assay available in the field	May not be related to sewage; correlation must be shown for each site type
Cryptosporidium (animal source pathogens)	Required for potential zoonoses, such as *Cryptosporidium* spp., where fecal indicator bacteria may have died out, or are not present	Expensive and specialized assay (e.g., Method 1622, USEPA) Human/animal speciation of serotypes not currently defined

Reproduced with permission from Bartram and Rees (2000).

The 'Annapolis Protocol' requires field-testing and improvement based upon the experience gained before application. Its application leads to a classification scheme through which a beach may be assigned to a class related to health risk. By enabling local management to respond to sporadic or limited areas of pollution (and thereby to upgrade the classification of a beach), it provides significant incentive for local management action as well as for pollution abatement. The protocol recognizes that a large number of factors can influence the safety of a given beach. In order to better reflect risk to public health, the classification scheme takes account of three aspects:

1. Counts of fecal indicator bacteria in samples collected from the water adjoining the beach.
2. An inspection-based assessment of the susceptibility of the area to direct influence from human fecal contamination.
3. Assessment of the effectiveness of management interventions if they are deployed to reduce human exposure at times or in places of increased risk.

The process of beach classification is undertaken in two phases:

1. Initial classification based upon the combination of inspection-based assessment and the results of microbiological monitoring.
2. Taking account of the management interventions.

Inspection-based assessment takes account of the three most important sources of human fecal contamination for public health: sewage (including CSO and storm water discharges); riverine discharges where the river is receiving water from sewage discharges and is used either directly for recreation or discharges near a coastal or lake area used for recreation; and bather-derived contamination. The result of assessment is an estimate of relative risk potential in bands as outlined in **Tables 5, 6** and **7** Use of microbial and nonmicrobial indicators of fecal pollution requires an understanding of their characteristics and properties and their applicability for different purposes. Some very basic indicators such as sanitary plastics and grease in marine environments may be used for some purposes under some circumstances. Some newer indicators are under extensive study, but conventional fecal indicator bacteria remain those of greatest importance. Indicators of fecal contamination and their principal uses are summarized in **Table 8**.

By combining the results of microbiological testing with those of inspection, it is possible to derive a primary beach classification using a simple lookup table of the type outlined in **Table 9**. This primary classification may be modified to take account of management interventions that reduce or prevent exposure at times when or in areas where pollution is unusually high. Such 'reclassification' requires a database adequate to describe the times or locations of elevated contamination and demonstration that management action is effective. Since this 'reclassification' may have significant economic importance, independent audit and verification may be appropriate.

Implementation of a monitoring and assessment scheme of the type envisaged in the Annapolis Protocol would be likely to have a significant impact upon the nature and cost of monitoring activities. In comparison with established practice, it would typically involve a greater emphasis on inspection and relatively less on sampling and analysis than

Table 9 Primary classification matrix

Sanitary inspection category (susceptibility to fecal influence)	Microbiological assessment category (indicator counts)				
	A	B	C	D	E
Very low	Excellent	Excellent	Good	Good (+)	Fair (+)
Low	Excellent	Good	Good	Fair	Fair (+)
Moderate	Good[a]	Good	Fair	Fair	Poor
High	Good[a]	Fair[a]	Fair	Poor	Very poor
Very high	Fair[a]	Fair[a]	Poor[a]	Very poor	Very poor

[a]Unexpected result requiring verification.
(+) implies non-sewage sources of fecal indicators (e.g., livestock) and this should be verified.
Reproduced with permission from Bartram and Rees (2000).

is presently common place. At the level of an administrative area with a number of diverse beaches, it would imply an increased short-term monitoring effort when beginning monitoring, but a decreased overall workloadin the medium to longterm.

Recreational use of the water environment provides benefits as well as potential dangers for human health andwell-being. It may also create economic benefits but can add tocompeting local demands upon a finite and sometimes already over-exploited local environment. Regulation, monitoring, and assessmentof areas of coastal recreational water use should be seen or undertaken not in isolation but within this broader context. Integrated approaches to management that take account of overlapping, competing, and sometimes incompatible uses ofthe coastal environment have been increasingly developed and applied in recent years. Extensive guidance concerning integrated coastal management is now available. However, recreational use of coastal areas is also significantly affected by river discharge and therefore upstream discharge and land use practice. While the need to integrate management around the water cycle is recognized, no substantial experience has yet accrued and tools for its implementation remain unavailable.

See also

Sandy Beaches, Biology of.

Further Reading

Bartram J and Rees G (eds.) (2000) *Monitoring Bathing Waters*. London: EFN Spon.

Bradford-Hill A (1965) The environment and disease: association or causation? *Proceedings of the Royal Society of Medicine* 58: 295–300.

Esrey S, Feachem R, and Hughes J (1985) Interventions for the control of diarrhoeal diseases among young children: improving water supplies and excreta disposal facilities. *Bulletin of the World Health Organization* 63(4): 757–772.

Ludwig RG (1988) *Environmental Impact Assessment. Siting and Design of Submarine Outfalls.* An EIA Guidance Document. MARC Report No. 43. Geneva: Monitoring and Assessment Research Centre/World Health Organization.

Mara D and Cairncross S (1989) *Guidelines for the Safe Use of Wastewater and Excreta in Agriculture and Aquaculture.* Geneva: WHO.

WHO (1998) *Guidelines for Safe Recreational-water Environments: Coastal and Freshwaters.* Draft for Consultation. Document EOS/DRAFT/98.14 Geneva: World Health Organization.

SEDIMENTS, SLIDES, SLUMPS AND CYCLING

OCEAN MARGIN SEDIMENTS

S. L. Goodbred Jr, State University of New York, Stony Brook, NY, USA

Introduction

Ocean margin sediments are largely detrital deposits of terrestrial origin that extend from the shoreline to the foot of the continental rise. Indeed, about 80% of the world's sediment is stored within margin systems, which cover about 14% of the Earth's surface (Table 1; Figure 1). Margin deposits typically consist of sand, silt, and clay-sized particles, the characteristics of which reflect the geology and climate of the adjacent continent. Some of the signals relevant to terrestrial conditions include sediment size, mineralogy, geochemistry, and isotopic signature. Upon entering the marine realm, however, sediments take on new characteristics indicative of coastal ocean processes that include waves, tides, currents, sea level, and biological productivity. Given that these numerous terrestrial and marine processes impart a signature to the sediments, ocean margins preserve an important record of Earth history, providing insights into past atmospheric, terrestrial, and marine conditions. Beyond their significance as environmental recorders ocean margin sediments support major petroleum and mineral resources. Currently, over 50% of the world's oil is recovered along ocean margins, and much of the remaining fraction is held within ancient margin deposits.

In the 1950s, early investigations of ocean margins focused on tectonic structure and how overlying sedimentary sequences developed on timescales of 10^5–10^7 years. Originally aimed at understanding plate tectonics and the nature of ocean–continent boundaries, these large-scale studies continued through the 1960s with specific interest in petroleum resources. Such efforts culminated in the publication of large scientific volumes such as Burk and Drake's *Geology of Continental Margins* (1974). At the time of these summary publications, new models of sedimentary margin systems were already being developed, has also been notably the approach of seismic stratigraphy established at the Exxon Production Research Company. Growing out of this approach was the more general model of sequence stratigraphy, which helped establish that margin strata could be grouped into discrete packages reflecting cycles of sea-level rise and fall. These concepts represent a general approach that has been applied to stratigraphic development and margin evolution in most of the world's ancient and modern sedimentary systems. Sequence stratigraphic data has also been mated with lithologic, magnetic, and biostratigraphic records, allowing researchers to establish the history of sea-level change since the Triassic era (260 million years ago). This historic record significantly advanced out understanding of ocean margin sedimentary records, because sea level is a major control on the distribution and accumulation of margin deposits, as well as an indicator of global climate.

At a much shorter time scale, a great deal of research in recent decades has focused on sediment dynamics along modern margin systems. Aimed at understanding the fate of sediments entering the marine realm, some of the issues driving margin-sediment research included coastal hazards, land loss and development, storm impacts, contaminant fate, and military interests. In part, the field has advanced with the development of new technology such as marine radioisotope geochronology, sonar seafloor mapping, and instruments for remote wave, current, and sediment concentration measures. Recently, research programs have sought to integrate long and short geologic timescales to understand how daily, seasonal, and annual sedimentary beds are ultimately preserved to form thick millennial and longer sedimentary sequences (e.g., US and European STRATA-FORM programs). Other newer initiatives are seeking to take integration a step further by recognizing that ocean margin sediments lie among a continuum of terrestrial and marine influences. Therefore, to better understand this critical accumulation zone, the

Table 1 Area of Earth's major physiographic provinces and the volume of stored sediments. Bracketed are ocean margin components with relative contributions shown

	Area[a] ($10^6 \times km^2$)	Volume[b] ($10^6 \times km^3$)
Land	148.1	45
Interior basins	8.7	35
Shelves	18.4 ⎫ 14%	75 ⎫ 80%
Slopes	28.7 ⎭	200 ⎭
Rises	25.0	150
Ocean basins	281.2	25
Totals	510.1	530

Data from [a]Burk and Drake (1974) and [b]Kennett (1982).

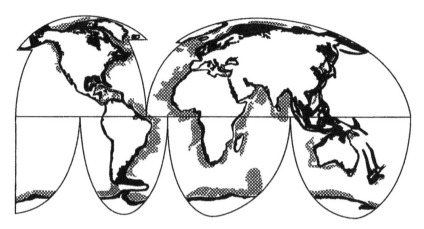

Figure 1 Distribution of major ocean margin deposits, showing extent of continental shelves (black) and rises (hatched). (Modified from Emery (1980).)

terrestrial processes that are responsible for sediment input must be considered in conjunction with the marine processes that are responsible for their redistribution (e.g., US and Japanese MARGINS programs). Such efforts to integrate over various spatial and temporal scales will likely govern the foreseeable future of margin sedimentary research.

Margin Structure

All margins represent past or present tectonic plate boundaries where crusts of varying age, density, and composition meet. In general these boundaries also comprise large-scale basins that trap sediments shed from the adjacent land surface. Most ocean margins specifically denote the boundary between oceanic and continental crusts, where different densities between these adjacent blocks give rise to a steep gradient at the margin. This tectonic structure controls the physiography of the margin as well, which comprises shelf, slope, and rise settings (**Figure 2**). The landward part of the margin, the shelf, overlies the buoyant continental-crust block and is thus the shallowest part of the margin, extending along a low gradient (0.1°) from the coast to the shelf break at 100–200 m water depth. Because of the low gradient and shallow depths, shelf environment are sensitive to sea-level change and isostatic movements (crustal buoyancy) causing the shelf to be periodically flooded and exposed. The seaward edge of the shelf, called the break, is identified by an increase in seafloor gradient (3–6°) and represents a transition between the shelf and continental slope. Roughly overlying the continental and oceanic crusts, sediment accumulation on the slope is partly controlled by on the shelf, because any slope sediments must first bypass this depositional region. Beyond the slope extends the rise, a

low gradient (0.3–1.4°) sedimentary apron that begins in 1500–3500 m water depth. Although the rise may extend hundreds of kilometers into the ocean basin, its sediment source largely derives from sediment originating on the slope and moves as turbidity currents. In addition to the major shelf, slope, and rise features, other regionally significant marginal features may include deltas and canyons, each playing a role in controlling or modifying the dispersal and deposition of sediments.

Tectonically, margins may be broadly divided into divergent (passive) and convergent (active) systems, each with a characteristic structure and physiography that are important to sediment accumulation and transport dynamics. Divergent margins are largely stable crustal boundaries, although the structure and movement of deep basement rocks remain a significant influence on sediment deposition and stratigraphy. Divergent margins are also largely constructive settings in which sediments shed from the continent form thick sedimentary sequences and a wide, low gradient shelf. This loading of sediment onto the margin also drives a slow downwarping of the crust, creating more accommodation space for the accumulation of thick sedimentary sequences. The resulting margin physiography is characterized by a broad shelf, slope, and rise. Margins along the Atlantic Ocean are among the best-studied examples of divergent systems.

In contrast, convergent margins are characterized by the subduction of oceanic crust beneath the adjacent continental block. Active mountain building in convergent settings frequently drives uplift of the margin, limiting the development of thick sedimentary sequences on the shelf. Thus, convergent margins tend to have narrow shelves that bypass most sediment to a steep and well-developed slope. The rise is generally not a significant depocenter because

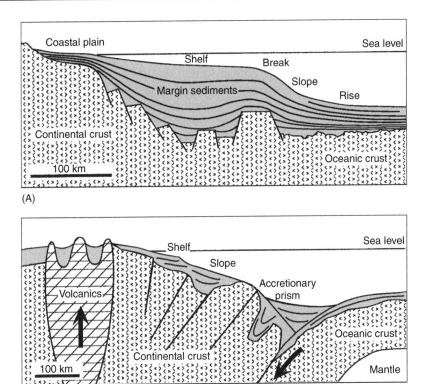

Figure 2 Generalized cross-sections of (A) divergent and (B) convergent-type continental margins. Note the different scales. The divergent margin shows thick sedimentary deposits that have prograded into the ocean basin. Margin deposits in the convergent setting are much thinner, and most deposition occurs below the shelf break in the accretionary prism.

sediment transport is intercepted by trenches and elevated topography that inhibit the long-distance movement of turbidity currents. Seaward of this, though, there is frequently a wedge of deformed marine sediments called an accretionary prism that derives from material scraped from the subducting ocean crust. This situation is often found along the eastern North Pacific margin. Along margins with major trench systems such as those of the western Pacific, the rise and accretionary prism may be altogether absent.

Margin Sediment Sources, Character, and Distribution

Ocean margin sediments generally comprise small particles ranging in size from fine sands (62–125 µm) to silts (4–62 µm) and clays (<4 µm). Larger sediments are occasionally found in ocean margin deposits, including coarse sands to boulders, but these particles are generally either relict (not actively transported), of glacial or ice-rafted origin, or biogenically precipitated (i.e. shell or coral). Overall patterns of grain-size distribution in margin sediments depend on the regional climate, geology, and sediment transport processes operating on the margin.

Three major mineral groups comprise ocean margin sediments, including siliciclastics, carbonates, and evaporites, or any combination of these. Worldwide, however, siliciclastics dominate ocean margin sediments. Siliciclastics are almost exclusively silicon-bearing minerals that derive from the physical and chemical weathering of continental rocks. Because most of the hundreds of rock-forming minerals are weathered before reaching the coast, margin siliciclastics dominantly comprise quartz and four clay species: kaolinite, chlorite, illite, and smectite. There is often a small component (several percent) of remnant feldspars, micas, and heavy minerals (sp. gr. >2.85) as well. Quartz, although an abundant and ubiquitous mineral, is often useful for interpreting margin sediments because different grain sizes reflect varying sources (e.g. eolian input) or energy regimes. For clay minerals, though, relative abundance of the major species in part reflects regional climate, thus making them a useful parameter for interpreting terrestrial signals preserved on the margin. Furthermore, within the heavy mineral fraction, more unique silicate species can help determine the specific source area (provenance) of margin sediments. This approach is often limited because of regionally common mineralogy, but researchers have begun to use isotope ratios to target more precisely margin sediment source areas. Since ocean margin sediments and deposits are closely tied to continental fluxes, such detailed studies are critical for an integrated understanding of the terrestrial-margin–marine sedimentary system.

Although nearly all siliciclastic margin sediments are derived from the continents, they may be delivered via different mechanisms such as glaciers, rivers, coastal erosion, and wind. Although the latter mechanism is important to deep-sea sedimentation, the others dominate transport to the margin. At high latitudes, glaciers are a major and sometimes sole sediment source. The characteristics of glacial margin sediments is highly variable because of the capacity of ice to transport sediments of all sizes, but in general glacial sediments are coarser and more poorly sorted near the ice front (e.g., till) and become progressively finer and better sorted with distance (e.g., proglacial clays). Examples of glacially dominated modern margins include south-east Alaska, Greenland, and Antarctica. Along most of the world's margins, rivers dominate sediment delivery to the shelf. The grain size of river sediments is typical of margin deposits, mostly including fine sand, silt, and clay-sized particles. However, the range of textures and the total amount of sediment delivered to the margin varies with many factors, including the size, elevation, and climate of the river basin. Because rivers are largely point sources, coastal processes such as waves and tides are important controls on redistributing fluvial sediments once they reach the ocean margin.

In contrast to siliciclastics, carbonate sediments are largely biogenic in origin and are precipitated *in situ* by various corals, algae, bryozoans, mollusks, barnacles, and serpulid worms. Carbonates can be a dominant or significant component of margin sediments where the flux of siliciclastics is not great, often in arid regions or on the outer shelf. Carbonates do not usually form on the slope because of unstable bottom surfaces and depths below the photic zone. In general the production of carbonate sediments is higher in warmer low-latitude waters, but they are found along margins throughout the world. Unlike siliciclastic sediments, carbonates are not widely transported and tend to be concentrated within local regions on the shelf, where the build-up of bioherms and reefs may comprise locally important structural features. Overall, areas of purely carbonate production are rare and limited to ocean margins detached from the continents, such as the Bahamas. In contrast, mixed siliciclastic–carbonate margins are not uncommon and include regions such as the Yucatan and Florida peninsulas, Great Barrier

Reef, and Indonesian islands. In very arid regions, margins along smaller marine basins may also support evaporite deposits. These minerals are precipitated directly from the water column to form thick salt deposits. Evaporites are also plastic, low-density sediments that frequently migrate upwards to form salt diapirs, causing major deformation of the overlying margin strata. Evaporite margin deposits are presently forming in the Red Sea and are also an important component of ancient margin sequences around the Mediterranean and Gulf of Mexico.

Along many modern margins, relict and recently eroded sediments comprise the bulk of material that is actively transported on the shelf. An example of a relict margin is the US East coast, where modern fluvial sediment is trapped in coastal estuaries and embayments, thus starving the shelf of significant input. Thus, the shelf there is characterized by a thin sheet of reworked sands (remnant from the Holocene transgression) that overlie much older, partly indurated deposits. Along mountainous high-energy coastlines where rivers are absent or very small, the erosion of headlands can be a major source of sediment to the margin. Often these are also convergent margins with steep shelves that aid in advecting eroded near-shore sediments to the outer shelf and slope. Examples include portions of the eastern Pacific margin and south-eastern Australia (divergent).

Margin Sediment Transport

Rivers are one of the major sources of sediment delivered to the shelf, and thus the fate of river plumes is an important control on the distribution of ocean margin sediments. The initial dispersal of the river plume and the subsequent sediment deposition are controlled by waves, tides, and three basic effluent properties that include the inertia and buoyancy of the plume, and its frictional interaction with the seafloor. In general, wave energy has the effect of keeping sediments close to shore, particularly sands that may be reworked onshore and/or transported alongshore by wave-driven currents. Margins that are wave dominated are typically steep and narrow and found in the high-latitudes, such as southern Australia, southern Africa and the margins of the north and south Pacific (i.e., high wind-stress regions). In contrast to waves, tides force a general onshore–offshore movement of sediment, most significantly along the coast where tidal energy is focused. Tide-dominated margin sediments typically occur on wide shelves and in shallow marginal basins such as the Yellow or North Seas. Compared with waves and tides, the effect of river plume dynamics on sediment dispersal is more varied, particularly as complex feedbacks exist between coastal geology and riverine, wave, and tidal processes.

The fate of sediments along margins is not a simple process of transport, deposition, and burial, but rather involves multiple cycles of erosion, transport, and deposition before being preserved and incorporated into ocean-margin strata (**Figure 3**). Thus, the triad of processes discussed above largely controls the initial phase of dispersal and deposition at the coast and inner shelf. The implication is that sediments undergo a succession of transport steps prior to preservation, and between each step the controlling processes, and thus sediment character, progressively change from riverine to coastal, shelf, and marine-dominated signals. Thus, beyond the coastal zone a different set of mechanisms is responsible for advecting sediments across the margin to the outer shelf, slope, and rise. On the shelf where seafloor gradients are often too low for failure (i.e., mass-wasting events), storms can be a major agent for cross-shelf transport. Two components of storms are involved in this process. First, strong winds generate steep, long-period waves that can resuspend sediment at much greater depths than fair-weather waves (e.g., a storm wave with a 12 s period will begin to affect the seafloor at ~110 m depth). Second, resuspended sediments may be transported offshore via a near-bottom return flow (set up by wind and wave stresses) or by group-bound infragravity (long period) waves. Although these mechanisms for seaward sediment transport are well recognized, their magnitude, frequency, and distribution are not precisely known.

A second and well-described mode for sediment transport across ocean margins is the suite of gravity-driven mass movements that includes slumps, slides, flows, and currents (**Figure 4**). This suite comprises nearly a continuum of properties from the movement of consolidated sediment blocks as slumps and slides to the superviscous non-Newtonian fluids of debris and sediment flows to the low-density suspended load of turbidity currents. Mass wasting events are common, widespread, and most frequently occur where slopes are oversteepened and/or sediments are underconsolidated (i.e., low shear strength). These last two factors are related since the threshold for oversteepening is partly a function of the sediment's shear strength, with possible failure occurring at gradients <1° in poorly consolidated sediments (e.g., 80–90% water content) and relative stability found on slopes of 20° for well-consolidated material (e.g., 20–30% water content). Many other factors also contribute to mass movements including groundwater pressure, sub-bottom gas production, diapirism, fault planes, and other zones of weakness. In addition to these intrinsic sedimentary characteristics, trigger mechanisms are an important control on mass movements. Triggers may include earthquakes, storms, and wave pumping, each of which may have the effect of inducing shear along a plane of weakness or disrupting the cohesion between sediment grains (i.e., liquefaction). Where trigger mechanisms are frequent and intense, such as along a tectonically active or high-energy margin, mass wasting may occur in deposits that would be stable in a passive, low-energy setting.

Although storm activity can be important on the shelf, mass movements are by far the most dominant mechanism for transferring sediment to the slope and rise. Indeed, estimates indicate that up to 90% of

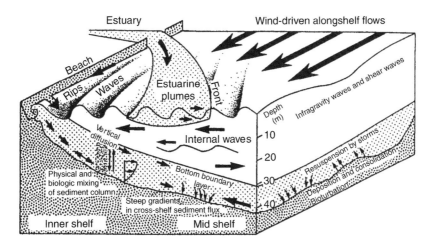

Figure 3 Conceptual diagram illustrating the major physical processes responsible for transporting sediment across the continental shelf. Note the many processes that contribute to the resuspension, erosion, and transport of sediment in this region. (Reproduced with permission from Nittrouer and Wright, 1994.)

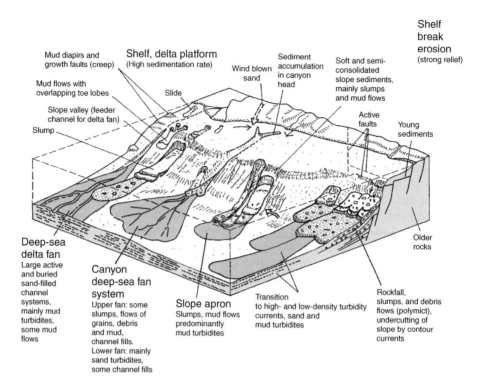

Figure 4   Summary diagram of sedimentary features and transport mechanisms typical of continental margins. Note the variety of mass-wasting events that originate on the slope and extend onto the rise. (Reproduced with permission from Einsele G (1991) Submarine mass flows, deposits and turbidities. In: Einsele G *et al.* (eds) (1991) *Cycles and Events in Stratigraphy*. New York: Springer-Verlag.)

continental rise deposits derive from turbidity currents. In terms of their occurrence, slumps (movement of consolidated sedimentary blocks) are most common along steep gradients such as the continental slope or along the channels and walls of submarine canyons. Slumps and slides may move over distances of meters to tens of kilometers, and at rates of centimeters per year to centimeters per second. Although slumps and slides do not frequently travel great distances, these energetic events often generate turbidity currents at their front, which can travel hundreds of kilometers across shallow gradients of the continental shelf. Although turbidity currents occur throughout the world, the development of thick and extensive turbidite deposits is largely a function of margin structure, where the trenches and seafloor topography of collision margins inhibit the propagation of these flows (see earlier discussion of **Figure 2**).

Margin Stratigraphy

Strata Formation

Because of the important role that margin sediments play in geochemical cycling (both natural and anthropogenic inputs), there has been a great deal of interest in how seafloor strata are formed and what

sort of physical, biological, and chemical impacts sediments undergo before being buried. These processes and controls on strata formation are also of great significance for interpreting the record of Earth's history preserved in ocean-margin deposits. Initially regarded as a one-way sink for organic carbon, dissolved metals, and coastal pollutants, the role of margin sediments has since been shown to be complicated by physical and biological processes that may alter chemical conditions (e.g., redox) and continue re-exposing sediments as deep as several meters to the water column and biogeochemical exchanges. Thus, through mixing processes, seafloor sediments may remain in contact with the water column for 10^1–10^3 years before being permanently buried. In order to quantify the effective importance of biological mixing relative to burial rate, the non-dimensional parameter (G) states that

$$G_b = \frac{D_b/L_b}{A}$$

where D_b represents the rate of biological (diffusive) mixing, L_b is depth of mixing, and A is the sedimentation rate. Field values of G are often 0.1–10, which falls within the range of codominance between mixing and sedimentation. Although the parameter

G_b reflects the intensity of biological effects, physical mixing (G_p) may be better represented by T_p/L_p for the upper term, where T_p and L_p are the recurrence interval (period) and depth of physical disturbance, respectively. A diffusive coefficient is not relevant here because physical mixing is primarily caused by resuspension, which results in a complete advective turnover of the affected sediments. The notion that natural mixing intensities fall in the range of codominance is shown by the frequency with which sedimentary sequences are found to be partially mixed (at various scales), displaying juxtaposed laminated and bioturbated strata. Such patterns indicate that mixing is more complex than represented here, but the idea that strata preservation and geochemical availability are a function of sedimentation rate (i.e., burial) versus the rate and depth (i.e., exhumation) of mixing remains a useful, if imperfect, manner in which to consider these processes.

In a spatial sense, the dynamics of strata formation varies systematically across the ocean margin with changes in the controlling processes. Near to shore, the relatively constant processes of waves, tides, and river discharge support the frequent deposition of very thin strata (mm to cm). Although short-term rates of sedimentation are often high, the thin nature of each deposit gives them a low chance of being preserved due to physical and biological mixing. In contrast, mean sedimentation rates on the outer margin are often relatively low, but deposition is dominated by stochastic events such as storms and mass wasting which generally produce thicker deposits. Therefore, mixing between sedimentation events may only affect the upper portion of the deposit and thus preserve lower strata. Another consideration is that the intensity of biological mixing is generally less in offshore regions than near the coast.

Sequences

One of the most difficult problems in sedimentary geology is scaling up from short-term strata formation as discussed above to longer-term preservation and the accumulation of thick ($10^2–10^3$ m) sediment sequences that comprise ocean margins. Because the majority of sedimentary deposits formed over a particular timescale have a low chance of being preserved, geologists can only model strata formation across two to three orders of magnitude (e.g., incorporation of a 1 m thick deposit into a 100 m thick sequence, or the chance of a seasonal flood layer being preserved for several decades). In reality six or more orders of magnitude are appropriate, as kilometer thick ancient-margin sequences comprise millimeter scale strata such as tidal

deposits, current ripples, and mud layers. Although stratigraphic reconstructions across this large scale remain too complex, geologists do have a good understanding of the incorporation of meter-scale strata into sedimentary packages that are many tens to hundreds of meters thick. Called *sequences*, these stratigraphic units are defined as a series of genetically related strata that are bound by unconformities (surfaces representing erosion or no deposition) (**Figure 5**).

Sequences are the major stratigraphic units that lead to continental margin growth, with deposition generally occurring on the shelf during periods of high sea level and sediments being dispersed to the slope and rise during sea-level lowstands. Forming over $10^4–10^7$ years, the major factors in sequence production are: (1) the input of sediments, (2) space in which to store them (accommodation), and (3) some function of cyclicity to produce unconformity-bound groups of deposits. These factors are largely controlled by tectonics, sea level, and sediment supply. In the long term, tectonics is the major control on accommodation, whereby space for sediment storage is created by isostatic subsidence in response to sediment loading and crustal cooling along the margin. At shorter time scales, sea level controls both the availability of accommodation and its location across the margin. Thus, it can be considered that sea level partly controls the formation of sequences, whereas sequence preservation is dependent on tectonic forcings. Sediment supply is perhaps the fundamental control on the size of a sequence, and thus partly affects its chance for preservation. Sediments of riverine and glacial origin are the major sources for sequence production, and changes in these sources with varying climatic conditions is one mechanism for generating the unconformity bounding surfaces. Sequence boundaries can also be

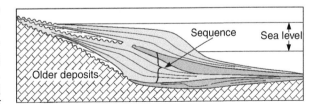

Figure 5 Generalized structure of continental margin sequences formed during cycles of sea-level change. Shown here are two sequences that reflect two periods of sea-level rise and fall. Transgressive deposits (□) are associated with rising seas and the highstand deposits (□) with a high, stable sea level. The lowstand deposits (■) form under falling and low phases of sea level. Reconstruction of these features from continental margins has provided a record of sea-level change over the past 200 million years.

generated by tectonics and eustatic (global) sea-level change, with the latter being significantly affected by the growth and retreat of continental ice sheets.

Conclusions

Although the study of ocean-margin sediments began more than a century ago, the field has grown most significantly in the past 50 years with the birth of marine geology and oceanography. Typical of geological systems, however, relevant research extends across a range of spatial and temporal scales and incorporates a variety of disciplines including geophysics, geochemistry, and sedimentary dynamics. The future of margin sedimentary research lies with the continued integration of these various scales and disciplines, moving toward the development of coherent models for the production and dispersal of margin sediments and their ultimate incorporation into margin sequences. The impacts of such capability would greatly benefit the world's growing population in the areas of energy and mineral resources, climate change, coastal hazards, sea-level rise, and marine pollution. Ongoing research initiatives are currently advancing the field of ocean-margin sediments toward these goals, and nascent programs are providing a promising lead to the future.

See also

Beaches, Physical Processes Affecting. Geomorphology. Glacial Crustal Rebound, Sea Levels, and Shorelines. River Inputs. Sea Level Change. Storm Surges. Tides.

Further Reading

Aller RC (1998) Mobile deltaic and continental shelf muds as fluidized bed reactors. *Marine Chemistry* 61: 143–155.

Burk CA and Drake CL (eds.) (1974) *The Geology of Continental Margins.* New York: Springer-Verlag.

Emery KO (1980) Continental margins – classification and petroleum prospects. *The American Association of Petroleum Geologists Bulletin* 64: 297–315.

Guinasso NL and Schink DR (1975) Quantitative estimates of biological mixing rates in abyssal sediments. *Journal of Geophysical Research* 80: 3032–3043.

Kennett JP (1982) *Marine Geology.* Englewood Cliffs, NJ: Prentice-Hall.

Nittrouer CA and Wright LD (1995) Transport of particles across continental shelves. *Reviews of Geophysics* 32: 85–113.

Payton CE (ed.) (1977) Seismic stratigraphy – applications to hydrocarbon exploration. *American Association of Petroleum Geologists Memoir* 2.

Sandford LP (1992) New sedimentation, resuspension, and burial. *Limnology and Oceanography* 37: 1164–1178.

Wilgus CW, Hastings BS, Kendall CGStC *et al.* (eds) (1988) Sea-level changes: an integrated approach. *Society of Economic Paleontologists and Mineralogists Special Publication No. 42.*

OFFSHORE SAND AND GRAVEL MINING

E. Garel, CIACOMAR, Algarve University,
Faro, Portugal
W. Bonne, Federal Public Service Health, Food Chain
Safety and Environment, Brussels, Belgium
M. B. Collins, National Oceanography Centre,
Southampton, UK

Introduction

Offshore sand and gravel extraction involves the abstraction of sediments from a bed which is always covered with seawater. This activity started in the early twentieth century (in the mid-1920s, in the UK), but did not reach a significant scale until the 1960s and 1970s, when markets for marine sand and gravel expanded and dredging technology improved (**Figure 1**). In the mining industry, the term 'aggregates' describes a variety of diverse particulate materials, which are provided in bulk, that is, sand and gravel, together with crushed rocks. The distinction between sand and gravel varies, on the basis of the classification adopted. Hence, the nomenclature may change between the end-users or countries (or even regions within a particular country). For example, in the UK industry, 'sand' refers to (noncohesive) minerals with a mean grain size lying between 0.63 and 5 mm, and 'gravel' is reserved for coarser material; in Belgium, the industry considers as gravel the sediment with a mean grain size larger than 4 mm. For comparison, within the scientific literature, the limit between sand and gravel is generally at 2 mm.

Resource Origin

Marine aggregate deposits from the continental shelf can be either relict or modern. Relict deposits have been formed under hydrodynamic and sedimentological regimes that no longer exist. They consist of river or coastal bank deposits, formed during the lower seawater stands, induced by the glacial periods affecting the Earth over the past 2 My. At that time, the rivers extended widely across the continental

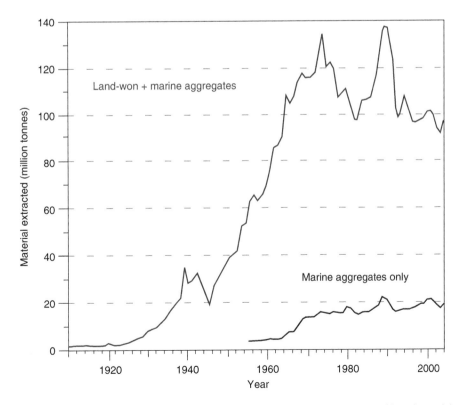

Figure 1 Production of aggregates in the UK between 1900 and 2004 (excluding fill contract and beach nourishment, for marine aggregates). Note the rapid growth of extraction during the late 1960s and the early 1970s, due to the boom in construction in southern England. Sources: Crown-Estate, British Geological Survey.

shelf, transporting and depositing large amounts of sand and gravel in their valleys and coastal waters. This material has been submerged and possibly re-worked, especially the mobile sandy deposits, during subsequent warmer interglacial periods, due to the retreating and melting of the ice. Concentration of deposits has resulted from the numerous repetitions of this warm–cold cycle. These sedimentary bodies have remained undisturbed since their formation and are immobile within the context of prevailing hydrodynamic conditions. Relict sand and gravel deposits include mostly thin sheets, banks, drowned beaches and spits, and infilled channel systems, such as river courses and terraces. In contrast, modern deposits have been formed and are controlled by the present prevailing hydrodynamic and sedimentologi-cal regime. Therefore, they are related to present-day coastal sediment budgets and sediment dynamics on the seabed. These deposits include, exclusively, sandbanks and sand bedform fields.

The composition, roughness, and grain size of marine aggregate particles vary considerably, de-pending upon their mode of formation, depositional processes, and history of reworking. Although dependent upon the intensity of post-erosional trans-port, the grains consist usually of fragments derived from rocks of the surrounding emerged regions. Due to their glacial origin, relict deposits have a com-position which is similar to those quarried on land (which are their upstream equivalent), having a low content of shell fragments. In comparison, modern deposits may include high concentration of shelly sediments, which affect the commercial quality of the deposit. In addition, marine aggregates tend to have an higher chloride content. As such, some objections have been raised on their use, in relation to potential alkali–silica reactions, which could affect concrete, or chloride attack on steel reinforcement. However, ex-perience and some experiments have shown that marine aggregates are as suitable as on-land quarried aggregates, for construction purposes.

Production and Usage

Marine aggregates are used mostly in the con-struction industry, but also for beach replenishment and shore protection, for land reclamation, and in other fill-related uses (such as drainage and capping material). The quality required is governed generally by the usage of the product. High-quality marine aggregates (i.e., well sorted and free from impurities such as clay) are used for mixing with a cementing agent, to produce hard building material (e.g., con-crete, mortar, and plaster), or are coated with bitu-men for road surfacing. For other usages, such as base material under foundations, roads, and rail-ways, or as the construction of embankments, lower-grade materials can be used. The quality of the ma-terial used for beach nourishment may also show strong variability, with location and local/regional policies. Nonetheless, not only should the grain size of the nourished material be similar to (or greater than) that eroded, but it should also not contain fresh biogenic material or contaminants.

Production and usage of marine aggregates vary considerably between countries, with the largest global producers being Japan, the Netherlands, Hong Kong, Korea, and the UK (**Figure 2**). Clusters of extraction areas lie in the vicinity of these coun-tries (e.g., in the North Sea, the Channel, and the Baltic Sea), although aggregate extraction takes place

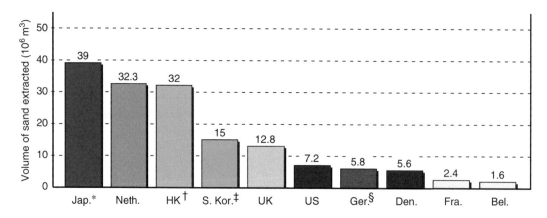

Figure 2 Overview of sand extracted from coastal waters around the world (2002). Annual volumes in million cubic meters per year, for different countries (Japan, the Netherlands, Hong Kong, Korea, United Kingdom, United States, Germany, Denmark, France, Belgium). Key: * data from 1998; † Hong Kong, averaged over 1990–98; ‡ averaged over 1993–95; and § from 2000. Source: Roos PC (2004) *Seabed Pattern Dynamics and Offshore Sand Extraction*, 166pp. PhD Thesis, University of Twente, The Netherlands (ISBN 90-365-2067-3).

in shallow continental seas all over the world. The largest producers present several – if not all – of the following characteristics: limited, or depleted, on-land aggregate resources, increasing environmental pressure on quarries, and large consumption of material. Besides, the yearly national productions may rise drastically as a result of major public works involving, temporarily, a huge volume of extracted material (e.g., Belgium, in 1997, for the construction of pipelines; and the Netherlands, in 2001, for major land reclamation schemes). Countries producing large quantities of marine aggregates extract different types of material, and for distinct purposes (e.g., concrete, filling, or beach recharge). In addition, they may import and export material to and from other countries. For example, in 1998, almost 90% of the marine aggregates production in the UK was for concrete (gravel and coarse sand), but a third of this (c. 7 Mt) was exported to other northwestern European countries. In Spain, by contrast, the dredging of marine aggregates is authorized only for beach replenishment.

Prospecting

The identification of marine aggregate resources is based upon both desk studies and offshore prospecting surveys. The purpose of the desk study is to locate suitable deposits, based upon a scientific literature review and other considerations. Prospecting surveys involve geophysical data acquisition and sediment sampling. In addition to an accurate vessel positioning system (e.g., GPS), the geophysical instrumentation includes usually: an echo sounder, to investigate the seafloor bathymetry; a high-resolution seismic system (e.g., Boomer, Pinger, and Sparker), to produce reflection profiles of the first (c. 10) few meters of the seabed layers; and, a side scan sonar, to characterize the seabed hardness and roughness (**Figure 3**). The collected data set provides information about the thickness, horizontal distribution, and surficial character of the deposit. Vertical and horizontal ground-truth information is provided by core and grab samples, respectively. In addition, grab sampling is useful to examine the surficial sediment grain-size distribution (and also to identify the benthic biological communities present within the area). The selection of the sampling equipment depends upon the difficulty of penetration into the various sediment layers. Large hydraulic grabs can be used to obtain surficial sediment samples, while vibrocorers (up to 10-m long) are capable of coring and recovering coarse-grained material. The level of accuracy of deposit recognition, dependent upon the resolution of both the instruments and the sampling spatial coverage, is governed by the cost of the surveys.

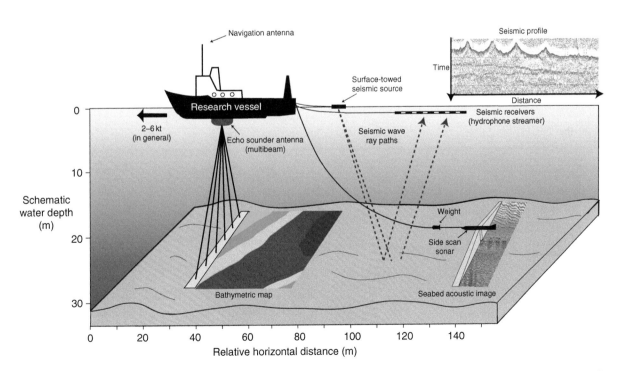

Figure 3 Schematic representation of a survey vessel undertaking geophysical prospecting operations: (multibeam) echo sounder mapping; side scan sonar imaging; and seismic reflection survey.

Extraction Process

Offshore sand and gravel extraction is performed by anchor (or static) suction dredging, and trailer suction dredging. Both techniques utilize powerful centrifugal pumps to draw up the seabed material into the hopper dredger, through pipes of up to 1 m diameter. The dredged material displaces seawater within the hold of the ship, loaded previously as ballast. Anchor suction dredging involves a vessel anchoring or remaining stationary over a deposit, together with forward suction of the bed material through the pipe. As such, the technique is effective for small spatially restricted or, locally, thick deposits. This abstraction procedure creates crater-like pits, or a series of pits, up to 25 m deep and 200 m in diameter, with a slope of about 5° (**Figure 4**). In contrast, trailer suction dredging requires the dredger to drag the lower end of its rear-facing pipe(s) slowly along the seabed, while the ship is underway. This technique permits relatively large areas with thin and more evenly distributed deposits to be worked. It is used also for thicker deposits, such as sandbanks, to limit the harmful environmental impacts to the superficial layers. At the bottom, the head of the pipe creates linear furrows, of 1–3 m in width and up to 50 cm in depth (**Figure 5**). However, repeated trailer suction hopper dredging over an area can lead to large dredged depressions (**Figure 6**).

The total dredged cargo contains generally a mix of different grain sizes. In some cases, the dredgers

(a)

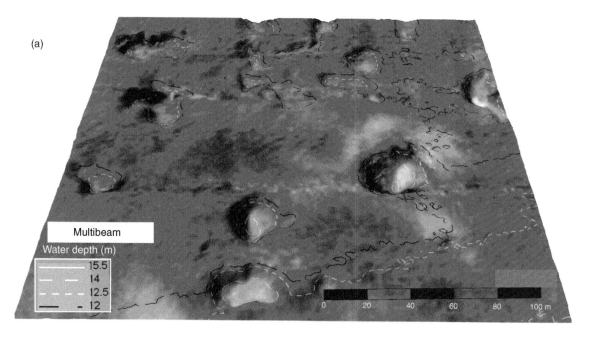

Multibeam

Water depth (m)

	15.5
	14
	12.5
	12

0   20   40   60   80   100 m

(b)

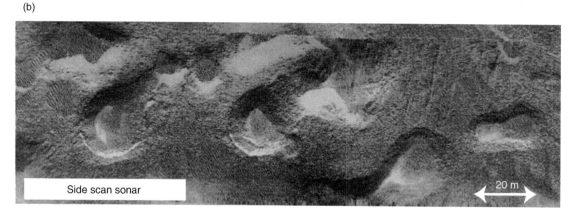

Side scan sonar

20 m

Figure 4 Examples of gravel pits created by anchor hopper dredging (Tromper Wiek Area, Baltic Sea): (a) bathymetry draped with backscatter (illuminated from the west), obtained from a multibeam system; (b) side scan sonar. Source: Manso F, Diesing M, Schwarzer K, and Garel E (2004) Monitoring of dredged areas by combination of multibeam echosounder and sidescan sonar data. *Conference Littoral 2004*, Aberdeen, UK, 20–22 September 2004.

(a)

(b)

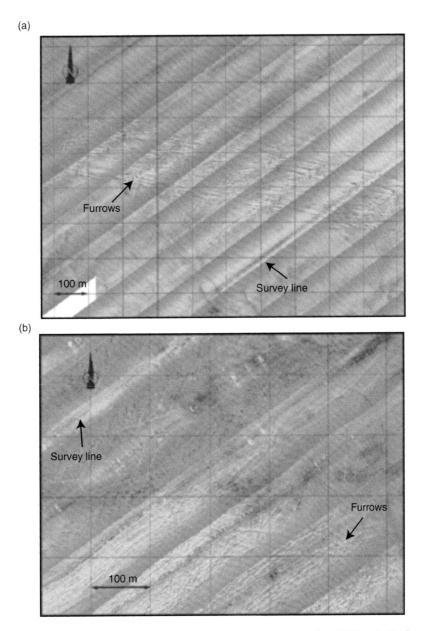

Figure 5 Side scan sonar mosaics showing sandy furrows, in two adjacent areas, at Graal Müritz, Baltic Sea, in December 2000 (with the survey lines in a NE–SW direction): (a) furrows are observed in the center of the image, with a subequatorial direction (note the reverse in direction of the operating dredger at the western tip of the tracks); (b) the furrows are oriented more randomly, although predominantly NE–SW in the lower part of the image. From Diesing M (2003) *Die Regeneration von Materialentnahmestellen in der südwestlichen Ostsee unter besonderer Berücksichtigung der rezenten Sedimentdynamik*, 158pp. PhD Thesis, Christian-Albrechts-Universität, Kiel.

retain all the dredged material on board, for discharge and processing ashore. However, dredging operations often target only a specific type of aggregate, defined by market demand. Since it is more cost-effective to load and transport only a cargo of the required type, dredgers often have the facility to screen onboard the dredged material, spilling back to the sea the unwanted fines. This screening process spills an important amount of material, up to 3–4 times the retained cargo load, generating sediment plumes that settle down and spread over the extraction zone (making subsequent extractions sometimes more difficult).

Nowadays, some dredgers with cargo capacities of up to 8500 t operate with pumps capable of drawing aggregates at 2600 t of material per hour, in water depths up to 50 m (even 90 m, for Japanese vessels). Discharging on the wharf is performed by a range of machinery, such as bucket wheels, scrapers, or wire-hoisted grabs, which place the aggregates on a

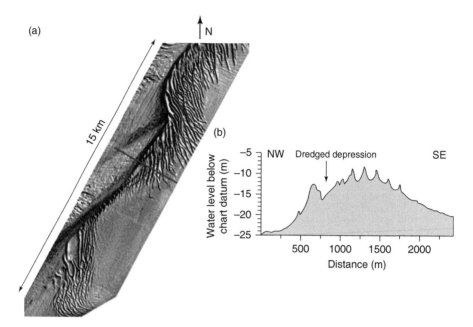

Figure 6 Example of depressions created by repeated trailer suction hopper dredging on a sandbank: the Kwintebank, offshore Belgium. (a) Bathymetric map of the Kwintebank, where the depression is located with the red dotted ellipse; the red straight line indicates the location of the cross section (b); the bank is 15 km in length, 1–2 km in width, and 10–20 m in height. (b) Cross section perpendicular to the bank long-axis, including the dredged depression, 300 m in width, 1 km in length, and up to 5 m deep. (a) Source: Belgian Fund for Sand Extraction.

conveyor belt. Hydraulic discharge is also used, mainly for beach-nourishment schemes.

Environmental Impacts

The environmental effects of sand and gravel extraction include physical effects, such as the modification of the sea bottom topography, the creation of turbidity plumes, substrate alteration, changes in the local wave and current patterns, and impacts on the coast. Biological effects include changes in the density, diversity, biomass, and community structure of the benthos or fish populations, as a consequence of the physical effects. One dredging activity may not have a significant direct or indirect environmental impact, but the cumulative effect of several (adjacent) dredging sites may induce significant changes.

Physical Impacts

The estimation of the regeneration time of dredged features is still very difficult and depends upon the extracted material, the geometry of the excavation, the water depth, and the hydrodynamic regime of the system. Typical timescales for the regeneration of dredged furrows in sandy dynamic substrates lie within the range of months. In very energetic shallow sandy areas, such as those found in estuaries, they may recover after just one (or a few) tidal cycle. In contrast,

in deep and low-energy areas of fine sand (e.g., the German Baltic Sea), recovery could take decades, even for small features such as furrows. In waters of the Netherlands, southern North Sea, dredged tracks in a gravelly substrate exposed to long-period waves were found to disappear completely within 8 months, at 38-m water depth. In contrast, in waters of SE England, dredged tracks in gravelly sediments can take from 3 to more than 7 years to recover. Dredged depressions or pits created by static dredging have also been reported to remain, as recognizable seabed features, for several years and up to decades. In some cases, pits have been observed to migrate slowly in the direction of the dominant current.

Surface turbid plumes are generated by the screening process and by the overflow of material from the hopper, during dredging. A further source of turbidity, with a far lesser quantity of suspended material, results from the mechanical disturbance of the bed sediment, by the head of the pipe. Large increases in suspended solid concentrations tend to be short-lived and localized, close to the operating dredger. However, turbid plumes with low suspended sediment concentrations can affect much larger areas of the seabed, over extended time periods (several days instead of several hours), especially when dredging is occurring simultaneously in adjacent extraction areas.

Changes in sediment composition, as a direct result of dredging, range from minor alterations of superficial granulometry to a large increase in the proportion of fines (sand or silt), or to an increase of gravel through the exposure of coarser sediments. In addition, dredged depressions or furrows, in gravel and coarse sand, can trap gravelly sand remobilized by storms, or finer sediment mobilized by tidal currents. This process may reduce significantly the volume of sediment bypassing in an area. In the case of relatively steep and deep dredged trough, a significant increase in silt and clay material, at the bottom, is sometimes observed following dredging, with long-term consequences (i.e., alteration) upon the *in situ* infaunal assemblage.

Local hydrodynamic changes (i.e., in wave heights and current patterns) may result from the modification of the bathymetry by dredging. In turn, the local erosional and depositional patterns may be affected (near-field effects). In general, the dimensions of dredged features are, in relative terms, so small, that there is little influence of the deepened area on the macroscale hydrodynamics. However, local hydrodynamic changes induced by dredging may result in wider environmental effects (far-field effects). The main issue concerns possible erosion at the coastline, according to various scenarios. Beach erosion may occur, as a result of offshore dredging, due to a reduction of the sediment supply to the coast. This effect is induced either by the direct extraction of material which would normally supply the coast, or by trapping these sediments within the dredged depressions. Any 'beach drawdown' is a particular example of this case: if the area of extraction is not located far enough from the shore, that is, on the subtidal part of a beach, beach sediment may slump down into the excavated area (e.g., during winter storms) and not be able to return subsequently to the beach (in the summer). Coastal erosion may take place also if the protection of the coast is reduced. In particular, dredging may lower significantly the crest heights of (shallow) banks that normally reduce wave breaking and hence, when removed, would lead to larger wave heights at the coast. In addition, if the alterations in the sea bottom topography affect the wave refraction patterns, erosion may result from modification of the wave directions at the coastline.

A substantial list of parameters has to be considered to evaluate whether dredging leads to erosion on the shore: water depth; the distance from the shore; the distribution of the banks; the degree of exposure; the direction of the prevailing waves and tidal currents; the frequency, direction, and severity of storms; the wave reflection and refraction pattern; and seabed type and sediment mobility. Predictions on the changes are performed generally using mathematical models in which input data are provided through a number of sources, including surveys, charts, and offshore buoys. The modeling results have often a wide error range due to the relative uncertainties in the estimation of processes, especially for sediment transport calculations.

Biological Impacts

The most obvious harmful direct effect of marine aggregate extraction results from the direct removal of substrata and associated benthic epifauna and inflora. Available evidence indicates that dredging causes an initial reduction in the abundance, species diversity, and biomass of a benthic community. Other impacts on benthic communities (and referred to as indirect effects) arise from the modification of abiotic factors, which determine the broad-scale benthic community patterns. These modifications concern the nature and stability of the sediment, through the exposition of underlying strata, and the tidal current strength, through the alteration of the local topography. The creation of relatively deep depressions with anoxic conditions is another source of concern. Sediment plumes arising from the dredging activities may also have adverse biological effects, by affecting water quality, increasing the sediment deposition, and modifying the sediment type. In the case of clean mobile sands, increased sedimentation and resuspension, as a consequence of dredging, are considered generally to be of less concern, as the fauna inhabiting such areas tend to be adapted to naturally high levels of suspended sediments, caused by wave and tidal current action. The effects of sediment deposition and resuspension may be more significant in gravelly habitats dominated by encrusting epifaunal taxa, due to the abrasive impacts of suspended sediments. In addition, turbidity plumes may have indirect impacts on fisheries, due to avoidance behavior by some species, the interruption of migratory pathways (e.g., lobsters), and the loss of access to traditional fishing grounds.

The importance of the impacts that affect a benthic community is investigated commonly on the basis of the number of species, abundance, biomass, and the age of the population, compared to an undisturbed similar baseline. The existence of negligible or significant impacts depends largely upon the spatial and temporal intensity of dredging and, likewise, upon the nature of the benthos. While the recolonization of a dredged site by benthic species is generally rapid, the complete recovery (also 'restoration' or 'readjustment') of a benthic community can take many years. The recovery rate, after dredging, depends primarily upon several factors: the community diversity and richness of the area prior to dredging; the life cycle and growth rate of species; the nature, magnitude, and duration of the dredging operations; the nature of the new sediment which is exposed, or subsequently accumulated at the extraction site; the larval and adult pool of potential new colonizers; the hydrodynamics and associated bed load transport processes; and the nature and intensity of the stresses (caused, for example, by storms) which the community normally withstands. Thus, among a number of studies addressing the consequences of long-term dredging operations on the recolonization of biota and the composition of sediment following cessation, some disparity appears in the findings. These results range from minimal effects of disturbance after dredging to significant changes in community structure, persisting over many years.

Supply and Demand: The Future of Offshore Sand and Gravel Mining

Offshore sand and gravel constitute a limited and nonrenewable resource. The notion of supply is linked strongly to the usage of the material. As such, a distinction has to be made between the (reserves of) primary aggregates, suitable for use in construction, and the (resources of) secondary aggregates, which may include contaminants, and used typically as in-filling material. For example, in-filling sand is available in relatively large quantities, whereas sand for construction, whose demand for quality is comparatively higher, is not so abundant. Gravel, which is also considered generally as a high-quality material, is in short supply. In the UK waters, the industry estimates that marine aggregate availability will last for 50 years, at present levels of extraction. However, such estimations are subject to important fluctuations, due to limited knowledge of the seafloor surface and subsurface. Furthermore, the level of accuracy of the identification of the resource varies greatly, from country to country (in general, data are far more complete in countries that have a greater reliance on marine sand and gravel). Another frequent difficulty toward a reliable assessment of the supply comes from the dispersal of the data relevant to the resource among governmental organizations, hydrographic departments, the dredging industry, and other commercial companies. In order to tackle this issue, a number of countries are undertaking seabed mapping programs, dedicated to the recognition of marine aggregates deposits. These programs range from detailed resource assessment, to reconnaissance level of the resource.

The precise estimation of the future demand for aggregates in general, including marine sand and gravel, is not an easy task. Difficulties arise from the highly variable and changing character of a number of parameters, such as the construction market and other economic factors, political strategies, environmental considerations, etc. In addition, the estimates can be distorted by extensive public work projects, involving large amounts of material. Moreover, the predictions should consider the exports and imports of material, and, therefore, take into account the demand at a much broader scale than the national one alone. Few countries have published estimates for the future demand of aggregate (only the UK and the Netherlands, in Europe).

However, in a number of countries, on-land aggregate mining takes place within the context of increasing environmental pressure and the depletion of the resource. At the same time, large parts of the coasts, on a worldwide scale, experience increasing coastal erosion. Therefore, it is expected that the future marine sand and gravel extraction is bound to increase, in order to supply high-quality material and to provide the large quantities of aggregates required for the realization of large-scale infrastructure projects designed to manage coastal retreat and accommodate the development of the coastal zones (i.e., beach replenishment, dune restoration, foreshore nourishment, land reclamation, and other coastal defense schemes). Hence, in order to find these new resources, it is expected that the activity will move farther offshore, within the next decade.

See also

Beaches, Physical Processes Affecting. Sandy Beaches, Biology of.

Further Reading

Bellamy AG (1998) The UK marine sand and gravel dredging industry: An application of quaternary geology. In: Latham J-P (ed.) *Engineering Geology Special Publications 13: Advances in Aggregates and Armourstone Evaluation*, pp. 33–45. London: Geological Society.

Boyd SE, Limpenny DS, Rees HL, and Cooper KM (2005) The effects of marine sand and gravel extraction on the macrobenthos at a commercial dredging site (results 6 years post-dredging). *ICES Journal of Marine Science* 62: 145–162.

Byrnes MR, Hammer RM, Thibaut TD, and Snyder DB (2004) Physical and biological effects of sand mining offshore Alabama, USA. *Journal of Coastal Research* 20(1): 6–24.

Desprez M (2000) Physical and biological impact of marine aggregate extraction along the French coast of the Eastern English Channel. Short and long-term post-dredging restoration. *ICES Journal of Marine Science* 57: 1428–1438.

Diesing M (2003) Die Regeneration von Material-entnahmestellen in der Südwestlichen Ostsee unter Besonderer Berücksichtigung der Rezenten Sediment-dynamik, 158pp. PhD Thesis, Christian-Albrechts-Universität, Kiel.

Diesing M, Schwarzer K, Zeiler M, and Kelon H (2006) *Special Issue: Comparison of Marine Sediment Extraction Sites by Mean of Shoreface Zonation. Journal of Coastal Research* 39: 783–788.

ICES (1992) Effects of extraction of marine sediments on fisheries. *Cooperative Research Report, No. 182.* Copenhagen: International Council for the Exploration of the Sea.

ICES (2001) Report of the ICES Working Group on the effects of extraction of marine sediments on the marine ecosystem. *Cooperative Research Report, No. 247.* Copenhagen: International Council for the Exploration of the Sea.

Kenny AJ and Rees HL (1996) The effects of marine gravel extraction on the macrobenthos: Results 2 years post-dredging. *Marine Pollution Bulletin* 32(8/9): 615–622.

Manso F, Diesing M, Schwarzer K, and Garel E (2004) Monitoring of dredged areas by combination of multibeam echosounder and sidescan sonar data. *Conference Littoral 2004*, Aberdeen, UK, 20–22 September 2004.

Newell RC, Seiderer LJ, and Hitchcock DR (1998) The impact of dredging works in coastal waters: A review of the sensitivity to disturbance and subsequent recovery of biological resources on the sea bed. *Oceanography and Marine Biology* 36: 127–178.

Roos PC (2004) *Seabed Pattern Dynamics and Offshore Sand Extraction*, 166pp. PhD Thesis, University of Twente, The Netherlands (ISBN 90-365-2067-3).

van Dalfsen JA, Essink K, Toxvig madsen H, *et al.* (2000) Differential response of macrozoobenthos to marine sand extraction in the North Sea and the Western Mediterranean. *ICES Journal of Marine Science* 57(5): 1439–1445.

van der Veer HW, Bergman MJN, and Beukema JJ (1985) Dredging activities in the Dutch Waddensea: Effects on macrobenthic infauna. *Netherlands Journal of Sea Research* 19: 183–190.

van Moorsel GWNM (1994) The Klaverbank (North Sea), geomorphology, macrobenthic ecology and the effect of gravel extraction. *Report No. 94.24.* Culemborg: Bureau Waardenburg BV.

Van Rijn LC, Soulsby RL, Hoekstra P, and Davies AG (eds.) (2005) *SANDPIT: Sand Transport and Morphology of Offshore Sand Mining Pits Process Knowledge and Guidelines for Coastal Management.* Amsterdam: Aqua Publications.

Relevant Websites

http://www.bmapa.org
– British Marine Aggregate Producers Association (BMAPA).

http://www.dredging.org
– Central Dredging Association (CEDA): Serving Africa, Europe, and the Middle East.

http://www.ciria.org
– Construction Industry Research and Information Association (CIRIA)

http://www.ciria.org
– European Marine Sand and Gravel Group (EMSAGG).

http://www.dvz.be
– ICES Working Group on the Effect of Extraction of Marine Sediments on the Marine Ecosystem (WGEXT).

http://www.iadc-dredging.com
– International Association of Dredging Companies (IADC).

http://www.ices.dk
– International Council for the Exploration of the Sea (ICES).

http://www.sandandgravel.com
– Marine Sand and Gravel Information Service (MAGIS).

http://www.westerndredging.org
– Western Dredging Association (WEDA): Serving the Americas.

http://www.woda.org
– World Organisation of Dredging Associations (WODA).

SLIDES, SLUMPS, DEBRIS FLOWS, AND TURBIDITY CURRENTS

G. Shanmugam, The University of Texas at Arlington, Arlington, TX, USA

Introduction

Since the birth of modern deep-sea exploration by the voyage of HMS *Challenger* (21 December 1872–24 May 1876), organized by the Royal Society of London and the Royal Navy, oceanographers have made considerable progress in understanding the world's oceans. However, the physical processes that are responsible for transporting sediment into the deep sea are still poorly understood. This is simply because the physics and hydrodynamics of these processes are difficult to observe and measure directly in deep-marine environments. This observational impediment has created a great challenge for communicating the mechanics of gravity processes with clarity. Thus a plethora of confusing concepts and related terminologies exist. The primary objective of this article is to bring some clarity to this issue by combining sound principles of fluid mechanics, laboratory experiments, study of modern deep-marine systems, and detailed examination of core and outcrop.

A clear understanding of deep-marine processes is critical because the petroleum industry is increasingly moving exploration into the deep-marine realm to meet the growing demand for oil and gas. Furthermore, mass movements on continental margins constitute major geohazards. Velocities of subaerial debris flows reached up to $500 \, \text{km} \, \text{h}^{-1}$. Submarine landslides, triggered by the 1929 Grand Banks earthquake (offshore Newfoundland, Canada), traveled at a speed of $67 \, \text{km} \, \text{h}^{-1}$. Such catastrophic events could destroy offshore oil-drilling platforms. Such events could result in oil spills and cost human lives. Therefore, numerous international scientific projects have been carried out to understand continental margins and related mass movements.

Terminology

The term 'deep marine' commonly refers to bathyal environments, occurring seaward of the continental shelf break (>200-m water depth), on the continental slope and the basin (**Figure 1**). The continental rise, which represents that part of the continental margin between continental slope and abyssal plain, is included here under the broad term 'basin'.

Two types of classifications and related terminologies exist for gravity-driven processes: (1) mechanics-based terms and (2) velocity-based terms.

Mechanics-based Terms

Continental margins provide an ideal setting for slope failure, which is the collapse of slope sediment (**Figure 2**). Following a failure, the failed sediment moves downslope under the pull of gravity. Such gravity-driven processes exhibit extreme variability in mechanics of sediment transport, ranging from mobility of kilometer-size solid blocks on the seafloor to transport of millimeter-size particles in suspension of dilute turbulent flows. Gravity-driven processes are broadly classified into two types based on the physics and hydrodynamics of sediment mobility: (1) mass transport and (2) sediment flows (**Table 1**).

Mass transport is a general term used for the failure, dislodgement, and downslope movement of sediment under the influence of gravity. Mass transport is composed of both slides and slumps. Sediment flow is an abbreviated term for sediment-gravity flow. It is composed of both debris flows and turbidity currents (**Figure 3**). In rare cases, debris flow may be classified both as mass transport and as sediment flow. Thus the term mass transport is used for slide, slump, and debris flow, but not for turbidity current. Mass transport can operate in both subaerial and subaqueous environments, but turbidity currents can operate only in subaqueous environments. The advantage of this classification is that physical features preserved in a deposit directly represent the physics of sediment movement that existed at the final moments of deposition. The link between the deposit and the physics of the depositional process can be established by practicing the principle of process sedimentology, which is detailed bed-by-bed description of sedimentary rocks and their process interpretation.

The terms slide and slump are used for both a process and a deposit. The term debrite refers to the deposit of debris flows, and the term turbidite represents the deposit of turbidity currents. These

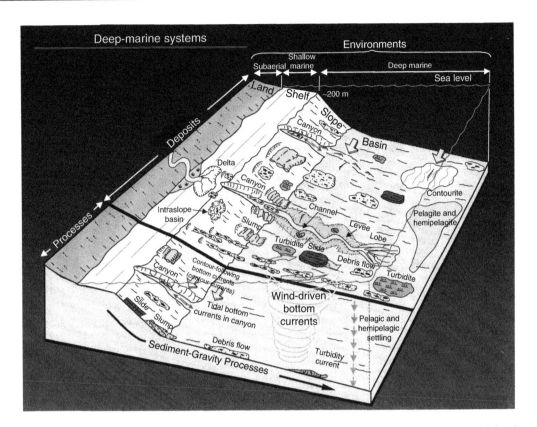

Figure 1 Schematic diagram showing complex deep-marine sedimentary environments occurring at water depths deeper than 200 m (shelf-slope break). In general, sediment transport in shallow-marine (shelf) environments is characterized by tides and waves, whereas sediment transport in deep-marine (slope and basin) environments is characterized by mass transport (i.e., slides, slumps, and debris flows), and sediment flows (i.e., debris flows and turbidity currents). Internal waves and up and down tidal bottom currents in submarine canyons (opposing arrows), and along-slope movement of contour-following bottom currents are important processes in deep-marine environments. Circular motion of wind-driven bottom currents can be explained by eddies induced by the Loop Current in the Gulf of Mexico or by the Gulf Stream Gyre in the North Atlantic. However, such bottom currents are not the focus of this article. From Shanmugam G (2003) Deep-marine tidal bottom currents and their reworked sands in modern and ancient submarine canyons. *Marine and Petroleum Geology* 20: 471–491.

Figure 2 Multibeam bathymetric image of the US Pacific margin, offshore Los Angeles (California), showing well-developed shelf, slope, and basin settings. The canyon head of the Redondo Canyon is at a water depth of 10 m near the shoreline. DB, debris blocks of debris flows; MT, mass-transport deposits. Vertical exaggeration is 6 ×. Modified from USGS (2007) US Geological Survey: Perspective view of Los Angeles Margin. http://wrgis.wr.usgs.gov/dds/dds-55/pacmaps/la_pers2.htm (accessed 11 May 2008).

Table 1   A classification of subaqueous gravity-driven processes

Major type	Nature of moving material	Nature of movement	Sediment concentration (vol. %)	Fluid rheology and flow state	Depositional process
Mass transport (also known as mass movement, mass wasting, or landslide)	Coherent Mass without internal deformation	Translational motion between stable ground and moving mass	Not applicable	Not applicable	Slide
	Coherent Mass with internal deformation	Rotational motion between stable ground and moving mass			Slump
Sediment flow (in cases, mass transport)	Incoherent Body (sediment–water slurry)	Movement of sediment–water slurry en masse	High (25–95)	Plastic rheology and laminar state	Debris flow (mass flow)
Sediment flow	Incoherent Body (water-supported particles in suspension)	Movement of individual particles within the flow	Low (1–23)	Newtonian rheology and turbulent state	Turbidity current

Reproduced from Shanmugam G (2006). *Deep-Water Processes and Facies Models: Implications for Sandstone Petroleum Reservoirs.* Amsterdam: Elsevier, with permission from Elsevier.

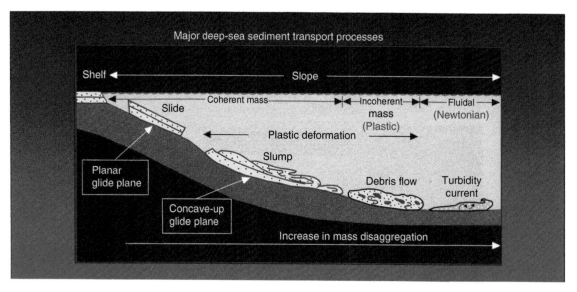

Figure 3   Schematic diagram showing four common types of gravity-driven downslope processes that transport sediment into deep-marine environments. A slide represents a coherent translational mass transport of a block or strata on a planar glide plane (shear surface) without internal deformation. A slide may be transformed into a slump, which represents a coherent rotational mass transport of a block or strata on a concave-up glide plane (shear surface) with internal deformation. Upon addition of fluid during downslope movement, slumped material may transform into a debris flow, which transports sediment as an incoherent body in which intergranular movements predominate over shear-surface movements. A debris flow behaves as a plastic laminar flow with strength. As fluid content increases in debris flow, the flow may evolve into Newtonian turbidity current. Not all turbidity currents, however, evolve from debris flows. Some turbidity currents may evolve directly from sediment failures. Turbidity currents can develop near the shelf edge, on the slope, or in distal basinal settings. From Shanmugam G, Lehtonen LR, Straume T, Syvertsen SE, Hodgkinson RJ, and Skibeli M (1994) Slump and debris flow dominated upper slope facies in the Cretaceous of the Norwegian and northern North Seas (61°–67° N): Implications for sand distribution. *American Association of Petroleum Geologists Bulletin* 78: 910–937.

process-product terms are also applicable to deep-water lakes. For example, turbidites have been recognized in the world's deepest lake (1637 m deep), Lake Baikal in south-central Siberia.

There are several synonymous terms in use:

1. rotational slide = slump;
2. mass transport = mass movement = mass wasting = landslide;
3. muddy debris flow = mud flow = slurry flow;
4. sandy debris flow = granular flow.

Velocity-Based Terms

Examples of velocity-based terms are:

1. rock fall: free falling of detached rock mass that increases in velocity with fall;
2. flow slide: fast-moving mass transport;
3. rock slide: fast-moving mass transport;
4. debris avalanche: fast-moving mass transport;
5. mud flow: fast-moving debris flow;
6. sturzstrom: fast-moving mass transport;
7. debris slide: slow-moving mass transport;
8. creep: slow-moving mass transport.

These velocity-based terms are subjective because the distinction between a fast-moving and a slow-moving mass has not been defined using a precise velocity value. It is also difficult to measure velocities because of common destruction of velocity meters by catastrophic mass transport events. Furthermore, there are no scientific criteria to determine the absolute velocities of sediment movement in the ancient rock record. This is because sedimentary features preserved in the deposit do not reflect absolute velocities. Therefore, velocity-based terms, although popular, are not practical.

Initiation of Movement

The initiation of mass movements and sediment flows is closely related to shelf-edge sediment failures. These failures along continental margins are triggered by one or more of the following causes:

1. *Earthquakes.* Earthquakes and related shaking can cause an increase in shear stress, which is the downslope component of the total stress. Submarine mass movements have been attributed to the 1929 Grand Banks earthquake off the US East Coast and Canada.
2. *Tectonic oversteepening.* Tectonic compression has elevated the northern flank of the Santa Barbara Basin and overturned the slope in southern California along the US Pacific margin.

Uplift (thrusting) along these slopes has led to oversteepening and sediment failure.
3. *Depositional oversteepening.* Mass movements have been attributed to oversteepening near the mouth of the Magdalena River, Colombia.
4. *Depositional loading.* Delta-front mass movements have been attributed to rapid sedimentation in the Mississippi River Delta, Gulf of Mexico.
5. *Hydrostatic loading.* Slope-failure deposits originated during the late Quaternary sea level rise on the eastern Tyrrhenian margin. These slope failures were attributed to rapid drowning of unconsolidated sediment, which resulted in increased hydrostatic loading.
6. *Glacial loading.* Submarine mass movements have been attributed to glacial loading and unloading along the Scotian margin in the North Atlantic.
7. *Eustatic changes in sea level.* Shelf-edge sediment failures and related gravity-driven processes have been attributed to worldwide fall in sea level, close to the shelf edge.
8. *Tsunamis.* The advancing wave front from a tsunami is capable of generating large hydrodynamic pressures on the seafloor that would produce soil movements and slope instabilities.
9. *Tropical cyclones.* Category 5 hurricane Katrina (2005) induced sediment failures and mudflows near the shelf edge in the Gulf of Mexico.
10. *Submarine volcanic activity.* Submarine mass movements have been attributed to volcanic activity in Hawaiian Islands.
11. *Salt movements.* Submarine mass movements along the flanks of intraslope basins have been attributed to mobilization of underlying salt masses in the Gulf of Mexico.
12. *Biologic erosion.* Submarine mass movements have been associated with erosion of the walls and floors of submarine canyons by invertebrates and fishes and boring by animals in offshore California.
13. *Gas hydrate decomposition.* Submarine mass movements have been associated with seepage of methane hydrate and related collapse of unconsolidated sediment along the US Atlantic Margin.

Mass Transport

Various methods are used to recognize mass transport deposits (slides and slumps) in modern submarine environments and their deposits in the ancient geologic record. These methods are (1) direct observations; (2) indirect velocity calculations; (3) remote-sensing technology (e.g., seismic profiling); and (4) examination of the rock (see Glossary).

Slides

A slide is a coherent mass of sediment that moves along a planar glide plane and shows no internal deformation (**Figure 3**). Slides represent translational shear-surface movements. Submarine slides can travel hundreds of kilometers on continental slopes (**Figure 4**). Long-runout distances of up to 800 km for slides have been documented (**Table 2**). Submarine slides are common in fiords because the submerged sides of glacial valleys are steep and because the rate of sedimentation is high due to sediment-laden rivers that drain glaciers into fiords. Submarine canyon walls are also prone to generate slides because of their steep gradients.

Multibeam bathymetric data show that the northern flank of the Santa Barbara Channel (Southern California) has experienced massive slope failures that resulted in the large (130 km²) Goleta landslide complex (**Figure 5(a)**). Approximately 1.75 km³ has been displaced by this slide during the Holocene (approximately the last 10 000 years). This complex has an upslope zone of evacuation and a downslope zone of accumulation (**Figure 5(b)**). It measures 14.6 km long, extending from a depth of 90 m to nearly 574 m, and is 10.5 km wide. It contains both surficial slump blocks and muddy debris flows in three distinct segments (**Figure 5(b)**).

Slides are capable of transporting gravel and coarse-grained sand because of their inherent strength. General characteristics of slides are:

- gravel to mud lithofacies;
- upslope areas with tensional faults;
- area of evacuation near the shelf edge (**Figures 4 and 5(a)**);
- occur commonly on slopes of 1–4°;
- long-runout distances of up to 800 km (**Table 2**);
- transported sandy slide blocks encased in deep-water muddy matrix (**Figure 6**);
- primary basal glide plane or décollement (core and outcrop) (**Figure 7(a)**);
- basal shear zone (core and outcrop) (**Figure 7(b)**);
- secondary internal glide planes (core and outcrop) (**Figure 7(a)**);
- associated slumps (core and outcrop) (**Figure 6**);
- transformation of slides into debris flows in frontal zone;
- associated clastic injections (core and outcrop) (**Figure 7(a)**);
- sheet-like geometry (seismic and outcrop) (**Figure 6**);
- common in areas of tectonic activity, earthquakes, steep gradients, salt movements, and rapid sedimentation.

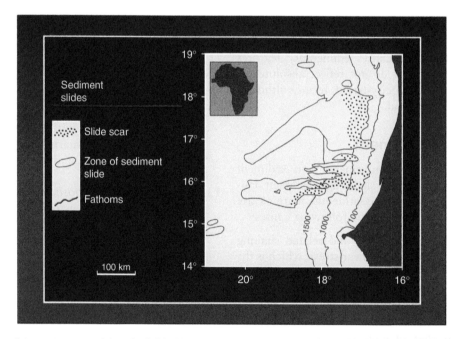

Figure 4 Long-distance transport of detached slide blocks from the shelf edge at about 100 fathom (i.e., 600 ft or 183 m) contour, offshore northwestern Africa. Note that two slide blocks near the 15° latitude marker have traveled nearly 300 km from the shelf edge (i.e., 100 fathom contour). Three contour lines (100, 1000, 1500 fathom) show increasing water depths to the west. From Jacobi RD (1976) Sediment slides on the northwestern continental margin of Africa. *Marine Geology* 22: 157–173.

Table 2 Long-runout distances of modern mass transport processes

Name and location	Runout distance (km)	Data	Process
Storegga, Norway	800	Seismic and GLORIA side-scan sonar images	Slide, slump, and debris flow
Agulhas, S Africa	750	Seismic	Slide and slump
Hatteras, US Atlantic margin	~500	Seismic and core	Slump and debris flow
Saharan NW African margin	>400	Seismic and core	Slump and debris flow
Mauritania–Senegal, NW African margin	~300	Seismic and core	Slump and debris flow
Nuuanu, NE Oahu (Hawaii)	235	GLORIA side-scan sonar images	Mass transport
Wailau, N Molakai (Hawaii)	<195	GLORIA side-scan sonar images	Mass transport
Rockall, NE Atlantic	160	Seismic	Mass transport
Clark, SW Maui (Hawaii)	150	GLORIA side-scan sonar images	Mass transport
N Kauai, N Kauai (Hawaii)	140	GLORIA side-scan sonar images	Mass transport
East Breaks (West), Gulf of Mexico	110	Seismic and core	Slump and debris flow
Grand Banks, Newfoundland	>100	Seismic and core	Mass transport
Ruatoria, New Zealand	100	Seismic	Mass transport
Alika-2, W Hawaii (Hawaii)	95	GLORIA side-scan sonar images	Mass transport
Kaena, NE Oahu (Hawaii)	80	GLORIA side-scan sonar images	Mass transport
Bassein, Bay of Bengal	55	Seismic	Slide and debris flow
Kidnappers, New Zealand	45	Seismic	Slump and slide
Munson–Nygren, New England	45	Seismic	Slump and debris flow
Ranger, Baja California	35	Seismic	Mass transport

On the modern Norwegian continental margin, large mass transport deposits occur on the northern and southern flanks of the Voring Plateau. The Storegga slide (offshore Norway), for example, has a maximum thickness of 430 m and a length of more than 800 km. The Storegga slide on the southern flank of the plateau exhibits mounded seismic patterns in sparker profiles. Although it is called a slide in publications, the core of this deposit is composed primarily of slumps and debrites. Unlike kilometers-wide modern slides that can be mapped using multibeam mapping systems and seismic reflection profiles, huge ancient slides are difficult to recognize in outcrops because of limited sizes of outcrops.

A summary of width:thickness ratio of modern and ancient slides is given in **Table 3**.

Slumps

A slump is a coherent mass of sediment that moves on a concave-up glide plane and undergoes rotational movements causing internal deformation (**Figure 3**). Slumps represent rotational shear-surface movements. In multibeam bathymetric data, distinguishing slides from slumps may be difficult because internal deformation cannot be resolved. In seismic profiles, however, slumps may be recognized because of their chaotic reflections. Therefore, a general term mass transport is preferred when interpreting bathymetric images. Slumps are capable of transporting gravel and coarse-grained sand because of their inherent strength. General characteristics of slumps are:

- gravel to mud lithofacies;
- basal zone of shearing (core and outcrop);
- upslope areas with tensional faults (**Figure 8**);
- downslope edges with compressional folding or thrusting (i.e., toe thrusts) (**Figure 8**);
- slump folds interbedded with undeformed layers (core and outcrop) (**Figure 9**);
- irregular upper contact (core and outcrop);
- chaotic bedding in heterolithic facies (core and outcrop);
- steeply dipping and truncated layers (core and outcrop) (**Figure 10**);
- associated slides (core and outcrop) (**Figure 6**);

(a)

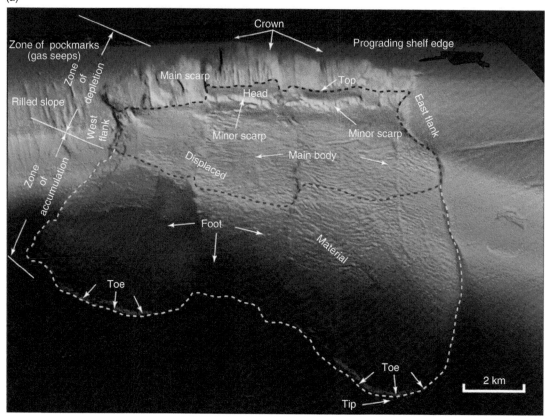

(b)

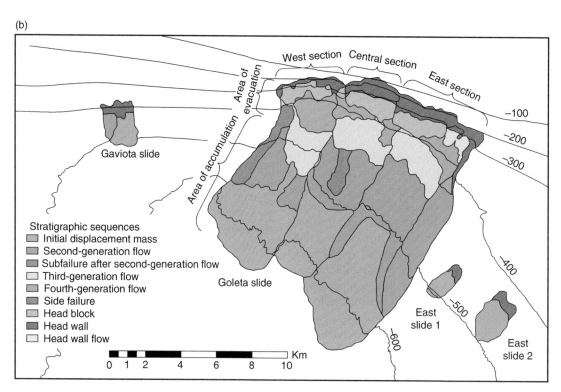

Figure 5 (a) Multibeam bathymetric image of the Goleta slide complex in the Santa Barbara Channel, Southern California. Note lobe-like (dashed line) distribution of displaced material that was apparently detached from the main scarp near the shelf edge. This mass transport complex is composed of multiple segments of failed material. (b) Sketch of the Goleta mass transport complex in the Santa Barbara Channel, Southern California (a) showing three distinct segments (i.e., west, central, and east). Contour intervals (−100, −200, −300, −400, −500, and −600) are in meters. (a) From Greene HG, Murai LY, Watts P, *et al.* (2006) Submarine landslides in the Santa Barbara Channel as potential tsunami sources. *Natural Hazards and Earth System Sciences* 6: 63–88. (b) From Greene HG, Murai LY, Watts P, *et al.* (2006) Submarine landslides in the Santa Barbara Channel as potential tsunami sources. *Natural Hazards and Earth System Sciences* 6: 63–88.

Figure 6 Outcrop photograph showing sheet-like geometry of an ancient sandy submarine slide (1000 m long and 50 m thick) encased in deep-water mudstone facies. Note the large sandstone sheet with rotated/slumped edge (left). Person (arrow): 1.8 m tall. Ablation Point Formation, Kimmeridgian (Jurassic), Alexander Island, Antarctica. Photo courtesy of D.J.M. Macdonald.

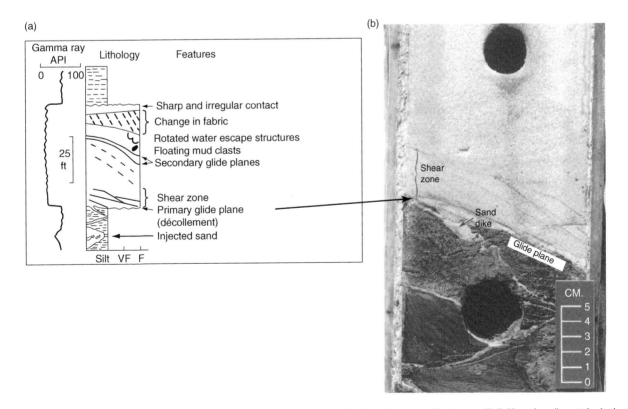

Figure 7   (a) Sketch of a cored interval of a sandy slide/slump unit showing blocky wireline log motif (left) and sedimentological details (right). (b) Core photograph showing the basal contact of the sandy slide (arrow). Note a sand dike (i.e., injectite) at the base of shear zone. Eocene, North Sea. Compare this small-scale slide (15 m thick) with a large-scale slide (50 m thick) in **Figure 6**. From Shanmugam G (2006) *Deep-Water Processes and Facies Models: Implications for Sandstone Petroleum Reservoirs*. Amsterdam: Elsevier.

Table 3 Dimensions of modern and ancient mass transport deposits

Example	Width:thickness ratio (observed dimensions)
Slide, Lower Carboniferous, England	7:1 (100 m wide/long, 15 m thick)
Slide, Cambrian–Ordovician, Nevada	30:1 (30 m wide/long, 1 m thick)
Slide, Jurassic, Antarctica	45:1 (20 km wide/long, 440 m thick)
Slide, Modern, US Atlantic margin	40–80:1 (2–4 km wide/long, 50 m thick)
Slide, Modern, Gulf of Alaska	130:1 (15 km wide/long, 115 m thick)
Slide, Middle Pliocene, Gulf of Mexico	250:1 (150 km wide/long, 600 m thick)
Slide/slump/debris flow/turbidite 5000–8000 BP, Norwegian continental margin	675:1 (290 km wide/long, 430 m thick)
Slump, Cambrian–Ordovician, Nevada	10:1 (100 m wide/long, 10 m thick)
Slump, Aptian–Albian, Antarctica	10:1 (3.5 km wide/long, 350 m thick)
Slump/slide/debris flow, Lower Eocene, Gryphon Field, UK	21:1 (2.6 km wide/long, 120 m thick)
Slump/slide/debris flow, Paleocene, Faeroe Basin, north of Shetland Islands	28:1 (7 km wide/long, 245 m thick)
Slump, Modern, SE Africa	171:1 (64 km wide/long, 374 m thick)
Slump, Carboniferous, England	500:1 (5 km wide/long, 10 m thick)
Slump, Lower Eocene, Spain	900–3600:1 (18 km wide/long, 5–20 m thick)
Debrite, Modern, British Columbia	12:1 (50 m wide/long, 4 m thick)
Debrite, Cambrian–Ordovician, Nevada	30:1 (300 m wide/long, 10 m thick)
Debrite, Modern, US Atlantic margin	500–5000:1 (10–100 km wide/long, 20 m thick)
Debrite, Quaternary, Baffin Bay	1250:1 (75 km wide/long, 60 m thick)
Turbidite (depositional lobe) Cretaceous, California	167:1 (10 km wide/long, 60 m thick)
Turbidite (depositional lobe) Lower Pliocene, Italy	1200:1 (30 km wide/long, 25 m thick)
Turbidite (basin plain) Miocene, Italy	11400:1 (57 km wide/long, 5 m thick)
Turbidite (basin plain) 16 000 BP Hatteras Abyssal Plain	125 000:1 (500 km wide/long, 4 m thick)

Reproduced from Shanmugam G (2006) *Deep-Water Processes and Facies Models: Implications for Sandstone Petroleum Reservoirs*. Amsterdam: Elsevier, with permission from Elsevier.

- associated sand injections (core and outcrop) (**Figure 10**);
- lenticular to sheet-like geometry with irregular thickness (seismic and outcrop);
- contorted bedding has been recognized in Formation MicroImager (FMI);
- chaotic facies on high-resolution seismic profiles.

In Hawaiian Islands, the Waianae Volcano comprises the western half of O'ahu Island. Several submersible dives and multibeam bathymetric imaging have confirmed the timing of seabed failure that formed the Waianae mass transport (slump?) complex. Multiple collapses and deformation events resulted in compound mass wasting features on the volcano's southwest flank. This complex is the largest in Hawaii, covering an area of about 5500 km².

Sediment Flows

Sediment flows (i.e., sediment-gravity flows) are composed of four types: (1) debris flow, (2) turbidity current, (3) fluidized sediment, and (4) grain flow. In this article, the focus is on debris flows and turbidity currents because of their importance. These two processes are distinguished from one another on the basis of fluid rheology and flow state. The rheology of fluids can be expressed as a relationship between applied shear stress and rate of shear strain (**Figure 11**). Newtonian fluids (i.e., fluids with no inherent strength), like water, will begin to deform the moment shear stress is applied, and the deformation is linearly proportional to stress. In contrast, some naturally occurring materials (i.e., fluids with strength) will not deform until their yield stress has been exceeded (**Figure 11**); once their yield stress is exceeded, deformation is linear. Such materials with strength (e.g., wet concrete) are considered to be Bingham plastics (**Figure 11**). For flows that exhibit plastic rheology, the term plastic flow is appropriate. Using rheology as the basis, deep-water sediment flows are divided into two broad groups, namely (1) Newtonian flows that represent turbidity currents and (2) plastic flows that represent debris flows.

In addition to fluid rheology, flow state is used in distinguishing laminar debris flows from turbulent turbidity currents. The difference between laminar and turbulent flows was demonstrated in 1883 by Osborne Reynolds, an Irish engineer, by injecting a thin stream of dye into the flow of water through a glass tube. At low rates of flow, the dye stream traveled in a straight path. This regular motion of

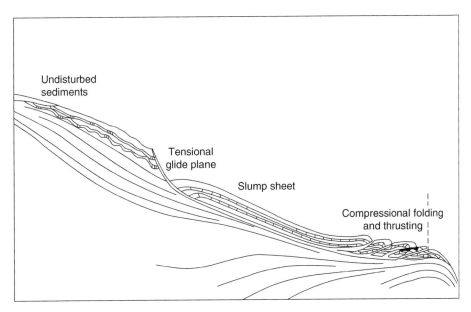

Figure 8 Sketch of a submarine slump sheet showing tensional glide plane in the updip detachment area and compressional folding and thrusting in the downdip frontal zone. From Lewis KB (1971) Slumping on a continental slope inclined at 1°–4°. *Sedimentology* 16: 97–110.

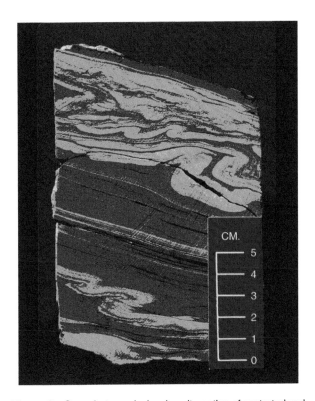

Figure 9 Core photograph showing alternation of contorted and uncontorted siltstone (light color) and claystone (dark color) layers of slump origin. This feature is called slump folding. Paleocene, North Sea. Reproduced from Shanmugam (2006). *Deep-Water Processes and Facies Models: Implications for Sandstone Petroleum Reservoirs.* Amsterdam: Elsevier, with permission from Elsevier.

fluid in parallel layers, without macroscopic mixing across the layers, is called a laminar flow. At higher flow rates, the dye stream broke up into chaotic eddies. Such an irregular fluid motion, with macroscopic mixing across the layers, is called a turbulent flow. The change from laminar to turbulent flow occurs at a critical Reynolds number (the ratio between inertia and viscous forces) of *c.* 2000 (**Figure 11**).

Debris Flows

A debris flow is a sediment flow with plastic rheology and laminar state from which deposition occurs through 'freezing' *en masse*. The terms debris flow and mass flow are used interchangeably because each exhibits plastic flow behavior with shear stress distributed throughout the mass. In debris flows, intergranular movements predominate over shear-surface movements. Although most debris flows move as incoherent material, some plastic flows may be transitional in behavior between coherent mass movements and incoherent sediment flows (**Table 1**). Debris flows may be mud-rich (i.e., muddy debris flows), sand-rich (i.e., sandy debris flows), or mixed types. Sandy debrites comprise important petroleum reservoirs in the North Sea, Norwegian Sea, Nigeria, Equatorial Guinea, Gabon, Gulf of Mexico, Brazil, and India. In multibeam bathymetric data, recognition of debrites is possible.

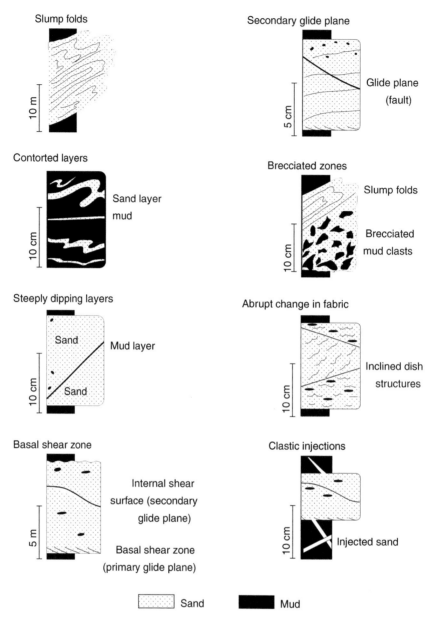

Figure 10 Summary of features associated with slump deposits observed in core and outcrop. Slump fold, an intraformational fold produced by deformation of soft sediment; contorted layer, deformed sediment layer; basal shear zone, the basal part of a rock unit that has been crushed and brecciated by many subparallel fractures due to shear strain; glide plane, slip surface along which major displacement occurs; brecciated zone, an interval that contains angular fragments caused by crushing of the rock; dish structures, concave – up (like a dish) structures caused by upward-escaping fluids in the sediment; clastic injections, natural injection of clastic (transported) sedimentary material (usually sand) into a host rock (usually mud). From Shanmugam G (2006) *Deep-Water Processes and Facies Models: Implications for Sandstone Petroleum Reservoirs*. Amsterdam: Elsevier.

Debris flows are capable of transporting gravel and coarse-grained sand because of their inherent strength. General characteristics of muddy and sandy debrites are:

- gravel to mud lithofacies;
- lobe-like distribution (map view) in the Gulf of Mexico (**Figure 12**);

- tongue-like distribution (map view) in the North Atlantic (**Figure 13**);
- floating or rafted mudstone clasts near the tops of sandy beds (core and outcrop) (**Figure 14**);
- projected clasts (core and outcrop);
- planar clast fabric (core and outcrop) (**Figure 14**);
- brecciated mudstone clasts in sandy matrix (core and outcrop);

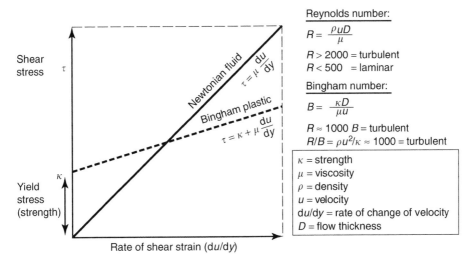

Figure 11 Graph showing rheology (stress–strain relationships) of Newtonian fluids and Bingham plastics. Note that the fundamental rheological difference between debris flows (Bingham plastics) and turbidity currents (Newtonian fluids) is that debris flows exhibit strength, whereas turbidity currents do not. Reynolds number is used for determining whether a flow is turbulent (turbidity current) or laminar (debris flow) in state. From Shanmugam G (1997) The Bouma sequence and the turbidite mind set. *Earth-Science Reviews* 42: 201–229.

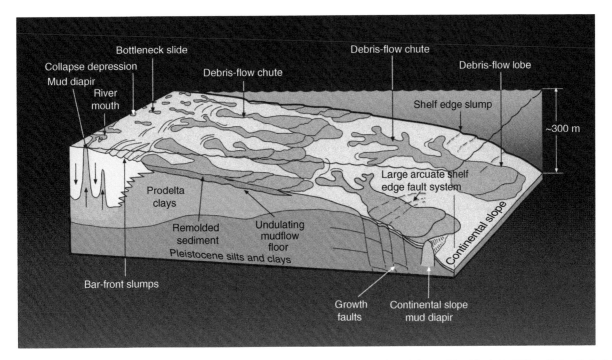

Figure 12 Schematic diagram showing lobe-like distribution of submarine debris flows in front of the Mississippi River Delta, Gulf of Mexico. This shelf-edge deltaic setting, associated with high sedimentation rate, is prone to develop ubiquitous mud diapers and contemporary faults. Mud diaper, intrusion of mud into overlying sediment causing dome-shaped structure; chute, channel. From Coleman JM and Prior DB (1982) Deltaic environments. In: Scholle PA and Spearing D (eds.) *American Association of Petroleum Geologists Memoir 31: Sandstone Depositional Environments*, pp. 139–178. Tulsa, OK: American Association of Petroleum Geologists.

- inverse grading of rock fragments (core and outcrop);
- inverse grading, normal grading, inverse to normal grading, and no grading of matrix (core and outcrop);
- floating quartz granules (core and outcrop);
- inverse grading of granules in sandy matrix (core and outcrop);
- pockets of gravels (core and outcrop);
- irregular, sharp upper contacts (core and outcrop);

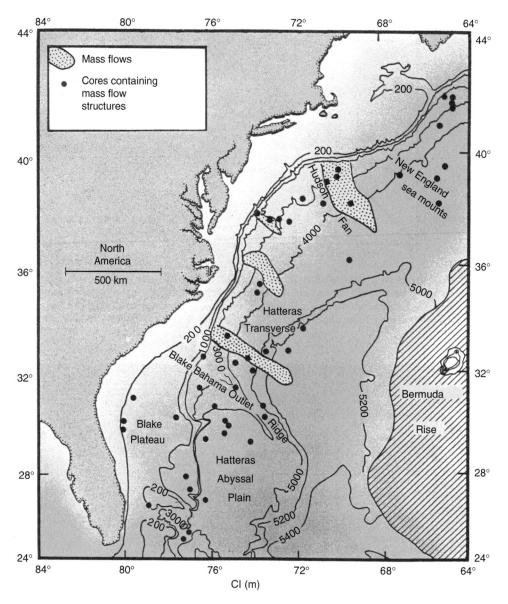

Figure 13 Tongue-like distribution of mass flows (i.e., debris flows) in the North Atlantic. Debrite units are about 500 km long, 10–100 km wide, and 20 m thick. Note a debrite tongue has traveled to a depth of about 5250 m water depth. CI, contour intervals. From Embley RW (1980) The role of mass transport in the distribution and character of deep-ocean sediments with special reference to the North Atlantic. *Marine Geology* 38: 23–50.

- side-by-side occurrence of garnet granules (density: 3.5–4.3) and quartz granules (density: 2.65) (core and outcrop);
- lenticular geometry (**Figure 15**).

The modern Amazon submarine channel has two major debrite deposits (east and west). The western debrite unit is about 250 km long, 100 km wide, and 125 m thick. In the US Atlantic margin, debrite units are about 500 km long, 10–100 km wide, and 20 m thick (**Figure 13**). In offshore northwest Africa, the Canary debrite is about 600 km long, 60–100 km wide, and 5–20 m thick. Submarine debris flows and

their flow-transformation induced turbidity currents have been reported to travel over 1500 km from their triggering point on the northwest African margin.

Turbidity Currents

A turbidity current is a sediment flow with Newtonian rheology and turbulent state in which sediment is supported by turbulence and from which deposition occurs through suspension settling. Turbidity currents exhibit unsteady and nonuniform flow behavior (**Figure 16**). Turbidity currents are surge-type waning flows. As they flow downslope,

Figure 14 Core photograph of massive fine-grained sandstone showing floating mudstone clasts (above the scale) of different sizes. Note planar clast fabric (i.e., long axis of clast is aligned parallel to bedding surface). Note sharp and irregular upper bedding contact (top of photo). Paleocene, North Sea. Reproduced from Shanmugam G (2006). *Deep-Water Processes and Facies Models: Implications for Sandstone Petroleum Reservoirs.* Amsterdam: Elsevier, with permission from Elsevier.

turbidity currents invariably entrain ambient fluid (seawater) in their frontal head portion due to turbulent mixing (**Figure 16**).

With increasing fluid content, plastic debris flows may tend to become Newtonian turbidity currents (**Figure 3**). However, not all turbidity currents evolve from debris flows. Some turbidity currents may evolve directly from sediment failures. Although turbidity currents may constitute a distal end member in basinal areas, they can occur in any part of the system (i.e., shelf edge, slope, and basin). In seismic profiles and multibeam bathymetric images, it is impossible to recognize turbidites.

Turbidity currents cannot transport gravel and coarse-grained sand in suspension because they do not possess the strength. General characteristics of turbidites are:

- fine-grained sand to mud;
- normal grading (core and outcrop) (**Figure 17**);
- sharp or erosional basal contact (core and outcrop) (**Figure 17**);
- gradational upper contact (core and outcrop) (**Figure 17**);
- thin layers, commonly centimeters thick (core and outcrop) (**Figure 18**);
- sheet-like geometry in basinal settings (outcrop) (**Figure 18**);
- lenticular geometry that may develop in channel-fill settings.

Figure 15 Outcrop photograph showing lenticular geometry (dashed line) of debrite with floating clasts (arrow heads). Cretaceous, Tourmaline Beach, California. Reproduced from Shanmugam G (2006). *Deep-Water Processes and Facies Models: Implications for Sandstone Petroleum Reservoirs.* Amsterdam: Elsevier, with permission from Elsevier.

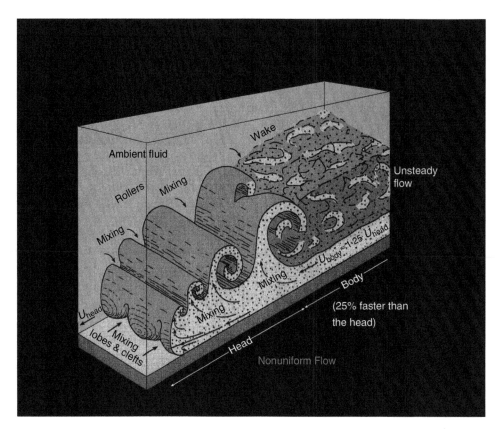

Figure 16 Schematic illustration showing the leading head portion of an unsteady, nonuniform, and turbulent turbidity current. Due to turbulent mixing, turbidity currents invariably entrain ambient fluid (seawater) at their head regions. Modified from Allen JRL (1985) Loose-boundary hydraulics and fluid mechanics: Selected advances since 1961. In: Brenchley PJ and Williams BPJ (eds.) *Sedimentology: Recent Developments and Applied Aspects*, pp. 7–28. Oxford, UK: Blackwell.

In the Hatteras Abyssal Plain (North Atlantic), turbidites have been estimated to be 500 km wide and up to 4 m thick (**Table 3**).

Selective emphasis of turbidity currents Turbidity currents have received a skewed emphasis in the literature. This may be attributed to existing myths about turbidity currents and turbidites. Myth no. 1: Some turbidity currents may be nonturbulent (i.e., laminar) flows. Reality: All turbidity currents are turbulent flows. Myth no. 2: Some turbidity currents may be waxing flows. Reality: All turbidity currents are waning flows. Myth no. 3: Some turbidity currents may be plastic in rheology. Reality: All turbidity currents are Newtonian in rheology. Myth no. 4: Some turbidity currents may be high in sediment concentration (25–95% by volume). Reality: All turbidity currents are low in sediment concentration (1–23% by volume). Because high sediment concentration damps turbulence, high-concentration (i.e., high-density) turbidity currents cannot exist. Myth no. 5: All turbidity currents are high-velocity flows and therefore they elude documentation. Reality:

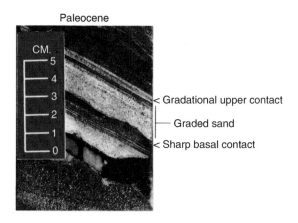

Figure 17 Core photograph showing a sandy unit with normal grading (i.e., grain size decreases upward), sharp basal contact, and gradational upper contact. These features are interpreted to be deposition from a waning turbidity current. Dark intervals are mudstone. Paleocene, North Sea.

Turbidity currents can operate under a wide range of velocity conditions. Myth no. 6: Turbidity currents are believed to operate in modern oceans. Reality: No one has ever documented turbidity

Figure 18 Outcrop photograph showing tilted thin-bedded turbidite sandstone beds with sheet-like geometry, Lower Eocene, Zumaya, northern Spain. Reproduced from Shanmugam G (2006). *Deep-Water Processes and Facies Models: Implications for Sandstone Petroleum Reservoirs.* Amsterdam: Elsevier, with permission from Elsevier.

currents in modern oceans based on the physics of the flow. Myth no. 7: Some turbidites are deposits of debris flows. Reality: All turbidites are the exclusive deposits of turbidity currents. Myth no. 8: Some turbidites exhibit inverse grading. Reality: All turbidites develop normal grading. Myth no. 9: Cross-bedding is a product of turbidity currents. Reality: Cross-bedding is a product of bottom currents. Myth no. 10: Turbidite beds can be recognized in seismic reflection profiles. Reality: Normally graded turbidite units, centimeters in thickness, cannot be resolved on seismic profiles. As a consequence, many debris flows and their deposits have been misclassified as turbidity currents and turbidites.

Sediment Transport via Submarine Canyons

Submarine canyons, which are steep-sided valleys incised into the continental shelf and slope, serve as major conduits for sediment transport from land and the shelf to the deep-sea environment. Although downslope sediment transport occurs both inside and outside of submarine canyons, canyons play a critical role because steep canyon walls are prone to slope failures. Submarine canyons are prominent erosional features along both the US Pacific (**Figure 2**) and the Atlantic margins. Many submarine canyons in the US Atlantic margin commence at a depth of about 200 m near the shelf edge, but heads of California canyons in the US Pacific margin begin at an average depth of about 35 m. The Redondo Canyon, for example, commences at a depth of 10 m near the shoreline (**Figure 2**). Such a scenario would allow for a quick transfer of sediment from shallow-marine into deep-marine environments. In the San Pedro Sea Valley, large debris blocks have been recognized as submarine landslides. Some researchers have proposed that these submarine landslides may have triggered local tsunamis. The significance of this relationship is that tsunamis can trigger submarine landslides, which in turn can trigger tsunamis. Such mutual triggering mechanisms can result in frequent sediment failures in deep-marine environments.

Tsunamis and tropical cyclones are important factors in transferring sediment into deep-marine environments via submarine canyons. For example, a rapid ($190\,cm\,s^{-1}$) sediment flow was recorded in the Scripps Submarine Canyon (La Jolla, California) during the passage of a storm front on 24 November 1968 over La Jolla. In the Scripps Canyon, a large slump mass of about $105\,m^3$ in size was triggered by the May 1975 storm.

Hurricane Hugo, which passed over St. Croix in the US Virgin Islands on 17 September 1989, had generated winds in excess of 110 knots ($204\,km\,h^{-1}$, category 3 in the Saffir–Simpson scale) and waves 6–7 m in height. In the Salt River submarine canyon ($>100\,m$ deep), offshore St. Croix, a current meter measured net downcanyon currents reaching velocities of $2\,m\,s^{-1}$ and oscillatory flows up to $4\,m\,s^{-1}$. Hugo had caused erosion of 2 m of sand in the Salt River Canyon at a depth of about 30 m. A minimum of 2 million kg of sediment were flushed down the Salt River Canyon into deep water. The transport rate associated with hurricane Hugo was 11 orders of magnitude greater than the rate measured during a fair-weather period. In the Salt River Canyon, much of the soft reef cover (e.g., sponges) had been eroded away by the power of the hurricane. Debris composed of palm fronds, trash, and pieces of boats found in the canyon were the evidence for storm-generated debris flows. Storm-induced sediment flows have also been reported in a submarine canyon off Bangladesh, in the Capbreton Canyon, Bay of Biscay in SW France, and in the Eel Canyon, Northern California, among others. In short, sediment transport in submarine canyons is accelerated by tropical cyclones and tsunamis in the world's oceans.

Glossary

Abyssal plain The deepest and flat part of the ocean floor that occupies depths between 2000 and 6000 m (6560 and 19 680 ft).

Ancient The term refers to deep-marine systems that are older than the Quaternary period, which began approximately 1.8 Ma.

Basal shear zone The basal part of a rock unit that has been crushed and brecciated by many subparallel fractures due to shear strain.

Bathyal Ocean floor that occupies depths between 200 (shelf edge) and 4000 m (656 and 13 120 ft). Note that abyssal plains may occur at bathyal depths.

Bathymetry The measurement of seafloor depth and the charting of seafloor topography.

Brecciated clasts Angular mudstone clasts in a rock due to crushing or other deformation.

Brecciated zone An interval that contains angular fragments caused by crushing or breakage of the rock.

Clastic sediment Solid fragmental material (unconsolidated) that originates from weathering and is transported and deposited by air, water, ice, or other processes (e.g., mass movements).

Continental margin The ocean floor that occupies between the shoreline and the abyssal plain. It consists of shelf, slope, and basin (**Figure 2**).

Contorted bedding Extremely disorganized, crumpled, convoluted, twisted, or folded bedding. Synonym: chaotic bedding.

Core A cylindrical sample of a rock type extracted from underground or seabed. It is obtained by drilling into the subsurface with a hollow steel tube called a corer. During the downward drilling and coring, the sample is pushed upward into the tube. After coring, the rock-filled tube is brought to the surface. In the laboratory, the core is slabbed perpendicular to bedding. Finally, the slabbed flat surface of the core is examined for geological bedding contacts, sedimentary structures, grain-size variations, deformation, fossil content, etc.

Dish structures Concave-up (like a dish) structures caused by upward-escaping water in the sediment.

Floating mud clasts Occurrence of mud clasts at some distance above the basal bedding contact of a rock unit.

Flow Continuous, irreversible deformation of sediment–water mixture that occurs in response to applied stress (**Figure 11**).

Fluid A material that flows.

Fluid dynamics A branch of fluid mechanics that deals with the study of fluids (liquids and gases) in motion.

Fluid mechanics Study of the properties and behaviors of fluids.

Geohazards Natural disasters (hazards), such as earthquakes, landslides, tsunamis, tropical cyclones, rogue (freak) waves, floods, volcanic events, sea level rise, karst-related subsidence (sink holes), geomagnetic storms; coastal upwelling; deep-ocean currents, etc.

Heterolithic facies Thinly interbedded (millimeter- to decimeter-scale) sandstones and mudstones.

Hydrodynamics A branch of fluid dynamics that deals with the study of liquids in motion.

Injectite Injected material (usually sand) into a host rock (usually mudstone). Injections are common in igneous rocks.

Inverse grading Upward increase in average grain size from the basal contact to the upper contact within a single depositional unit.

Lithofacies A rock unit that is distinguished from adjacent rock units based on its lithologic (i.e., physical, chemical, and biological) properties (see Rock).

Lobe A rounded, protruded, wide frontal part of a deposit in map view.

Methodology Four methods are in use for recognizing slides, slumps, debris flows, and turbidity currents and their deposits. *Method 1*: Direct observations – Deep-sea diving by a diver allows direct observations of submarine mass movements. The technique has limitations in terms of diving depth and diving time. These constraints can be overcome by using a remotely operated deep submergence vehicle, which would allow observations at greater depths and for longer time. Both remotely operated vehicles (ROVs) and manned submersibles are used for underwater photographic and video documentation of submarine processes. *Method 2*: Indirect velocity calculations – A standard practice has been to calculate velocity of catastrophic submarine events based on the timing of submarine cable breaks. The best example of this method is the 1929 Grand Banks earthquake (Canada) and related cable breaks. This method is not useful for recognizing individual type of mass movement (e.g., slide vs. slump). *Method 3*: Remote sensing technology – In the 1950s, conventional echo sounding was used to construct seafloor profiles. This was done by emitting sound pulses from a ship and by recording return echos from the sea bottom. Today, several types of seismic profiling techniques are available depending on the desired degree of resolutions. Although popular in the petroleum industry and academia, seismic profiles cannot resolve subtle sedimentological features that are required to distinguish turbidites from debrites. In the 1970s, the most significant progress in mapping the seafloor was made by adopting multibeam side-scan sonar survey. The Sea MARC 1 (Seafloor Mapping and Remote Characterization) system uses up to 5-km-broad swath of the seafloor. The GLORIA (Geological Long Range Inclined Asdic) system uses up to 45-km-broad swath of the seafloor. The advantage of GLORIA is that it can map an area of $27\,700\,\text{km}^2\,\text{day}^{-1}$. In the 1990s, multibeam mapping systems were adopted to map the seafloor. This system utilizes hull-mounted sonar arrays that collect bathymetric soundings. The ship's position is determined by Global Positioning System (GPS). Because the transducer arrays are hull mounted rather than towed in a vehicle behind the ship, the data are gathered with navigational accuracy of about 1 m and depth resolution of 50 cm. Two of the types of data collected are bathymetry (seafloor depth) and backscatter (data that can provide insight into the geologic makeup of the seafloor). An example is a bathymetric image of the US Pacific margin with mass transport deposits (**Figure 2**). The US National Geophysical Data Center (NGDC) maintains a website of bathymetric images of continental margins. Although morphological features seen on bathymetric images are useful for recognizing mass transport as a general mechanism, these images may not be useful for distinguishing slides from slumps. Such a distinction requires direct examination of the rock in detail. *Method 4*: Examination of the rock – Direct examination of core and outcrop is the most reliable method for recognizing individual deposits of slide, slump, debris flow, and turbidity current. This method known as process sedimentology, is the foundation for reconstructing ancient depositional environments and for understanding sandstone petroleum reservoirs.

Modern The term refers to present-day deep-marine systems that are still active or that have been active since the Quaternary period that began approximately 1.8 Ma.

Nonuniform flow Spatial changes in velocity at a moment in time.

Normal grading Upward decrease in average grain size from the basal contact to the upper contact within a single depositional unit composed of a single rock type. It should not contain any floating mudstone clasts or outsized quartz granules. In turbidity currents, waning flows deposit successively finer and finer sediment, resulting in a normal grading (see waning flows).

Outcrop A natural exposure of the bedrock without soil capping (e.g., along river-cut subaerial canyon walls or submarine canyon walls) or an artificial exposure of the bedrock due to excavation for roads, tunnels, or quarries.

Planar clast fabric Alignment of long axis of clasts parallel to bedding (i.e., horizontal). This fabric implies laminar flow at the time of deposition.

Primary basal glide plane (or décollement) The basal slip surface along which major displacement occurs.

Projected clasts Upward projection of mudstone clasts above the bedding surface of host rock (e.g., sand). This feature implies freezing from a laminar flow at the time of deposition.

Rock The term is used for (1) an aggregate of one or more minerals (e.g., sandstone); (2) a body of undifferentiated mineral matter (e.g., obsidian); and (3) solid organic matter (e.g., coal).

Scarp A relatively straight, cliff-like face or slope of considerable linear extent, breaking the continuity of the land by failure or faulting. Scarp is an abbreviated form of the term escarpment.

Secondary glide plane Internal slip surface within the rock unit along which minor displacement occurs.

Sediment flows They represent sediment-gravity flows. They are classified into four types based on sediment-support mechanisms: (1) turbidity current with turbulence; (2) fluidized sediment flow with upward moving intergranular flow; (3) grain flow with grain interaction (i.e., dispersive pressure); and (4) debris flow with matrix strength. Although all turbidity currents are turbulent in state, not all turbulent flows are turbidity currents. For example, subaerial river currents are turbulent, but they are not turbidity currents. River currents are fluid-gravity flows in which fluid is directly driven by gravity. In sediment-gravity flows, however, the interstitial fluid is driven by the grains moving down slope under the influence of gravity. Thus turbidity currents cannot operate without their entrained sediment, whereas river currents can do so. River currents are subaerial flows, whereas turbidity currents are subaqueous flows.

Sediment flux (1) A flowing sediment–water mixture. (2) Transfer of sediment.

Sedimentology Scientific study of sediments (unconsolidated) and sedimentary rocks (consolidated) in terms of their description, classification, origin, and diagenesis. It is concerned with physical, chemical, and biological processes and products. This article deals with physical sedimentology and its branch, process sedimentology.

Submarine canyon A steep-sided valley that incises into the continental shelf and slope. Canyons serve as major conduits for sediment transport from land and the shelf to the deep-sea environment. Smaller erosional features on the continental slope are commonly termed gullies; however, there are no standardized criteria to distinguish canyons from gullies. Similarly, the distinction between submarine canyons and submarine erosional channels is not straightforward. Thus, alternative terms, such as gullies, channels, troughs, trenches, fault valleys, and sea valleys, are in use for submarine canyons in the published literature.

Tropical cyclone It is a meteorological phenomenon characterized by a closed circulation system around a center of low pressure, driven by heat energy released as moist air drawn in over warm ocean waters rises and condenses. Structurally, it is a large, rotating system of clouds, wind, and thunderstorms. The name underscores their origin in the Tropics and their cyclonic nature. Worldwide, formation of tropical cyclones peaks in late summer months when water temperatures are warmest. In the Bay of Bengal, tropical cyclone activity has double peaks; one in April and May before the onset of the monsoon, and another in October and November just after. Cyclone is a broader category that includes both storms and hurricanes as members. Cyclones in the Northern Hemisphere represent closed counterclockwise circulation. They are classified based on maximum sustained wind velocity as follows:

- tropical depression: 37–61 km h^{-1};
- tropical storm: 62–119 km h^{-1};
- tropical hurricane (Atlantic Ocean): $>119\,\text{km h}^{-1}$;
- tropical typhoon (Pacific or Indian Ocean): $>119\,\text{km h}^{-1}$.

The Saffir–Simpson hurricane scale:

- category 1: 119–153 km h^{-1};
- category 2: 154–177 km h^{-1};
- category 3: 178–209 km h^{-1};
- category 4: 210–249 km h^{-1};
- category 5: $>249\,\text{km h}^{-1}$.

Tsunami Oceanographic phenomena that are characterized by a water wave or series of waves with long wavelengths and long periods. They are caused by an impulsive vertical displacement of the body of water by earthquakes, landslides, volcanic explosions, or extraterrestrial (meteorite) impacts. The link between tsunamis and sediment flux in the world's oceans involves four stages: (1) triggering stage, (2) tsunami stage, (3) transformation stage, and (4) depositional stage. During the triggering stage, earthquakes, volcanic explosions, undersea landslides, and meteorite impacts can trigger displacement of the sea surface, causing tsunami waves. During the tsunami stage, tsunami waves carry energy traveling through the water, but these waves do not move the water. The incoming wave is depleted in entrained sediment. This stage is one of energy transfer, and it does not involve sediment transport. During the transformation stage, the incoming tsunami waves tend to erode and incorporate sediment into waves near the coast. This sediment-entrainment process transforms sediment-depleted waves into outgoing mass transport processes and sediment flows. During the depositional stage, deposition from slides, slumps, debris flows, and turbidity currents would occur.

Unsteady flow   Temporal changes in velocity through a fixed point in space.

Waning flow   Unsteady flow in which velocity becomes slower and slower at a fixed point through time. As a result, waning flows would deposit successively finer and finer sediment, resulting in a normal grading.

Waxing flow   Unsteady flow in which velocity becomes faster and faster at a fixed point through time.

See also

Ocean Margin Sediments. Organic Carbon Cycling in Continental Margin Environments. Tsunami.

Further Reading

Allen JRL (1985) Loose-boundary hydraulics and fluid mechanics: Selected advances since 1961. In: Brenchley PJ and Williams BPJ (eds.) *Sedimentology: Recent Developments and Applied Aspects*, pp. 7–28. Oxford, UK: Blackwell.

Coleman JM and Prior DB (1982) Deltaic environments. In: Scholle PA and Spearing D (eds.) *American Association of Petroleum Geologists Memoir 31: Sandstone Depositional Environments*, pp. 139–178. Tulsa, OK: American Association of Petroleum Geologists.

Dingle RV (1977) The anatomy of a large submarine slump on a sheared continental margin (SE Africa). *Journal of Geological Society of London* 134: 293–310.

Dott RH Jr. (1963) Dynamics of subaqueous gravity depositional processes. *American Association of Petroleum Geologists Bulletin* 47: 104–128.

Embley RW (1980) The role of mass transport in the distribution and character of deep-ocean sediments with special reference to the North Atlantic. *Marine Geology* 38: 23–50.

Greene HG, Murai LY, Watts P, *et al.* (2006) Submarine landslides in the Santa Barbara Channel as potential tsunami sources. *Natural Hazards and Earth System Sciences* 6: 63–88.

Hampton MA, Lee HJ, and Locat J (1996) Submarine landslides. *Reviews of Geophysics* 34: 33–59.

Hubbard DK (1992) Hurricane-induced sediment transport in open shelf tropical systems – an example from St. Croix, US Virgin Islands. *Journal of Sedimentary Petrology* 62: 946–960.

Jacobi RD (1976) Sediment slides on the northwestern continental margin of Africa. *Marine Geology* 22: 157–173.

Lewis KB (1971) Slumping on a continental slope inclined at 1°–4°. *Sedimentology* 16: 97–110.

Locat J and Mienert J (eds.) (2003) *Submarine Mass Movements and Their Consequences*. Dordrecht: Kluwer.

Middleton GV and Hampton MA (1973) Sediment gravity flows: Mechanics of flow and deposition. In: Middleton GV and Bouma AH (eds.) *Turbidites and Deep-Water Sedimentation*, pp. 1–38. Los Angeles, CA: Pacific Section Society of Economic Paleontologists and Mineralogists.

Sanders JE (1965) Primary sedimentary structures formed by turbidity currents and related resedimentation mechanisms. In: Middleton GV (ed.) *Society of Economic Paleontologists and Mineralogists Special Publiation 12: Primary Sedimentary Structures and Their Hydrodynamic Interpretation*, 192–219.

Schwab WC, Lee HJ, and Twichell DC, (eds.) (1993) Submarine Landslides: Selected Studies in the US Exclusive Economic Zone. *US Geological Survey Bulletin 2002*.

Shanmugam G (1996) High-density turbidity currents: Are they sandy debris flows? *Journal of Sedimentary Research* 66: 2–10.

Shanmugam G (1997) The Bouma sequence and the turbidite mind set. *Earth-Science Reviews* 42: 201–229.

Shanmugam G (2002) Ten turbidite myths. *Earth-Science Reviews* 58: 311–341.

Shanmugam G (2003) Deep-marine tidal bottom currents and their reworked sands in modern and ancient submarine canyons. *Marine and Petroleum Geology* 20: 471–491.

Shanmugam G (2006) *Deep-Water Processes and Facies Models: Implications for Sandstone Petroleum Reservoirs*. Amsterdam: Elsevier.

Shanmugam G (2006) The tsunamite problem. *Journal of Sedimentary Research* 76: 718–730.

Shanmugam G (2008) The constructive functions of tropical cyclones and tsunamis on deep-water sand deposition during sea level highstand: Implications for petroleum exploration. *American Association of Petroleum Geologists Bulletin* 92: 443–471.

Shanmugam G, Lehtonen LR, Straume T, Syvertsen SE, Hodgkinson RJ, and Skibeli M (1994) Slump and debris flow dominated upper slope facies in the Cretaceous of the Norwegian and northern North Seas (61°–67° N): Implications for sand distribution. *American Association of Petroleum Geologists Bulletin* 78: 910–937.

Shepard FP and Dill RF (1966) *Submarine Canyons and Other Sea Valleys*. Chicago: Rand McNally.

Talling PJ, Wynn RB, Masson DG, *et al.* (2007) Onset of submarine debris flow deposition far from original giant landslide. *Nature* 450: 541–544.

USGS (2007) US Geological Survey: Perspective view of Los Angeles Margin. http://wrgis.wr.usgs.gov/dds/dds-55/pacmaps/la_pers2.htm (accesed May 11, 2008).

Varnes DJ (1978) Slope movement types and processes. In: Schuster RL and Krizek RJ (eds.) *Transportation Research Board Special Report 176: Landslides:*

Analysis and Control, pp. 11–33. Washington, DC: National Academy of Science.

Relevant Websites

http://www.ngdc.noaa.gov
– NGDC Coastal Relief Model (images and data), NGDC (National Geophysical Data Center).

http://www.mbari.org
– Submarine Volcanism: Hawaiian Landslides, MBARI (Monterey Bay Aquarium Research Institute)

http://www.nhc.noaa.gov
– The Saffir–Simpson Hurricane Scale, NOAA/National Weather Service.

ORGANIC CARBON CYCLING IN CONTINENTAL MARGIN ENVIRONMENTS

R. A. Jahnke, Skidaway Institute of Oceanography, Savannah, GA, USA

Introduction

Continental margin environments, including estuarine, salt marsh, mangrove forest, coral reef, continental shelf, and continental slope ecosystems, are the interface zone between oceanic and terrestrial realms with gateways also to the atmosphere and solid Earth (sediment) reservoirs. Transports between these carbon reservoirs exert critical control on the global carbon cycle and will be a major factor in mitigating anthropogenically driven climate change. This interface is also where man most directly interacts with and exploits the oceans. Much of the global economy's goods and services are transported along and across coastal zones; most of the world's fisheries are found within continental margin ecosystems; and much of the waste and materials mobilized from humans and their industrial and agricultural activities are delivered to the coastal zone. In addition, three-fourths of the world's human population will reside on the coastal plain by 2025. Rising population not only increases human pressures on and interactions with coastal ecosystems but also increases the susceptibility of humans and their property to ocean-related natural hazards such as inundation by tsunamis and hurricanes. Advancing understanding of coastal marine processes is, therefore, of fundamental importance. The following section focuses specifically on the cycling of organic carbon in this critically important environment.

Processes that Characterize Coastal Ecosystems

Many transport and exchange processes are either intensified within or unique to ocean boundaries and exert considerable influence on the structure, composition, and characteristics of continental margin habitats. Examples of some of these processes are schematically depicted in **Figure 1**. Many of the processes highlighted in the figure emphasize interactions with the seafloor which rises to intersect the sea surface at the shoreline. This is not to minimize the importance of processes in the water column,

some of which are included as well, but rather reflects the recognition that at the most fundamental level, it is the presence of the seafloor that differentiates continental margin ecosystems from their oceanic counterparts. The important inputs to the margin depicted in **Figure 1** include surface freshwater inputs; groundwater exchanges; atmospheric inputs, which while not unique to coastal systems are generally intensified due to proximity to terrestrial sources of airborne materials; and oceanic inputs represented by wind-driven upwelling but also including inputs by currents and upwelling driven by other processes such as eddies associated with boundary currents.

Important to the overall character of coastal systems are processes that internally exchange and transform materials, promoting the biogeochemical cycling of carbon, nutrient elements, and associated trace elements. Examples include benthic solute exchange; sediment resuspension; intensified water column mixing due to friction with the seafloor; internal wave–seafloor interactions; turbulent interactions with the surface and bottom boundary layers; and water column and benthic interactions with surface gravity waves. These material exchanges support transformative processes such as biological production and respiration and abiotic processes such as mineral precipitation and dissolution. Processes within margin systems also provide efficient conduits for the transport of materials to the open ocean surface, intermediate and deep water column, and the seafloor. Contributing to these important fluxes are transports associated with offshore surface and subsurface advective plumes; cascading downslope high-density flows; particle fluxes associated with the above advective transports; gravitational settling; and downslope benthic boundary layer lateral transport.

As complex as this figure is, numerous processes are not displayed, such as ice-driven exchanges and organism migrations, and **Figure 1** is admittedly still a simplistic view of the margin system. The processes depicted in the figure structure margin ecosystems and support high rates of metabolic activity and vertical and lateral material fluxes. Furthermore, it must be recognized that the intensity, mix, and effectiveness of the processes vary temporally in response to atmospheric and hydrographic forcings. The magnitudes and interactions of these processes and timing of responses are not well known.

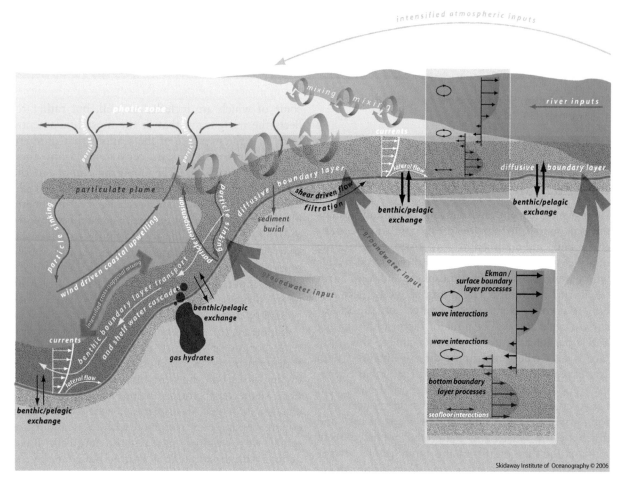

Figure 1 Examples of important transport processes that are unique or intensified along continental margins. Reproduced with permission, © Skidaway Institute of Oceanography.

The magnitude and mix of dominant processes vary across this coastal/margin interface zone between the land and ocean system. In most instances, variations are gradual and dividing this system into subsystems is a difficult and occasionally arbitrary process. Nevertheless, for discussion and quantification purposes, it is useful to divide this complex system into simpler subsystems. In the following, the coastal ecosystems closest to and with the greatest interaction with the shoreline are discussed as a group and include estuarine, salt marsh, and mangrove forest ecosystems. This section is then followed by a discussion of continental shelf and slope ecosystems.

Shore Zone Ecosystems (Estuaries, Salt Marshes, and Mangrove Forests)

Perhaps the most common and best-studied coastal ecosystem is the estuary. An estuary is typically considered to be a semi-enclosed region where fresh water, predominantly from a river but possibly also including groundwater input, and salt water mix.

The generalized pattern of circulation is that of fresh and brackish waters advecting off shore at the surface, saline waters being drawn shoreward along the bottom and mixing of these waters at the boundary between them. The thickness of the saline water along the bottom thins progressively upstream, eventually pinching out and resulting in the commonly described 'salt wedge' estuary (depicted in **Figure 1**). This circulation concentrates nutrients within the estuary and maintains water-column stability, thereby supporting high rates of primary production. While this simple picture describes most estuaries, there are natural variations. For example, for some large rivers, the freshwater discharge is sufficient to 'push' the salt wedge out onto the open continental shelf.

Estuaries are important pathways for transferring materials from the land to the ocean. The efficiency with which materials such as suspended sediments and nutrients are passed through the estuary and delivered to the continental margin depends on a variety of factors including the magnitude of river flow, morphology of the estuary, and residence time of waters within the estuarine system. In addition, driven by the strong physical-chemical gradients and intense metabolic rates, important transformations of materials may occur within the estuary. One critically important example is denitrification, conversion of nitrogen from a 'fixed' form such as nitrate or ammonium that is biologically available to nitrogen gas which can be utilized by only a very select group of microorganisms, the nitrogen fixers. Because nitrogen is a required nutrient element for life and overall biological primary production is limited by the availability of fixed nitrogen in many regions, the rate of denitrification in estuaries can be a major factor controlling estuarine biological production.

Because estuaries are often excellent harbors, providing access to ocean- and river-based transportation and supporting productive fisheries, many of the largest cities are also located adjacent to estuaries. Human influences on the coastal environment are often concentrated and accentuated within estuarine systems. Influences include altering ecosystems by removing specific top predators through fishing activities, introducing exotic species through transportation and dumping of ballast waters in ships, eutrophication of the estuarine system by increased nutrient loading from sewage, farm runoff, and other nonpoint sources of nutrients, and the introduction of contaminants (e.g., trace metals and pesticides). Through these activities and interactions, the cycling of carbon within many, or most, estuarine systems has already been altered from natural rates.

Other ecosystems at the shore zone include mangrove forests and salt marshes. Mangroves are found along sheltered tropical and subtropical shorelines. While the mangrove trees are the obvious characteristic of these ecosystems and a major contributor to total biological primary production, other plants such as macroalgae, periphyton, and phytoplankton also contribute significantly. The proportion of the contribution from each is controlled by such factors as the total surface area covered by each plant type, water turbidity, and nutrient delivery pathways and magnitude. Supported by contributions from these different plants, mangrove ecosystems exhibit some of the highest rates of primary production observed for any ecosystem on earth (**Table 1**).

Salt marshes occur over a wider latitudinal range than mangroves, extending from the subpolar regions to the tropics. Marshes, sometimes called tidal marshes, occur between the high- and low-tide levels and require a significant tidal range. In general, they occur in low-energy, depositional environments, along estuaries and on the protected side of barrier islands. The most obvious characteristic of salt

Table 1 Total global area and primary production of shore zone ecosystems

Environment	Surface area $(10^6 km^2)$	Primary production rate $(mol\,C\,m^{-2}\,yr^{-1})$	Total production $(10^{12}\,mol\,C\,yr^{-1})$
Estuaries	1.4	22	31
Mangroves	0.2	232	46
Salt marshes	0.4	185	74

marshes is the dominance of cord grass, the most common being *Spartina alterniflora* and *Spartina patens*, and the black needle rush, *Juncus roemerianus*. These unique plants can tolerate large variations in salinity and account for much of the productivity of the marsh system although other plants, such as benthic diatoms living at the sediment surface, also contribute significantly to total productivity (**Table 1**).

Together, these shoreline ecosystems provide critical habitats for many life forms including juvenile forms of fish, shellfish, and birds. As such, understanding, managing, and maintaining the overall health of these ecosystems are priorities of local municipalities and governmental agencies.

The contribution of these ecosystems to the cycling of organic carbon within the local continental margin system depends critically on the balance between the production and respiration rates. Because of the magnitude of the rates, even a small difference would represent a significant net carbon transfer. The difficulty of determining the difference between production and respiration is compounded by natural variability; added complexity because organic carbon can be temporarily stored on the seafloor and then periodically transported by large storms; and the fact that human activities continue to alter the system. Based on numerous comparisons, it is currently thought that in most estuarine systems respiration is greater than production because of added organic input from the river while nutrient inputs to salt marshes and mangrove forests support production that exceeds respiration. Therefore, these latter ecosystems may export organic matter to adjacent continental shelves and other ecosystems. As shown in the next section, however, while the impacts of this carbon cycling are important to local ecosystems, they are relatively small when compared to transfers on the global scale.

Continental Shelf and Slope Ecosystems (The Continental Margin)

Continental shelf and slope environments (hereafter referred collectively as the continental margin) occur everywhere at the continent–ocean boundary. To facilitate the discussion, these environments can be divided into broad categories such as eastern and western boundary current, polar, subpolar, monsoonal, and tropical margins. Where there is significant exchange, inland seas are linked to the margin across which the major material exchange occurs. In the following, these broad categories will be used to generalize rates of organic carbon cycling and exchange.

In addition to the regional distinctions noted above, local geomorphology of the margin also influences the magnitude and mixture of processes controlling carbon and associated bioactive element transfers. In the broadest sense, margins can be classified into two fundamental types, slope-dominated and shelf-dominated margins (**Figure 2**). These margin types are differentiated by (1) the location of the majority of primary production (seaward of the shelf break for slope-dominated, shoreward for shelf-dominated), (2)

the relative importance of nutrient input from upwelling versus benthic nutrient regeneration, and (3) the extent to which turbulence and water column mixing are supported by friction with the seafloor. These margin types are punctuated with other features such as river plumes and canyons while variations in factors such as local meteorological forcing and tidal ranges and currents add spatial variability, but nevertheless, all margin regions can be classified within this simple two end-member framework.

At slope-dominated margins (**Figure 2(a)**), the majority of the primary production occurs seaward of the shelf break. Generally, but not exclusively, these margins exhibit relatively narrow continental shelves. Examples include eastern margins of the Pacific from Vancouver Island to central Chile, and the eastern margin of the Atlantic from the Iberian Coast to Cape Town. Residence time of water on the narrow, open shelf is relatively short. When upwelling, often wind-driven, occurs, a significant portion of upwelled waters is driven offshore, supporting high production rates seaward of the shelf break, over the continental slope and rise. In addition, benthic boundary layer processes act to resuspend particles facilitating downward lateral transport of materials off the shelf. Combined, these processes result in large transfers of particulate organic matter seaward of the shelf break, often for distances of 200–300 km. This may occur within the benthic boundary layer, within surface plumes extending offshore or in subducted intermediate layers in offshore plumes. It is difficult to generalize the width of this margin zone but 150–200 km seaward of the shelf break in the surface and 200–300 km in the benthic boundary layer are reasonable estimates.

At shelf-dominated margins, on the other hand, the majority of primary production occurs shoreward of the shelf break. In general, these margins have broad shallow shelves (**Figure 2(b)**) although when bounded by western boundary currents, some narrowed-shelf systems may be classified as shelf-dominated margins. Examples are the eastern seaboard of the US south of Cape Cod, the broad shelf adjacent to southern Brazil, Uruguay, and Argentina, East China Sea, and North Sea. Due to their shallow nature, sufficient energy is transmitted to the seafloor to frequently resuspend sediments. Many of these systems are also characterized by low relief and low sediment flux from the adjacent land. The combination of winnowing away of the fine-grained sediments and slow replenishment leaves the majority of these shelves with nonaccumulating, coarse-grained sandy sediments. Recent studies have revealed, however, that while these sandy shelf sediments are generally depauperate in organic matter, high metabolic rates

(a)

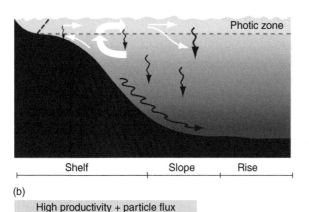

(b)

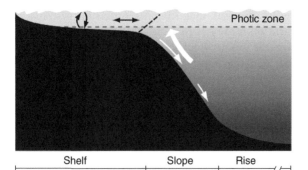

Figure 2 Generalized representation of major distinctions between (a) a slope-dominated margin and (b) a shelf-dominated margin. From Jahnke RA (in press) Global synthesis. In: Lui K-K, Atkinson L, Quinones R, and Talaue-McManus L (eds.) *Global Change, The IGBP Series: Carbon and Nutrient Fluxes in Continental Margins: A Global Synthesis*, 450pp. New York: Springer.

are observed. This seeming paradox is most likely the result of efficient advective transport of substrates and metabolites into and out of these high permeability sediments.

Friction with the seafloor and the shallow nature of these systems enhance water column mixing with fronts commonly found near the shelf edge. Nutrient inputs are dominated by dynamic processes at the shelf break that bring nutrient-rich ocean waters onto the shelf but may also be augmented via atmospheric deposition, river outflow, and groundwater discharge. On the western side of ocean basins, these systems are often bounded by western boundary currents such as the Gulf Stream, Kuroshio, and Brazil Currents and eddy interactions induce upwelling of nutrient-rich waters onto the shelf. Once upwelled, nutrient-rich waters either move onto the shelf via lateral exchange

processes or are subducted, restricting high biological productivity to the shelf break and shoreward.

The high productivity and particle flux region in these systems is shoreward of the shelf break and sinking particles are intercepted by the seafloor and efficiently remineralized. Because shelf-dominated systems are generally wide with a shallow water column, exchange between ocean waters and shelf waters is reduced while shelf water residence times are relatively long, further promoting remineralization of particulate organic matter in these environments.

The overall length and areas characterizing each major region type are listed in **Table 2**. The total length of the shelf break as shown in **Figure 3** is 1.5×10^5 km. The focus here is on margin–open ocean exchange and the shelf break is defined as the

Table 2 Summary of areas of major margin regions

Region	Margin type	Shelf width (km)	Shelf break length (km)	Shelf area (10^6 km^2)	Slope area (10^6 km^2)	Rise area (10^6 km^2)	Surface expression (10^6 km^2)	Slope and rise area (10^6 km^2)
Arctic – P	Shelf	253	13 897	3.51	2.78	1.39	3.51	4.17
Antarctic – P	Slope	134	16 332	2.19	3.27	1.63	5.46	4.90
NW Atlantic – SP	Shelf	417	5 393	2.25	1.08	0.54	2.25	2.16
NW Atlantic – WBC	Shelf	136	11 332	1.54	2.27	1.13	1.54	3.40
W Atlantic – T	Shelf	180	3 413	0.62	0.68	0.34	0.62	1.02
SW Atlantic – WBC	Shelf	305	5 501	1.68	1.10	0.55	1.68	1.65
SW Atlantic – SP	Shelf	693	1 731	2.33	0.35	0.17	2.33	0.52
NE Atlantic – SP	Shelf	536	4 362	2.34	0.87	0.44	2.34	1.31
NE Atlantic – EBC	Slope	89	5 201	1.68	1.04	0.52	2.72	1.56
E Atlantic – T	Shelf	43	4 179	0.18	0.84	0.42	0.18	1.25
SE Atlantic – EBC	Slope	69	3 161	0.22	0.63	0.32	0.85	0.95
Atlantic subtotal			44 273	12.83	8.85	4.43	14.51	13.82
W Indian – M	Slope	65	7 708	0.50	1.54	0.77	2.04	2.31
W Indian – T	Slope	34	2 236	0.08	0.45	0.22	0.52	0.67
W Indian – WBC	Shelf	50	3 713	0.18	0.74	0.37	0.18	1.11
E Indian – M	Shelf	113	5 475	0.62	1.10	0.55	0.62	1.64
E Indian – T	Slope	53	4 405	0.23	0.88	0.44	1.11	1.32
E Indian – EBC	Slope	96	2 624	0.25	0.52	0.26	0.78	0.79
E Indian – SP	Shelf	112	3 363	0.38	0.67	0.34	0.38	1.01
Indian subtotal			29 524	2.24	5.90	2.95	5.64	8.86
NW Pacific – SP	Shelf	471	6 180	2.91	1.24	0.62	2.91	1.85
NW Pacific – WBC	Shelf	269	5 069	1.36	1.01	0.51	1.36	1.52
W Pacific – T	Shelf	226	9 514	2.15	1.90	0.95	2.15	2.85
SW Pacific – WBC	Shelf	377	5 340	2.01	1.07	0.53	2.01	1.60
NE Pacific – SP	Slope	57	3 815	0.22	0.76	0.38	0.98	1.14
NE Pacific – EBC	Slope	47	8 356	0.40	1.67	0.84	2.07	2.51
E Pacific – T	Slope	48	2 122	0.10	0.42	0.21	0.53	0.64
SE Pacific – EBC	Slope	35	5 672	0.20	1.13	0.57	1.33	1.70
SE Pacific – SP	Slope	65	539	0.03	0.11	0.05	0.14	0.16
Pacific subtotal			46 607	9.38	9.32	4.66	13.48	13.98
Global totals			150 633	30.15	30.13	15.06	42.59	45.19

For abbreviations, see **Figure 3**.

From Jahnke RA (in press) Global synthesis. In: Lui K-K, Atkinson L, Quinones R, and Talaue-McManus L (eds.) *Global Change, The IGBP Series: Carbon and Nutrient Fluxes in Continental Margins: A Global Synthesis*, 450pp. New York: Springer.

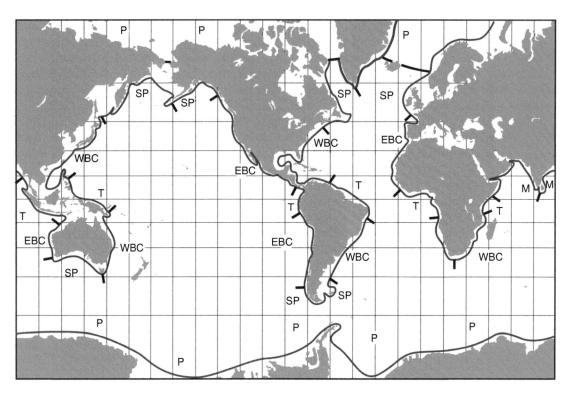

Figure 3   Location of shelf break and distribution of major region designations for global integration. WBC, western boundary current; EBC, eastern boundary current; SP, subpolar; P, poloar; T, tropical; M, monsoonal. From Jahnke RA (in press) Global synthesis. In: Lui K-K, Atkinson L, Quinones R, and Talaue-McManus L (eds.) *Global Change, The IGBP Series: Carbon and Nutrient Fluxes in Continental Margins: A Global Synthesis*, 450pp. New York: Springer.

top of the continental slope that is contiguous with the main abyssal ocean basins. This results in a shelf break length that is considerably shorter than the $3–4 \times 10^5$ km length one would estimate if a higher-resolution coastline and bathymetry were used but probably better reflects the length of boundary across which ocean–margin exchange must occur. Note that fully enclosed inland seas such as the Baltic Sea, Mediterranean Sea, Hudson Bay, Arabian Sea, Red Sea, Persian Gulf, and the Sea of Japan are shoreward of the shelf break and are, therefore, included in the shelf area as defined in this simple, two-end-member framework. The total shelf area estimated from mean shelf width and shelf break length is 30.2×10^6 km^2.

The surface expression of the margin ecosystem is defined to be the shelf area for shelf-dominated margins and the shelf plus slope areas for slope-dominated margins and is considerably larger than the shelf area alone (42.6×10^6 km^2). With this definition, 12.6×10^6 km^2 or 30% of the surface expression of the margin is in water depths greater than 200 m. The inclusion of a significant deep-water area within the margin system is an important difference between the description here and previous efforts but is consistent with the processes displayed in **Figures 1** and 2. Note that all of these values are thought to be

minimum estimates for several reasons. At many margin regions such as eastern boundary current systems and the Argentine shelf, satellite imagery reveals elevated chlorophyll and presumed productivity at much greater distances from the continent than are used here. There is no definitive value or threshold between high coastal productivity and low gyre productivity at which to insert a boundary. The point is to identify regions that are sufficiently different from the open ocean because of the dominance of margin processes that they require specific study. The definition of the surface expression adopted here meets this criterion. Another reason these values are believed to be minimum estimates is that islands and the Canadian and Indonesian archipelagos have been excluded. Various estimates suggest that there are roughly 18 000 islands in the oceans. Many of these occur adjacent to shorelines and would not provide significant additional continental shelf break length or margin area. Additionally, many islands are clustered together and therefore share shelf and slope areas and shelf-break length. However, even if only 180 or 1% are considered separate islands or groups of islands protruding to the ocean's surface and if it is assumed as a minimum, each is an infinitely small point surrounded by a 30-km-wide shelf and a

200-km-wide slope, these 180 islands would contribute 33 840 km of shelf break length, 0.59×10^6 km^2 of shelf area, and 29.4×10^6 km^2 of slope area. Thus, omission of islands is a serious deficiency in the present estimates of coastal carbon cycling that future efforts must address.

All of the major ecosystem types contribute significantly to shelf break length (**Table 3**). Values range from $c.~3.1 \times 10^4$ km for western boundary current systems to 1.3×10^4 km for monsoonal systems. A greater range is observed in the contribution of each to total global continental shelf area. Because of the preponderance of shallow marginal seas (e.g., North Sea, Bering Sea, and Sea of Okhotsk) in regions classified as subpolar, the subpolar margins account for 1.05×10^7 km^2 or 35% of total shelf area. Globally, the polar and subpolar regions contain 54% of continental shelf area. Eastern boundary current and monsoonal systems contain only a small portion of the global shelf area (9.1% and 3.7%, respectively). A slightly different pattern is observed in the relative ranking of the surface areas of the margin types. Polar and subpolar systems still contain the largest amounts of sea surface area, 21% and 26%, respectively. However, now eastern boundary current and monsoonal systems provide 17.8% and 6.1%, respectively. This increase in the relative sea surface area reflects the inclusion of the offshore area that is still dominated by margin processes.

Shelf- and slope-dominated margins each contribute significantly to total shelf break length (88 462 km and 62 171 km, respectively) while shelf-dominated margins account for 80% of global shelf area. However, because surface processes that extend over adjacent deeper waters are considered in defining margin boundaries, both margin classifications contribute relatively equally to margin sea surface area (24.1×10^6 km^2 (shelf-dominated); 18.5×10^6 km^2 (slope-dominated)).

A total of 9.7 Pg C yr^{-1} (1 Pg $= 10^{15}$ g) is estimated for continental margin production (**Table 4**). This is $c.~21$% of the current estimates of global production ($c.~48$ Pg C yr^{-1}) and is significantly larger than the total contributed by estuaries, salt marshes, and mangrove forests. Highest productivities are reported for the monsoonal and eastern boundary current regions (**Table 4**) with the lowest values attributed to the polar region. Interestingly, when productivities are integrated over each major ecosystem type, the contributions to total production from subpolar, western boundary current, eastern boundary current, and monsoonal regions are all roughly similar, between 1.9 and 2.7 Pg yr^{-1} (**Table 5**). The western boundary current system is lowest of the group but still significant due to contributions from broad shelves while the eastern boundary current system is highest. Also, similar contributions to total production are estimated for shelf-dominated and slope-dominated systems (5.0 vs. 4.6 Pg C yr^{-1}, respectively). Of the estimated productivity, 3.5 Pg C yr^{-1} or 36% occurs over the slope where water depths are greater than 200 m. High export efficiency over the slope could provide an important conduit for the transfer of carbon to the deep sea via the biological pump.

Direct exchange of CO_2 with the atmosphere suggests a net uptake of -0.29 Pg C yr^{-1} (-2.4×10^{13} mol yr^{-1}; note that fluxes from the atmosphere to the ocean are expressed as negative fluxes). Consistent trends among the major ecosystem types are observed and summarized in **Table 5**. The majority of the influx is estimated to occur in the polar and

Table 3 Summary of areas by major ecosystem types

Ecosystem type	Shelf break length (km)	Shelf area (10^6 km^2)	Slope area (10^6 km^2)	Rise area (10^6 km^2)	Surface expression (10^6 km^2)	Slope and rise area (10^6 km^2)
Polar	30 229	5.70	6.05	3.02	8.97	9.07
Subpolar	25 583	10.46	5.08	2.54	11.33	8.15
Western boundary current	30 955	6.77	6.19	3.10	6.77	9.27
Eastern boundary current	25 014	2.74	5.00	2.50	7.74	7.50
Tropical	25 869	3.36	5.17	2.59	5.11	7.76
Monsoonal	13 183	1.12	2.64	1.32	2.66	3.95
Shelf-dominated	88 462	24.06	17.69	8.85	24.06	26.53
Slope-dominated	62 171	6.10	12.43	6.22	18.53	18.65
Total	150 633	30.15	30.13	15.06	42.59	45.19

From Jahnke RA (in press) Global synthesis. In: Lui K-K, Atkinson L, Quinones R, and Talaue-McManus L (eds.) *Global Change, The IGBP Series: Carbon and Nutrient Fluxes in Continental Margins: A Global Synthesis*, 450pp. New York: Springer.

Table 4 Primary production, air–sea and margin–ocean carbon exchange, and slope and rise of organic carbon deposition for margin regions

Region	Primary production $(gCm^{-2}yr^{-1})$	Total production $(10^{12}gCyr^{-1})$	Air–sea exchange $(molCO_2m^{-2}yr^{-1})$	Total air–sea exchange $(10^{12}molCO_2 yr^{-1})$	Deposition rate $(molm^{-2}yr^{-1})$	Total deposition $(10^{12}molOCyr^{-1})$
Arctic – P	59	207.1	− 2.4	− 8.49	0.11	0.46
Antarctic – P	80	436.5	− 0.8	− 4.53	0.56	2.74
NW Atlantic – SP	175	393.8	− 1.7	− 3.83	0.34	0.55
NW Atlantic – WBC	350	539.0	+ 2.5	+ 3.85	0.17	0.58
W Atlantic – T	350	215.6	0.0	0.00	0.30	0.31
SW Atlantic – WBC	350	588.0	+ 1.0	+ 1.68	0.25	0.41
SW Atlantic – SP	292	680.4	− 1.5	− 3.50	0.39	0.20
NE Atlantic – SP	175	409.5	− 1.3	− 3.04	0.22	0.29
NE Atlantic – EBC	300	816.1	− 0.4	− 1.09	0.44	0.69
E Atlantic – T	150	27.0	0.0	0.00	0.51	0.64
SE Atlantic – EBC	450	382.6	− 0.4	− 0.34	0.51	0.48
Atlantic subtotal		4051.9		− 6.26		4.14
W Indian – M	365	745.9	+ 1.4	+ 2.86	0.59	1.36
W Indian – T	150	78.6	0.0	0.00	0.30	0.20
W Indian – WBC	175	32.2	+ 1.0	+ 0.18	0.30	0.33
E Indian – M	300	186.0	− 0.4	0.25	0.59	0.97
E Indian – T	150	167.1	0.0	0.00	0.30	0.40
E Indian – EBC	225	174.6	0.0	0.00	0.23	0.18
E Indian – SP	175	65.6	− 1.8	− 0.68	0.21	0.21
Indian subtotal		1450.0		+ 2.12		3.66
NW Pacific – SP	175	509.3	− 2.0	− 5.82	0.30	0.56
NW Pacific – WBC	350	476.0	− 1.0	1.36	0.17	0.26
W Pacific – T	188	404.2	+ 0.1	+ 0.22	0.25	0.71
SW Pacific – WBC	150	301.5	+ 1.0	+ 2.01	0.12	0.19
NE Pacific – SP	300	294.6	− 1.8	− 1.77	0.15	0.17
NE Pacific – EBC	345	713.2	0.0	0.00	0.51	1.28
E Pacific – T	225	118.4	0.0	0.00	0.59	0.38
SE Pacific – EBC	500	665.2	0.0	0.00	0.59	1.00
SE Pacific – SP	300	42.8	− 1.8	− 0.26	0.22	0.04
Pacific subtotal		3525.2		− 6.98		4.59
Grand total		9670.6		− 24.14		15.59

For abbreviations, see **Figure 3**.
From Jahnke RA (in press) Global synthesis. In: Lui K-K, Atkinson L, Quinones R, and Talaue-McManus L (eds.) *Global Change, The IGBP Series: Carbon and Nutrient Fluxes in Continental Margins: A Global Synthesis*, 450pp. New York: Springer.

Table 5 Summary of primary production, air–sea exchange, and slope and rise of organic carbon deposition by major ecosystem type

Ecosystem type	Total primary production $(10^{12}gCyr^{-1})$	Total air–sea exchange $(10^{12}molCO_2yr^{-1})$	Total slope and rise deposition $(10^{12}molOCyr^{-1})$
Polar	644	− 13.02	3.20
Subpolar	2396	− 18.88	2.02
Western boundary current	1937	+ 6.36	1.78
Eastern boundary current	2751	− 1.43	3.63
Tropical	1011	+ 0.22	2.63
Monsoonal	932	+ 2.61	2.33
Shelf-dominated	5035	− 19.02	6.67
Slope-dominated	4636	− 5.12	8.92
Total	9671	− 24.14	15.59

From Jahnke RA (in press) Global synthesis. In: Lui K-K, Atkinson L, Quinones R, and Talaue-McManus L (eds.) *Global Change, The IGBP Series: Carbon and Nutrient Fluxes in Continental Margins: A Global Synthesis*, 450pp. New York: Springer.

subpolar regions. Contrastingly, near-zero net exchange rates are generally reported for tropical and eastern boundary current regions and effluxes are reported for most western boundary current regions. Based on the values reported here, the majority (79%) of the air–sea exchange at margins is attributed to shelf-dominated margins. An important caveat for these conclusions, however, is to note that for many of the regions, no direct measurements have been reported and the values used here have been extrapolated from other, presumably similar, locations. The observed trends must be verified by future studies.

In principle, the above estimate of CO_2 uptake from the atmosphere defines the major role margins may play in mitigating the impacts of the release of anthropogenic CO_2 to the atmosphere. However, along the margin, lateral exchange with terrestrial pools may provide another pathway by which CO_2 input to the ocean may be augmented. Simply, productive coastal zone ecosystems such as salt marshes and mangrove forests may take up carbon directly from the atmosphere and release it to coastal waters as dissolved organic and inorganic carbon. Water-exchange processes may transfer it offshore, significantly enhancing the coastal–open ocean carbon flux. Recent studies at shelf-dominated margins, such as the North Sea, Georgia Bight, and East China Sea, in fact suggest that the net carbon transfer from the margin to the open ocean is significantly enhanced relative to what would be predicted from air–sea exchange alone. More measurements are required to yield a global-scale estimate but it is likely that the transfer of carbon from the margins to the open ocean is significantly larger than that estimated from air–sea exchange alone listed above.

Integrating over the regions defined here suggests that 15.6×10^{12} mol of organic carbon is deposited on the adjacent slope and rise sediments. This value accounts for 47% of the estimated total deep-seafloor carbon deposition (3.3×10^{13} mol organic carbon yr^{-1}). Assuming that there is no dramatic difference in the transfer efficiency of sinking organic matter between the base of the main thermocline and the seafloor, this pattern of deposition should reflect the relative strength of the biological pump transfer of carbon to the deep ocean. Thus, continental margin ecosystems likely supply roughly one-half of the carbon transferred by the biological pump to the deep ocean. This conclusion is consistent with the global productivity assessment provided earlier if the margin export efficiency averages 3 times the open ocean average. As noted earlier, because a significant portion of the margin production occurs over deep water where sinking particles will not be intercepted by the seafloor, higher export efficiencies are likely. Studies of how the biological pump may change with future climates must, therefore, include these quantitatively important environments.

Summary and Concluding Remarks

Intense carbon cycling in coastal ecosystems is driven by a complex mixture of physical and biological processes. The impact of ocean boundaries on the global carbon budget is not well known but likely provides major pathways for transferring carbon among the important, climatically relevant reservoirs. With the caveat that significant uncertainty remains in the estimates of many of the important carbon transfers, results to date imply that 21% of global primary production, a likely higher proportion of the export production, a significant proportion of the net oceanic uptake of CO_2, and nearly half of the biological pump transfer of carbon to the deep sea reservoir occur at the ocean boundaries, driven by coastal ecosystems. Ocean margins must be a priority in future marine research given the magnitude of these values, the concentration of human-induced pressures on coastal systems, and the role margins may play in the mitigation of and oceanic response to climate change.

Further Reading

Gattuso J-P, Frankignoulle M, and Wollast R (1998) Carbon and carbonate metabolism in coastal aquatic ecosystems. *Annual Review of Ecology and Systematics* 29: 405–434.

Jahnke RA (in press) Global synthesis. In: Lui K-K, Atkinson L, Quinones R, and Talaue-McManus L (eds.) *Global Change, The IGBP Series: Carbon and Nutrient Fluxes in Continental Margins: A Global Synthesis*, 450pp. New York: Springer.

Robinson AR and Brink K (eds.) (2005) *The Sea, Vol. 13: The Global Coastal Ocean: Multiscale Interdisciplinary Processes*. Cambridge, MA: Harvard University Press.

Walsh JJ (1988) *On the Nature of Continental Shelves*, 520pp. London: Academic Press.

Wollast R (1998) Evaluation and comparison of the global carbon cycle in the coastal zone and in the open ocean. In: Brink KH and Robinson AR (eds.) *The Sea*, vol. 10, pp. 213–252. New York: Wiley.

PHOSPHORUS CYCLE

K. C. Ruttenberg, Woods Hole Oceanographic
Institution, Woods Hole, MA, USA

Introduction

The global phosphorus cycle has four major components: (i) tectonic uplift and exposure of phosphorus-bearing rocks to the forces of weathering; (ii) physical erosion and chemical weathering of rocks producing soils and providing dissolved and particulate phosphorus to rivers; (iii) riverine transport of phosphorus to lakes and the ocean; and (iv) sedimentation of phosphorus associated with organic and mineral matter and burial in sediments (**Figure 1**). The cycle begins anew with uplift of sediments into the weathering regime.

Phosphorus is an essential nutrient for all life forms. It is a key player in fundamental biochemical reactions involving genetic material (DNA, RNA) and energy transfer (adenosine triphosphate, ATP), and in structural support of organisms provided by membranes (phospholipids) and bone (the biomineral hydroxyapatite). Photosynthetic organisms utilize dissolved phosphorus, carbon, and other essential nutrients to build their tissues using energy from the sun. Biological productivity is contingent upon the availability of phosphorus to these organisms, which constitute the base of the food chain in both terrestrial and aquatic systems.

Phosphorus locked up in bedrock, soils, and sediments is not directly available to organisms. Conversion of unavailable forms to dissolved orthophosphate, which can be directly assimilated, occurs through geochemical and biochemical reactions at various stages in the global phosphorus cycle. Production of biomass fueled by phosphorus bioavailability results in the deposition of organic matter in soil and sediments, where it acts as a source of fuel and nutrients to microbial communities. Microbial activity in soils and sediments, in turn, strongly influences the concentration and chemical form of phosphorus incorporated into the geological record.

This article begins with a brief overview of the various components of the global phosphorus cycle. Estimates of the mass of important phosphorus reservoirs, transport rates (fluxes) between reservoirs, and residence times are given in **Tables 1** and **2**. As is

clear from the large uncertainties associated with these estimates of reservoir size and flux, there remain many aspects of the global phosphorus cycle that are poorly understood. The second half of the article describes current efforts underway to advance our understanding of the global phosphorus cycle. These include (i) the use of phosphate oxygen isotopes ($\delta^{18}O$-PO_4) as a tool for identifying the role of microbes in the transformer of phosphate from one reservoir to another; (ii) the use of naturally occurring cosmogenic isotopes of phosphorus (^{32}P and ^{33}P) to provide insight into phosphorus-cycling pathways in the surface ocean; (iii) critical evaluation of the potential role of phosphate limitation in coastal and open ocean ecosystems; (iv) reevaluation of the oceanic residence time of phosphorus; and (v) rethinking the global phosphorus-cycle on geological timescales, with implications for atmospheric oxygen and phosphorus limitation primary productivity in the ocean.

The Global Phosphorus Cycle: Overview

The Terrestrial Phosphorus Cycle

In terrestrial systems, phosphorus resides in three pools: bedrock, soil, and living organisms (biomass) (**Table 1**). Weathering of continental bedrock is the principal source of phosphorus to the soils that support continental vegetation (F_{12}); atmospheric deposition is relatively unimportant (F_{82}). Phosphorus is weathered from bedrock by dissolution of phosphorus-bearing minerals such as apatite ($Ca_{10}(PO_4)_6(OH, F, Cl)_2$), the most abundant primary phosphorus mineral in crustal rocks. Weathering reactions are driven by exposure of minerals to naturally occurring acids derived mainly from microbial activity. Phosphate solubilized during weathering is available for uptake by terrestrial plants, and is returned to the soil by decay of litterfall (**Figure 1**).

Soil solution phosphate concentrations are maintained at low levels as a result of absorption of phosphorus by various soil constituents, particularly ferric iron and aluminum oxyhydroxides. Sorption is considered the most important process controlling terrestrial phosphorus bioavailability. Plants have different physiological strategies for obtaining phosphorus despite low soil solution concentrations. For example, some plants can increase root volume

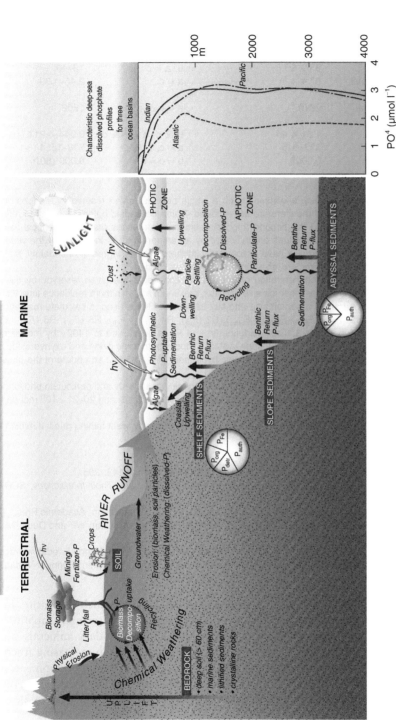

Figure 1 Cartoon illustrating the major reservoirs and fluxes of phosphorus described in the text and summarized in Tables **Table 1** and **2**. The oceanic photic zone, idealized in the cartoon, is typically thinner in coastal environments owing to turbidity from continental terrigenous input, and deepens as the water column clarifies with distance away from the continental margins. The distribution of phosphorus among different chemical/mineral forms in marine sediments is given in the pie diagrams, where the abbreviations used are: P_{org}, organic phosphorus; P_{Fe}, iron-bound phosphorus; P_{detr}, detrital apatite; P_{auth}, authigenic/biogenic apatite. The P_{org}, P_{Fe}, and P_{auth} reservoirs represent potentially reactive phosphorus pools (see text and **Tables 2** and **5** for discussion), whereas the P_{detr} pool reflects mainly detrital apatite weathered off the continents and passively deposited in marine sediments (note that P_{detr} is not an important sedimentary phosphorus component in abyssal sediments, far from continents). Continental margin phosphorus speciation data were compiled from Louchouarn P, Lucotte M, Duchemin E and de Vernal A (1997) Early diagenetic processes in recent sediments of the Gulf of St-Lawrence: Phosphorus, carbon and iron burial rates. *Marine Geology* 139(1/4): 181–200, and Ruttenberg KC and Berner RA (1993) Authigenic apatite formation and burial in sediments from non-upwelling continental margin environments. *Geochimica et Cosmochimica Acta* 57: 991–1007. Abyssal sediment phosphorus speciation data were compiled from Filippelli GM and Delaney ML (1996) Phosphorus geochemistry of equatorial Pacific sediments. *Geochimica et Cosmochimica Acta* 60: 1479)1495, and Ruttenberg KC (1990) *Diagenesis and burial of phosphorus in marine sediments: implications for the marine phosphorus budget.* PhD thesis, Yale University. The global phosphorus cycle cartoon is from Ruttenberg (2000). The global phosphorus cycle. In: *The Encyclopedia of Global Change*, Oxford University Press. (in Press), with permission. The vertical water column distributions of phosphate typically observed in the three ocean basins are shown in the panel to the right of the global phosphorus cycle cartoon, and are from Sverdrup HV, Johnson MW and Fleming RH (1942) *The Oceans, Their Physics, Chemistry and General Biology.* New York: Prentice Hall; used with permission. © 1942 Prentice Hall; used with permission.

Table 1 Major reservoirs active in the global phosphorus cycle and associated residence times

Reservoir no.	Reservoir description	Reservoir size (10^{12} mol P)	Reference	Residence time τ (years)
R1	Sediments (crustal rocks and soil > 60 cm deep and marine sediments)	0.27×10^8–1.3×10^8	b, a = c = d	42–201×10^6
R2	Land ($\approx$ total soil < 60 cm deep: organic + inorganic)	3100–6450	b, a = c = d	425–2311
R3	Land biota	83.9–96.8	b, a = c = d	13–48
R4	Surface ocean, 0–300 m (total dissolved P)	87.4	a = c	2.46–4.39
R5	Deep sea, 300–3300 m (total dissolved P)	2810	a = c & d	1502
R6	Oceanic biota	1.61–4.45	b & d, a = c & d	0.044–0.217 (16–78 d)
R7	Minable P	323–645	a = c, b & d	718–1654
R8	Atmospheric P	0.0009	b = c = d	0.009 (80 h)

^{a}Notes

(1) Ranges are reported for those reservoirs for which a consensus on a single best estimated reservoir size does not exist. Maximum and minimum estimates found in a survey of the literature are reported. References cited before the comma refer to the first (lowest) estimate, those after the comma refer to the second (higher) estimate. References that give identical values are designated by an equality sign, references giving similar values are indicated by an ampersand. As indicated by the wide ranges reported for some reservoirs, all calculations of reservoir size have associated with them a large degree of uncertainty. Methods of calculation, underlying assumptions, and sources of error are given in the references cited.

(2) Residence times are calculated by dividing the concentration of phosphorus contained in a given reservoir by the sum of fluxes out of the reservoir. Where ranges are reported for reservoir size and flux, maximum and minimum residence time values are given; these ranges reflect the uncertainties inherent in reservoir size and flux estimates. Fluxes used to calculate residence times for each reservoir are as follows: R1 (F_{12}), R2 ($F_{23} + F_{28} + F_{24(d)} + F_{24(p)}$), R3 ($F_{32}$), R4 ($F_{45} + F_{46}$), R5 ($F_{54}$), R6 ($F_{64} + F_{65}$), R7 ($F_{72}$), R8 ($F_{82} + F_{84}$). Flux estimates are given in **Table 2**. The residence time of R5 is decreased to 1492 y by inclusion of the scavenged flux of deep-sea phosphate at hydrothermal mid-ocean ridge systems, mostly onto ferric oxide and oxyhydroxide phases (Wheat CG, Feely RA and Mottl MJ (1996). Phosphate removal by oceanic hydrothermal processes: an update of the phosphorus budget in the oceans. *Geochimica et Cosmochimica Acta* 60(19): 3593–3608).

(3) Estimates for the partitioning of the oceanic reservoir between dissolved inorganic phosphorus and particulate phosphorus are given in references b and d as follows: 2581–2600×10^{12} mol dissolved inorganic phosphorus (b, d) and 20–21×10^{12} mol particulate phosphorus (d, b).

(4) The residence times estimated for the minable phosphorus reservoir reflect estimates of current mining rates; if mining activity increases or diminishes the residence time will change accordingly.

References

(a) Lerman A, Mackenzie FT and Garrels RM (1975) *Geological Society of America Memoir* 142: 205.

(b) Richey JE (1983). In: Bolin B and Cook RB (eds) *The Major Biogeochemical Cycles and Their Interactions*, SCOPE 21, pp. 51–56. Chichester: Wiley.

(c) Jahnke RA (1992) In: Butcher SS *et al.* (eds) *Global Geochemical Cycles*, pp. 301–315. San Diego: Academic Press. (Values identical to Lerman *et al.*, except the inclusion of the atmospheric reservoir estimate taken from Graham WF and Duce RA (1979) *Geochimica et Cosmochimica Acta* 43: 1195.)

(d) Mackenzie FT, Ver LM, Sabine C, Lane M and Lerman A (1993) In: Wollast R, Mackenzie FT and Chou L (eds) *Interactions of C, N, P and S Biogeochemical Cycles and Global Change*. NATO ASI Series 1, vol. 4, pp. 1–61. Berlin: Springer-Verlag.

and surface area to optimize uptake potential. Alternatively, plant roots and/or associated fungi can produce chelating compounds that solubilize ferric iron and calcium-bound phosphorus, enzymes and/or acids that solubilize phosphate in the root vicinity. Plants also minimize phosphorus loss by resorbing much of their phosphorus prior to litterfall, and by efficient recycling from fallen litter. In extremely unfertile soils (e.g., in tropical rain forests) phosphorus recycling is so efficient that topsoil contains virtually no phosphorus; it is all tied up in biomass.

Systematic changes in the total amount and chemical form of phosphorus occur during soil development. In initial stages, phosphorus is present mainly as primary minerals such as apatite. In mid-stage soils, the reservoir of primary apatite is diminished; less-soluble secondary minerals and organic phosphorus make up an increasing fraction of soil phosphorus. Late in soil development, phosphorus is partitioned mainly between refractory minerals and organic phosphorus (**Figure 2**).

Transport of Phosphorus from Continents to the Ocean

Phosphorus is transferred from the continental to the oceanic reservoir primarily by rivers (F_{24}). Deposition of atmospheric aerosols (F_{84}) is a minor

Table 2 Fluxes between the major phosphorus reservoirs[a]

Flux no.	Description of flux	Flux (10^{12} $molP$ y^{-1})	References and comments
Reservoir fluxes			
F_{12}	Rocks/sediments → soils (erosion/weathering, soil accumulation)	0.645	a = c & d
F_{21}	Soils → rocks/sediments (deep burial, lithification)	0.301–0.603	d, a = c
F_{23}	Soils → land biota	2.03–6.45	a = c, b & d
F_{32}	Land biota → soils	2.03–6.45	a = c, b & d
$F_{24(d)}$	Soil → surface ocean (river total dissolved P flux)	0.032–0.058	e, a = c; ~ > 50% of TDP is DOP (e)
$F_{24(p)}$	Soil → surface ocean (river particulate P flux)	0.59–0.65	d, e; ~ 40% of RSPM-P (Riverine Suspended Particulate Matter-Phosphorus) is organic P (e); it is estimated that between 25–45% is reactive once it enters the ocean (f).
F_{46}	Surface ocean → oceanic biota	19.35–35	b, d; a = c = 33.5, b reports upper limit of 32.3; d reports lower limit of 28.2
F_{64}	Oceanic biota → surface ocean	19.35–35	b, d; a & c = 32.2, b reports upper limit of 32.3, d reports lower limit of 28.2
F_{65}	Oceanic biota → deep sea (particulate rain)	1.13–1.35	d, a = c
F_{45}	Surface ocean → deep sea (downwelling)	0.581	a = c
F_{54}	Deep sea → surface ocean (upwelling)	1.87	a = c
F_{42}	Surface ocean → land (fisheries)	0.01	d
F_{72}	Minable P → land (soil)	0.39–0.45	a = c = d, b
F_{28}	Land (soil) → atmosphere	0.14	b = c = d
F_{82}	Atmosphere → land (soil)	0.1	b = c = d
F_{48}	Surface ocean → atmosphere	0.01	b = c = d
F_{84}	Atmosphere → surface ocean	0.02–0.05	c, b; d gives 0.04; ~ 30% of atmospheric aerosol P is soluble (g)
Subreservoir fluxes: marine sediments			
sF_{ms}	Marine sediment accumulation (total)	0.265–0.280	i, j; for higher estimate (j), use of sediment P concentration below the diagenesis zone implicitly accounts for P loss via benthic remineralization flux and yields pre-anthropogenic net burial flux. For estimates of reactive P burial see note (j).
sF_{cs}	Continental margin ocean sediments → burial	0.150–0.223	j, i; values reported reflect total P, reactive P burial constitutes from 40–75% of total P (h). These values reflect pre-agricultural fluxes, modern value estimated as 0.33 (d).
sF_{as}	Abyssal (deep sea) sediments → burial	0.042–0.130	i, j; a = c gives a value of 0.055. It is estimated that 90–100% of this flux is reactive P (h). These values reflect pre-agricultural fluxes, modern value estimates range from 0.32 (d) to 0.419 (b).

(Continued)

Table 2 *Continued*

Flux no.	Description of flux	Flux (10^{12} $mol\,P\,y^{-1}$)	References and comments
sF_{cbf}	Coastal sediments → coastal waters (remineralization, benthic flux)	0.51–0.84	d, k; these values reflect pre-agricultural fluxes, modern value estimated as 1.21 with uncertainties $\pm40\%$ (k)
sF_{abf}	Abyssal sediments → deep sea (remineralization, benthic flux)	0.41	k; this value reflects pre-agricultural fluxes, modern value estimated as 0.52, uncertainty $\pm30\%$ (k)

[a] *Notes*

(1) Reservoir fluxes (F) represent the P-flux between reservoirs #R1–R8 defined in **Table 1**. The subreservoir fluxes (sF) refer to the flux of phosphorus into the marine sediment portion of reservoir #1 via sediment burial, and the flux of diagenetically mobilized phosphorus out of marine sediments via benthic return flux. These subfluxes have been calculated as described in references h–k. Note that the large magnitude of these sub-fluxes relative to those into and out of reservoir #1 as a whole, and the short oceanic-phosphorus residence time they imply (**Tables 1** and **5**), highlight the dynamic nature of the marine phosphorus cycle.

(2) Ranges are reported where consensus on a single best estimate does not exist. References cited before the comma refer to the first (lowest) estimate, those after the comma refer to the second (higher) estimate. References that give identical values are designated by an equality sign, references giving similar values are indicated by an ampersand. Maximum and minimum estimates found in a survey of the literature are reported. In some cases this range subsumes ranges reported in the primary references. As indicated by the wide ranges reported, all flux calculations have associated with them a large degree of uncertainty. Methods of calculation, underlying assumptions, and sources of error are given in the references cited.

References

(a) Lerman A, Mackenzie FT and Garrels RM (1975) *Geological Society of America Memoir* 142: 205.

(b) Richey JE (1983). In: Bolin B and Cook RB (eds), *The Major Biogeochemical Cycles and Their Interactions*, SCOPE 21, pp. 51–56. Chichester: Wiley.

(c) Jahnke RA (1992) In: Butcher SS *et al.* (eds), *Global Geochemical Cycles*, pp. 301–315. San Diego: Academic Press. (Values are identical to those found in Lerman *et al.* (1975) except for atmospheric phosphorus fluxes taken from Graham WF and Duce RA (1979) *Geochimica et Cosmochimica Acta* 43: 1195.)

(d) Mackenzie FT, Ver LM, Sabine C, Lane M and Lerman A (1993) In: Wollast R, Mackenzie FT and Chou L (eds), *Interactions of C, N, P and S Biogeochemical Cycles and Global Change*, NATO ASI Series 1, vol. 4, pp. 1–61. Berlin: Springer-Verlag.

(e) Meybeck M. (1982). *American Journal of Science* 282(4): 401.

(f) The range of riverine suspended particulate matter that may be solubilized once it enters the marine realm (e.g. so-called 'reactive phosphorus') is derived from three sources. Colman AS and Holland HD ((2000). In: Glenn, C.R., Prévôt-Lucas L and Lucas J (eds) *Marine Authigenesis: From Global to Microbial*, SEPM Special Publication No. 66. pp. 53–75) estimate that 45% may be reactive, based on RSPM-P compositional data from a number of rivers and estimated burial efficiency of this material in marine sediments. Berner RA and Rao J-L ((1994) *Geochimica et Cosmochimica Acta* 58: 2333) and Ruttenberg KC and Canfield DE ((1994) *EOS, Transactions of the American Geophysical Union* 75: 110) estimate that 35% and 31% of RSPM-P is released upon entering the ocean, based on comparison of RSPM-P and adjacent deltaic surface sediment phosphorus in the Amazon and Mississippi systems, respectively. Lower estimates have been published: 8% (Ramirez AJ and Rose AW (1992) *American Journal of Science* 292: 421); 18% (Froelich PN (1988) *Limnology and Oceanography* 33: 649); 18%: (Compton J, Mallinson D, Glenn CR *et al.* (2000) In: Glenn CR, Prévôt-Lucas L and Lucas J (eds) *Marine Authigenesis: From Global to Microbial*, SEPM Special Publication No. 66, pp. 21–33). Higher estimates have also been published: 69% (Howarth RW, Jensen HS, Marine R and Postma H (1995) In: Tiessen H (ed) *Phosphorus in the Global Environment*, SCOPE 54, pp. 323–345. Chichester: Wiley). Howarth *et al.* (1995) also estimate the total flux of riverine particulate P to the oceans at 0.23×10^{12} moles $P\,y^{-1}$, an estimate likely too low because it uses the suspended sediment flux from Milliman JD and Meade RH ((1983) *Journal of Geology* 91: 1), which does not include the high sediment flux rivers from tropical mountainous terranes (Milliman JD and Syvitski JPM (1992) *Journal of Geology* 100: 525).

(g) Duce RA, Liss PS, Merrill JT *et al.* (1991) *Global Biogeochemical Cycles* 5: 193.

(h) Ruttenberg KC (1993) *Chemical Geology* 107: 405.

(i) Howarth RW, Jensen HS, Marino R and Postma H (1995) In: Tiessen H (ed.) *Phosphorus in the Global Environment*, SCOPE 54, pp. 323–345, Chichester: Wiley.

(j) Phosphorus-burial flux estimates as reported in Ruttenberg ((1993) *Chemical Geology* 107: 405) modified using pre-agricultural sediment fluxes updated by Colman AS and Holland HD ((2000) In: Glenn CR, Prévôt-Lucas L and Lucas J (eds) *Marine Authigenesis: From Global to Microbial*, SEPM Special Publication No. 66, pp. 53–75). Using these total phosphorus burial fluxes and the ranges of likely reactive phosphorus given in the table, the best estimate for reactive phosphorus burial flux in the oceans lies between 0.177 and 0.242×10^{12} moles $P\,y^{-1}$. Other estimates of whole-ocean reactive phosphorus burial fluxes range from, at the low end, 0.032–0.081×10^{12} moles $P\,y^{-1}$ (Compton J, Mallinson D, Glenn CR *et al.* (2000) In: Glenn CR, Prévôt-Lucas L and Lucas J (eds) *Marine Authigenesis: From Global to Microbial*, SEPM Special Publication No. 66, pp. 21–33), and 0.09×10^{12} moles $P\,y^{-1}$ (Wheat CG, Feely RA and Mottl MJ (1996) *Geochimica et Cosmochimica Acta* 60: 3593); to values more comparable to those derived from the table above (0.21×10^{12} moles $P\,y^{-1}$: Filippelli GM and Delaney ML (1996) *Geochimica Cosmochimica Acta* 60: 1479)

(k) Colman AS and Holland HD (2000) In: Glenn CR, Prévôt-Lucas L and Lucas J (eds) *Marine Authigenesis: From Global to Microbial*, SEPM Special Publication No. 66, pp. 53–75.

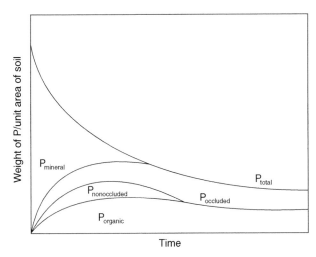

Figure 2 The fate of phosphorus during soil formation can be viewed as the progressive dissolution of primary mineral P (dominantly apatite), some of which is lost from the system by leaching (decrease in P_{total}), and some of which is reincorporated into nonoccluded, occluded and organic fractions within the soil. Nonoccluded P is defined as phosphate sorbed to surface of hydrous oxides of iron and aluminum, and calcium carbonate. Occluded P refers to P present within the mineral matrix of discrete mineral phases. The initial build-up in organic P results from organic matter return to soil from vegetation supported by the soil. The subsequent decline in $P_{organic}$ is due to progressive mineralization and soil leaching. The time-scale over which these transformations occur depends upon the initial soil composition, topographic, and climatic factors. Figure is after Walker and Syers (1976). The fate of phosphorus during pedogenesis. *Geoderma* 15: 1–19, with permission.

flux. Groundwater seepage to the coastal ocean is a potentially important but undocumented flux.

Riverine phosphorus derives from weathered continental rocks and soils. Because phosphorus is particle-reactive, most riverine phosphorus is associated with particulate matter. By most estimates, over 90% of the phosphorus delivered by rivers to the ocean is as particulate phosphorus ($F_{24(p)}$). Dissolved phosphorus in rivers occurs in both inorganic and organic forms. The scant data on dissolved organic phosphorus suggest that it may account for 50% or more of dissolved riverine phosphorus. The chemical form of phosphorus associated with riverine particles is variable and depends upon the drainage basin geology, on the extent of weathering of the substrate, and on the nature of the river itself. Available data suggest that approximately 20–40% of phosphorus in suspended particulate matter is organic. Inorganic forms are partitioned mainly between ferric oxyhydroxides and apatite. Aluminum oxyhydroxides and clays may also be significant carriers of phosphorus.

The fate of phosphorus entering the ocean via rivers is variable. Dissolved phosphorus in estuaries at the continent–ocean interface typically displays nonconservative behavior. Both negative and positive deviations from conservative mixing can occur, sometimes changing seasonally within the same estuary. Net removal of phosphorus in estuaries is typically driven by flocculation of humic-iron complexes and biological uptake. Net phosphorus release is due to a combination of desorption from fresh water particles entering high-ionic-strength marine waters, and flux of diagenetically mobilized phosphorus from benthic sediments. Accurate estimates of bioavailable riverine phosphorus flux to the ocean must take into account, in addition to dissolved forms, the fraction of riverine particulate phosphorus released to solution upon entering the ocean.

Human impacts on the global phosphorus cycle   The mining of phosphate rock (mostly from marine phosphorite deposits) for use as agricultural fertilizer (F_{72}) increased dramatically in the latter half of the twentieth century. In addition to fertilizer use, deforestation, increased cultivation, and urban and industrial waste disposal all have enhanced phosphorus transport from terrestrial to aquatic systems, often with deleterious results. For example, elevated phosphorus concentrations in rivers resulting from these activities have resulted in eutrophication in some lakes and coastal areas, stimulating nuisance algal blooms and promoting hypoxic or anoxic conditions that are harmful or lethal to natural populations.

Increased erosion due to forest clear-cutting and widespread cultivation has increased riverine suspended matter concentrations, and thus increased the riverine particulate phosphorus flux. Dams, in contrast, decrease sediment loads in rivers and therefore diminish phosphorus-flux to the oceans. However, increased erosion below dams and diagenetic mobilization of phosphorus in sediments trapped behind dams moderates this effect. The overall effect has been a 50–300% increase in riverine phosphorus flux to the oceans above pre-agricultural levels.

The Marine Phosphorus Cycle

Phosphorus in its simplest form, dissolved orthophosphate, is taken up by photosynthetic organisms at the base of the marine food web. When phosphate is exhausted, organisms may utilize more complex forms by converting them to orthophosphate via enzymatic and microbiological reactions. In the open ocean most phosphorus associated with biogenic particles is recycled within the upper water column. Efficient stripping of phosphate from surface waters by photosynthesis combined with build-up at depth due to respiration of biogenic particles results in the classic oceanic dissolved nutrient profile. The progressive accumulation of respiration-derived phosphate at depth along the deep-water circulation trajectory results in higher phosphate concentrations in Pacific Ocean deep waters at the end of the trajectory than in the North Atlantic where deep water originates (**Figure 1**).

The sole means of phosphorus removal from the oceans is burial with marine sediments. The phosphorus flux to shelf and slope sediments is larger than the phosphorus flux to the deep sea (**Table 2**) for several reasons. Coastal waters receive continentally derived nutrients via rivers (including phosphorus, nitrogen, silicon, and iron), which stimulate high rates of primary productivity relative to the deep sea and result in a higher flux of organic matter to continental margin sediments. Organic matter is an important, perhaps primary, carrier of phosphorus to marine sediments. Owing to the shorter water column in coastal waters, less respiration occurs prior to deposition. The larger flux of marine organic phosphorus to margin sediments is accompanied by a larger direct terrigenous flux of particulate phosphorus (organic and inorganic), and higher sedimentation rates overall. These factors combine to enhance retention of sedimentary phosphorus. During high sea level stands, the sedimentary phosphorus reservoir on continental margins expands, increasing the phosphorus removal flux and therefore shortening the oceanic phosphorus residence time.

Terrigenous-dominated shelf and slope (hemipelagic) sediments and abyssal (pelagic) sediments have distinct phosphorus distributions. While both are dominated by authigenic Ca-P (mostly carbonate fluorapatite), this reservoir is more important in pelagic sediments. The remaining phosphorus in hemipelagic sediments is partitioned between ferric iron-bound phosphorus (mostly oxyhydroxides), detrital apatite, and organic phosphorus; in pelagic sediments detrital apatite is unimportant. Certain coastal environments characterized by extremely high, upwelling-driven biological productivity and low terrigenous input are enriched in authigenic apatite; these are proto-phosphorite deposits. A unique process contributing to the pelagic sedimentary Fe-P reservoir is sorptive removal of phosphate onto ferric oxyhydroxides in mid-ocean ridge hydrothermal systems.

Mobilization of sedimentary phosphorus by microbial activity during diagenesis causes dissolved phosphate build-up in sediment pore waters, promoting benthic efflux of phosphate to bottom waters or incorporation in secondary authigenic minerals.

The combined benthic flux from coastal (sF_{cbf}) and abyssal (sF_{abf}) sediments is estimated to exceed the total riverine phosphorus flux ($F_{24(d+p)}$) to the ocean. Reprecipitation of diagenetically mobilized phosphorus in secondary phases significantly enhances phosphorus burial efficiency, impeding return of phosphate to the water column. Both processes impact the marine phosphorus cycle by affecting the primary productivity potential of surface waters.

Topics of Special Interest in Marine Phosphorus Research

Phosphate Oxygen Isotopes (δ^{18}O-PO$_4$)

Use of the oxygen isotopic composition of phosphate in biogenic hydroxyapatite (bones, teeth) as a paleotemperature and climate indicator was pioneered by Longinelli in the late 1960s–early 1970s, and has since been fairly widely and successfully applied. A novel application of the oxygen isotope system in phosphates is its use as a tracer of biological turnover of phosphorus during metabolic processes. Phosphorus has only one stable isotope (^{31}P) and occurs almost exclusively as orthophosphate (PO_4^{3-}) under Earth surface conditions. The phosphorus–oxygen bond in phosphate is highly resistant to nonenzymatic oxygen isotope exchange reactions, but when phosphate is metabolized by living organisms, oxygen isotopic exchange is rapid and extensive. Such exchange results in temperature-dependent fractionations between phosphate and ambient water. This property renders phosphate oxygen isotopes useful as indicators of present or past metabolic activity of organisms, and allows distinction of biotic from abiotic processes operating in the cycling of phosphorus through the environment.

Currently, the δ^{18}O-PO$_4$ system is being applied in a number of studies of marine phosphorus cycling, including (i) application to dissolved sea water inorganic phosphate as a tracer of phosphate source, water mass mixing, and biological productivity; (ii) use in phosphates associated with ferric iron oxyhydroxide precipitates in submarine ocean ridge sediments, where the δ^{18}O-PO$_4$ indicates microbial phosphate turnover at elevated temperatures. This latter observation suggests that phosphate oxygen isotopes may be useful biomarkers for fossil hydrothermal vent systems.

Reevaluating the Role of Phosphorus as a Limiting Nutrient in the Ocean

In terrestrial soils and in the euphotic zone of lakes and the ocean, the concentration of dissolved orthophosphate is typically low. When bioavailable phosphorus is exhausted prior to more abundant nutrients, it limits the amount of sustainable biological productivity. Phosphorus limitation in lakes is widely accepted, and terrestrial soils are often phosphorus-limited. In the oceans, however, phosphorus limitation is the subject of controversy and debate.

The prevailing wisdom has favored nitrogen as the limiting macronutrient in the oceans. However, a growing body of literature convincingly demonstrates that phosphate limitation of marine primary productivity can and does occur in some marine systems. In the oligotrophic gyres of both the western North Atlantic and subtropical North Pacific, evidence in the form of dissolved nitrogen : phosphorous (N:P) ratios has been used to argue convincingly that these systems are currently phosphate-limited. The N(:)P ratio of phytoplankton during nutrient-sufficient conditions is 16N:1P (the Redfield ratio). A positive deviation from this ratio indicates probable phosphate limitation, while a negative deviation indicates probable nitrogen limitation. In the North Pacific at the Hawaiian Ocean Time Series (HOT) site, there has been a shift since the 1988 inception of the time series to N:P ratios exceeding the Redfield ratio in both particulate and surface ocean dissolved nitrogen and phosphorus (**Figure 3**). Coincident with this shift has been an increase in the prevalence of the nitrogen-fixing cyanobacterium *Trichodesmium* (**Table 3**). Currently, it appears as though the supply of new nitrogen has shifted from a limiting flux of upwelled nitrate from below the euphotic zone to an unlimited pool of atmospheric N_2 rendered bioavailable by the action of nitrogen fixers. This shift is believed to result from climatic changes that promote water column stratification, a condition that selects for N_2-fixing microorganisms, thus driving the system to phosphate limitation. A similar situation exists in the subtropical Sargasso Sea at the Bermuda Ocean Time Series (BATS) site, where currently the dissolved phosphorus concentrations (especially dissolved inorganic phosphorus (DIP)) are significantly lower than at the HOT site, indicating even more severe phosphate limitation (**Table 3**).

A number of coastal systems also display evidence of phosphate limitation, sometimes shifting seasonally from nitrogen to phosphate limitation in concert with changes in environmental features such as upwelling and river runoff. On the Louisiana Shelf in the Gulf of Mexico, the Eel River Shelf of northern California (USA), the upper Chesapeake Bay (USA), and portions of the Baltic Sea, surface water column dissolved inorganic N:P ratios indicate seasonal phosphate limitation. The suggestion of phosphate

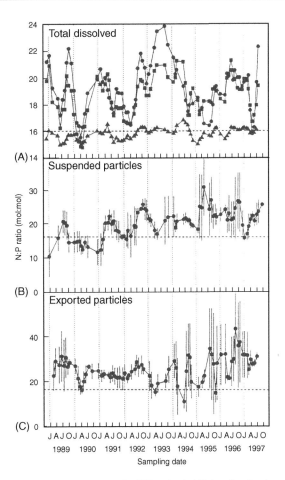

Figure 3 Time-series molar N:P ratios in (A) the dissolved pool, (B) suspended particulate matter, and (C) exported particulate matter from the HOT time series site at station ALOHA in the subtropical North Pacific near Hawaii. (A) The 3-point running mean N:P ratios for 0–100 m (circles), 100–200 m (squares), and 200–500 m (triangles). (B) The 3-point running mean (± 1 SD) for the average suspended particulate matter in the upper water column (0–100 m). (C) The 3-point running mean (± 1 SD) for the average N:P ratio of sediment trap-collected particulate matter at 150 m. The Redfield ratio (N:P16) is represented by a dashed line in all three panels. Particulate and upper water column dissolved pools show an increasing N:P ratio throughout the time-series, with a preponderance of values in excess of the Redfield ratio. (After Karl DM, Letelier R, Tupas L *et al.* (1997) The role of nitrogen fixation in the biogeochemical cycling in the subtropical North Pacific Ocean. *Nature* 388: 533–538, with permission.)

limitation is reinforced in the Louisiana Shelf and Eel River Shelf studies by the occurrence or presence of alkaline phosphatase activity, an enzyme induced only under phosphate-limiting conditions. Alkaline phosphatase has also been observed seasonally in Narragansett Bay. Although these coastal sites are recipients of anthropogenically derived nutrients (nitrogen and phosphate) that stimulate primary productivity above 'natural' levels, the processes that result in shifts in the limiting nutrient are not

Table 3 Parameters affecting nutrient limitation: comparison between North Atlantic and North Pacific Gyres

Parameter	Sargasso Sea	Pacific HOT site
DIP	0.48 ± 0.27^a	$9-40^b$
TDN (n mol l^{-1})	4512 ± 430	5680 ± 620^b
TDP (n mol l^{-1})	75 ± 42	222 ± 14^b
TDN : TDP	60 ± 7	26 ± 3^b
N$_2$-fixation rate (mmol N m^{-2} y^{-1})	72^c	$31-51^c$

After Wu J, Sunda W, Boyle EA and Karl DM (2000) *Science* 289: 759, with permission.

[a]Average DIP between 26° and 31°N in Sargasso Sea surface waters in March 1998.

[b]North Pacific near Hawaii at station ALOHA (the HOT site) during 1991–1997.

[c]See Wu *et al.* (2000) for method of calculation or measurement.

necessarily related to anthropogenic effects. Other oceanic sites where phosphorus limitation of primary productivity has been documented include the Mediterranean Sea and Florida Bay (USA).

One key question that studies of nitrogen and phosphorus limitation must address before meaningful conclusions may be drawn about phosphorus versus nitrogen limitation of marine primary productivity is the extent to which the dissolved organic nutrient pools are accessible to phytoplankton. In brief, this is the question of bioavailability. Many studies of nutrient cycling and nutrient limitation do not include measurement of these quantitatively important nutrient pools, even though there is indisputable evidence that at least some portion of the dissolved organic nitrogen (DON) and dissolved organic phosphorus (DOP) pools is bioavailable. An important direction for future research is to characterize the DON and DOP pools at the molecular level, and to evaluate what fraction of these are bioavailable. The analytical challenge of identifying the molecular composition of the DOP pool is significant. Recent advances in ^{31}P-nuclear magnetic resonance (NMR) spectroscopy have permitted a look at the high molecular weight (>1 nm) fraction of the DOP pool. However, this fraction represents only one-third of the total DOP pool; the other two-thirds of this pool, made up of smaller-molecular-weight DOP compounds, remains outside our current window of analytical accessibility.

Debate continues among oceanographers about the most probable limiting nutrient on recent and on long, geological timescales. While there is an abundant reservoir of nitrogen (gaseous N$_2$) in the atmosphere that can be rendered bioavailable by nitrogen-fixing photosynthetic organisms, phosphorus supply to the ocean is limited to that

weathered off the continents and delivered by rivers, with some minor atmospheric input. As a consequence of continental weathering control on phosphorus supply to the oceans, phosphorus limitation has been considered more likely than nitrogen limitation on geological timescales. As the recent studies reviewed in this section suggest, phosphorus limitation may be an important phenomenon in the modern ocean as well. Overall, current literature indicates that there is a growing appreciation of the complexities of nutrient limitation in general, and the role of phosphorus limitation in particular, in both fresh water and marine systems.

Cosmogenic ^{32}P and ^{33}P as Tracers of Phosphorus Cycling in Surface Waters

There are two radioactive isotopes of phosphorus, ^{32}P (half-life $= 14.3$ days) and ^{33}P (half-life $= 25.3$ days). Both have been widely used in the study of biologically mediated phosphorus cycling in aquatic systems. Until very recently, these experiments have been conducted by artificially introducing radiophosphorus into laboratory incubations or, far more rarely, by direct introduction into natural waters under controlled circumstances. Such experiments necessarily involve significant perturbation of the system, which can complicate interpretation of results. Recent advances in phosphorus sampling and radioisotope measurement have made it possible to use naturally produced ^{32}P and ^{33}P as *in situ* tracers of phosphorus recycling in surface waters. This advance has permitted studies of net phosphorus recycling in the absence of experimental perturbation caused by addition of artificially introduced radiophosphorus.

^{32}P and ^{33}P are produced naturally in the atmosphere by interaction of cosmic rays with atmospheric argon nuclei. They are then quickly scavenged onto aerosol particles and delivered to the ocean surface predominantly in rain. The ratio of ^{33}P/^{32}P introduced to the oceans by rainfall remains relatively constant, despite the fact that absolute concentrations can vary from one precipitation event to another. Once the dissolved phosphorus is incorporated into a given surface water phosphorus pool (e.g., by uptake by phytoplankton or bacteria, grazing of phytoplankton or bacteria by zooplankton, or abiotic sorption), the ^{33}P/^{32}P ratio will increase in a systematic way as a given pool ages. This increase in the ^{33}P/^{32}P ratio with time results from the different half-lives of the two phosphorus radioisotopes. By measuring the ^{33}P/^{32}P ratio in rain and in different marine phosphorus pools – e.g., DIP, DOP (sometimes called soluble nonreactive phosphorus, or SNP), and particulate phosphorus of various size classes corresponding to different levels in the food chain – the net age of phosphorus in any of these reservoirs can be determined (**Table 4**). New insights into phosphorus cycling in oceanic surface waters derived from recent work using the cosmogenically produced $^{33}P/^{32}P$ ratio include the following. (1) Turnover rates of dissolved inorganic phosphorus in coastal and oligotrophic oceanic surface waters ranges from 1 to 20 days. (2) Variable turnover rates in the DOP pool range from <1 week to >100 days, suggesting differences in either the demand for DOP, or the lability of DOP toward enzymatic breakdown. (3) In the Gulf of Maine, DOP turnover times vary seasonally, increasing from 28 days in July to >100 days in August, suggesting that the DOP pool may evolve compositionally during the growing seasons. (4) Comparison of the $^{33}P/^{32}P$ ratio in different particulate size classes indicates that the age of phosphorus generally increases at successive levels in the food chain. (5) Under some circumstances, the $^{33}P/^{32}P$ ratio can reveal which dissolved pool is being ingested by a particular size class of organisms. Utilization of this new tool highlights the dynamic nature of phosphorus cycling in surface waters by revealing the rapid rates and temporal variability of phosphorus turnover. It further stands to provide new insights into ecosystem nutrient dynamics by revealing, for example, that low phosphorus concentrations can support high primary productivity through rapid turnover rates, and that there is preferential utilization of particular dissolved phosphorus pools by certain classes of organisms.

The Oceanic Residence Time of Phosphorus

As phosphorus is the most likely limiting nutrient on geological timescales, an accurate determination of its oceanic residence time is crucial to understanding how levels of primary productivity may have varied in the Earth's past. Residence time provides a means of evaluating how rapidly the oceanic phosphorus inventory may have changed in response to variations in either input (e.g., continental weathering, dust flux) or output (e.g., burial with sediments). In its role as limiting nutrient, phosphorus will dictate the amount of surface-ocean net primary productivity, and hence atmospheric CO_2 drawdown that will occur by photosynthetic biomass production. It has been suggested that this so-called 'nutrient–CO_2' connection might link the oceanic phosphorus cycle to climate change due to reductions or increases in the atmospheric greenhouse gas inventory. Oceanic phosphorus residence time and response time (the inverse of residence time) will dictate the timescales over which such a phosphorus-induced climate effect may operate.

Over the past decade there have been several reevaluations of the marine phosphorus cycle in the literature, reflecting changes in our understanding of the identities and magnitudes of important phosphorus sources and sinks. One quantitatively important newly identified marine phosphorus sink is precipitation of disseminated authigenic carbonate fluoroapatite (CFA) in sediments in nonupwelling environments. CFA is the dominant phosphorus mineral in economic phosphorite deposits. Its presence in terrigenous-dominated continental margin environments is detectable only by indirect methods (coupled pore water/solid-phase chemical analyses), because dilution by terrigenous debris renders it below detection limits of direct methods (e.g. X-ray diffraction). Disseminated CFA has now been found in numerous continental margin environments, bearing out early proposals that this is an important marine phosphorus sink. A second class of authigenic phosphorus minerals identified as a quantitatively significant phosphorus sink in sandy continental margin sediments are aluminophosphates. Continental margins in general are quantitatively important sinks for organic and ferric iron-bound phosphorus, as well. When newly calculated phosphorus burial fluxes in continental margins, including the newly identified CFA and aluminophosphate (dominantly aluminum rare-earth phosphate) sinks are combined with older estimates of phosphorus burial fluxes in the deep sea, the overall burial flux results in a much shorter residence time than the canonical value of 100 000 years found in most text books (**Table 5**). This reduced residence time suggests that the oceanic phosphorus-cycle is subject to perturbations on shorter timescales than has previously been believed.

The revised, larger burial flux cannot be balanced by the dissolved riverine input alone. However, when the fraction of riverine particulate phosphorus that is believed to be released upon entering the marine realm is taken into account, the possibility of a balance between inputs and outputs becomes more feasible. Residence times estimated on the basis of phosphorus inputs that include this 'releasable' riverine particulate phosphorus fall within the range of residence time estimates derived from phosphorus burial fluxes (**Table 5**). Despite the large uncertainties associated with these numbers, as evidenced by the maximum and minimum values derived from both input and removal fluxes, these updated residence times are all significantly shorter than the canonical value of 100 000 years. Revised residence times on the order of 10 000–17 000 y make phosphorus-perturbations of the ocean–atmosphere CO_2 reservoir on the timescale of glacial–interglacial climate change feasible.

Table 4 Turnover rates of dissolved inorganic phosphorus (DIP) [a] and dissolved organic phosphorus (DOP) [b] in surface sea water

| Phosphorus pool | Phosphorus turnover rate | | References |
	Coastal	Open Ocean	
DIP	< 1 h to 10 d (> 1000 d in Bedford Basin)	Weeks to months	c, d, e, f, g, h, i, j, k, l, m
Total DOP	3 to > 90 d	50 to 300 d	l, m, n, o, p, q, r
Bioavailable DOP (model compounds)	2 to 30 d	1 to 4 d	k, s, t, u
Microplankton (< 1 μm)	> 1 to 3 d	NA	m
Phytoplankton (> 1 μm)	< 1 to 8 d	< 1 week	m, v
Macrozooplankton (> 280 μm)	14 to 40 d	30 to 80 d	m, p, q, v, w

After Benitez-Nelson CR (2000) *Earth-Science Reviews* 51: 109, with permission.

[a] DIP is equivalent to the soluble reactive phosphorus (SRP) pool, which may include some phosphate derived from hydrolysis of DOP (e.g., see Monaghan EJ and Ruttenberg KC (1998) *Limnology and Oceanography* 44(7): 1702).

[b] DOP is equivalent to the soluble nonreactive phosphorus (SNP) pool which may include dissolved inorganic polyphosphates (e.g., see Karl DM and Yanagi K (1997) *Limnology and Oceanography* 42: 1398).

[c] Pomeroy LR (1960) *Science* 131: 1731.

[d] Duerden CF (1973) PhD thesis, Dalhousie University, Halifax.

[e] Taft JL, Taylor WR and McCarthy JJ (1975) *Marine Biology* 33: 21.

[f] Harrison WG, Azam F, Renger EH and Eppley RW (1977) *Marine Biology* 40: 9.

[g] Perry MJ and Eppley RW (1981) *Deep-Sea Research* 28: 39.

[h] Smith RE, Harrison WG and Harris L (1985) *Marine Biology* 86: 75.

[i] Sorokin YI (1985) *Marine Ecology Progress Series* 27: 87.

[j] Harrison WG and Harris LR (1986) *Marine Ecology Progress Series* 27: 253.

[k] Björkman K and Karl DM (1994) *Marine Ecology Progress Series* 111: 265.

[l] Björkman K, Thomson-Bulldis AL and Karl DM (2000) *Aquatic and Microbial Ecology* 22: 185.

[m] Benitez-Nelson CR and Buesseler KO (1999) *Nature* 398: 502.

[n] Jackson GA and Williams PM (1985) *Deep-Sea Research* 32: 223.

[o] Orrett K and Karl DM (1987) *Limnology and Oceanography* 32: 383.

[p] Lal D and Lee T (1988) *Nature* 333: 752.

[q] Lee T, Barg E and Lal D (1992) *Analytica Chimica Acta* 260: 113.

[r] Karl DM and Yanagi K (1997) *Limnology and Oceanography* 42: 1398.

[s] Ammerman JW and Azam F (1985) *Limnology and Oceanography* 36: 1437.

[t] Nawrocki MP and Karl DM (1989) *Marine Ecology Progress Series* 57: 35.

[u] Björkman K and Karl DM (1999) *Marine Ecology Progress Series* (submitted).

[v] Waser NAD, Bacon MP and Michaels AF (1996) *Deep-Sea Research* 43(2–3): 421.

[w] Lee T, Barg E and Lal D (1991) *Limnology and Oceanography* 36: 1044.

Long Time-scale Phosphorus Cycling, and Links to Other Biogeochemical Cycles

The biogeochemical cycles of phosphorus and carbon are linked through photosynthetic uptake and release during respiration. During times of elevated marine biological productivity, enhanced uptake of surface water CO_2 by photosynthetic organisms results in increased CO_2 evasion from the atmosphere, which persists until the supply of the least abundant nutrient is exhausted. On geological timescales, phosphorus is likely to function as the limiting nutrient and thus play a role in atmospheric CO_2 regulation by limiting CO_2 drawdown by oceanic photosynthetic activity. This connection between nutrients and atmospheric CO_2 could have played a role in triggering or enhancing the global cooling that resulted in glacial episodes in the geological past. It has recently been proposed that tectonics may play the ultimate role in controlling the exogenic phosphorus mass, resulting in long-term phosphorus-limited productivity in the ocean. In this formulation, the balance between subduction of phosphorus bound up in marine sediments and underlying crust and creation of new crystalline rock sets the mass of exogenic phosphorus.

Phosphorus and oxygen cycles are linked through the redox chemistry of iron. Ferrous iron is unstable at the Earth's surface in the presence of oxygen, and oxidizes to form ferric iron oxyhydroxide precipitates, which are extremely efficient scavengers of dissolved phosphate. Resupply of phosphate to surface waters where it can fertilize biological productivity is reduced when oceanic bottom waters are well oxygenated owing to scavenging of phosphate by ferric oxyhydroxides. In contrast, during times in Earth's history when oxygen was not abundant in the atmosphere (Precambrian), and when expanses of the deep ocean were anoxic (e.g., Cretaceous), the potential for a larger oceanic dissolved phosphate inventory could have been realized due to the reduced importance of sequestering with ferric

Table 5 Revised oceanic phosphorus input fluxes, removal fluxes, and estimated oceanic residence time

	Flux description[a]	Flux (10^{12} mol P y^{-1})	Residence time (y)[e]
Input fluxes			
F_{84}	atmosphere → surface ocean	0.02–0.05	
$F_{24(d)}$	soil → surface ocean (river dissolved P flux)[b]	0.032–0.058	
$F_{24(p)}$	soil → surface ocean (river particulate P flux)[b]	0.59–0.65	
	Minimum reactive-P input flux	0.245	12 000
	Maximum reactive-P input flux	0.301	10 000
Removal fluxes			
sF_{cs}	Best estimate of total-P burial in continental margin marine sediments (**Table 2**, note j)[c]	0.150	
sF_{as}	Best estimate of total-P burial in abyssal marine sediments (**Table 2**, note j)[c]	0.130	
	Minimum estimate of reactive-P burial in marine sediments[d]	0.177	17 000
	Maximum estimate of reactive-P burial in marine sediments[d]	0.242	12 000

[a]All fluxes are from **Table 2**.
[b]As noted in **Table 2**, 30% of atmospheric aerosol phosphorus (Duce *et al.* (1991) *Global Biogeochemical Cycles* 5: 193) and 25–45% of the river particulate flux (see note (f) in **Table 2**) is believed to be mobilized upon entering the ocean. The reactive phosphorus input flux was calculated as the sum of $0.3(F_{84}) + F_{24(d)} + 0.35(F_{24(p)})$, where the mean value of the fraction of riverine particulate phosphorus flux estimated as reactive phosphorus (35%) was used. Reactive phosphorus is defined as that which passes through the dissolved oceanic phosphorus reservoir, and thus is available for biological uptake.
[c]These estimates are favored by the author, and reflect the minimum sF_{cs} and maximum sF_{as} fluxes given in **Table 2**. Because the reactive phosphorus contents of continental margin and abyssal sediments differ (see **Table 2** and note d, below), these fluxes must be listed separately in order to calculate the whole-ocean reactive phosphorus burial flux. See note (j) in **Table 2** for other published estimates of reactive-phosphorus burial flux.
[d]As noted in **Table 2**, between 40% and 75% of phosphorus buried in continental margin sediments is potentially reactive, and 90% to 100% of phosphorus buried in abyssal sediments is potentially reactive. The reactive phosphorus fraction of the total sedimentary phosphorus reservoir represents that which may have passed through the dissolved state in oceanic waters, and thus represents a true phosphorus sink from the ocean. The minimum reactive phosphorus burial flux was calculated as the sum of $0.4(sF_{cs}) + 0.9 (sF_{as})$; the maximum reactive phosphorus burial flux was calculated as the sum of $0.75(sF_{cs}) + 1(sF_{as})$. Both the flux estimates and the % reactive phosphorus estimates have large uncertainties associated with them.
[e]Residence time estimates are calculated as the oceanic phosphorus inventory (reservoirs #4 and 5 (**Table 1**) $= 3 \times 10^{15}$ mol P) divided by the minimum and maximum input and removal fluxes.

oxyhydroxides. This iron–phosphorus–oxygen coupling produces a negative feedback, which may have kept atmospheric O_2 within equable levels throughout the Phanerozoic. Thus, it is in the oceans that the role of phosphorus as limiting nutrient has the greatest repercussions for the global carbon and oxygen cycles.

Summary

The global cycle of phosphorus is truly a biogeochemical cycle, owing to the involvement of phosphorus in both biochemical and geochemical reactions and pathways. There have been marked advances in the last decade on numerous fronts of phosphorus research, resulting from application of new methods as well as rethinking of old assumptions and paradigms. An oceanic phosphorus residence time on the order of 10 000–20 000 y, a factor of 5–10 shorter than previously cited values, casts phosphorus in the role of a potential player in climate change through the nutrient–CO_2 connection. This possibility is bolstered by findings in a number of recent studies that phosphorus does function as the limiting nutrient in some modern oceanic settings. Both oxygen isotopes in phosphate ($\delta^{18}O$-PO_4) and *in situ*-produced radiophosphorus isotopes (^{33}P and ^{32}P) are providing new insights into how phosphorus is cycled through metabolic pathways in the marine environment. Finally, new ideas about global phosphorus cycling on long, geological timescales include a possible role for phosphorus in regulating atmospheric oxygen levels via the coupled iron–phosphorus–oxygen cycles, and the potential role of tectonics in setting the exogenic mass of phosphorus. The interplay of new findings in each of these areas is providing us with a fresh look at the marine phosphorus cycle, one that is sure to evolve further as these new areas are explored in more depth by future studies.

Further Reading

Blake RE, Alt JC, and Martini AM (2001) Oxygen isotope ratios of PO_4: an inorganic indicator of enzymatic activity and P metabolism and a new biomarker in the search for life. *Proceedings of the National Academy of Sciences of the USA* 98: 2148–2153.

Benitez-Nelson CR (2001) The biogeochemical cycling of phosphorus in marine systems. *Earth-Science Reviews* 51: 109–135.

Clark LL, Ingall ED, and Benner R (1999) Marine organic phosphorus cycling: novel insights from nuclear magnetic resonance. *American Journal of Science* 299: 724–737.

Colman AS, Holland HD, and Mackenzie FT (1996) Redox stabilization of the atmosphere and oceans by phosphorus limited marine productivity: discussion and reply. *Science* 276: 406–408.

Colman AS, Karl DM, Fogel ML, and Blake RE (2000) A new technique for the measurement of phosphate oxygen isotopes of dissolved inorganic phosphate in natural waters. *EOS, Transactions of the American Geophysical Union* F176.

Delaney ML (1998) Phosphorus accumulation in marine sediments and the oceanic phosphorus cycle. *Global Biogeochemical Cycles* 12(4): 563–572.

Falkowski PG (1997) Evolution of the nitrogen cycle and its influence on the biological sequestration of CO_2 in the ocean. *Nature* 387: 272–275.

Föllmi KB (1996) The phosphorus cycle, phosphogenesis and marine phosphate-rich deposits. *Earth-Science Reviews* 40: 55–124.

Guidry MW, Mackenzie FT and Arvidson RS (2000) Role of tectonics in phosphorus distribution and cycling. In: Glenn CR, Prévt-Lucas L and Lucas J (eds) *Marine Authigenesis: From Global to Microbial.* SEPM Special Publication No. 66, pp. 35–51. (See also in this volume Compton *et al.* (pp. 21–33), Colman and Holland (pp. 53–75), Rasmussen (pp. 89–101), and others.)

Howarth RW, Jensen HS, Marino R and Postma H (1995) Transport to and processing of P in near-shore and oceanic waters. In: Tiessen H (ed.) *Phosphorus in the Global Environment.* SCOPE 54, pp. 232–345. Chichester: Wiley. (See other chapters in this volume for additional information on P-cycling.)

Karl DM (1999) A sea of change: biogeochemical variability in the North Pacific Subtropical Gyre. *Ecosystems* 2: 181–214.

Longinelli A and Nuti S (1973) Revised phosphate–water isotopic temperature scale. *Earth and Planetary Science Letters* 19: 373–376.

Palenik B and Dyhrman ST (1998) Recent progress in understanding the regulation of marine primary productivity by phosphorus. In: Lynch JP and Deikman J (eds.) *Phosphorus in Plant Biology: Regulatory Roles in Molecular, Cellular, Organismic, and Ecosystem Processes*, pp. 26–38. American Society of Plant Physiologists

Ruttenberg KC (1993) Reassessment of the oceanic residence time of phosphorus. *Chemical Geology* 107: 405–409.

Tyrell T (1999) The relative influences of nitrogen and phosphorus on oceanic productivity. *Nature* 400: 525–531.

LAND–SEA GLOBAL TRANSFERS

F. T. Mackenzie and L. M. Ver, University of Hawaii, Honolulu, HI, USA

Introduction

The interface between the land and the sea is an important boundary connecting processes operating on land with those in the ocean. It is a site of rapid population growth, industrial and agricultural practices, and urban development. Large river drainage basins connect the vast interiors of continents with the coastal zone through river and groundwater discharges. The atmosphere is a medium of transport of substances from the land to the sea surface and from that surface back to the land. During the past several centuries, the activities of humankind have significantly modified the exchange of materials between the land and sea on a global scale – humans have become, along with natural processes, agents of environmental change. For example, because of the combustion of the fossil fuels coal, oil, and gas and changes in land-use practices including deforestation, shifting cultivation, and urbanization, the direction of net atmospheric transport of carbon (C) and sulfur (S) gases between the land and sea has been reversed. The global ocean prior to these human activities was a source of the gas carbon dioxide (CO_2) to the atmosphere and hence to the continents. The ocean now absorbs more CO_2 than it releases. In pre-industrial time, the flux of reduced sulfur gases to the atmosphere and their subsequent oxidation and deposition on the continental surface exceeded the transport of oxidized sulfur to the ocean via the atmosphere. The situation is now reversed because of the emissions of sulfur to the atmosphere from the burning of fossil fuels and biomass on land. In addition, river and groundwater fluxes of the bioessential elements carbon, nitrogen (N), and phosphorus (P), and certain trace elements have increased because of human activities on land. For example, the increased global riverine (and atmospheric) transport of lead (Pb) corresponds to its increased industrial use. Also, recent changes in the concentration of lead in coastal sediments appear to be directly related to changes in the use of leaded gasoline in internal combustion engines. Synthetic substances manufactured by modern society, such as pesticides and pharmaceutical products, are now appearing in river and groundwater flows, thus moving toward the sea in a greater degree than before.

The Changing Picture of Land–Sea Global Transfers

Although the exchange through the atmosphere of certain trace metals and gases between the land and the sea surface is important, rivers are the main purveyors of materials to the ocean. The total water discharge of the major rivers of the world to the ocean is $36\,000\,\text{km}^3\,\text{yr}^{-1}$. At any time, the world's rivers contain about 0.000 1% of the total water volume of $1459 \times 10^6\,\text{km}^3$ near the surface of the Earth and have a total dissolved and suspended solid concentration of $c.$ 110 and $540\,\text{ppm}\,\text{l}^{-1}$, respectively. The residence time of water in the world's rivers calculated with respect to total net precipitation on land is only 18 days. Thus the water in the world's rivers is replaced every 18 days by precipitation. The global annual direct discharge to the ocean of groundwater is about 10% of the surface flow, with a recent estimate of $2400\,\text{km}^3\,\text{yr}^{-1}$. The dissolved constituent content of groundwater is poorly known, but one recent estimate of the dissolved salt groundwater flux is $1300 \times 10^6\,\text{t}\,\text{yr}^{-1}$.

The chemical composition of average river water is shown in **Table 1**. Note that the major anion in river water is bicarbonate, HCO_3^-; the major cation is calcium, Ca^{2+}, and that even the major constituent concentrations of river water on a global scale are influenced by human activities. The dissolved load of the major constituents of the world's rivers is derived from the following sources: about 7% from beds of salt and disseminated salt in sedimentary rocks, 10% from gypsum beds and sulfate salts disseminated in rocks, 38% from carbonates, and 45% from the weathering of silicate minerals. Two-thirds of the HCO_3^- in river waters are derived from atmospheric CO_2 via the respiration and decomposition of organic matter and subsequent conversion to HCO_3^- through the chemical weathering of silicate ($\sim$30% of total) or carbonate ($\sim$70% of total) minerals. The other third of the river HCO_3^- comes directly from carbonate minerals undergoing weathering.

It is estimated that only about 20% of the world's drainage basins have pristine water quality. The organic productivity of coastal aquatic environments has been heavily impacted by changes in the

Table 1 Chemical composition of average river water

By continent	River water concentration[a] (mg l⁻¹)												Water discharge (10³ km³ yr⁻¹)	Runoff ratio[c]
	Ca^{2+}	Mg^{2+}	Na^+	K^+	Cl^-	SO_4^{2-}	HCO_3^-	SO_2	TDS[b]	TOC[b]	TDN[b]	TDP[b]		
Africa														
Actual	5.7	2.2	4.4	1.4	4.1	4.2	36.9	12.0	60.5				3.41	0.28
Natural	5.3	2.2	3.8	1.4	3.4	3.2	26.7	12.0	57.8					
Asia														
Actual	17.8	4.6	8.7	1.7	10.0	13.3	67.1	11.0	134.6				12.47	0.54
Natural	16.6	4.3	6.6	1.6	7.6	9.7	66.2	11.0	123.5					
S. America														
Actual	6.3	1.4	3.3	1.0	4.1	3.8	24.4	10.3	54.6				11.04	0.41
Natural	6.3	1.4	3.3	1.0	4.1	3.5	24.4	10.3	54.3					
N. America														
Actual	21.2	4.9	8.4	1.5	9.2	18.0	72.3	7.2	142.6				5.53	0.38
Natural	20.1	4.9	6.5	1.5	7.0	14.9	71.4	7.2	133.5					
Europe														
Actual	31.7	6.7	16.5	1.8	20.0	35.5	86.0	6.8	212.8				2.56	0.42
Natural	24.2	5.2	3.2	1.1	4.7	15.1	80.1	6.8	140.3					
Oceania														
Actual	15.2	3.8	7.6	1.1	6.8	7.7	65.6	16.3	125.3				2.40	
Natural	15.0	3.8	7.0	1.1	5.9	6.5	65.1	16.3	120.3					
World average														
Actual	14.7	3.7	7.2	1.4	8.3	11.5	53.0	10.4	110.1	12.57	0.574	0.053	37.40	0.46
Natural (unpolluted)	13.4	3.4	5.2	1.3	5.8	5.3	52.0	10.4	99.6	9.89	0.386	0.027	37.40	0.46
Pollution	1.3	0.3	2.0	0.1	2.5	6.2	1.0	0.0	10.5	2.67	0.187	0.027		
World % pollutive	9%	8%	28%	7%	30%	54%	2%	0%	10%	21%	33%	50%		

[a] Actual concentrations include pollution. Natural concentrations are corrected for pollution.
[b] TDS, total dissolved solids; TOC, total organic carbon; TDN, total dissolved nitrogen; TDP, total dissolved phosphorus.
[c] Runoff ratio = average runoff per unit area/average rainfall.

Revised after Meybeck M (1979) Concentrations des eaux fluviales en elements majeurs et apports en solution aux oceans. *Revue de Geologie Dynamique et de Geographie Physique* 21: 215–246; Meybeck M (1982) Carbon, nitrogen, and phosphorus transport by world rivers. *American Journal of Science* 282: 401–450; Meybeck M (1983) C, N, P and S in rivers: From sources to global inputs. In: Wollast R, Mackenzie FT, and Chou L (eds.) *Interactions of C, N, P and S Biogeochemical Cycles and Global Change*, pp. 163–193. Berlin: Springer.

dissolved and particulate river fluxes of three of the major bioessential elements found in organic matter, C, N, and P (the other three are S, hydrogen (H), and oxygen (O)). Although these elements are considered minor constituents of river water, their fluxes may have doubled over their pristine values on a global scale because of human activities. Excessive river-borne nutrients and the cultural eutrophication of freshwater and coastal marine ecosystems go hand in hand. In turn, these fluxes have become sensitive indicators of the broader global change issues of population growth and land-use change (including water resources engineering works) in the coastal zone and upland drainage basins, climatic change, and sea level rise.

In contrast to the situation for the major elements, delivery of some trace elements from land to the oceans via the atmosphere can rival riverine inputs. The strength of the atmospheric sources strongly depends on geography and meteorology. Hence the North Atlantic, western North Pacific, and Indian Oceans, and their inland seas, are subjected to large atmospheric inputs because of their proximity to both deserts and industrial sources. Crustal dust is the primary terrestrial source of these atmospheric inputs to the ocean. Because of the low solubility of dust in both atmospheric precipitation and seawater and the overwhelming inputs from river sources, dissolved sources of the elements are generally less important. However, because the oceans contain only trace amounts of iron (Fe), aluminum (Al), and manganese (Mn) (concentrations are in the ppb level), even the small amount of dissolution in seawater ($\sim$10% of the element in the solid phase) results in eolian dust being the primary source for the dissolved transport of these elements to remote areas of the ocean. Atmospheric transport of the major nutrients N, silicon (Si), and Fe to the ocean has been hypothesized to affect and perhaps limit primary productivity in certain regions of the ocean at certain times. Modern processes of fossil fuel combustion and biomass burning have significantly modified the atmospheric transport from land to the ocean of trace metals like Pb, copper (Cu), and zinc (Zn), C in elemental and organic forms, and nutrient N.

As an example of global land–sea transfers involving gases and the effect of human activities on the exchange, consider the behavior of CO_2 gas. Prior to human influence on the system, there was a net flux of CO_2 out of the ocean owing to organic metabolism (net heterotrophy). This flux was mainly supported by the decay of organic matter produced by phytoplankton in the oceans and part of that transported by rivers to the oceans. An example

overall reaction is:

$$C_{106}H_{263}O_{110}N_{16}S_2P + 141O_2 \Rightarrow 106CO_2$$
$$+ 6HNO_3 + 2H_2SO_4 + H_3PO_4 + 120H_2O \quad [1]$$

Carbon dioxide was also released to the atmosphere due to the precipitation of carbonate minerals in the oceans. The reaction is:

$$Ca^{2+} + 2HCO_3^- \Rightarrow CaCO_3 + CO_2 + H_2O \quad [2]$$

The CO_2 in both reactions initially entered the dissolved inorganic carbon (DIC) pool of seawater and was subsequently released to the atmosphere at an annual rate of about 0.2×10^9 t of carbon as CO_2 gas. It should be recognized that this is a small number compared with the 200×10^9 t of carbon that exchanges between the ocean and atmosphere each year because of primary production of organic matter and its subsequent respiration.

Despite the maintenance of the net heterotrophic status of the ocean and the continued release of CO_2 to the ocean–atmosphere owing to the formation of calcium carbonate in the ocean, the modern ocean and the atmosphere have become net sinks of anthropogenic CO_2 from the burning of fossil fuels and the practice of deforestation. Over the past 200 years, as CO_2 has accumulated in the atmosphere, the gradient of CO_2 concentration across the atmosphere–ocean interface has changed, favoring uptake of anthropogenic CO_2 into the ocean. The average oceanic carbon uptake for the decade of the 1990s was c. 2×10^9 t annually. The waters of the ocean have accumulated about 130×10^9 t of anthropogenic CO_2 over the past 300 years.

The Coastal Zone and Land–Sea Exchange Fluxes

The global coastal zone environment is an important depositional and recycling site for terrigenous and oceanic biogenic materials. The past three centuries have been the time of well-documented human activities that have become an important geological factor affecting the continental and oceanic surface environment. In particular, historical increases in the global population in the areas of the major river drainage basins and close to oceanic coastlines have been responsible for increasing changes in land-use practices and discharges of various substances into oceanic coastal waters. As a consequence, the global C cycle and the cycles of N and P that closely interact with the carbon cycle have been greatly affected. Several major perturbations of the past three centuries of the industrial age have affected the processes of transport from land, deposition of terrigenous materials, and *in situ* production of organic matter in coastal zone environments. In addition, potential future changes in oceanic circulation may have significant effects on the biogeochemistry and CO_2 exchange in the coastal zone.

The coastal zone is that environment of continental shelves, including bays, lagoons, estuaries, and near-shore banks that occupy 7% of the surface area of the ocean ($36 \times 10^{12}\,m^2$) or *c*. 9% of the volume of the surface mixed layer of the ocean ($3 \times 10^{15}\,m^3$). The continental shelves average 75 km in width, with a bottom slope of $1.7\,m\,km^{-1}$. They are generally viewed as divisible into the interior or proximal shelf, and the exterior or distal shelf. The mean depth of the global continental shelf is usually taken as the depth of the break between the continental shelf and slope at *c*. 200 m, although this depth varies considerably throughout the world's oceans. In the Atlantic, the median depth of the shelf-slope break is at 120 m, with a range from 80 to 180 m. The depths of the continental shelf are near 200 m in the European section of the Atlantic, but they are close to 100 m on the African and North American coasts. Coastal zone environments that have high sedimentation rates, as great as 30–$60\,cm\,ky^{-1}$ in active depositional areas, act as traps and filters of natural and human-generated materials transported from continents to the oceans via river and groundwater flows and through the atmosphere. At present a large fraction ($\sim 80\%$) of the land-derived organic and inorganic materials that are transported to the oceans is trapped on the proximal continental shelves. The coastal zone also accounts for 30–50% of total carbonate and 80% of organic carbon accumulation in the ocean. Coastal zone environments are also regions of higher biological production relative to that of average oceanic surface waters, making them an important factor in the global carbon cycle. The higher primary production is variably attributable to the nutrient inflows from land as well as from coastal upwelling of deeper ocean waters.

Fluvial and atmospheric transport links the coastal zone to the land; gas exchange and deposition are its links with the atmosphere; net advective transport of water, dissolved solids and particles, and coastal upwelling connect it with the open ocean. In addition, coastal marine sediments are repositories for much of the material delivered to the coastal zone. In the last several centuries, human activities on land have become a geologically important agent affecting the land–sea exchange of materials. In particular, river and groundwater flows and atmospheric transport of materials to the coastal zone have been substantially altered.

Bioessential Elements

Continuous increase in the global population and its industrial and agricultural activities have created four major perturbations on the coupled system of the biogeochemical cycles of the bioessential elements C, N, P, and S. These changes have led to major alterations in the exchanges of these elements between the land and sea. The perturbations are: (1) emissions of C, N, and S to the atmosphere from fossil-fuel burning and subsequent partitioning of anthropogenic C and deposition of N and S; (2) changes in land-use practices that affect the recycling of C, N, P, and S on land, their uptake by plants and release from decaying organic matter, and the rates of land surface denudation; (3) additions of N and P in chemical fertilizers to cultivated land area; and (4) releases of organic wastes containing highly reactive C, N, and P that ultimately enter the coastal zone. A fifth major perturbation is a climatic one: (5) the rise in mean global temperature of the lower atmosphere of about 1 °C in the past 300 years, with a projected increase of about 1.4–5.8 °C relative to 1990 by the year 2100. **Figure 1** shows how the fluxes associated with these activities have changed during the past three centuries with projections to the year 2040.

Partially as a result of these activities on land, the fluxes of materials to the coastal zone have changed historically. **Figure 2** shows the historical and projected future changes in the river fluxes of dissolved inorganic and organic carbon (DIC, DOC), nitrogen (DIN, DON), and phosphorus (DIP, DOP), and fluxes associated with the atmospheric deposition and denitrification of N, and accumulation of C in organic matter in coastal marine sediments. It can be seen in **Figure 2** that the riverine fluxes of C, N, and P all increase in the dissolved inorganic and organic phases from about 1850 projected to 2040. For example, for carbon, the total flux (organic + inorganic) increases by about 35% during this period. These increased fluxes are mainly due to changes in land-use practices, including deforestation, conversion of forest to grassland, pastureland, and urban centers, and regrowth of forests, and application of fertilizers to croplands and the subsequent leaching of N and P into aquatic systems.

Inputs of nutrient N and P to the coastal zone which support new primary production are from the land by riverine and groundwater flows, from the open ocean by coastal upwelling and onwelling, and to a lesser extent by atmospheric deposition of nitrogen. New primary production depends on the

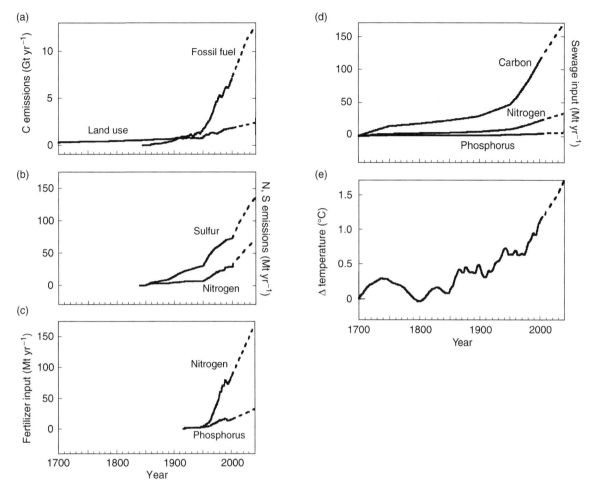

Figure 1 Major perturbations on the Earth system over the past 300 years and projections for the future: (a) emissions of CO_2 and (b) gaseous N and S from fossil-fuel burning and land-use activities; (c) application of inorganic N and P in chemical fertilizers to cultivated land; (d) loading of highly reactive C, N, and P into rivers and the coastal ocean from municipal sewage and wastewater disposal; and (e) rise in mean global temperature of the lower atmosphere relative to 1700. Revised after Ver LM, Mackenzie FT, and Lerman A (1999) Biogeochemical responses of the carbon cycle to natural and human perturbations: Past, present, and future. *American Journal of Science* 299: 762–801.

availability of nutrients from these external inputs, without consideration of internal recycling of nutrients. Thus any changes in the supply of nutrients to the coastal zone owing to changes in the magnitude of these source fluxes are likely to affect the cycling pathways and balances of the nutrient elements. In particular, input of nutrients from the open ocean by coastal upwelling is quantitatively greater than the combined inputs from land and the atmosphere. This makes it likely that there could be significant effects on coastal primary production because of changes in ocean circulation. For example, because of global warming, the oceans could become more strongly stratified owing to freshening of polar oceanic waters and warming of the ocean in the tropical zone. This could lead to a reduction in the intensity of the oceanic thermohaline circulation (oceanic circulation owing to differences in density of water masses, also

popularly known as the 'conveyor belt') and hence the rate at which nutrient-rich waters upwell into coastal environments.

Another potential consequence of the reduction in the rate of nutrient inputs to the coastal zone by upwelling is the change in the CO_2 balance of coastal waters: reduction in the input of DIC to the coastal zone from the deeper ocean means less dissolved CO_2, HCO_3^-, and CO_3^{2-} coming from that source. With increasing accumulation of anthropogenic CO_2 in the atmosphere, the increased dissolution of atmospheric CO_2 in coastal water is favored. The combined result of a decrease in the upwelling flux of DIC and an enhancement in the transfer of atmospheric CO_2 across the air–sea interface of coastal waters is a lower saturation state for coastal waters with respect to the carbonate minerals calcite, aragonite, and a variety of magnesian calcites. The

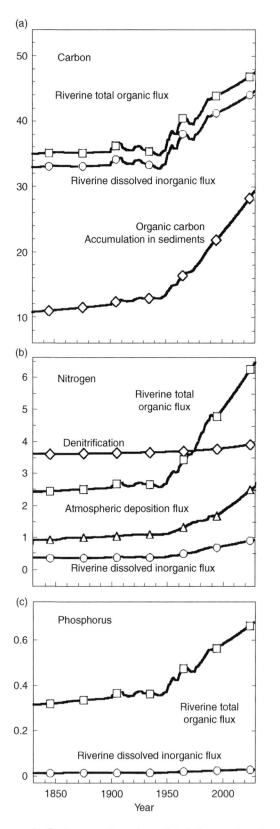

Figure 2 Past, present, and predicted fluxes of carbon, nitrogen, and phosphorus into or out of the global coastal margin, in 10^{12} mol yr^{-1}.

lower saturation state in turn leads to the likelihood of lower rates of inorganic and biological precipitation of carbonate and hence deposition and accumulation of sedimentary carbonate.

In addition, the present-day burial rate of organic carbon in the ocean may be about double that of the late Holocene flux, supported by increased fluxes of organic carbon to the ocean via rivers and groundwater flows and increased *in situ* new primary production supported by increased inputs of inorganic N and P from land and of N deposited from the atmosphere. The organic carbon flux into sediments may constitute a sink of anthropogenic CO_2 and a minor negative feedback on accumulation of CO_2 in the atmosphere.

The increased flux of land-derived organic carbon delivered to the ocean by rivers may accumulate there or be respired, with subsequent emission of CO_2 back to the atmosphere. This release flux of CO_2 may be great enough to offset the increased burial flux of organic carbon to the seafloor due to enhanced fertilization of the ocean by nutrients derived from human activities. The magnitude of the CO_2 exchange is a poorly constrained flux today. One area for which there is a substantial lack of knowledge is the Asian Pacific region. This is an area of several large seas, a region of important river inputs to the ocean of N, P, organic carbon, and sediments from land, and a region of important CO_2 exchange between the ocean and the atmosphere.

Anticipated Response to Global Warming

From 1850 to modern times, the direction of the net flux of CO_2 between coastal zone waters and the atmosphere due to organic metabolism and calcium carbonate accumulation in coastal marine sediments was from the coastal surface ocean to the atmosphere (negative flux, **Figure 3**). This flux in 1850 was on the order of -0.2×10^9 t yr^{-1}. In a condition not disturbed by changes in the stratification and thermohaline circulation of the ocean brought about by a global warming of the Earth, the direction of this flux is projected to remain negative (flux out of the coastal ocean to the atmosphere) until early in the twenty-first century. The increasing partial pressure of CO_2 in the atmosphere because of emissions from anthropogenic sources leads to a reversal in the gradient of CO_2 across the air–sea interface of coastal zone waters and, hence, invasion of CO_2 into the coastal waters. From that time on the coastal ocean will begin to operate as a net sink (positive flux) of atmospheric CO_2 (**Figure 3**). The role of the open ocean as a sink for anthropogenic CO_2 is slightly reduced while that of the coastal oceans

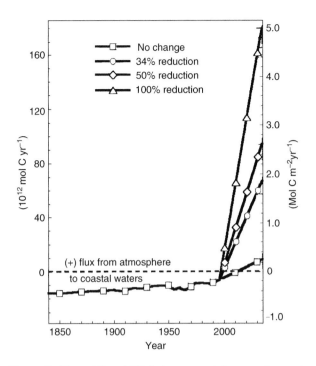

Figure 3 The net flux of CO$_2$ between coastal zone waters and the atmosphere due to organic metabolism and calcium carbonate accumulation in coastal marine sediments, under three scenarios of changing thermohaline circulation rate compared to a business-as-usual scenario, in units of 10^{12} mol C yr^{-1}.

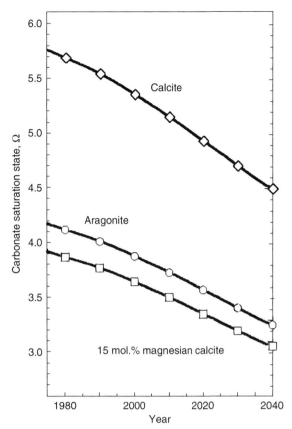

Figure 4 Changes in saturation state with respect to carbonate minerals of surface waters of the coastal ocean projected from 1999 to 2040. Calculations are for a temperature of 25 °C.

increases. The net result is the maintenance of the role of the global oceans as a net sink for anthropogenic CO$_2$.

The saturation index (Ω) for calcite or aragonite (both CaCO$_3$) is the ratio of the ion activity product IAP in coastal waters to the respective equilibrium constant K at the *in situ* temperature. For aqueous species, IAP $= a$Ca$^{2+} \times a$CO$_3^{2-}$ (where a is the activity; note that for 15 mol.% magnesian calcite, the IAP also includes the activity of the magnesium cation, aMg^{2+}). Most coastal waters and open-ocean surface waters currently are supersaturated with respect to aragonite, calcite, and magnesian calcite containing 15 mol.% Mg, that is, Ω_{calcite}, $\Omega_{\text{aragonite}}$, and $\Omega_{15\%\text{magnesian calcite}}$ are >1. Because of global warming and the increasing land-to-atmosphere-to-seawater transport of CO$_2$ (due to the continuing combustion of fossil fuels and burning of biomass), the concentration of the aqueous CO$_2$ species in seawater increases and the pH of the water decreases slightly. This results in a decrease in the concentration of the carbonate ion, CO$_3^{2-}$, resulting in a decrease in the degree of supersaturation of coastal zone waters. **Figure 4** shows how the degree of saturation might change into the next century because of rising atmospheric CO$_2$ concentrations. The overall reduction in the saturation state of coastal

zone waters with respect to aragonite from 1997 projected to 2040 is about 16%, from 3.89 to 3.26.

Modern carbonate sediments deposited in shoal-water ('shallow-water') marine environments (including shelves, banks, lagoons, and coral reef tracts) are predominantly biogenic in origin derived from the skeletons and tests of benthic and pelagic organisms, such as corals, foraminifera, echinoids, mollusks, algae, and sponges. One exception to this statement is some aragonitic muds that may, at least in part, result from the abiotic precipitation of aragonite from seawater in the form of whitings. Another exception is the sand-sized, carbonate oöids composed of either aragonite with a laminated internal structure or magnesian calcite with a radial internal structure. In addition, early diagenetic carbonate cements found in shoal-water marine sediments and in reefs are principally aragonite or magnesian calcite. Thus carbonate production and accumulation in shoal-water environments are dominated by a range of metastable carbonate minerals associated with skeletogenesis and abiotic processes, including calcite, aragonite, and a variety of magnesian calcite compositions.

With little doubt, as has been documented in a number of observational and experimental studies, a reduction in the saturation state of ocean waters will lead to a reduction in the rate of precipitation of both inorganic and skeletal calcium carbonate. Conversely, increases in the degree of supersaturation and temperature will increase the precipitation rates of calcite and aragonite from seawater. During global warming, rising sea surface temperatures and declining carbonate saturation states due to the absorption of anthropogenic CO_2 by surface ocean waters result in opposing effects. However, experimental evidence suggests that within the range of temperature change predicted for the next century due to global warming, the effect of changes in saturation state will be the predominant factor affecting precipitation rate. Thus decreases in precipitation rates should lead to a decrease in the production and accumulation of shallow-water carbonate sediments and perhaps changes in the types and distribution of calcifying biotic species found in shallow-water environments.

Anticipated Response to Heightened Human Perturbation: The Asian Scenario

In the preceding sections it was shown that the fluxes of C, N, and P from land to ocean have increased because of human activities (refer to **Table 1** for data comparing the actual and natural concentrations of C, N, P, and other elements in average river water). During the industrial era, these fluxes mainly had their origin in the present industrialized and developed countries. This is changing as the industrializing and developing countries move into the twenty-first century. A case in point is the countries of Asia.

Asia is a continent of potentially increasing contributions to the loading of the environment owing to a combination of such factors as its increasing population, increasing industrialization dependent on fossil fuels, concentration of its population along the major river drainage basins and in coastal urban centers, and expansion of land-use practices. It is anticipated that Asia will experience similar, possibly even greater, loss of storage of C and nutrient N and P on land and increased storage in coastal marine sediments per unit area than was shown by the developed countries during their period of industrialization. The relatively rapid growth of Asia's population along the oceanic coastal zone indicates that higher inputs of both dissolved and particulate organic nutrients may be expected to enter coastal waters.

A similar trend of increasing population concentration in agricultural areas inland, within the drainage basins of the main rivers flowing into the ocean, is also expected to result in increased dissolved and particulate organic nutrient loads that may eventually reach the ocean. Inputs from inland regions to the ocean would be relatively more important if no entrapment or depositional storage occurred en route, such as in the dammed sections of rivers or in alluvial plains. In the case of many of China's rivers, the decline in sediment discharge from large rivers such as the Yangtze and the Yellow Rivers is expected to continue due to the increased construction of dams. The average decadal sediment discharge from the Yellow River, for example, has decreased by 50% from the 1950s to the 1980s. If the evidence proposed for the continental United States applies to Asia, the damming of major rivers would not effectively reduce the suspended material flow to the ocean because of the changes in the erosional patterns on land that accompany river damming and more intensive land-use practices. These flows on land and into coastal ocean waters are contributing factors to the relative importance of autotrophic and heterotrophic processes, competition between the two, and the consequences for carbon exchange between the atmosphere and land, and the atmosphere and ocean water. The change from the practices of land fertilization by manure to the more recent usage of chemical fertilizers in Asia suggests a shift away from solid organic nutrients and therefore a reduced flow of materials that might promote heterotrophy in coastal environments.

Sulfur is an excellent example of how parts of Asia can play an important role in changing land–sea transfers of materials. Prior to extensive human interference in the global cycle of sulfur, biogenically produced sulfur was emitted from the sea surface mainly in the form of the reduced gas dimethyl sulfide (DMS). DMS was the major global natural source of sulfur for the atmosphere, excluding sulfur in sea salt and soil dust. Some of this gas traveled far from its source of origin. During transport the reduced gas was oxidized to micrometer-size sulfate aerosol particles and rained out of the atmosphere onto the sea and continental surface. The global sulfur cycle has been dramatically perturbed by the industrial and biomass burning activities of human society. The flux of gaseous sulfur dioxide to the atmosphere from the combustion of fossil fuels in some regions of the world and its conversion to sulfate aerosol greatly exceeds natural fluxes of sulfur gases from the land surface. It is estimated that this flux for the year 1990 was equivalent to 73×10^6 $t\,yr^{-1}$, nearly 4 times the natural DMS flux from the ocean. This has led to a net transport of sulfur from the land to the ocean via the atmosphere, completely reversing the flow direction in preindustrial times. In addition, the sulfate aerosol content of the atmosphere derived from human activities has increased. Sulfate aerosols affect global climate directly as particles that scatter incoming solar radiation and indirectly as cloud condensation nuclei (CCNs), which lead to an increased number of cloud droplets and an increase in the solar reflectance of clouds. Both effects cause the cooling of the planetary surface. As can be seen in **Figure 5** the eastern Asian

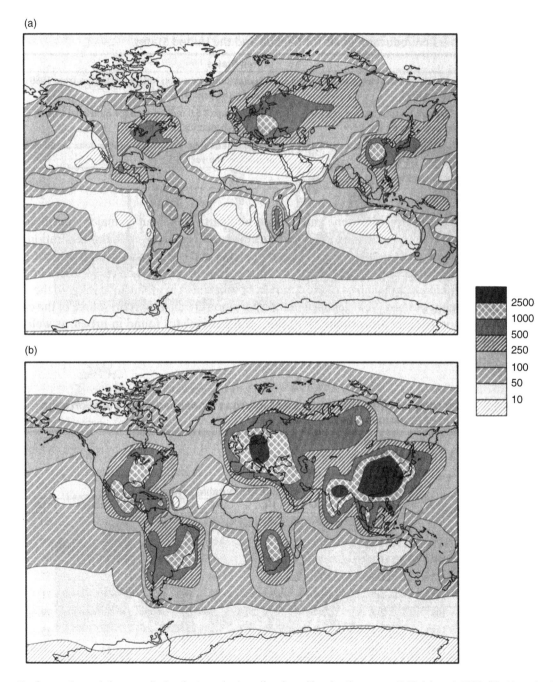

Figure 5 Comparison of the magnitude of atmospheric sulfur deposition for the years 1990 (a) and 2050 (b). Note the large increases in both spatial extent and intensity of sulfur deposition in both hemispheres and the increase in importance of Asia, Africa, and South America as sites of sulfur deposition between 1990 and 2050. The values on the diagrams are in units of kg $S m^{-2} yr^{-1}$. Revised after Mackenzie FT (1998) *Our Changing Planet: An Introduction to Earth System Science and Global Environmental Change*. Upper Saddle River, NJ: Prentice Hall; Rodhe H, Langner J, Gallardo L, and Kjellström E (1995) Global transport of acidifying pollutants. *Water, Air and Soil Pollution* 85: 37–50.

region is an important regional source of sulfate aerosol because of the combustion of fossil fuels, particularly coal. This source is predicted to grow in strength during the early- to mid-twenty-first century (**Figure 5**).

Conclusion

Land–sea exchange processes and fluxes of the bioessential elements are critical to life. In several cases documented above, these exchanges have been substantially modified by human activities. These modifications have led to a number of environmental issues including global warming, acid deposition, excess atmospheric nitrogen deposition, and production of photochemical smog. All these issues have consequences for the biosphere – some well known, others not so well known. It is likely that the developing world, with increasing population pressure and industrial development and with no major changes in agricultural technology and energy consumption rates, will become a more important source of airborne gases and aerosols and materials for river and groundwater systems in the future. This will lead to further modification of land–sea global transfers. The region of southern and eastern Asia is particularly well poised to influence significantly these global transfers.

See also

Coastal Topography, Human Impact on. Phosphorus Cycle.

Further Reading

Berner EA and Berner RA (1996) *Global Environment: Water, Air and Geochemical Cycles.* Upper Saddle River, NJ: Prentice Hall.

Galloway JN and Melillo JM (eds.) (1998) *Asian Change in the Context of Global Change.* Cambridge, MA: Cambridge University Press.

Mackenzie FT (1998) *Our Changing Planet: An Introduction to Earth System Science and Global Environmental Change.* Upper Saddle River, NJ: Prentice Hall.

Mackenzie FT and Lerman A (2006) *Carbon in the Geobiosphere – Earth's Outer Shell.* Dordrecht: Springer.

Meybeck M (1979) Concentrations des eaux fluviales en elements majeurs et apports en solution aux oceans. *Revue de Geologie Dynamique et de Geographie Physique* 21: 215–246.

Meybeck M (1982) Carbon, nitrogen, and phosphorus transport by world rivers. *American Journal of Science* 282: 401–450.

Meybeck M (1983) C, N, P and S in rivers: From sources to global inputs. In: Wollast R, Mackenzie FT, and Chou L (eds.) *Interactions of C, N, P and S Biogeochemical Cycles and Global Change*, pp. 163–193. Berlin: Springer.

Rodhe H, Langner J, Gallardo L, and Kjellström E (1995) Global transport of acidifying pollutants. *Water, Air and Soil Pollution* 85: 37–50.

Schlesinger WH (1997) *Biogeochemistry: An Analysis of Global Change.* San Diego, CA: Academic Press.

Smith SV and Mackenzie FT (1987) The ocean as a net heterotrophic system: Implications from the carbon biogeochemical cycle. *Global Biogeochemical Cycles* 1: 187–198.

Ver LM, Mackenzie FT, and Lerman A (1999) Biogeochemical responses of the carbon cycle to natural and human perturbations: Past, present, and future. *American Journal of Science* 299: 762–801.

Vitousek PM, Aber JD, and Howarth RW (1997) Human alteration of the global nitrogen cycle: Sources and consequences. *Ecological Applications* 7(3): 737–750.

Wollast R and Mackenzie FT (1989) Global biogeochemical cycles and climate. In: Berger A, Schneider S, and Duplessy JC (eds.) *Climate and Geo-Sciences*, pp. 453–473. Dordrecht: Kluwer Academic Publishers.

CIRCULATION AND MODELS

SHELF SEA AND SHELF SLOPE FRONTS

J. Sharples, Proudman Oceanographic Laboratory, Liverpool, UK
J. H. Simpson, Bangor University, Bangor, UK

Introduction

Horizontal gradients in temperature and salinity in the ocean are generally very weak. Regions of enhanced horizontal gradient are referred to as fronts. The scalar gradients across a front indicate concomitant changes in the physical processes that determine water column structure. Fronts are important oceanographic features because, corresponding to the physical gradients, they are also sites of rapid chemical and biological changes. In particular, fronts are often sites of enhanced standing stock of primary producers and primary production, with related increases in zooplankton biomass, fish, and foraging seabirds. These frontal aggregations of fish are also an important marine resource, targeted specifically by fishing vessels.

Here we will discuss three types of fronts. In shelf seas, fronts can be generated by the influence of freshwater runoff, or by surface heat fluxes interacting with horizontal variations of tidal mixing. At the edge of the continental shelves, fronts are often seen separating the inherently contrasting temperature–salinity characteristics of shelf and open ocean water. These three types are illustrated schematically in **Figure 1**.

Freshwater Fronts in Shelf Seas

The lateral input of fresh water from rivers results in coastal waters having a lower salinity than the ambient shelf water. In the absence of any vertical mixing, this low-salinity water would spread out above the saltier shelf water as a density-driven current, eventually turning anticyclonically (i.e., to the right in the Northern Hemisphere, left in the Southern Hemisphere) to form a buoyancy current. Parallel to the coastline there will be a thermohaline front separating the low-salinity water from the shelf water. If there is stronger vertical mixing, supplied by tidal currents or wind stress, or if the freshwater input is very strong, then the front can extend from the surface to the seabed.

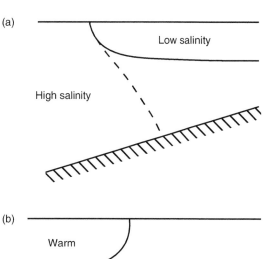

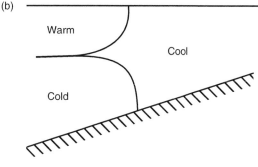

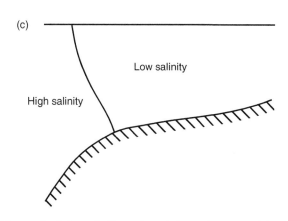

Figure 1 Schematic illustration of the main types of fronts found in shelf seas. (a) A freshwater front, caused by the input of fresher estuarine water into the coastal zone. The front can either be confined to the surface (weak vertical mixing) or extend from the surface to the seabed (strong vertical mixing). (b) A shelf sea tidal mixing front, caused by competition between surface heating and tidal mixing. The stratified water on the left occurs because of weak tidal mixing being unable to counter the stratification generated by surface heating. The mixed water on the right is the result of strong tidal mixing being able to prevent thermal stratification. (c) A shelf break front. The low-salinity water on the shelf results from the combination of all the estuarine inputs from the coast. There can also be a cross-shelf edge contrast in temperature.

Determining the Position of a Freshwater Front

The distance offshore at which this front lies depends on whether or not the buoyancy current 'feels' the seabed. If the buoyancy flow is confined to the surface, then the coast-parallel region containing the low-salinity surface layer will be approximately one internal Rossby radius thick (i.e., the distance traveled seaward by the surface buoyancy current before the effect of the Earth's rotation drives it parallel to the coastline). Typically, in temperate regions, this distance will be up to 10 km.

When the buoyancy current is in contact with the seabed, the situation is altered by the breakdown of geostrophy within the bottom Ekman layer of the flow. Within this Ekman layer there is a component of transport perpendicular to the front, pushing the bottom front offshore and driving low-density water beneath the higher-density shelf water. Thus the frontal region becomes convectively unstable and overturns rapidly. The effect of this is to shift the position of the front further offshore. Numerical modeling studies have suggested that this continual offshore movement of the front is halted as a result of the vertical shear in the along-shore buoyancy current. As the water deepens, this vertical shear results in a reduction of the offshore bottom flow. Eventually the offshore flow is reversed, so that low-density water is no longer transported underneath the shelf water, and the front becomes fixed at that particular isobath.

Mixing and Frontogenesis in ROFIs

Fronts associated with the lateral buoyancy flux from rivers are affected by the amount of vertical mixing. In regions of freshwater influence (ROFIs) the modulation of tidal mixing over the spring–neap cycle has a dramatic effect on frontal dynamics. Strong vertical mixing at spring tides results in a vertically mixed water column, with salinity increasing offshore and often only a weak horizontal front. As the mixing then decreases toward neap tides, a point is reached when the vertical homogeneity of the water column cannot be maintained against the tendency for the low-density coastal water to flow offshore above the denser shelf water. This surface density-driven offshore current then rapidly establishes vertical stratification, with the offshore progression eventually being halted by the Earth's rotation.

Fluid dynamics experiments have shown that such a relaxation of the initially vertically mixed density structure will produce a strong front within any nonlinear region of the initial horizontal density gradient. Furthermore, a periodic modulation of the mixing about the level required to prevent this

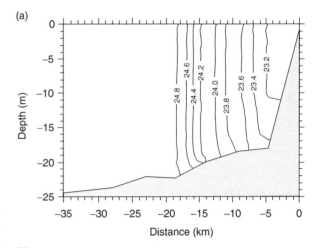

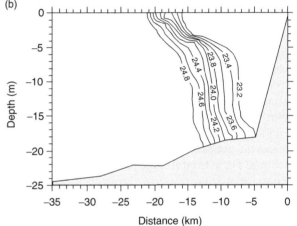

Figure 2 Density sections (σ_t, kg m^{-3}), normal to the coastline through the Rhine outflow. (a) Spring tide section, showing vertically mixed water with a freshwater-induced horizontal density gradient. (b) Neap tide section, showing the relaxation of the horizontal gradient and stratification caused by the reduction in mixing. Reproduced from Souza AJ and Simpson JH (1997) Controls on stratification in the RHINE ROFI system. *Journal of Marine Systems* 12: 311–323, with permission from Elsevier.

frontogenesis will result in a similar periodic variation in the density-driven mass flux. Spring–neap control of frontogenesis has been observed in a number of shelf seas; for instance, Liverpool Bay (eastern Irish Sea), the Rhine outflow (southern North Sea), and Spencer Gulf (South Australia). The physical switching between vertically mixed and stratified conditions is well established (**Figure 2**), though the biological responses within these dynamic environments are yet to be determined.

Tidal Mixing Fronts in Shelf Seas

In summer, away from sources of fresh water, temperate shelf seas are partitioned into thermally

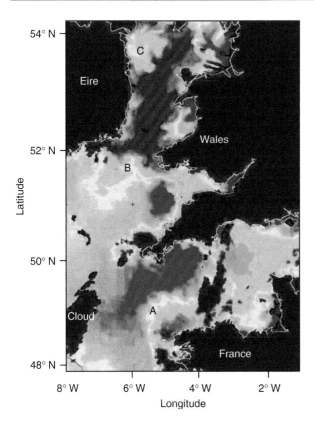

Figure 3 SST image from the advanced very high resolution radiometer (AVHRR). The image was taken at 04.19 GMT on 12 July 1999. Violet/blue represents a temperature of 13–14 °C, and shows regions of shelf sea that are vertically mixed. Red represents 18–19 °C, indicating the surface temperature of strongly stratified water. Green/yellow represents 16–17 °C, and shows the regions of weak stratification at the tidal mixing fronts. The regions of strong horizontal temperature gradient separating the mixed and stratified areas are the tidal mixing fronts. A, Ushant front; B, Celtic Sea front; C, Western Irish Sea front. Image courtesy of the Dundee Satellite Receiving Station, and the Remote Sensing Group, Plymouth Marine Laboratory.

stratified and vertically well-mixed regions. Such partitioning is clearly visible in satellite remote sensing images of sea surface temperature (SST; see **Figure 3**). Warm SST indicates the temperature of the surface mixed layer of a stratified water column, while cool SST shows the temperature of the entire, vertically homogeneous water column. The transition region between these stratified and well-mixed regions, with horizontal temperature gradients of typically $1\,°C\,km^{-1}$, are the shelf sea tidal mixing fronts.

Physical Control of Fronts and h/u^3

The suggestion that the intensity of tidal mixing was responsible for controlling the vertical structure of shelf seas was first made by Bigelow in the late 1920s, with reference to the variations in vertical temperature structure on and off Georges Bank. The first quantitative link between shelf sea fronts and tidal mixing was made by Simpson and Hunter in 1974. Surface heating, which is absorbed rapidly within the upper few meters of the ocean, acts to stabilize the water column by expanding the near-surface water and thus reducing its density. Friction between tidal currents and the seabed generates turbulence. Most of this turbulence is dissipated as heat, but a small fraction of it (typically 0.3%) is available for working against the thermal stratification near the sea surface. This seemingly low conversion rate of turbulence to mixing arises because turbulence is dissipated very close to where it was generated (the 'local equilibrium hypothesis'). Most of the vertical current shear (and hence turbulence production and dissipation) in a tidal flow is close to the seabed. Current shear higher in the water column near the thermocline (where turbulence can work against stratification) is much weaker, leading to a low overall efficiency.

Thus, there is a competition between the rate at which the water column is being stratified by the surface heating, and the ability of the tidal turbulence to erode and prevent stratification. If the magnitude of the heating component exceeds that of the tidal mixing term, then the water column will stratify. Alternatively, a stronger tidal current, and therefore more mixing, results in a situation where the heat input is continuously being well distributed through the entire depth and the water column is kept vertically mixed. A shelf sea front marks the narrow transition between these two conditions, with equal contributions from the heating and tidal mixing. This simple analysis led Simpson and Hunter to predict that tidal mixing fronts should follow lines of a critical value of h/u^3, with h (m) the total water depth and u (m s^{-1}) a measure of the amplitude of the tidal currents. Subsequent analysis of satellite SST images in comparison with maps of h/u^3 confirmed the remarkable power of this simple theory: shelf sea front positions are controlled by a local balance between the vertical physical processes of tidal mixing and sea surface heating (**Figure 4**).

Modifications to h/u^3

A prediction of the Simpson–Hunter h/u^3 hypothesis is that a shelf sea front should change position periodically with the spring–neap tidal cycle, due to the fortnightly variation in tidal currents (**Figure 5**).

In NW European shelf seas, spring tidal current amplitudes are typically twice those at neap tides. However, predicting the horizontal displacement of a shelf sea front using h/u^3_{Springs} and h/u^3_{Neaps} leads to

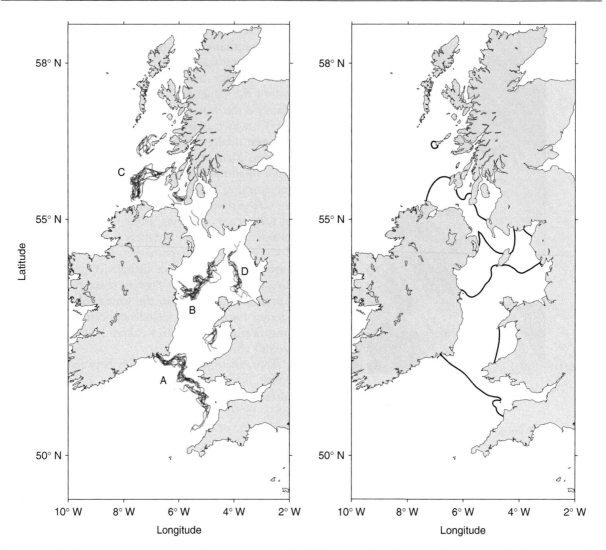

Figure 4 (Left) Mean positions of tidal mixing fronts observed in SST images, May 1978. A, Celtic Sea front; B, Western Irish Sea front; C, Islay front. (Right) Contours of $\log_{10}(h/u^3) = 2.7$. Note the correspondence between fronts A, B, and C, and the contours of constant h/u^3. Front D is caused by freshwater inputs from the estuaries of NW England, and so does not conform to the h/u^3 hypothesis. Adapted from Simpson JH and James ID (1986) Coastal and estuarine fronts. In: Mooers CNK (ed.) *Baroclinic Processes on Continental Shelves*, pp. 63–93. Washington, DC: American Geophysical Union. Courtesy of the American Geophysical Union.

a substantial over-estimate, typically suggesting a transition distance of 40–50 km, compared to satellite-derived observations of only 2–4 km. Two modifications to the theory were subsequently made. First, as tidal turbulence increases from neap to spring tides, the mixing not only has to counteract the instantaneous heat supply, but it must also break down the existing stratification that had developed as a result of the previous neap tide. Incorporating this behavior into the theory reduced the amplitude of the adjustment region to 10–20 km. Second, stratification inhibits vertical mixing, so the mixing efficiency would be expected to be lower as the existing stratification was being eroded. Simpson and Bowers used a simple parametrization linking mixing efficiency to the strength of the stratification, and showed that the

predicted spring–neap adjustment was then similar to that observed. This feedback is also responsible for constraining the frontal transition to a narrow range of h/u^3. The use of a turbulence closure model, providing a less arbitrary link between stability and mixing, has provided further confirmation of the need to include variable mixing efficiency.

The only source of mixing accounted for in the h/u^3 theory is tidal friction with the seabed, and the success of the theory in NW European shelf seas is arguably a result of the dominance of the tides in these regions. A better prediction of frontal position can be made by incorporating wind-driven mixing, so that the competition becomes one of surface heating versus the sum of tidal + wind mixing. Again, only a small fraction of the wind-driven

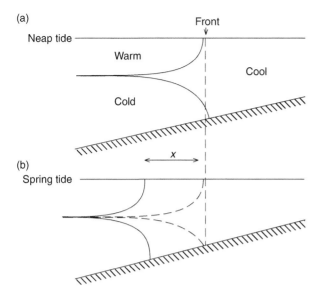

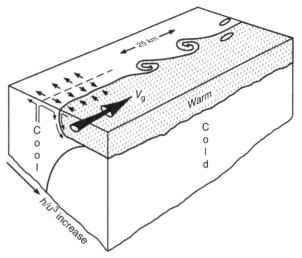

Figure 5 Schematic illustration of the adjustment of a tidal mixing front as a consequence of the spring–neap variation of tidal currents. At neap tides (a), the weaker tidal mixing allows stratification to develop in shallower water. At spring tides (b), the stronger tidal currents remix the shallow stratification. The adjustment distance, x, is typically 2–4 km.

Figure 6 Pattern of circulation at a tidal mixing front. V_g is an along-front surface geostrophic jet, typically about $10 \, \mathrm{cm \, s^{-1}}$. There is a surface current convergence at the front, followed by downwelling, with a compensatory upwelling and surface divergence in the mixed region. Baroclinic eddies develop along the front, with typical wavelengths of around 25 km. Adapted from Simpson JH and James ID (1986) Coastal and estuarine fronts. In: Mooers CNK (ed.) *Baroclinic Processes on Continental Shelves*, pp. 63–93. Washington, DC: American Geophysical Union. Courtesy of the American Geophysical Union.

turbulence is available to work against the stratification, about 2–3%. This is significantly larger than the tidal mixing efficiency because the thermocline is generally nearer to the sea surface than the seabed, and hence in a region of wind-driven current shear.

A debate arose in the late 1980s concerning the validity of the h/u^3 theory. Loder and Greenberg, and subsequently Stigebrandt, put forward the alternative hypothesis based on a more realistic description of tidal turbulence that includes the effect of the bottom rotational boundary layer as a control on the vertical extent of tidal turbulence away from the seabed. The position of the shelf sea front would, in this theory, simply reflect the position at which the tidal boundary layer was thick enough to reach over the entire depth, and that fronts should follow a critical value of h/u. In particular, Soulsby noted that in temperate latitudes the similar values of Coriolis and tidal frequencies suggests that there should be a very significant rotational constraint on vertical mixing as the tidal currents become cyclonically polarized.

Observations of frontal positions were not precise enough to determine which of the two theories was correct. However, use of a numerical model has showed that both mechanisms contributed. For anticyclonically polarized tidal currents boundary layer limitation is not a significant factor, and the frontal position is well described by the h/u^3 theory.

As currents become more cyclonic the vertical limitation of turbulence due to the reducing thickness of the boundary layer does alter the frontal position away from that predicted using h/u^3, but by less than predicted using the boundary layer theory alone.

Circulation at Shelf Sea Fronts

The density gradients associated with shelf sea fronts drive a weak residual circulation, superimposed on the dominant tidal flows (**Figure 6**). A surface convergence of flow at the front often leads to an accumulation of buoyant debris. This can form a clear visual indicator of a front. The convergence is associated with a downwelling, predicted by models to be around $4 \, \mathrm{cm \, s^{-1}}$. On the stratified side of the front a surface, geostrophic jet is predicted to flow parallel to the front. Models have predicted this flow to be of the order of $10 \, \mathrm{cm \, s^{-1}}$. Direct observations of such flows against the background of strong tidal currents is difficult, but both drogued buoys and high-frequency radar have been used successfully to observe along-front speeds of $10–15 \, \mathrm{cm \, s^{-1}}$, highlighting the dominance of the heating–stirring competition in these regions. These frontal jets are prone to baroclinic instability, with meanders forming along the front, growing, and eventually producing baroclinic eddies that transfer water between the two sides of the front.

Biological Implications

The physical structure of a shelf sea front controls associated biochemical gradients. From the mid-1970s, alongside the physical oceanographic studies of fronts, it was recognized that enhanced levels of chlorophyll (phytoplankton biomass) were often seen in the frontal surface water (**Figure 7(a)**). The recent availability of satellite remote sensing of surface chlorophyll (in particular the Sea-viewing Wide Field-of-view Sensor (SeaWIFS)) now allows dramatic evidence of these frontal accumulations of phytoplankton (**Figure 7(b)**). It is conceivable that the convergence of flow at a front could lead to enhancement of surface chlorophyll, by concentrating the biomass from the mixed and stratified water on either side. However, the spatial extent of the observed frontal chlorophyll (~ 1–10 km) is typically at least an order of magnitude greater than the horizontal extent of the convergence region ($\sim O(100$ m$)$). More recently, at the Georges Bank frontal system, the enhanced frontal chlorophyll has been observed directly associated with an increase in rates of primary production, compared to the waters on either side of the front. Thus, it appears that locally enhanced concentrations of frontal phytoplankton biomass is a result of locally increased production, and so requires some source of nitrate to be mixed into the region.

For primary production the shelf sea front marks the transition between a nutrient-replete, but light-limited, environment, and a stratified water column with a well-lit but nutrient-deficient surface layer (**Figure 8**). Highest nutrient levels are usually found in the bottom mixed layer on the stratified side of the front, due to negligible utilization and the contribution from detritus sinking down from the surface layer. Enhanced primary production in the frontal surface waters requires a mechanism to transport nutrients into the region, from the deep, high-nutrient water (vertical nutrient flux), and/or from the moderate nutrient-containing waters on the mixed side of the front (horizontal nutrient flux).

Four supply mechanisms have been suggested (**Figure 8**). First, the surface outcropping of the front is a region of gradually reducing vertical stratification, and thus a region where the inhibition of vertical mixing is reduced. The increased turbulent flux of nutrients will be available for surface primary production, as long as the residence time of the phytoplankton cells in the photic zone is still sufficient to allow net growth. Second, the spring–neap adjustment of a front's position results in the 2–4 km adjustment region undergoing a fortnightly mixing–stratification cycle. Thus, toward spring tides the region becomes vertically mixed and replenished with nutrients throughout the water column, and as the water restratifies toward neap tides the new surface nutrients become available for primary production. The predicted fortnightly pulses in surface frontal biomass have been reported at the Ushant shelf sea front, in the Western English Channel. A third nutrient supply mechanism is via baroclinic eddies that will transfer pools of water from the mixed side into the stratified side, though at the cost

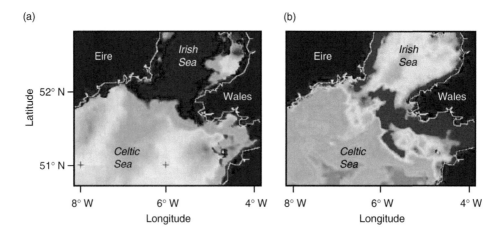

Figure 7 (a) SST image of the Celtic Sea front, 12 July 1999. (b) Sea surface chlorophyll concentration on 12 July in the same region, derived from the SeaWIFS sensor on NASA's SeaStar satellite. The mixed water of the Irish Sea is associated with chlorophyll concentrations of 1–2 mg m^{-3}. The strongly stratified water in the Celtic Sea has surface chlorophyll concentrations of less than 0.5 mg m^{-3}. At the front there is a clear signature of enhanced chlorophyll concentration, reaching about 5 mg m^{-3}. Image courtesy of the Dundee Satellite Receiving Station, and the Remote Sensing Group, Plymouth Marine Laboratory.

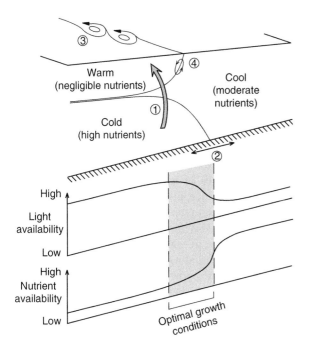

Figure 8 Nutrient supply mechanisms that are thought to fuel primary production at a tidal mixing front. (1) Vertical turbulent nutrient flux through the weaker stratification at the front. (2) The spring–neap adjustment of the front, causing a fortnightly replenishment of surface nutrients within the adjustment region. (3) Baroclinic eddy shedding along the front, transferring nutrient-rich water from the mixed side of the front into the stratified side. (4) A weak cross-frontal circulation caused by friction between the residual flows within the front (see **Figure 6**). In the surface layer on the stratified side of the front, the algae receive plenty of light but are prevented from growing because new nutrients cannot be supplied through the strong thermocline. On the mixed side of the front, nutrients are plentiful but growth is limited by a lack of light as the algae are mixed throughout the entire depth of the water column. The problem of lack of light on the mixed side is compounded by tidal resuspension of bed sediments, acting to increase the opacity of the water. As the transition zone between these two extremes, the front provides suitable conditions for algal growth. Processes (1) and (2) are thought to be capable of supplying about 80% of the nutrient requirements at a typical front.

of a similar flux of water containing phytoplankton in the opposite direction. Finally, weak transverse circulation associated with frictional boundary layers at the front will transfer water from the mixed side of the front into the surface frontal water.

Shelf Slope Fronts

Typical seabed slopes in shelf seas are about 0.5–0.8°. At the edge of the shelf seas this slope increases to 1.3–3.2°, a transition that occurs at a depth typically between 100 and 200 m. This region of steeper bathymetry, just seaward of the shelf edge, is the shelf slope, and is often associated with sharp

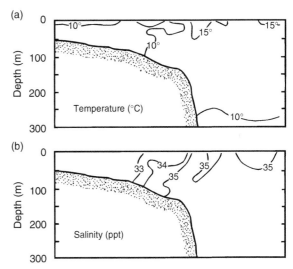

Figure 9 (a) Temperature and (b) salinity structure across the shelf break south of Cape Cod, eastern North America. Reproduced from Wright WR (1976) The limits of shelf water south of Cape Cod, 1941–1972. *Journal of Marine Research* 34: 1–14, with permission from the Sears Foundation for Marine Research.

horizontal gradients in temperature and/or salinity (e.g., see **Figure 9**). The difference in the water characteristics of shelf seas and the open ocean arises as the result of several mechanisms. Coastal and shelf waters tend to have lower salinity that the open ocean, due to the input of fresh water from land runoff. The freshwater input also alters the temperature of the shelf water, as does the seasonal heating/cooling cycle which will generate more pronounced temperature fluctuations within the shallow water. Offshore, the open ocean is part of a larger, basin-scale circulation that, for instance, brings much warmer water from equatorial regions past the shelf edge (e.g., consider the along-slope circulation of the world's western boundary currents). There is often a marked seasonality in the form of the shelf break front. Shelf waters can be more buoyant than oceanic waters during summer, but surface cooling in winter can reverse this to leave a denser water mass on the shelf that has the potential for cascading off the shelf and down the shelf slope.

The Position of a Shelf Slope Front

While the reasons for the contrast in water characteristics are straightforward, an explanation is required concerning why these differences between shelf and oceanic waters are maintained across such sharp fronts at the shelf slope. This is not as straightforward as for the case of the tidal mixing front. The limited number of processes governing the vertical structure of shelf seas resulted in a testable prediction for the position of a tidal mixing front in terms of water depth and tidal current amplitude. At the shelf break there are a number of potential controlling factors on a front's position, and a corresponding difficulty in producing an unambiguous, testable hypothesis. Numerical modeling provides the best technique for investigating frontal dynamics, allowing simultaneous consideration of several physical processes. However, a major problem with the assessment of any description of controls on shelf slope fronts is that there are considerable logistic difficulties in collecting current and scalar observations of sufficient quality and resolution to compare with the model outputs. The following arguments are based on both analytical and numerical models of shelf slope fronts.

A fundamental dynamic constraint on the exchange of water masses across the shelf slope lies with the geostrophic behavior of the oceanic flows. Geostrophic currents cannot cross steep bathymetry. Instead, they are forced by the Earth's rotation to flow along isobaths, parallel to the topography of the shelf slope and shelf edge. Both the oceanic flow seaward of the shelf edge, and the buoyancy-driven flows of shelf water close to the shelf edge, behave geostrophically. This basic topographic constraint on these geostrophic flows often forms the basis of descriptions of shelf slope fronts (e.g., work by Csanady, Ou, and Hsueh). The breakdown of geostrophy in the bottom boundary layer of the frontal along-slope flow is thought to be critical in controlling cross-slope transports and determining the stable position of the front (e.g., work by Chapman and Lentz).

Such gravitational relaxation of the horizontal density structure across the shelf slope only explains the formation of the front. The resulting strong along-slope flows and current shear suggest that the frontal signature should be rapidly mixed and dissipated, and yet observations clearly show that the fronts exist for prolonged periods. This implies that the dynamics of the fronts must also act to maintain frontal structure, in addition to causing its initial formation. A suggestion by Ou is that the front can be maintained, paradoxically, by the action of wind stress at the sea surface. This wind mixing generates a surface mixed layer, which still contains a cross-shelf horizontal density gradient and so continues to relax under gravity and feed the along-slope current. Chapman and Lentz have shown how a surface to bottom slope front is formed and maintained when the shelf water has the lower density. There the offshore flux of shelf water in the bottom Ekman layer of the along-slope flow drives local convective mixing and pushes the front offshore. As the front moves into deeper water, vertical shear in the along-slope flow results in cessation and eventual reversal of the nearbed cross-shore flow, thus halting the frontal movement and setting the stable position of the slope front.

The above mechanisms for frontal formation and maintenance explicitly use a cross-shore density gradient as the pivotal dynamical process. However, it has been noted (e.g., by Chapman), that in the Middle Atlantic Bight in summer the combined frontal structures of temperature and salinity compensate to produce no horizontal density gradient. In other words, a shelf slope front can exist in the scalar fields without any apparent horizontal density structure to maintain them. Chapman, again utilizing a numerical model, showed that such a situation could be supported if there is a strong alongshore flow on the shelf and a distinct shelf break. Friction with the seabed in the shallower shelf water causes a cross-shelf component of the flow. Above the shelf slope, in deeper water, the effect of friction is reduced, and so there is a convergence of the cross-shelf flow close to the shelf break. The existence of the temperature and/or salinity front is then dependent on the relative contributions of advection and diffusion. Seaward of the shelf edge diffusion is the dominant process, smoothing out any horizontal gradients. The convergence at the shelf edge concentrates the cross-shelf scalar gradients into a front, and the dominance of advection moves this structure along the shelf edge faster than diffusive processes can erode it. Thus, the front can be visible along several hundred kilometers of the Middle Atlantic Bight. For both scalar and density fronts it seems that the breakdown of geostrophy in the bottom Ekman layer of the along-slope current is a key aspect of frontal formation and positioning.

Implications of Shelf Slope Fronts

As with the tidal mixing fronts, shelf slope fronts in summer are often associated with concentrations of relatively high chlorophyll biomass, compared to the oceanic and shelf surface waters on either side. Fundamentally, this is again likely to be due to the diffusion of bottom water nutrients through the weaker stratification just at the surface front. Evidence from some shelf edge regions indicates the areas to be influenced by energetic internal waves on the thermocline, driven by the dissipation of the internal tidal wave (which is itself generated on the steep slope bathymetry seaward of the shelf edge). There has been some suggestion that secondary production can be more clearly linked to the primary production at shelf slope fronts than at shelf sea tidal mixing fronts, due to the temporal variability of the tidal fronts (e.g., spring–neap adjustment). Certainly, many shelf slope regions are places of intense fishing activity.

The shelf slope is recognized as a key region to the global cycling of carbon. The shelf seas and slope areas are highly productive, and thus have a high capacity for uptake of atmospheric carbon. Atmospheric carbon is drawn into the ocean as the result of algal growth extracting carbon from the seawater. The fate of some of this carbon uptake is to sink when the algae die, and become buried in the shelf and slope sediments. This flux of carbon to the seabed is an important carbon removal process, with shelf sea and slope regions currently thought to be responsible for up to 90% of the global oceanic removal of organic carbon. Thus, one of the important questions in oceanography concerns the transfer of water across the shelf edge, between the shelf seas and the open ocean. This transfer controls both the rate at which carbon is transferred to the open ocean from the shelf seas, and the rate at which new nutrients from the slope waters are supplied to the shelf waters ready to fuel new carbon uptake. The breakdown of slope fronts when the shelf water is denser than the ocean is of increasing interest in this context of cross-slope transports. Intense cooling or evaporation on the shelf can lead to a dynamically unstable slope front, with a downslope cascade of near-bed shelf water potentially triggered by, for instance, the bottom Ekman layer of the along-slope flow. A recent catalog of such cascades (Ivanov and co-workers) has suggested a mean downslope flow of 0.05–$0.08 \times 10^6 \, \mathrm{m^3 \, s^{-1}}$ per $100 \, \mathrm{km}$ of shelf length during a cascade event (which could last a few months). Canyons at the shelf edge may act as conduits for such transports.

Summary

Locally enhanced regions of horizontal salinity, temperature, and density gradient occur across the coastal and shelf seas, driven by a variety of mechanisms. The dynamics controlling the structure and position of freshwater fronts and shelf sea tidal mixing fronts are relatively well understood. Freshwater fronts result from the relaxation of a horizontal density gradient, arrested by either the diversion of flow caused by the Earth's rotation, or by the interaction between near-bed cross-frontal flows and a sloping seabed. Tidal mixing fronts are controlled by the competition between the rate of supply of mixing energy (supplied by either tidal current stress against the seabed, or by wind stress against the sea surface) and the rate of stratification (produced by surface heating). For fronts at the shelf edge/slope region the change in the slope of the seabed must play a pivotal role, but the full dynamics controlling the fronts are less clear. This is partially due to the difficulty in collecting observations of sufficient resolution, in both time and space, against which to test hypotheses. Also, in particular contrast with the tidal mixing fronts, there appears to be no dominant process controlling these fronts, though cross-shore, near-bed flows arising from the bottom Ekman layer of the along-slope current are often suggested to play an important role.

All fronts are observed to be regions of enhanced surface primary production. The common feature causing this production is likely to be the reduced stability close to surface fronts allowing increased vertical turbulent mixing of nutrients into the well-lit surface water. Fronts close to the shelf edge, or other regions of steep bathymetry, have the additional feature of locally generated internal waves providing enhanced mixing across the shallowing pycnocline. This increased primary production is often seen to be associated with increases in zooplankton and larger fish, ultimately supporting populations of seabirds and providing an important fisheries resource for people.

There are still important questions that remain to be answered concerning the physics of fronts. For instance, direct measurements of turbulent mixing have only recently become possible, so the potential for horizontal gradients in rates of vertical turbulent exchange still needs to be addressed. Shelf slope fronts are perhaps the most lacking in terms of a coherent theory of their dynamics (assuming such a general approach is possible), and have particular questions related to cross-frontal transfers that still require attention. The link between the physics of fronts and the closely coupled biology and chemistry is perhaps the area of greatest research potential. The shelf edge and slope region is an area driving substantial development in numerical modeling techniques. The steep topography there has always been a serious challenge, requiring very high resolution in order to correctly simulate the shelf slope physics. Combined with the computer processing power available, this often places constraints on the size of region or length of time that can be modeled. The move toward unstructured, finite element grids (from the traditional structured, finite difference grids) is possibly the most important recent development in modeling transports between the open ocean and the shelf.

See also

Regional and Shelf Sea Models.

Further Reading

Chapman DC and Lentz SJ (1994) Trapping of a coastal density front by the bottom boundary layer. *Journal of Physical Oceanography* 24: 1464–1479.

Hill AE (1998) Buoyancy effects in coastal and shelf seas. In: Brink KH and Robinson AR (eds.) *The Sea*, vol. 10, pp. 21–62. New York: Wiley.

Holligan PM (1981) Biological implications of fronts on the northwest European continental shelf. *Philosophical Transactions of the Royal Society of London, A* 302: 547–562.

Ivanov VV, Shapiro GI, Huthnance JM, Aleynik DL, and Golovin PN (2004) Cascades of dense water around the world ocean. *Progress in Oceanography* 60(1): 47–98.

Mann KH and Lazier JRN (1996) *Dynamics of Marine Ecosystems*, 2nd edn. 394pp. New York: Blackwell Science.

Sharples J (2008) Potential impacts of the spring–neap tidal cycle on shelf sea primary production. *Journal of Plankton Research* 30(2): 183–197.

Sharples J and Holligan PM (2006) Interdisciplinary processes in the Celtic seas. In: Robinson AR and Brink KH (eds.) *The Sea*, vol. 14B. Boston, MA: Harvard University Press.

Simpson JH (1998) Tidal processes in shelf seas. In: Brink KH and Robinson AR (eds.) *The Sea*, vol. 10, pp. 113–150. New York: Wiley.

Simpson JH and Hunter JR (1974) Fronts in the Irish Sea. *Nature* 250: 404–406.

Simpson JH and James ID (1986) Coastal and estuarine fronts. In: Mooers CNK (ed.) *Baroclinic Processes on Continental Shelves*, pp. 63–93. Washington, DC: American Geophysical Union.

Souza AJ and Simpson JH (1997) Controls on stratification in the Rhine ROFI system. *Journal of Marine Systems* 12: 311–323.

Wright WR (1976) The limits of shelf water south of Cape Cod, 1941–1972. *Journal of Marine Research* 34: 1–14.

Relevant Websites

http://amcg.ese.ic.ac.uk
    – ICOM, Imperial College Ocean Modelling, The Applied Modelling and Computation Group (AMCG).
http://www.sos.bangor.ac.uk
    – Physical Oceanography Department, Bangor University.
http://www.pol.ac.uk
    – Proudman Oceanographic Laboratory.
http://www.es.flinders.edu.au/~mattom/
    – Shelf and Coastal Oceanography at Matthias Tomczak's oceanography website, Flinders University.
http://www.whoi.edu
    – Woods Hole Physical Oceanography Group.

COASTAL CIRCULATION MODELS

F. E. Werner and B. O. Blanton, The University of North Carolina at Chapel Hill, Chapel Hill, NC, USA

Introduction

Coastal environments are among the most complex regions of the world's oceans. They are the transition zone between the open ocean and terrestrial watersheds with important and disparate spatial and temporal scales occurring in the physical as well as biogeochemical processes. Coastal oceans have three major components, the estuarine and nearshore areas, the continental shelf, and the continental slope. The water column depth ranges from areas where flooding and drying of topography occurs over a tidal cycle at the landward boundary, to depths of thousands of meters seaward of the shelf break. The offshore extent of coastal environments can range from a few kilometers (off the Peru/Chile coast), to hundreds of kilometers (over the European or Patagonian shelf). The topography in coastal oceans can be relatively featureless, or it can be complex and include river deltas, canyons, submerged banks, and sand ridges.

Coastlines and coastal oceans span the globe, from near the North Pole to the Antarctic, and thus are subject to a full range of climatic conditions. Circulation in coastal regions is forced locally (for example by winds, freshwater discharges, formation of ice) or remotely (through interactions with the neighboring deep ocean, terrestrial watersheds, or large-scale atmospheric disturbances). Resulting motions include tides, waves, mean currents, jets, plumes, eddies, fronts, instabilities, and mixing events. Vertical and horizontal spatial scales of motion range from centimeters to hundreds of kilometers, and timescales range from seconds to interannual and longer.

Coastal regions can be very productive biologically and they support the world's largest fisheries. These regions are also preferred as recreational and dwelling sites for our increasing human population. There is evidence suggesting that changes in the coastal environment, such as degradation in habitat, water quality, as well as changes in the structure and abundance of fisheries have resulted from increases in commercial and residential development, agriculture, livestock, soil, and sediment loss. Therefore, although natural phenomena shaped the coastal environment in the past, in the future it will be defined jointly by natural and anthropogenic processes. To understand the coastal oceans, predict their future states, and reduce the human impact on the region through management strategies, it is necessary to develop a quantitative understanding of the processes that define the state of the coastal ocean. Coastal circulation models are tools rooted in mathematical and computational science formalism that allow the integration of measurements, theory and computational capability in our attempt to quantify the above processes.

Governing Equations and State Variables

The starting point for coastal circulation models is modified versions of the Navier–Stokes equations derived for the study of classical fluid mechanics. The fundamental differences are the inclusion of the Coriolis force associated with the Earth's rotation, and the inclusion of hydrostatic and Boussinesq approximations appropriate for a thin layer of stratified fluid on a sphere for circulation features of hundreds of meters and larger. Certain smaller-scale motions, such as convection and mixing, are not admitted by these approximations as they may be possibly non-hydrostatic. Additional departures from descriptions of other fluid motions are the consideration of temperature and salinity as thermodynamic variables, and a nonlinear equation of state.

Coastal ocean domains are subsets of the global ocean basins and are typically defined by a solid wall (landward) boundary and open (wet) boundaries which connect the region of interest to neighboring bodies of water. Islands inside the model domains are also considered as solid wall boundaries. The open boundaries are generally of two types: offshore boundaries along which the coastal domain exchanges information with the neighboring deep ocean, and cross-shelf boundaries along which the coastal domain receives and/or radiates information to regions up- or downstream of the study site. The sea surface and the bottom complete the definition of the model domain. With the model domain defined, solution of the governing equations is sought subject to specified initial and boundary conditions.

The simplest initial conditions specify the fluid to be at rest (no motion), and the sea level, temperature and salinity fields to be flat (no horizontal gradients). Boundary conditions are more problematic and must be specified on all model boundaries. They include stress conditions at the free surface where atmospheric winds are imposed and input energy into the coastal ocean, and at the bottom where frictional forces extract energy from the overlying motions. Heating and cooling are imposed through prescribed flux conditions at the surface (where the coastal ocean is in contact with the atmosphere), or along the model's lateral boundaries (where it is in contact with offshore regions). Additional buoyancy fluxes due to variations in salinity are imposed through prescribed evaporation or precipitation fluxes at the surface, along the open boundaries as a result of exchanges with the offshore and up- and downstream regions, and from either point- or line-sources representing riverine or larger watershed (terrestrial) inputs.

The proper specification of boundary conditions is one of the more difficult aspects of modeling coastal circulation. Although for most cases the mathematical approaches are well established, the data required to quantitatively specify the mass and momentum fluxes across the model boundaries are lacking. As a result, boundary conditions are generally idealized. In practice boundary conditions are a mixture of imposed observed quantities, derived values from larger-domain models, and conditions that minimize the uncertainties associated with the artificial nature of open (wet) boundaries.

The solution of the governing equations consists of the time-history in three-dimensions of the velocity field, the temperature, salinity, and density, and additional derived quantities describing mixing rates of mass, momentum, and other tracers. Analytic solutions, also known as closed form solutions, are not possible except for highly idealized cases. For example, topography and forcing must be simplified and certain nonlinear processes must be ignored. Nevertheless, even in these limiting cases, analytic approaches are desirable as they include in a single statement the solutions' dependence on a wide range of parameters over the entire model domain, allowing for a comprehensive understanding of the interaction between the fundamental processes. In the early 1980s, analytic solutions developed for coastal-trapped waves presented a breakthrough in the study of remotely forced currents in coastal regions.

Numerical approaches offer the possibility of retaining full dynamic and topographic complexity in the study of coastal circulation. In these approaches, the governing equations are discretized in space and time and the resulting algebraic discrete equations are solved using methods of numerical analysis. Spatial discretization in the horizontal is accomplished using the finite difference method with either structured regular grids, or, in cases where the shape of the coastline is highly irregular, curvilinear grids (**Figure 1**). The latter allow some degree of resolution and geometric flexibility. The finite element method uses unstructured grids and allows for greatest flexibility in capturing spatial heterogeneity and geometric complexity (**Figure 2**). Horizontal spatial discretization is important as the convergence of the models' solution is dependent on proper refinement of topography and flow structures associated with the presence of stratification, among others. The relative merits of structured versus unstructured meshes has not been fully addressed by the research community.

In the vertical, three approaches are commonly used in computing the depth-dependent structure of the circulation. The 'z'-coordinate computes the vertical structure along constant geopotential levels, the 'sigma' or σ-coordinate is bottom- or terrain-following and the solution is computed at the same number of points in the vertical regardless of the water column depth, and the isopycnal or density-coordinates in which the vertical structure is computed along the time-dependent location of the density surfaces. As in the case of horizontal discretization approaches, there is no optimum choice of vertical coordinate systems.

There are many algorithmic questions and mathematical formulations that are still not fully answered. Assuming smoothness in forcing functions, initial data, topography, etc., the choice of discretization method to the solution should not matter in the continuum limit of the equations. However, errors arising from solving the approximate forms of the governing equations display different behaviors due to discretization methods, and in some cases these solutions are spurious. Higher-order discretization schemes that reduce the truncation error although not significantly increasing the computational effort continue to be investigated. Similarly, many physical processes are not well understood, such as vertical mixing near the free surface, flow instabilities and horizontal mixing rates. The scales of these processes are frequently too small to be resolved in models and it is necessary to represent them by what is often 'ad hoc' parametrization. Thus, the development of coastal circulation models continues to be a specialized undertaking, with several approaches being developed by teams of investigators worldwide.

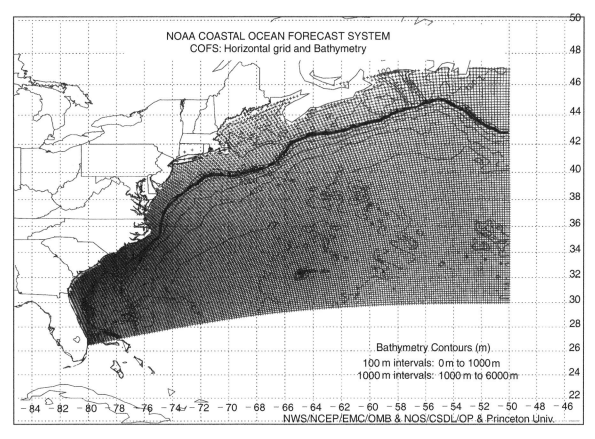

Figure 1 National Oceanic and Atmospheric Administration's Coastal Ocean Forecast System (COFS) curvilinear finite difference grid. Provided courtesy of National Weather Service's Environmental Modeling Center.

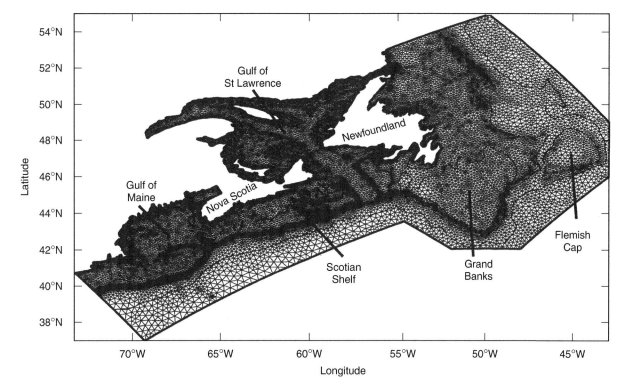

Figure 2 Northwest Atlantic shelf finite element domain. Flexibility of method increases resolution near sharp bathymetric gradients at the shelf break, deep channels, submerged banks, and along the coast.

However, the advent of significant computational capabilities (readily accessible on present-day desktop and laptop computers) has enabled coastal ocean models to become increasingly complex, and are now based on the fully stratified, nonlinear equations of motion. These advances coupled with the importance and interest in understanding coastal ocean processes has resulted in expansive growth and applications in some of the areas discussed next.

Applications

Applications of coastal circulation models can be broadly classified as process studies or regional studies. Process studies seek to identify the fundamental physical mechanisms responsible for observed features of the coastal ocean by idealizing complications of irregular shoreline geometry, time-dependent stochastic boundary forcing, and possibly simplifying the governing equations. Typically, these models retain effects such as the earth's rotation, idealized stratification and topography, idealized boundary conditions of heat flux and wind stress, and simplified turbulence closure. Early studies in the 1970s and 1980s focused on understanding large-scale wind-forced response of coastal regions including upwelling, and the nonlinear propagation of tides. Recently, with the increase in computing capabilities and improved mathematical formulations, the spatial and temporal resolution of process studies has also increased. The result has been a greater understanding of the detailed structure of phenomena such as the interaction of coastally trapped waves with bathymetric features and irregular coastlines (e.g., canyons, ridges, and capes), the generation of instabilities in the currents and formation of upwelling filaments, the formation of temperature and/or salinity fronts along the continental shelf break, and of river plume dynamics in the near-shore coastal and estuarine regions.

Regional coastal circulation studies attempt to include as much realism as possible into the numerical simulation of a specific region, including geometry and boundary conditions. These studies include the estimation of climatological circulation and tracer (e.g., temperature and salinity) distribution, fine resolution tidal simulations, storm surge analysis and prediction, transport of dissolved and particulate matter, coastal ocean prediction and forecasting, and coupled effects between estuaries, tidal inlets and the coastal ocean.

Sea Level

Many of the world's coastal regions are affected by large variations in sea level. The ability of coastal circulation models to accurately simulate coastal sea level has enabled the quantitative study of the impact of large tidal amplitudes and storm surges on low-lying coastal areas. Regional coastal sea-level models have become *de facto* components of emergency management systems in areas sensitive to sea-level variations. Robust and very good predictions of sea level can be obtained with horizontal two-dimensional models provided that accurate predictions of the surface wind field are available. The simplification from fully three-dimensional approaches is accomplished by averaging the governing equations along the vertical coordinate. Usually these models will also ignore effects of stratification, but include very high-resolution bottom topography and coastline features. The simplifications allow for significant speed-up of computations, which is necessary when issuing real-time forecasts. The rise of sea level associated with the passage of storm systems is known as storm surge. Accurate prediction of sea level during a storm surge (**Figure 3**) and its timing relative to the time of high tide are essential for the protection of property and life in low-lying coastal areas, such as the Dutch coast, the Gulf of Mexico, the east coast of the United States, and the southern Asian continent.

Engineering

Coastal circulation models are also used in engineering applications, such as in the design of ports, offshore platforms, and in the dredging of shipping lanes. These applications usually deal with the impact of the circulation on the structure being built. However, there are instances where the structure

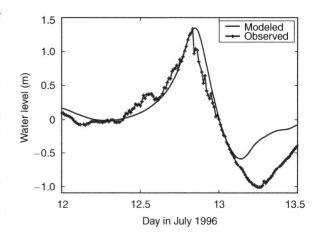

Figure 3 Observed and modeled water levels of New Bern, North Carolina, during the passage of Hurricane Bertha in July 1996. The modeled water level was computed using a two-dimensional circulation model, and accurately captures the magnitude and timing of the sea level response to the storm.

itself can have a significant effect on the coastal ocean, and circulation models are used in the quantitative assessment of its impact. The Bay of Fundy in the Gulf of Maine is known for its extreme (over 6 meter) tidal amplitudes, a result of the region's near-resonance with the principal lunar M_2 tide with period 12.4 h. The natural resonant period of the gulf-bay system is about 13.3 h. The large amplitude tide offers the opportunity to harness the tidal elevation and resulting potential energy to generate hydroelectric power by constructing dams. In 1987, a two-dimensional circulation model of the Gulf of Maine was used to investigate the tides, the effects of building tidal power plants, and their potential impact on the natural resonant period of the gulf-bay system. These studies showed that the tidal amplitude near the proposed barriers would decrease by about 25 cm due to the shortening of the bay's natural length and consequent decrease in the bay's resonant period. Furthermore, and perhaps somewhat unexpectedly the results also indicated that increases in tidal amplitude of 15–20 cm would occur in remote coastal areas, some of which are potentially sensitive to sea-level fluctuations and flooding. Predicted changes in circulation also suggested changes in sedimentation rates.

Coastal Ecosystems

The study of marine ecosystems requires that models of different systems be coupled to properly capture biological, geochemical, and hydrodynamic interactions across a wide range of temporal and spatial scales. Important biological processes are affected by transport mechanisms that can occur over hundreds of kilometers as well as turbulent mixing events that can occur on scales of several meters or less. Coastal circulation models have now achieved a level of sophistication and realism where new and significant opportunities for scientific progress in studying coupled physical–biological simulations are within reach. We are close to being able to construct spatially and temporally explicit models of the coastal physical environment, including the specification of velocity, hydrography, and turbulent fields, on scales relevant to biological processes. The investigation of ecosystem-level questions involving the role of hydrodynamics in determining the variability and regulation of planktonic and fish populations (**Figures 4** and **5**) are now being attempted. There are now many case studies that have coupled the growth and feeding environment of planktonic and larval fish species with coastal circulation models. Examples include: the study of retention, survival, and

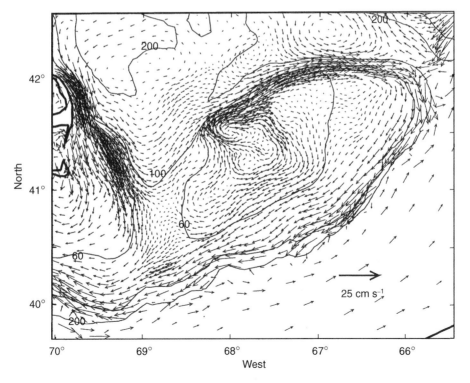

Figure 4 Modeled depth and time-averaged circulation on Georges Bank forced by tides, winds and upstream inflow. Adapted from Werner *et al* (1996).

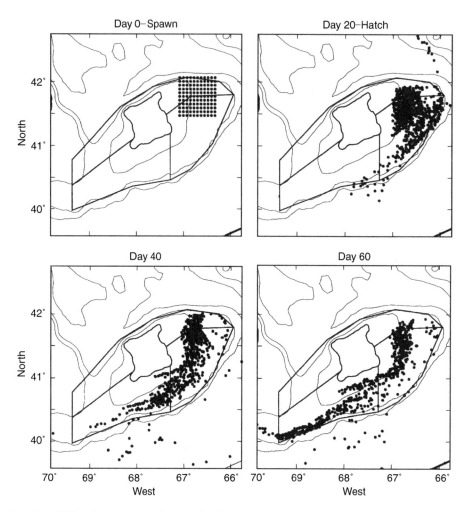

Figure 5 Simulated larval fish trajectories over Georges Bank at 20, 40, and 60 days post-spawn using flow fields from **Figure 4**. These trajectories are used to evaluate the on-bank retention versus off-bank loss. Adapted from Werner *et al.* (1996).

dispersal of larval cod, haddock and their prey in the Northwest Atlantic and North Sea; the transport of estuarine dependent fish from offshore coastal spawning regions to estuarine nursery habitats on the eastern US coast; the interannual recruitment variability of pollock in the Gulf of Alaska; and the dispersal of coral reef species. The development of management strategies used in the definition of marine sanctuaries or marine reserves can now look to circulation models for guidance in estimating population exchanges within and among neighboring coastal regions.

Operational Forecast Systems

A recent and evolving application of coastal circulation models is in operational coastal ocean prediction and forecasting systems. These systems are used to estimate in real-time the state of a particular region of the coastal ocean for the purposes of navigation, naval operations, search-and-rescue, oil spill impact assessment, or commercial and recreational fishing. The coastal circulation model is driven in part by forecasts of heat, moisture, and momentum from weather, tidal or large-scale ocean circulation models. However, to partially correct for erroneous (or imperfectly known) boundary and initial conditions and the resulting continuous accumulations of these errors, algorithms have been developed that allow assimilation of observed currents, water level, and/or hydrography within the domain and thereby improve model forecasts in real-time. The US National Oceanic and Atmospheric Administration's Coastal Ocean Forecast System (COFS) provides real-time forecasts of the coastal and open ocean state for the eastern US coast (**Figure 6**) by taking advantage of recent advances in coastal circulation models and observational systems. The coastal circulation model is forced at the surface by forecast surface flux fields of momentum, heat, and

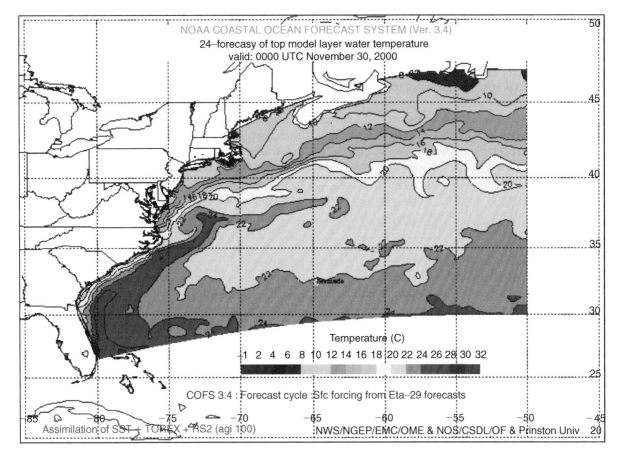

Figure 6 ECOFS 24-hour sea surface temperature forecast for 30 November 2000, computed on the grid in **Figure 1**. Note that a large portion of the deep water adjacent to the east coast shelf needs to be included in such a forecasting system. Provided courtesy of National Weather Service's Environmental Modeling Center.

moisture from a high-resolution weather forecast model. An assimilation system that incorporates both *in situ* and remotely sensed observations of surface and sysubsurface temperatures and sea-surface heights enables ECOFS to make relatively accurate 24-hour forecasts of Gulf Stream frontal position, water levels, three-dimensional currents, temperature, and salinity on a daily basis.

Sampling Design

Intense observational efforts focus on sampling physical and biogeochemical fields in the coastal ocean. The design of field sampling programs through Observational System Simulation Experiments (OSSEs) is a challenging modeling opportunity with extremely valuable results. Sampling strategies at sea can be difficult due to the evolving nature of the circulation and OSSEs provide a realistic site-specific simulation that can affect field protocols. Additionally, real-time limited-area forecast systems have been implemented on board research vessels to predict the transport of physical or biological tracers

at sea. Thus, using a coastal circulation forecasting system the likely path of the tracer of interest can be predicted and help researchers in the field develop appropriate sampling schemes.

Conclusions and Future Directions

The ocean science community is currently presented with unprecedented opportunities and advanced technologies for understanding and managing coastal ecosystems. Rapid advancement of computer resources, observational systems and instruments, and numerical techniques are converging to enable real-time coastal observation systems for coastal monitoring and marine forecasting. In the past two decades, a variety of models for simulating the coastal ocean have emerged as significant tools for investigating processes and mechanisms as well as regional coastal ecosystems and environmental questions and issues. The application of these models to almost any regionin the world represents a re-markable scientific achievement. The state of the art of coastal oceancirculation models and computer technology is such that a comprehensive and quantitative description of the hydrodynamics in a specific region can be obtained relatively easily by coastal oceanographers in general. Their application no longer requires expertise in numerical techniques and mathematics.

However, many fundamental research questions still remain. There are several areas of active investigation for coastal circulation models that include formal development issues as well as applications. As computational power increases, larger-scaleproblems requiring more memory and faster computer speed will enable higher resolution regional studies as well as faster longer-term integrations for the purposes of climate studies. Advanced numerical methods for discretizing the model domain in both thehorizontal and vertical are being developed, particularly regarding mass conservation and the algorithms that transport scalar properties of the fluid volume like salt and heat, as well as nonconservative tracers like oxygen and nutrients.

Advances in observational systems that include satellite and radar remote sensing, fixed instrumented platforms, remotely operated vehicles, and moored instruments, are currently being harnessed to provide as much near real-time information as possible to use in data assimilation schemes for oceanic numerical models. The modeling community in general is striving to provide forecasted global ocean circulation fields in nereal-time that resolve basin-scale to coastal-scale features. Coastal circulation models will play an important role in: (1) communicating the open-ocean information to coastal and near-shore regions; (2) providing extensions to the basin-scale models to regions that are typically underresolved by the larger-scale models; and (3) providing realistic cross-shelf fluxes of mass and momentum to the bain-scale models.

Operational forecasting systems are being developed for site-specific, limited-area predictions of the coastal ocean. *In situ* and remotesensed data are being assimilated by these systems, driven by forecasting results frommeteorological and basin-scale ocean models. Mesoscale weather models are being used to provide spatially dense estimates of surface flux parameters to the coastal circulation models but this coupling is largely one-way; the forecasted weather parameters affect the coastal hydrodynamic evolution. However, it is well known that the ocean affects the atmosphere. Observations have shown, for example, that the surface heat flux between the ocean and atmosphere over Gulf Stream waters significantly affects the development and evolution of extratropical cyclones that routinely pass along the eastern United States seaboard. Effective two-way coupling that communicates surface fluxes from the coastal ocean to overlying atmosphere in coupled coastal ocean and regional weather forecasting models is currently being developed and will provide a significant enhancement to regional meteorological forecasting skill.

The open-water boundaries of coastal circulation models require specification of either the velocity or the water level. For realistic regional simulations, there exists uncertainty in these boundary conditions related to the sparsity of observations on which to directly deduce them. The further development of schemes for assimilating observations into model integrations to provide optimal boundary conditions for forecasting is critical. Obtaining the open water boundary conditions from a larger basin-scale prediction model is another method for specifying open boundary conditions in operational limited-area coastal prediction models. This, in effect, generates the smaller domain boundary conditions from the larger model.

Recognizing that the entire ocean functions as a single unit from global to estuarine scales, the coupling of coastal- and basin-scale ocean models will represent a significant advance toward global ocean forecasting. Since the formulations of the coastal and basin models are usually quite different, this poses an unsolved question of the communication between the two models.

The ability of coastal circulation models to integrate the governing equations and boundary conditions for the coastal environment is a powerful tool for exploring both questions of process and mechanism and for addressing realistic regional problems that include forecasting of the coastal ocean analogous to atmospheric weather prediction. As the need to understand and address the growing list of environmental concerns accelerates, broad interdisciplinary efforts that couple models of different physical, biological, chemical, and geological systems will be critical in addressing these issues.

See also

Coastal Trapped Waves. Oil Pollution. Storm Surges. Tidal Energy. Tides.

Further Reading

Brink KH and Robinson AR (eds.) (1999) *The Sea*, vols 10 and 11. New York: John Wiley.

Crowder LB and Werner FE (eds.) (1999) Fisheries oceanography of the estuarine-dependent fishes in the South Atlantic Bight. *Fisheries Oceanography* 8: 242.

Gill AE (1982) *Atmosphere–Ocean Dynamics*. New York: Academic Press.

Greenberg DA (1987) Modeling Tidal Power. *Scientific American* November: 128–131.

Haidvogel DB and Beckmann A (1998) Numerical models of the coastal ocean. In: Brink KH and Robinson AR (eds.) *The Sea*, vol 10, pp. 457–482. New York: John Wiley.

Heaps NS (ed.) (1987) *Three-dimensional Coastal Ocean Models*. Washington, DC: American Geophysical Union.

Lynch DR and Davies AM (eds.) (1995) *Quantitative Skill Assessment for Coastal Ocean Models*. Washington, DC: American Geophysical Union.

Malanotte-Rizzoli P (ed.) (1996) *Modern Approaches to Data Assimilation in Ocean Modeling*. New York: Elsevier.

Mooers CNK (ed.) (1998) *Coastal Ocean Prediction*. Washington, DC: Amercian Geophysical Union.

Werner FE, Perry RI, Lough G, and Naimie CE (1996) Trophodynamic amd advective influences on Georges Bank larval cod and haddock. *Deep-Sea Research II* 43: 1793–1822.

Werner FE, Quinlan JA, Blanton BO, and Luettich RA Jr (1997) The role of hydrodynamics in explaining variability in fish populations. *Journal of Sea Research* 37: 195–212.

Westerink JJ, Luettich RA Jr, Baptista AM, Scheffner NW, and Farrar P (1992) Tide and storm surge predictions using a finite element model. *ASCE Journal of Hydraulic Engineering* 118: 1373–1390.

REGIONAL AND SHELF SEA MODELS

J. J. Walsh, University of South Florida,
St. Petersburg, FL, USA

Introduction

Simple biological models of nutrient (N), phytoplankton (P), and zooplankton (Z) state variables were formulated more than 50 years ago (Riley *et al.*, 1949), with minimal physics, to explore different facets of marine plankton dynamics. These N–P–Z models, usually in units of nitrogen or carbon are still used in coupled biophysical models of ocean basins, where computer constraints preclude the use of more complex ecological formulations of global biogeochemical budgets. At the regional scale (**Table 1**) of individual continental shelves (Steele, 1974; Walsh, 1988), however, pressing questions no longer focus on the amount of biomass of the total phytoplankton community, but are instead concerned with the functional types of algal species that may continue to support fisheries, generate anoxia, fix nitrogen, or form toxic red tides.

Regional models that presently address the outcome of such plankton competition must of course specify the rules of engagement among distinct functional groups of P, Z, and larval fish (F) living and dying within time-dependent physical (e.g., light, temperature, and current) and chemical (e.g., N) habitats. Successful ecological models are data-driven, distilling qualitative hypotheses and aliased field observations into simple analogs of the real world in a continuing cycle of model testing and revision. They are usually formulated as part of multidisciplinary field studies, in which the temporal and spatial distributions of the model's state variables, from water motion to plankton abundances, are measured to provide validation data. Here, examples of some state-of-the-art, complex regional biophysical models are drawn from specific field programs designed to validate them. Depending upon the questions asked of regional models, processes at smaller and larger time/space scales (**Table 1**) are variously ignored, parametrized, or specified as boundary conditions.

Statistical models elicit noncausal relationships among presumed independent variables with poorly known probability density functions. Deterministic simulation models in contrast assign cause and effect among the state variables and forcing functions described by a set of usually nonlinear ordinary or partial differential equations, whose solutions are obtained numerically. Within a fluid subjected to such external forcing, the vagaries of population changes of the embedded plankton, along measured spatial gradients, must be described in relation to both water motion and biotic processes.

For example, in a Lagrangian sense – following the motion of a parcel of fluid – the time dependence of larval fish (F) can be written simply as an ordinary differential equation (eqn [1]), where the total derivative is given by eqn [2].

$$\frac{dF}{dt} = (b - d)F \qquad [1]$$

$$\frac{d}{dt} = \frac{\partial}{\partial t} + u\frac{\partial}{\partial x} + v\frac{\partial}{\partial y} + w\frac{\partial}{\partial z} \qquad [2]$$

Mixing is ignored and biotic factors are just the linear birth, b, and death, d, rates of the fish. Because drogues have difficulty following plankton patches,

Table 1 Domains of ecological and physical processes within a hierarchy of coupled models

Model	Time domain	Space domain	Ecological focus	Physical focus
Small scale	Seconds	mm–cm	Cellular metabolism	Dissipation, nutrient supply
Turbulent	Minutes	dm–m	Photoadaptation, cell quotas	Waves, Langmuir cells
Local	Hours	m	Plankton migration, cell division	Tides, inertial oscillations
Regional	Days–weeks	km	Food web interactions, populations, resuspension events, nutrient depletion, grazing controls	Trapped waves, storms, upwelling, frontal currents
Basin	Months	degree (latitude)	Seasonal succession, mammal migration, carbon sequestration	Gyre flows, changes of mixed layer depths
Global	Years	Planetary	Stock fluctuations	Thermohaline circulation
Geological	Centuries	–	Evolution, sedimentation	Climate change

and current meters measure flows at a few fixed locations, it is usually more convenient to model circulation and the coupled biological fields in an Eulerian sense – over a grid mesh of spatial cells.

The underlying physical models (Csanady, 1982; Heaps, 1987; Mooers, 1998) have a rich history, since the development of calculus by Isaac Newton and Gottfried Leibnitz *c.* 1675, allowing expression of the rates of change, or derivatives, of properties over both time and space. After 1800, for example, Newton's second law of motion became transformed into the time (*t*)-dependent Navier–Stokes equations for the horizontal (*u*, *v*) and vertical (*w*) motions of a fluid, as a function of density (*ρ*), pressure (*p*), gravitational (*gρ/ρ*$_0$), where *ρ*$_0$ is a reference density and Coriolis (*fv*, −*fu*, 0) forces, in a Cartesian co-ordinate system (*x*, *y*, *z*) as eqns [3]–[5].

$$\frac{\partial u}{\partial t} + u\frac{\partial u}{\partial x} + v\frac{\partial u}{\partial y} + w\frac{\partial u}{\partial z} = \frac{\partial}{\partial x}\left(K_x\frac{\partial u}{\partial x}\right)$$
$$+ \frac{\partial}{\partial y}\left(K_y\frac{\partial u}{\partial y}\right) + \frac{\partial}{\partial z}\left(K_z\frac{\partial u}{\partial z}\right) - \frac{1}{\rho}\frac{\partial \varphi_e}{\partial x}$$
$$- \frac{1}{\rho}\frac{\partial \varphi_\rho}{\partial x} + fv \qquad [3]$$

$$\frac{\partial v}{\partial t} + u\frac{\partial v}{\partial x} + v\frac{\partial v}{\partial y} + w\frac{\partial v}{\partial z} = \frac{\partial}{\partial x}\left(K_x\frac{\partial v}{\partial x}\right)$$
$$+ \frac{\partial}{\partial y}\left(K_y\frac{\partial v}{\partial y}\right) + \frac{\partial}{\partial z}\left(K_z\frac{\partial v}{\partial z}\right) - \frac{1}{\rho}\frac{\partial \varphi_e}{\partial y}$$
$$- \frac{1}{\rho}\frac{\partial \varphi_\rho}{\partial y} - fu \qquad [4]$$

$$\frac{\partial w}{\partial t} + u\frac{\partial w}{\partial x} + v\frac{\partial w}{\partial y} + w\frac{\partial w}{\partial z} = \frac{\partial}{\partial x}\left(K_x\frac{\partial w}{\partial x}\right)$$
$$+ \frac{\partial}{\partial y}\left(K_y\frac{\partial w}{\partial y}\right) + \frac{\partial}{\partial z}\left(K_z\frac{\partial w}{\partial z}\right) - \frac{1}{\rho}\frac{\partial \varphi_\rho}{\partial z} - \frac{g\rho}{\rho_0} \qquad [5]$$

subject to appropriate boundary conditions at the surface (s), bottom (b), and sides of the ocean.

For example, the surface wind and bottom friction stresses, *τ*$_{s,b}$, are part of the boundary conditions for the third term of internal vertical Reynolds stresses, involving *K*$_z$ on the right side of [3]–[5]. No water moves across the land interfaces of the first two terms, representing horizontal turbulent mixing at smaller length scales than those of the flow components (*u*, *v*, *w*) over longer timescales. The last three terms on the left side of eqns [3]–[5] are thus the nonlinear advection of momentum in the horizontal and vertical directions, while the first one is the local time change.

In terms of other forces exerting acceleration of the fluid on the right side of eqns [3]–[5], their last terms represent the impact of Coriolis (*f*) and gravitational (*g*) effects. The fourth and fifth terms of eqns [3] and [4] are the respective pressure forces exerted by the piled-up mass of sea water (*φ*$_e$) (i.e., the barotropic force from slope of the sea surface, *e*, and the baroclinic one (*φ*$_\rho$) from the density field, *ρ*, which in turn is a function of temperature, (*T*), salinity (*S*), and pressure. At the surface of the sea, it is given simply by eqn [6], where *ρ*$_0$ is a reference density of pure water and *γ* is a polynomial function of *S* and *T*.

$$\rho = \rho_0 + \gamma(S, T) \qquad [6]$$

As in the case of momentum, the time-dependent spatial distributions of these conservative parameters of temperature and salinity are described by equations [7] and [8].

$$\frac{\partial T}{\partial t} + u\frac{\partial T}{\partial x} + v\frac{\partial T}{\partial y} + w\frac{\partial T}{\partial z} = \frac{\partial}{\partial x}\left(K_x\frac{\partial T}{\partial x}\right)$$
$$+ \frac{\partial}{\partial y}\left(K_y\frac{\partial T}{\partial y}\right) + \frac{\partial}{\partial z}\left(K_z\frac{\partial T}{\partial z}\right) \qquad [7]$$

$$\frac{\partial S}{\partial t} + u\frac{\partial S}{\partial x} + v\frac{\partial S}{\partial y} + w\frac{\partial S}{\partial z} = \frac{\partial}{\partial x}\left(K_x\frac{\partial S}{\partial x}\right)$$
$$+ \frac{\partial}{\partial y}\left(K_y\frac{\partial S}{\partial y}\right) + \frac{\partial}{\partial z}\left(K_z\frac{\partial S}{\partial z}\right) \qquad [8]$$

The different forcings of net incident radiation, latent/sensible heat losses, precipitation, and evaporation at the sea surface are boundary conditions for the last terms on the right side of eqns [7] and [8]. Depending upon the time and space scales of interest (**Table 1**), the coefficients *K*$_x$, *K*$_y$, and *K*$_z$ may represent either molecular diffusion, turbulence, viscosity, tidal stirring, or eddy mixing processes, while *u*, *v*, and *w* are obtained from solutions of eqns [3]–[5] and [9] (below).

Finally, to complete description of the physical habitat, conservation of momentum is invoked as the continuity equation [9] by assuming that sea water is an incompressible fluid.

$$\frac{\partial u}{\partial x} + \frac{\partial v}{\partial y} + \frac{\partial w}{\partial z} = 0 \qquad [9]$$

Thomas Malthus's consideration of exponential population growth in *c.* 1800 was placed in a marine context of eqn [1] during the 1920s, after recognition of the sources of variation of stock recruitment. Since then, quantitative description of

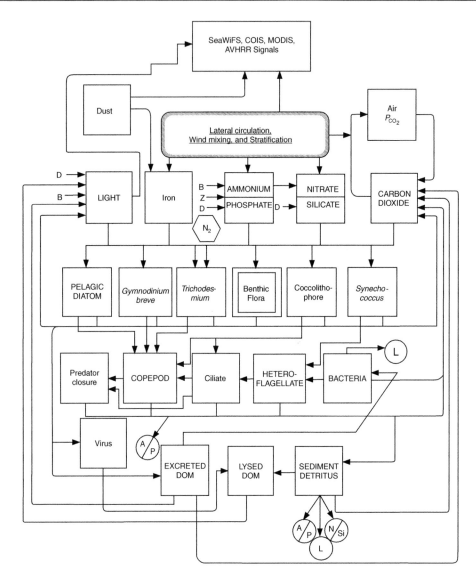

Figure 1 State variables of an ecologically complex numerical food web, coupled to a physical model of water circulation, for analysis of phytoplankton competition leading both to harmful algal blooms of *Gymnodinium breve* and to larval fish predators on the West Florida shelf.

the fluctuations of larval fish abundance have expanded to simple marine food webs (**Figure 1**) that support these vertebrates. State equation [10] for fish now has mixing and a nonlinear term for zooplankton prey, which are linked to N, P, bacteria (B), dissolved organic matter (D), and benthic microbiota (M) in eqns [11]–[16], where the advective and diffusive terms of eqns [10]–[16] are analogous to those of eqns [3]–[5].

$$\frac{\partial F}{\partial t} + u\frac{\partial F}{\partial x} + v\frac{\partial F}{\partial y} + w\frac{\partial F}{\partial z} = \frac{\partial}{\partial x}\left(K_x\frac{\partial F}{\partial x}\right)$$
$$+ \frac{\partial}{\partial y}\left(K_y\frac{\partial F}{\partial y}\right) + \frac{\partial}{\partial z}\left(K_z\frac{\partial F}{\partial z}\right) + (bZ - d)F \quad [10]$$

$$\frac{\partial Z}{\partial t} + u\frac{\partial Z}{\partial x} + v\frac{\partial Z}{\partial y} + w\frac{\partial Z}{\partial z} = \frac{\partial}{\partial x}\left(K_x\frac{\partial Z}{\partial x}\right)$$
$$+ \frac{\partial}{\partial y}\left(K_y\frac{\partial Z}{\partial x}\right) + \frac{\partial}{\partial z}\left(K_z\frac{\partial Z}{\partial z}\right)$$
$$+ (\alpha aP + cB - e - \varepsilon_1 - \delta_1 - bF)Z \quad [11]$$

$$\frac{\partial P}{\partial t} + u\frac{\partial P}{\partial x} + v\frac{\partial P}{\partial y} + w\frac{\partial P}{\partial z} = \frac{\partial}{\partial x}\left(K_x\frac{\partial P}{\partial x}\right)$$
$$+ \frac{\partial}{\partial y}\left(K_y\frac{\partial P}{\partial x}\right) + \frac{\partial}{\partial z}\left(K_z\frac{\partial P}{\partial z}\right)$$
$$+ (gN - \varepsilon_2 - \delta_2 - aZ)P + w_p\frac{\partial P}{\partial z} \quad [12]$$

$$\frac{\partial B}{\partial t} + u\frac{\partial B}{\partial x} + v\frac{\partial B}{\partial y} + w\frac{\partial B}{\partial z} = \frac{\partial}{\partial x}\left(K_x\frac{\partial B}{\partial x}\right)$$
$$+ \frac{\partial}{\partial y}\left(K_y\frac{\partial B}{\partial y}\right) + \frac{\partial}{\partial z}\left(K_z\frac{\partial B}{\partial Z}\right)$$
$$+ (bD - \delta_3 - cZ)B \qquad [13]$$

$$\frac{\partial N}{\partial t} + u\frac{\partial N}{\partial x} + v\frac{\partial N}{\partial y} + w\frac{\partial N}{\partial z} = \frac{\partial}{\partial x}\left(K_x\frac{\partial N}{\partial x}\right)$$
$$+ \frac{\partial}{\partial y}\left(K_y\frac{\partial N}{\partial y}\right) + \frac{\partial}{\partial z}\left(K_z\frac{\partial N}{\partial Z}\right) + \frac{\varepsilon_1}{\delta_1}Z$$
$$+ \delta_2 P + \delta_3 B + \frac{\partial}{\partial x}\left(K_x\frac{\partial M}{\partial z}\right) - gNP \qquad [14]$$

$$\frac{\partial D}{\partial t} + u\frac{\partial D}{\partial x} + v\frac{\partial D}{\partial y} + w\frac{\partial D}{\partial z} = \frac{\partial}{\partial x}\left(K_x\frac{\partial D}{\partial x}\right)$$
$$+ \frac{\partial}{\partial y}\left(K_y\frac{\partial D}{\partial y}\right) + \frac{\partial}{\partial z}\left(K_z\frac{\partial D}{\partial z}\right)$$
$$+ (1 - \alpha aP) + \varepsilon_2 P + \frac{\partial}{\partial z}\left(K_z\frac{\partial M}{\partial z}\right) - bDB \quad [15]$$

$$\frac{\partial M}{\partial t} + w_z\frac{\partial eZ}{\partial z} + w_p\frac{\partial P}{\partial z} - \frac{\partial}{\partial z}\left(K_z\frac{\partial M}{\partial z}\right) \qquad [16]$$

Thus far, F has represented various larval fish species subjected to the vagaries of the currents, of their supply of zooplankton prey (bZF), and of losses (dF) to predators. The simulated Z are a more diverse zooplankton group of protozoans, salps, copepods, and euphausiids, however, while P are numerous functional groups of light-regulated phytoplankton, e.g., the silicate-requiring diatoms, nitrogen-fixing diazotrophs, calcium-using coccolithophores, sulfur-releasing prymnesiophytes, and heavily-grazed chlorophytes, cyanophytes, and cryptophytes of the microbial food web.

Accordingly, the types of inorganic nutrients that have been modeled with eqn [14] are total carbon dioxide, nitrate, ammonium, dinitrogen gas, iron, silicate, calcium, and phosphate, while the dissolved organic matter (DOM) pools have been those of carbon, nitrogen, phosphorus, and sulfur, to provide sufficient algal niches. The particulate pools of the phytoplankton are usually chlorophyll (chl) biomass, or their elemental equivalents. Finally, B are ammonifying and nitrifying bacteria, and M represents both microflora and microfauna, as well as particulate and dissolved debris of the sediments.

The other biological terms of eqns [11]–[16] involve non-linear rate processes, in units of reciprocal time and associated with the various coefficients, such as in the temperature-dependent grazing of phytoplankton by zooplankton (aPZ). Their sloppy feeding is represented by α such that both the herbivores and the phytoplankton excretion ($\varepsilon_2 P$) are

sources of the DOM consumed by the bacteria (bDB), who in turn are eaten by other smaller zooplankton (cBZ). Some of the ingested food is released as egested fecal pellets (eZ) to sink (w_z), together with settling phytodetritus ($w_p\partial P/\partial z$) to the seafloor. Ammonium and phosphorus are excreted ($\varepsilon_1 Z$) by the respiring ($\delta_1 Z$) grazers as one source of recycled nutrients (including carbon dioxide), as well as those derived from respiring phytoplankton ($\delta_2 P$) and bacteria ($\delta_3 B$) and sediment releases $\partial/\partial z(K_z\,\partial M/\partial z)$. Finally, the temperature-dependent, light-regulated uptake of nutrients and subsequent growth (gNP) of the phytoplankton is usually a Liebig's law of the minimum formulation, involving saturation kinetics of the available resources, as function of cell quota and a ratio of the required elements.

The respective exchanges of biogenic gases, dissolved solutes, and bioturbated particulate matter across either the air–sea or water–sediment interfaces have been treated as part of the boundary conditions for the vertical diffusive term of these ecological state variables, similarly to the physical formulations of wind stress, bottom friction, and buoyancy fluxes. Further, since few elements escape the planet, a conservation of mass within marine food webs is also invoked as eqn [17], like that of eqn [9] for momentum.

$$\frac{dF}{dt} + \frac{dZ}{dt} + \frac{dP}{dt} + \frac{dB}{dt} + \frac{dN}{dt} + \frac{dD}{dt} + \frac{dM}{dt} = 0 \quad [17]$$

Model Hierarchies

Despite extensive *in situ* and satellite validation data for today's nonlinear biophysical models, based on some or all of state equations [3]–[17], the models still have numerical problems, since any set of coupled partial differential equations is solved iteratively at each individual grid point of a spatial three-dimensional mesh (**Figure 2**), using time-dependent forcing functions. The smaller the spatial grid mesh and the larger the number of state variables (**Figure 1**), the more computer memory is required, such that any one model must be formulated to provide only a few answers to a hierarchy of possible questions (**Table 1**) – *ergo*, the reason for construction of regional models.

For example, 50 years after the pioneer studies of Riley and co-workers, the present general circulation models (GCMs) still have minimal ecological realism with just nitrate and the total phytoplankton community to allow spatial meshes of $\sim 1/3°$ resolution of important physical processes of eddy mixing within basin simulations over ~ 37 depth levels.

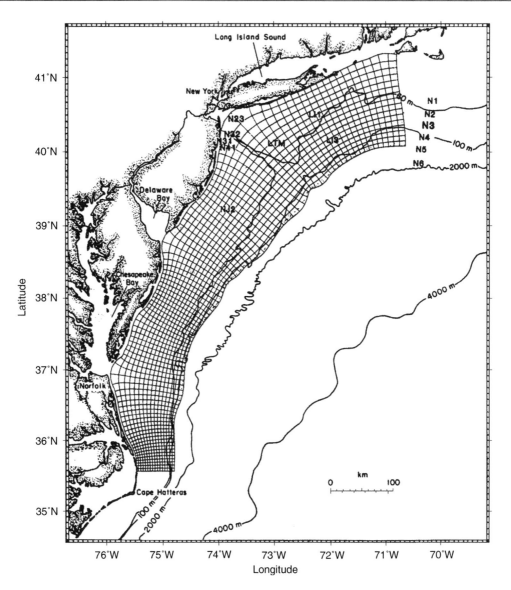

Figure 2   A curvilinear grid, showing every other grid line, of the first coupled physical-biochemical model of the Mid-Atlantic Bight, in relation to upstream (N1–N6) and interior (LI 1, LI 3, LTM, N23, N31–32. N41, and NJ2) current meter moorings.

Ideally, future GCMs may include more biogeochemical state variables to explicitly link carbon uptake during net photosynthesis with associated element cycles, e.g., nitrogen fixation, dimethyl sulfide evasion, organic phosphorus lability, calcification, and silica depletion in global budgets of climate change. But present questions of predicting the onset of harmful algal blooms, of the consequences of overfishing regional stocks, and of the fate of eutrophied coastal ecosystems require circulation models at smaller spatial resolution and biological models of greater ecological realism than those of GCMs (**Table 1** and **Figure 1**).

Regional models both incorporate local processes at the high end of their time/space scales of hours and decimeters and extend into the low end of

basin models at resolution of months and degrees (**Table 1**). At a residual flow of $5\,\mathrm{cm\,s^{-1}}$, or $\sim 5\,\mathrm{km\,d^{-1}}$, it would take a plankton population about 8 weeks to drift $\sim 300\,\mathrm{km}$ across a region of $\sim 1 \times 10^5\,\mathrm{km^2}$ area, about the size of the Mid-Atlantic Bight, the southern Caribbean Sea, or the West Florida shelf. With a large set of ecological variables (**Figure 1**), however, present computer resources barely allow computation of the terms of [3]–[17] over the $\sim 4\,\mathrm{km}$ resolution of a regional circulation model (**Figure 2**), with ~ 10 depth layers. Thus the complexities of turbulent flows and cellular metabolism must be glossed over, while the results of the regional model cannot be extrapolated over $\sim 3 \times 10^7\,\mathrm{km^2}$ of an adjacent basin without the loss of additional realism.

At the global ocean level of $\sim 3.7 \times 10^8$ km^2, lateral boundary conditions of a GCM are not a problem, because land boundaries are specified and flows are computed throughout all sectors of the simulated ocean; surface boundary conditions are another story, requiring input from some atmospheric model! In a regional model of the shelf seas, however, three open boundary conditions must be specified, which determine the solution over the grid mesh of the region of interest. Various options are (1) to use data at some of the boundaries, (2) to move them far way with a telescoping grid mesh, or (3) to make some assumptions about the water flows and fluxes of elements across the boundaries from other models. Examples of results from coupled biophysical models of three shelf regions are shown, using each of these options.

Regional Case Studies

Mid-Atlantic Bight

The first set of coupled biophysical models of the continental shelf and slope of the Mid-Atlantic Bight (MAB) between Martha's Vineyard and Cape Hatteras used the same circulation model, which was wind-forced, with current meter data specifying flows at the upstream boundary (**Figure 2**). The horizontal grid mesh was an average of ~ 4 km, with analytical solutions yielding continuous vertical profiles of the barotropic flow fields, from the shallow water formulation of the depth-averaged Navier–Stokes equations. During spring conditions of minimal density structure, the barotropic flow is $\sim 90\%$ of the total transport, and the simulated flows of the model exhibited a vector error of 1–2 cm s^{-1} speed and 10–30° direction, compared to the observed currents at interior moorings (**Figure 2**). During a strong upwelling event (**Table 1**), when nutrient-rich subsurface water moves up into the sunlit euphotic zone, the model's vertical velocity, w from eqn [9], was 9–14 m d^{-1} for effecting both supply of unutilized nutrients in eqn [14] and resuspension of settled-out phytoplankton (**Figure 3A**) in eqn [12].

In the first version of a coupled biological model of the MAB, only three vertical levels of ecological interactions were computed within the spatial mesh of the circulation model, with a minimal realism of just nitrate and the whole phytoplankton community as state variables, forced in turn by incident light and time-dependent grazing stress. Within parts of the MAB, this study of 'new' nitrate-based production yielded similar chlorophyll stocks of phytoplankton on 10 April 1979 (**Figure 3B**), after such an upwelling event, to those seen concurrently by a color-sensing satellite (**Figure 3A**) (the Coastal Zone Color Scanner (CZCS)) and an *in situ* moored fluorometer, moored south of Long Island, i.e., between New Jersey and Martha's Vineyard, as well as off Cape Hatteras. At mid-shelf in the wider, central region of the MAB, however, the simulated surface stocks of the first model overestimated those seen by the CZCS.

In another version of the coupled model of both shelf and slope domains, temperature-dependent growth of two groups of phytoplankton (net plankton diatoms and nanoplankton flagellates) utilized both nitrate and ammonium over 10 depth intervals, allowing calculation of the total ('new' + 'recycled') primary production. With the more heavily grazed nanoplankton dominating on the outer shelf and slope, the simulated stocks on 10 April 1979 (**Figure 3C**) now more closely approximated those estimated from the CZCS (**Figure 3A**). The model's total net photosynthesis over March–April was then within 28–97% of the ^{14}C measurements of productivity, depending upon the area and month. A mean of 0.5 g C m^{-2} d^{-1} suggested that only $\sim 13\%$ of the total spring production was exported to the deep sea, compared to prior estimates of $\sim 50\%$.

Southern Caribbean Sea

The second biophysical model of the Venezuelan shelf was again wind-forced with a mesh of ~ 4 km within 200 km of the land boundary, but the seaward boundary now extended ~ 1400 km out from the coast, intersecting Puerto Rico and Cuba, with a telescoping mesh of maximal size of ~ 110 km. A western boundary current was assumed to flow east–west along the model's shelf-break at a depth of ~ 100 m (**Figure 4**). Both barotropic and baroclinic flow fields were computed from a two-layered, linearized circulation model – i.e., again without advection of momentum and ignoring tidal forces. The ecological formulation over 30 depth intervals is also more complex, with the additional state variables of carbon dioxide, labile and refractory dissolved organic matter, bacteria, and zooplankton fecal pellets, as well as nitrate, ammonium, and light-regulated diatoms.

During spring upwelling of 8 m d^{-1} near the coast and 3 m d^{-1} offshore at a time-series site ('+' on **Figure 4**), the thermal fields of the physical model match satellite and ship estimates of surface temperature in February–April 1996–1997. Within the three-dimensional flow field, the steady solutions of detrital effluxes from a simple food web of diatoms, adult calanoid copepods, and the ammonifying/nitrifying

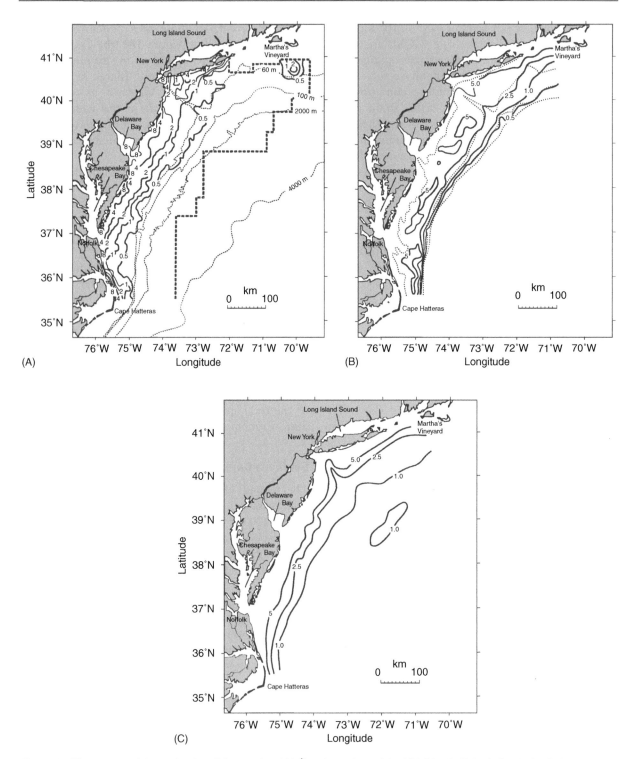

Figure 3 The observed (A) and simulated biomass (μg chl l^{-1}) at the surface of the Mid-Atlantic Bight during 10 April 1979, as seen by the Coastal Zone Color Scanner (CZCS) and modeled by simple food webs of (B) one and (C) two functional groups of phytoplankton within the same barotropic circulation model, under wind forcing measured at John F. Kennedy Airport.

bacteria are also ∼91% of the mean spring observations of settling fluxes of particulate carbon caught by a sediment trap at a depth of ∼240 m, moored in the Cariaco Basin (+). At this station, the other state variables of the coupled models are within ∼71% of the chlorophyll biomass (**Figure 4A**), ∼89% of the average ^{14}C net primary production of 2.0 g C m^{-2} d^{-1} (**Figure 4B**), ∼80% of the light

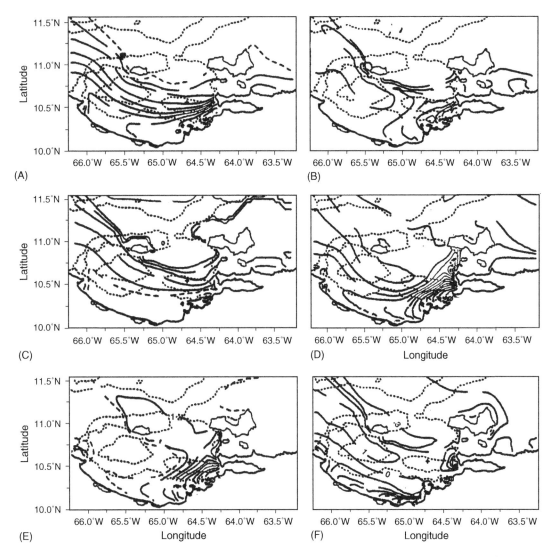

Figure 4 The simulated spring distributions within a two-layered baroclinic circulation model, under wind forcing measured during February–April 1996–1997 at Margarita Island, of (A) near-surface chlorophyll ($\mu g\,l^{-1}$), (B) water column net primary production ($mg\,C\,m^{-2}\,d^{-1}$), (C) 1% light depth (m), and, at $\sim 25\,m$ depth of (D) total CO_2 ($\mu\,mol\,DIC\,kg^{-1}$), (E) nitrate ($\mu\,mol\,NO_3\,kg^{-1}$), and (F) ammonium ($\mu\,mol\,NH_4\,kg^{-1}$) around the CARIACO time-series site ($+$) on the Venezuelan shelf.

penetration (**Figure 4C**), and 97% of the total CO_2 stocks (**Figure 4D**). No data were then available for nitrate, ammonium, or satellite color.

When the Intertropical Convergence Zone of the North Atlantic winds moves north in summer, local winds along the Venezuelan coast slacken. Within one summer case of the model with weaker wind forcing, the simulated net primary production is only 14% of that measured in August–September, while the predicted detrital flux is then 30% of the observed. Addition of a diazotroph state variable (**Figure 1**), with another source of 'new' nitrogen via nitrogen fixation rather than upwelled nitrate, would remedy the seasonal deficiencies of the biological model, attributed to use of a single phytoplankton groups. We explore such a scenario in the last case.

West Florida Shelf

All of the terms eqns [3]–[5] and [9] were maintained in a recent series of wind-forced barotropic circulation models of the West Florida shelf, with horizontal grid meshes of $\sim 4–9\,km$ and 16–21 vertical levels over the water column. Open boundary conditions of flow at a depth, H, of $\sim 1500\,m$ in the Gulf of Mexico basin, as well as on the shelf, were of the radiation form [18], where C is the speed of gravity waves (i.e., $(gH)^{1/2}$) and u is determined from the adjacent interior grid points of the model.

$$\frac{\partial u}{\partial t} = C\frac{\partial u}{\partial x} \qquad [18]$$

When the additional constraint is imposed that the depth-integrated transport across these boundaries is zero, as in the shelf-break condition of the Mid-Atlantic Bight (**Figure 2**), fluxes of water within the surface and bottom Ekman layers cancel each other. Since biochemical variables exhibit nonuniform vertical profiles, however, the net influx of nutrient within the bottom layer of the embedded ecological model may instead be balanced by a net efflux of plankton within the surface layer. Prior one-dimensional models of element cycling by eight functional groups of phytoplankton on the West Florida shelf concerned most of the state variables of **Figure 1**, while the three-dimensional results, shown here, consider just the transport of a harmful algal bloom (**Figure 5**), with diel migration of the dinoflagellate *Gymnodinium breve*.

Once a red tide of *G. breve* is formed, after DON supplied from nitrogen-fixers of our one-dimensional model, its simulated trajectory over 16 vertical levels during December 1979 (**Figure 5A**) matches repeated shipboard and helicopter observations (**Figure 5B**) of this dinoflagellate bloom at the surface of the West Florida shelf, if one samples the model at sunrise after nocturnal convective mixing: at noon, the simulated red tide instead aggregates in a subsurface maximum, as observed during additional time-series studies.

Under the predominantly upwelling-favorable winds of fall/winter, the circulation model yields a positive w of ~ 0.5–$1.0 \, \mathrm{m \, d^{-1}}$ within the red tide patch and of 1.0–$2.0 \, \mathrm{m \, d^{-1}}$ at the coast. In another model case, without the vertical downward migration of *G. breve* at a speed of $\sim 1 \, \mathrm{m \, h^{-1}}$ to avoid bright light, the model's surface populations then did not replicate the data; they were instead advected farther offshore than the *in situ* populations. It appears that in the 'real world' *G. breve* spent most of their time in the lower layers of the water column, before ascending to be sampled by ship and helicopters during daylight at the sea surface.

Furthermore, within the bottom Ekman layer, the simulated red tide is advected onshore, mimicking observations of shellfish bed closures on the barrier islands. Thus, the coupled models suggest that, upon maturation of a red tide from successful competition among functional groups of the phytoplankton community (**Figure 1**), vertical migration of *G. breve* in relation to seasonal changes of summer downwelling and fall/winter upwelling flow fields then determines the duration and intensity of red tide landfalls along the beaches of the west coast of Florida.

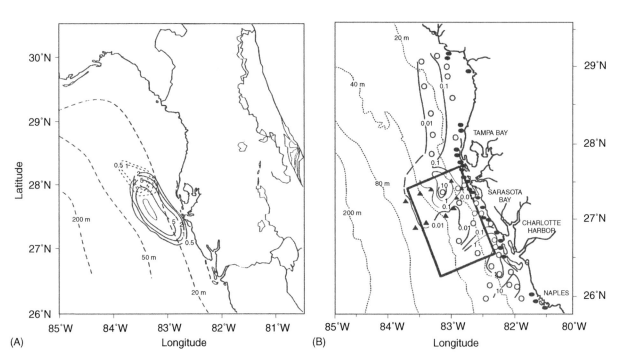

Figure 5   The (A) simulated and (B) observed daily trajectories on the West Florida shelf of a near-surface red tide (10^5 cells $l^{-1} = 1 \, \mu g \, chl \, l^{-1}$) of *Gymnodinium breve*. The model's population is growing at a maximal rate of $\sim 0.15 \, d^{-1}$ with vertical migration and nocturnal convective mixing, during sunrise (06:00) on 20 December 1979, after initiation on 25 November (dashed contours), and under forcing of a barotropic circulation model by winds measured at the Tampa International Airport. The offshore observations were made during 10–11 December 1979 (solid triangles) and 19–21 December 1979 (open circles) in relation to landfalls of red tides sampled (solid circles) on piers and bridges during 11/25/79–2/8/80.

Prospectus

Other regional models of varying ecological and physical realism have been constructed for numerous shelf regions. They are mainly classic N–P–Z formulations, however, such that they may be improved with inclusion of a larger number of ecological state variables. Simply adding biochemical and physical variables for the next generation of coupled regional models is not sufficient, because the initial and boundary conditions will always be poorly known. Like models of the weather on land, such predictive models must be continually validated with data to correct for the poor knowledge of these conditions.

Given the expense of shipboard monitoring programs, a few bio-optical moorings (e.g., fluorometers or remote sensors (**Figure 1**)), are the most likely sources of such updates for the ecological models. Furthermore, the veracity of the underlying circulation models must be maintained with a complete suite of buoyancy flux measurements at the same moorings, to derive the baroclinic contributions important to the regional flow fields. For example, the barotropic calculations of the West Florida shelf case did not match current meter observations during summer on the outer shelf. The bio-optical implications of regional physical/ecological models driven by time-dependent density fields must be included in future simulation analyses.

Further Reading

Csanady GT (1982) *Circulation in the Coastal Ocean.* Dordrecht: Riedel.

Heaps NS (1987) *Three-Dimensional Coastal Ocean Models.* Coastal and Estuarine Series 4. Washington, DC: American Geophysical Union.

Mooers CN (1998) *Coastal Ocean Prediction.* Coastal and Estuarine Series 56. Washington, DC: American Geophysical Union.

Riley GA, Stommel H, and Bumpus DF (1949) Quantitative ecology of the plankton of the western North Atlantic. *Bulletin of Bingham Oceanographic College* 12: 1–169.

Steele JH (1974) *The Structure of Marine Ecosystems.* Cambridge, MA: Harvard University Press.

Walsh JJ (1988) *On the Nature of Continental Shelves.* San Diego, CA: Academic Press.

REMOTE SENSING BY ACOUSTICS, AIRCRAFT AND SATELLITES

ACOUSTICS, SHALLOW WATER

F. B. Jensen, SACLANT Undersea Research Centre, La Spezia, Italy

Introduction

Using the commonly accepted definition of shallow water to mean coastal waters with depth up to 200 m, the shallow-water regions of the world constitute around 8% of all oceans and seas. These regions are particularly important since they are national economic zones and also more assessible.

Sound waves in the sea play the role of light in the atmosphere, i.e. acoustics is the only means of 'seeing' objects at distances beyond a few hundred meters in seawater. All forms of electromagnetic waves (light, radar) are rapidly attenuated in seawater. Low-frequency acoustic signals, on the other hand, propagate with little attenuation and can be heard over thousands of kilometers in the deep ocean.

The use of sound in the sea is ubiquitous. It is employed by the military to detect mines and submarines, and ship-mounted sonars measure water depth, ship speed and the presence of fish shoals. Side-scan systems are used to map bottom topography, sub-bottom profilers for getting information about the deeper layering, and other sonar systems for locating pipelines and cables on and beneath the seafloor. Sound is also used for navigating submerged vehicles, for underwater communications and for tracking marine mammals. In an inverse sense sound is used for measuring physical parameters of the ocean environment and for monitoring oceanic processes through the techniques of acoustical oceanography and ocean acoustic tomography.

Optimal sonar design for this great variety of applications demands using a wide range of acoustic frequencies. Practical shallow-water systems cover a frequency range from 50 Hz to 500 kHz, which, with a mean sound speed of 1500 m s^{-1}, correspond to acoustic wavelengths from 30 m down to 3 mm.

The principal characteristic of shallow-water propagation is that the sound-speed profile is nearly constant over depth or downward refracting, meaning that long-range propagation takes place exclusively via lossy bottom-interacting paths. This is very different from deep-water scenarios, where the sound-speed structure is such that sound is refracted away from the bottom and therefore can propagate to long ranges with little attenuation. Moreover, the environmental variability is much higher in coastal regions than in the deep ocean, with the result that there is much more acoustic variability in shallow water than in deep water.

The Ocean Acoustic Environment

The ocean is an acoustic waveguide limited above by the sea surface and below by the seafloor. The speed of sound in the waveguide plays the same role as the index of refraction does in optics. Sound speed is normally related to density and compressibility. In the ocean, density is related to static pressure, salinity and temperature. The sound speed in the ocean is an increasing function of temperature, salinity, and pressure, the latter being a function of depth. It is customary to express sound speed (c) as an empirical function of three independent variables: temperature (T) in degrees centigrade, salinity (S) in parts per thousand (‰), and depth (D) in meters. A simplified expression for this dependence is

$$c = 1449.2 + 4.6T - 0.055T^2 + 0.00029T^3 \\ + (1.34 - 0.010T)(S - 35) + 0.016D \quad [1]$$

In shallow water, where the depth effect on sound speed is small, the primary contributor to sound speed variations is the temperature. Thus, for a salinity of 35‰, the sound speed in seawater varies between 1450 m s^{-1} at 0°C and 1545 m s^{-1} at 30°C.

Seasonal and diurnal changes affect the oceanographic parameters in the upper ocean. In addition, all of these parameters are a function of geography. In a warmer season (or warmer part of the day) in shallow seas where tidal mixing is weak, the temperature increases near the surface and hence the sound speed increases toward the sea surface. This near-surface heating (and subsequent cooling) has a profound effect on surface-ship sonars. Thus the diurnal heating causes poorer sonar performance in the afternoon – a phenomenon known as the afternoon effect. The seasonal variability, however, is much greater and therefore more important acoustically.

A ray picture of propagation in a 100-m deep shallow water duct is shown in **Figure 1**. The sound-speed profile in the upper panel is typical of the Mediterranean in the summer. There is a warm surface layer causing downward refraction and hence repeated bottom interaction for all ray paths. Since the seafloor is a lossy boundary, propagation in

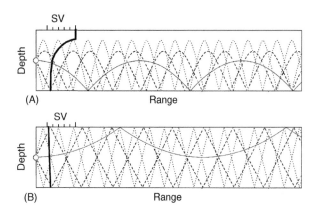

Figure 1 Ray paths in shallow water for typical Mediterranean summer and winter profiles. (A) In summer sound interacts repeatedly with the seabed but not with the sea surface. (B) In winter sound interacts with both the sea surface and the seabed, except for shallow rays emitted near the horizontal.

shallow water is dominated by bottom reflection loss at low and intermediate frequencies (<1 kHz) and scattering losses at high frequencies. The seasonal variation in sound-speed structure is significant with winter conditions being nearly iso-speed (**Figure 1B**). The result is that there is less bottom interaction in winter than in summer, which again means that propagation conditions are generally better in winter than in summer.

Of course, the ocean sound-speed structure is neither frozen in time nor space. On the contrary, the ocean has its own weather system. There are currents, internal waves and thermal microstructure present in most shallow-water areas. **Figure 2** illustrates the sound speed variability along a 15 km-long track in the Mediterranean Sea. The data were recorded on a towed thermistor chain covering depths between 5 and 90 m. In general, this type of time-varying oceanographic structure has an effect on sound propagation, both as a source of attenuation (acoustic energy being scattered into steeperangle propagation paths suffers increased bottom

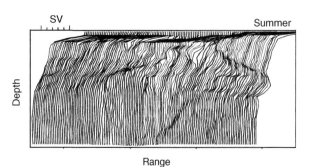

Figure 2 Spatial variability of sound speed in shallow-water area of the Mediterranean. The depth covered is around 100 m and the range 15 km.

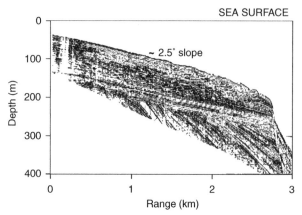

Figure 3 Seismic profile of bottom layering in coastal-water area of the Mediterranean.

reflection loss) and of acoustic signal fluctuations with time.

Turning to the upper and lower boundaries of the ocean waveguide, the sea surface is a simple horizontal boundary and a nearly perfect reflector. The seafloor, on the other hand, is a lossy boundary with varying topography. Both boundaries have small-scale roughness associated with them which causes scattering and hence attenuation of sound due to the increased bottom reflection loss associated with steep-angle propagation paths. In terms of propagation physics, the seafloor is definitely the most complex boundary, exhibiting vastly different reflectivity characteristics in different geographical locations.

The structure of the ocean bottom in shallow water generally consists of a thin stratification of sediments overlying the continental crust. The nature of the stratification is dependent on many factors, including geological age and local geological activity. Thus, relatively recent sediments will be characterized by plane stratification parallel to the sea bed, whereas older sediments and sediments close to the crustal plate boundaries may have undergone significant deformation. An example of a complicated bottom layering is given in **Figure 3**, which displays a seismic section from the coastal Mediterranean. The upper stratification here is almost parallel to the seafloor, whereas deeper layers are strongly inclined.

Transmission Loss

The decibel (dB) is the dominant unit in ocean acoustics and denotes a ratio of intensities (not pressures) expressed on a $\log_{10}$ scale.

An acoustic signal traveling through the ocean becomes distorted due to multipath effects and weakened due to various loss mechanisms. The standard measure in underwater acoustics of the change in signal strength with range is transmission loss defined as the ratio in decibels between the acoustic intensity $I(r,z)$ at a field point and the intensity I_0 at 1 m distance from the source, i.e.

$$\begin{aligned} TL &= -10\log\frac{I(r,z)}{I_0} \\ &= -20\log\frac{|p(r,z)|}{|p_0|} \quad \text{[dB re 1 m]}. \end{aligned} \quad [2]$$

Here use has been made of the fact that the intensity of a plane wave is proportional to the square of the pressure amplitude. The major contributors to transmission loss in shallow water are: geometrical spreading loss, water volume attenuation, bottom reflection loss, and various scattering losses.

Geometrical Spreading

The spreading loss is simply a measure of the signal weakening as it propagates outward from the source. **Figure 4** shows the two geometries of importance in underwater acoustics. First consider a point source in an unbounded homogeneous medium (**Figure 4A**). For this simple case the power radiated by the source is equally distributed over the surface area of a sphere surrounding the source. If the medium is assumed to be lossless, the intensity is inversely proportional to the surface of the sphere, i.e. $I \propto 1/(4\pi R^2)$. Then from eqn [2] the spherical spreading loss is given by

$$TL = 20\log r \quad \text{[dB re 1 m]} \quad [3]$$

where r is the horizontal range in meters.

When the medium has plane upper and lower boundaries as in the waveguide case in **Figure 4B**, the farfield intensity change with horizontal range becomes inversely proportional to the surface of a cylinder of radius R and depth D, i.e. $I \propto 1/(2\pi RD)$. The cylindrical spreading loss is therefore given by

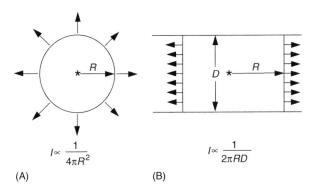

$$I \propto \frac{1}{4\pi R^2}$$

$$I \propto \frac{1}{2\pi RD}$$

(A)   (B)

Figure 4 Geometrical spreading laws. (A) Spherical spreading; (B) Cylindrical spreading.

$$TL = 10\log r \quad \text{[dB re 1 m]} \quad [4]$$

Note that for a point source in a waveguide, there is spherical spreading in the nearfield ($r \leq D$) followed by a transition region toward cylindrical spreading which applies only at longer ranges ($r \gg D$).

As an example consider propagation in a shallow-water waveguide to a range of 20 km with spherical spreading applying on the first 100 m. The total propagation loss (neglecting attenuation) then becomes: 40 dB + 23 dB = 63 dB. This figure represents the minimum loss to be expected at 20 km. In practice, the total loss will be higher due both to the attenuation of sound in seawater, and to various reflection and scattering losses.

Sound Attenuation in Seawater

When sound propagates in the ocean, part of the acoustic energy is continuously absorbed, i.e. the energy is transformed into heat. Moreover, sound is scattered by different kinds of inhomogeneities, also resulting in a decay of sound intensity with range. As a rule, it is not possible in real ocean experiments to distinguish between absorption and scattering effects; they both contribute to sound attenuation in seawater.

A simplified expression for the frequency dependence (f in kHz) of the attenuation is given by Thorp's formula,

$$\alpha = \frac{0.11f^2}{1+f^2} + \frac{44f^2}{4100+f^2} \quad \text{[dB km}^{-1}\text{]}, \quad [5]$$

where the two terms describe absorption due to chemical relaxations of boric acid, $B(OH)_3$, and magnesium sulphate, $MgSO_4$, respectively.

According to eqn [5] the attenuation of low-frequency sound in seawater is indeed very small. For instance, at 100 Hz a tenfold reduction in sound intensity (-10 dB) occurs over a distance of around 8300 km. Even though attenuation increases with frequency ($r_{-10\,dB} \simeq 150$ km at 1 kHz and $\simeq 9$ km at 10 kHz), no other kind of radiation can compete with sound waves for long-range propagation in the ocean.

Bottom Reflection Loss

Reflectivity, the ratio of the amplitudes of a reflected plane wave to a plane wave incident on an interface separating two media, is an important measure of the effect of the bottom on sound propagation. Ocean bottom sediments are often modeled as fluids which means that they support only one type of sound wave – a compressional wave.

The expression for reflectivity at an interface separating two homogeneous fluid media with density ρ_i and sound speed c_i, $i = 1, 2$, was first worked out by Rayleigh as

$$R(\theta_1) = \frac{(\rho_2/\rho_1)\sin\theta_1 - \sqrt{\left((c_1/c_2)^2 - \cos^2\theta_1\right)}}{(\rho_2/\rho_1)\sin\theta_1 + \sqrt{\left((c_1/c_2)^2 - \cos^2\theta_1\right)}} \quad [6]$$

where θ_1 denotes the grazing angle of the incident plane wave of unit amplitude.

The reflection coefficient has unit magnitude, meaning perfect reflection, when the numerator and denominator of eqn [6] are complex conjugates. This can only occur when the square root is purely imaginary, i.e. for $\cos\theta_1 > c_1/c_2$ (total internal reflection). The associated critical grazing angle below which there is perfect reflection is found to be

$$\theta_c = \arccos\left(\frac{c_1}{c_2}\right) \quad [7]$$

Note that a critical angle only exists when the sound speed of the second medium is higher than that of the first.

A closer look at eqn [6] shows that the reflection coefficient for lossless media is real for $\theta_c > \theta_c$, which means that there is loss ($|R| = 1$) but no phase shift associated with the reflection process. On the other hand, for $\theta_c < \theta_c$ we have perfect reflection ($|R| = 1$) but with an angle-dependent phase shift. In the general case of lossy media (c_i complex), the reflection coefficient is complex, and, consequently, there is both a loss and a phase shift associated with each reflection.

The critical-angle concept is very important for understanding the waveguide nature of shallow-water propagation. **Figure 5** shows bottom loss curves (BL $= -10\log|R|^2$) for a few simple fluid bottoms with different compressional wave speeds (C_p), densities and attenuations. Note that for a lossy bottom we never get perfect reflection. However, there is in all cases an apparent critical angle ($\theta_C \simeq 33°$ for $c_p = 1800\,\text{m s}^{-1}$ in **Figure 5**), below which the reflection loss is much smaller than for supercritical incidence. With paths involving many bottom bounces such as in shallow-water propagation, bottom losses even as small as a few tenths of a decibel per bounce accumulate to significant total losses since the propagation path may involve many tens or even hundreds of bounces.

Real ocean bottoms are complex layered structures of spatially varying material composition. A geoacoustic model is defined as a model of the real

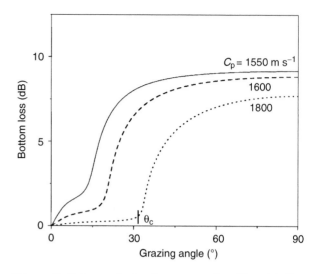

Figure 5 Bottom reflection loss curves for different bottom types. Note that low-speed bottoms (clay, silt) are more lossy than high-speed bottoms (sand, gravel). C_p is the compressional wave speed.

seafloor with emphasis on measured, extrapolated, and predicted values of those material properties important for the modeling of sound transmission. In general, a geoacoustic model details the true thicknesses and properties of sediment and rock layers within the seabed to a depth termed the effective acoustic penetration depth. Thus, at high frequencies (>1 kHz), details of the bottom composition are required only in the upper few meters of sediment, whereas at low frequencies (<100 Hz) information must be provided on the whole sediment column and on properties of the underlying rocks.

The information required for a complete geoacoustic model should include the following depth-dependent material properties: the compressional wave speed, C_p; the shear wave speed, C_p; the compressional wave attenuation, C_p; the shear wave attenuation, C_p; and the density, C_p. Moreover, information on the variation of all of these parameters with geographical position is required.

The amount of literature dealing with acoustic properties of seafloor materials is vast. **Table 1** lists the geoacoustic properties of some typical seafloor materials, as an indication of the many different types of materials encountered just in continental shelf and slope environments.

Boundary and Volume Scattering Losses

Scattering is a mechanism for loss, interference and fluctuation. A rough sea surface or seafloor causes attenuation of the mean acoustic field propagating in the ocean waveguide. The attenuation increases with increasing frequency. The field scattered away from

Table 1 Geoacoustic properties of continental shelf environments

Bottom type	p (%)	ρ_b/ρ_w –	C_p/C_w –	C_p (m s^{-1})	C_s (m s^{-1})	α_p ($\alpha B \lambda_p^{-1}$)	α_s ($\alpha B \lambda_s^{-1}$)
Clay	70	1.5	1.00	1500	< 100	0.2	1.0
Silt	55	1.7	1.05	1575	$C_s{}^a$	1.0	1.5
Sand	45	1.9	1.1	1650	$C_s{}^b$	0.8	2.5
Gravel	35	2.0	1.2	1800	$C_s{}^c$	0.6	1.5
Moraine	25	2.1	1.3	1950	600	0.4	1.0
Chalk	–	2.2	1.6	2400	1000	0.2	0.5
Limestone	–	2.4	2.0	3000	1500	0.1	0.2
Basalt	–	2.7	3.5	5250	2500	0.1	0.2

$^a C_s = 80\ \bar{z}^{0.3}$
$^b C_s = 110\ \bar{z}^{0.3}$
$^c C_s = 180\ \bar{z}^{0.3}$
$C_w = 1500\ \mathrm{m s}^{-1}$, $\rho_w = 1000\ \mathrm{kg\ m}^{-3}$.

the specular direction, and, in particular, the back-scattered field (called reverberation) acts as interference for active sonar systems. Because the ocean surface moves, it will also generate acoustic fluctuations. Bottom roughness can also generate fluctuations when the sound source or receiver is moving. The importance of boundary roughness depends on the sound-speed profile which determines the degree of interaction of sound with the rough boundaries.

Often the effect of scattering from a rough surface is thought of as simply an additional loss to the specularly reflected (coherent) component resulting from the scattering of energy away from the specular direction. If the ocean bottom or surface can be modeled as a randomly rough surface, and if the roughness is small with respect to the acoustic wavelength, the reflection loss can be considered to be modified in a simple fashion by the scattering process. A formula often used to describe reflectivity from a rough boundary is:

$$R'(\theta) = R(\theta)e^{-0.5\Gamma^2} \qquad [8]$$

where $R'(\theta)$ is the new reflection coefficient, reduced because of scattering at the randomly rough interface. Γ is the Rayleigh roughness parameter defined as

$$\Gamma = 2k\sigma \sin\theta \qquad [9]$$

where $k = 2\pi/\lambda$ is the acoustic wavenumber and σ is the *rms* roughness. Note that the reflection coefficient for the smooth ocean surface is simply -1 (the pressure-release condition is obtained from eqn [6] by setting $\rho_2 = 0$) so that the rough-sea-surface reflection coefficient for the coherent field is $R'(\theta) = -\exp(-0.5\Gamma^2)$. For the ocean bottom, the appropriate geoacoustic parameters (see **Table 1**) are used for evaluating $R'(\theta)$, and the rough-bottom reflection coefficient is then obtained from eqn [8].

Volume scattering is thought to arise primarily from biological organisms. For lower frequencies (less than 10 kHz), fish with air-filled swim bladders are the main scatterers whereas above 20 kHz, zooplankton or smaller animals that feed on the phytoplankton, and the associated biological food chain, are the scatterers. Many of the organisms undergo a diurnal migration rising towards the sea surface at sunset and descending to depth at sunrise. Since the composition and density of the populations vary with the environmental conditions, the scattering characteristics depend on geographical location, time of day, season and frequency. As an example, data from the Mediterranean Sea for volume scattering losses due to fish shoals show excess losses of 10–15 dB for a propagation range of 12 km and frequencies between 1 and 3 kHz.

Finally, scattering off bubbles near the surface is sometimes referred to as either a volume- or surface-scattering mechanism. These bubbles arise not only from sea surface action, but also from biological origins and from ship wakes. Furthermore, bubbles are not the only scattering mechanism, but bubble clouds may have significantly different sound speed than plain seawater thereby altering local refraction conditions. At the sea surface, the relative importance of roughness versus bubble effects is not yet resolved.

Transmission-loss Data

Figure 6 gives an example of transmission-loss variability in shallow water. The graph displays a collection of experimental data from different shallow-water areas (100–200 m deep) all over the world. The data refer to downward-refracting summer conditions in the frequency band 0.5–1.5 kHz. Two features are of immediate interest. One is the

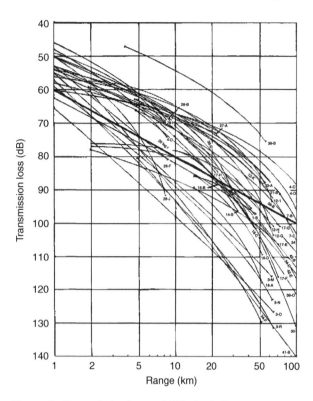

Figure 6 Transmission loss variability in shallow water.

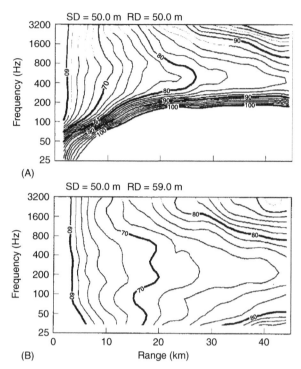

Figure 7 Examples of frequency-dependent propagation losses measured in two shallow-water areas: (A) Barents Sea, (B) English Channel. Note the presence of an optimum frequency of propagation between 200 and 400 Hz.

spread of the data amounting to around 50 dB at 100 km and caused primarily by the varying bottom-loss conditions in different areas of the world. The second feature is the fact that transmission is generally better than free-field propagation (20 log r) at short and intermediate ranges but worse at longer ranges. This peculiarity is due to the trapping of energy in the shallow-water duct, which improves transmission at shorter ranges (cylindrical versus spherical spreading), but, at the same time, causes increased boundary interaction, which degrades transmission at longer ranges.

A second example of transmission-loss variability in shallow water is given in **Figure 7**, where broadband data from two different geographical areas are compared. The data set in **Figure 7A**, was collected in the Barents Sea in 60 m water depth. Note the high transmission losses recorded below 200 Hz, where energy levels fall off rapidly indicating that most of the acoustic energy emitted by the source is lost to the seabed. It is believed that this excess attenuation is caused by the coupling of acoustic energy into shear waves in the seabed. In contrast to the high-loss environment in the Barents Sea, **Figure 7B** shows a data set from the English Channel in 90 m water depth. Here propagation conditions are excellent over the entire frequency band. This second data set represents typical propagation conditions for thick sandy sediments with negligible shear-wave effects.

Cutoff Frequency and Optimum Frequency

A common feature of all acoustic ducts is the existence of a low-frequency cutoff. Hence, there is a critical frequency below which the shallow-water channel ceases to act as a waveguide, causing energy radiated by the source to propagate directly into the bottom. The cutoff frequency is given by,

$$f_0 = \frac{c_w}{4D\sqrt{\left(1 - (c_w/c_b)^2\right)}} \qquad [10]$$

This expression is exact only for a homogeneous water column of depth D and sound speed c_w overlying a homogeneous bottom of sound speed c_b. As an example, let us take $D = 100$ m, $c_w = 1500$ m s^{-1}, and $c_b = 1600$ m s^{-1} (sand–silt), which yields $f_0 \simeq 11$ Hz.

Sound transmission in shallow water has the characteristic frequency-dependent behavior shown in **Figure 7**, i.e. there is an optimum frequency of propagation at longer ranges. Thus the 80 dB contour line extends farthest in range for frequencies around 400 Hz in **Figure 7A** and around 200 Hz in **Figure 7B**, implying that transmission is best at these frequencies – the optimum frequencies of propagation for the two sites.

Optimum frequency is a general feature of ducted propagation in the ocean. It occurs as a result of competing propagation and attenuation mechanisms at high and low frequencies. In the high-frequency regime we have increasing volume and scattering loss with increasing frequency. At lower frequencies the efficiency of the duct to confine sound decreases (the cutoff phenomenon). Hence propagation and attenuation mechanisms outside the duct (in the seabed) become important. In fact, the increased penetration of sound into a lossy seabed with decreasing frequency causes the overall attenuation of waterborne sound to increase with decreasing frequency. Thus we get high attenuation at both high and low frequencies, whereas intermediate frequencies have the lowest attenuation. It can be shown that the optimum frequency for shallow-water propagation is strongly dependent on water depth ($f_{opt} \propto D^{-1}$), has some dependence on the sound-speed profile, but is only weakly dependent on the bottom type. Typically, the optimum frequency is in the range 200–800 Hz for a water depth of 100 m.

Signal Transmission in the Time Domain

Even though underwater acousticians have traditionally favored spectral analysis techniques for gaining information about the band-averaged energy distribution within a shallow-water waveguide, additional insight into the complication of multipath propagation can be obtained by looking at signal transmission in the time domain.

Figure 8 indicates that the signal structure measured downrange will consist of a number of arrivals with time delays determined by the pathlength differences, and individual pulse shapes being modified due to frequency-dependent amplitude and phase changes associated with each boundary reflection.

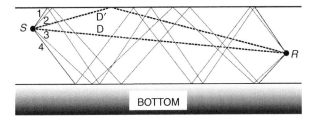

Figure 8 Schematic of ray arrivals in shallow-water waveguide. A series of multipath arrivals are expected, represented by eigenrays connecting source (S) and receiver (R). The shortest path is the direct arrival D followed by the surface-reflected arrival D'. Next comes a series of four arrivals all with a single bottom bounce; then four rays with two bottom bounces as illustrated in the figure, followed by four rays with three bottom bounces, etc.

From simple geometrical considerations, the time dispersion is found to be

$$\Delta_\tau \simeq \frac{R}{\bar{c}}\left(\frac{1}{\cos\theta} - 1\right) \qquad [11]$$

where R is the range between source and receiver, $\bar{c}$ is the mean sound speed in the channel, and θ is the maximum propagation angle with respect to the horizontal. This angle will be determined either by the source beamwidth or by the critical angle at the bottom (the smaller of the two). Since the dispersion considered here is solely due to the geometry of the waveguide, it is called geometrical dispersion.

An example of measured pulse arrivals over a 10 h period in the Mediterranean is given in **Figure 9**. Note that the time-varying ocean (internal waves, currents, tides) causes strong signal fluctuations with time, particularly in the earlier part of the signal. The time dispersion is 15–20 ms and at least four main energy packets, each consisting of several ray arrivals can be identified.

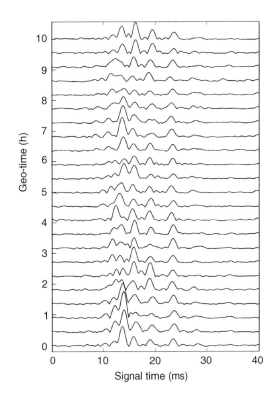

Figure 9 Measured pulse arrivals versus geo-time over a 10 km shallow-water propagation track in the Mediterranean Sea. The bandwidth is 200–800 Hz.

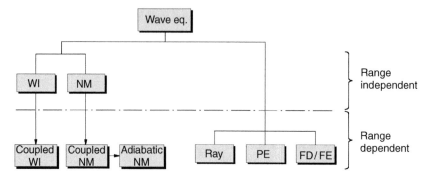

Figure 10 Hierarchy of numerical models in underwater acoustics. WI, wavenumber integration; NM, normal modes; PE, parabolic equation; FD, finite difference; FE, finite element.

Numerical Modeling

The advent of computers has resulted in an explosive growth in the development and use of numerical models since the mid-1970s. Numerical models have become standard research tools in acoustic laboratories, and computational acoustics is becoming an evermore important branch of the ocean acoustic science. Only the numerical approach permits an analysis of the full complexity of the acoustic problem.

An assortment of models has been developed over the past 25 years to compute the acoustic field in shallow-water environments in both the frequency and time domains. Entire textbooks are dedicated to the development of theoretical and numerical formalisms which can provide quantitative acoustic predictions for arbitrary ocean environments. Sound propagation is mathematically described by the wave equation, whose parameters and boundary conditions are descriptive of the ocean environment. As shown in **Figure 10**, there are essentially five types of models (computer solutions to the wave equation) to describe sound propagation in the sea: wavenumber integration (WI); normal mode (NM); ray; parabolic equation (PE) and direct finite-difference (FD) or finite-element (FE) solutions of the full wave equation. All of these models permit the ocean environment to vary with depth. A model that also permits horizontal variations in the environment, i.e. sloping bottom or spatially varying oceanography, is termed range dependent.

As shown in **Figure 10**, an a priori assumption about the environment being range independent, leads to solutions based on spectral techniques (WI) or normal modes (NM); both of these techniques can, however, be extended to treat range dependence. Ray, PE and FD/FE solutions are applied directly to range-varying environments. For high frequencies (a few kilohertz or above), ray theory, the infinite frequency approximation, is still the most practical, whereas the other five model types become more and more applicable below, say, a kilohertz in shallow water. Models that handle ocean variability in three spatial dimensions have also been developed, but these models are used less frequently than two-dimensional versions because of the computational cost involved.

Conclusions

The acoustics of shallow water has been thoroughly studied both experimentally and theoretically since World War II. Today the propagation physics is well understood and sophisticated numerical models permit accurate simulations of all processes (reflection, refraction, scattering) that contribute to the complexity of the shallow-water problem. Sonar performance predictability, however, is limited by knowledge of the controlling environmental inputs. The current challenge is therefore how best to collect relevant environmental data from the world's enormously variable shallow-water areas.

See also

Acoustics in Marine Sediments.

Further Reading

Brekhovskikh LM and Lysanov YP (1990) *Fundamentals of Ocean Acoustics*, 2nd edn. New York: Springer-Verlag.

Etter PC (1996) *Underwater Acoustic Modeling*, 2nd edn. London: E & FN Spon.

Jensen FB, Kuperman WA, Porter MB, and Schmidt H (2000) *Computational Ocean Acoustics*. New York: Springer-Verlag.

Medwin H and Clay CS (1998) *Fundamentals of Acoustical Oceanography*. San Diego: Academic Press.

Urick RJ (1996) *Principles of Underwater Sound*, 3rd edn. Los Altos: Peninsula Publishing.

ACOUSTIC MEASUREMENT OF NEAR-BED SEDIMENT TRANSPORT PROCESSES

P. D. Thorne and P. S. Bell, Proudman Oceanographic Laboratory, Liverpool, UK

Introduction: Sediments and Why Sound Is Used

Marine sediment systems are complex, frequently comprising mixtures of different particles with non-cohesive (sands) and cohesive (clays and muds) properties. The movement of sediments in coastal waters impacts on many marine processes. Through the actions of accretion, erosion, and transport, sediments define most of our coastline. Their deposition and resuspension by waves and tidal currents in estuarine and nearshore environments control seabed morphology. Fine sediments, which act as reservoirs for nutrients and contaminants and as regulators of light transmission through the water column, have significant impact on water chemistry and on primary production. Therefore, an improved understanding of sediment dynamics in coastal waters has relevance to a broad spectrum of marine science ranging from physical and chemical processes, to the complex biological and ecological structures supported by sedimentary environments. However, it is commonly acknowledged that our capability to describe the coupled system of the bed, the hydrodynamics, and the sediments themselves is still relatively primitive and often based on empiricism. Recent advances in observational technologies now allow sediment processes to be investigated with greater detail and precision than has previously been the case. The combined use of acoustics, laser and radar, both at large and small scales, is facilitating exciting measurement opportunities. The enhancement of computing capabilities also allows us to make use of more complex coupled sediment–hydrodynamic models, which, linked with the emerging observations, provide new openings for model development.

It is readily acknowledged by sedimentologists that the presently available commercial instrumentation does not satisfy a number of requirements for near-bed sediment transport processes studies. Here we focus on the development of acoustics to fulfill some of these needs. The question could be asked: Why use acoustics for such studies? Sediment transport can be thought of as dynamic interactions between: (1) the seabed morphology, (2) the sediment field, and (3) the hydrodynamics. These three components interrelate with each other in complex ways, being mutually interactive and interdependent as illustrated schematically as a triad in **Figure 1**. The objective therefore is to measure this interacting triad with sufficient resolution to study the dominate mechanisms. Acoustics uniquely offers the prospect of being able to nonintrusively provide profiles of the flow, the suspension field, and the bed topography. This exclusive combination of being able to measure all three components of the sediment triad, co-located and simultaneously, has been and is the driving force for applying acoustics to sediment transport processes.

The idea of using sound to study fundamental sediment processes in the marine environment is attractive, and, in concept, straightforward. A pulse of high-frequency sound, typically in the range 0.5–5 MHz in frequency, and centimetric in length, is transmitted downward from a directional sound source usually mounted a meter or two above the bed. As the pulse propagates down toward the bed, sediment in suspension backscatters a proportion of the sound and the bed generally returns a strong echo. The amplitude of the signal backscattered from the suspended sediments can be used to obtain vertical profiles of the suspended concentration and particle size. Utilizing the rate of change of phase of the backscattered signal provides profiles of the three orthogonal components of flow. The strong echo from the bed can be used to measure the bed forms.

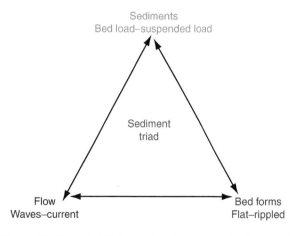

Figure 1 Illustration of the sediment processes triad and their interactions.

Acoustics therefore has the potentiality to provide profile measurements of near-bed sediment processes, with sufficient spatial and temporal resolution to allow turbulence and intrawave processes to be probed; this coupled with the bedform morphology observations provides sedimentologists and coastal engineers with an extremely powerful tool to advance understanding of sediment entrainment and transport. All of this is delivered with almost no influence on the processes being measured, because sound is the instrument of measurement.

Some Historical Background

For over two decades the vision of a number of people involved in studying small-scale sediment processes in the coastal zone has been to attempt to utilize the potential of acoustics to simultaneously and nonintrusively measure seabed morphology, suspended sediment particle size and concentration profiles, and profiles of the three components of flow, with the required resolution to observe the perceived dominant sediment processes. The capabilities to measure the three components have developed at different rates, and it is only in the past few years that the potentiality of an integrated acoustic approach for measuring the triad has become realizable. A schematic of the vision is shown in **Figure 2**.

The figure shows a visualization of the application of acoustics to sediment transport processes. 'A' is a multifrequency acoustic backscatter system (ABS), consisting, in this case, of three downward-looking narrow beam transceivers. The differential scattering characteristic of the suspended particles with frequency is used to obtain profiles of suspended sediment particle size and concentration profiles. 'B' is a three-axis coherent Doppler velocity profiler (CDVP) for measuring co-located profiles of the three orthogonal components of flow velocity; two horizontal and one vertical. It consists of an active narrow beam transceiver pointing vertically downward and two passive receivers having a wide beam width in the vertical and a narrow beam width in the horizontal. This system uses the rate of change of phase from the backscattered signal to obtain the three velocity components. 'C' is a pencil beam transceiver which rotates about a horizontal axis and functions as an acoustic ripple profiler (ARP). This is used to extract the bed echo and provide profiles of the bed morphology along a transect. These measurements are used to obtain, for example, ripple height and wavelength, and assess bed roughness. 'D' is a high-resolution acoustic bed (ripple) sector scanner (ARS) for imaging the local bed features. Although the ARS does not provide quantitative measurements of bed form height, it does provide the spatial distribution and this can be very useful when used in conjunction with the ARP. 'E' is a rapid backscatter ripple profiler system (BSARP) for measuring the instantaneous relationship between bed forms and the suspended sediments above it.

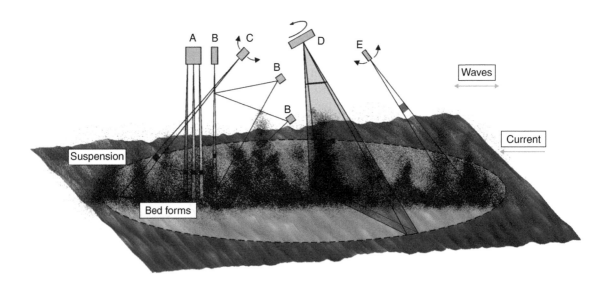

Figure 2 A vision of the application of acoustics to sediment transport processes. A, multifrequency acoustic backscatter for measuring suspended sediment particle size and concentration profiles. B, coherent Doppler velocity profiler for measuring the three orthogonal components of flow velocity. C, bed ripple profiler for measuring the bed morphology along a transect. D, high-resolution sector scanner for imaging the local bed features. E, backscatter scanning system for measuring the relationship between bed form morphology and suspended sediments.

What Can Be Measured?

The Bed

Whether the bed is rippled or flat has a profound influence on the mechanism of sediment entrainment into the water column. Steep ripples are associated with vortices lifting sediment well away from the bed, while for flat beds sediments primarily move in a confined thin layer within several grain diameters of the bed. Therefore knowing the form of the bed is a central component in understanding sediment transport processes. The development of the ARP and the ARS has had a significant impact on how we interpret sediment transport observations. These specifically designed systems typically either provide quantitative measurements of the evolution of a bed profile with time, the ARP, or generate an image of the local bed features over an area, the ARS.

Figure 3 shows data collected with a 2-MHz narrow pencil beam ARP in a marine setting. The figure shows the variability of a bed form profile, over nominally a 3-m transect, covering a 24-h period. Over this period the bed was subject to both tidal currents and waves and the figure shows the complex evolution of the bed with periods of ripples and less regular bed forms. The figure clearly shows the detailed quantitative measurements of the bed that can be obtained with the ARP.

Acoustic ripple scanners, ARSs, are based on sector-scanning technology, which has been specifically adapted for high-resolution images of bed form morphology. They typically have a frequency of around 1–2 MHz with beam widths of about 1° in the horizontal and 30° in the vertical. As the pulse is backscattered from the bed, the envelope of the signal is measured and usually displayed as image intensity.

An example of the data collected by an ARS is shown in Figure 4. As can be seen, this provides an aerial image of the bed, clearly showing the main bed features. The advantage of the ARS is the area coverage that is obtained, as opposed to a single line profile with the ARP; however, direct information on the height of features within the image cannot readily be extracted.

Ideally one would like to combine the two instruments and recently such systems have become available. This is essentially an ARP which also rotates horizontally through 180° and therefore allows a three-dimensional (3-D) measurements of the bed. The sonar gathers a single swath of data in the vertical plane and then rotates the transducer around the vertical axis and repeats the process until a circular area underneath the sonar has been scanned in a sequence of radial spokes. An example of data collected by a 3-D ARP operating at a frequency of 1.1 MHz is shown in Figure 5.

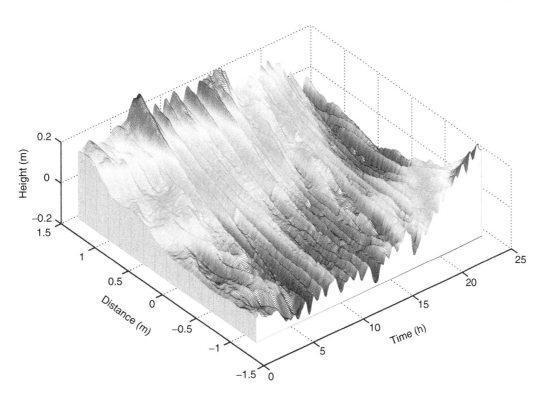

Figure 3   ARP measurements of the temporal evolution of a 3-m profile on a rippled bed, covering a 24-h period.

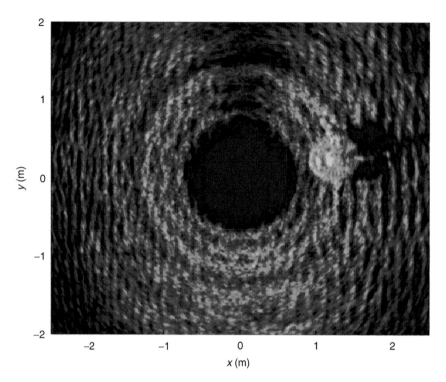

Figure 4 Image of ripples on the bed in a large-scale flume facility, the Delta Flume, collected using an acoustic ripple sector scanner, ARS. The 0.5-m-diameter circle at the center right of the image is one of the feet of the instrumented tripod used to collect the measurements, while the dark area at the center of the image is a blind spot directly beneath the sonar.

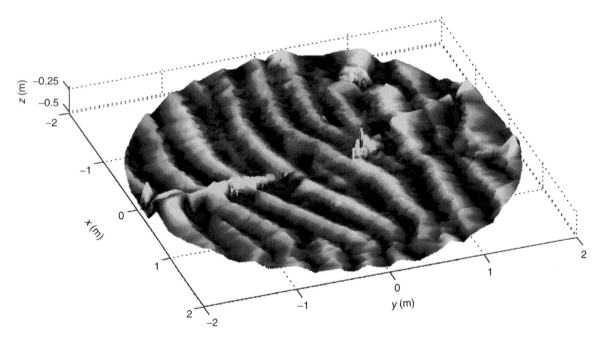

Figure 5 Three-dimensional measurements of a rippled bed collected in the marine environment using a 3-D ripple scanner, 3-D ARP. The artifacts in the image are associated with reflections from the instrumented tripod used to collect the measurements.

The data shown in the figure were collected during a recent marine deployment. The bed surface shown is based on a 3-D scan, with a 100 vertical swaths at 1.8° intervals, each of which comprised 200 acoustic samples at 0.9° intervals and spanning 180°. The plot clearly shows a rippled bed. There are one or two artifacts in the plot; these are associated with reflection from the rig on which the 3-D ARP was mounted. However, the figure plainly contains information on both the horizontal and vertical dimensions of the ripples and can therefore be used to precisely define quantitatively the features of the bed over an area. The 3-D ARP is a substantial advance on both the ARP and the ARS.

The Flow

The success of acoustic Doppler current profilers (ADCPs), which typically provide mean current profiles with decimeter spatial resolution, and more recently the acoustic Doppler velocimeter (ADV), which measures, subcentimetric, subsecond, three velocity components at a single height, has stimulated interest in using acoustics to measure near-bed velocity profiles. The objective is to use the same backscattered signal as used by the ABSs, but process the rate of change of phase of the signal (rather than the amplitude as used by ABSs) to obtain velocity profiles with comparable spatial and temporal resolution to ABSs. The phase technique is utilized in CDVPs and the phase approach has been the preferred method for obtaining high spatial and temporal resolution velocity profiles. An illustration of a three-axis CDVP is shown in **Figure 6**. A narrow-beam, downwardly pointing transceiver, Tz, transmits a pulse of sound. The scattered signal is picked up by Tz, and two passive receivers Rx and Ry which are orthogonal to each other and have a wide beam in the vertical and narrow in the horizontal.

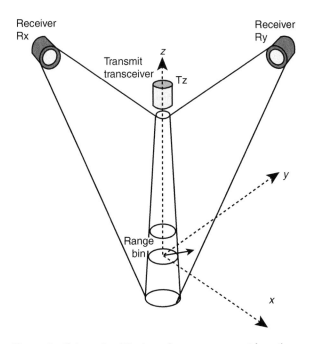

Figure 6 Schematic of the transducer arrangement for a three-axis CDVP.

To examine the capability of such a system, measurements from a three-axis CDVP have been compared with a commercially available ADV. The system had a spatial resolution of 0.04 m, operated over a range of 1.28 m, and provided 16-Hz velocity measurements of the vertical and two horizontal components of the flow. **Figure 7** shows a typical example of CDVP and ADV time series, power spectral density, and probability density function plots for **u**, the streamwise flow. The velocities presented in **Figure 7** show CDVP results that compare very favorably with the ADV measurements; having time series, spectra, and probability distributions in general agreement. There are differences in the spectrum; the CDVP spectra begins to depart from the ADV above about 4 Hz, with the CDVP measuring larger spectral components at the higher frequencies. This trend was common to all the records and is a limitation of the CDVP system used to obtain the data, rather than an intrinsic limitation to the technique. **Figure 8** illustrates the capability of the CDVP for flow visualization in the marine environment. The figure shows mean zeroed velocity vectors, **u**–**w**, **v**–**w**, and **u**–**v**, plotted over a 5-s time period, between 0.05 and 0.7 m above the bed. The length of the velocity vectors is indicated in the figure. A single-point measurement instrument such as an ADV can provide the time-varying velocity vectors at a single height above the bed; however, the spatial profiling which is achievable with the three-axis CDVP provides a capability to visualize structures in the flow. The structures seen in **Figure 8** are associated with combined turbulent and wave flows. This type of plot exemplifies the value of developing a three-axis CDVP with co-located measurement volumes, since it clearly illustrates the fine-scale temporal and spatial flow structures which can be measured in the near-bed flow regime. Linking such measurements with ABS profiles of particle size and concentration will provide a very powerful tool for studying near-bed fluxes and sediment transport processes.

The Suspended Sediments

Multifrequency acoustic backscattering (ABS) can be used to obtain profiles of mean particle size and concentration. The ABS is the only system available that profiles both parameters rapidly and simultaneously. Also the bed echo references the profile to the local bed position. This is important because all sediment transport formulas use the bed as the reference point and predict profiles of suspension parameters relative to the bed location. Examples of the results that can be obtained are shown

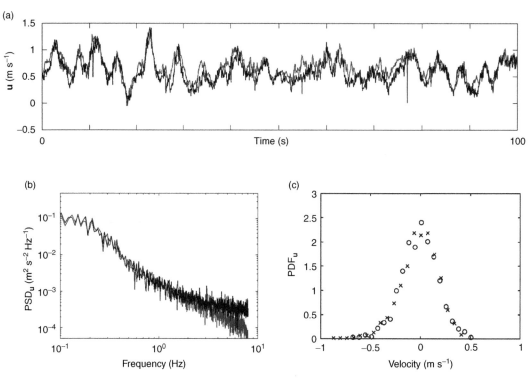

Figure 7 Comparison of the streamwise flow, **u**, measured by an ADV (red) and a CDVP (black). (a) The velocities measured at 16 Hz over 100 s. (b) The power spectra of the zero-mean velocities. (c) The probability density functions of the zero-mean velocities for the ADV (open circles) and the CDVP (crosses).

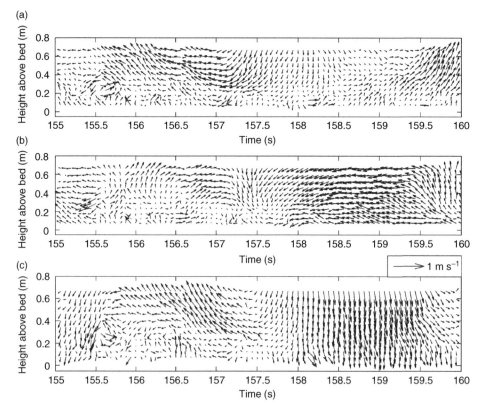

Figure 8 Demonstration of the capability of the triple-axis CDVP to provide visualizations of intrawave and turbulent flow. Plots in (a)–(c) show a time series over a 5 s period of the zero-mean velocities displayed as vectors **u–w**, **v–w**, and **u–v**, respectively.

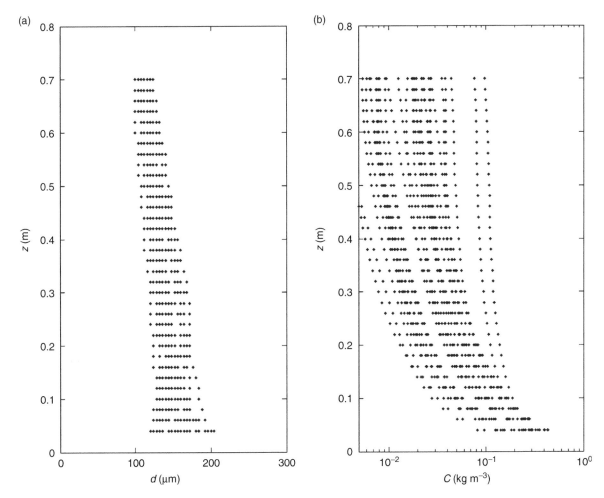

Figure 9 Profiles of suspended sediments: (a) particle size and (b) concentration.

in **Figures 9** and **10**. These observations were collected off Santa Cruz Pier, California, as a seasonal storm passed through the area over the period of a couple of days. **Figure 9** shows profiles of the suspended sediment particle size and concentration. The profiles of particle size are seen to be relatively consistent with a mean diameter of c. 180 μm near the bed and reducing to c. 120 μm at 0.7 m above the bed. The variability in size is relatively limited and due to changing bed and hydrodynamic conditions as the storm passed by.

In **Figure 9(b)** the suspended concentrations are comparable in their form, although they have absolute values that vary by greater than an order of magnitude. This variation in suspended concentration is associated with the changing conditions as the storm passed through the observational area. It is interesting to note that over the period even though there is a large variation in concentration, the particle size remains nominally consistent. **Figure 10** shows the temporal variation in particle size and

concentration with height above the bed. It can be clearly seen that some of the periods of increased particle size are associated with substantial suspended sediment events, as one might expect, however, there are one or two events where the correlation is not as clear. **Figures 9** and **10** clearly illustrate the capability of ABS to simultaneously measure profiles of concentration and particle size and the combination of both significantly adds to the assessment and development of sediment transport formulas.

A Case Study of Waves over a Rippled Bed

Here the use of acoustics is illustrated by application to a specific experimental study. Over large areas of the continental shelf outside the surf zone, sandy seabeds are covered with wave-formed ripples. If the ripples are steep, the entrainment of sediments into the water column, due to the waves, is considered

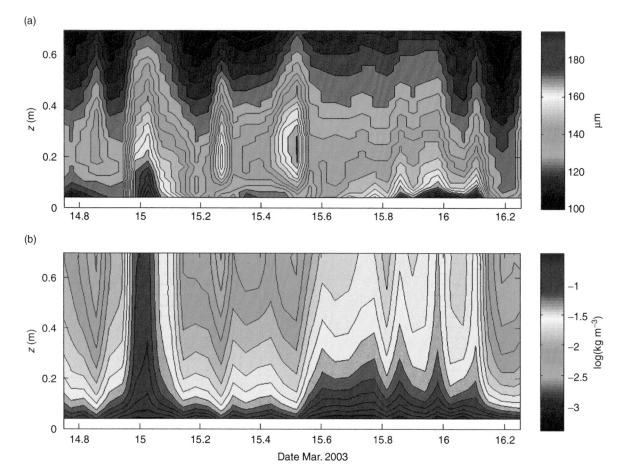

Figure 10 Measurements of the temporal variations with height above the bed, of: (a) particle size with the color bar scaled in microns and (b) logarithmic concentration with the color bar scaled relative to $1.0\,\mathrm{kg\,m^{-3}}$.

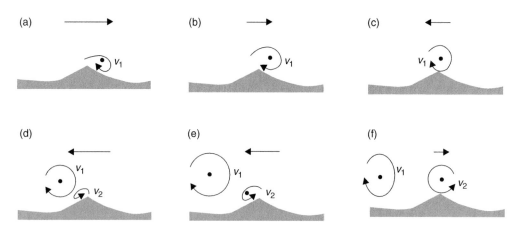

Figure 11 A schematic of vortex sediment entrainment by waves over a steeply rippled bed. The arrows show the direction and relative magnitude of the near-bed wave velocity. v_1 and v_2 are the lee slope-generated vortices.

to be primarily associated with the generation of vortices. This process is illustrated in **Figure 11**. As shown in **Figures 11(a)** and **11(b)**, a spinning parcel of sediment-laden water, v_1, is formed on the leeside of the ripple at the peak positive velocity in the wave

cycle. This sediment-rich vortex is then thrown up into the water column at around flow reversal (**Figures 11(c)** and **11(d)**), carrying sediment well away from the bed and allowing it to be transported by the flow. At the same time, a sediment-rich vortex,

v_2, is being formed on the opposite side of the ripple due to the reversed flow. As shown in **Figures 11(d)–11(f)**, v_2 grows, entrains sediment, becomes detached, and moves over the crest at the next flow reversal carrying sediments into suspension. The main feature of the vortex mechanism is that sediment is carried up into the water column twice per wave cycle at flow reversal. This mechanism is completely different to the flat bed case where maximum near-bed concentration is at about the time of maximum flow velocity.

To study this fundamental process of sediment entrainment, experiments were conducted in one of the world's largest man-made channels, specifically constructed for such sediment transport studies, the Delta Flume; this is located at the De Voorst Laboratory of the Delft Hydraulics in the north of the Netherlands. The flume is shown in **Figure 12(a)**; it is 230 m in length, 5 m in width, and 7 m in depth, and it allows waves and sediment transport to be studied at full scale. A large wave generator at one end of the flume produced waves that propagated along the flume, over a sandy bed, and dissipated on a beach at the opposite end. The bed was comprised of coarse sand, with a mean grain diameter by weight of 330 μm, and this was located approximately halfway along the flume in a layer of thickness 0.5 m and length 30 m. In order to make the acoustic and other auxiliary measurements, an instrumented tripod platform was developed and is shown in **Figure 12(b)**. The tripod, STABLE II (Sediment Transport and Boundary Layer Equipment II), used an ABS to measure profiles of particle size and concentration, a pencil beam ARP to measure the bed forms, and, in this case, electromagnetic current meters (ECMs) to measure the horizontal and vertical flow components. **Figure 12(c)** shows a wave propagating along the flume with STABLE II submerged in water of depth 4.5 m, typical of coastal zone conditions.

To investigate and then model the vortex entrainment process it was necessary to establish at the outset whether or not the surface waves were generating ripples on the bed in the Delta Flume. Using an ARP a 3-m transect of the bed was measured over time. The results of the observations over a 90-min recording period are shown in **Figure 13**. Clearly ripples were formed on the bed and the ripples were mobile. To obtain flow separation and hence vortex formation requires a ripple steepness (ripple height/

(a)   (b)   (c)

Figure 12 (a) Photograph of the Delta Flume showing the sand bed at approximately midway along the flume and the wave generator at the far end of the flume. (b) The instrumented tripod platform, STABLE II, used to make the measurements. (c) A surface wave propagating along the flume.

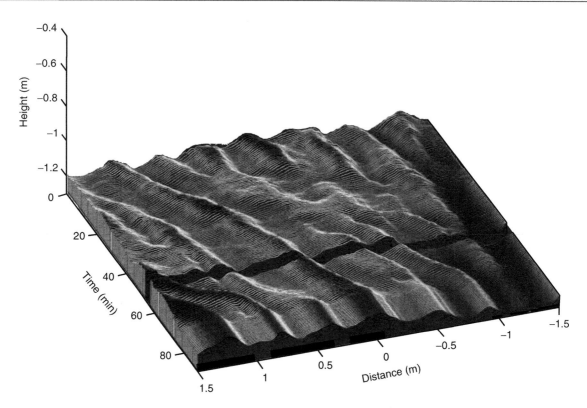

Figure 13 Acoustic ripple profiler measurements of a transect of the bed, over time, in the Delta Flume.

ripple wavelength) of the order of 0.1 or greater; an analysis of the observations showed that this was indeed the case.

Using the ABS, some of the most detailed full scale measurements of sediment transport over a rippled bed under waves were captured. These measurements from the Delta Flume are shown in **Figure 14**. The images shown were constructed over a 20-min period as a ripple passed beneath the ABS. The suspended concentrations over a ripple, at the same velocity instants during the wave cycle, were combined to generate a sequence of images of the concentration over the ripple with the phase of the wave. Four images from the sequence have been shown to illustrate the measured vortex entrainment. The length and direction of the arrows in the figure give the magnitude and direction of the wave velocity, respectively. Comparison of **Figure 14** with **Figure 11** shows substantial similarities. In **Figure 14(a)**, there can be observed the development of a high-concentration event at high flow velocity above the lee slope of the ripple, v_1. In **Figure 14(b)**, as the flow reduced in strength, the near-bed sediment-laden parcel of fluid travels up the leeside of the ripple toward the crest. As the flow reverses, this sediment-laden fluid parcel, v_1, travels over the crest and expands. As the reverse flow increases in strength (**Figure 14(d)**), the parcel v_1 begins to lift

away from the bed and a new sediment-laden lee vortex, v_2, is initiated on the lee slope of the ripple.

In order to capture the essential features of these data within a relatively simple, and hence practical, 1-DV (one-dimensional in the vertical) model, the data has first been horizontally averaged over one ripple wavelength at each phase instant during the wave cycle. The resulting pattern of sediment suspension contours is shown in the central panel of **Figure 15**, while the upper panel shows the oscillating velocity field measured at a height of 0.3 m above the bed. The concentration contours shown here are relative to the ripple crest level, the mean (undisturbed) bed level being at height $z = 0$. The measured concentration contours presented in **Figure 15** show two high concentration peaks near the bed that propagate rapidly upward through a layer of thickness corresponding to several ripple heights. The first, and the strongest, of these peaks occurs slightly ahead of flow reversal, while the second, weaker and more dispersed peak, is centered on flow reversal. The difference in the strengths of the two peaks reflects the greater positive velocity that can be seen to occur beneath the wave crest (time $= 0$ s) than beneath the wave trough (time $= 2.5$ s). Between the two concentration peaks the sediment settles rapidly to the bed. Maybe rather unexpectedly this settling effect occurs at the times of strong forward and backward velocity at measurement

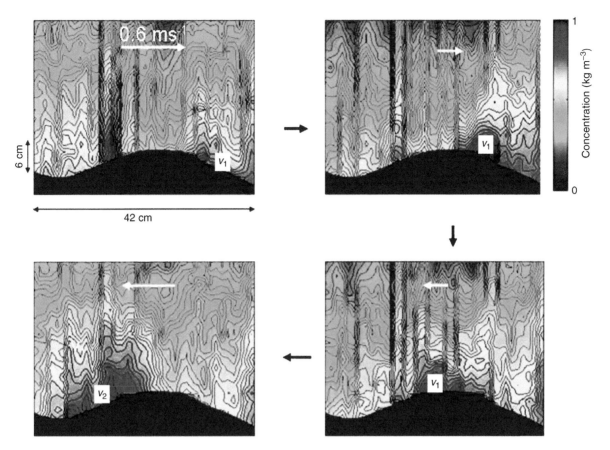

Figure 14 Acoustic imaging of suspended sand entrainment over a rippled bed due to waves, at four phases of the wave velocity. The length of the white arrow in each plot gives the magnitude and direction of the near-bed wave velocity.

levels well above the bed. The underlying mechanism of sediment entrainment by vortices shed at or near flow reversal is clearly evident in the spatially averaged measurements shown in **Figure 15**.

Any conventional 'flat rough bed' model that attempts to represent the above sequence of events in the suspension layer runs into immediate and severe difficulties, since such models predict maximum near-bed concentration at about the time of maximum flow velocity, and not at flow reversal. Here therefore, for the first time in a 1-DV model, it has been attempted to capture these effects realistically through the use of a strongly time-varying eddy viscosity that represents the timing and strength of the upward mixing events due to vortex shedding. The model initially predicts the size of the wave-induced ripples and the size of the grains found in suspension, and then goes on to solve numerically the equations governing the upward diffusion and downward settling of the suspended sediment. The resulting concentration contours in the present case are shown on the lower panel of **Figure 15**. The essential two-peak structure of the eddy shedding process can be seen to be represented rather well, with the initial

concentration peak being dominant. The decay rate of the concentration peaks as they go upward is also represented quite well, though a phase lag develops with height that is not seen to the same extent in the data. Essentially, the detailed acoustic observations of sediment entrainment under waves over ripples of moderate steepness have begun to establish a new type of 1-DV modeling, thereby allowing the model to go on to be used for practical prediction purposes in the rippled regime, which is the bed form regime of most importance over wide offshore areas in the coastal seas.

Discussion and Conclusions

The aim of this article has been to illustrate the application of acoustics in the study of near-bed sediment processes. It was not to detail the theoretical aspects of the work, which can be found elsewhere. To this end, measurements of bed forms, the hydro-dynamics, and the movement of sediments have been used. These results show that acoustics is progressively approaching the stage where it

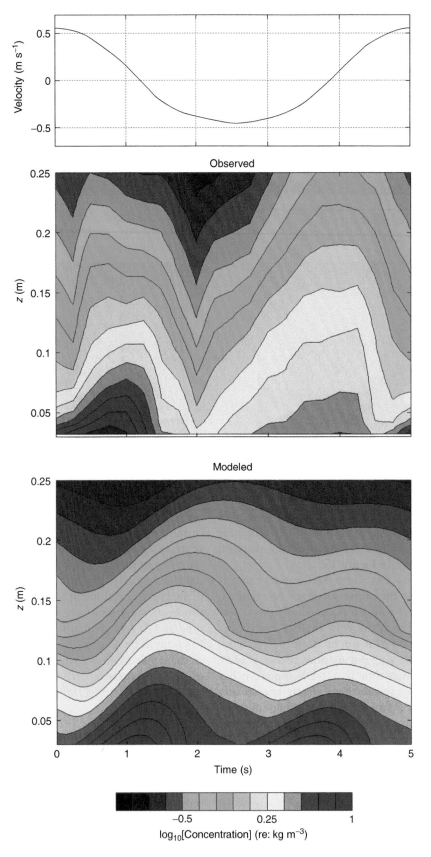

Figure 15 Measurement and modeling of suspended sediments with height z above a rippled bed under a 5-s period wave.

can measure nonintrusively, co-located and simultaneously, with high temporal and spatial resolution, all three components of the interacting sediment triad. **Figure 16** shows the instrumentation used in a further recent deployment in the Delta Flume. This shows the convergence of the instrumentation and also its use in conjunctions with instruments such as laser *in situ* scattering transmissometry (LISST).

Although substantial advances have been made in the past two decades, the application of acoustics to sediment transport processes is still in an ongoing developmental phase and there are limitations and shortcomings that need to be overcome, and further applications explored. Although there have been few reports to date on data collected using 3-D ARP, such systems are now becoming available. The 3-D ARP is a substantial development over the single-line ARP and the ARS, and should make a considerable contribution to the measurement and understanding of the formation and development of bed forms. There have been a number of reports on single-axis CDVP; however, it is the three-axis CDVP which is the way ahead. Again, these instruments are coming online, though they are still very much a research tool. However, the concept has now essentially been

proven and its use with a vengeance in sediment studies will begin to make an important impact in the next couple of years. It was with the concept of using acoustics to measure suspended sediment concentration that its application to the other components of the small-scale sedimentation processes triad followed. To date the use of sound to measure suspended sediment concentration and particle size has been successful when systems have been deployed over nominally homogeneous sandy beds. However, all who use acoustics recognize that the marine sedimentary environment is frequently much more complicated, and suspensions of cohesive sediments and combined cohesive and noncohesive sediments are common. To employ acoustics quantitatively in mixed and cohesive environments requires the development of a description of the scattering properties of suspensions of cohesive sediments and sediment mixtures. This would be interesting and very valuable work, and should significantly extend the deployment regime over which acoustic backscattering can be employed quantitatively.

In conclusion, the objective of this article has been to describe the role of acoustics, in near-bed sediment transport studies. It is clearly acknowledged

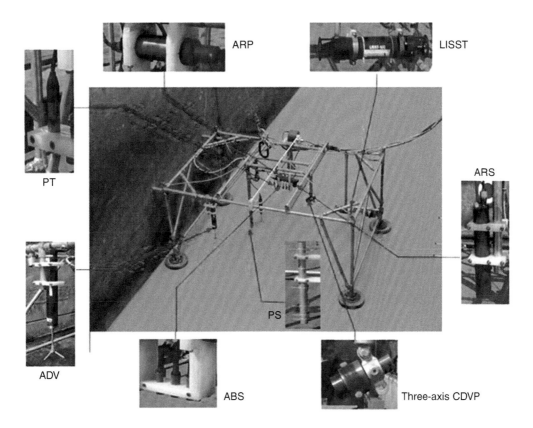

Figure 16 The instrumentation used in a recent Delta Flume experiment. LISST, laser *in situ* scattering transmissometry; PS, pumped samples; PT, pressure transducer.

that acoustics is one of a number of technologies advancing our capabilities to probe sediment processes. However, its nonintrusive profiling ability, coupled with its capability to measure all three components of the sediment dynamics triad, make it a unique and very powerful tool for studying the fundamental mechanisms of sediment transport.

See also

Acoustics in Marine Sediments. Estuarine Circulation. Offshore Sand and Gravel Mining.

Further Reading

Crawford AM and Hay AE (1993) Determining suspended sand size and concentration from multifrequency acoustic backscatter. *Journal of the Acoustical Society of America* 94(6): 3312–3324.

Davies AG and Thorne PD (2005) Modelling and measurement of sediment transport by waves in the vortex ripple regime. *Journal of Geophysical Research* 110: C05017 (doi:1029/2004JC002468).

Hay AE and Mudge T (2005) Principal bed states during SandyDuck97: Occurrence, spectral anisotropy, and the bed storm cycle. *Journal of Geophysical Research* 110: C03013 (doi:10.1029/2004JC002451).

Thorne PD and Hanes DM (2002) A review of acoustic measurements of small-scale sediment processes. *Continental Shelf Research* 22: 603–632.

Vincent CE and Hanes DM (2002) The accumulation and decay of near-bed suspended sand concentration due to waves and wave groups. *Continental Shelf Research* 22: 1987–2000.

Zedel L and Hay AE (1999) A coherent Doppler profiler for high resolution particle velocimetry in the ocean: Laboratory measurements of turbulence and particle flux. *Journal of Atmosphere and Ocean Technology* 16: 1102–1117.

Relevant Websites

http://www.aquatecgroup.com
  – Aquatec Group Ltd.
http://www.marine-electronics.co.uk
  – Marine Electronics Ltd.
http://www.pol.ac.uk
  – POL Research, Proudman Oceanographic Laboratory.

ACOUSTICS IN MARINE SEDIMENTS

T. Akal, NATO SACLANT Undersea Research Centre, La Spezia, Italy

Introduction

Because of the ease with which sound can be transmitted in sea water, acoustic techniques have provided a very powerful means for accumulating knowledge of the environment below the ocean surface. Consequently, the fields of underwater acoustics and marine seismology have both used sound (seismo-acoustic) waves for research purposes.

The ocean and its boundaries form a composite medium, which support the propagation of acoustic energy. In the course of this propagation there is often interaction with ocean bottom. As the lower boundary, the ocean bottom is a multilayered structure composed of sediments, where acoustic energy can be reflected from the interface formed by the bottom and subbottom layers or transmitted and absorbed. At low grazing angles, wave guide phenomena become significant and the ocean bottom, covered with sediments of different physical characteristics, becomes effectively part of the wave guide. Depending on the frequency of the acoustic energy, there is a need to know the acoustically relevant physical properties of the sediments from a few centimeters to hundreds of meters below the water/sediment interface.

Underwater acousticians and civil engineers are continuously searching for practical and economical means of determining the physical parameters of the marine sediments for applications in environmental and geological research, engineering, and underwater acoustics. Over the past three decades much effort has been put into this field both theoretically and experimentally, to determine the physical properties of the marine sediments. Experimental and forward/inverse modeling techniques indicate that the acoustic wave field in the water column and seismo-acoustic wave field in the seafloor can be utilized for remote sensing of the physical characteristics of the marine sediments.

Sediment Structure as an Acoustic medium

Much of the floor of the oceans is covered with a mixture of particles of sediments range in size from boulder, gravels, coarse and fine sand to silt and clay, including materials deposited from chemical and biological products of the ocean, all being saturated with sea water. Marine sediments are generally a combination of several components, most of them coming from the particles eroded from the land and the biological and chemical processes taking place in sea water. Most of the mineral particles found in shallow and deep-water areas, have been transported by runoff, wind, and ice and subsequently distributed by waves and currents.

After these particles have been formed, transported, and transferred, they are deposited to form the marine sediments where the physical factors such as currents, dimensions and shapes of particles and deposition rate influence the spatial arrangements and especially sediment layering. Particles settle to the ocean floor and remain in place when physical forces are not sufficiently strong to move them. In areas with strong physical forces (tidal and ocean currents, surf zones etc.) large particles dominate, whereas in low motion energy areas (ocean basins, enclosed bays) small particles dominate.

During the sedimentation process these particles, based on the physical and chemical interparticle forces between them, form the sedimentary acoustic medium: larger particles (e.g., sands) by direct contact forces; small particles (e.g., clays and fine silts) by attractive electrochemical forces; and silts, remaining between sands and clays are formed by the combination of these two forces. The amount of the space between these particles is the result of different factors, mainly size, shape, mineral content and the packing of the particles determined by currents and the overburden pressure present on the ocean bottom.

Figure 1 is an example of a core taken very carefully by divers, to ensure an undisturbed internal structure of the sediment sample. Sediment structures have been quantified by using X-ray computed tomography to obtain values of density with a millimeter resolution for the full three-dimensional volume. The image shown in **Figure 1** is a false color 3D reconstruction of a core sample at a site where sediments consist of sandy silt (75% sand, 15% silt, and 10% clay) and shell pieces.

The results of the X-ray tomography of the same core, can also be shown on an X-ray cross-section slice along the center of the core (**Figure 2A**) and the corresponding two-dimensional spectral density levels for that cross-section (**Figure 2B**). The complex

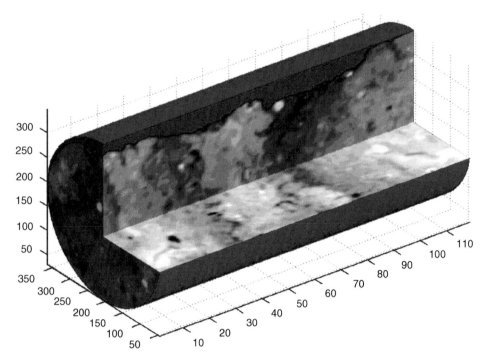

Figure 1   Three-dimensional reconstruction of a sediment core sample containing 75% sand, 15% silt and 10% clay. Scale in mm.

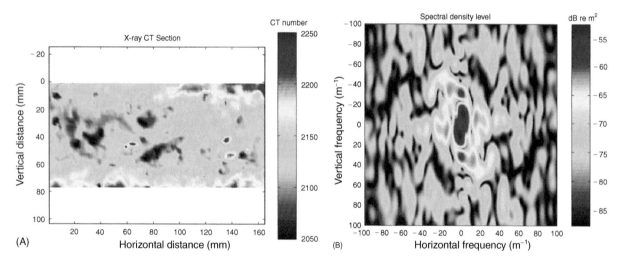

Figure 2   (A) X-ray cross-section slice along the center of the core shown in **Figure 1**. (B) 2-D spectral density levels for the cross-section shown in (A).

structure of the sediments can be with seen strong heterogeneity (local density fluctuation) of the medium that controls the interaction of the seismo-acoustic energy. In addition to the complex fine structure described above, the seafloor can also show complex layering (**Figure 3A**) or a simpler structure (**Figure 3B**). These structures result from the lowering of sea levels during the glaciation of the Pleistocene epoch during which sand was deposited over wide areas of the continental shelves. Unconsolidated

sediments subsequently covered the shelves as the sea level are in postglacial times.

Biot–Stoll Model

Various theories have been developed to describe the geoacoustic response of marine sediments. The most comprehensive theory is based on the Biot model as elaborated by Stoll. This model takes into account various loss mechanisms that affect the response of

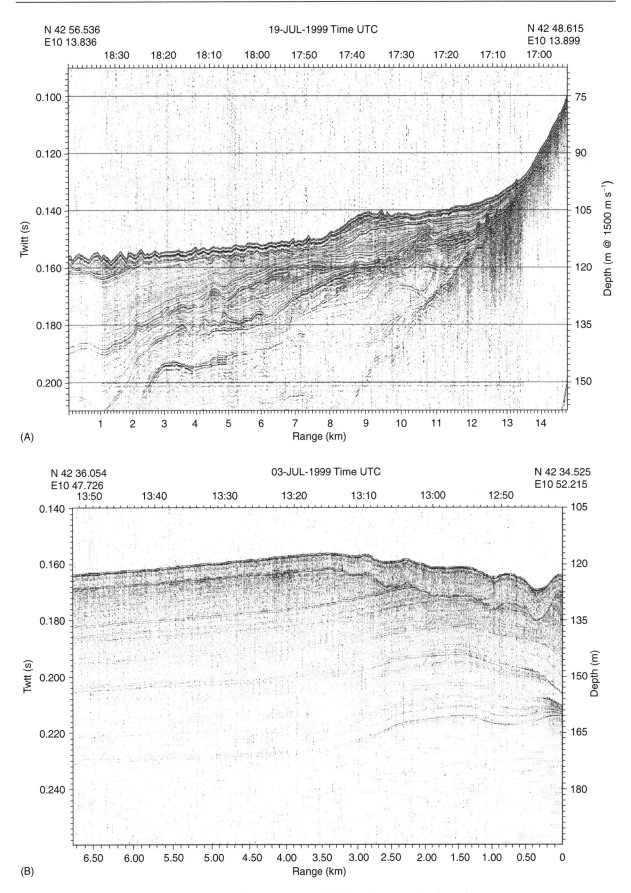

Figure 3 Seismic reflection profiles showing (A) complex and (B) simple structures of sediment layers.

porous sediments that are saturated with fluid. The Biot–Stoll theory shows that acoustic wave velocity and attenuation in porous, fluid-saturated sediments depend on a number of parameters including porosity, mean grain size, permeability, and the properties of the skeletal frame.

According to the Biot–Stoll theory, in an unbounded, fluid-saturated porous medium, there are three types of body wave. Two of these are dilatational (compressional) and one is rotational (shear). One of the compressional waves ('first kind') and the shear wave are similar to body waves in an elastic medium. In a compressional wave of the 'first kind', the skeletal frame and the sea water filling the pore space move nearly in phase so that the attenuation due to viscous losses becomes relatively small. In contrast, due to out-of-phase movement of the frame and the pore fluid, the compressional wave of the 'second kind' becomes highly attenuated. The Biot theory and its extensions by Stoll have been used by many researchers for detailed description of the acoustic wave–sediment interaction when basic input

parameters such as those shown in **Table 1** are available.

Seismo-acoustic Waves in the Vicinity of the Water–Sediment Interface

As mentioned above, when acoustic energy interacts with the seafloor, the energy creates two basic types of deformation: translational (compressional) and rotational (shear). Solution of the equations of the wave motion shows that each of these types of deformation travels outward from the source with its own velocity. Wave type, velocity, and propagation direction vary in accordance with the physical properties and dimensions of the medium.

The ability of seafloor sediments to support the seismo-acoustic energy depends on the elastic properties of the sediment, mainly the bulk modulus (incompressibility, K) and the shear modulus (rigidity, G). These parameters are related to the compressional and shear velocities C_P and C_S respectively, by:

$$C_P = [(K + 4G/3)/\rho]^{1/2}$$

$$C_S = (G/\rho)^{1/2}$$

where, ρ is the bulk density. **Table 2** shows basic seismic–acoustic wave types and their velocities related to elastic parameters

These two types of deformations (compressional and shear) belong to a group of waves (body waves) that propagate in an unbounded homogeneous medium. However, in nature the seafloor is bounded and stratified with layers of different physical properties. Under these conditions, propagating energy undergoes characteristic conversions every time it interacts with an interface: propagation velocity,

Table 1 Basic input parameters to Biot–Stoll model

Frequency-independent Variables	
Porosity (%)	P
Mass density of grains	ρ_r
Mass density of pore fluid	ρ_f
Bulk modulus of sediment grains	K_r
Bulk modulus of pore fluid	K_f
Variables affecting global fluid motion	
Permeability	k
Viscosity of pore fluid	η
Pore-size parameter	a
Structure factor	α
Variables controlling frequency-dependent response of frame	
Shear modulus of skeletal frame	$\bar{\mu} = \mu_r(\omega) + i\mu_i(\omega)$
Bulk modulus of skeletal frame	$\bar{K}_b = K_{br}(\omega) + iK_{bi}(\omega)$

Table 2 Basic seismo-acoustic wave types and elastic parameters

Basic wave type		Wave velocity
Body wave	Compressional	$C_P = [(K + 4G/3)/\rho]^{1/2}$
	Shear	$C_S = (G/\rho)^{1/2}$
Ducted wave	Love	$C_L = (G/\rho)^{1/2}$
Surface wave	Scholte	$C_{SCH} = (G/\rho)^{1/2}$
Elastic parameters in terms of wave velocities (C) and bulk density (ρ)		
Bulk modulus (incompressibility)	$K = \rho(C_P^2 - 4C_S^2/3)$	
Compressibility	$\beta = 1/K$	
Young's modulus	$E = 2C_S^2\rho(1 + \sigma)$	
Poisson's ratio (transv./long. strain)	$\sigma = (3E - \rho C_P^2)/(3E + 2\rho C_P^2)$	
Shear modulus (rigidity)	$G = \rho C_S^2$	
Lame's constant	$\lambda = \rho(C_P^2 - 2C_S^2)$	

energy content, and spectral structure and propagation direction changes. In addition, other types of waves, i.e., ducted waves and surface waves, may be generated. These basic types of waves and their characteristics together with their arrival structure as synthetic seismograms for orthogonal directions are illustrated in **Figure 4**.

Body Waves

These waves propagate within the body of the material, as opposed to surface waves. External forces can distort solids in two different ways. The first involves the compression of the material without changing its shape; the second implies a change in shape without changing its volume (distortion). From earthquake seismology, these compressional and distortion waves are called primus (P) and secoundus (S), respectively, for their arrival sequence on earthquake records. However, the distortional waves are very often called shear.

Compressional waves Compressional waves involve compression of the material in such a manner that the particles move in the direction of propagation.

Shear waves Shear waves are those in which the particle motion is perpendicular to the direction of propagation. These waves can be generated at a layer interface by the incidence of compressional waves at other than normal incidence. Shear wave energy is polarized in the vertical or horizontal planes, resulting in vertically polarized shear waves (SV) and horizontally polarized shear waves (SH). However, if the interfaces are close (relative to a wavelength), one cannot distinguish between body and surface waves.

Ducted Waves

Love waves Love waves are seismic surface waves associated with layering; they are characterized by horizontal motion perpendicular to the direction of propagation (SH wave). These waves can be considered as ducted shear waves traveling within the duct of the upper sedimentary layer where total reflection occurs at the boundaries; thus the waves represent energy traveling by multiple reflections.

Interface Waves

Seismic interface waves travel along or near an interface. The existence of these waves demands the combined action of compressional and shear waves. Thus at least one of the media must be solid, whereas

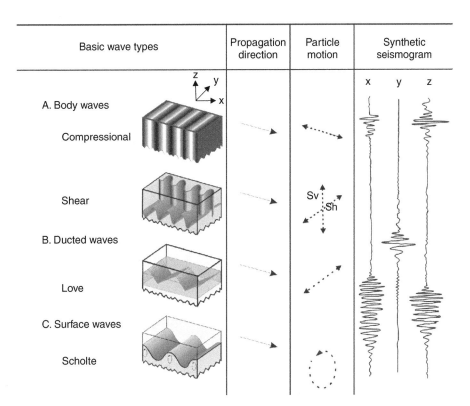

Figure 4   Basic seismo-acoustic waves in the vicinity of a water–sediment interface.

the other may be a solid (Stoneley wave), a liquid (Scholte wave), or a vacuum (Rayleigh wave). For two homogeneous half-spaces, interface waves are characterized by elliptical particle motion confined to the radial/vertical plane and by a constant velocity that is always smaller than the shear wave.

When different types of seismic waves propagate and interact with the layered sediments they are partly converted into each other and their coupling may create mixed wave types in the vicinity of the interface. **Figure 4** shows the basic seismo-acoustic waves in the vicinity of a water/sediment interface. Basic characteristics of these waves together with their particle motion and synthetic seismograms at three orthogonal directions (x, y and z) are also illustrated in the same Figure. Under realistic conditions, in which the seafloor cannot be considered to be homogeneous, isotropic, nor a half-space, some of these waves become highly attenuated or travel together, making identification very difficult. In fact, the interface waves shown in **Figure 4** are for a layered seafloor, where the dispersion of the signal is evident (homogeneous half-space would not give any dispersion).

Under realistic conditions, i.e., for an inhomogeneous, bounded and anisotropic seafloor, some of these waves convert from one to another. The different wave types may travel with different speeds or together, and they generally have different attenuation. As an example, **Figure 5** shows signals from an explosive source (0.5 kg trinitrotoluene (TNT))

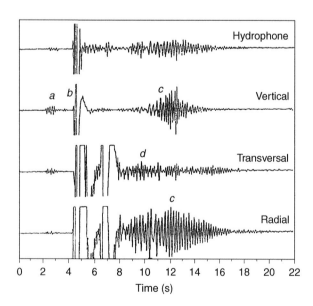

Figure 5 Signals from an explosive source of 0.5 kg trinitrotoluene received by a hydrophone and three orthogonal geophones at 1.5 km distance (a, Head wave; b, Compressional wave; c, Interface wave; d, Love wave).

received by a hydrophone and three orthogonal geophones placed on the seafloor, at a distance of 1.5 km from the source. The broadband signal of a few milliseconds duration generated by the explosive source is dispersed over nearly 18 s demonstrating the arrival structure of different types of waves. The characteristics of these waves are indicated in the figure for four different sensors. They can be identified in order of their arrival time as: (a) head wave; (b) water arrival (compressional wave); (c) interface wave; (d) Love waves.

Seafloor Roughness

The roughness of the water–sediment interface and layers below is another important parameter that needs to be considered in sediment acoustics. The seafloor contains a wide spectrum of topographic roughness, from features of the order of tens of kilometers, to those of the order of millimeters. The shape of the seafloor and its scattering effect on acoustic signals is be covered here.

Techniques to Measure Geoacoustic Parameters of Marine Sediments

The geoacoustic properties of the seafloor defined by the compressional and shear wave velocities, their attenuation, together with the knowledge of the material bulk density, and their variation as a function of depth, are the main parameters needed to solve the acoustic wave equation. To be able to determine these properties of the seabed, different techniques have been developed using samples taken from the seafloor, instruments and divers conducting measurements *in situ*, and remote techniques measuring seismo-acoustic waves and inverting this information with realistic models into sediment properties. Some of the current methods of obtaining geoacoustic parameters of the marine sediments are briefly described here.

Laboratory Measurements on Sediment Core Samples

Most of our knowledge of the physical properties of sediments is acquired through core sampling. A large number of measurements on marine sediments have been made in the past. In undisturbed sediment core samples, under laboratory conditions, density and compressional velocity can be measured with accuracy, and having measured values of density and compressional velocity, the bulk modulus can be selected as the third parameter, where it is can be calculated (**Table 1**).

There are several laboratory techniques available to measure some of the sediment properties. However, the reliability of such measurements can be degraded by sample disturbance and temperature and pressure changes. In particular, the acoustic properties are highly affected by the deterioration of the chemical and mechanical bindings caused by the differences in temperature and pressure between the sampling and the laboratory measurements. Controlling the relationships between various physical parameters can be used to check the accuracy of measured parameters of sediment properties. It has been shown that the density and porosity of sediments have a relationship with compressional velocity, and different empirical equations between them have been established.

Over the past three decades, at the NATO Undersea Research Centre (SACLANTCEN), a large number of laboratory and *in situ* measurements have been made of the physical properties of the seafloor sediments. These measurements have been conducted on the samples with the same techniques as when the same laboratory methods were applied. This data set with a great consistency of the hardware and measurement technique, has been used to demonstrate the physical characteristics of the sediment that affect acoustic waves.

The relationship between measured physical parameters From 300 available cores, 20 000 measured data samples for the density and porosity, and 10 000 samples for the compressional velocity were obtained. To be able to handle this large data set taken from different oceans at different water depths, all bulk density and compressional velocity data were converted into relative density (ρ) and relative velocity (C) with respect to the *in situ* water values:

$$\rho = \rho_S/\rho_W$$
$$C = C_S/C_W$$

where, ρ_S is sediment bulk density, ρ_W is water density, C_S is sediment compressional velocity and C_W is water compressional velocity. The data are not only from the water–sediment interface but cover sedimentary layers of up to 10 m deep.

Relative density and porosity The density and porosity of the marine sediment are least affected during coring and laboratory handling of the samples. Porosity is given by the percentage volume of the porous space and sediment bulk density by the weight of the sample per unit volume. The relationship between porosity and bulk density has been investigated by many authors with fewer data

than used here and shown to have a strong linear correlation. Theoretically this linearity only exists if the dry densities of the mineral particles are the same for all marine sediments. The density of the sediment would then be the same as the density of the solid material at zero porosity, and the same as the density of the water at 100% porosity. **Figure 6** shows this relationship.

Porosity and relative compressional wave velocity The relationship between porosity and compressional wave velocity has received much attention in the literature because porosity can be measured easily and accurately. Data from the SACLANTCEN sediment cores giving the relationship between porosity and compressional wave velocity are shown in **Figure 7**. As shown in **Figure 6** due to the linear relation between density and porosity, the relationship between density and compressional wave velocity is similar to the porosity compressional wave velocity relation.

In situ Techniques

There are several *in situ* techniques available that use instruments lowered on to the seafloor mainly by means of submersibles, remotely operated vehicles (ROVs), and autonomous underwater vehicles AUVs and divers. The first deep-water, *in situ* measurements of sediment properties were made from the bathyscaph *Trieste* in 1962. These measurements provided accurate results due to the minimum disturbance of temperature and pressure changes compared to bringing the sample to the surface and for

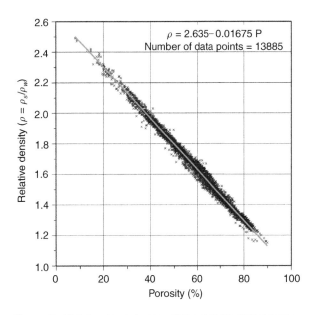

Figure 6 Relationship between relative density and porosity.

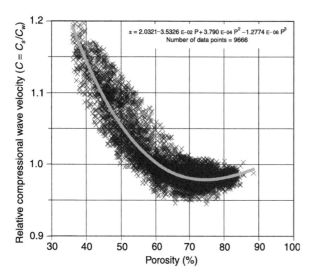

Figure 7   Relationship between relative compressional wave velocity and porosity.

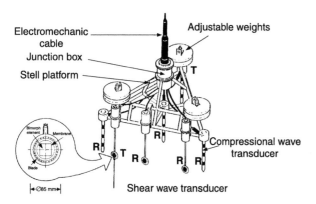

Figure 8   *In Situ* Sediment Acoustic Measurement System (ISSAMS) for near-surface *in situ* geoacoustic measurements.

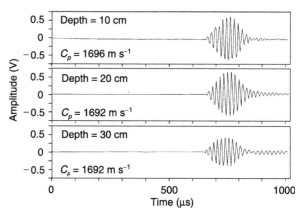

Figure 9   Examples of received compressional wave signals from fine sand sediment.

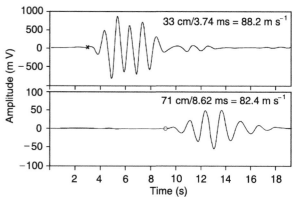

Figure 10   Examples of signals recorded from two shear wave receivers in hard-packed fine sand sediments.

analysis in the laboratory. The most reliable direct geoacoustic measurement techniques for marine sediments are *in situ* techniques. Some of the approaches used in recent years are discussed below.

Near-surface method   A system has been developed to measure sediment geoacoustic parameters, including compressional and shear wave velocities and their attenuation at tens of centimeters below the sediment–water interface. **Figure 8** shows the main features of the *In situ* Sediment Acoustic Measurement System (ISSAMS). Shear and compressional wave probes are attached to a triangular frame that uses weights to force the probes into the sediment. In very shallow water, divers can be used to insert the probes into the sediment, whereas in deeper water, a sleeve system (not shown) allows the ISSAMS to penetrate into the seafloor.

Using the compressional and shear wave transducers measurements are made with a continuous wave (cw) pulse technique where the ratio of measured transducer separation and pulse arrival time yields the wave velocity. Samples of compressional and shear wave data are shown in **Figure 9** and **10** respectively. **Table 3** gives a comparison of laboratory and *in situ* values of compressional and shear wave velocities from two different types of Adriatic Sea sediment.

Cross-hole method   Measurements as a function of depth in sediments can be made with boreholes, using either single or cross-hole techniques. Boreholes are made by divers using water–air jets to penetrate thin-walled plastic tubes for cross-hole measurements. **Figure 11** shows the experimental set up for the cross-hole measurements. The source is in the form of an electromagnetic mallet securely coupled to the inner wall of one of the plastic tubes with a hydraulic clamping device.

Table 3 Comparison of laboratory and *in situ* measurements

Sediment type	Porosity (%)	Mean grain size (ϕ)	Wave velocity (ms^{-1})			
			In situ (C_P)	Laboratory (C_P)	In situ (C_S)	Laboratory (C_S)
Sand	37	3.5	1557–1568	1580–1604	78–82	50
Mud	68	8.6	1467–1488	1468–1487	27–31	15

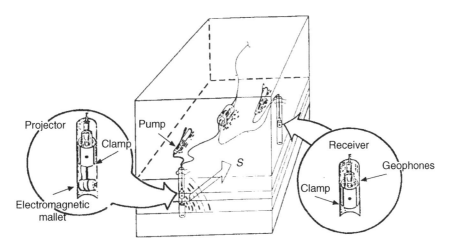

Figure 11 The experimental setup for cross-hole measurements.

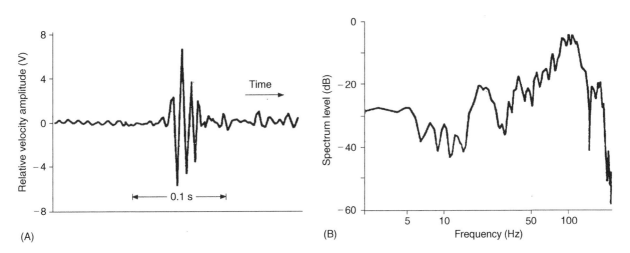

Figure 12 Cross-hole shear wave signal (A) and its spectrum (B) for a silty-clay bottom in the Ligurian Sea.

With a separation range of 2.9 m, moving-coil geophone receivers are also coupled to the inner wall of the second plastic tube with a hydraulic clamping device. The electromagnetic mallet generates a point source on the thin plastic tube wall, which results in a multipolarized transient signal to the sediment. Depending on the orientation of receiving sensors, vector components of the propagating compressional and shear waves are received and analyzed for velocity and attenuation parameters. Examples of a time series and a frequency spectrum for a shear wave signal received by a geophone with a natural frequency of 4.5 Hz are shown in **Figure 12**.

Figure 13 shows shear wave velocity as a function of depth obtained in the Ligurian Sea. It can be seen that the shear wave velocities are around $60 \, m^{-1} s$ at the sediment interface and increase with depth.

Remote Sensing Techniques

Even though *in situ* techniques provide the most reliable data, they are usually more time consuming and expensive to make and they are limited to small areas. Remote sensing and inversion to obtain geoacoustic parameters can cover larger areas in less time and provide reliable information. These techniques are based on the use of a seismo-acoustic signal received by sensors on the seafloor and/or in the water column. **Figure 14** illustrates a characteristic shallow water signal from an explosion received by a hydrophone close to the sea bottom. Three different techniques to extract information relative to bottom parameters from these signals are described briefly below.

Reflected waves

The half-space seafloor. When the seafloor consists of soft unconsolidated sediments, due to its very low shear modulus it can be treated as fluid.

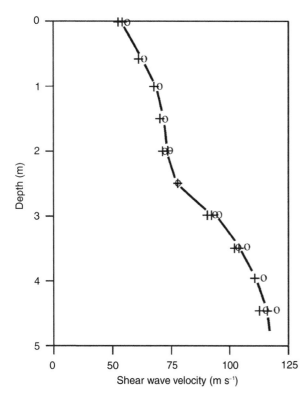

Figure 13 Shear wave velocity profile obtained from cross-hole measurements.

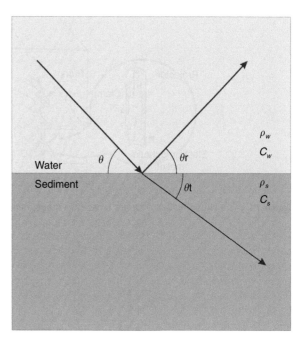

Figure 15 Geometry and notations for a simple half-space water–sediment interface.

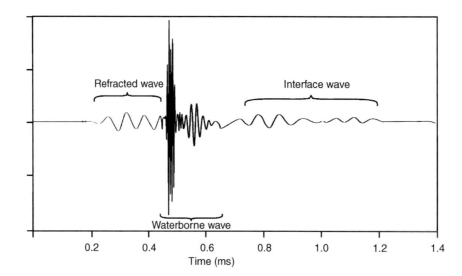

Figure 14 A characteristic shallow-water signal.

Reflection may occur whenever an acoustic wave is incident on an interface between two media. The amount of energy reflected, and its phase relative to the incident wave, depend on the ratio between the physical properties on the opposite sides of the interface. It is possible to calculate frequency independent reflection loss as the ratio between the amplitude of the shock pulse and the peak of the first reflection (after correcting for phase shift, absorption in the water column, and differences in spreading loss). If the relative density $\rho = \rho_S/\rho_W$, and the relative compressional wave velocity

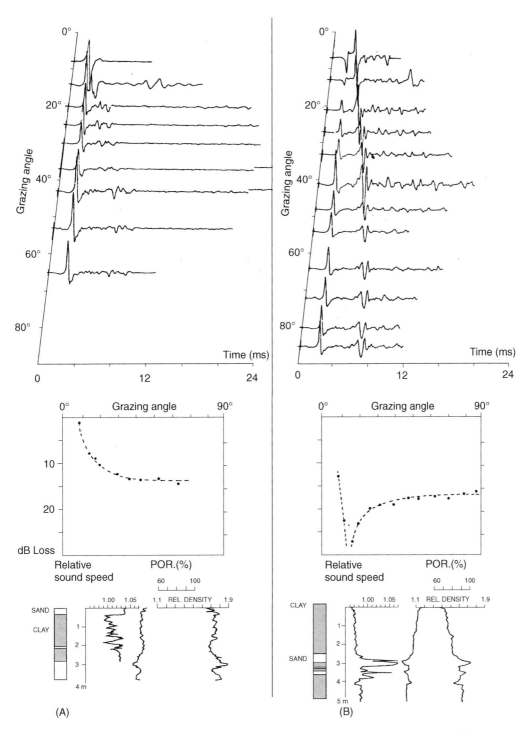

Figure 16 Acoustic signals and reflection loss as a function of grazing angle and sediment core properties: (A) for critical angle; (B) for angle of incidence case.

$C = C_S/C_W$, are used to present the contrast between the two media, reflection coefficient for such a simple environmental condition can be written as:

$$R = \frac{\rho\sin\Theta - \sqrt{(1/C^2 - \cos^2\Theta)}}{\rho\sin\Theta - \sqrt{(1/C^2 - \cos^2\Theta)}}$$

Where Θ is the grazing angle with respect to the interface as shown in **Figure 15**. For an incident path normal to a reflecting horizon, i.e., $\Theta = 90°$, the reflection coefficient is:

$$R = \rho C - 1/\rho C + 1$$

Critical angle case. When the velocity of compressional wave velocity is greater in the sediment layer ($C > 1$), as the grazing angle is decreased, a unique value is reached at which the acoustic energy totally reflects back to the water column. This is known as the critical angle and is given by:

$$\Theta_{cr} = \arccos(1/C)$$

When the grazing angle is less than this critical angle, all the incident acoustic energy is reflected.

However, the phase of the reflected wave is then shifted relative to the phase of the incidence wave by an angle varying from 0° to 180° and is given as:

$$\Phi = -2\arctan\frac{\sqrt{(\cos^2\Theta - 1/C^2)}}{\rho\sin\Theta}$$

Figure 16 shows measured and calculated reflection losses (20 log R) for this simple condition together with the basic physical properties of the core sample taken in the same area for explosive signals at different grazing angles.

Angle of intromission case. Especially in deep-water sediments the sound velocity in the top layer of the bottom is generally less than in the water above ($C < 1$). In such conditions there is an angle of incidence at which all of the incident energy is transmitted into the sedimentary layer and the reflection coefficient becomes zero:

$$\cos\Theta_i = \sqrt{\left(\frac{\rho^2 - 1/C^2}{\rho^2 - 1}\right)}$$

The phase shift is 0° when the ray angle is greater than the intromission angle and 180° when it is smaller. Thus, acoustical characteristics of the bottom, such as the critical angle, the angle of incidence, the phase shift,

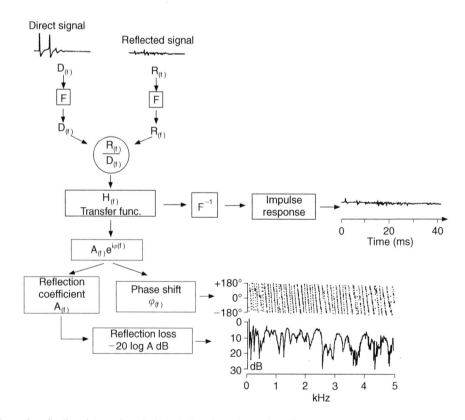

Figure 17 Acoustic reflection data and analysis technique for a layered seafloor.

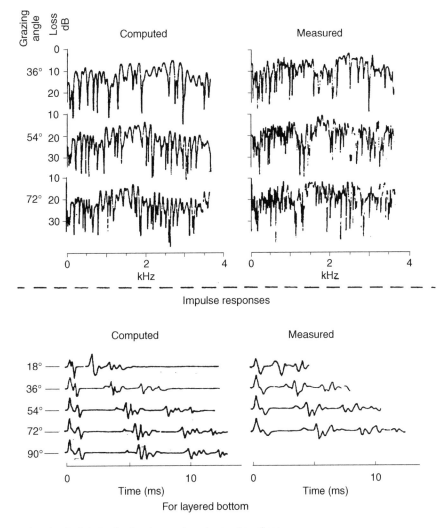

Figure 18 Measured and calculated reflection losses for a layered seafloor.

and the reflection coefficient, are primarily influenced by the relative density and relative sound velocity of the environment as a function of the ray angle.

The layered seafloor. Since the seafloor is generally layered, a simple peak amplitude approximation cannot be implemented because of the frequency dependence. In this case one calculates the transfer function (or reflection coefficient) from the convolution of the direct reference signal with the reflected signal. Examples of the phase shift and reflection loss as a function of frequency are shown in **Figure 17**. The reflectivity can also be described in the time domain by the impulse response, which is the inverse Fourier transform of the transfer function as shown in the same figure. This type of data can be utilized for inverse modeling to obtain the unknown parameters of the sediments. **Figure 18** is an example of measured and calculated reflection losses and impulse responses for a layered seafloor.

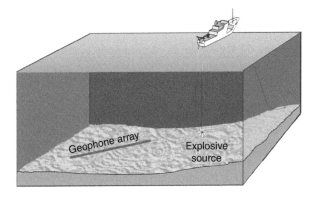

Figure 19 Experimental setup to measure seismo-acoustic waves on the seafloor.

Refracted waves Techniques developed for remote sensing of the uppermost sediments (25–50 m below the seafloor) utilize broad-band sources (small

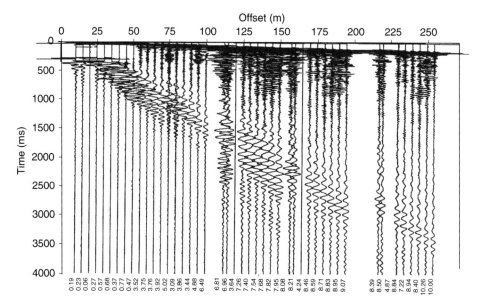

Figure 20 Signals received by geophone array.

explosives) and an array of geophones deployed on the seafloor. To obtain estimates of the bottom properties as a function of depth, both refracted compressional and shear waves as well as interface waves are analyzed. Inversion of the data is carried out using modified versions of techniques developed by earthquake seismologists to study dispersed Rayleigh-waves and refracted waves. **Figure 19** illustrates the basic experimental setup and **Figure 20** the signals received by an array of 24 geophones that permit studies of both interface and refracted waves.

These data are analyzed and inverted to obtain both compressional and shear wave velocities as a function of depth in the seafloor. Studies of attenuation and lateral variability are also possible using the same data set.

Figure 21 shows the expanded early portion of the geophone data shown in **Figure 20**, where, the first arriving energy out to a range of about 250 m has been fitted with a curve showing that compressional waves refracted through the sediments just beneath the seafloor travel faster as they penetrate more deeply into the sediment. At zero offset, the slope indicates a velocity of $1505 \, \text{m s}^{-1}$ whereas at a range offset at 250 m the slope corresponds to a velocity of $1573 \, \text{m s}^{-1}$. At ranges over 250 m, a strong head wave becomes the first arrival and, the interpretation would be that there is an underlying rock layer with compressional wave velocity of about $4577 \, \text{m s}^{-1}$.

A compressional wave velocity–depth curve for the upper part of the seafloor can be derived from the first arrivals shown in **Figure 21** using the classical Herglotz–Bateman–Wieckert integration method.

The slope of the travel-time curve (fitted parabola) gives the rate of change of the range with respect to time that is also the velocity of propagation of the diving compressional wave at the level of its deepest penetration (turning point) into the sediment. At each range Δ, the depth corresponding to the deepest penetration is then calculated using the following integral

$$z(V) = 1/\pi \int_0^\Delta \cosh^{-1}(V(\mathrm{d}t/\mathrm{d}x))\mathrm{d}x$$

where

$$1/V = (\mathrm{d}t/\mathrm{d}x)_{x = \Delta}$$

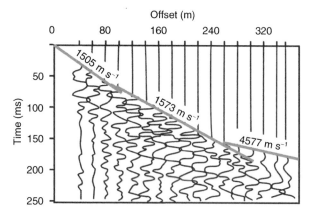

Figure 21 Expanded early portion of the data shown in **Figure 21**.

The result is the solid velocity–depth curve shown in **Figure 22**.

Interface waves In order to obtain a shear wave velocity–depth profile from the data, later arrivals corresponding to dispersed interface waves may be utilized (**Figure 20**). The portion of each individual signal corresponding to the interface-wave arrival can be processed using multiple filter analysis to create a group velocity dispersion diagram (Gabor diagram). The result of applying this technique to a dispersed signal is a filtered time signal whose envelope reaches a maximum at the group velocity arrival time for a selected frequency. The envelope is computed by taking the quadrature components of the inverse Fourier transform of the filtered signal. Filtering is carried out at many discrete frequencies over selected frequency bands. Once the arrival times are converted in to velocity, the envelopes are arranged in a matrix and contoured and dispersion curves are obtained by connecting the maximum values of the contour diagram (**Figure 23**).

Having obtained the dispersion characteristics of the interface waves, the geoacoustic model, made of a stack of homogeneous layers with different compressional and shear wave velocities for each layer that predict the measured dispersion curve is determined. **Figure 24** illustrates a number of examples from the Mediterranean sea covering data from soft clays to hard sands.

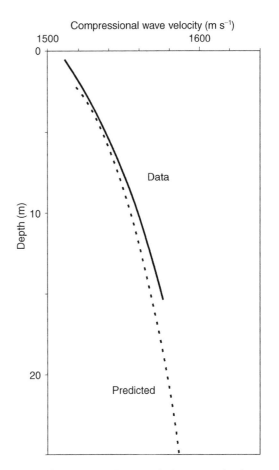

Figure 22 Compressional wave velocity versus depth curves derived from data and predicted (dashed line) by the model.

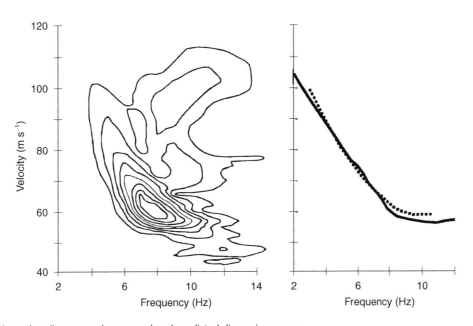

Figure 23 Dispersion diagram and measured and predicted dispersion curves.

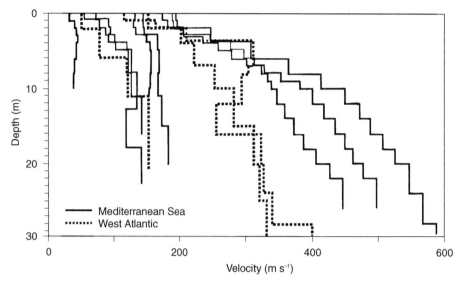

Figure 24   Summary of shear wave velocity profiles from the Mediterranean Sea.

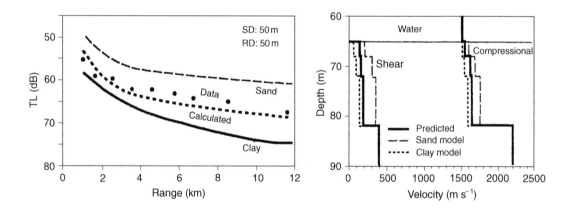

Figure 25   Comparison of transmission loss (TL) data and SAFARI prediction. Effects of changing bottom parameters from sand to clay is also shown together with the input profiles utilized for SAFARI predictions.

Transmission Loss Technique

The seafloor is known to be the controlling factor in low-frequency shallow water acoustic propagation. Forward modeling is performed with models giving exact solutions to the wave equation, i.e., SAFARI where, compressional and shear wave velocities, the attenuation factors associated with these waves, and the sediment density as a function of depth are the main input parameters. Acoustic energy propagating through a shallow water channel interacts with the seafloor causing partitioning of waterborne energy into different types of seismic and acoustic waves. The propagation and attenuation of these waves observed in such an environment are strongly dependent on the physical characteristics of the sea bottom. Transmission loss (TL), representing the amount of energy lost along an acoustic propagation path, carries the information relative to the environment through which the wave is propagating. **Figure 25** shows a comparison of TL data and model predictions together with the input parameters used at 400 Hz. The effects of changes in bottom parameters are also shown in the figure. This technique becomes extremely useful when seafloor information is sparse.

Conclusions

Acoustic/seismic characteristics of the marine sediments have been of interest to a wide range of activities covering commercial operations involving trenching and cable lying, construction of offshore foundations, studies of slope stability, dredging and military applications like mine and submarine detection. Experimental and theoretical work over the years has shown that it is possible to determine the geoacoustic properties of sediments by different techniques. The characteristics of the marine sediments and techniques to obtain information about these characteristics have been briefly described. The studies so far conducted indicate that direct and indirect methods developed over the last four decades may give sufficient information to deduce some of the fundamental characteristics of the marine sediments. It is evident that still more research needs to be done to develop these techniques for fast and reliable results.

See also

Acoustics, Shallow Water. Ocean Margin Sediments.

Further Reading

Akal T and Berkson JM (eds.) (1986) *Ocean Seismo-Acoustics*. New York and London: Plenum press.

Biot MA (1962) Generalized theory of acoustic propagation in porous dissipative media. *Journal of Acoustical Society of America* 34: 1254–1264.

Brekhovskikh LM (1980) *Waves in Layered Media*. New York: Academic Press.

Cagniard L (1962) *Reflection and Refraction of Progressive Seismic Waves*. New York: McGraw-Hill.

Grant FS and West GF (1965) *Interpretation and Theory in Applied Geophysics*. New York: McGraw Hill.

Hampton L (ed.) (1974) *Physics of Sound in Marine Sediments*. New York and London: Plenum Press.

Hovem JM, Richardson MD, and Stoll RD (eds.) (1992) *Shear Waves in Marine Sediments*. Dordrecht: Kluwer Academic.

Jensen FB, Kuperman WA, Porter MB, and Schmidt H (1999) *Computational Ocean Acoustics*. New York: Springer-Verlag.

Kuperman WA and Jensen FB (eds.) (1980) *Bottom Interacting Ocean Acoustics*. New York and London: Plenum Press.

Lara-Saenz A, Ranz-Guerra C and Carbo-Fite C (eds) (1987) *Acoustics and Ocean Bottom*. II F.A.S.E. Specialized Conference. Inst. De Acustica, Madrid.

Pace NG (ed.) (1983) *Acoustics and the Sea-Bed*. Bath: Bath University Press.

Pouliquen E, Lyons AP, Pace NG *et al.* (2000) *Backscattering from unconsolidated sediments above 100 kHz*. In: Chevret P and Zakhario ME (ed.) Proceedings of the fifth European Conference on Under water Acoustics, ECUA 2000 Lyon, France.

Stoll RD (1989) *Lecture Notes in Earth Sciences*. New York: Springer-Verlag.

AIRCRAFT REMOTE SENSING

L. W. Harding, Jr and W. D. Miller, University of
Maryland, College Park, MD, USA
R. N. Swift, and C. W. Wright, NASA Goddard
Space Flight Center, Wallops Island, VA, USA

Introduction

The use of aircraft for remote sensing has steadily
grown since the beginnings of aviation in the early
twentieth century and today there are many applications in the Earth sciences. A diverse set of remote
sensing uses in oceanography developed in parallel
with advances in aviation, following increased aircraft capabilities and the development of instrumentation for studying ocean properties. Aircraft
improvements include a greatly expanded range of
operational altitudes, development of the Global
Positioning System (GPS) enabling precision navigation, increased availability of power for instruments, and longer range and duration of missions.
Instrumentation developments include new sensor
technologies made possible by microelectronics,
small, high-speed computers, improved optics, and
increased accuracy of digital conversion of electronic
signals. Advances in these areas have contributed
significantly to the maturation of aircraft remote
sensing as an oceanographic tool.

Many different types of aircraft are currently used
for remote sensing of the oceans, ranging from bal-
loons to helicopters, and from light, single engine
piston-powered airplanes to jets. The data and information collected on these platforms are commonly used to enhance sampling by traditional
oceanographic methods, giving increased spatial and
temporal resolution for a number of important
properties. Contemporary applications of aircraft
remote sensing to oceanography can be grouped into
several areas, among them ocean color, sea surface
temperature (SST), sea surface salinity (SSS), wave
properties, near-shore topography, and bathymetry.
Prominent examples include thermal mapping using
infrared (IR) sensors in both coastal and open ocean
studies, lidar and visible radiometers for ocean color
measurements of phytoplankton distributions and
'algal blooms', and passive microwave sensors to
make observations of surface salinity structure of
estuarine plumes in the coastal ocean. These topics
will be discussed in more detail in subsequent

sections. Other important uses of aircraft remote
sensing are to test instruments slated for deployment
on satellites, to calibrate and validate space-based
sensors using aircraft-borne counterparts, and to
make 'under-flights' of satellite instruments and assess the efficacy of atmospheric corrections applied
to data from space-based observations.

Aircraft have some advantages over satellites for
oceanography, including the ability to gather data
under cloud cover, high spatial resolution, flexibility
of operations that enables rapid responses to 'events',
and less influence of atmospheric effects that complicate the processing of satellite data. Aircraft remote sensing provides nearly synoptic data and
information on important oceanographic properties
at higher spatial resolution than can be achieved by
most satellite-borne instruments. Perhaps the greatest advantage of aircraft remote sensing is the ability
to provide consistent, high-resolution coverage at
larger spatial scales and more frequent intervals than
are practical with ships, making it feasible to use
aircraft for monitoring change.

Disadvantages of aircraft remote sensing include
the relatively limited spatial coverage that can be
obtained compared with the global coverage available from satellite instruments, the repeated expense
of deploying multiple flights, weather restrictions on
operations, and lack of synopticity over large scales.
Combination of the large-scale, synoptic data that are
accessible from space with higher resolution aircraft
surveys of specific locations is increasingly recognized
as an important and useful marriage that takes advantages of the strengths of both approaches.

This article begins with a discussion of sensors that
use lasers (also called active sensors), including airborne laser fluorosensors that have been used to
measure chlorophyll (chl-a) and other properties;
continues with discussions of lidar sensors used for
topographic and bathymetric mapping; describes
passive (sensors that do not transmit or illuminate,
but view naturally occurring reflections and emissions) ocean color remote sensing directed at quantifying phytoplankton biomass and productivity;
moves to available or planned hyper-spectral aircraft
instruments; briefly describes synthetic aperture
radar applications for waves and wind, and closes
with a discussion of passive microwave measurements of salinity. Readers are directed to the Further
Reading section if they desire additional information
on individual topics.

Active Systems

Airborne Laser Fluorosensing

The concentrations of certain waterborne constituents, such as chl-a, can be measured from their fluorescence, a relationship that is exploited in shipboard sensors such as standard fluorometers and flow cytometers that are discussed elsewhere in this encyclopedia. NASA first demonstrated the measurement of laser-induced chl-a fluorescence from a low-flying aircraft in the mid-1970s. Airborne laser fluorosensors were developed shortly thereafter in the USA, Canada, Germany, Italy, and Russia, and used for measuring laser-induced fluorescence of a number of marine constituents in addition to chl-a. Oceanic constituents amenable to laser fluorosensing include phycoerythrin (photosynthetic pigment in some phytoplankton taxa), chromophoric dissolved organic matter (CDOM), and oil films. Airborne laser fluorosensors have also been used to follow dyes such as fluorescein and rhodamine that are introduced into water masses to trace their movement.

The NASA Airborne Oceanographic Lidar (AOL) is the most advanced airborne laser fluorosensor. The transmitter portion features a dichroic optical device to spatially separate the temporally concurrent 355 and 532 nm pulsed-laser radiation, followed by individual steering mirrors to direct the separated beams to respective oceanic targets separated by ~1 m when flown at the AOL's nominal 150 m operational altitude. The receiver focal plane containing both laser-illuminated targets is focused onto the input slits of the monochromator. The monochromator output focal planes are viewed by custom-made optical fibers that transport signal photons from the focal planes to the photo-cathode of each photo-multiplier module (PMM) where the conversion from photons to electrons takes place with a substantial gain. Time-resolved waveforms are collected in channels centered at 404 nm (water Raman) and 450 nm (CDOM) from 355 nm laser excitation, and at 560 nm and 590 nm (phycoerythrin), 650 (water Raman), and 685 nm (chl-a) from the 532 nm laser excitation. The water Raman from the respective lasers is the red-shifted emission from the OH^- bonds of water molecules resulting from radiation with the laser pulse. The strength of the water Raman signal is directly proportional to the number of OH^- molecules accessed by the laser pulse. Thus, the water Raman signal is used to normalize the fluorescence signals to correct for variations in water attenuation properties in the surface layer of the ocean. The AOL is described in more detail on http://lidar.wff.nasa.gov.

The AOL has supported major oceanographic studies throughout the 1980s and 1990s, extending the usefulness of shipboard measurements over wide areas to permit improved interpretation of the ship-derived results. Examples of data from the AOL show horizontal structure of laser-induced fluorescence converted to chl-a concentration (**Figure 1**). Prominent oceanographic expeditions that have benefited from aircraft coverage with the AOL include the North Atlantic Bloom Experiment (NABE) of the Joint Global Ocean Flux Study (JGOFS), and the Iron Enrichment Experiment (IRONEX) of the Equatorial Pacific near the Galapagos Islands. This system has been flown on NASA P-3B aircraft in open-ocean missions that often exceeded 6 h in duration and collected hundreds of thousands of spectra. Each 'experiment' is able to generate both active and passive data in 'pairs' that are used for determining ocean color and recovering chl-a and other constituents, and that are also useful in the development of algorithms for measuring these constituents from oceanic radiance spectra.

Airborne Lidar Coastal Mapping

The use of airborne lidar (light detection and ranging) sensors for meeting coastal mapping requirements is a relatively new and promising application of laser-ranging technology. These applications include high-density surveying of coastline and beach morphology and shallow water bathymetry. The capability to measure distance accurately with lidar sensors has been available since the early 1970s, but their application to airborne surveying of terrestrial features was seriously hampered by the lack of knowledge of the position of the aircraft from which the measurement was made. The implementation of the Department of Defense GPS constellation of satellites in the late 1980s, coupled with the development of GPS receiver technology, has resulted in the capability to provide the position of a GPS antenna located on an aircraft fuselage in flight to an accuracy approaching 5 cm using kinematic differential methodology. These methods involve the use of a fixed receiver (generally located at the staging airport) and a mobile receiver that is fixed to the aircraft fuselage. The distance between mobile and fixed receivers, referred to as the baseline, is typically on the order of tens of kilometers, and can be extended to hundreds of kilometers by using dual frequency survey grade GPS receivers aided by tracking the phase code of the carrier from each frequency.

Modern airborne lidars are capable of acquiring 5000 or more discrete range measurements per second. Aircraft attitude and heading information are

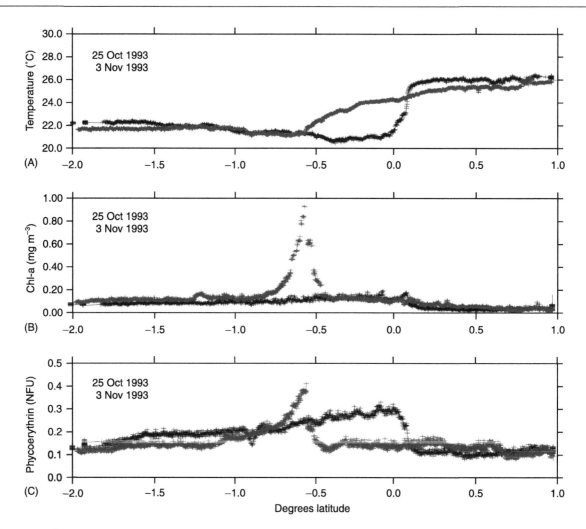

Figure 1   Cross-section profiles flown across a large oceanic front west of the Galapagos Islands on 25 October and 3 November 1993. (A) A dramatic horizontal displacement of the front as measured with an infrared radiometer; (B) and (C) show corresponding changes in laser-induced fluorescence of chl-a and phycoerythrin. This flight was made as part of the original IRONEX investigation in late 1993. NFU, normalized fluorescence units.

used along with the GPS-determined platform position to locate the position of the laser pulse on the Earth's surface to a vertical accuracy approaching 10 cm with some highly accurate systems and <30 cm for most of these sensors. The horizontal accuracy is generally 50–100 cm. Depending on the pulse repetition rate of the laser transmitter, the off-nadir pointing angle, and the speed of the aircraft platform, the density of survey points can exceed one sample per square meter.

NASA's Airborne Topographic Mapper (ATM) is an example of a topographic mapping lidar used for coastal surveying applications. An example of data from ATM shows shoreline features off the east coast of the USA (**Figure 2**). ATM was originally developed to measure changes in the elevation of Arctic ice sheets in response to global warming. The sensor was applied to measurements of changes in coastal morphology beginning in 1995. Presently, baseline

topographic surveys exist for most of the Atlantic and Gulf coasts between central Maine and Texas and for large sections of the Pacific coast. Affected sections of coastline are re-occupied following major coastal storms, such as hurricanes and 'Nor'easters', to determine the extent of erosion and depositional patterns resulting from the storms. Additional details on the ATM and some results of investigations on coastal morphology can be found on websites (http://lidar.wff.nasa.gov and http://aol.wff.nasa.gov/aoltm/projects/beachmap/98results/).

Other airborne lidar systems have been used to survey coastal morphology, including Optec lidar systems by Florida State University and the University of Texas at Austin. Beyond these airborne lidar sensors, there are considerably more instruments with this capability that are currently in use in the commercial sector for surveying metropolitan areas, flood plains, and for other terrestrial

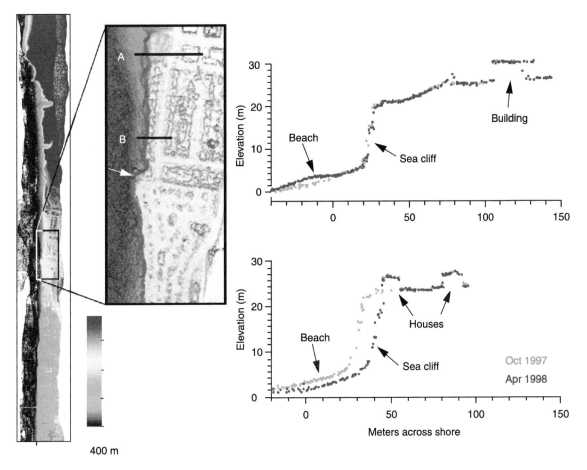

Figure 2 (Left) Map of coastal topography around Pacifica, CA, USA, derived from lidar data obtained in April 1998, after a winter of severe storms associated with El Niño. Map insert shows the Esplanade Drive area of Pacifica rendered from lidar data gridded at 2 m resolution and colored according to elevation. (Right) Cross-sections derived from lidar data of October 1997 and April 1998 at locations marked in the map inset. The profiles in (A) show a stable cliff and accreting beach, whereas about 200 m to the south the profiles in (B) show erosion of the sea cliff and adjacent beach resulting in undermining of houses. Each profile shows individual laser spot elevations that fall within a 2 m wide strip oriented approximately normal to shoreline. (Reproduced with permission from Sallenger *et al.*, 1999.)

applications. At the last count (early 2000) there were approximately 60 airborne lidars in operation worldwide, with most engaged in a variety of survey applications generally outside the field of coastal mapping.

Pump and Probe Fluorometry

Several sections in this chapter describe recoveries of phytoplankton biomass as chl-a by active and passive measurements. Another recent accomplishment is an active, airborne laser measurement intended to aid in remote detection of photosynthetic performance, an important ingredient of primary productivity computations. Fluorometric techniques, such as fast repetition rate (FRR) fluorometry (explained in another article of this encyclopedia), have provided an alternative approach to ^{14}C assimilation and O_2 evolution in the measurement of primary productivity. This

technology has matured with the commercial availability of FRR instruments that can give vertical profiles or operate in a continuous mode while underway. There have been several attempts to develop airborne lidar instruments to determine phytoplankton photosynthetic characteristics from aircraft in the past decade. NASA scientists have deployed a pump and probe fluorometer from aircraft, wherein the AOL laser (described above) acts as the pump and a second laser with variable power options and rapid pulsing capabilities (10 ns) functions as the probe.

Passive Systems

Multichannel Ocean Color Sensor (MOCS)

Passive ocean color measurements using visible radiometers to measure reflected natural sunlight from the ocean have been made with a number of instruments in the past two decades. These instruments include the Multichannel Ocean Color Sensor (MOCS) that was flown in studies of Nantucket Shoals in the early 1980s, the passive sensors of the AOL suite that have been used in many locations around the world, and more recently, simple radiometers that have been deployed on light aircraft in regional studies of Chesapeake Bay (see below). MOCS was one of the earliest ocean color sensors used on aircraft. It provided mesoscale data on shelf and slope chl-a in conjunction with shipboard studies of physical structure, nutrient inputs, and phytoplankton primary productivity.

Ocean Data Acquisition System (ODAS) and SeaWiFS Aircraft Simulator (SAS)

Few aircraft studies have obtained long time-series sufficient to quantify variability and detect secular trends. An example is ocean color measurements made from light aircraft in the Chesapeake Bay region for over a decade, providing data on chl-a and SST from >250 flights. Aircraft over-flights of the Bay using the Ocean Data Acquisition System (ODAS) developed at NASA's Goddard Space Flight Center commenced in 1989. ODAS was a nadir-viewing, line-of-flight, three-band radiometer with spectral coverage in the blue-green region of the visible spectrum (460–520 nm), a narrow 1.5° field-of-view, and a 10 Hz sampling rate. The ODAS instrument package included an IR temperature sensor (PRT-5, Pyrometrics, Inc.) for measuring SST. The system was flown for ~7 years over Chesapeake Bay on a regular set of tracks to determine chl-a and SST. Over 150 flights were made with ODAS between 1989 and 1996, coordinated with *in situ* observations from a multi-jurisdictional monitoring program and other cruises of opportunity.

ODAS was flown together with the SeaWiFS Aircraft Simulator (SAS II, III, Satlantic, Inc., Halifax, Canada) beginning in 1995 and was retired soon thereafter and replaced with the SAS units. SAS III is a multi-spectral (13-band, 380–865 nm), line-of-flight, nadir viewing, 10 Hz, passive radiometer with a 3.5° field-of-view that has the same wavebands as the SeaWiFS satellite instrument, and several additional bands in the visible, near IR, and UV. The SAS systems include an IR temperature sensor (Heimann Instruments, Inc.). Chl-a estimates are obtained using a curvature algorithm applied to water-leaving radiances at wavebands in the blue-green portion of the visible spectrum with validation from concurrent shipboard measurements. Flights are conducted at ~50–60 m s^{-1} (100–120 knots), giving an along-track profile with a resolution of 5–6 m averaged to

50 m in processing, and interpolated to 1 km^2 for visualization. Imagery derived from ODAS and SAS flights is available on a web site of the NOAA Chesapeake Bay Office for the main stem of the Bay (http://noaa.chesapeakebay.net), and for two contrasting tributaries, the Choptank and Patuxent Rivers on a web site of the Coastal Intensive Sites Network (CISNet) (http://www.cisnet-choptank.org).

Data from ODAS and SAS have provided detailed information on the timing, position, and magnitude of blooms in Chesapeake Bay, particularly the spring diatom bloom that dominates the annual phytoplankton cycle. This April–May peak of chl-a represents the largest accumulation of phytoplankton biomass in the Bay and is a proximal indicator of over-enrichment by nutrients. Data from SeaWiFS for spring 2000 show the coast-wide chl-a distribution for context, while SAS III data illustrate the high-resolution chl-a maps that are obtained regionally (**Figure 3**). A well-developed spring bloom corresponding to a year of relatively high freshwater flow from the Susquehanna River, the main tributary feeding the estuary, is apparent in the main stem Bay chl-a distribution. Estimates of primary productivity are now being derived from shipboard observations of key variables combined with high-resolution aircraft measurements of chl-a and SST for the Bay.

Hyper-spectral Systems

Airborne Visible/Infrared Imaging Spectrometer (AVIRIS)

The Airborne Visible/Infrared Imaging Spectrometer (AVIRIS) was originally designed in the late 1980s by NASA at the Jet Propulsion Laboratory (JPL) to collect data of high spectral and spatial resolution, anticipating a space-based high-resolution imaging spectrometer (HIRIS) that was planned for launch in the mid-1990s.

Because the sensor was designed to provide data similar to satellite data, flight specifications called for both high altitude and high speed. AVIRIS flies almost exclusively on a NASA ER-2 research aircraft at an altitude of 20 km and an airspeed of 732 km h^{-1}. At this altitude and a 30° field of view, the swath width is almost 11 km. The instantaneous field of view is 1 mrad, which creates individual pixels at a resolution of 20 m^2. The sensor samples in a whisk-broom fashion, so a mirror scans back and forth, perpendicular to the line-of-flight, at a rate of 12 times per second to provide continuous spatial coverage. Each pixel is then sent to four separate spectrometers by a fiber optic cable. The spectrometers are arranged so that they each cover a part of the

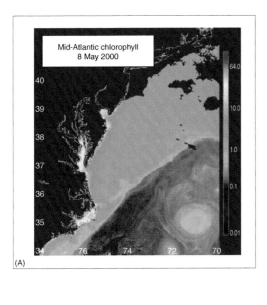

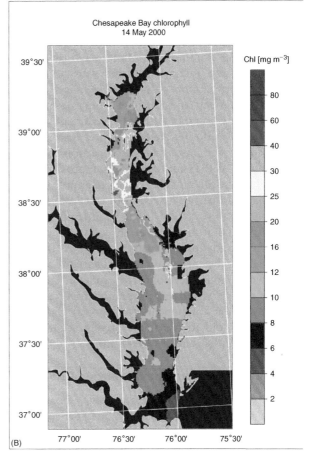

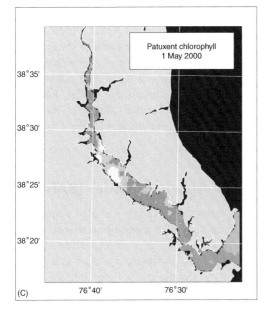

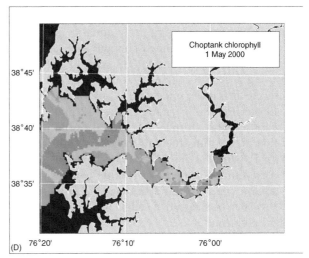

Figure 3 Spring chl-a (mg m^{-3}) in: (A) the mid-Atlantic region from SeaWiFS; (B) Chesapeake Bay; (C) Patuxent R; (D) Choptank R from SAS III.

spectrum from 0.40 to 2.4 μm, providing continuous spectral coverage at 10 nm intervals over the entire spectrum from visible to near IR. Data are recorded to tape cassettes for storage until rectification, atmospheric correction, and processing at JPL. A typical AVIRIS 'scene' is a 40 min flight line. At ER-2 flight parameters, this creates an image roughly 500 km long and 11 km wide. Data are encoded at 12-bits for a high degree of discrimination. The physical dimensions of AVIRIS are quite large, 84 cm wide × 160 cm long × 117 cm tall at a weight of 720 pounds.

Data collected with AVIRIS have been used for terrestrial, marine, and atmospheric applications. Accomplishments of AVIRIS include separation of the chl-a signature from bottom reflectance for clear lake waters of Lake Tahoe and turbid waters near Tampa Bay, interpretation of spectral signals from resuspended sediment and dissolved organic materials in W. Florida, and of suspended sediment and kelp beds in S. California. Recent efforts have focused on improving atmospheric correction procedures for both AVIRIS and satellite data, providing inputs for bio-optical models which determine inherent optical properties (IOPs) from reflectance, algorithm development, and sporadic attempts at water quality monitoring (e.g., chl-a, suspended sediment, diffuse attenuation coefficient, k_d). AVIRIS data have recently been used as an input variable to a neural network model developed to estimate water depth. The model was able to separate the contributions of different components to the total water-leaving radiance and to provide relatively accurate estimates of depth (rms error = 0.48 m).

Compact Airborne Spectrographic Imager (CASI)

The Compact Airborne Spectrographic Imager (CASI) is a relatively small, lightweight hyper-spectral sensor that has been used on a variety of light aircraft. CASI was developed by Itres Research Ltd (Alberta, Canada) in 1988 and was designed for a variety of remote sensing applications in forestry, agriculture, land-use planning, and aquatic monitoring. By allowing user-defined configurations, the 12-bit, push-broom-type sensor (333 scan lines per second) using a charge-coupled detector (CCD) can be adapted to maximize either spatial (37.8° across track field of view, 0.077° along-track, 512 pixels – pixel size varies with altitude) or spectral resolution (288 bands at 1.9 nm intervals between 400 and 1000 nm). Experiments have been conducted using CASI to determine bottom type, benthic cover, submerged aquatic vegetation, marsh type, and in-water constituents such as suspended sediments, chl-a, and other algal pigments.

Portable Hyper-spectral Imager for Low-Light Spectroscopy (PHILLS)

The Portable Hyper-spectral Imager for Low-Light Spectroscopy (PHILLS) has been constructed by the US Navy (Naval Research Laboratory) for imaging the coastal ocean. PHILLS uses a backside-illuminated CCD for high sensitivity, and an all-reflective spectrograph with a convex grating in an Offner configuration to produce a distortion-free image. The instrument benefits from improvements in large-format detector arrays that have enabled increased spectral resolution and higher signal-to-noise ratios for imaging spectrographs, extending the use of this technology in low-albedo coastal waters. The ocean PHILLS operates in a push-broom scanned mode whereby cross-track ground pixels are imaged with a camera lens onto the entrance slit of the spectrometer, and new lines of the along-track ground pixels are attained by aircraft motion. The Navy's interest in hyper-spectral imagers for coastal applications centers on the development of methods for determining shallow water bathymetry, topography, bottom type composition, underwater hazards, and visibility. PHILLS precedes a planned hyper-spectral satellite instrument, the Coastal Ocean Imaging Spectrometer (COIS) that is planned to launch on the Naval Earth Map Observer (NEMO) spacecraft.

Radar Altimetry

Ocean applications of airborne radar altimetry systems include several sensors that retrieve information on wave properties. Two examples are the Radar Ocean Wave Spectrometer (ROWS), and the Scanning Radar Altimeter (SRA), systems designed to measure long-wave directional spectra and near-surface wind speed. ROWS is a K_u-band system developed at NASA's Goddard Space Flight Center in support of present and future satellite radar missions. Data obtained from ROWS in a spectrometer mode are used to derive two-dimensional ocean spectral wave estimates and directional radar backscatter. Data from the pulse-limited altimeter mode radar yield estimates of significant wave height and surface wind speed.

Synthetic Aperture Radar (SAR)

Synthetic Aperture Radar (SAR) systems emit microwave radiation in several bands and collect the reflected radiation to gain information about sea surface conditions. Synthetic aperture is a technique that is used to synthesize a long antenna by combining signals, or echoes, received by the radar as it moves along a flight track. Aperture refers to the opening that is used to collect reflected energy and form an image. The analogous feature of a camera to the aperture would be the shutter opening. A synthetic aperture is constructed by moving a real aperture or antenna through a series of positions along a flight track.

NASA's Jet Propulsion Laboratory and Ames Research Center have operated the Airborne SAR (AIRSAR) on a DC-8 since the late 1980s. The radar of AIRSAR illuminates the ocean at three microwave wavelengths: C-band (6 cm), L-band (24 cm), and P-band (68 cm). Brightness of the ocean (the amount of energy reflected back to the antenna) depends on the roughness of the surface at the length scale of the microwave (Bragg scattering). The primary source of roughness, and hence brightness, at the wavelengths used is capillary waves associated with wind. Oceanographic applications derive from the responsiveness of capillary wave amplitude to factors that affect surface tension, such as swell, atmospheric stability, and the presence of biological films. For example, the backscatter characteristics of the ocean are affected by surface oil and slicks can be observed in SAR imagery as a decrease of radar backscatter; SAR imagery appears dark in an area affected by an oil spill, surface slick, or biofilm, as compared with areas without these constituents.

Microwave Salinometers

Passive microwave radiometry (L-band) has been tested for the recovery of SSS from aircraft, and it may be possible to make these measurements from space. Salinity affects the natural emission of EM radiation from the ocean, and the microwave signature can be used to quantify SSS. Two examples of aircraft instruments that have been used to measure SSS in the coastal ocean are the Scanning Low-Frequency

Microwave Radiometer (SLFMR), and the Electronically Thinned Array Radiometer (ESTAR).

SLFMR was used recently in the estuarine plume of Chesapeake Bay on the east coast of the USA to follow the buoyant outflow that dominates the nearshore density structure and constitutes an important tracer of water mass movement. SLFMR is able to recover SSS at an accuracy of about 1 PSU (**Figure 4**). This resolution is too coarse for the open ocean, but is quite suitable for coastal applications where significant gradients occur in regions influenced by freshwater inputs. SLFMR has a bandwidth of 25 MHz, a frequency of 1.413 GHz, and a single antenna with a beam width of approximately $16°$ and six across-track positions at $\pm 6°$, $\pm 22°$ and $\pm 39°$. Tests of SLFMR off the Chesapeake Bay demonstrated its effectiveness as a 'salinity mapper' by characterizing the trajectory of the Bay plume from surveys using light aircraft in joint operations with ships. Flights were conducted at an altitude of 2.6 km, giving a resolution of about 1 km. The accuracy of SSS in this example is ~ 0.5 PSU.

ESTAR is an aircraft instrument that is the prototype of a proposed space instrument for measuring SSS. This instrument relies on an interferometric technique termed 'aperture synthesis' in the across-track dimension that can reduce the size of the antenna aperture needed to monitor SSS from space. It has been described as a 'hybrid of a real and a synthetic aperture radiometer.' Aircraft surveys of

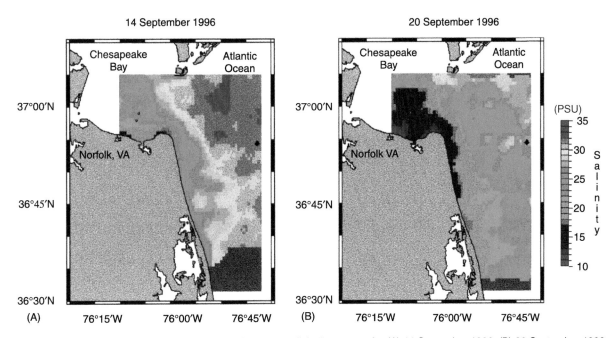

Figure 4 Sea surface salinity from an airborne microwave salinity instrument for (A) 14 September 1996; (B) 20 September 1996. Images reveal strong onshore-offshore gradients in salinity from the mouth of Chesapeake Bay to the plume and shelf, and the effect of high rainfall and freshwater input on the salinity distribution over a 1-week interval. (Adapted with permission from Miller *et al.*, 1998.)

SSS using ESTAR in the coastal current off Maryland and Delaware showed good agreement with thermosalinograph measurements from ships in the range of 29–31 PSU.

See also

Beaches, Physical Processes Affecting. River Inputs.

Further Reading

Blume H-JC, Kendall BM, and Fedors JC (1978) Measurement of ocean temperature and salinity via microwave radiometry. *Boundary Layer Meteorology* 13: 295–308.

Campbell JW and Esaias WE (1985) Spatial patterns in temperature and chlorophyll on Nantucket Shoals from airborne remote sensing data, May 7–9, 1981. *Journal of Marine Research* 43: 139–161.

Harding LW Jr, Itsweire EC, and Esaias WE (1994) Estimates of phytoplankton biomass in the Chesapeake Bay from aircraft remote sensing of chlorophyll concentrations, 1989–92. *Remote Sensing Environment* 49: 41–56.

Le Vine DM, Zaitzeff JB, D'Sa EJ, *et al.* (2000) Sea surface salinity: toward an operational remote-sensing system. In: Halpern D (ed.) *Satellites, Oceanography and Society*, pp. 321–335. Elsevier Science.

Miller JL, Goodberlet MA, and Zaitzeff JB (1998) Airborne salinity mapper makes debut in coastal zone. *EOS Transactions of the American Geophysical Union* 79: 173–177.

Sallenger AH Jr, Krabill W, Brock J, *et al.* (1999) *EOS Transactions of the American Geophysical Union* 80: 89–93.

Sandidge JC and Holyer RJ (1998) Coastal bathymetry from hyper-spectral observations of water radiance. *Remote Sensing Environment* 65: 341–352.

REMOTE SENSING OF COASTAL WATERS

N. Hoepffner and G. Zibordi, Institute for
Environment and Sustainability, Ispra, Italy

Introductory Comments

Coastal waters occupy at the most 8–10% of the
ocean surface and only 0.5% of its volume, but
represent an important fraction in terms of economic, social, and ecological value. With growing
concerns about the rapid and negative changes of the
coastal areas and marine resources, there has been
over the last decade a pressing request for the development of quantitative and cost-effective methodologies to detect and characterize both long-term
changes and short-term events in the coastal environment. The challenge, however, is that the
properties of coastal waters are controlled by complex interactions and fluxes of material between
land, ocean, and atmosphere. As a result, coastal
zones are among the most changeable environment
on Earth, typically involving phenomena which vary
along time and space scales shorter than those in the
open ocean.

In this view, remote sensing techniques have a key
role to play, providing consistent products for a wide
range of coastal applications over scales otherwise
inaccessible from standard ship surveys. In spite of
this advantage, remote sensing techniques have also
their limitations and should not be seen as a substitute to field measurements but rather complementary
to them in terms of both sampling scales and variables to be measured.

In this article, particular emphasis is given to the
Space-borne sensors measuring radiances in the
visible and near-infrared, enabling the observation of
biogeochemical processes in the marine environment.
Coastal waters are optically more complex than the
open oceanic waters because of the large influence by
the land system and catchment areas delivering significant amount of dissolved and particulate material. Thus, relative to other Earth-observing
techniques, optical remote sensing of coastal waters
requires more specificity in the treatment of the signal, and in the choice of the algorithms, to
support any quantitative assessment of relevant
biogeochemical variables and their associated
processes.

Coastal Water Optics

The radiance leaving the water at a given wavelength,
λ, $L_w(\lambda)$, in the direction of the remote sensor results
from three major processes of unequal magnitude: (1)
the reflected sunlight at the surface of the water; (2) the
light that has entered the sea surface and retransmitted
back to the atmosphere through scattering; and (3) the
fluorescence of some specific material in suspension in
the upper water column. In addition, in shallow and
relatively clear waters, remote sensing techniques are
often affected by the ocean bottom.

As the photons enter the water column, the spectral signature of the sunlight is altered depending on
the optical properties of the medium itself and its
constituents. Optically complex waters are commonly labeled as 'case 2' waters in a bipartite classification scheme where 'case 1' refers to marine
waters in which phytoplankton and covarying materials are the principal components responsible for
the changes in the optical properties. Optical properties of case 2 waters are influenced by substances
that vary independently from phytoplankton. Such
substances are inorganic particles in suspension (e.g.,
sediment), and some colored or chromophoric dissolved organic materials (CDOMs), also called yellow substances.

The water-type classification is based on the relative contribution of these three types of substances
affecting the light field, irrespectively of possible
changes in the concentration of each substance. For
example, case 1 waters can range from clear, oligotrophic open ocean waters with chlorophyll concentration less than $0.1\,\mathrm{mg\,m^{-3}}$, to some eutrophic
conditions where chlorophyll can reach a concentration of $10\,\mathrm{mg\,m^{-3}}$.

Accordingly, all coastal water samples can be optically charted in a triangular diagram (**Figure 1**) in
which the axes represent the fractional contributions
due to each of the three types of optically active
components (AOCs). As the data point moves
toward one of the three apices, the sample will be
classified as either case 1 water, case 2 'yellow-
substance' dominated water, or case 2 'suspended
material' dominated water. This classification may
appear rather restricted considering the huge variety
of living and nonliving compounds that contributes
to the optical variability of marine waters, but it is
important in remote sensing of coastal waters to select the type of algorithms which are best suited to
retrieve biogeochemical variables.

The Color of Coastal Waters

As for any water body, the color of coastal waters is defined by the spectral variations in reflectance, $R(\lambda)$, at the sea surface. This is an apparent optical

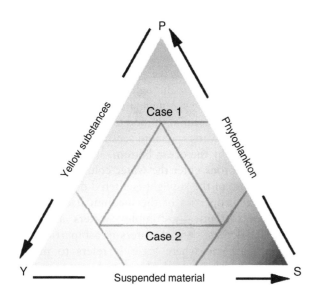

Figure 1 Ternary or triangular diagram illustrating the relative contribution of phytoplankton, dissolved organic matter (yellow substances) and suspended material (nonalgal particles), and the division between case 1 and case 2 waters. Copyright: IOCCG.

property (AOP) determined by the ratio between the upwelling and downwelling irradiances at a given wavelength. This quantity is strictly related to the normalized water-leaving radiance, $L_{WN}(\lambda)$, which can be determined from the total radiance measured by a space-borne sensor. The reflectance varies depending on the illumination and viewing geometry, as well as on the inherent optical properties (IOPs) of the water constituents. In the context of remote sensing, the IOPs of relevance are the absorption coefficient, $a(\lambda)$, and the backscattering coefficient, $b_b(\lambda)$, which represents the integral of the volume scattering function, $\beta(\lambda,\varphi)$, in the backward direction from 90° to 180°. Accordingly, the reflectance at the sea surface is formulated as

$$R(\lambda) = f[b_b(\lambda)/(a(\lambda) + b_b(\lambda))] \qquad [1]$$

where f is a function of the solar zenith angle and depends also on the shape of the volume scattering function.

In coastal waters, the optically active constituents may vary in space and time, independently of each other, and consequently, this variation affects the color (i.e., reflectance) of the water (**Figure 2**).

Optical remote sensing in coastal waters thus necessitates a thorough assessment of the water bulk

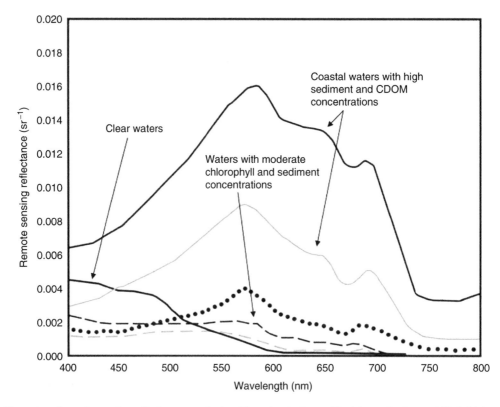

Figure 2 Examples of remote sensing reflectance spectra for different water types. Maximum reflectance shifts to higher wavelength in coastal waters with high concentration of nonalgal particles and CDOMs. Adapted from IOCCG.

optical properties, quantifying the role of each individual component. To this end, the total spectral absorption and backscattering coefficients are divided into as many additive components:

$$a(\lambda) = a_{w}(\lambda) + a_{phy}(\lambda) + a_{NAP}(\lambda) + a_{CDOM}(\lambda) \quad [2]$$

$$b_{b}(\lambda) = b_{b\,w}(\lambda) + b_{b\,phy}(\lambda) + b_{b\,NAP}(\lambda) \quad [3]$$

where the subscript 'w' refers to water molecules (pure water), 'phy' to phytoplankton, 'NAP' to nonalgal particle (e.g., sediment), and 'CDOM' to colored dissolved organic matter (i.e., yellow substances, assumed to have negligible scattering properties).

The basic theory developed for open ocean to parametrize the wavelength dependence of each term in eqns [2] and [3] is directly applicable in coastal waters. The implementation of these models require, however, different strategies to account for unrelated distributions between the various optically active components as well as differences in the values of some model parameters. A notable difference, for example, is in the contribution of nonalgal particles which can be significantly higher in coastal waters due to a supply of sediment either from land or resuspension at the bottom.

The spectral absorption by nonalgal particles and CDOMs obeys similar exponential shape defined by their respective absorption coefficients at a reference wavelength and the slope of the absorption curve in the lower part of the spectrum (400–440 nm). In coastal areas influenced by river runoff, CDOMs mainly consist of terrestrial humic acids with lower slope values than those observed in open ocean and in coastal waters not directly connected with terrestrial sources. In the case of nonalgal particles, the variability in the exponential slope in marine waters is restricted to a narrow range of values (~ 0.01 to < 0.02 nm^{-1}). A potential distinction between coastal and open ocean waters remains therefore difficult, even though assemblage of nonalgal particles may be dominated by mineral sediments or organic debris.

The analysis of the backscattering coefficient is still advancing because of the difficulty in measuring it directly. Our knowledge largely benefits from theoretical studies, which require a good estimate of the particle-size distribution and their refractive index. In coastal waters, the large variety of particle type and size increases the complexity of the calculation. In addition, the presence of fine mineral particles with high refractive index represents there a crucial factor controlling the reflectance variability

when compared with open ocean waters. Specific blooms of phytoplankton (e.g., coccolithophores) and bubbles entrained by breaking waves can also play a significant role in shaping the backscattering coefficient of coastal waters.

Bottom Reflectance

In shallow and clear coastal waters, the ocean color signal is affected by the light reflected off the bottom. The influence of the bottom depends on the depth of the water body, the bottom type, and, of course, the optical state of the water column above. In clear waters, a 30-m-deep bottom feature can affect the remote sensing signal over a narrow spectral band in the blue part of the spectrum. As the bottom becomes shallower, the surface water reflectance is affected over a wider spectral region (assuming the optical state of the water column above remains unchanged).

To account for the bottom, the water-leaving radiance in shallow waters is usually modeled using a two-flow system where the upwelled irradiance is made of a water column and a bottom component of known reflectance. The seabed, however, may be made of sand, rocks, mud, or covered with benthic organisms such as algae, seagrasses, and coral reefs. All of these reflect light in different ways, which add to the complexity in using remote sensing technique in nearshore waters, albeit open to new challenges to apply satellite data for mapping benthic features.

Influence of Inelastic Processes

In marine waters, the water reflectance can also be affected by some transpectral (or inelastic) processes such as those associated with solar-stimulated fluorescence by organic compounds, and Raman (molecular) scattering. Fluorescence is mostly generated by phytoplankton chlorophyll pigments, and by phycoerythrin-containing cells. At high concentration (typically, Chl > 1 mg m^{-3}), chlorophyll can have a significant effect on the reflectance signal within a narrow band in the red part of the spectrum centered at 685 nm, whereas phycoerythrin fluorescence occurs over a larger band centered at 585 nm.

In coastal 'CDOM-dominated' waters, fluorescence by the dissolved organic matter can also influence the blue part of the spectrum, c. 430–450 nm. In all cases, the integration of fluorescence into bio-optical models requires a good knowledge of the absorption coefficients of each component at the excitation wavelengths, as well as their fluorescence quantum yield. Note that in coastal waters, the effect of Raman scattering on the surface reflectance is reduced as the water turbidity increases.

Algorithms for Coastal Waters

Established algorithms are currently on a routine basis by the space agencies and others to generate multi-sensors archives of satellite biogeochemical products over the global ocean. These algorithms were developed to perform as efficiently as possible in the open ocean. The complexity of coastal waters, as well as the atmosphere above, requires more complex algorithms capable of handling nonlinear multivariate bio-optical systems. In addition, the large variability in time and space of the bulk properties of coastal waters hampers any implementation of a single and global algorithm that would work with the same efficiency in all regions.

Atmospheric Correction

The atmospheric correction process is applied to remove the effects of the atmosphere that contribute to the signal measured by a satellite sensor. The objective of this process is the discrimination, from top-of-atmosphere radiance, of the signal emerging from the sea carrying information on the materials suspended and dissolved in seawater. The atmospheric correction of coastal data is challenged by the presence of continental aerosols, bottom reflectance, and adjacency of land. Minimizing these perturbing effects, which generally are site-specific, requires knowledge of the regional aerosol and bottom optical properties. Specifically, effects of continental aerosols are minimized by properly accounting for their scattering phase function and single scattering albedo; the increase in radiance due to bottom reflectance can be removed through iterative processes knowing the water depth and spectral reflectance of the seabed; and the adjacency effects are minimized by determining the reduction in image contrast as a function of the aerosol and of the land reflectance.

Coastal Water Algorithms

In optical remote sensing of marine waters, the algorithms used to retrieve the concentrations of water constituents are either analytical (or semi-) or empirical. Empirical algorithms (so-called blue–green ratio algorithms; have been implemented in the processing of satellite data over open ocean with various degrees of success. But the performance of these algorithms remains limited to the range of input values that have been used for their development. In a highly changeable environment such as coastal waters, these algorithms are clearly restricted to specific regions, and even specific periods. Some improvements can be made to increase the efficiency of these algorithms in turbid waters, by using more than one band ratio, within spectral regions that are more sensitive to the optical conditions observed locally. For example, additional wavelengths in the blue part of the spectrum could be used to separate the dissolved material from the particulate matter, whereas other bands in the green and orange part may provide information on specific phytoplankton blooms often observed in coastal waters (see Table 1). In reality, the spectral characteristics of combined seawater constituents are not unique, suggesting that no single wavelength can really be representing information on just one constituent. Under these conditions, more complex modeling approaches, and optimization techniques are envisaged to solve the problem.

Analytical (or semi-) algorithms are based on the inversion of a forward radiance model describing both the relationship between water constituents and the water reflectance at the surface (typically as in eqn [1]), and the light propagation through the water and atmosphere. This approach makes use of known optical properties for the specific components, even though the spectral characteristics of these components are usually determined from empirical formulation, thus justifying the term 'semi-analytical' to describe these algorithms. In coastal waters, the number of unknowns in each set of equations may increase rapidly with the number of independent variables (i.e., water constituents), causing difficulties in solving the system without applying some approximation that eventually leads to a lower accuracy of the algorithms.

Table 1 Example of spectral range for coastal waters applications

Product name	Critical spectral range (nm)
Chlorophyll, and other pigment concentration	443–445 (max. abs.), 550–560 (min. abs.) 490, 548, 640 (for other pigments)
Red tides	510
Sediment concentration	530–650, NIR
CDOM	410–420
Seabed reflectance	490–580
Chlorophyll fluorescence	680–685
Aerosol optical properties Atmospheric corrections	700–NIR
Sea surface temperature (SST)	TIR

NIR, near-infrared; TIR, thermal infrared.

Nonlinear optimization techniques and other machine learning methods such as artificial neural networks provide an interesting alternative to examine complex coastal waters and to handle multivariate systems. Models based on these methods determine the output values (e.g., the concentrations of optically active components) from input data (e.g., water reflectance at various wavelengths) through nonlinear multidimensional parametric functions (**Figure 3**). The determination of the model parameters, as well as the assessment of the model performance rely on a reference data set.

Remote sensing of coastal waters and algorithm development are still topics of ongoing research in spite of substantial effort and promising results that have taken place within the last few years. Improvements remain to be achieved on several aspects of both the atmospheric correction and in-water bio-optical modeling to ensure an operational retrieval of the products at a reasonable accuracy. These investigations rely on field experiments and the collection of high-quality optical data for a wide range of water bodies.

With the rapid changes occurring in coastal waters, a unique algorithm that will serve all optical conditions with the same accuracy is doubtful. On the other hand, the implementation of more than one regional algorithm into a unique processing scheme remains a challenge, as that approach could result in significant discontinuities at the region boundaries.

Data Validation and Measurement Protocols

The normalized water-leaving radiance, $L_{WN}(\lambda)$, is the primary ocean color remote sensing product obtained from top-of-atmosphere data corrected for the atmospheric perturbations. Higher-level marine products, like the concentration of phytoplankton pigments or seawater inherent optical properties (i.e., absorption, scattering, and backscattering coefficients), are determined from $L_{WN}(\lambda)$. The accuracy of coastal remote sensing products may be affected by the presence of turbid waters or absorbing aerosols. Because of this, the validation of remote sensing $L_{WN}(\lambda)$ and aerosol optical thickness data, resulting from the atmospheric correction process, is a first fundamental step in assessing the accuracy of ocean color products. Additional steps include the validation of derived products to assess the consistency of the whole process leading to the determination of higher-level products.

Validation measurements can result from oceanographic campaigns, or from continuous data collection by autonomous systems deployed on fixed platforms or ships of opportunity. Oceanographic campaigns ensure a wide spatial data collection with a comprehensive optical and biogeochemical characterization of each sampling site, but with a poor time resolution. Autonomous systems enable continuous a highly repeated collection of a restricted set of parameters at a specific site or along a

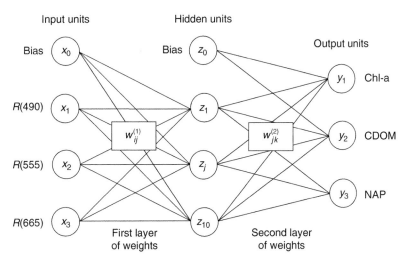

Figure 3 Representation of a multilayer perceptron (MLP) neural network adopted for the retrieval of the optically active seawater components. In this particular case, the network architecture consists of three inputs (reflectances at three satellite wavelengths), 10 hidden units, and three outputs (chlorophyll *a*, colored dissolved organic matter, and nonalgal particulate matter). An additional input is implemented for the bias parameter. Note that the MLP architecture presented here has only an illustrative purpose. Adapted from D'Alimonte D and Zibordi G (2003) Phytoplankton determination in an optically complex coastal region using a multilayer perceptron neural network. *IEEE Transactions on Geoscience and Remote Sensing* 41(12): 2861–2868.

given route. Obviously, measurement campaigns and autonomous systems are complementary and both contribute to the objective of validating remote sensing products.

Within the framework of *in situ* optical technologies commonly applied for the determination of the fundamental quantity $L_{WN}(\lambda)$, both in- and above-water radiometry can produce individual measurements or time series of optical data. In this context, various in-water autonomous systems have been widely exploited in open ocean regions where the biofouling perturbation is low. Differently, the use of autonomous above-water radiometry is more suitable to produce long-term measurements for the validation of ocean color data in optically complex coastal waters.

Offshore fixed platforms like lighthouses, oceanographic towers, and oil rigs, generally located at several nautical miles from the main land in coastal areas (i.e., regions permanently or occasionally affected by bottom resuspension, coastal erosion, river inputs, or by relevant anthropogenic impact), are suitable for the deployment of autonomous above-water radiometers provided that (1) their distance from the coast makes negligible the adjacency effects in the remote sensing data; and (2) the bottom structure and type have no effect (**Figure 4**).

Time, spatial, and spectral coincident pairs of satellite and *in situ* data, so-called matchups, are the basis for any validation analysis. Due to the intrinsic higher variability exhibited by coastal waters, when compared to oceanic waters, particular care must be given to the spatial variability of satellite data and to the temporal variability of *in situ* observations (**Figure 5**).

Remote Sensing Coastal Applications

Remote sensing of coastal waters satisfies on its own a large number of applications, owing to substantial progress in sensor technology, as well as algorithm performance, made over the last few years. Important contributions have been accomplished in various disciplines of marine sciences, like marine physics and biogeochemistry, process modeling, as well as fisheries and coastal management. More importantly, satellites provide continuous long time series of bio-optical and geophysical variables, thus representing a cost-effective tool to monitor changes in coastal waters that might occur as a direct or indirect result of external pressures (e.g., climate change and human activities). In parallel, the scientific and technical community is progressing in the search for

Figure 4 Lighthouse tower in the Baltic Sea used for the deployment of above-water radiometer systems supporting the validation of ocean color products in coastal areas.

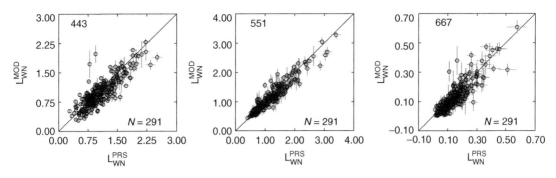

Figure 5 Scatter plots of satellite-derived L_{WN}^{MOD} vs. *in situ* L_{WN}^{MOD} normalized water-leaving radiances in units of mW cm^{-2} μm^{-1}s^{-1} at the 443, 551, and 667 nm MODIS center-wavelengths.

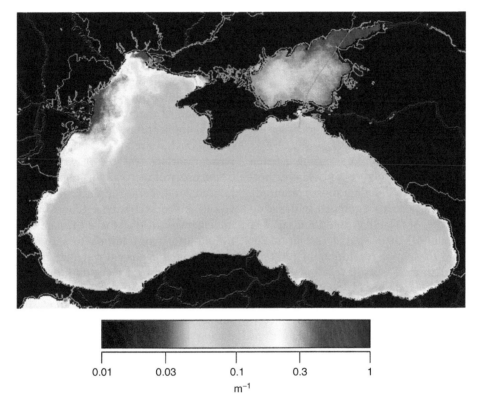

Figure 6 Light attenuation coefficient K_d in units of m^{-1} for the Black Sea derived from SeaWiFS monthly composite (Jan. 2003). The coefficient is calculated using a semi-analytical algorithm to retrieve the IOPs of the water constituents. Source: Joint Research Centre of the European Commission © Joint Research Centre, 2006.

the optimal model to be applied in these complex waters, taking benefit from advanced instruments with higher spectral resolution and better calibration. Recent development of hyperspectral sensor prototypes, providing the full spectral coverage over the visible wavelengths, offers new opportunities to discriminate optical disparities, and thus to classify turbid waters and quantify individual seawater components down to phytoplankton species.

Optical remote sensing in coastal waters typically provides quantitative estimates of phytoplankton biomass (indexed by chlorophyll concentrations),

total suspended matter, and dissolved organic substances, with accuracies as good as in open ocean (*c.* 30–40% in biomass) when using regional dedicated algorithms. These components, in turn, influence the light attenuation coefficient in the water which is also directly retrieved from satellite (**Figure 6**), providing information on water transparency. The latter represents an important quality parameter to assess the ecological status of lakes and coastal waters. Remotely sensed chlorophyll concentration and light attenuation are also required to estimate water-column primary production, which represents a critical indicator to assess coastal eutrophication, and to analyze the carbon budget directly associated with the continental shelf pump.

The concentrations of suspended sediment and dissolved organic matter can be used to track river plumes and other coastal dynamics, and to calibrate and validate sediment transport models. Recently, the availability of IOPs, that is, absorption and backscattering coefficients, as direct products of satellite ocean color, has given an alternative approach to address changes in the water constituents, as well as water transparency, further extending satellite applications to particle size and composition characteristics.

Several algal species can be present in coastal waters as intense blooms (e.g., coccolithophorid species, red tides, *Trichodesmium*), which can be readily distinguished on the basis of their differences in pigment composition and optical properties (**Figure 7**). In that case, remote sensing acts as an efficient technique to monitor these anomalous blooms; some are associated with harmful consequences on the surrounding ecosystem, causing mass mortalities of marine organisms.

In shallow coastal waters, remote sensing is being used to monitor benthic structure and classify bottom types. At present, satellite data most commonly exploited for that purpose are equipped with radiometers, such as the SPOT-HRV and Landsat TM series with typically 10–30 m spatial resolution, or even IKONOS with spatial resolution better than 10 m. Even though these sensors have a limited set of wavelengths (three to four broad spectral bands in the visible), the spectral contrast between various bottom structures (e.g., corals and noncoral objects) can be used to construct classification algorithms based on differential reflectance (**Figure 8**).

Surface features simultaneously observed from ocean color and thermal imagery have assisted coastal fisheries to locate target areas with high fish densities. Many pelagic species aggregate along frontal structures and upwelling areas that bear distinctive signals on the sea surface temperature and primary production (or phytoplankton biomass) maps. Moreover, many aspects of the life cycle of fishes (e.g., reproduction and early-stage feeding) depends on physical and biological processes, and discerning these from remote sensing is important for marine conservation and sustainable management of the marine coastal resources.

Satellite instruments, other than optical sensors, can provide additional information of importance to coastal water management. Microwave sensors collect data either passively or actively in that part of the electromagnetic spectrum ranging from 3 mm to 1 m wavelength. These instruments provide significant advantages in viewing the Earth under all-weather conditions, even through clouds; but their spatial resolution, which depends on the physical dimension of the antenna, is often not adequate for coastal applications. Sophisticated synthetic aperture radars (SARs) are an exception to that rule and can retrieve surface features at extremely fine resolution of the order of meters. This remote sensing technique is used for fisheries applications to monitor fishing vessels in coastal waters, contributing thus to the management of the stocks through evaluation of the fishing effort. In addition to ship survey, SAR data are exploited in coastal waters to monitor sea surface roughness and to detect pollution at the surface. Oil spill detection is of particular importance to coastal management. The presence of oil slicks on the sea surface increases the surface tension of seawater, reducing its natural roughness which appears as a black signature on a SAR image. Multiple techniques using, for example, ocean color and SAR can be helpful to discriminate between dense algal blooms

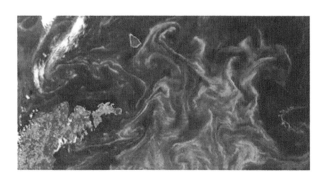
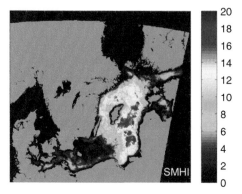

Figure 7   Cyanobacteria bloom in the Baltic Sea. Left: MODIS-Aqua 30 Jul. 2003. True color image of the Baltic Sea showing a bloom of *Nodularia spumigena*. Source: http://oceancolor.gsfc.nasa.gov. Right: Occurrence of cyanobacterial accumulations in the Baltic during 2003 expressed as number of days (data based on satellite images). Source: http://www.helcom.fi

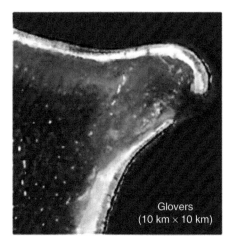

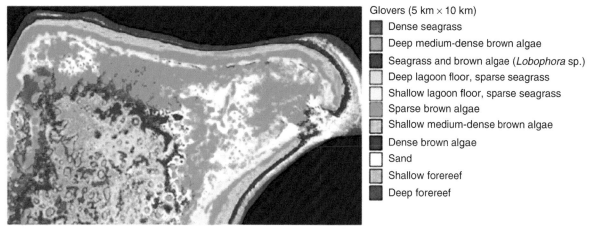

Glovers (5 km × 10 km)
- Dense seagrass
- Deep medium-dense brown algae
- Seagrass and brown algae (*Lobophora* sp.)
- Deep lagoon floor, sparse seagrass
- Shallow lagoon floor, sparse seagrass
- Sparse brown algae
- Shallow medium-dense brown algae
- Dense brown algae
- Sand
- Shallow forereef
- Deep forereef

Figure 8 Classification of tropical coral reef environment (Belize, Caribbean Sea). Top: RGB color composite based on the red, green, and blue bands of IKONOS sensor. Bottom: Eleven-class scheme classification of the benthic features. Reproduced from Andréfouët S, Kramer P, Torres-Pulliza D, *et al.* (2003) Multi-site evaluation of IKONOS data for classification of tropical coral reef environments. *Remote Sensing of Environment* 88: 128–143, with permission from Elsevier.

and surface slicks. The low revisit cycle (e.g., 35 days for the European Remote Sensing satellites ERS-1 and ERS-2 instruments) limits, however, the use of SAR to monitor rapidly changing events.

In conclusion, there is a large and ever-increasing interest in applying remote sensing technique in coastal waters. It offers a unique solution to monitor this environment at scales that are compatible with many biological and physical processes. At the same time, the number of satellite products is growing rapidly with the advent of new sensors, better algorithms and validation experiments, providing a range of key indicators to support scientific investigations, as well as operations for coastal management.

Further Reading

Andréfouët S, Kramer P, Torres-Pulliza D, *et al.* (2003) Multi-site evaluation of IKONOS data for classification of tropical coral reef environments. *Remote Sensing of Environment* 88: 128–143.

Babin M, Stramski D, Ferrari GM, *et al.* (2003) Variations in the light absorption coefficients of phytoplankton, nonalgal particles, and dissolved organic matter in coastal waters around Europe. *Journal of Geophysical Research* 108(C7): 3211 (doi:10.1029/2001JC000882).

Berthon J-F, Mélin F, and Zibordi G (2007) Ocean colour remote sensing of the optically complex European seas. In: Barale V and Gade M (eds.) *Remote Sensing of European Seas*, pp. 35–52. Dordrecht: Springer.

Bukata RP, Jerome JH, Kondratyev KY, and Pozdnyakov DV (eds.) (1995) *Optical Properties and Remote Sensing of Inland and Coastal Waters*. Boca Raton, FL: CRC Press.

D'Alimonte D and Zibordi G (2003) Phytoplankton determination in an optically complex coastal region using a multilayer perceptron neural network. *IEEE Transactions on Geoscience and Remote Sensing* 41(12): 2861–2868.

Gordon HR (2002) Inverse methods in hydrologic optics. *Oceanologia* 44: 9–58.

Hansson M (2004) Cyanobacterial blooms in the Baltic Sea. HELCOM Indicator Fact Sheets 2004. Online (2008-05-06) http://www.helcom.fi/environment2/ifs/archive/ifs2004/enGB/cyanobacteria (accessed July 2008).

IOCCG (2000) Remote sensing of ocean colour in coastal, and other optically-complex waters. In: Sathyendranath S (ed.) *Report of the International Ocean-Colour Coordinating Group, No. 3.* Dartmouth, NS: IOCCG.

IOCCG (2006) Remote sensing of inherent optical properties: Fundamentals, tests of algorithms and applications. In: Lee ZP (ed.) *Report of the International Ocean-Colour Coordinating Group, No. 5.* Dartmouth, NS: IOCCG.

Miller RL, Del Castillo CE, and McKee BA (eds.) (2005) *Remote Sensing of Coastal Aquatic Environments: Technologies, Techniques and Applications.* Dordrecht: Springer.

Oceanography Magazine (2004) *Special Issue: Coastal Ocean Optics and Dynamics. The Official Magazine of the Oceanography Society* 17(2).

Zibordi G, Holben B, Hooker SB, *et al.* (2006) A network for standardized ocean color validation measurements. *EOS Transactions, American Geophysical Union* 87(30): 293–297.

RIGS, STRUCTURES AND SHIPPING

RIGS AND OFFSHORE STRUCTURES

C. A. Wilson III, Department of Oceanography and
Coastal Sciences, and Coastal Fisheries Institute,
CCEER Louisiana State University, Baton Rouge, LA,
USA
J. W. Heath, Coastal Fisheries Institute,
CCEER Louisiana State University, Baton Rouge, LA,
USA

Introduction

The rapid advance in offshore engineering and
physical oceanography has provided mankind with
the ability to emplace and maintain structures that
support what are effectively small cities in some of
the ocean's harshest environments. The majority of
these offshore structures are associated with the development of hydrocarbon extraction off the world's
continents. Offshore structures have been focal
points for controversies concerning pollution and
hazard to navigation, yet serve as a basis for many
coastal economies. Oil spills, drilling muds, and
produced water have been continuous public and
regulatory issues. Ironically, oil and gas platforms
have also been the target of the locally popular 'rigs
to reefs' program since the early 1980s. State and
federal governments of the United States have encouraged states and private sector organizations to
maintain these structures as artificial reefs as it is
perceived that they improve fishing on a local scale.
In addition to their ecological value as 'artificial

reefs' these platforms are now recognized as excellent
facilities for research as we explore our Earth's final
frontier.

History

Since the mid 1950s approximately 7000 structures
associated with oil and gas development have been
emplaced on the continental shelf and slope, and are
now approaching the ocean basins (**Figure 1**). The
majority of these structures (>65%) are along United
States Gulf Coast, and the remainder are concentrated in the North Sea, Middle East, Africa, Australia, Asia, and South America.

The oldest, most well developed, and largest concentration of structures is in the US Gulf of Mexico
(GOM). There are currently about 4000 oil and gas
platforms off the coasts of Louisiana and Texas, and
industry is now challenging depths in excess of 900 m
and over 220 km from shore. The GOM now hosts
over 99% of structures associated with US offshore
petroleum production. The earliest of these structures
were installed in the late 1920s, and were wooden
structures erected with steam power and human
muscle in the shallow bays and near-shore coastal
waters. They were later followed by single piled
caissons and multiple leg concrete structures. As
technology improved, large steel towers supported by
multiple, pipe-like legs approaching 200 cm in
diameter were used on the continental shelf by the
1950s. Today, engineers use wire-supported towers

Figure 1 Distribution of the world's offshore platforms in the year 2000. (Information provided by William Griffin.)

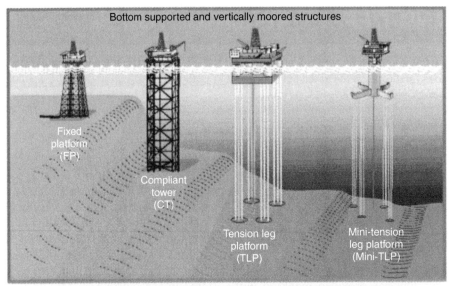

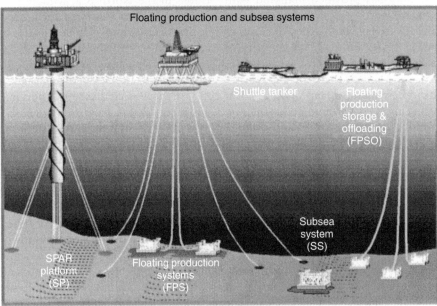

Figure 2   Various shapes and sizes of common offshore platforms (modified from the http://www.gomr.mnms.gov/homepg/offshore/deepwater/options.html/deepwater info).

and floating platforms in water depths not suitable for standard steel frame structures.

Oil and gas structures come in a variety of shapes and sizes (**Figure 2**) that reflect engineering demands depending on water depth, range of sea state conditions, advancements in engineering design, metallurgy, and ocean physics (e.g. currents). Several of the larger platforms have set engineering records for their time, and current structures reach 3.5 km tall. The largest of these structures are found in the North Sea, where ocean conditions are amongst the harshest in the world – winter storm waves can exceed 45 m in height and last for many days. These types of

environments demand tremendous feats in engineering. For example, the North Sea 'Troll' weighs over 1 million tons, is over 450 m tall, and is expected to have 70 year lifetimes to exhaust the mineral resources beneath its legs. North Sea structures dwarf those found in the rest of the world.

Environmental Issues

Public perception of oil and gas development has been greatly influenced by oil spills, visual pollution, drilling discharges, and unrelated but dramatic oil tanker accidents. The greatest wealth of information concerning oil and gas environmental issues comes from regulations enforced by the Environmental Protection Agency of the United States and regulations of the UK government for operations in the North Sea. Otherwise, international treaties guide the environmental conscience of industry such as the International Maritime Organization and the Law of the Treaty of the Sea. During the early exploratory years of offshore drilling there were several recorded instances of oil spills and the occasional 'blow out'. However, over the past 25 years strict regulations, put in place by the developed countries (e.g. the US Minerals Management Service and the Environmental Protection Agency), have provided safeguards to prevent such ecological disasters.

Environmental concern of the 1980s focused on the fate of discharged fluids and drilling mud. Several books have been written about produced waters; these are the hydrocarbon-permeated waters pumped to the surface with oil and gas. Recently US regulations were enacted that require operators of platforms located in 60 m of water or less to re-inject produced waters back into the wells. Farther offshore these fluids are discharged in association with large compressor-driven bubblers that mix produced water with the surrounding water. The fate of the re-injected produced waters is unknown. Drilling muds have also caused a major amount of concern due to the heavy metals present in them. A number of studies have been funded by the US Mineral Management Service to investigate the impact of drilling muds. The general conclusions of these studies were that environmental impacts of drilling muds are confined to the area immediately around the drilling site.

The oil and gas industry in the Gulf of Mexico has enjoyed a much different public relationship with the community than off California and elsewhere. Since the first platform was placed off California there has been strong public objection concerning visual pollution and the potential for chemical pollution. However, the oil and gas industries of Louisiana and Texas have notably had a dramatic impact on coastal economies, and enjoyed a very positive relationship with the recreational and commercial fisherman. These user groups have come to depend upon the industry for infrastructure support as well as fishing opportunity. The oil and gas industry has been friendly toward helping fishermen in distress, providing weather updates, and serving as a visual reference point for navigation prior to the advent of Loran or GPS technology. Most noteworthy is the common knowledge that fishing around platforms is exceptional and that the Gulf Coast oil and gas industry does not object to fishermen mooring to a structure.

The appearance of oil and gas platforms has inadvertently impacted on historic or developing fishing fleets along the world's coastlines. High coastal primary and secondary production serves as the basis for production of commercial and recreational fisheries. Coincident with the large oil and gas fields, the crustacean fishery in the Gulf of Mexico is amongst the world's largest. Likewise, the North Sea has historically been one of the richest trawling fisheries in the world. Offshore structures interfere with net fisheries through their physical presence and the materials that were discarded from the oil and gas platforms during the earlier years of development. These issues have plagued Gulf of Mexico shrimpers for years. In the United States there are federal and state regulations and efforts by major oil and gas companies to totally eliminate all debris associated with oil- and gas-related activities. However, fishermen are still catching artifacts which are damaging to their gear, and oil and gas companies have gone to great lengths to provide funding mechanisms to replace the various gear that are damaged. Nevertheless, industries still coexist in most areas.

'Rigs to Reefs'

It is the observation of many federal and state fishery managers throughout the world that fisheries are in a serious state of decline. The last assessment by the FAO indicated that well over 60% of the world's fisheries were in a state of overfishing or overfished and in decline and that the world's fishery harvest is at maximum. As mentioned above many of these fisheries are located in areas of oil and gas development. It has been suggested by some that the presence of oil and gas platforms actually provides a 'safe haven' for many of these fish species that prevents them from being captured. On the contrary, scientists and managers have also proposed that platforms in fact make fish easier to catch for the hook and line fisherman.

Since the first platforms were emplaced in the Gulf of Mexico, off California, and other parts of the world where hook and line fishing activities occur, fishers have realized that these tall steel structures serve as fish havens or 'artificial reefs'. Scientists have documented that catch rates around platforms in the Gulf of Mexico (number of fish caught per hour per person by hook and line) are amongst the highest in the world. Popular recreational and commercial species such as red snapper, gray snapper, amberjack, mackerel, tuna, marlin, sheepshead, and triggerfish are abundant around the platforms on the continental shelf of the Gulf of Mexico. Charter boat fisheries and commercial fisheries now depend upon many of these structures for income. Pursuant to the public's perception of the fishery value of oil and gas platforms, several states including Louisiana, Texas, Mississippi, and Florida have established artificial reef programs that encourage the use of oil and gas platforms as artificial reefs. In most cases these are extremely positive and cooperative efforts between state regulators and industry to establish specific areas for the long-term maintenance of fishing opportunity.

However, these programs have their opponents. Scientists have yet to establish whether artificial reefs in general, and specifically oil and gas platforms, concentrate fish and make them easier to catch or if artificial reefs increase productivity and benefit some life stages of a fish species. Most scientists have come to an agreement that artificial reefs and specifically platforms increase productivity for some fish species and concentrate others. However, the extent and value of that habitat has yet to be determined. A major concern for species such as red snapper is that the benefit of increased production and survival may be outweighed by increased vulnerability to human capture. Due to the hook and line vulnerability of some fish species, such as red snapper, regulators continue to pursue the use of obsolete platforms as artificial reefs, 'rigs to reefs', with caution.

The Minerals Management Service of the United States has funded a number of ecological investigations off California and in the Gulf of Mexico to explore the ecology and habitat value of platforms. Scientists at Louisiana State University have established that an average of 25 000 fish 12 cm in length inhabit the area immediately around and under oil and gas platforms; such density is 50–1000 times greater than adjacent soft bottom sediment. The fish community associated with offshore platforms includes a number of important recreational and commercial species. Research in the Gulf of Mexico indicates that species composition for near-shore platforms (typically within 50 km offshore) is dominated mostly by estuarine species, while platforms farther offshore contain more pelagic and tropical species. However, this is not the case in the North Sea, where species composition is dominated predominantly by pelagic species. Divers in the GOM have also reported large schools of larval and juvenile species of fish that obviously depend upon the hard steel substrate for protection, orientation, and perhaps food. Scientists in the North Sea report fish densities around large steel towers to be two to three times higher than adjacent water bottoms. Hence, there is little debate regarding the higher numbers of fish living around offshore structures than adjacent water bottoms.

The Rigs to Reef issue is fairly unique to the United States, although there has been limited interest in other parts of the world such as Australia, Japan, and in the Caribbean. In reality the utility of oil and gas platforms as artificial reefs will be limited to the most developed countries where recreational and commercial fishing command attention.

Removal

Federal law in the United States, law in the UK, and EC, as well as international treaty through the International Maritime Organization, requires that once an oil and gas company has completed extraction of oil and gas hydrocarbon reserves the platform must be removed and the bottom returned to its original state. This law was developed through international agreements to protect navigation and military interests, and to preserve ocean ecology.

During the early years of oil and gas development, when the industry was operating in shallow water, technology existed for the removal of these massive structures. Since then the semi-permanent placement of offshore structures has challenged engineering and led to several new disciplines, the most formidable of which is decommissioning. Currently the technology does not exist to remove deep-water structures like the Troll, Mars, and others (**Table 1**). The majority of the structures in shallower waters will be subject to removal over the next 50 years.

The oldest platforms, particularly those in the Gulf of Mexico and off California, are 20–30 years old and are nearing the depletion of their hydrocarbon reserves. Several international organizations continue to debate removal technology, and to explore ocean dumping as an alternative. Most of the environmental groups throughout the world strongly oppose this activity, as the concept of ocean dumping is not palatable to most people. Shell's North Sea Brent Spar incident in 1995 illustrates the public's sensitivity to ocean dumping. Greenpeace blocked the towing of the Brent Spar out to sea, and gas stations were burned in Germany in association with this highly publicized event. The British Government conceded to public pressure and reversed its decision to allow ocean dumping, and the fate of the Brent Spar is yet undecided. It is however a reality that the oil and gas industry will be faced with some serious engineering questions as these deep-water platforms are retired.

Table 1 Company, structure name, water depth, and date of installation of deep-water structures in the Gulf of Mexico, USA[a]

Operator	Structure name	Water depth (m)	Date installed
Shell Offshore Inc.	COGNAC	312	1 Jan. 1978
Shell Offshore Inc.	BULLWINKLE	412	1 Jan. 1988
Conoco Inc.	JOLLIET	536	1 Jan. 1989
BP Exploration & Oil Inc.	A	335	1 Jan. 1991
Exxon Mobil Corporation	A	423	3 Oct. 1991
Shell Deepwater Production Inc.	A	872	5 Feb. 1994
BP Exploration & Oil Inc.	A	393	19 Aug. 1994
EEX Corporation	COOPER	639	7 Jul. 1995
Shell Deepwater Production Inc.	MARS	894	18 Jul. 1996
Kerr-McGee Oil & Gas Corporation	NEPTUNE	588	19 Nov. 1996
Shell Deepwater Production Inc.	RAM-POWELL	980	21 May 1997
Amerada Hess Corporation	BALDPATE	502	31 May 1998
Chevron USA Inc.	GENESIS	789	21 July 1998
British-Borneo Exploration, Inc.	MORPETH EAST	518	10 Aug. 1998
British-Borneo Exploration, Inc.	ALLEGHENY SEA	1004	19 Aug. 1999
Elf Exploration, Inc.	VIRGO	344	17 Sep. 1999

[a]From http://www.gomr.mms.gov/homepg/offshore/deepwater/options.html.

The accepted removal technology along the continental shelf involves the use of explosives where the platform legs are severed 5 m below the mud line. Not only are explosives extremely damaging to the marine environment and the associated explosion kills most of the 10 000–25 000 fish that live around a platform, but also the activity involves risk of human life as divers are frequently lowered down inside the legs to place the explosives. More recently mechanical cutting and cryogenics have been used. Cryogenics is still improving and mechanical cutting is practical only to depths of about 60 m. There are many challenges to removing a large steel platform that is some 3.5 km long. The engineers of deep-water platforms of this size need to provide designs or methods for removal when the structure depletes the petroleum stores.

Research

Not only do the oil and gas platforms serve as a support structure for operations involved in the extraction and transmission in oil and gas; these platforms also serve as excellent research stations. The ocean is a final frontier for exploration, as we know more about the surface of the moon than we do about our continental slopes and ocean basins. To aid in the quest for knowledge, many major and minor oil and gas companies have allowed scientists on their platforms for the purpose of research over the past 20 years.

The US Minerals Management Service has funded a number of ecological investigations to explore the physical, chemical, biological, and meteorological events that occur around oil and gas platforms. These permanent steel structures provide us with the opportunity to collect long-term databases on changes in weather pattern and shallow and deep water currents, to track bird migration, and to monitor the aquatic biological community in many parts of the world. This information will be invaluable in predicting the path and strength of hurricanes, to follow changes in sea life due to global warming, and improve general knowledge of the world's oceans and resolve the impacts of offshore structures and the marine environment. This research activity is carried out at a fraction of the cost that it would be if larger vessels were used to collect this data, and under weather conditions that prevent research vessel operations. Oil and gas companies are often cooperative in providing logistical support (lodging and meals) and transportation for scientists who visit these platforms for such investigations. Obviously information such as hurricane path and current patterns are of extreme interest to the platform engineers as well as to coastal resource managers in state and federal government. The advent of microwave, satellite, and cellular technology allow for the transmission of real-time data collected at these platforms and much of this information is being made online to students around the world, in addition to the scientists that use the data.

Summary

Offshore structures are concentrated in areas of oil and gas development off most coastal countries of the world. Platforms will be a common sight along the world's coastlines for as long as hydrocarbon resources are present. Although the industry provides the energy to fuel the world, the presence of offshore structures has both positive and negative impacts concerning pollution, navigation, and fisheries production. Current development and research is being implemented for the best use of offshore structures concerning these issues. Offshore structures are the result of remarkable feats of engineering, and engineers are continually challenged in the design and in the removal guidelines within the constraints of current technology.

See also

Coral Reef Fishes. Coral Reefs. Oil Pollution.

Further Reading

American Petroleum Institute (1995) *Impact on U.S. Companies of a Worldwide Ban on Disposal of Offshore Oil and Gas Platforms at Sea*. IFC Resources Inc.

Boesch DF and Rabalais NW (1987) *Long-term Environmental Effects of Offshore Petroleum Development*. New York: Elsevier Applied Science.

Love MS, Nishimoto M, Schroeder D, and Caselle J (1999) *The Ecological Role of Natural Reefs and Oil and Gas Production Platforms on Rocky Reef Fishes in Southern California: Final Interim Report*. US Department of the Interior, US Geological Survey, Biological Resources, Division, USGS/BRD/CR-1999-007.

Pulsipher AG (ed.) (1996) *Proceedings: An International Workshop on Offshore Lease Abandonment and Platform Disposal: Technology, Regulation and Environmental Effects*. MMS contract 14-35-0001-30794. Baton Rouge, LA: Center for Energy Studies, Louisiana State University.

Reggio VC Jr (1987) *Rigs to Reefs: The Use of Obsolete Petroleum Structures as Artificial Reefs*. OCS Report/MMS 87-0015. New Orleans, LA: US Department of the Interior. Minerals Management Service. Gulf of Mexico OCS Region.

Reggio VC Jr (1989) *Petroleum Structures as Artificial Reefs: A Compendium*. Fourth International Conference on Artificial Habitats for Fisheries, Rigs-to-Reefs Special Session, Miami, FL, November 4, 1987. New Orleans, LA: US Department of the Interior, Minerals Management Service, Gulf of Mexico OCS Regional Office.

Wilson CA and Stanley DR (1990) A fishery-dependent based study of fish species composition and associated catch rates around oil and gas structures off Louisiana. *Fishery Bulletin US* 88: 719–730.

Wilson CA, Stickle VR and Pope DL (1987) *Louisiana Artificial Reef Plan*. Louisiana Department of Wildlife and Fisheries Technical Bulletin No. 41. Baton Rouge, LA Louisiana Sea Grant College Program.

SHIPPING AND PORTS

H. L. Kite-Powell, Woods Hole Oceanographic
Institution, Woods Hole, MA, USA

Introduction

Ships and ports have been an important medium for trade and commerce for thousands of years. Today's maritime shipping industry carries 90% of the world's 5.1 billion tons of international trade. Total seaborne cargo movements exceeded 21 trillion ton-miles in 1998. Shipping is an efficient means of transport, and becoming more so. World freight payments as a fraction of total import value were 5.24% in 1997, down from 6.64% in 1980. Modern container ships carry a 40-foot container for < 10 cents per mile – a fraction of the cost of surface transport.

Segments of the shipping industry exhibit varying degrees of technological sophistication and economic efficiency. Bulk shipping technology has changed little in recent decades, and the bulk industry is economically efficient and competitive. Container shipping has advanced technologically, but economic stability remains elusive. The shipping industry has a history of national protectionist measures and is governed by a patchwork of national and international regulations to ensure safety and guard against environmental damage.

Units

Two standard measures of ship size are deadweight tonnage (dwt) and gross tonnage (gt). Deadweight describes the vessel's cargo capacity; gross tonnage describes the vessel's enclosed volume, where 100 cubic feet equal one gross ton. In container shipping, the standard unit of capacity is the twenty-foot equivalent unit (TEU), which is the cargo volume available in a container 20 feet long, 8 feet wide, and 8 feet (or more) high. These containers are capable of carrying 18 metric tons but more typically are filled to 13 tons or less, depending on the density of the cargo.

World Seaborne Trade

Shipping distinguishes between two main types of cargos: bulk (usually shiploads of a single commodity) and general cargos (everything else). Major dry bulk trades include iron ore, coal, grain, bauxite, sand and gravel, and scrap metal. Liquid bulk or tanker cargos include crude oil and petroleum products, chemicals, liquefied natural gas (LNG), and vegetable oil. General cargo may be containerized, break-bulk (non-container packaging), or 'neo-bulk' (automobiles, paper, lumber, etc.). **Table 1** shows that global cargo weight has grown by a factor of 10 during the second half of the twentieth century, and by about 4% per year since 1986. Tanker cargos (mostly oil and oil products) make up about 40% of all cargo movements by weight.

Both the dry bulk and tanker trades have grown at an average rate of about 2% per year since 1980. Current bulk trades total some 9 trillion ton-miles for dry bulk and 10 trillion ton-miles for tanker cargos.

Before containerization, general cargo was transported on pallets or in cartons, crates, or sacks (in general cargo or break-bulk vessels). The modern container was first used in US domestic trades in the 1950s. By the 1970s, intermodal services were developed, integrating maritime and land-based modes (trains and trucks) of moving containers. Under the 'mini land-bridge' concept, for example, containers bound from Asia to a US east coast port might travel across the Pacific by ship and then across the USA by train. In the 1980s, intermodal services were streamlined further with the introduction of double-stack trains, container transportation became more

Table 1 World seaborne dry cargo and tanker trade volume (million tons) 1950–1998

Year	Dry cargo	Tanker cargo	Total	% change
1950	330	130	460	—
1960	540	744	1284	—
1970	1165	1440	2605	—
1986	2122	1263	3385	3
1987	2178	1283	3461	2
1988	2308	1367	3675	6
1989	2400	1460	3860	5
1990	2451	1526	3977	3
1991	2537	1573	4110	3
1992	2573	1648	4221	3
1993	2625	1714	4339	3
1994	2735	1771	4506	4
1995	2891	1796	4687	4
1996	2989	1870	4859	4
1997	3163	1944	5107	5
1998	3125	1945	5070	− 1

Table 2 Cargo volume on major shipping routes (million tons) 1998 (for containers: millions of TEU, 1999)

Cargo type	Route	Cargo volume	% of total for cargo
Petroleum	Middle East–Asia	525.0	26
	Intra-Americas	326.5	16
	Middle East–Europe	229.5	12
	Intra-Europe	142.7	7
	Africa–Europe	141.4	7
	Middle East–Americas	137.7	7
	Africa–Americas	106.1	5
Dry bulk: iron ore	Australia–Asia	116.8	28
	Americas–Europe	87.1	21
	Americas–Asia	63.3	15
Dry bulk: coal	Australia–Asia	118.4	25
	Americas–Europe	61.3	13
	S. Africa–Europe	37.8	8
	Americas–Asia	35.2	7
	Intra-Asia	29.3	6
	Australia–Europe	27.3	6
Dry bulk: grain	Americas–Asia	60.2	31
	Americas–Europe	20.6	10
	Intra-Americas	33.1	17
	Americas–Africa	15.2	8
	Australia–Asia	15.1	8
Container	Trans-Pacific (Asia/N. America)	7.3	14
	Asia/Europe	4.5	8
	Trans-Atlantic (N. America/Europe)	4.0	8

standardized, and shippers began to treat container shipping services more like a commodity. Today, >70% of general cargo moves in containers. Some 58.3 million TEU of export cargo moved worldwide in 1999; including empty and domestic movement, the total was over 170 million TEU. The current total container trade volume is some 300 billion TEU-miles. Container traffic has grown at about 7% per year since 1980.

Table 2 lists the major shipping trade routes for the most significant cargo types, and again illustrates the dominance of petroleum in total global cargo volume. The most important petroleum trade routes are from the Middle East to Asia and to Europe, and within the Americas. Iron ore and coal trades are dominated by Australian exports to Asia, followed by exports from the Americas to Europe and Asia. The grain trades are dominated by American exports to Asia and Europe, along with intra-American routes. Container traffic is more fragmented; some of the largest routes connect Asia with North America to the east and with Europe to the west, and Europe and North America across the Atlantic.

The World Fleet

The world's seaborne trade is carried by an international fleet of about 25 000 ocean-going commercial cargo ships of more than 1000 gt displacement. As trade volumes grew, economies of scale (lower cost per ton-mile) led to the development of larger ships: ultra-large crude oil carriers (ULCCs) of >550 000 dwt were launched in the 1970s. Structural considerations, and constraints on draft (notably in US ports) and beam (notably in the Panama Canal) have curbed the move toward larger bulk vessels. Container ships are still growing in size. The first true container ships carried around 200 TEU; in the late 1980s, they reached 4000 TEU; today they are close to 8000 TEU. The world container fleet capacity was 4.1 million TEU in 1999.

Table 3 shows the development of the world fleet in dwt terms since 1976. Total fleet capacity has grown by an average of 1.5% per year, although it contracted during the 1980s. The significant growth in the 'other vessel' category is due largely to container vessels; container vessel dwt increased faster than any other category (nearly 9% annually) from 1985 to 1999. Specialized tankers and auto carriers have also increased rapidly, by around 5% annually. Combos (designed for a mix of bulk and general cargo) and general cargo ships have decreased by >5% per year.

Four standard vessel classes dominate both the dry bulk and tanker fleets, as shown in Table 4. Panamax bulkers and Suezmax tankers are so named because they fall just within the maximum dimensions for the

Table 3   World fleet development (million dwt) 1976–1998

Year	Oil tankers	Dry bulk carriers	Other vessels	Total world fleet	% change
1976	320.0	158.1	130.3	608.4	—
1977	335.3	174.4	139.1	648.8	7
1978	339.1	184.5	146.8	670.4	3
1979	338.3	188.5	154.7	681.5	2
1980	339.8	191.0	160.1	690.9	1
1981	335.5	199.5	162.2	697.2	1
1982	325.2	211.2	165.6	702.0	1
1983	306.1	220.6	167.8	694.5	− 1
1984	286.8	228.4	168.1	683.3	− 2
1985	268.4	237.3	168.0	673.7	− 1
1986	247.5	235.2	164.9	647.6	− 4
1987	245.5	231.8	163.5	640.8	− 1
1988	245.0	230.1	162.0	637.1	− 1
1989	248.4	231.4	166.9	646.7	2
1990	257.4	238.9	170.5	666.8	3
1991	264.2	244.0	176.1	684.3	3
1992	270.6	245.7	178.3	694.6	2
1993	270.2	251.3	194.4	715.9	3
1994	269.4	254.3	220.3	744.0	4
1995	265.8	259.8	241.5	767.1	3
1996	271.2	269.6	252.2	793.0	4
1997	272.5	277.8	263.0	813.3	3
1998	279.7	271.6	274.0	825.3	2

Table 4   Major vessel categories in the world's ocean-going cargo ship fleet, 2000

Cargo type	Category	Typical size	Number in world fleet
Dry bulk	Handysize	27 000 dwt	2700
	Handymax	43 000 dwt	1000
	Panamax	69 000 dwt	950
	Capesize	150 000 dwt	550
Tanker	Product tanker	45 000 dwt	1400
	Aframax	90 000 dwt	700
	Suezmax	140 000 dwt	300
	VLCC	280 000 dwt	450
General cargo	Container	>2000 TEU	800
	Container feeder	<2000 TEU	1800
	Ro/Ro	<2000 TEU	900
	Semi-container	<1000 TEU	2800

VLCC, very large crude carrier; RoRo, roll-on/roll-off.

Panama and Suez Canals, respectively. Capesize bulkers and VLCCs (very large crude carriers) are constrained in size mostly by the limitations of port facilities (draft restrictions).

In addition to self-propelled vessels, ocean-going barges pulled by tugs carry both bulk and container cargos, primarily on coastal routes.

Dry Bulk Carriers

The dry bulk fleet is characterized by fragmented ownership; few operators own more than 100 vessels. Most owners charter their vessels to shippers through a network of brokers, under either long-term or single-voyage ('spot') charter arrangements (see **Table 5**). The top 20 dry bulk charterers account for about 30% of the market. The dry bulk charter market is, in general, highly competitive.

Capesize bulkers are used primarily in the Australian and South American bulk export trades. Grain shipments from the USA to Asia move mainly on Panamax vessels. Some bulk vessels, such as oil/bulk/ore (OBO) carriers, are designed for a variety of cargos. Others are 'neo-bulk' carriers, specifically

Table 5 Average daily time charter rates, 1980–2000

Cargo type	Vessel type	$/day
Dry bulk	Handysize	6000
	Handymax	8000
	Panamax	9500
	Capesize	14 000
Tanker	Product tanker	12 000
	Aframax	13 000
	Suezmax	16 500
	VLCC	22 000
Container	400 TEU geared	5000
	1000 TEU geared	9000
	1500 TEU geared	13 500
	2000 TEU gearless	18 000

VLCC, very large crude carrier.

designed to carry goods such as automobiles, lumber, or paper.

Tankers

Ownership of the tanker fleet is also fragmented; the average tanker owner controls fewer than three vessels; 70% of the fleet is owned by independent owners and 25% by oil companies. Like dry bulk ships, tankers are chartered to shippers on longer time charters or spot voyage charters (see **Table 5**). The top five liquid bulk charterers account for about 25% of the market. Like the dry bulk market, the tanker charter market is highly competitive.

VLCCs are used primarily for Middle East exports to Asia and the USA. In addition to liquid tankers, a fleet of gas carriers (world capacity: 17 million gt) moves liquefied natural gas (LNG) and petroleum gas (LPG).

General Cargo/Container Ships

Ownership is more concentrated in the container or 'liner' fleet. ('Liner' carriers operate regularly scheduled service between specific sets of ports.) Of some 600 liner carriers, the top 20 account for 55% of world TEU capacity. The industry is expected to consolidate further. Despite this concentration, liner shipping is competitive; barriers to entry are low, and niche players serving specialized cargos or routes have been among the most profitable. Outside ownership of container vessels and chartering to liner companies increased during the 1990s.

Large ships (>4000 TEU) are used on long ocean transits between major ports, while smaller 'feeder' vessels typically carry coastwise cargo or serve niche markets between particular port pairs. In addition to container vessels, the general cargo trades are served

by roll-on/roll-off (RoRo) ships (world fleet capacity: 23 million gt) that carry cargo in wheeled vehicles, by refrigerated ships for perishable goods, and by a diminishing number of multi-purpose general cargo ships.

Passenger Vessels

The world's fleet of ferries carries passengers and automobiles on routes ranging from urban harbor crossings to hundreds of miles of open ocean. Today's ferry fleet is increasingly composed of so-called fast ferries, >1200 of which were active in 1997. Many of these are aluminum-hull catamarans, the largest capable of carrying 900 passengers and 240 vehicles at speeds between 40 and 50 knots. Fast ferries provide an inexpensive alternative to land and air travel along coastal routes in parts of Asia, Europe, and South America.

The cruise ship industry was born in the 1960s, when jet aircraft replaced ships as the primary means of crossing oceans, and underutilized passenger liners were converted for recreational travel. The most popular early cruises ran from North America to the Caribbean. Today, the Caribbean remains the main cruise destination, with more than half of all passengers; others are Alaska and the Mediterranean. The US cruise market grew from 500 000 passengers per year in 1970 to 1.4 million in 1980 and 3.5 million in 1990. Today, more than 25 cruise lines serve the North American cruise market. The three largest lines control over 50% of the market, and further consolidation is expected; 149 ships carried 6 million passengers and generated $18 billion in turnover in 1999. The global market saw 9 million passengers. Cruise vessels continue to grow larger due to scale economies; the largest ships today carry more than 2000 passengers. Most cruise lines are European-owned and fly either European or open registry flags (see below).

Financial Performance

Shipping markets display cyclical behavior. Charter rates (see **Table 5**) are set by the market mechanism of supply and demand, and are most closely correlated with fleet utilization (rates rise sharply when utilization exceeds 95%). Shipping supply (capacity) is driven by ordering and scrapping decisions, and sometimes by regulatory events such as the double hull requirement for tankers (see below). Demand (trade volume) is determined by national and global business cycles, economic development, international trade policies, and trade disruptions such as wars/ conflicts, canal closures, or oil price shocks. New vessel deliveries typically lag orders by 1–2 years.

Excessive ordering of new vessels in times of high freight rates eventually leads to overcapacity and a downturn in rates; rates remain low until fleet contraction or trade growth improves utilization, and the cycle repeats.

Average annual returns on shipping investments ranged from 8% (dry bulk) to 11% (container feeder ships) and 13% (tankers) during 1980–2000, with considerable volatility. Second-hand ship asset prices fluctuate with charter rates, and many ship owners try to improve their returns through buy-low and sell-high strategies.

Law and Regulation

Legal Regimes

International Law of the Sea The United Nations Convention on the Law of the Sea (UNCLOS) sets out an international legal framework governing the oceans, including shipping. UNCLOS codifies the rules underlying the nationality (registry) of ships, the right of innocent passage for merchant vessels through other nations' territorial waters, etc. (see the entry in this volume on the Law of the Sea for details).

International Maritime Organization The International Maritime Organization (IMO) was established in 1948 and became active in 1959. Its 158 present member states have adopted some 40 IMO conventions and protocols governing international shipping. Major topics include maritime safety, marine pollution, and liability and compensation for third-party claims. Enforcement of these conventions is the responsibility of member governments and, in particular, of port states. The principle of port state control allows national authorities to inspect foreign ships for compliance and, if necessary, detain them until violations are addressed.

In 1960, IMO adopted a new version of the International Convention for the Safety of Life at Sea (SOLAS), the most important of all treaties dealing with maritime safety. IMO next addressed such matters as the facilitation of international maritime traffic, load lines, and the carriage of dangerous goods. It then turned to the prevention and mitigation of maritime accidents, and the reduction of environmental effects from cargo tank washing and the disposal of engine-room waste.

The most important of all these measures was the Marine Pollution (MARPOL) treaty, adopted in two stages in 1973 and 1978. It covers accidental and operational oil pollution as well as pollution by chemicals, goods in packaged form, sewage, and garbage. In the 1990s, IMO adopted a requirement for all new tankers and existing tankers over 25 years of age to be fitted with double hulls or a design that provides equivalent cargo protection in the event of a collision or grounding.

IMO has also dealt with liability and compensation for pollution damage. Two treaties adopted in 1969 and 1971 established a system to provide compensation to those who suffer financially as a result of pollution.

IMO introduced major improvements to the maritime distress communications system. A global search and rescue system using satellite communications has been in place since the 1970s. In 1992, the Global Maritime Distress and Safety System (GMDSS) became operative. Under GMDSS, distress messages are transmitted automatically in the event of an accident, without intervention by the crew.

An International Convention on Standards of Training, Certification and Watchkeeping (STCW), adopted in 1978 and amended in 1995, requires each participating nation to develop training and certification guidelines for mariners on vessels sailing under its flag.

National Control and Admiralty Law The body of private law governing navigation and shipping in each country is known as admiralty or maritime law. Under admiralty, a ship's flag (or registry) determines the source of law. For example, a ship flying the American flag in European waters is subject to American admiralty law. This also applies to criminal law governing the ship's crew.

By offering advantageous tax regimes and relatively lax vessel ownership, inspection, and crewing requirements, so-called 'flags of convenience' or 'open registries' have attracted about half of the world's tonnage (see **Table 6**). The open registries include Antigua and Barbuda, Bahamas, Bermuda, Cayman Islands, Cyprus, Gibraltar, Honduras, Lebanon, Liberia, Malta, Mauritius, Oman, Panama, Saint Vincent, and Vanuatu. Open registries are the flags of choice for low-cost vessel operation, but in some instances they have the disadvantage of poor reputation for safety.

Protectionism/Subsidies

Most maritime nations have long pursued policies that protect their domestic flag fleet from foreign competition. The objective of this protectionism is usually to maintain a domestic flag fleet and cadre of seamen for national security, employment, and increased trade.

Table 6   Fleets of principal registries (in 1000s of gt) 1998 (includes ships of 100 gt)

Country of registry	Tankers	Dry bulk	General cargo	Other	Total	% of world
Panama	22 680	40 319	26 616	8608	98 223	18.5
Liberia	26 361	16 739	8803	8590	60 493	11.4
Bahamas	11 982	4990	7143	3601	27 716	5.2
Greece	12 587	8771	1943	1924	25 225	4.7
Malta	9848	8616	4658	952	24 074	4.5
Cyprus	3848	11 090	6981	1383	23 302	4.4
Norway	8994	4041	4153	5949	23 137	4.4
Singapore	8781	4585	5596	1408	20 370	3.8
Japan	5434	3869	3111	5366	17 780	3.3
China	2029	6833	6199	1443	16 504	3.1
USA	3436	1268	3861	3286	11 851	2.2
Russia	1608	1031	3665	4786	11 090	2.1
Others	33 448	46 414	57 350	34 917	172 129	32.4
World total	151 036	158 566	140 079	82 213	531 894	100.0

Laws that reserve domestic waterborne cargo (cabotage) for domestic flag ships are common in many countries, including most developed economies (although they are now being phased out within the European Union). The USA has a particularly restrictive cabotage system. Cargo and passengers moving by water between points within the USA, as well as certain US government cargos, must be carried in US-built, US-flag vessels owned by US citizens. Also, US-flag ships must be crewed by US citizens.

Vessels operating under restrictive crewing and safety standards may not be competitive in the international trades with open registry vessels due to high operating costs. In some cases, nations have provided operating cost subsidies to such vessels. The US operating subsidy program was phased out in the 1990s.

Additional subsidies have been available to the shipping industry through government support of shipbuilding (see below).

Liner Conferences

The general cargo trades have traditionally been served by liner companies that operate vessels between fixed sets of ports on a regular, published schedule. To regulate competition among liner companies and ensure fair and consistent service to shippers, a system of 'liner conferences' has regulated liner services and allowed liner companies, within limits, to coordinate their services.

Traditionally, so-called 'open' conferences have been open to all liner operators, published a single rate for port-to-port carriage of a specific commodity, and allowed operators within the conference to compete for cargo on service. The US foreign liner trade has operated largely under open conference rules. By contrast, many other liner trades have been governed by closed conferences, participation in which is restricted by governments. In 1974, the United Nations Conference on Trade and Development (UNCTAD) adopted its Code of Conduct for Liner Conferences, which went into effect in 1983. Under this code, up to 40% of a nation's trade can be reserved for its domestic fleet.

In the 1970s, price competition began within conferences as carriers quoted lower rates for intermodal mini land-bridge routes than the common published port-to-port tariffs. By the 1980s, intermodal rates were incorporated fully in the liner conference scheme. The 1990s saw growing deregulation of liner trades (for example, the US 1998 Ocean Shipping Reform Act), which allowed carriers to negotiate confidential rates separately with shippers. The effect has been, in part, increased competition and a drive for consolidation among liner companies that is expected to continue in the future.

Environmental Issues

With the increased carriage of large volumes of hazardous cargos (crude oil and petroleum products, other chemicals) by ships in the course of the twentieth century, and especially in the wake of several tanker accidents resulting in large oil spills, attention has focused increasingly on the environmental effects of shipping. The response so far has concentrated on the operational and accidental discharge of oil into the sea. Most notably, national and international regulations (the Oil Pollution Act (OPA) of 1990 in the US; IMO MARPOL 1992 Amendments, Reg. 13F and G) are now forcing conversion of the world's tanker fleet to double hulls. Earlier, MARPOL 1978 required segregated ballast tanks on tankers to avoid the discharge of oil residue following the use of cargo tanks to hold ballast water.

Recently, ballast and bilge water has also been identified as a medium for the transport of non-indigenous species to new host countries. A prominent example is the zebra mussel, which was brought to the USA from Europe in this way.

Liability and Insurance

In addition to design standards such as double hulls, national and international policies have addressed the compensation of victims of pollution from shipping accidents through rules governing liability and insurance. The liability of ship owners and operators for third-party claims arising from shipping accidents historically has been limited as a matter of policy to encourage shipping and trade. With the increased risk associated with large tankers, the international limits were raised gradually in the course of the twentieth century. Today, tanker operators in US waters face effectively unlimited liability under OPA 90 and a range of state laws.

Most liability insurance for the commercial shipping fleet is provided through a number of mutual self-insurance schemes known as P&I (protection and indemnity) clubs. The International Group of P&I Clubs includes 19 of the largest clubs and collectively provides insurance to 90% of the world's ocean-going fleet. Large tankers operating in US waters today routinely carry liability insurance coverage upward of $2 billion.

Shipbuilding

Major technical developments in twentieth century shipbuilding began with the introduction of welding in the mid-1930s, and with it, prefabrication of steel sections. Prefabrication of larger sections, improved welding techniques, and improved logistics were introduced in the 1950s. Mechanized steel prefabrication and numerically controlled machines began to appear in the 1970s. From 1900 to 2000, average labor hours per ton of steel decreased from 400–500 to fewer than 100; and assembly time in dock/berth decreased from 3–4 years to less than 6 months.

South Korea and Japan are the most important commercial shipbuilding nations today. In the late 1990s, each had about 30% of world gross tonnage on order; China, Germany, Italy, and Poland each accounted for between about 4 and 6%; and Taiwan and Romania around 2%. The US share of world shipbuilding output fell from 9.5% in 1979 to near zero in 1989.

Shipbuilding is extensively subsidized in many countries. Objectives of subsidies include national security goals (maintenance of a shipbuilding base), employment, and industrial policy goals (shipyards are a major user of steel industry output). Direct subsidies are estimated globally around $5–10 billion per year. Indirect subsidies include government loan guarantees and domestic-build requirements (for example, in the US cabotage trade).

An agreement on shipbuilding support was developed under the auspices of the Organization for Economic Cooperation and Development (OECD) and signed in December 1994 by Japan, South Korea, the USA, Norway, and the members of the European Union. The agreement would limit loan guarantees to 80% of vessel cost and 12 years, ban most direct subsidy practices, and limit government R&D support for shipyards. Its entry into force, planned for 1996, has been delayed by the United States' failure to ratify the agreement.

Ports

Commercial ships call on thousands of ports around the world, but global cargo movement is heavily concentrated in fewer than 100 major bulk cargo ports and about 24 major container ports. The top 20 general cargo ports handled 51% of all TEU movements in 1998 (see **Table 7**).

Most current port development activity is in container rather than bulk terminals. Port throughput efficiency varies greatly. In 1997, some Asian container ports handled 8800 TEU per acre per year, while European ports averaged 3000 and US ports only 2100 TEU per acre per year. The differences are due largely to the more effective use of automation and lesser influence of organized labor in Asian ports. Because container traffic is growing rapidly and container ships continue to increase in size, most port investment today is focused on the improvement and new development of container terminals.

In many countries, the public sector plays a more significant role in port planning and development than it does in shipping. Most general cargo or 'commercial' ports are operated as publicly controlled or semi-public entities, which may lease space for terminal facilities to private terminal operators. Bulk cargo or 'industrial' ports for the export or import/processing of raw materials traditionally have been built by and for a specific industry, often with extensive public assistance. The degree of national coordination of port policy and planning varies considerably. The USA has one of the greatest commercial port densities in the world, particularly along its east coast. US commercial ports generally compete among each other as semi-private entities run in the interest of local economic development objectives. The US federal government's primary role in

Table 7 Top 20 container ports by volume, 1998

Port	Country	Throughput (1000s TEU)
Singapore	Singapore	15 100
Hong Kong	China	14 582
Kaohsiung	Taiwan	6271
Rotterdam	Netherlands	6011
Busan	South Korea	5946
Long Beach	USA	4098
Hamburg	Germany	3547
Los Angeles	USA	3378
Antwerp	Belgium	3266
Shanghai	China	3066
Dubai	UAE	2804
Manila	Philippines	2690
Felixstowe	UK	2524
New York/New Jersey	USA	2500
Tokyo	Japan	2169
Tanjung Priok	Indonesia	2131
Gioia Tauro	Italy	2126
Yokohama	Japan	2091
San Juan	Puerto Rico	2071
Kobe	Japan	1901

port development is in the improvement and maintenance of navigation channels, since most US ports (apart from some on the west coast) are shallow and subject to extensive siltation.

Future Developments

A gradual shift of the 'centroid' of Asian manufacturing centers westward from Japan and Korea may in the future shift more Asia–America container cargo flows from trans-Pacific to Suez/trans-Atlantic routes. Overall container cargo flows are expected to increase by 4–7% per year. Bulk cargo volumes are expected to increase as well, but at a slower pace.

Container ships will continue to increase in size, reaching perhaps 12 000 TEU capacity in the course of the next decade. Smaller, faster cargo ships may be introduced to compete for high-value cargo on certain routes. One concept calls for a 1400 TEU vessel capable of 36–40 knots (twice the speed of a conventional container ship), making scheduled Atlantic crossings in 4 days.

Port developments will center on improved container terminals to handle larger ships more efficiently, including berths that allow working the ship from both sides, and the further automation and streamlining of moving containers to/from ship and rail/truck terminus.

Conclusions

Today's maritime shipping industry is an essential transportation medium in a world where prosperity is often tied to international trade. Ships and ports handle 90% of the world's cargo and provide a highly efficient and flexible means of transport for a variety of goods. Despite a history of protectionist regulation, the industry as a whole is reasonably efficient and becoming more so. The safety of vessels and their crews, and protection of the marine environment from the results of maritime accidents, continue to receive increasing international attention. Although often conservative and slow to adopt new technologies, the shipping industry is poised to adapt and grow to support the world's transportation needs in the twenty-first century.

Further Reading

Containerization Yearbook 2000. London: National Magazine Co.

Shipping Statistics Yearbook. Bremen: Institute of Shipping Economics and Logistics.

Lloyd's Shipping Economist. London: Lloyd's of London Press.

United Nations Conference on Trade and Development (UNCTAD) (Annual) *Review of Maritime Transport.* New York: United Nations.

INDEX

Notes

Cross-reference terms in italics are general cross-references, or refer to subentry terms within the main entry (the main entry is not repeated to save space). Readers are also advised to refer to the end of each article for additional cross-references - not all of these cross-references have been included in the index cross-references.

The index is arranged in set-out style with a maximum of three levels of heading. Major discussion of a subject is indicated by bold page numbers. Page numbers suffixed by T and F refer to Tables and Figures respectively. vs. indicates a comparison.

This index is in letter-by-letter order, whereby hyphens and spaces within index headings are ignored in the alphabetization. For example, 'oceanography' is alphabetized before 'ocean optics.' Prefixes and terms in parentheses are excluded from the initial alphabetization.

Where index subentries and sub-subentries pertaining to a subject have the same page number, they have been listed to indicate the comprehensiveness of the text.

Abbreviations used in subentries

DIC - dissolved inorganic carbon
DOC - dissolved organic carbon
ENSO - El Niño Southern Oscillation
ROV - remotely operated vehicle
SAR - synthetic aperture radar
SST - sea surface temperature

Additional abbreviations are to be found within the index.

A

Abalone (*Haliotis* spp.)
  harvesting, 379–380
  mariculture, 380–381, 381, 382F
Abiki, 201, 206
Absorption (optical) coefficient, 670
  components, 670–671
Absorption spectra
  gelbstoff, 671
Abyssal gigantism, 316
Abyssal plain, 550
Abyssal zone, 312T, 315–316, 316
Acanthaster planci (crown-of-thorns starfish), 114
Acartia spp. copepods, 303
Accelerator mass spectrometry (AMS), 482
Acorn barnacle (*Semibalanus balanoides*), 96
Acoustic backscatter system (ABS), 630, 630F
  STABLE II, 637, 638
Acoustic Doppler current profiler (ADCP), 633
Acoustic Doppler velocimeter (ADV), 633

Acoustic measurement, sediment transport, **629–642**
Acoustic remote sensing, 651–654
  characteristic shallow water signal, 652, 652F
  data inversion, 655–656
  experiment setup, 655F
  geophone arrays, 655–656, 655F, 656F
  interface waves, 652F, 657
    dispersion diagram, 657, 657F
  reflected waves, 652–654
    angle of intromission case, 654–655
    critical angle case, 653F, 654
    grazing angle, 643, 652–654, 652F, 653F
    half-space seafloor, 652–654, 652F
    layered seafloor, 654F, 655, 655F
    phase shift, 654–655, 654F
  reflection coefficient, 652–654, 654–655, 654F
  reflection loss, 654F, 655, 655F
  refracted waves, 652F, 655–657
    Herglotz-Bateman-Wieckert integration, 656–657657F
Acoustic ripple profiler (ARP), 630, 630F, 631, 631F, 637, 641F

3-D, 631, 632F, 641
  bed profile, 638F
Acoustic ripple scanner (ARS), 629–630, 630F, 631, 632F
Acoustics, marine sediments, **643–659**
  attenuation, 644–646, 648, 650, 658,
  bulk density, 646, 646T, 648
  bulk modulus, 646, 646T, 648
  forward modeling, 658, 658F
  measurement of geoacoustic parameters, 648–649
    laboratory, sediment core samples *see* Sediment core samples
    transmission loss technique, 657–658, 658F
  relationship between geoacoustic parameters, 649
    porosity and relative compressional wave velocity, 649, 650F
    relative density and porosity, 649, 649F
  SAFARI, 658, 658F
  seafloor roughness, 648
  sedimentary acoustic medium, 643, 646
  sediment structure, 643, 643–644
    composition, 643, 644F

Printed and bound by CPI Group (UK) Ltd, Croydon, CR0 4YY

03/10/2024

01040311-0008